W9-CSA-929

Superconductor Applications: SQUIDs and Machines

NATO ADVANCED STUDY INSTITUTES SERIES

A series of edited volumes comprising multifaceted studies of contemporary scientific issues by some of the best scientific minds in the world, assembled in cooperation with NATO Scientific Affairs Division.

Series B: Physics

SOME VOLUMES IN THIS SERIES

Volume 1 – Superconducting Machines and Devices: Large Systems Applications edited by Simon Foner and Brian B. Schwartz

Volume 2 – Elementary Excitations in Solids, Molecules, and Atoms (Parts A and B) edited by J. Devreese, A. B. Kunz, and T. C. Collins

Volume 9 – Electronic Structure of Polymers and Molecular Crystals edited by Jean-Marie André and János Ladik

Volume 11 – Fluctuations, Instabilities, and Phase Transitions edited by Tormod Riste

Volume 13 – Weak and Electromagnetic Interactions at High Energies – *Cargèse 1975* (Parts A and B) edited by Maurice Lévy, Jean-Louis Basdevant, David Speiser, and Raymond Gastmans

Volume 16 – Electronic Structure and Reactivity of Metal Surfaces edited by E. G. Derouane and A. A. Lucas

Volume 17 – Linear and Nonlinear Electron Transport in Solids edited by J. T. Devreese and V. van Doren

Volume 19 – Defects and Their Structure in Nonmetallic Solids edited by B. Henderson and A. E. Hughes

Volume 20 – Physics of Structurally Disordered Solids edited by Shashanka S. Mitra

Volume 21 – Superconductor Applications: SQUIDs and Machines edited by Brian B. Schwartz and Simon Foner

The series is published by an international board of publishers in conjunction with NATO Scientific Affairs Division

A	Life Sciences	Plenum Publishing Corporation
B	Physics	New York and London
C	Mathematical and Physical Sciences	D. Reidel Publishing Company Dordrecht and Boston
D	Behavioral and Social Sciences	Sijthoff International Publishing Company Leiden
E	Applied Sciences	Noordhoff International Publishing Leiden

Nato Advanced Study Institute on
" Small-Scale Applications of
Superconductivity

Superconductor Applications: SQUIDs and Machines

Edited by

Brian B. Schwartz and Simon Foner

Francis Bitter National Magnet Laboratory
M. I. T.
Cambridge, Massachusetts

PLENUM PRESS • NEW YORK AND LONDON
Published in cooperation with NATO Scientific Affairs Division

Library of Congress Cataloging in Publication Data

Nato Advanced Study Institute on Small-Scale Applications of Superconductivity, Gardone Riviera, Italy, 1976.
Superconductor applications: SQUIDs and machines.

(Nato Advanced Study Institute series; Series B, Physics; v. 21)
Includes index.
1. Superconductors–Congresses. I. Schwartz, Brian B., 1938- II. Foner, Simon. III. Title. IV. Series.
TK7872.S8N2 1976 621.39 76-51750
ISBN 0-306-35721-6

The Francis Bitter National Magnet Laboratory is sponsored by the National Science Foundation

Proceedings of a NATO Advanced Study Institute on Small-Scale Applications of Superconductivity held in Gardone Riviera (Lago di Garda), Italy, September 1–10, 1976

A Division of Plenum Publishing Corporation
227 West 17th Street, New York, N.Y. 10011

Printed in the United States of America

PREFACE

This book includes small and large scale applications of superconductivity. Part I, SQUIDs, comprises about 75% of this volume, and is devoted to small scale applications, mainly Superconducting QUantum Interference Devices (SQUIDs), and the remainder, Part II, Machines, presents an updated review of large scale applications of superconductivity. The present book combined with the previous book Superconducting Machines and Devices: Large Systems Applications edited by S. Foner and B.B. Schwartz, Plenum Press, New York (1974) represents a detailed and most up-to-date review of the applications of superconducting technology. The text of the current book is suitable for advanced undergraduates or graduate students in applied physics and engineering courses. The book should be valuable to scientists, engineers and technologists interested in the current status and future applications of superconductivity technology. The last 7 chapters in Part I review the major national efforts on small scale technology and should prove useful for industrial and government planners as well as scientists and engineers.

This book is based on a NATO Advanced Study Institute entitled "Small Scale Applications of Superconductivity" which was held from 1 September to 10 September 1976 in Gardone Riviera on Lake Garda in Northern Italy. This Study Institute complements a previous NATO Advanced Study Institute held in 1973 on Large Scale Superconducting Devices. As with the previous Institute, the focus of the lecturers and the present book involves both applications as well as the scientific principles. Part I, the major part of the present book, treats small-scale applications of superconductivity; and Part II contains an updated review of large scale applications. Part II, together with the previous NATO Institute book (Superconducting Machines and Devices: Large Systems Applications) gives a thorough coverage of large scale applications of superconductivity up to 1977.

Part I provides reviews of all the major principles and devices of small scale superconductivity. The opening chapter by Sir Brian Pippard gives a personal review of some of the historical highlights of superconductivity from the 1930's through the 1960's. Professor Pippard was Brian Josephson's thesis advisor at Cambridge and places Josephson's discovery within the context of the rapid theoretical and experimental developments in superconductivity in the late 1950's and early 1960's.

Chapters 2-6 discuss the basic principles of macroscopic quantum interference phenomena and weak links, superconducting quantum interference devices (SQUIDs), equivalent circuits and analogs, superconducting devices for metrology and standards, and high frequency applications of Josephson Junctions. Chapters 7 - 12 involve special topics including reviews of junction fabrication techniques, biomagnetism applications, status of commercial instruments in the United States, resistive superconducting devices, nonequilibrium properties of superconductors, and application of SQUIDs to computers. The last 7 chapters of Part I (13 - 17) in this book continue an innovation we introduced in the 1973 NATO Institute proceedings; reviews of national efforts. These reviews for Canada, France, Germany, Italy, The Netherlands, United Kingdom and the United States give an up-to-date summary of the current programs in the area of small scale devices. Part II, Chapter 20 presents a thorough review of Large Scale Applications of Superconductivity. (For convenience we have subdivided the subject index into two parts, Index Part I – SQUIDs, Index Part II – Machines.)

The 1976 NATO Institute which resulted in the present volume involved planning which dates back to the 1973 NATO Institute. We were fortunate in having a very effective advisory committee which helped us with the planning. Dr. Edelsack of the Naval Research Laboratory helped us throughout the planning of the Institute. We also profitted from many useful suggestions by S. Shapiro, University of Rochester, A. Baratoff, IBM Zurich, and A.D. Appleton, IRD, England, throughout the planning of the Institute. Members of our International Advisory Committee included R. Adde, G. Bogner, I. Giaever, B. Josephson, R.A. Kamper, C. Rizzuto, and K. Saermark.

We wish to thank Dr. T. Kester and Dr. M. Di Lullo, from the NATO Scientific Affairs Division for their continued interest and encouragement, and the NATO Science Council for their support of the Advanced Study Institute. We also wish to thank the National Science Foundation for Travel Grants to three students.

In addition to the lecturers, the NATO Institute had approximately 95 participants from 23 countries. Professor Carlo Rizzuto was the Local Chairman. He and his associates at the University of Genoa gave continuous help in all aspects of the planning and operation of the Institute. Professor Cerdonio of the University of Rome with Professor Rizzuto helped us to choose the site of the Institute. We would like to thank the Gardone Riviera region, the town of Salò, and the Lake region tourist agency for their hospitality.

We received excellent cooperation from all the lecturers, and we wish to thank them for their excellent talks and prompt completion of the manuscripts. Their cooperation in meeting our deadline dates has allowed us to adhere to a very tight publication schedule. We apologize to all the lecturers for the many demands we made, and wish to thank each of the lecturers for their dedication to the Institute. The success of the Institute required the continued cooperation of each lecturer before, during, and after the Institute. In addition to our personal thanks, we hope that the response of the students at the Institute and the present volume justifies their efforts. In return, we have attempted to set a record in rapidly publishing these contributions.

We would especially like to thank Dr. G. Bogner, Siemens for his review lectures on Large Scale Applications which comprise Part II of these Proceedings. His lectures and notes gave the participants a broad perspective of many additional areas of superconductor applications which are not normally encountered in Small Scale Applications.

We would like to thank R.B. Frankel for his help with the Institute. We would also like to thank Mary Filoso, Michael McDowell, Delphine Radcliffe and Nancy Brandon for helping with the typing and correcting of these manuscripts. We particularly wish to thank Mary Filoso who, as in 1973, was closely involved with the planning and execution of the Institute and Proceedings. Her experience and continued attention to the Institute arrangements and the present book were invaluable.

Cambridge, Massachusetts

October 1976

Brian B. Schwartz
Simon Foner

CONTENTS

PART I - SQUIDS

CHAPTER 3: SUPERCONDUCTING QUANTUM INTERFERENCE DEVICES FOR LOW FREQUENCY MEASUREMENTS

J. Clarke

CHAPTER 4: EQUIVALENT CIRCUITS AND ANALOGS OF THE JOSEPHSON EFFECT

T. A. Fulton

CHAPTER 5: SUPERCONDUCTING DEVICES FOR METROLOGY AND STANDARDS
R. A. Kamper

CHAPTER 6: HIGH FREQUENCY PROPERTIES AND APPLICATIONS OF JOSEPHSON JUNCTIONS FROM MICROWAVES TO FAR-INFRARED

R. Adde and G. Vernet

PART II - MACHINES

CHAPTER 20: LARGE-SCALE APPLICATIONS OF SUPERCONDUCTIVITY

G. Bogner

Part I
SQUIDs

Chapter 1

THE HISTORICAL CONTEXT OF JOSEPHSON'S DISCOVERY

A. B. Pippard

Cavendish Laboratory

Cambridge, England CB3 OHE

In an introductory talk like this you might expect a complete history of superconductivity from 1911 onward. I was not a physicist in 1911; I never met Kamerlingh Onnes and I cannot tell you how it was then, nor am I going to try. I am going to start considerably later, about the time of the B. C. S. theory. I am not a historian, and when one starts to think about writing history, one realizes that historians may not be skilled in physics, but they certainly have expertise of their own. It is quite difficult to both write history and be involved in discovery. The thing which normally saves historians from being completely bogged down in detail is that they let enough time elapse between the events and the writing for almost everything to be forgotten; then it is possible to reconstruct the bits you care about and let the trivial details fade into insignificance. Because we are not talking about that sort of time lapse here it is very hard to separate the details from the important things. We feel the urge to find out what people were really thinking as the discoveries were made, but this is an extremely hard task.

Most of you probably have read Watson's book "The Double Helix" and his account of the discovery of the form of the DNA

Note: This paper is based on the opening talk presented by Sir Brian Pippard to the participants at the NATO Advanced Study Institute on Small-Scale Superconducting Devices. The talk was recorded and it represents essentially the transcript. The editors have requested that Professor Pippard allow us to maintain the informal nature of his talk in order to give the reader the flavor of his remarks. We encouraged only slight editing for clarity.

molecule. Watson was basing his story on copious notes and letters which he wrote at the time and was able to do a magnificent job of reconstruction. I was in Cambridge at the time that this work on the double helix was going on in the Cavendish laboratory. I can assure you that the flavor which comes through in Watson's book is very close to what I remember it to have been. To tell an equally good story of superconductivity would require someone with almost perfect recall and also with very good notes to make sure that his memory is not falsified by subsequent interpretations. I cannot promise you that this is what I can manage in this paper. The double helix is the only example that I know which gets at all close to the feel of events. One of the troubles one has in reconstructing important scientific events is that anyone working on a difficult problem goes through a great many fanciful interpretations. A lot of nonsense goes on in his mind until the moment when suddenly the right idea appears, and that usually comes through as a moment of enlightenment - sudden enough to drive out completely any feeling for what it was like before, so that almost immediately it is impossible to reconstruct the process of thought which led to that inspiration. Already history has begun to be blurred. One can try to go back and plot out the main lines which led to the new insight. However, once having seen the truth it is impossible to give an exact description of what it is like not to be able to see the truth. All I promise then is that I have done my best to find out what happened, but I cannot guarantee that any of the details are right.

I was, as far as the Josephson effect was concerned, an observer of the events but a very poor observer. One reason is that you could hardly find two people whose minds worked on more different lines than Brian Josephson and myself, so that I was never able to understand what he was talking about. The other thing is that one is not told at the time that something important is happening. It is only afterwards - years afterwards - that people will say, "You ought to have taken more notice of what was happening, because it was obviously important." All I can say is that it was not obviously important at the time - it was obviously interesting, but that is quite another matter. If something is interesting but you don't actually understand it, you can hardly recall it very closely. Apart from that, I was interested in other things at the time. It was only a few months since I had commissioned a high magnetic field laboratory. I was concerned with the behavior of electrons in high magnetic fields and had no deep interest in superconductivity at that time, apart from certain outstanding problems that I had a few students working on. My mind was going on other things, so I didn't take all that much notice. This is why I cannot present you with the evidence of a notebook in which I say "On January the third, Josephson told me that he had solved the Hamiltonian."

Well, where shall we start? I think the best thing is to go back a little into history. It is very easy to overestimate the importance of the Josephson effect when the major part of this NATO Institute is devoted to its applications. I am not saying that the Josephson effect is not important, but it is not the only important thing that happened in superconductivity about that time, and I think it is wise to get some perspective on it. How one gets perspective depends on taste, and one way (a bit curious perhaps) is through numerical research. I have studied Physics Abstracts and counted the number of papers published year by year on superconductivity and compared their total with the total number of papers published. This (as illustrated in Fig. 1) shows how the subject of superconductivity was far from dying as a result of B. C. S. theory. Normalization to the total number of papers is to allow for the overall five-fold expansion in the number of papers published in physics annually in this period. Far from superconductivity coming to an end in 1957, the B. C. S. theory represents the moment when it took off both in terms of the proportion of physicists actually publishing papers and (even more striking) in absolute numbers; in ten years there is a twelve-fold increase in the number of published papers, from 66 to 830. Some of this represents the rise of letter journals. (It may not represent more work done but just shorter papers

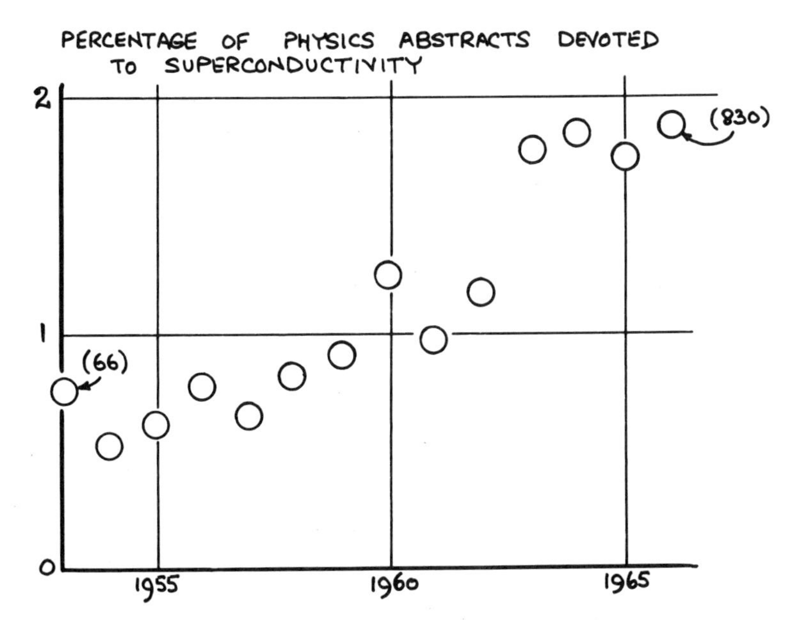

Fig. 1 A graph of the number of papers on superconductivity in Physics Abstracts from 1953 to 1966.

being published, with still less in each paper even than before.) The data in Fig. 1 stops in 1965-66 since it was quite clear looking at later Physics Abstracts that the curve was still increasing and my patience was exhausted. There is an enormous amount of applications work in superconductivity nowadays. Although 1957 was the year of the B.C.S. theory, this rise in papers is not associated just with B.C.S. nor with the Josephson effect which was discovered in 1962. Some of the bump in 1960 is perhaps the result of Giaever's discovery of tunneling, but there is a general rise of interest at this time in all areas of superconductivity. Another very important factor, of course, was the emergence of technical superconductivity; high-field alloys and devices must account for a great deal of the increase. Before we leave the simple numerology of the subject, there was a conference at Colgate on the science of superconductivity in August 1963, which was published in the Reviews of Modern Physics. The published proceedings ran to 330 pages. Although the conference took place one and a half years after Josephson, only 40 pages were devoted to tunneling - just over 10% - whereas 130 pages, more than one-third of the work, was concerned with high-field superconductors and with transition metal alloys. Nowadays, technical applications of superconductivity are separately listed in Physics Abstracts and they overwhelm, by quite a large factor, the papers published on small-scale devices. Let us not then be too parochial when we are talking about small-scale devices. We are not the only pebble on the beach, there is quite a large rock standing quite close by.

Bearing in mind that many factors were responsible for the large increase of papers in superconductivity, I have traced historically four rather closely related fields, and have listed in Table I, against the year, some of the milestones. I hope nobody will be offended at being left out, for it is not meant to be anything like a complete list, but it includes some of the more significant developments. Let us start with the third column, dealing with type II superconductors which are possibly less intimately related to the main theme than the others. I start in 1935, when there was already quite a lot of experimental work by Mendelssohn, Shubnikov and others showing that alloy superconductors behave quite differently from pure superconductors. In 1935, Gorter specifically suggests that it is a short mean-free-path effect which is responsible for the peculiar properties of alloy superconductors. He doesn't know why, however. His ideas are in contrast to Mendelssohn's that alloys are inhomogeneous and form a superconducting sponge which can trap the flux. The war came along and the next significant event is the Ginsburg and Landau theory in 1950. I have put my own name there in 1951. These two developments are quite independent (it takes time for Russian papers to filter through). I revived Gorter's idea (without knowing he had had it already) in a more specific

Table I: Milestones in Superconductivity

	From Energy Gap to Tunneling	Weak Links	Type II Supercon-ductors	Fundamental Theory
1935			Gorter	
|				
1946	Ginsburg Daunt and Mendelssohn			
|				
1950			Ginsburg and Landau	Frölich Maxwell
51			Pippard	
52				
53	Goodman			
54				
55				
56	Glover and Tinkham			Cooper
57			Abrikosov	Bardeen, Cooper Schrieffer
58		Meissner		Anderson
59			Gorkov	
1960	Giaever Nicol, Shapiro and Smith	Parmenter		
61			Kunzler, Buehler, Hsu & Wernick [Goodman]	
62	Cohen, Phillips and Falicov			
63	Josephson Anderson & Rowell Shapiro			
64	Zimmerman and Silver	De Gennes		

REFERENCES FOR TABLE I

FROM ENERGY GAP TO TUNNELING

1946 Ginsberg: Superconductivity (book; in Russian).
Daunt & Mendelssohn: Proc. Roy. Soc. A185, 225 (1946).
First suggestions of energy gap to account for London equations (G), and absence of Thomson heat (D & M).

1953 Goodman: Proc. Phys. Soc. A66, 217 (1953).
Experimental evidence for energy gap from thermal conductivity.

1956 Glover and Tinkham: Phys. Rev. 104, 844 (1956).
Absorption of far infrared reveals gap directly.

1960 Giaever: Phys. Rev. Lett. 5, 147 (1960).
Discovery of N-S tunneling through oxide layer.
Nicol, Shapiro & Smith: Phys. Rev. Lett. 5, 461 (1960).
Giaever: Phys. Rev. Lett. 5, 464 (1960).
S-S tunneling; explicit exhibition of zero-voltage tunneling current, discussed as metallic bridge (G) or uncommented upon (N, S & S).

1962 Cohen, Phillips and Falicov: Phys. Rev. Lett. 8, 316(1962)
Tunneling Hamiltonian.
Josephson: Phys. Lett. 1, 251 (1962).
Prediction of pair tunneling through oxide layer separating superconductors.

1963 Anderson & Rowell: Phys. Rev. Lett. 10, 230 (1963).
Observation and recognition of Josephson tunneling.
Shapiro: Phys. Rev. Lett. 11, 80 (1963).
Stepped characteristic with microwave irradiation.

1964 Zimmermann and Silver: Phys. Lett. 10, 47 (1964).
Double-junction device as forerunner of fabricated interferometers.

WEAK LINKS

1958 Meissner: Phys. Rev. 109, 686 (1958).
Evidence of supercurrents through normal metals not caused by superconducting bridges.
Parmenter: Phys. Rev. 118, 1173 (1958).
Application of BCS theory to NS sandwiches.

1964 De Gennes: Rev. Mod. Phys. 36, 225 (1964).
Refinement of earlier theories of proximity effect.
Anderson and Dayem: Phys. Rev. Lett. 13, 195 (1964).
Supercurrents through narrow bridges.

REFERENCES FOR TABLE I (CONTINUED)

TYPE II SUPERCONDUCTORS

1935 Gorter: Physica 2, 449 (1935).
Difference between dirty and pure superconductors tentatively related to mean free path.

1950 Ginsburg & Landau: Zh. Eksperim. i Teor. Fiz. 20, 1064 (1950).
Phenomenological theory.

1951 Pippard: Proc. Camb. Phil. Soc. 47, 617 (1951).
Coherence length shortened by collisions, leading to negative surface energy in impure superconductors.

1957 Abrikosov: Soviet Physics J.E.T.P. 5, 1174 (1957).
Flux-lattice solution of Ginsburg-Landau equations.

1959 Gorkov: Soviet Physics J.E.T.P. 9, 1364 (1959).
Derivation of Ginsburg-Landau equations from microscopic theory.

1961 Kunzler, Buehler, Hsu and Wernick: Phys. Rev. Lett. 6, 89 (1961).
Discovery of very high critical field in Nb_3Sn.

[Goodman: see IBM Journal 6, 63 (1962).
Draws attention of Western world to Abrikosov (1957).]

FUNDAMENTAL THEORY

1950 Fröhlich: Phys. Rev. 79, 845 (1950).
Electron-phonon interaction proposed as cause of electronic phase change.
Maxwell: Phys. Rev. 78, 477 (1950).
Reynolds, Serin, Wright and Nesbitt: Phys. Rev. 78 487 (1950).)
Isotope effect strongly supporting Fröhlich's proposal.

1956 Cooper: Phys. Rev. 104, 1189 (1956).
Pairing of electron states lowers energy.
Bardeen, Cooper and Schrieffer: Phys. Rev. 108, 1175 (1956).

1958 Anderson: Phys. Rev. 112, 1900 (1958).
Pseudo-spin formulation of BCS theory.

form and tried to give some reason why mean-free-paths, when they are short, can lead to negative surface energies and the possibility of high field superconductivity. Mine was essentially a qualitative idea, in contrast to Ginsburg and Landau who, as we all know, produced a phenomenological theory in which for the first time we get the famous concept of a wavefunction ψ describing the collective state of the superconductor. The (imaginary) gradient of the wavefunction $\nabla\psi$ is responsible for the current, in accordance with normal quantum mechanical principles. The gradient of the wavefunction squared, $|\nabla\psi|^2$, enters the Ginsburg-Landau equation in a way which represents in energy terms the difficulty of confining the wave function. You can't change the spatial distribution of particles without affecting their kinetic energy.

Here I am going to touch on personal history, because I think it illustrates how physics is not always as simple as it ought to be. We all know now that the Ginsburg-Landau theory has been established microscopically. Why in 1950 did it take such a long time for the Ginsburg-Landau theory to be recognized? Why, bearing in mind the extreme lucidity and clarity of their paper, didn't everybody accept it? Why, I ask in particular, did I not accept it, since I soon came to know it perfectly well? Well, there is quite a good reason for this. It wasn't merely the dislike of someone else's ideas, though we all suffer from that. It was that in the Ginsburg-Landau theory the parameter $\varkappa$, whose size controls whether the superconductor is type I or type II, is determined by the penetration depth and by certain other parameters such as the transition temperature. The penetration depth in London theory, which the Ginsburg-Landau theory incorporates, is fixed by the number of superconducting electrons and their mass. In other words, the penetration depth is a fundamental parameter according to London. What made me skeptical, at that time, was that in the early 1950's we knew that the penetration depth was changed by scattering. When the mean-free path was made shorter, the penetration depth increased, as could be explained easily by a non-local equation. One didn't have to infer that the number of superconducting electrons changes because of scattering. But Ginsburg and Landau implied that when you alloy a superconductor, making the mean-free-path shorter, the penetration depth increases and $\varkappa$ changes because the fundamental parameters which go into the theory change. I found that quite unacceptable. I couldn't understand how the number of superconducting electrons could be altered by alloying without, shall we say, altering the transition temperature. Of course all this was cleared up satisfactorily later by Gorkov, but for some years there was a feeling that the Ginsburg-Landau theory was altogether too arbitrary. You read it now, post-1957, and see how beautifully it ties onto the B.C.S. theory. Believe me, however, it's quite different when you haven't got a theory of superconductivity and chaps come out suddenly saying, "Let's write down a wavefunction ψ which will behave just like an ordinary wavefunction behaves in quantum mechanics, except that we will fudge up the boundary conditions because it is not convenient to have normal boundary conditions ...". You can say, "Well, Ginsburg and Landau are very clever chaps and they know much more than I do about it, but it doesn't necessarily mean they are right.". And it took quite a long time for Ginsburg and Landau to be accepted. It was only, I think, when Gorkov produced from the Green's function treatment of the B.C.S. theory an explicit demonstration of how the microscopic parameters could be interpreted, that the Ginsburg-Landau theory fell into place. So in the early 1950's there was a certain amount of conflict which wasn't helped, incidentally, by the fact that Ginsburg kept on writing small papers in which he said it would be much better if we interpreted the electronic charge as not being exactly e, but e times a small numerical factor which might be as large as 2! He didn't say it was exactly 2; instead he wanted to introduce a fudge factor of (say) 1.6, and Landau kept on telling him he couldn't just put in arbitrary numbers, and muttered darkly about gauge invariance going wrong if you did. So even Landau wasn't supporting Ginsburg, and the Soviet Union seemed to be falling apart at that time.

The same thing happened when Abrikosov developed the Ginsburg-Landau theory to produce the theory of vortices in type II superconductors. As he told the story at the time he received the Simon prize, Abrikosov was discouraged by Landau who loathed the theory and would have nothing at all to do with it for some years, so that it wasn't published till 1957. In this now classic paper Abrikosov shows how vortices (flux lines) can appear in the theory, which nobody took any notice of. I don't know when the Russians began to think seriously of it, but in the West we owe a debt to Goodman. He is on my list because at the IBM Conference in 1961 he drew attention to the existence of Abrikosov's paper. He said it would explain high field superconductors and flux penetration far better than any ideas that he (Goodman) and I had been developing. So there is a gap of about 7 years between Abrikosov's solution of the equations and the idea of type II superconductivity really becoming respectable. Meanwhile Kunzler and his colleagues at Bell Labs had found that Nb_3Sn has a critical field higher than 80 kG. This discovery represents the sudden recognition of the technical importance of superconductivity, and opens up the development of high-field superconducting magnets. At this point a new story begins, which is not relevant to my theme.

I thought it worthwhile going into this matter in some detail simply because, although it was high-field superconductivity that acted as the prime stimulus to the development and application of Ginsburg-Landau, yet once Gorkov had firmly based the G-L equations, they were soon recognized as the natural way of discussing weak links and many other things. In making applications of superconductivity one hopes to be allowed to forget fundamental theory and rely on the simpler G-L phenomenological theory.

Turning to the fundamental theory, I propose to take a moment to talk a little more about the Russian connection. This is a personal recollection of an incident I found instructive. In the early summer of 1957 I went to Moscow for the first and only time. Up to then, ever since the War, there had been extremely poor contact between Russia and the Western world. This was one of the first moments when it was possible for scientists working on superconductivity in Russia and the West to talk, and we all found it extremely revealing to discover each other's misapprehensions about work on the other side. In the course of a few days spent in Moscow, I spent a lot of time arguing fiercely with Landau, Ginsburg and Lifschitz. They are excellent people to argue with; there's no one like a Russian for having a fight with - no holds barred, intellectually speaking - and we all had a good time. But I was seriously assailed by Landau particularly, who would have nothing to do with non-local electrodynamics. The London equation was good enough for superconductivity, and it was a monstrous thing to throw this beautiful London equation out and replace it by an ugly non-local equation which was quite unnecessary. I did my best to explain why it was necessary, but Landau was not a man to be easily convinced once his mind was made up. Eventually I promised that I would do an experiment to show, once and for all, that the penetration depth varied with the direction of current flow in a non-tensorial manner. This, they said, was the only thing which would persuade them that the London equations were wrong. And so we parted.

Round about the end of the same year, I was rather astonished to receive one morning in the mail a letter from the BBC Monitor Service. They listen to foreign broadcasts and hand on the information to anyone interested. They wrote to say that a few days before Professor Lifschitz had been giving a general talk on science, and had mentioned particularly the pleasure the Russian physicists had in talking with people from the West. He went on to say "... and if Dr. Pippard should happen to be listening to this broadcoast, I would like him to know that in the discussions which took place earlier in the summer, it was he who was entirely right." Now I hadn't done any of the promised experiments, so this was a surprise, and I was gratified to find that these great scientists could change their minds so readily. Of course what had happened was quite simple. Cooper's letter on electron pairs had reached Russia in the Physical Review and immediately Bogoliubov had worked out his theory of superconductivity* (Bogoliubov is not post-B.C.S., but contemporary with B.C.S., having a powerful formalism absolutely ready from his earlier work on superfluid helium). As soon as the essential idea of Cooper pairs was put to him he lost no time in developing his extremely tidy and mathematically satisfying formulation of superconductivity. It was this theory that convinced Landau and Ginsburg - they really didn't care any more about experimental evidence; what they wanted was a nice tidy theory that looked convincing, and they would then happily accept its consequences, even non-local electrodynamics. This is a lesson then on how to persuade your critics. Don't give them what they think they need, give them what you know they need (if you possibly can).

Brian Schwartz remarked in introducing me that when the B.C.S. theory came out, it was obviously correct and it seemed as though there was little more theory to be done. Now this raises an interesting historical point. He is quite right in saying that the majority of people working on superconductivity, once they realized what was in the B.C.S. paper, recognized its worth. It took a month or two, but not much more than that, for the majority to appreciate that this was, if not the right answer, so near that it really was a fundamental breakthrough. But when I say the majority, I don't mean all. I must exclude from the consensus a considerable proportion of the leading theoretical physicists of the world who disliked the theory very much! This was somewhat similar to what happened in 1950, when Fröhlich proposed a theory in which the electron-phonon interaction was responsible for superconductivity. It was supported by experimental evidence from the isotope effect discovered by Maxwell at NBS and the Rutgers group. Most of us working in the field then were convinced that the isotope effect substantiated the idea of electron-phonon interaction, but the reaction from many leading theorists was hostile. Why was there this adverse reaction to Fröhlich and to Bardeen, Cooper and Schrieffer? To some degree the criticism of Fröhlich's paper was justified; he quite correctly traced superconductivity to the electron-phonon interaction, but the detailed model he constructed is not right. But that wasn't what those who disagreed objected to. The real trouble lay in the fact that Fröhlich's electron-phonon interaction is something they had almost all thought of themselves and rejected for what seemed very good reasons. Fröhlich's criterion was that in superconductors the electron-phonon interaction is strong enough to cause electron-electron attraction. Earlier theorists who had discovered this attraction had, I think,

---

*This may not be correct. B.C.S. published a Letter in Physical Review early in 1957, and it was perhaps this, and not Cooper's Letter, that did the trick. At all events, Bogoliubov's paper was submitted to Il Nuovo Cimento before the main B.C.S. paper was published.

concluded that in these circumstances the lattice is unstable and a crystal modification occurs. Perhaps they were right in general, and superconductivity is the last resort of a metal which cannot find a better way of eliminating the embarrassing interactions between the electron and lattice. Anyone who had travelled that road and decided it led nowhere was understandably irritated by the suggestion that he had missed discovering the theory of superconductivity; and it was easy enough for the expert to discover flaws in the working out that enabled him to overlook any possible merits. I think by the time B. C. S. appeared, seven years later, the iron had entered into the soul of these eminent men. They had already been had once on the electron-phonon interaction, and when B. C. S. showed that this very mechanism which they had rejected was in fact capable of leading to the right theory, their response was distinctly ill-natured. For two or three years after the B. C. S. theory, many of the leading theoreticians were saying, "It's all very well, but the theory is not manifestly gauge invariant." I don't know quite what those words stand for, but what they really meant was that they weren't having the theory at any price. It didn't make the slightest difference, of course; everybody else had accepted it.

The more one had been involved in superconductivity, the less happily one accepted B. C. S. I confess I muttered away for six months or so after B. C. S. before I was finally convinced that muttering did no good, and it was better to join the majority. But my disinclination in no way reflected a private disappointment at having missed getting the answer first - I could never have hit on anything so clever as B. C. S. My reaction sprang from regret for the end of an era, when superconductivity as an unsolved mystery posed the sort of problem that keeps an experimental physicist happy. From now on the subject was basically different. I need not pursue the history of fundamental theory any further, but must note that my inclusion of Anderson's pseudo-spin formulation of the theory is not intended to imply that it was, in itself, a major advance; it did, however, provide the framework which inspired Josephson's first thoughts about tunneling.

Let's go on to tunneling, and approach it through the idea of an energy gap. Now we are in deep water. Who first thought there was an energy gap in superconductors? Mendelssohn tells us that this is an idea which was being bandied around at Oxford before the War,* but he and Daunt never had time to write it up

---

*Since delivering the talk I have looked at some of the pre-War literature, especially the abortive models of superconductivity developed by, among others, Slater and Welker. It is clear that the energy gap was part of the mental furniture of the time, even if its precise role was ill-defined.

until after the War, in 1946, as Table I records. Ginsburg published a book on superconductivity in 1946, in Russian of course, which seems to have passed out of recollection. I suspect the ideas in the book go back several years, before 1946. Ginsburg is quite explicit that the London equations can come out of an energy gap model, and he stands out as the one who has the most cogent physical reasoning behind why he thinks there ought to be a gap.

The gap was then lost sight of for a few years, until Goodman revived it, to interpret his measurements on thermal conductivity well below the transition temperature. He found that at very low temperatures the electronic conductivity was exponentially dependent on temperature, and suggested an energy gap. Now you might expect him to have referred back to Daunt and Mendelssohn, but he doesn't; instead, he refers back to Koppe who isn't even in Table I. This is because it is difficult to give fair credit to an incorrect theory, even when it provides helpful clues. I refer to Heisenberg's theory of superconductivity which, like so many theories in the pre-BCS days, was based on the wrong mechanism and worked out wrongly. Nevertheless, we in Cambridge had cause to remember it with gratitude, because when Heisenberg went wrong he did not cease to be enlightening. In fact, in my own experience, two things came out of Heisenberg's theory. One is that Goodman knew about it because Heisenberg had lectured in Cambridge. We had discussed his theory in great detail in Cambridge, and had not forgotten it even though we were unconvinced. Goodman knew an energy gap came out of the Heisenberg theory, as developed by Koppe, and took it over as possibly the only correct thing to emerge from this work.

The other valuable outcome that I recollect took place when, in the early 1950's, I found compelling evidence for the non-local character of the supercurrent equation (about which I had the argument with Landau). I needed to formulate the non-local theory, and remembered how Heisenberg had explained the supercurrent as arising by a curious take-over mechanism from the normal current. So if the normal current obeyed a local equation, so did the supercurrent, and that gave the London equation; but if the normal current was non-local, as in the anomalous skin effect (my own baby), so too was the supercurrent. So I wrote down what the Heisenberg theory would have given in the non-local case, and (hurray!) it fitted the experiments. Here again, although the reasoning behind Heisenberg's theory is totally wrong, it provided the incentive and the formalism which was needed. This is something which crops up whenever one traces the development of ideas; almost always they start wrong and finish up right. There can be very few creative scientists who do not remember the debt they owe to other people's bad ideas which set their mind working in a new direction. The debt is greater still when we take another's good idea and improve it in a way he could not have imagined, and the apportioning of credit can then be difficult. I should not have broached this rather awkward topic but that it plays a part in the Josephson story, and I feel moved to try to resolve some of the tensions. But this comes later - let's get back to the energy gap.

Between 1953 and 1960 the energy gap gained acceptance steadily, and the matter was clinched by Giaever's tunneling experiments. He first found tunneling between a normal metal and a superconductor, and then a few months later between superconductors. His second paper jostles that of Nicol, Shapiro and Smith in Physical Review Letters, and both of them produce evidence of zero-voltage currents. Giaever remarks on this explicitly and says it is due to a superconducting bridge. Nicol, Shapiro and Smith publish a beautiful oscillograph showing the current at zero voltage, but they do not draw attention to it. Here is another instance of researchers having the data to make an important discovery, but unable to break away from their traditional notions. We had to wait until 1962 for Brian Josephson to take a good result and make it even better.

Since we have at last reached the central theme, I'll go into a little more detail of this particular discovery. Josephson started as a research student under my direction in October 1961. The reason for this choice is worth recounting. He had had a brilliant undergraduate career, first in mathematics and then in physics. At that point he felt that his understanding of practical matters was deficient, and he would therefore do a thesis in experimental physics in order to balance up his expertise and try to compensate for what was easiest for him - the mathematical formulation of problems. But for that, he would have started as a theoretical research student under John Ziman, probably, and I don't know where he would have gone - not into superconductivity at any rate. If you enjoy arid speculation you might try to guess if, and when, the Josephson effects would have been discovered; and you might also try guessing what it is we don't know now that he might have discovered in different circumstances. But the simple fact is that he came to me to do an experimental thesis. I suggested to him that he should study the variation of penetration depth with magnetic field in superconductors, as measured by microwaves. This is a problem which still remains unsolved. Josephson studied this and obtained experimental results, but did not add anything to the theoretical understanding. He did a perfectly satisfactory thesis, ultimately, but was rather side-tracked in the middle by his independent theoretical work. I think it must have been early in 1962 that he began this, and not long afterwards came to tell me about it. In retrospect I feel no shame at not understanding him then. I don't know how many people here have read his 1962 paper in Physics Letters. If you have, read it again, and remember that the paper is the result of systematic pressure on my part and Phil Anderson's to get the ideas into an intelligible form. I still find it a very difficult paper, but the first versions were really in a class of their own, since he was a new research student with almost no experience of technical writing. I hope you will be sympathetic to my reaction, which was to tell him that I just didn't have a clue what he was talking about, and he had better go and talk to Phil Anderson, who was in Cambridge as a visitor at the time. Phil is a much more clever man than I am, of course, and he did understand, though I think he too had to work for it. A few months later John Bardeen was very doubtful about Josephson's ideas, and he was far from being alone in this. I think this shows how hard it is for someone with a new idea to visualize what other peoples' difficulties will be in accepting it. It was so clear to Josephson in his particular way of looking at it, that he could not understand why, for example, I should fail to see the point. He remarks in his Nobel lecture, "In 1961 Pippard had considered the possibility that a Cooper pair could tunnel through an insulating barrier such as that which Giaever used, but argued that the probability of two electrons tunneling simultaneously would be very small so that any effects would be unobservable. This plausible argument is now known not to be valid. However, in view of it, I turned my attention to a different possibility, that the normal currents might be affected by phase difference." Let's look at this point. If a single electron only has a probability of 1 in 10^{10} of getting through, the probability for two electrons to tunnel simultaneously is 1 in 10^{20} - as near impossible as no matter. It is quite easy therefore to conclude that no current flows in pairs. I put this argument to Brian Josephson early on, and I am quite sure that this criticism of his theory was in the minds of almost everybody who was thinking about pair tunneling. I asked, "How can the pairs get through?" and he explained that the wave-functions of the electrons in the pair are phase-coherent, so that you have to add the amplitudes before you square. The electrons do not tunnel independently, but more like a single particle, and the probability of a pair going through is comparable to the probability for a single electron. It is like interference in optics with phase-coherent waves mixing, and if he had only said that in his early publication, he would have had no difficulty in carrying the world with him. But can you find it said in his 1962 paper? Of course you can't; because he didn't see the existence of a difficulty. So for some months a lot of people (especially those who had been thinking about tunneling before) thought he might well be talking nonsense because his orders of magnitude for the effect were completely astray.

While I am trying to correct the record, let me make reference to Phil Anderson's account of the early days of the Josephson paper. In one paragraph he manages to create two obscurities, one important and one not important. He says that "we were all - Josephson, Pippard and myself as well as various other people who habitually sat at the Mond tea table and participated in the discussions of the next few weeks - very much puzzled by the meaning of the fact that the current depends on the phase (this is the famous formula $J = J_1 \sin \phi$ relation). I think," he continues, "that it was residual uneasiness on this score that caused the two Brians (Pippard and Josephson) to decide to send the paper to Physics Letters, which was just then starting publication, rather than to Physical Review Letters." I disregard the insult that I don't mind publishing wrong papers, if they are published obscurely; in fact the reason for the choice of journal is quite simple - Physical Review Letters has a page charge which is charged in dollars. At that time exchange regulations in England made it very difficult to get dollars. Therefore it was not published in Physical Review Letters, but in Volume 1 of the new European journal, Physics Letters. This is a trivial point; more important is Anderson's statement about the famous $J = J_1 \sin \phi$ relation. Here I have some difficulty because my recollection on this is quite at variance with what Anderson says. The interesting thing is that $J = J_1 \sin \phi$ is not to be found in Josephson's original paper, nor is it to be found in the fellowship thesis which he wrote later in the year. I believe that the equation was derived by Phil Anderson either at the time of these discussions or within the following months; he quotes it openly in his lectures given in the spring of 1963. There is no question but that Phil has been generous to a fault in what he has written about Josephson. He did an enormous amount to help Josephson get his ideas across, and some of the suggestions which are associated with Josephson are really Phil's. Although Josephson may have been aware, at the back of his mind, of the simple sine relationship, I believe it was Anderson who recognized its importance as a statement of the tunneling law, and also the related expression for the coupling energy of two superconductors as a cosine of the phase. I cannot find these formulae in anything Josephson wrote at that time. By 1963, though, in his article for the Colgate Conference, he quotes the sine relation in a way that leaves one uncertain whether he means it is actually in his original paper or implicit in it (which it certainly is). I'm sure the ambiguity is unintentional. This is not the only example of undue modesty on Anderson's part; he deserves great credit for recognizing two obstacles in the way of detecting Josephson tunneling: (1) The deleterious effect of stray magnetic fields such as the Earth's, and (2) much more subtle, room temperature noise getting down the leads. When Phil suggested they should be eliminated, John Rowell was able to exhibit a clear demonstration of Josephson tunneling.

I must try to make a few remarks about Josephson's own approach to the problem. I can't say much of value because, as I have already indicated, his habit of thought was (and remains) alien to mine. But I can paraphrase his own account. Anderson is on the scene from the start, but his role in this case is unconscious. Being on leave from Bell Labs and spending a year in Cambridge, he gave a course of lectures on solid state physics. At the end he devoted some time to the question of broken symmetry which was intriguing him at that time. Let me explain broken symmetry briefly. The lattice of a crystal has certain symmetry properties, say cubic symmetry, described by a set of rotations that transform it into itself. It is not isotropic, for any other rotation leaves it looking different. However, if you write down the Hamiltonian describing the particles that make up the block of solid, the Hamiltonian itself contains only isotropic central forces. There are no symmetry elements corresponding to the symmetry of the solid crystal which results as the ground state solution. If the particles had been non-interacting, the ground state would have been isotropic, the lowest state of a quantum gas. As you start increasing the interaction between the particles there will come a time when there will precipitate out, as the lowest state, a new ground state which contains certain symmetry properties not present in the original Hamiltonian - a gas-to-solid phase transition has occurred. In a sense the ground state is still isotropic, for the solid lattice can be oriented in any direction, and an isotropic wave-function can be constructed as the superposition of all the degenerate states describing orientations. But this is physically unrealistic - what we observe is one particular orientation at a time.

Now this phenomenon, which is widespread and not confined to solid lattices, has excited the interest of many physicists, including Anderson, and has been a source of inspiration for general ideas about how physics works. He was talking about it in his lectures and used his pseudo-spin formulation of B. C. S. theory to show how the phase transition into the superconducting state could also be seen as a symmetry transition. This intrigued Josephson very much because when a symmetry is changed a new parameter enters, which in general we may refer to as the order-parameter. In superconductors the energy gap plays the role of order-parameter, and as Josephson pondered this in the light of Anderson's general discussion, he recognized that the energy gap alone was an incomplete specification - a phase must be associated with it; moreover (and all this is still consistent with general broken-symmetry theory) this phase was a variable which was non-commutative with respect to the number of electrons in the sample. The Heisenberg uncertainty principle applies to phase and number - if you know exactly how many electrons you have in the superconductor you can say nothing about the phase of the wavefunction. This is the case for two pieces of superconductor quite separate from one another so that you can count the number of particles in each. On the other hand, Josephson saw if you take a single piece of superconductor and imagine dividing it into two, it is easy to transfer particles from one half to the other and correspondingly, you can have complete knowledge of the phase. Therefore you have a superconducting junction between one half and the other, because the Ginsburg-Landau equations tell one how the supercurrent is related to phase. Thus Josephson recognized early on that the uncertainty relation of phase and number distinguished clearly between superconductors which were separate and superconductors which were joined together. He then asked himself what happened in the intermediate case when they are weakly joined together. I find it particularly interesting to see the weak link appearing as the primary concept, with the oxide barrier coming later as a way of realizing it, and not the other way round - a theory of oxide barriers later being generalized to other sorts of weak link. Josephson recognized that if it was possible to transfer electrons, though with difficulty, from one superconductor to another; you could know something about the number and something about the phase. And since electron number is closely related to the current between the two, and so also is phase, through the G-L theory, he was on the way to describing the properties of a weak link. But, as he points out in the passage I quoted, he was also convinced by the argument against pair tunneling.

At precisely the right moment Cohen, Falicov and Phillips provided the formalism he needed, with their tunneling Hamiltonian, and first he looked at it to see what effect a knowledge of phase would have on the normal current. Not finding what he knew he needed there, he concentrated on the supercurrent and found pair tunneling as one element of the solution. It may be worth recording that a complete analysis of the tunneling Hamiltonian is an extremely formidable undertaking; Cohen, Falicov and Phillips tackled the simpler case of superconductor-normal tunneling, but were daunted by the superconductor-superconductor case. The thorough way they had approached the formulation of the problem prevented them in the end from seeing the wood for the trees. Josephson's advantage lay in his having started from a very general conception of what he wanted, so that he could seize on those features of the tunneling theory that appealed to his physical sensibilities. This is why I lay stress on the idea of weak links rather than tunneling.

Now weak links, though not so named, had been around for some time. In 1958 Meissner published a lot of careful work on the junction of two superconductors with a thin normal film between them. He provided a clear demonstration of the proximity effect (the normal metal between two superconductors carrying pairs and enabling supercurrents to pass from one superconductor to another). Two years before Giaever, and four before Josephson, then, he was talking about superconductors being joined together by weak non-superconducting links. But his conception of what was going on in the weak links is not radical like Josephson's; nor is that of Parmenter who, following Meissner's results, wrote a very solid and systematic account of the proximity effect based on the B. C. S. theory. There were many of us at that time wondering, more or less deeply, about the proximity effect, and we agreed that it was perfectly possible for this superconductor to infect that normal metal with superconducting pairs so that the supercurrent could pass from one side to the other. But our thought was strictly limited to the idea of the normal metal becoming a sort of dilute superconductor, passing current according to G-L, that is, proportional to the gradient of the wavefunction just as in a normal superconductor. It never entered any of our minds that the current could be a periodic function of the phase difference, sin ϕ; the essential idea of a weak link was missing. Since Josephson, we recognize the weak link as one in which the behavior of the electrons between the two superconductors is determined by what is going on in the superconductors themselves, and not by the local state of affairs at each point between. And the difference is crucial - it permits the current to be periodic in the phase difference - but it took several years for many of us to see just how significant the difference is in practical terms.

I could really stop here, but perhaps should make a comment on the last entries in Table I, marking the beginning of the device era. I have put Zimmerman and Silver here as the key contributors and that may not be fair. I know other people, Mercereau and others, whose names should also be mentioned. But Zimmerman and Silver published the first description I can find of a device with two Josephson junctions separated by a macroscopic gap. What they did was to take a strip and a V-shaped wire, both of Nb, and they bent the wire over the strip so that it contacted the two sides and left a gap in between. They then showed interference fringes in the critical current as a function of magnetic flux through the loop. I think this is the progenitor of the macroscopic devices based on the Josephson effect.

Well, to sum up, I have tried to show some of the windings in the interconnected research paths leading to the Josephson effects, to bring out the mixture of logicality and illogicality, inspiration and desperation, and how we get things wrong. And of course this is still the way things are done. John Clarke will tell you that his discovery of the Slug was the result of imperfectly formulated ideas providing the inspiration for an elegant and useful end-product, and almost everyone who believes himself to have made a significant discovery will admit that the first inklings had their origin in something read or heard, and very likely misunderstood. The final result is none the worse because it was reached by stumbling - it is only our pride which is hurt when we fail to measure up to that perfection of progress that the great men of the past always seemed to achieve. Or did they? If we wish to boast of our achievements, let us not point to the unerring pursuit of truth by a logically faultless thinking-machine, but to the even more astonishing way in which truth can be caused to emerge from the toils of error and stupidity.

Chapter 2

MACROSCOPIC QUANTUM PHENOMENA IN SUPERCONDUCTORS

R. de Bruyn Ouboter

Kamerlingh Onnes Laboratory

Leiden, the Netherlands

I. INTRODUCTION – dc QUANTUM EFFECTS

A. Meissner Effect and Flux Quantization

The essential feature of superfluidity (superconductivity) is according to F. London [1] a condensation of a macroscopic number of particles (bound-electron pairs, first described by Cooper [2]) in the same single quasi-particle quantum state. Such a condensation can be described, as usual, by an internal order parameter. According to the phenomenological theory of Ginzburg and Landau [3] the order parameter Ψ_s is a complex quantity, with properties similar to those of the wave function of the macroscopically occupied single quasi-particle quantum state. This complex order parameter Ψ_s has an amplitude $\Psi_o(\vec{r}, t, T)$ and a phase $\phi(\vec{r}, t)$

$$\Psi_s = \Psi_o(\vec{r}, t, T)\, e^{i\phi(\vec{r}, t)} \quad . \tag{1}$$

In the Ginzburg-Landau theory $|\Psi_s|^2$ is interpreted as the internal order parameter of the original two fluid model of Gorter and Casimir [4]. The charge density $\rho_s(T)$ of the superfluid particles at a given temperature T is equal to the square of the amplitude $\rho_s = |\Psi_s|^2$ and hence,

$$\Psi_s = \sqrt{\rho_s(\vec{r}, t, T)}\; e^{i\phi(\vec{r}, t)} \quad , \tag{2}$$

in which $\rho_s \to 0$ if $T \to T_c$. The absolute phase is not observable, but we will show that phase differences are directly observable quantities. In the Ginzburg-Landau theory, the current density $\vec{I}_s$ is related to the probability current density of quantum mechanics in terms of the wave function Ψ_s of this single quasi-particle quantum state,

$$\vec{I}_s = \frac{-i\hbar}{(2(2m))} (\Psi_s^* \vec{\nabla} \Psi_s - \Psi_s \vec{\nabla} \Psi_s^*) - \frac{(2e)}{(2m)} |\Psi_s|^2 \vec{A}$$

$$= |\Psi_s|^2 \frac{\hbar}{2m} [\vec{\nabla} \phi - \frac{2e}{\hbar} \vec{A}] = |\Psi_s|^2 \vec{v}_s = \rho_s \vec{v}_s \quad . \quad (3)$$

The charge (2e) which enters in this expression is equal to twice the charge of a free electron, since according to the microscopic theory of Bardeen, Cooper and Schrieffer [2] bound electron pairs are involved in the ordering process. The mass (2m) is very nearly equal to twice the mass of a free electron. Furthermore, Ginzburg and Landau [3] derived a differential equation for Ψ_s by using an expansion of the free energy in powers of Ψ_s and $\vec{\nabla}\Psi_s$ analogous to the time independent Schrödinger equation, but modified with a nonlinear term,

$$\xi^2(T) (\vec{\nabla} - \frac{2ie}{\hbar} \vec{A})^2 \Psi_s + (1 - \eta|\Psi_s|^2) \Psi_s = 0, \quad (4)$$

in which $\xi(T)$ is the characteristic temperature-dependent coherence length. From Eq. (3) one obtains the fundamental relation for the generalized dynamical momentum $\vec{P}_s$ of the superconducting pairs,

$$\vec{P}_s \equiv (2m)\vec{v}_s + (2e)\vec{A} = \hbar\vec{\nabla}\phi \quad . \quad (5)$$

Taking the curl of Eq. (5) the London relation [5] is obtained,

$$\vec{\nabla} \times \vec{P}_s = 0 \quad \text{or:} \; \vec{\nabla} \times \vec{v}_s = - \frac{(2e)}{(2m)} \vec{\nabla} \times \vec{A} = - \frac{e}{m} \vec{B} \quad . \quad (6)$$

Combining Eq. (6) with the Maxwell equations: $\varepsilon_o c^2 \vec{\nabla} \times \vec{B} = \vec{I}_s = \rho_s \vec{v}_s$ and $\vec{\nabla} \cdot \vec{B} = 0$ gives the following relations:

$$\vec{\nabla}^2 \vec{B} = \vec{B}/\lambda^2 \quad \text{and} \quad \vec{\nabla}^2 \vec{I}_s = \vec{I}_s/\lambda^2 \quad (7)$$

in which $\lambda = (\varepsilon_0 mc^2/\rho_s e)^{1/2}$ is the London-Ginzburg-Landau penetration depth, explaining the Meissner-Ochsenfeld [6] effect. The magnetic field (the magnetic induction) $\vec{B}$ vanishes completely in the bulk of a superconductor ($\vec{B} = 0$). Even if the magnetic field was already within the metal before cooling through the critical temperature, the magnetic field is expelled from the interior of the superconductor below the critical temperature.

We should like to remark that the Eqs. (3) and (5) are gauge invariant under the following transformations for the vector potential $\vec{A}$, the scalar potential φ, the phase ϕ and the generalized dynamical momentum $\vec{P}_s$

$$\vec{A}' = \vec{A} + \vec{\nabla}\chi ; \; \varphi' = \varphi - \frac{\partial\chi}{\partial t} ; \; \phi' = \phi + \frac{2e}{\hbar}\chi ; \; \vec{P}_s' = \vec{P}_s + 2e\vec{\nabla}\chi \tag{8}$$

and we introduce the gauge invariant phase ϕ^* defined by

$$\vec{\nabla}\phi^* \equiv \vec{\nabla}\phi - \frac{2e}{\hbar}\vec{A} \quad . \tag{9}$$

If one considers a multiply-connected superconductive region instead of a simply-connected one, for example a hollow superconducting cylinder, one finds that although everywhere in the superconductor $\vec{\nabla} \times \vec{P}_s = 0$, the circulation of the generalized dynamical momentum (and hence the generalized angular momentum) around the hole of the cylinder is quantized, $\oint \vec{P}_s \cdot d\vec{s} = nh$, with n an integer (Bohr-Sommerfeld quantum condition). This is a consequence of the fact that for each point in the superconductor there can be only one value of the wave function Ψ_s. Thus the phase ϕ cannot change arbitrarily in the superconductor. If one adds the phase changes in a closed loop around the cylinder, the wave function must stay single valued. No matter how the phase ϕ changes as one goes around the cylinder the phase must return to the same value for the wave function Ψ_s when one returns to the starting point. Hence $\oint_s d\phi = 2\pi n$.

Inside the bulk of a thick-walled superconducting cylinder $\vec{v}_s = 0$, hence Eq. (5) gives,

$$\vec{P}_s = 2e\vec{A} = \hbar\vec{\nabla}\phi \tag{10}$$

and the magnetic flux Φ_m enclosed by the hole with area a is equal to:

$$\Phi_m = \iint_a B_n da = \oint_a A_s ds = \frac{1}{2e}\oint_s P_s ds = \frac{h}{2e}\oint_s d\phi = n\frac{h}{2e} \tag{11}$$

In 1948 F. London [1] made the prediction above that the magnetic flux Φ_m enclosed in a superconducting cylinder should be quantized. This has been verified by Deaver and Fairbank [7] and by Doll and Näbauer [8] in 1961 and Φ_m is found to be quantized in units $h/2e$.

For a closed path in a region of non-zero current ($\vec{v}_s \neq 0$) one has fluxoid quantization,

$$\frac{1}{2e}\oint_s P_s ds = \frac{m}{e}\oint_s v_s ds + \oint_s A_s ds = \frac{m}{e}\oint_s v_s ds + \Phi_m = n\frac{h}{2e} \quad . \tag{12}$$

One starts with the material in the normal state. A magnetic flux is applied through the hole of the cylinder with area O equal to $B_\perp O$. Subsequently the cylinder is cooled into the superconducting state. A persistent circulating current i_{circ} is induced in such a way that the total magnetic flux enclosed by the cylinder is quantized,

$$\Phi_m = B_\perp O + L\, i_{circ} = n \frac{h}{2e} \quad , \tag{13}$$

in which L is the self-inductance of the cylinder. When the applied magnetic flux is equal to an integer number of magnetic flux quanta, $B_\perp O = n\frac{h}{2e}$, the circulating current is equal to zero. When the applied magnetic flux is equal to a half-integer number of flux quanta, $B_\perp O = (n+\frac{1}{2})\frac{h}{2e}$, the circulating current is maximum. At these values of flux, the quantum state changes from the state with n quanta to the state with (n + 1) quanta. A schematic plot of the total enclosed magnetic flux Φ_m through the area O and the self-induced flux $\Phi_s = L i_{circ}$ as a function of the applied flux $B_\perp O$ is given in Fig. 1 for the case that the cylinder is cooled through its transition temperature in an applied magnetic field $B_\perp$ (see also References [9-13]).

We investigate the problem of the dependence of the critical temperature T_c of a very thin-walled superconducting cylinder on the applied magnetic flux $\Phi_a = B_\perp O$. In this case no distinction need by made between Φ_a and Φ_m since at $T_c(B_\perp)$ the superfluid density $\rho_s = |\Psi_s|^2$ goes to zero and the supercurrent $I_s = \rho_s v_s = 0$. If the cylinder is cooled in an applied magnetic field from the normal to the superconductive state it passes into that quantum state with number n which gives the lowest Gibbs free energy. This quantum state is conserved during the further cooling process even if the external field is switched off.

From Eq. (12) one can solve v_s for a circular hole of inner radius r. In the penetration depth region with non-zero current, the velocity v_s is equal to

$$v_s = (\oint_s v_s ds)/2\pi r = (\frac{h}{2m})(n - \Phi_a/\frac{h}{2e})/2\pi r \quad . \tag{14}$$

The kinetic-energy density can be calculated as

$$\frac{1}{2}\left(\frac{n_s}{2}\right)(2m)\, v_s^2 = \frac{1}{2} n_s m v_s^2 = (n_s h^2/32\pi^2 m r^2)(n - \Phi_a/\frac{h}{2e})^2 \quad , \tag{15}$$

and is shown in Fig. 2. For a superconducting ring, cooled down in a magnetic field, the kinetic-energy term and the magnetic energy term $\frac{1}{2}L(i_{circ})^2$ are the most important contribution to the Gibbs free energy. The resulting energy has a quasi-periodic dependence upon the applied magnetic flux, since the quantum number n switches in such a way that

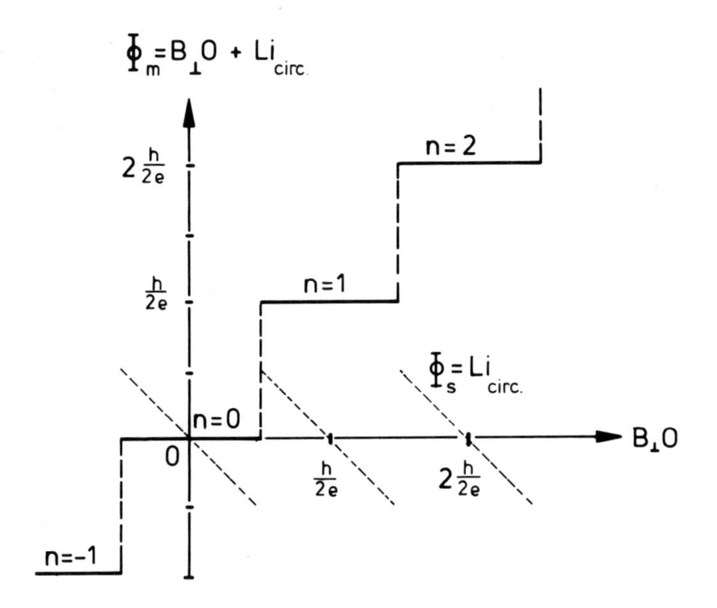

Fig. 1 The total flux Φ_m in the area O as a function of the applied flux $B_{\perp}O$.

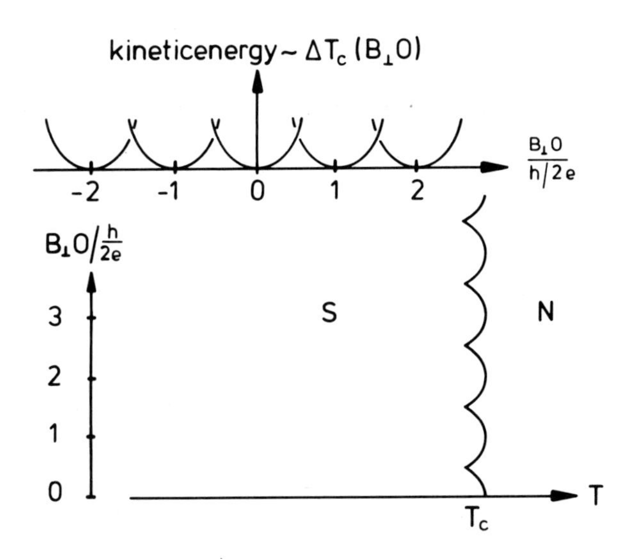

Fig. 2 The threshold curve of a superconducting cylinder.

the kinetic and magnetic energy is minimized. The transition temperature is determined by the difference in the Gibbs free-energy between the superconductive and the normal state and should therefore be periodic in the applied magnetic field. Little and Parks [14] have shown that the transition temperature of a thin ring is indeed periodic in the applied magnetic field (see Fig. 2).

B. The dc Josephson Effect

Macroscopic quantum "interference" effects are observed when two superconductors are weakly coupled, either by (a) a very thin oxide layer, (b) a superconducting point contact, (c) a narrow constriction or (d) a super-normal-super sandwich junction. Josephson [15] predicted that a dc superconducting tunnel current of limited magnitude can flow through such a junction in the absence of a voltage difference (V = 0) between the two coupled superconductors, Fig. 3. A simple explanation of the dc Josephson effect follows from Eq. (5), $\vec{v}_s = (\hbar/2m)\vec{\nabla}\phi - (e/m)\vec{A}$. We see that a supercurrent requires a phase difference. We note that at any point the vector potential $\vec{A}$ can be made to vanish by an appropriate choice of gauge. In fact the most general expression for the supercurrent at the junction must depend on the quantities which might change across it, the phase, the voltage, and the temperature. Hence to lowest order $v_s = \alpha f(\Delta\phi) + \alpha_1 \Delta T + \alpha_2 \Delta V$. The dependence of v_s on temperature or voltage is a dissipative correction. To lowest order we

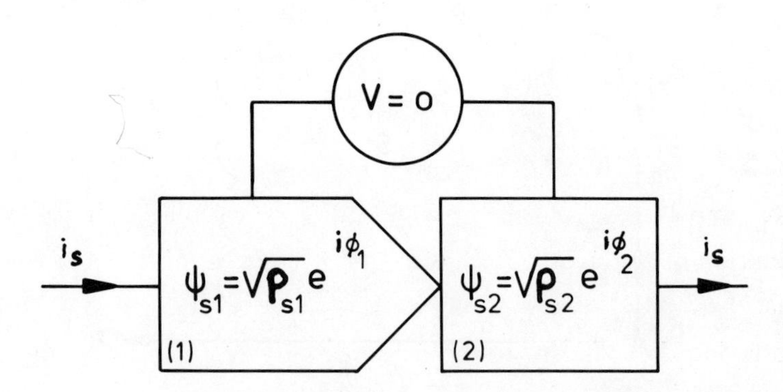

Fig. 3 A schematic diagram of a single junction.

can ignore the corrections. If the link is very weak we suppose the supercurrent in the link is proportional to a function of the phase difference,

$$i_s = i_1 f(\Delta\phi) \quad . \tag{16}$$

Eq. (5) is useful for a long wire, in which the phase $\phi(x)$ can be followed as a continuous variable of x along the wire. It is not satisfactory for a localized weak link in which only the total phase difference $\Delta\phi$ can be defined up to differences of $2\pi n$. Thus the function $f(\Delta\phi)$ must be a periodic function of 2π,

$$f(\Delta\phi) = f(\Delta\phi + n2\pi) \quad . \tag{17}$$

Any function satisfying Eq. (17) can be expanded as a series of harmonics of $\sin n\Delta\phi$ and $\cos n\Delta\phi$. If we consider only the $n = 1$ term we obtain the Josephson relation for the supercurrent through the junction,

$$i_s = i_i \sin \Delta\phi \quad . \tag{18}$$

This result can be derived using the equations in Fig. 4. The wave functions on both sides of the barrier are $\Psi_1 - \sqrt{\rho_s}\, e^{i\phi_1}$ and $\Psi_2 = \sqrt{\rho_s}\, e^{i\phi_2}$. Inside the insulating or normal barrier of thickness 2a, the wave functions Ψ_1 and Ψ_2 decay exponentially. The total wave function $\Psi(x)$ in the barrier is given by

$$\Psi_s(x) = \sqrt{\rho_s}\left[e^{i\phi_1} e^{-\frac{x+a}{\lambda}} + e^{i\phi_2} e^{\frac{x-a}{\lambda}}\right] = \Psi_o(x)\, e^{i\phi(x)} \quad . \tag{19}$$

in which $\Psi_o(x)$ and $\phi(x)$ are the real values for the amplitude and phase in the barrier and λ is a characteristic length of the barrier. From Eq. (3) it follows directly that the current density is equal to

$$I_s = \frac{-i\hbar}{2(2m)}[\Psi_s^* \nabla \Psi_s - \Psi_s \nabla \Psi_s^*] = \Psi_o^2(x)\frac{\hbar}{2m}\nabla\phi(x) = I_1 \sin \Delta\phi \tag{20}$$

in which $I_1 \equiv \frac{\hbar\rho_s}{m\lambda} e^{-\frac{2a}{\lambda}}$ and $\Delta\phi \equiv \phi_2 - \phi_1$. If we take into account the contribution of the vector potential A, it follows quite generally from Eq. (3) that $I_s = \rho_s \frac{\hbar}{2m}[\nabla\phi - \frac{2e}{\hbar}A] = \rho_s \frac{\hbar}{2m}\nabla\phi^*$. The gauge invariant phase difference $\Delta\phi^*$ enters in the dc Josephson relation Eq. (20),

$$i_s = i_1 \sin \Delta\phi^* \quad . \tag{21}$$

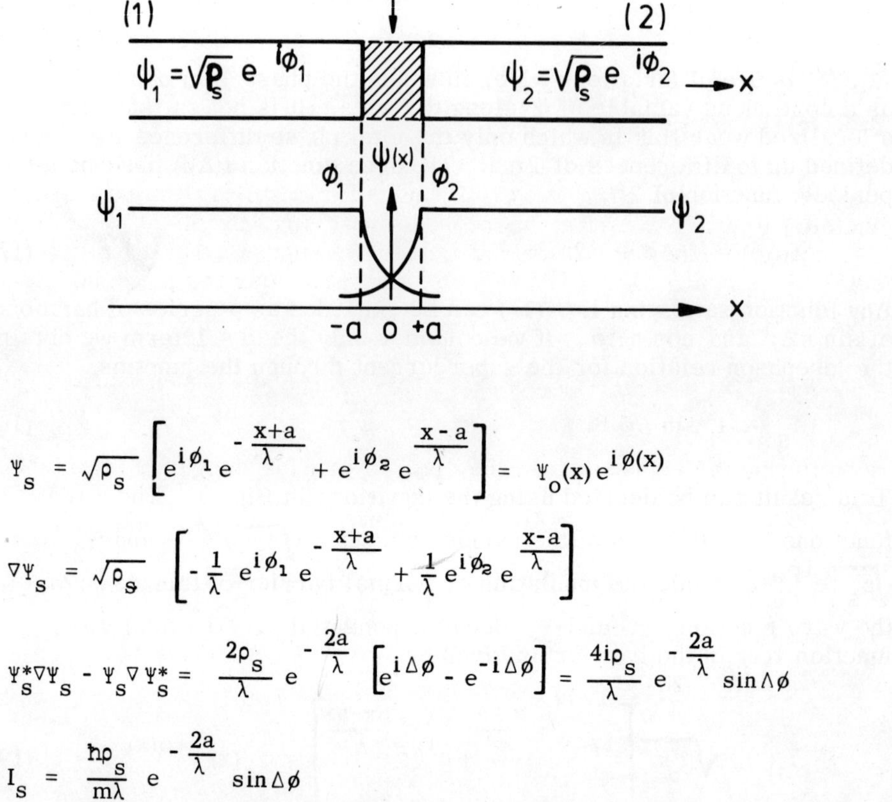

$$\Psi_s = \sqrt{\rho_s}\left[e^{i\phi_1} e^{-\frac{x+a}{\lambda}} + e^{i\phi_2} e^{\frac{x-a}{\lambda}}\right] = \Psi_o(x)\, e^{i\phi(x)}$$

$$\nabla\Psi_s = \sqrt{\rho_s}\left[-\frac{1}{\lambda} e^{i\phi_1} e^{-\frac{x+a}{\lambda}} + \frac{1}{\lambda} e^{i\phi_2} e^{\frac{x-a}{\lambda}}\right]$$

$$\Psi_s^*\nabla\Psi_s - \Psi_s\nabla\Psi_s^* = \frac{2\rho_s}{\lambda} e^{-\frac{2a}{\lambda}}\left[e^{i\Delta\phi} - e^{-i\Delta\phi}\right] = \frac{4i\rho_s}{\lambda} e^{-\frac{2a}{\lambda}} \sin\Delta\phi$$

$$I_s = \frac{\hbar\rho_s}{m\lambda} e^{-\frac{2a}{\lambda}} \sin\Delta\phi$$

$$\Psi_s(x)\Psi_s^*(x) = \rho_s\left[e^{-2\frac{x+a}{\lambda}} + e^{2\frac{x-a}{\lambda}} + e^{-\frac{2a}{\lambda}}\, 2\cos\Delta\phi\right]$$

$$|\Psi_s(x=0, \Delta\phi=0)|^2 = 2\rho_s e^{-\frac{2a}{\lambda}}(1+\cos\Delta\phi) = |\Psi_s(x=0, \Delta\phi=0)|^2\ \frac{(1+\cos\Delta\phi)}{2}$$

$$|\Psi_s\ (x=0,\ \Delta\phi=\tfrac{\pi}{2})|^2\ /\ |\Psi_s(x=0,\ \Delta\phi=0)|^2 = 1/2$$

Fig. 4 Derivation of the current-phase relation.

The supercurrent i_s flowing through the junction reaches its maximal (critical) value i_1 when the phase difference $\Delta\phi^* = \pi/2$. The dc Josephson effect was demonstrated experimentally for the first time by Anderson and Rowell [16]. If the weak contact area has a finite extension the supercurrent maximum as a function of applied magnetic field displays a diffraction pattern [17]. For a loop with two weak links, the applied magnetic field leads to interference effects [18]. The behavior of the supercurrent in Josephson junctions and weak links has been discussed in References [19-24].

As an example of the effect of an applied magnetic field consider a single superconducting weak link shunted in parallel with a superconductor. Figure 5 shows a superconducting ring containing a weak link. The phase change $\Delta\phi_L$ measured around the superconducting loop with self-inductance L, must be the same as that across the junction, $\Delta\phi_J$. From Eq. (10) we obtain

$$\phi_2 - \phi_1 = \Delta\phi_J = \Delta\phi_L + n2\pi = \frac{2e}{\hbar}\int_1^2 {}_L A_s ds + 2\pi n \quad . \tag{22}$$

Substituting the expression in Eq. (22) in that for the gauge invariant phase difference across the junction gives

$$\begin{aligned}\Delta\phi_J^* &= \Delta\phi_J - \frac{2e}{\hbar}\int_1^2 {}_J A_s ds = -\frac{2e}{\hbar}\left\{\int_1^2 {}_J A_s ds + \int_2^1 {}_L A_s ds\right\} + 2\pi n \\ &= -\frac{2e}{\hbar}\oint A_s ds + 2\pi n = -\frac{2e}{\hbar}\Phi_m + 2\pi n \\ &= -2\pi\Phi_m / \left(\frac{h}{2e}\right) + 2\pi n \quad .\end{aligned} \tag{23}$$

The total magnetic flux Φ_m through the hole is equal to

$$\begin{aligned}\Phi_m &= B_\perp O + L i_{circ} = B_\perp O + L i_1 \sin\Delta\phi_J^* \\ &= B_\perp O - L i_1 \sin\left(2\pi\Phi_m / \left(\frac{h}{2e}\right)\right) .\end{aligned} \tag{24}$$

Figure 6 illustrates for two L values that Φ_m oscillates periodically around the straight line $\Phi_m = B_\perp O$ with a period equal to one flux quantum. Φ_m is equal to $B_\perp O$ for values of Φ_m equal to 0, $\frac{1}{2}\left(\frac{h}{2e}\right)$, $\left(\frac{h}{2e}\right)$, $\frac{3}{2}\left(\frac{h}{2e}\right)$, ... independent of the value L since no circulating current is present at these values of Φ_m.

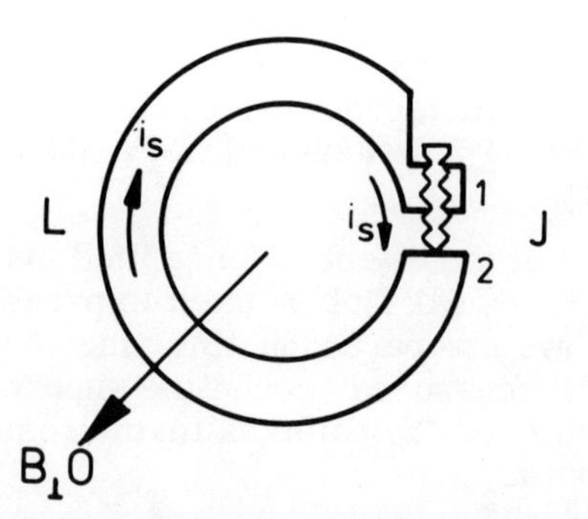

Fig. 5 A superconducting ring (self inductance L) containing a weak junction J.

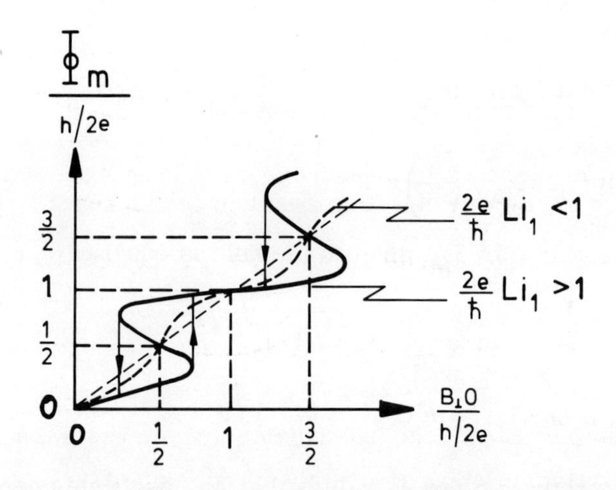

Fig. 6 The total enclosed magnetic flux Φ_m versus the applied magnetic field $B_\perp$ for situations in which $(2e/\hbar)Li_1$ is <1 and >1.

If we take the derivative of Φ_m with respect to $B_\perp O$ we obtain

$$\frac{d\Phi_m}{dB_\perp O} = \frac{1}{1 + 2\pi\left[Li_1/\frac{h}{2e}\right]\cos\left[2\pi\Phi_m/\frac{h}{2e}\right]} \quad . \tag{25}$$

$\frac{d\Phi_m}{dB_\perp O}$ is positive definite for all values of $B_\perp O$ for all values of inductance L such that

$$2\pi(Li_1/\frac{h}{2e}) = \frac{2e}{\hbar}Li_1 < 1 \quad . \tag{26}$$

If $\frac{2e}{\hbar}Li_1 > 1$, $\frac{d\Phi_m}{dB_\perp O}$ is not a single valued function. In the limit of a very weak superconducting junction $\frac{2e}{\hbar}Li_1 << 1$, it follow from Eq. (25) that the wavy line in Fig. 6 for Φ_m approaches the straight line $\Phi_m = B_\perp O$. In such a situation the external flux penetrates the hole almost completely (very incomplete flux quantization). In the limit of very strongly superconducting junction $\frac{2e}{\hbar}Li_1 >> 1$, the system behaves as a bulk superconductor. The external flux does not penetrate the hole at all. Compare Fig. 6 with Fig. 1. For intermediate situations when $\frac{2e}{\hbar}Li_1 \gtrsim 1$, the system becomes unstable at points where $d\Phi_m/dB_\perp O = \infty$. The flux Φ_m jumps discontinuously by changing from one flux quantum state to the next state. This process is indicated in Fig. 6 by an arrow.

An important application, the periodicity of the flux in superconducting loops with Josephson junctions, was developed by Zimmerman and Silver [25, 26] in order to measure small changes in external magnetic flux (ac SQUID, Superconducting Quantum Interference Device). In a dynamic situation such a device is coupled to a resonant circuit and the voltage across the tuned circuit is amplified and detected. Zimmerman and Silver observed that in the regime $2\pi(Li_1/\frac{h}{2e}) > 1$ the quantum states are discrete, and the transitions between states are well defined and irreversible. If $2\pi(Li_1/\frac{h}{2e}) \gtrsim 1$ as shown in Fig. 6, the transitions generally occur only between adjacent states ($\Delta n = \pm 1$). If $2\pi(Li_1/\frac{h}{2e}) >> 1$ then multiple quantum jumps can be observed. Whereas if $2\pi(Li_1/\frac{h}{2e}) < 1$ the quantum states merge into one another continuously and reversibly.

C. The Critical Current Through a Double Point Contact as a Function of the Applied Magnetic Field

The magnetic behavior of a symmetrical double point contact, Fig. 7, with zero applied current, should be different when compared with a single superconducting point contact shunted in parallel with a superconducting ring, Fig. 5. The total magnetic flux through the hole is now equal to $\Phi_m = B_\perp O \pm Li_1 \sin(\pi \Phi_m/(h/2e))$. By taking into account the contributions to the free energy of the barrier, $-2\,(\frac{h}{2e}\, i_1 \cos \Delta\phi^*)$ and the magnetic energy one finds that even for $\pi L i_1/(\frac{h}{2e}) < 1$, the system should jump into the adjacent most stable state at $\Phi_m = (n + \frac{1}{2})\frac{h}{2e}$.

The critical current $i_c(B_\perp)$ for a double point contact in a superconducting loop is defined as the maximum current that can flow through the point contacts without any voltage appearing across the junctions. The critical current i_c is a function of the magnetic field $B_\perp$ applied perpendicular to the enclosed area O between the contacts. This is a consequence of the fact that the phase differences across both weak contacts are related to the enclosed magnetic flux in the following way,

$$\oint d\phi = 2\pi n = 0 = \Delta\phi_a - \Delta\phi_b + \frac{2e}{\hbar}\oint{}^{*} A_s ds \quad . \tag{27}$$

$\oint^*$ is the line integral with the junctions excluded. $\Delta\phi_a$ and $\Delta\phi_b$ are the phase differences across the junctions a and b: $\Delta\phi_a = (\phi_2 - \phi_1)_a$ and $\Delta\phi_b = (\phi_2 - \phi_1)_b$. The

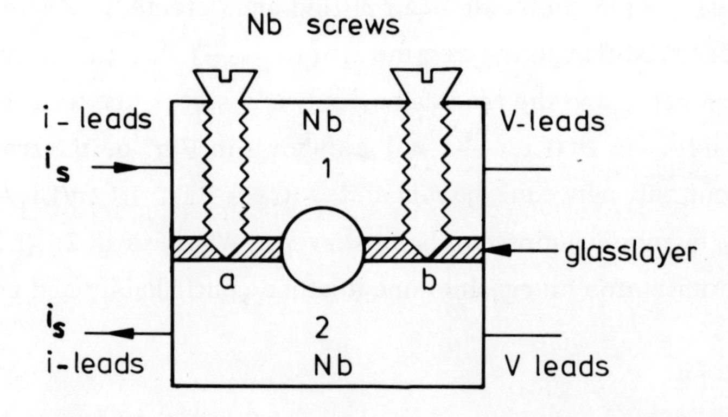

Fig. 7 Cross section of a double point contact.

phase differences inside the bulk superconductors (1) and (2) are found by integrating Eq. (10) around the hole excluding the junctions. The wave function must be single valued, so that the line integral of $\vec{\nabla}\phi$ along the closed loop is equal to $2\pi n$. Without any loss of generality we have taken $n=0$. The gauge-invariant phase differences across the junctions are defined by $\Delta\phi^* \equiv \Delta\phi - \frac{2e}{\hbar}\int_{1\,J}^{2} A_s ds$, and therefore differ by $\frac{2e}{\hbar}$ times the total enclosed magnetic flux

$$\Delta\phi_b^* - \Delta\phi_a^* = \frac{2e}{\hbar}\oint_S A_s ds = \frac{2e}{\hbar}\Phi_m = \frac{2e}{\hbar}(B_\perp O + Li_{circ}) \tag{28}$$

in which the total enclosed flux Φ_m consists of two parts, the applied magnetic flux $B_\perp O$ and a self-induced flux $\Phi_S = Li_{circ}$. In general Φ_m is not equal to an integer number of magnetic flux quanta.

In order to show some of the essential features of the double point contact device we first consider the situation when the self-inductance of the enclosed area is ignored. Setting $L = 0$ we obtain from Eq. (28)

$$\Delta\phi_b^* = \Delta\phi_a^* + \frac{2e}{\hbar}B_\perp O \quad . \tag{29}$$

The total current through the double junction is equal to the sum of the currents through the individual contacts and is a function of $B_\perp$ and $\Delta\phi_a^*$,

$$\begin{aligned} i_s(B_\perp, \Delta\phi_a^*) &= i_a \sin\Delta\phi_a^* + i_b \sin\Delta\phi_b^* \\ &= i_a \sin\Delta\phi_a^* + i_b \sin(\Delta\phi_a^* + \frac{2e}{\hbar}B_\perp O) \quad . \end{aligned} \tag{30}$$

The value of $i_s(B_\perp, \Delta\phi_a^*)$ can be maximized with respect to $\Delta\phi_a^*$ and yields the critical current of the device,

$$i_c(B_\perp) = \left[(i_a - i_b)^2 + 4 i_a i_b \cos^2(\frac{e}{\hbar}B_\perp O)\right]^{1/2} \quad . \tag{31}$$

When $i_a = i_b = i_1$ Eq. (31) reduces to the well-known interference relation,

$$i_c(B_\perp) = 2i_1 \left|\cos\frac{e}{\hbar}B_\perp O\right| \quad . \tag{32}$$

This relationship is presented in Fig. 8 for a double point contact with identical junction parameters $i_1 = i_2$ [19]. One of the significant features of this i_c vs $B_\perp$ dependence is that $i_c = 0$ when $B_\perp O = (n+1/2)(h/2e)$. However, this result $i_c = 0$ is difficult to observe experimentally. This

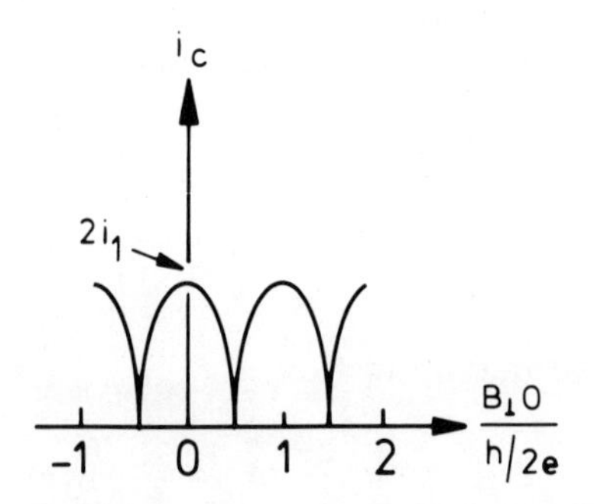

Fig. 8 The critical current i_c as a function of the applied magnetic flux in reduced units $(B_\perp O/(\hbar/2e))$ when the self induced flux $\Phi_s = 0$.

might partly be due to an asymmetry $i_1 \neq i_2$, Eq. (31). More importantly, the self-induced flux Φ_s in the enclosed area O cannot be neglected.

We consider the situation of two superconductors weakly coupled by a symmetrical double point contact and take into account the self-induced flux in the enclosed area. Half the externally applied supercurrent, $i_s/2$, will flow through each of the symmetrical double point contacts. In addition to this current in each junction, we must take into account the circulating supercurrent i_{circ} associated with the self-inductance L of the double contact. In one of the contacts both $i_s/2$ and i_{circ} have the same direction, and in the other they are opposite. The total current through each contact is given by the equations

$$i_s/2 - i_{circ} = i_1 \sin \Delta\phi_a^* \tag{33}$$

$$i_s/2 + i_{circ} = i_1 \sin \Delta\phi_b^* \quad . \tag{34}$$

By adding and subtracting Eqs. (33) and (34) and using Eq. (28) one obtains

$$i_s = i_1 [\sin \{\Delta\phi_a^* + \frac{2e}{\hbar} (B_\perp O + Li_{circ})\} + \sin \Delta\phi_a^*] \tag{35}$$

$$2i_{circ} = i_1 [\sin \{\Delta\phi_a^* + \frac{2e}{\hbar}(B_\perp O + Li_{circ})\} - \sin\Delta\phi_a^*] \quad . \qquad (36)$$

These two equations can be solved together by means of numerical or graphical methods. The critical current is determined as the maximum of $i_s(B_\perp, \Delta\phi_a^*)$ at a given magnetic field $B_\perp$ with respect to $\Delta\phi_a^*$. The result obtained in this way has been verified experimentally by De Waele and De Bruyn Ouboter [27] and is presented in Fig. 9. In the upper part of Fig. 9 the experimental critical current-magnetic field curve is plotted for a symmetrical double point contact with a relatively small enclosed area. From the period $\Delta B_\perp = 0.86$ gauss the area O is 24.10^{-8} cm^2. The lines representing the currents through the individual contacts (a) and (b) have been calculating using Eqs. (35) and (36), with $\pi L i_1/(\frac{h}{2e}) \approx \frac{1}{2} < 1$ and $i_1 = 185\ \mu A$. When a half integer number of flux quanta are applied in the area O, the critical current is a minimum $(i_{c,min})$. This critical current is shown in Fig. 10 as a function of $(\pi L i_1)/(h/2e)$. The currents through the individual contacts (a) and (b) are also shown. When $(\pi L i_1)/(h/2e)$ approaches 0, the minimum critical current is equal to zero $(i_a + i_b = 0)$ and the circulating current equal to $(i_a - i_b)/2 = i_1/\sqrt{2}$.

The main interest for the construction of dc double junction magneto interferometers (dc SQUID) is when $\pi L i_1/(h/2e) >> 1$. Equations (35) and (36) show that the total magnetic flux is nearly quantized in units $h/2e$. The maximum in the modulation of the critical current is in this limit equal to $(h/2e)/L$, which goes to zero if $L \to \infty$, see Fig. 11. This can easily be derived if one realises that in this limit the total magnetic flux is nearly quantized, $B_\perp O + Li_{circ} = n\frac{h}{2e}$. Due to flux conservation, the applied current i_s divides equally between both contacts of the symmetrical double junction. The circulating current adds to the applied current in one of the contacts (a) and subtracts from the other (b). When the applied current is increased from zero, the critical current is reached when the total current $i_c/2 + i_{circ}$ through one of the contacts (a) reaches its critical value i_1. From the equation $i_c/2 + i_{circ} = i_1$ we obtain

$$i_c = 2i_1 - 2|i_{circ}| = 2i_1 - \frac{2}{L}\left|n\left(\frac{h}{2e}\right) - B_\perp O\right| \qquad (37)$$

in which n is an integer determined by the condition that i_c is a maximum with respect to n but smaller than or equal to $2i_1$. The i_c vs $B_\perp$ dependence described by Eq. (37) is shown in Fig. 11 together with the total flux Φ_m in the area O as a function of $B_\perp$. Figure 12 shows a situation for Li_1 in between the cases of Fig. 9 and Fig. 11. In Fig. 12 $(\pi L i_1)/(h/2e) = 5$. For further details see References [18, 28, 29, 30].

Finally we consider a junction with N point contacts in parallel arranged on a straight line at equal distance from one another in the

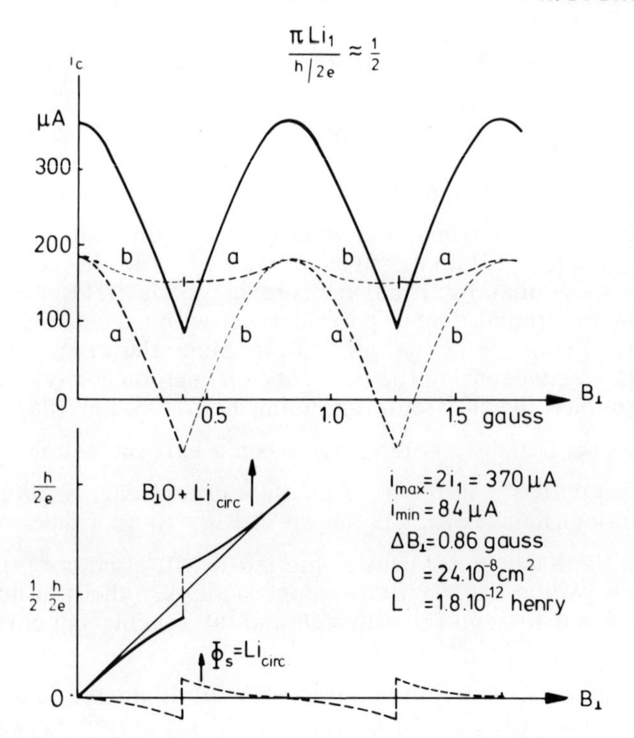

Fig. 9 The upper part of the figure gives the experimentally determined critical current-magnetic field curve. The lines a and b are the calculated currents through the individual point contacts.

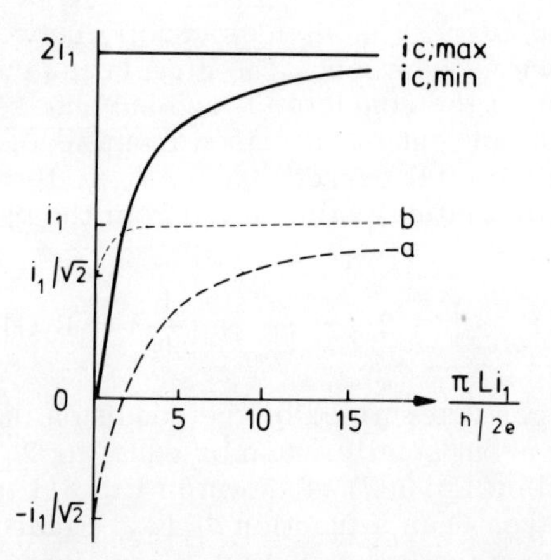

Fig. 10 When a half integer number of flux quanta is applied on 0, the critical current is a minimum. This critical current is shown as a function of $\pi Li_1/(h/2e)$.

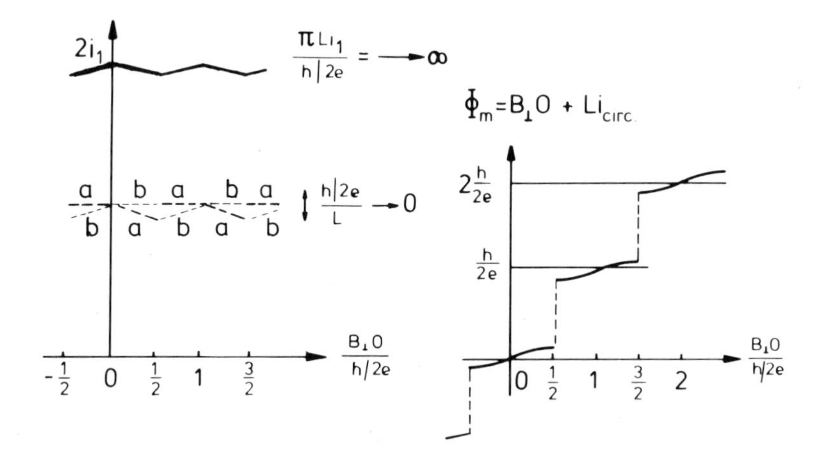

Fig. 11 The critical current of a weakly coupled double junction as a function of the applied flux in the limit $\pi L i_1/(\hbar/2e) \to \infty$ The total embraced magnetic flux is now nearly quantized in units $\hbar/2e$.

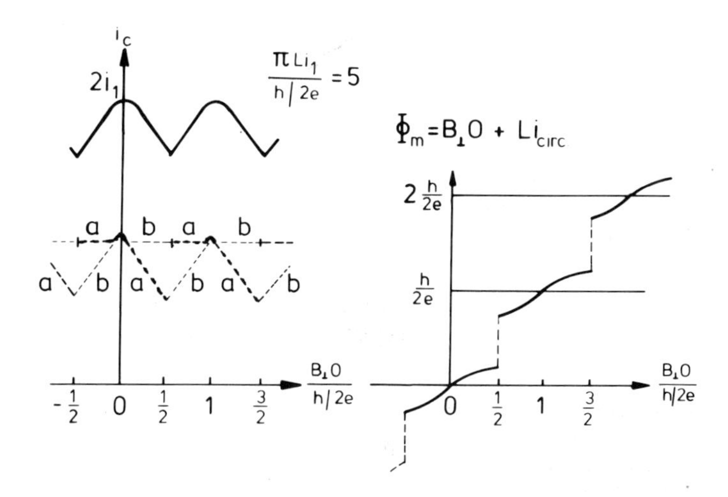

Fig. 12 The critical current of a weakly coupled double junction as a function of the applied flux. In this figure $\pi L i_1/(\hbar/2e) = 5$.

form of an interference grating. For simplicity we assume that all the contacts and holes are identical and we ignore the flux in the holes due to self-inductance and mutual-inductance. The total flux in the holes between contact 1 and contact k is then equal to $(k - 1)B_\perp O$ and analogous to Eq. (29). We obtain:

$$\Delta\phi_k^* = \Delta\phi_1^* + \frac{2e}{\hbar}(k - 1)\, B_\perp O \qquad (k = 2, 3, 4, \ldots, N) \quad . \tag{38}$$

The total current through the junction is equal to

$$i(B_\perp, \Delta\phi_1^*) = i_1 \sum_{k=1}^{N} \sin \Delta\phi_k^*$$

$$= i_1 \sin(\Delta\phi_1^* + (N - 1)\frac{e}{\hbar} B_\perp O) \; \frac{\sin N(\frac{e}{\hbar} B_\perp O)}{\sin(\frac{e}{\hbar} B_\perp O)} \quad . \tag{39}$$

At the critical current, $\Delta\phi_1^*$ will be such that $\sin(\Delta\phi_1^* + (N - 1)\frac{e}{\hbar} B_\perp O)$ $= \pm 1$ which results in the critical current as a function of $B_\perp$ given by

$$i_c(B_\perp) = i_1 \left| \frac{\sin N(\frac{e}{\hbar} B_\perp O)}{\sin(\frac{e}{\hbar} B_\perp O)} \right| = \frac{i_{max}}{N} \left| \frac{\sin \frac{N}{N-1}(\frac{e}{\hbar} B_\perp O_{tot})}{\sin \frac{1}{N-1}(\frac{e}{\hbar} B_\perp O_{tot})} \right| \tag{40}$$

in which the total area of the junction $O_{tot} = (N - 1)O$ and the maximum critical current of the junction $i_{max} = N i_1$. In the limit $N \to \infty$, keeping i_{max} constant, we obtain

$$\lim_{N\to\infty} i_c(B_\perp) = i_{max} \left| \frac{\sin(\frac{e}{\hbar} B_\perp O_{tot})}{\frac{e}{\hbar} B_\perp O_{tot}} \right| \quad . \tag{41}$$

This is identical to the Fraunhofer diffraction formula in optics and describes the famous diffraction experiment of Rowell [17] for a "single-slit" as shown in Fig. 13.

The "single-slit" diffraction pattern is obtained when the supercurrent flows between two superconductors separated by a very thin oxide layer in a magnetic field parallel to the oxice layer. Historically Rowell's experiment was the first demonstration of the dc Josephson effect. Subsequently, in 1964, Jaklevic, Lambe, Mercereau and Silver [18] constructed a quantum interferometer consisting of two parallel Josephson oxide tunnel junctions. Using such a quantum interferometer, they demonstrated that phase coherence of the wave function exists in superconductors over distances of at least a few centimeters.

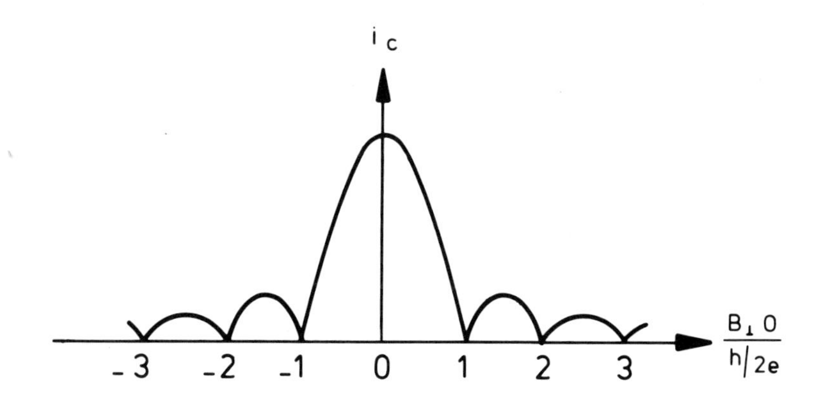

Fig. 13 The "single-slit" diffraction pattern for a single junction of finite extension.

The close agreement between the calculated and the measured i_c vs $B_\perp$ dependence, Figs. 9 and 13, is strong evidence that the relation $i_s = i_1 \sin \Delta\phi^*$ is correct in static conditions. Waldram and Lumley [31] used an ingeneous method to investigate the current-phase relation by inserting the weak link in a superconducting loop as shown in Fig. 14. The phase change around the loop must be the same as that across the junction. From Eq. (23) it follows that the gauge invariant phase difference $\Delta\phi^* = -2\pi\Phi_m / (\frac{h}{2e}) + 2\pi n$ and the current i_1 in Fig. 14 is given by

$$i_1 = i_c \sin \left(-2\pi \Phi_m / (\frac{h}{2e})\right) \qquad . \qquad (42)$$

The procedure of the measurement is as follows:

(a) Insert current i_1. This current will divide so that equal phases are produced around the loop and across the junction.

(b) Insert current i_2. Similarly, this current will divide and if inserted with the correct polarity, the value of i_0 will be reduced.

(c) When i_0 is zeroed (by means of a second double contact detector), the circuit is balanced. In this condition i_1 flows through the weak link and i_2 flows around the end of the loop. The total flux in the loop is proportional to i_2.

(d) A plot of i_1 vs i_2 at balance will yield the current-phase relation.

Figure 15 illustrates a typical result obtained by Waldram and Lumley [31]. The critical current i_c is adjusted so that $\frac{2e}{\hbar} Li_c < 1$, Eq. (26). At this value of Li_c flux jumps in the loop do not occur. In this situation, the whole current-phase curve can be plotted out, including those regions for which $\partial i_s / \partial \Delta\phi^*$ is negative, which are unstable when the weak link is simply connected to a source of current.

Returning to the i_c vs $B_\perp$ curves for a double junction (Figs. 9 and 12), which are obtained by solving Eqs. (35) and (36), we point out that for a given $B_\perp O$ and for a given applied subcritical current i_s, there are several (meta-)stable solutions for the gauge invariant phase difference $\Delta\phi_a^*$ and the corresponding circulating current i_{circ}. By increasing the applied current, the critical value will be reached when $\Delta\phi_a^*$ can no longer adjust itself to satisfy Eqs. (35) and (36). In general $\Delta\phi_a^*$ at the critical current changes continuously with the applied flux $B_\perp O$. However, when $B_\perp O = (n + \frac{1}{2}) \frac{h}{2e}$, $\Delta\phi_a^*$ changes discontinuously in order to permit the largest possible i_s. This change leads to the kinks in the i_c vs. $B_\perp$ curve shown in Figs. 9, 10 and 12.

In the region below the i_c vs $B_\perp$ curves of Figs. 9 and 12, both Zappe [32] (1974) and Guéret [33] succeeded in observing single flux quantum transitions for an undamped double Josephson oxide-junction memory device. By applying a magnetic field by means of a control current i_x, "vortex" modes transitions were induced and detected by a small Josephson junction located nearby. The small Josephson sense-junction placed in the vicinity of the storage cell is biased close to its maximum Josephson current. This junction is then normally in its superconducting state and very little energy is needed to trigger it into the voltage state. The storage cell is a double junction Josephson device with a small value for $(\pi Li_1 / \frac{h}{2e})$. If the device is initially empty a vortex will not be produced if the operating point remains inside the "zero-vortex" curve, (0). By crossing the vortex curve to the left in Fig. 16 a (+) vortex is generated and remains stable as long as the operating point remains inside the (+) vortex curve. Thus a single flux quantum can be stored at $i_x = i_y = 0$. By crossing the threshold curves with small i_y, the device changes simply from one quantum state to another. The existence of a switching threshold below the envelope of the critical current curves is crucial for the operation of the memory. If a (+) vortex was stored, the solid line of the (+) vortex curve can be crossed and the device switches to the gap voltage, after which the system can be reset to the zero voltage state.

In conclusion, it can be said with confidence that all these macroscopic quantum phenomena in superconductors form a nice demonstration that the wave function Ψ_s of the electron pairs is indeed always single valued and that the eigenstates depend on the vector potential. In accordance with the Aharonov-Bohm effect [11], even if the electron pair

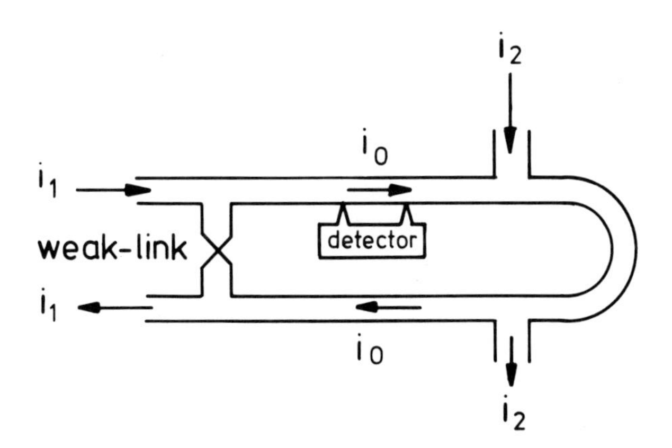

Fig. 14 Method used by Waldram and Lumley to investigate the current-phase relation of a weak link.

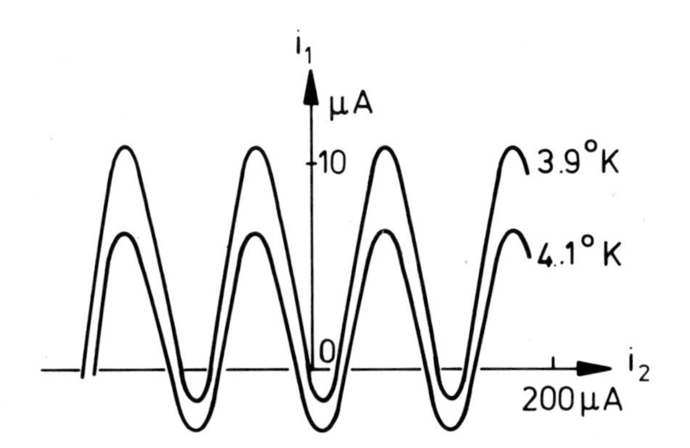

Fig. 15 A typical result obtained by Waldram and Lumley with the device of Fig. 14 for the current phase relation.

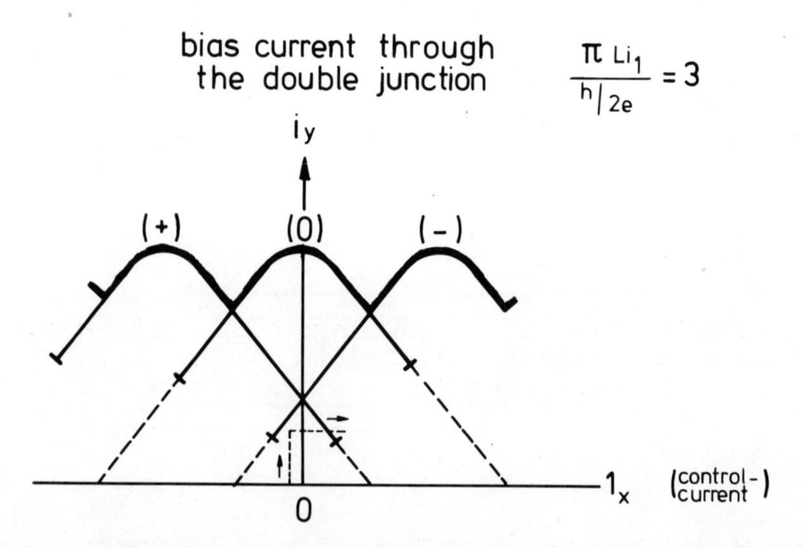

Fig. 16 Single flux quantum transitions observed in an undamped double Josephson oxide-junction memory device.

themselves do not find themselves in a region of magnetic field. The electrons still sense the vector potential of the enclosed magnetic field and their quantum state and therefore the current is dependent upon the vector potential.

II. AC QUANTUM EFFECTS

A. Extension of the Two-Fluid Interpretation of the London Theory

A comparison of superconductors with the superfluid state of helium II leads in first approximation to the Euler-Landau [1, 5] equation of motion for superconductivity which becomes

$$\frac{d_s \vec{v}_s}{dt} = \frac{\partial \vec{v}_s}{\partial t} - \vec{v}_s \times (\vec{\nabla} \times \vec{v}_s) + \vec{\nabla} \frac{1}{2} \vec{v}_s^{\,2}$$

$$= - \vec{\nabla} \mu_e + \frac{2e}{2m} (\vec{E} + \vec{v}_s \times \vec{B}) \quad . \tag{43}$$

According to the Landau two-fluid theory, the gradient of the chemical potential μ_e of the electrons per unit of mass depends on the velocity according to

$$\vec{\nabla}\mu_e = \vec{\nabla}\mu_e^o - \frac{\rho_n}{2\rho}\vec{\nabla}(\vec{v}_n - \vec{v}_s)^2 \quad . \tag{44}$$

The equation of motion Eq. (43) together with the London Eq. (6) leads to the equation

$$\frac{\partial \vec{v}_s}{\partial t} = \frac{e}{m}\vec{E} - \vec{\nabla}(\mu_e + \frac{1}{2}\vec{v}_s^2) \quad . \tag{45}$$

Following Khalatnikov [34], in the next approximation, a dissipative term $-\vec{\nabla}\mathcal{H}$ is added to the superfluid equation [35,36]. This term does not affect the validity of the London Eq. (6). Hence one obtains

$$\frac{\partial \vec{v}_s}{\partial t} = \frac{e}{m}\vec{E} - \vec{\nabla}(\mu_e + \mathcal{H} + \frac{1}{2}\vec{v}_s^2) = \frac{e}{m}\vec{E} - \vec{\nabla}\mu_c \tag{46}$$

in which μ_c is the generalized chemical potential of the superconducting condensate. Besides the Maxwell equations, the following relations for the dissipative fluxes which become in linear approximation must be added

$$\vec{I}_n = \sigma[\vec{E} - \frac{m}{e}\vec{\nabla}(\mu_e + \frac{\vec{v}_s^2}{2})] + \sigma\alpha\vec{\nabla}T = -\frac{\sigma m}{e}\vec{\nabla}\mu_e + \sigma\alpha\vec{\nabla}T \tag{47}$$

$$\mathcal{H} = -\zeta\nabla\cdot(\rho_s\vec{v}_s) = -\zeta\vec{\nabla}\cdot\vec{I}_s \quad . \tag{48}$$

Note that in the stationary state the generalized electro-chemical potential of the superfluid condensate

$$\bar{\mu}_c \equiv \mu_c + \frac{e}{m}\varphi \quad . \tag{49}$$

is a constant, as follows from Eq. (46). Thus there can be no e.m.f. (Seebeck effect) even if there exists a $\vec{\nabla}T \neq 0$ [35]. There is, however, the possibility of a thermoelectric current

$$\vec{I}_n = -\vec{I}_s = \sigma\alpha\vec{\nabla}T \neq 0 \quad . \tag{50}$$

Equation (50) was proposed by Ginzburg [37] in 1944 and very recently experimental indications were found by Zavaritskii [38], and by Pegrum, Guénault and Pickett [39], and by Falco [39]. Equation (46) is based on the early experiments of Casimir and Rademakers [40] on the Seebeck effect and of Daunt and Mendelssohn [41] on the Thompson heat, which show that in the usual stationary-state arrangement these effects vanish in the superconducting state.

The quantity $\mathcal{H}$ introduced by Khalatnikov [34] can be interpreted as the difference in the electrochemical potential of the normal- and superconducting electrons due to a non-equilibrium between the two fluids. The presence of this difference can in certain situations lead to interesting consequences, which have recently been observed by Yu and Mercereau [42] and by Clarke [43, 44] (see also: Tinkham [44, 45] and Waldram [46]). If $\frac{\partial v_s}{\partial t} = -(\vec{\nabla}\,\bar{\mu}_c) = 0$ and $\vec{\nabla} T = 0$, one obtains from Eqs. (48) and (49) that $\vec{I}_n = \sigma\zeta\frac{m}{e}\vec{\nabla}^2 I_n$. This effect due to the unbalance between both fluids penetrates over a characteristic distance, the quasi-particle diffusion length $\delta = [\sigma\zeta\frac{m}{e}]^{1/2}$.

By substituting Eq. (5) for $\vec{v}_s = \frac{\hbar}{2m}\vec{\nabla}\phi - \frac{e}{m}\vec{A}$ and $\vec{E} = -\vec{\nabla}\varphi - \frac{\partial\vec{A}}{\partial t}$ in Eq. (46) one obtains for the equation of motion

$$\frac{\hbar}{2m}\vec{\nabla}\frac{\partial\phi}{\partial t} = -\vec{\nabla}(\mu_c + \frac{e}{m}\varphi) = \vec{\nabla}\bar{\mu}_c \quad (51)$$

or

$$\hbar\vec{\nabla}\frac{\partial\phi}{\partial t} = -2\vec{\nabla}\bar{\mu} \quad (52)$$

in which $\bar{\mu} = m\bar{\mu}_c$ is the electro-chemical potential per superconducting electron. Integrating Eq. (52) between two points (1) and (2) in a superconductor gives an equation for the change of phase with respect to time,

$$\hbar\frac{\partial\Delta\phi}{\partial t} = -\Delta 2\bar{\mu} \quad (53)$$

in which $\Delta\phi(t) \equiv \phi_2 - \phi_1$ and $\Delta\bar{\mu} \equiv \bar{\mu}_2 - \bar{\mu}_1$. Equation (52) corresponds to a Schrödinger type equation for the condensate,

$$i\hbar\frac{\partial\Psi_s}{\partial t} = 2\bar{\mu}\Psi_s \quad . \quad (54)$$

Substituting Eq. (1) into Eq. (54) yields Eq. (52) for the real part.

In order to see how Eqs. (52) and (53) are related to experiments, we introduce the gauge-invariant electrochemical potential $\bar{\mu}^*$ [47] defined by,

$$\vec{\nabla}\bar{\mu}^* \equiv \vec{\nabla}\bar{\mu} + e\frac{\partial\vec{A}}{\partial t} = \vec{\nabla}\mu + e\vec{\nabla}\varphi + e\frac{\partial\vec{A}}{\partial t} = \vec{\nabla}\mu - e\vec{E} \quad . \quad (55)$$

The gauge-invariant phase ϕ^* is defined by Eq. (9). Using these definitions Eqs. (52), (53) and (46) become respectively

$$\hbar \vec{\nabla} \frac{\partial \phi^*}{\partial t} = -2 \vec{\nabla} \bar{\mu}^* \tag{52a}$$

$$\hbar \frac{\partial \Delta \phi^*}{\partial t} = -\Delta 2 \bar{\mu}^* \tag{53a}$$

$$m \frac{\partial \vec{v}_s}{\partial t} = e\vec{E} - \vec{\nabla} m\mu_c = -\vec{\nabla}\bar{\mu}^* \tag{46a}$$

From Eq. (46a) it follows that it is the difference of $\bar{\mu}^*$ from one point to another which determines the overall energy change when an electron is transferred. Since all ordinary voltmeters (except electrostatic ones) involve electric contact allowing electron transfer, they really measure differences in $\bar{\mu}^*$. In conclusion, we find for the difference in voltage V as measured by a voltmeter between the points (1) and (2)

$$V = \Delta \frac{\bar{\mu}^*}{e} \tag{56}$$

and Eq. (53a) leads to the well known second Josephson relation [15]

$$V = -\frac{\hbar}{2e} \frac{\partial \Delta \phi^*}{\partial t} \tag{57}$$

which describes the ac Josephson effect [5].

A different approach to the relations discussed here was adopted by Rieger, Scalapino and Mercereau [48] by generalizing the Ginzburg-Landau theory [3], Eq. (4), into a time-dependent version as was developed by Gor'kov and Eliashberg [49-51]. A relaxation equation, with a characteristic relaxation pairing time constant τ, is introduced in order to describe the dynamics of the order parameter Ψ, Eq. (1),

$$\left(\frac{\partial}{\partial t} + \frac{2i(m\bar{\mu}_e)}{\hbar}\right)\Psi_s = \frac{1}{\tau}\,[(1 - \eta|\Psi_s|^2) + \xi^2(T)(\vec{\nabla} - \frac{2ie}{\hbar}\vec{A})^2]\Psi_s \,. \tag{58}$$

ξ^2/τ appears to be proportional to the electron diffusion constant. The supercurrent is again given by Eq. (3). A superconducting weak link is essentially a one-dimensional object, which implies that the vector-potential can be ignored. An applied magnetic field depresses the magnitude of the order parameter, but plays no essential role in the dynamics of the system. Hence by setting $\vec{A} = 0$, Eq. (58) reduces to:

$$\left(\frac{\partial}{\partial t} + \frac{2i(m\bar{\mu}_e)}{\hbar}\right)\Psi_s = \frac{1}{\tau}\left[\xi^2(T)\,\vec{\nabla}^2 + 1 - \eta|\Psi_s|^2\right]\Psi_s \tag{59}$$

By substituting Eq. (1) into Eq. (59), and equating the real and imaginary parts we obtain

$$\frac{\partial \Psi_o}{\partial t} = \frac{1}{\tau}\left[\xi^2(T)\left(\vec{\nabla}^2 - (\vec{\nabla}\phi)^2\right) + 1 - \eta\Psi_o^2\right]\Psi_o \tag{60}$$

$$\hbar\frac{\partial \phi}{\partial t} = -2\,(m\bar{\mu}_e) + \frac{\hbar\xi^2(T)}{\tau}\left(\vec{\nabla}^2\phi + \frac{2}{\Psi_o}\vec{\nabla}\Psi_o\cdot\vec{\nabla}\phi\right) \tag{61}$$

$$= -2\,(m\bar{\mu}_e) + \frac{m\,\xi^2(T)}{\tau\,\Psi_o^2}\,\vec{\nabla}\cdot\vec{I}_s \quad .$$

If $\tau\,\frac{\partial \Psi_o}{\partial t} = 0$, Eq. (60) reduces to the time-independent Ginzburg-Landau theory, Eq. (4)

$$\left[\xi^2(T)\left\{\vec{\nabla}^2 - (\vec{\nabla}\phi)^2\right\} + 1 - \eta\Psi_o^2\right]\Psi_o = 0 \quad . \tag{62}$$

Equation (61) can be reduced by noting that,

$$\hbar\frac{\partial \phi}{\partial t} = -2(m\bar{\mu}_c) \tag{63}$$

and $\rho_s = \Psi_o^2$ to the Khalatnikov Eq. (48),

$$\bar{\mu}_c - \bar{\mu}_e = \mathcal{H} = -\frac{\xi^2(T)}{2\tau\rho_s}\,\vec{\nabla}\cdot\vec{I}_s = -\zeta\,\vec{\nabla}\cdot\vec{I}_s \tag{64}$$

in which

$$\zeta = -\frac{\xi^2(T)}{2\tau\rho_s} \qquad \text{and} \qquad \delta^2 = \frac{\sigma\, m\,\zeta^2}{2e\rho_s\tau} \quad . \tag{65}$$

B. The ac Josephson Effect

The ac Josephson effect [15] follows from the basic Eqs. (21) and (57),

$$i_s = i_1 \sin\Delta\phi^*(t) \tag{66}$$

$$V = -\frac{\hbar}{2e}\,\frac{\partial\,\Delta\phi^*(t)}{\partial t} \quad . \tag{67}$$

When a voltage difference V is established across the weak link the supercurrent oscillates with an amplitude i_1 and a frequency f equal to

$$f = \frac{V}{h/2e} \tag{68}$$

These oscillations are usually accompanied by electro-magnetic radiation with the same frequency. The occurrence of the ac Josephson effect was first demonstrated by Shapiro [52] in 1963 and then experimentally investigated by Fiske [53], Yanson, Svistuno and Dmitrenko [54] and, very intensively by Eck, Scalapino, Parker, Taylor and Langenberg [55-57]. Finnegan, Denenstein and Langenberg [58, 59] confirmed the relation Eq. (68) within 0.12 p.p.m. and found experimentally that f = 483.59372(6) MHz/$\mu V_{NBS'69}$ or $h/2e = 2.067851(2) \times 10^{-15}$ Weber. Hence a dc voltage of 1 μVolt across the weak link corresponds to a frequency of 483.6 MHz.

There is considerable interest in the intrinsic accuracy of the Josephson frequency-voltage relation Eq. (67). The ac Josephson effect establishes a relationship between the frequency and the electro-chemical potential difference $\Delta\bar{\mu}^*$ across the junction, Eq. (53a). The charge of the free electron e enters in the equations when $\Delta\bar{\mu}^*$ across the junction is compared with $\overline{\Delta\mu}^*$ across the voltage standard cell. When these two are equal, no current is flowing through the standard cell. In this case $\Delta\bar{\mu}^*$ is defined as eV, Eq. (56). These considerations are of relevance with respect to the reliability of the new experimental determination of the atomic constants h and e and the fine structure constant α ($\alpha^{-1} = 2\varepsilon_0 hc/e^2 = 137.036,1(2)$) by means of the Josephson effect.

As a final remark we note that in writing Eq. (67), we have implicitly assumed that the electrochemical potential of the condensate $\mu^* = m\bar{\mu}_c^*$ was a well-defined unique quantity for each superconductor. This is not the case in the immediate vicinity of the junction in which supercurrent must be converted to normal current and back again, as follows from the Eqs. (48), (64) and (65). In this nonequilibrium region the electrochemical potential of the superconducting condensate differs slightly from the electrochemical potential of the "normal"-electrons (quasi-particles). The Josephson frequency is determined by the difference in $\bar{\mu}_c$. Since $\bar{\mu}_c$ is constant in space in each bulk superconductor, and since $\bar{\mu}_c \neq \bar{\mu}_e$ only in the region where $\vec{\nabla}\cdot\vec{I}_s \neq 0$, a possible experimental error can only be made if the voltage leads are attached in the region where $\vec{\nabla}\cdot\vec{I}_s \neq 0$.

The ac Josephson effect was detected by Shapiro [52] (1963), by observing that current steps at constant voltage could be induced in the dc current-voltage characteristic of a junction at values $V = n(h/2e)$ (n = integer) when microwave radiation of proper frequency is supplied to the junction. These steps can be simply understood by considering a similar situation in which a small ac voltage with amplitude v is superimposed on the dc voltage V_0 (and $v \ll V_0$). The total time dependent voltage V(t) across the junction can be written,

$$V(t) = V_o + v \cos \omega t \tag{69}$$

The time dependence of the phase difference $\Delta\phi^*(t)$ is given by,

$$\Delta\phi^*(t) = \Delta\phi^*(0) - \frac{2e}{\hbar}\int_o^t V(t')dt'$$

$$= \Delta\phi^*(0) - \frac{2e}{\hbar}\int_o^t (V_o + v \cos \omega t')dt'$$

$$= \Delta\phi^*(0) - \frac{2e}{\hbar} V_o t - \frac{2e}{\hbar}\frac{v}{\omega} \sin \omega t \tag{70}$$

and the corresponding supercurrent by,

$$i_s(t) = i_1 \sin\left[\Delta\phi^*(0) - \frac{2e}{\hbar} V_o t - \frac{2e}{\hbar}\frac{v}{w} \sin \omega t\right] \quad . \tag{71}$$

When $v \ll V_o$ one can expand Eq. (71) to obtain

$$i_s(t) = i_1 \sin(\Delta\phi^*(0) - \frac{2e}{\hbar} V_o t) - i_1 \frac{2e}{\hbar}\frac{v}{w} \sin \omega t \cos(\Delta\phi^*(0) - \frac{2e}{\hbar} V_o t) \quad . \tag{72}$$

The time average of the first term is equal to zero, and the time average of the second term is not equal to zero only when

$$\frac{2e}{\hbar} V_o = \omega \quad . \tag{73}$$

The magnitude of the average supercurrent over a period is given by

$$\langle i_s(t) \rangle_t = -\frac{1}{2}\frac{v}{V_o} i_1 \sin \Delta\phi^*(0) \quad . \tag{74}$$

At constant $\overline{V} = V_o$, $\langle i_s(t) \rangle_t$ can have different values by varying $\Delta\phi^*(0)$. The current steps at constant voltage are explained from Eq. (74). Figure 17 shows another example of detection of radiation by a Josephson junction. A point contact is placed in the center of the bottom of a coaxial cavity. When the voltage matches a resonant frequency f_r of the cavity, $V = \frac{n}{m}\left(\frac{h}{2e}\right) f_r$ (n and m are integers) [60-62], steps in the I-V characteristics of the weak link are obtained.

III. RESISTIVE STATES IN WEAK LINK JUNCTIONS

A. The Current-Voltage Characteristic and the Resistive-Superconductive Region of a Single Superconducting Weak Link

In 1963 the Josephson effect [15] was observed for the first time by Anderson and Rowell [16] in a superconducting tunnel junction with an oxide barrier of a few atomic layers thickness, using evaporated film techniques, see Fig. 18. Subsequently Zimmerman and Silver [63] made weak contacts by pressing two mutually perpendicular niobium wires together in such a way that the contact resistance was of the order of one Ohm at room temperature. Clarke [64] developed another technique by dipping a thin niobium wire in molten tin-lead solder. The small droplet of solder encircling the niobium wire results in a two junction contact between the niobium and solder [65]. Another technique is due to Anderson and Dayem [66] and makes use of narrow constrictions in thin bridge devices (Dayem bridges). Notarys and Mercereau [67-69] weakened the narrow constriction by utilizing the proximity effect of a normal metal evaporated on top of the superconductor ("overlay" bridges). A stable point contact can be obtained by using a niobium screw and niobium blocks [70, 71] separated by a layer (150 μm thick) of Schott sealing glass C15, see Fig. 7. This type of glass has the same coefficient of thermal expansion as niobium. The niobium blocks and the glass are baked together in order to form a mechanically solid unit. Single-point contact diffraction like effects (Fig. 13, Eq. (41)) are not observed due to the very small size of the contact areas. Zimmerman and Silver [63] observed that interference phenomena in double point contacts persist undiminished up to fields as large as 3000 gauss and even at these fields no diffraction like attenuation was observed.

The system of two superconductors connected by a single point contact is the most simple of all possible weak connections between two superconductors. Nevertheless, in the resistive state, it is a complicated system because very small inductances and capacitances that usually can be neglected play an important role even in its dc current-voltage characteristic, owing to the high frequency related to the ac Josephson effect.

For a junction brought into the resistive-superconductive region a two fluid model will be assumed in which the current through the junction is the sum of an ideal Josephson supercurrent, $i_s(t) = i_1 \sin \Delta\phi^*(t) = i_1 \sin \{\Delta\phi^*_0 - \frac{2e}{\hbar} \int_0^t V(t')dt'\}$ and a normal current $i_n(t) = V(t)/R_n = -\frac{\hbar}{2e} \frac{\partial \Delta\phi^*(t)}{\partial t} / R_n$, in which R_n is the ideal ohmic resistance of the junction. Through the current leads of the junction one has in addition to these two contributions $i_s(t)$ and $i_n(t)$ a third contribution $C \frac{dV(t)}{dt} = -\frac{\hbar}{2e} C \frac{\partial^2 \Delta\phi}{\partial t^2}$ due to the intrinsice capacitive coupling between both

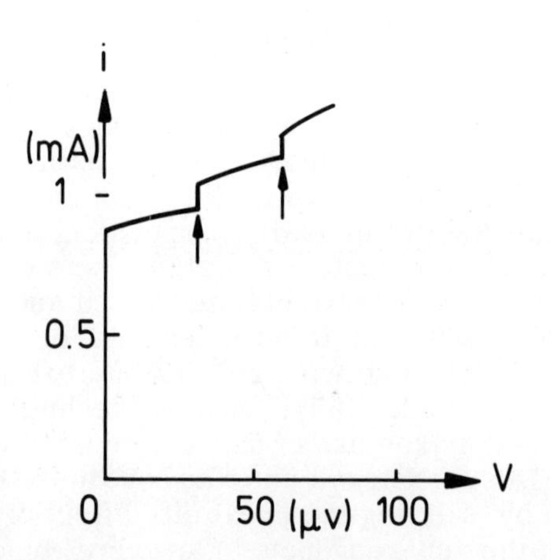

Fig. 17 The current-voltage characteristic for a point contact placed in the center of a coaxial cavity.

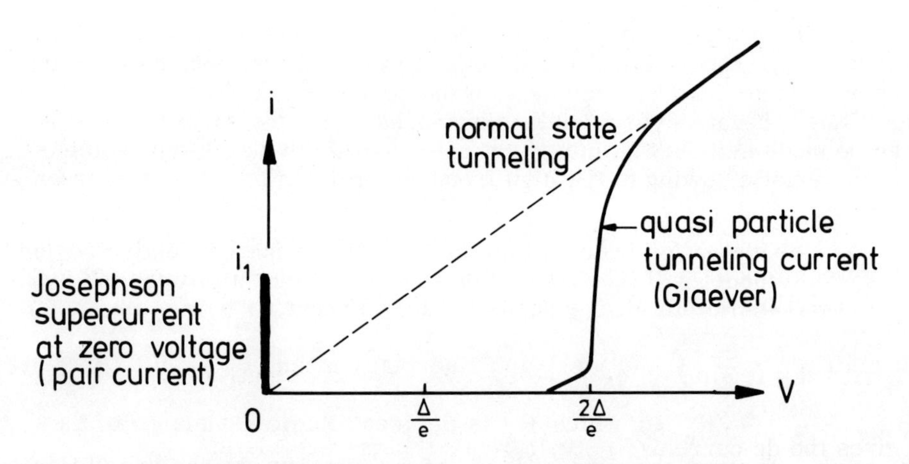

Fig. 18 The current voltage characteristic of an evaporated oxide tunnel junction.

superconductors (1) and (2). The total current through the current leads is equal to [72, 73],

$$i(t) = i_1 \sin \Delta\phi^*(t) - \frac{\hbar}{2e} \frac{\partial \Delta\phi^*(t)}{\partial t} / R_n - \frac{\hbar}{2e} C \frac{\partial^2 \Delta\phi^*(t)}{\partial t^2} \quad . \tag{75}$$

This equation was proposed for the first time by Stewart [72] and McCumber [73]. All these contributions greatly complicate the analysis of the dc current-voltage characteristic of a weak link and in only the very simplest cases can analytical solutions be obtained.

We start with the following approach [27]. At first both the super- and normal-current through the contact will be taken into account and we ignore the capacitive coupling between the superconductors as well as the self inductance of the point-contact. Furthermore we apply certain limitations to the differential equations of the contact. Primarily, only the two cases for which either the current or the voltage are constants of time will be discussed. In this case the total current through the junction is equal to,

$$i(t) = i_1 \sin \Delta\phi^*(t) - \frac{\hbar}{2e} \frac{\partial \Delta\phi^*(t)}{\partial t} / R_n \quad . \tag{76}$$

We first take the voltage as a constant in time. If $V = 0$, and dc supercurrent between $-i_1$ and $+i_1$ can flow through the junction. If $V \neq 0$ then the sinusoidal ac Josephson current oscillates at a frequency $f = (2e/h) V$ and with an amplitude i_1. The time average of the supercurrent is equal to zero and only the normal component contributes to the dc current, see Fig. 19,

$$\langle i(t) \rangle_t = i_n = V/R_n \quad . \tag{77}$$

Voltage biasing can be achieved by shunting the junction with a small parallel ideal resistance R ($R \ll R_n$, Zimmerman, Cowen and Silver [74]), or an ideal capacitance C (with $1/(2\pi\nu C) \ll R_n$, Stewart [72] and McCumber [73, 75]).

Secondly, we can take the total current $i = i_s(t) + i_n(t)$ through the weak link as a constant in time and calculate the mean voltage $\overline{V(t)}^t$. The solution which follows from the differential equation,

$$i = i_1 \sin \Delta\phi^* - \frac{\hbar}{2e} \frac{\partial \Delta\phi^*}{\partial t} / R_n = \text{constant} \tag{78}$$

gives the dc current (i) - voltage $(\overline{V(t)})$ characteristic

$$\langle V(t) \rangle_t = -\left\langle \frac{\hbar}{2e} \frac{\partial \Delta\phi^*(t)}{\partial t} \right\rangle_t = R_n \sqrt{i^2 - i_1^2} \tag{79}$$

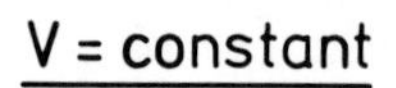

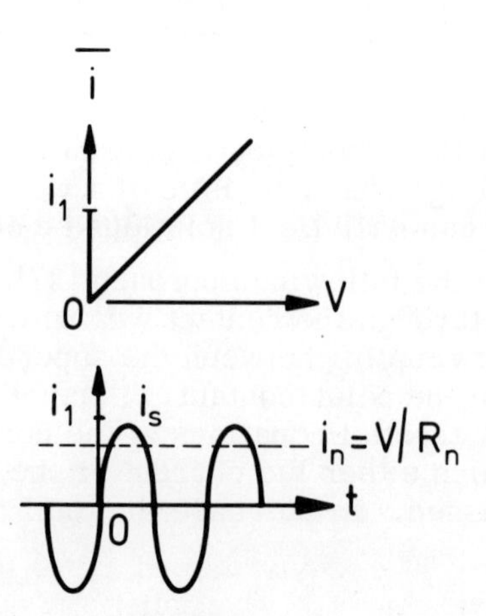

Fig. 19 The current-voltage characteristic of a single junction when the voltage is constant in time.

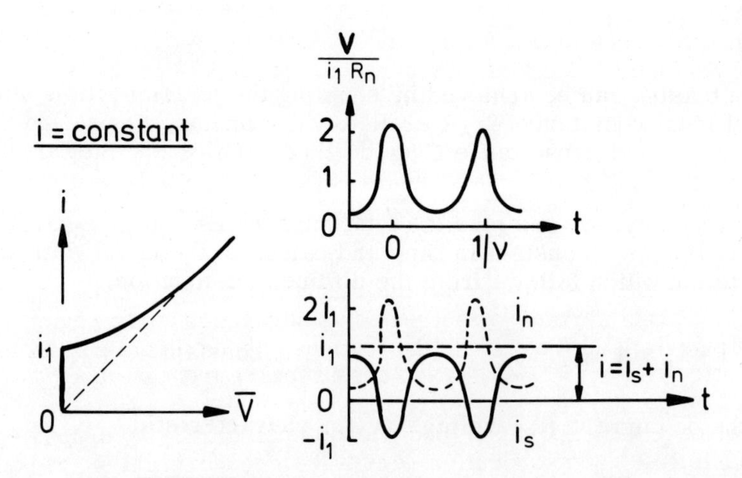

Fig. 20 The current-voltage characteristic of a single junction when the current is constant in time.

which is shown in Fig. 20. This analytical solution for the time averaged voltage was given by Aslamazov and Larkin [76] in 1969. The energy dissipation $\left\langle \frac{dw}{dt} \right\rangle_t = \langle iV \rangle_t = \langle V^2 \rangle_t / R_n = R_n \, i \sqrt{i^2 - i_1^2}$ is completely due to the normal current. When $i > i_1$, the voltage $V(t)$ is periodic and a sharply peaked function of time. When i is only slightly larger than i_1, the peak-height is equal to $(i + i_1)R_n$ and the peak width of the order $(\frac{h}{2e})/(i + i_1)R_n$. The time dependence of the voltage $V(t)$, the normal current $i_n(t) = V(t)/R_n$ and the supercurrent $i_s(t) = i - i_n(t)$ are plotted in Fig. 20. The supercurrent spends more time in the forward than in the backward direction, hence a dc supercurrent (averaged) persists at finite voltages. This dc supercurrent disappears gradually as the total current i rises. When $i \gg i_1$, the current-voltage characteristic approaches Ohm's law $i = V/R_n$. In this case the supercurrent is nearly sinusoidal in time, so that $\langle i_s(t) \rangle_t = 0$.

A constant current can be realized experimentally in a junction with a very small capacitance if one applies the current in a circuit having a high impedance. This approach appears to be essential for understanding the observed modulation in the voltage $\langle V(B_\perp) \rangle_t$ as a function of an external magnetic field $B_\perp$ at constant applied current i for a double point contact as will be shown in the next section. However, if one tries to measure the current-voltage characteristic with a constant current circuit, it is nearly impossible to avoid capacitive coupling between the two superconductors of order of 10^{-10} Farad. This capacitance plays a significant role for large values of $V(t)$. When $\langle V(t) \rangle$ is increased the frequency of the ac currents increases and the capacitance becomes a short for ac currents between the superconductors, $(R_n \gg 1/2\pi\nu C)$. In this limit we can take V as a constant in time. We remark that McCumber [73] performed calculations on a point contact in series with a self-inductance while the system is voltage biased. A typical result is shown in Fig. 21.

If we compare the current-voltage characteristic of a current biased ideal Josephson junction (see Fig. 20 and Eq. (79)), with experiments on narrow "overlay" bridges by Notarys and Mercereau [67-69] or on point contacts by De Bruyn Ouboter, Omar and De Waele [27,65, 71], in which macroscopic quantum-interference phenomena are observed, we conclude that in the resistive superconductive region considerable deviations from the ideal Josephson behavior are often found. Asymptotically for $i \gg i_{crit}$, an excess supercurrent is observed (see Fig. 23 [65]). The current-voltage characteristic of such a current biased weak link can be described phenomenologically by supposing that in the resistive-superconductive region the supercurrent i_s can be written as,

$$i_s = i_o + i_1 \sin \Delta\phi^*(t) \tag{80}$$

and

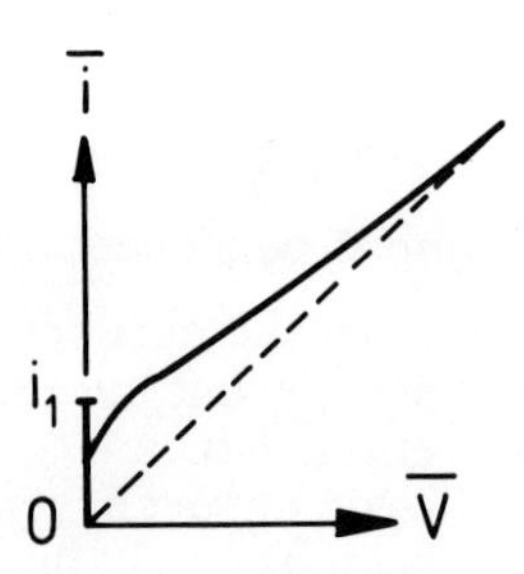

Fig. 21 The current voltage behavior for a point contact in series with a self-inductance [73].

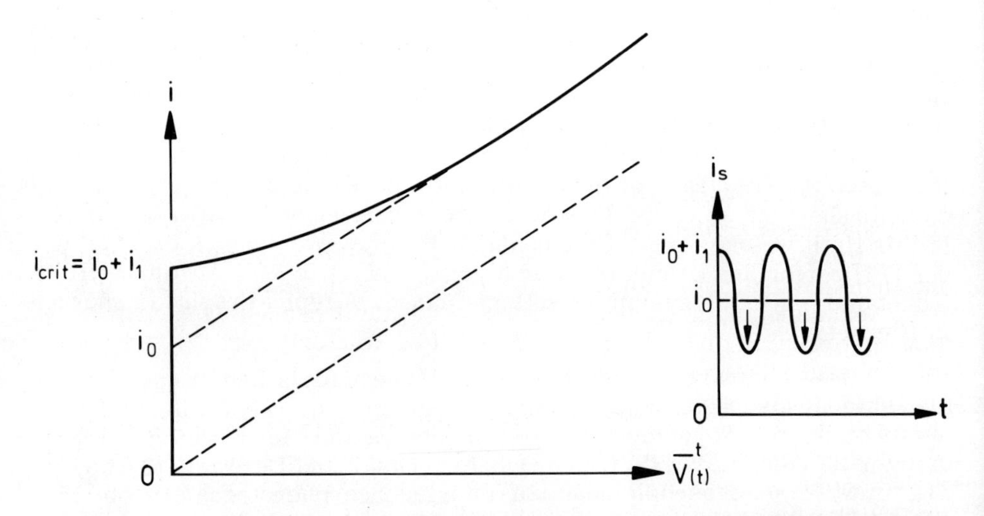

Fig. 22 A current-voltage characteristic for a junction showing an excess supercurrent i_o. Furthermore is indicated the dependence of i_s vs t according to the theory of Rieger, Scalapino and Mercereau.

$$i = i_s + i_n = i_o + i_1 \sin\Delta\phi^*(t) - \frac{\hbar}{2e}\frac{\partial\Delta\phi^*}{\partial t} / R_n \tag{81}$$

where i equals a constant in time and i_o and i_1 are constants. In the same way we find for the current-voltage characteristic, which follows from Eq. (81) in analogue to Eq. (79),

$$\langle V(t)\rangle_t = R_n \sqrt{(i - i_o)^2 - i_1^2} \quad . \tag{82}$$

For $i \gg i_{crit} = i_o + i_1$ the excess supercurrent asymtotically approaches

$$\langle i_s(t)\rangle_t = i_o \tag{83}$$

and

$$\langle V(t)\rangle_t = R_n (i - i_o) \quad . \tag{84}$$

It should be remarked that there is an addiational energy dissipation connected to the excess supercurrent in the weak link. The energy dissipation due to the normal-current is equal to $\langle i_n^2 \rangle_t = (i - i_o)\langle V\rangle_t$, thus the superconducting energy loss must equal $i_o V$. A nice example of such a behavior is found in the current-voltage characteristic of the double-point contact given in Fig. 23. In the "overlay" bridges Notarys and Mercereau [67] observed that the excess supercurrent is equal to $i_o = i_1/2$.

Such an excess supercurrent $\langle i_s\rangle_t = i_o$ is in conflict with the ideal Josephson model in which $\langle i_s\rangle_t = i_1 \langle \sin \Delta\phi^*\rangle_t = 0$ if $i \gg i_{crit}$, where the evolution of the relative phase is a continuous process with no intrinsic dissipation associated with the oscillating supercurrent. Rieger, Scalapino and Mercereau [48] used a time dependent Ginzburg-Landau theory in order to calculate the time varying currents in a one-dimensional phase slip model and introduced the concept of superconducting dissipation. There is good evidence that a weak link is narrower than the size of an Abrikosov vortex, and then a three dimensional phase slip due to vortex motion has to be excluded. For instance, a model based on the concept of the creation of a vortex ring at the edges of the weak link which moves inward and annihilates at the center cannot be considered. No diffraction-like attenuation effects are observed in single superconducting point-contacts. More realistic is the assumption that the amplitude of the order parameter goes periodically through zero across the entire weak link. Rieger et al. [48] consider the time evolution of a one-dimensional superconductor of a time scale $\Delta t = 1/\nu = \frac{h}{2e} V$ which is slow with respect to the pairing time τ. If the superconducting electrons in a weak link are accelerated due to an external electric field, they eventually reach a velocity, the Landau critical velocity

Δ/P_F, at which superconductivity is unstable relative to the normal state. The amplitude of the superconducting state begins to decay toward zero in a region one quasi-particle diffusion length δ long on a time scale comparable to the pairing time τ ($\tau \ll \Delta t$). The assumption of phase slip is that the system subsequently recovers into a phase coherent superconducting state in which the relative phase difference between any two points separated by the slip region has changed by 2π. The new state appears after the phase-slip region transition. The superelectrons again accelerate to their critical velocity to repeat the phase slip process. A small amount of condensation energy is lost during each cycle and there exists a dissipative process in the superconductor connected with the time variation of the order parameter. There may be conditions under which the superconducting dissipation exceeds the normal dissipation. Rieger, Scalapino and Mercereau [48] numerically solved the Eqs. (60), (61), (3) and (47) for a one dimensional weak link one coherence length long. Their results are compatible with Fig. 22 with $i_o = i_1/2$. The arrows indicate the time of phase slip.

Very recently Jackel et al. [77, 78] investigated the supercurrent-phase relations of weak links consisting of a Sn uniform-thickness microbridges in which a distinctly non sinusoidal behavior is observed. A linear increase of i_s as a function of $\Delta\phi$ is observed until a critical phase difference $\Delta\phi_c$ is reached, where an irreversible transition is made to an adjacent fluxoid state with a lower supercurrent. Furthermore, these authors measured for the same junction its dc current-voltage characteristic. They showed that these linear multivalued current phase relationships account for the observed excess supercurrent i_o. The equation for the current-voltage characteristic with such a linear supercurrent phase relation is given by,

$$V = -\left(\frac{2\pi R_n i_c}{\Delta\phi_c}\right)\left[\ell n\left(\frac{(i/i_c) - 1}{(i/i_c) - 1 + 2\pi/\Delta\phi_c}\right)^{-1}\right] \tag{85}$$

where R_n and i_c are again the normal resistance and the critical current respectively. This seems to be a very attractive alternative explanation for the excess supercurrent, for which they find the following expression,

$$i_o = i_c\,[(\Delta\phi_c - \pi)/\Delta\phi_c] \quad . \tag{86}$$

The non sinusoidal behavior of the current phase relation is attributed to the extra phase gradient in the electrodes leading to the bridge. The total phase difference across the entire weak link structure is often dominated by the latter contribution.

B. The Double Point Contact in the Resistive Superconductive Region, the dc SQUID

In Section I C, we demonstrated that the critical current of a double point contact is an oscillating function of the applied magnetic field.

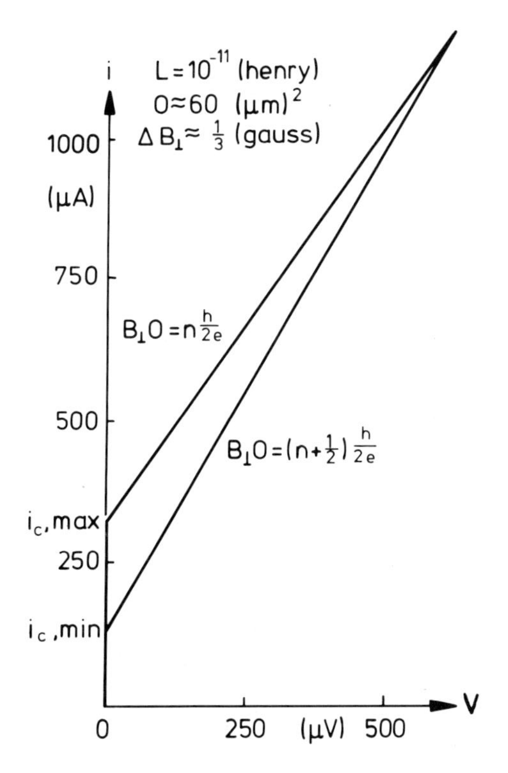

Fig. 23 X-Y recorder plots of two i-V characteristics of a double junction. One at $B_{\perp}O = n\frac{h}{2e}$ and one at $B_{\perp}O = (n+\frac{1}{2})\ \frac{h}{2e}$.

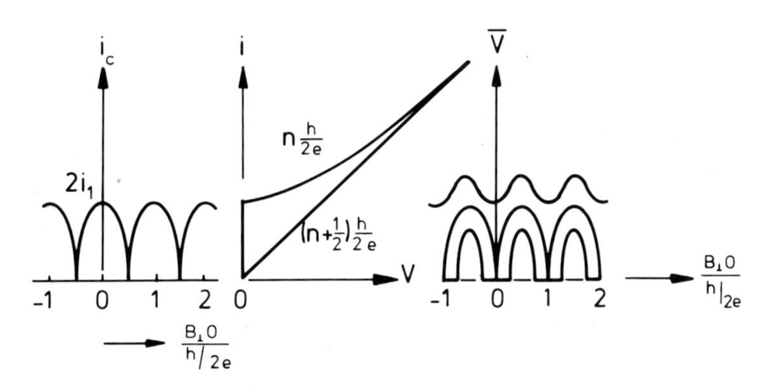

Fig. 24 Description of the i-V-$B_{\perp}$ dependence of a symmetrical double junction when the self inductance $L = 0$.

It is also observed experimentally that the dc voltage across a double point contact oscillates as a function of the applied magnetic field when a constant current is applied [27, 63, 65, 79, 80] to the junction, Fig. 23 [65]. The critical current i_c and the dc voltage $\overline{V}$ (at constant applied current) are both periodic functions of the applied magnetic field with the same period $\Delta B_\perp = (\frac{h}{2e})/O$. A maximum in the critical current corresponds to a minimum in the voltage. The amplitudes of both the i_c and $\overline{V}$ oscillations decrease when the enclosed area O between the contacts is increased. Larger areas up to $\approx 1\ \mathrm{cm}^2$ with $L \approx 10^{-8}$ henry are made in which oscillations were measured. It is observed that for different self-inductances the amplitude in the voltage oscillations can vary between $0.1\ \mu V$ and $1\ mV$.

For simplicity, we start by ignoring the self inductance ($L = 0$) of the enclosed area O of the symmetrical double contact. This has the advantage that a simple analytical solution can be obtained. The total current through the double junction is equal to the sum of the supercurrents through the individual contacts and the normal current. Again the total current is taken as a constant in time,

$$i = i_1[\sin\{\Delta\phi_a^* + \frac{2e}{\hbar}B_\perp O\} + \sin\Delta\phi_a^*] - \frac{\hbar}{2e}\frac{\partial\Delta\phi_a^*}{\partial t}/R_n = \tag{87}$$

$$= 2i_1 \sin\{\Delta\phi_a^*(t) + \frac{e}{\hbar}B_\perp O\}\cos(\frac{e}{\hbar}B_\perp O) - \frac{\hbar}{2e}\frac{\partial\Delta\phi_a^*}{\partial t}/R_n = \text{constant},$$

which is essentially the same as Eq. (78) for a single contact when i_1 is replaced by $i_c(B_\perp)$ given by Eq. (32),

$$i_c = 2i_1 \left|\cos\frac{e}{\hbar}B_\perp O\right| \quad . \tag{88}$$

Hence the solutions of the equations are the same:

$$\langle V(t)\rangle_t = -\frac{\hbar}{2e}\left\langle\frac{\partial\Delta\phi^*(t)}{\partial t}\right\rangle_t = R_n\sqrt{i^2 - |2i_1\cos(\frac{e}{\hbar}B_\perp O)|^2} \ . \tag{89}$$

This result already explains the observed voltage oscillations in the resistive superconductive region as a function of the external magnetic field $B_\perp$, for the case that the applied current (i) is constant and exceeds the critical value ($i > i_c(B_\perp)$). The current-voltage characteristic, i vs $\langle V\rangle_t$ for $i_c = 2i_1$ (when $B_\perp O = n\frac{h}{2e}$) and for $i_c = 0$ when $B_\perp O = (n + \frac{1}{2})\frac{h}{2e}$) are presented in Fig. 24 [27]. In the same figure we have also shown the voltage oscillations ($\overline{V}$) as a function of the applied magnetic flux $B_\perp O$ at several values of the constant applied current.

In double point contacts, ac properties can be observed as in single point contacts. The electromagnetic radiation, emitted by the two point contacts, interfere with each other. One can control the phase difference of the radiation emitted by the two contacts by means of the applied static magnetic field. The behavior of a double point contact, with a small enclosed area, in its own resonant electromagnetic radiation field can be observed by putting it in the center of a microwave cavity. De Bruyn Ouboter, De Waele and Omar [81] observed that in that case the current-voltage curve shows current steps at constant voltage, corresponding to a resonant frequency of the cavity. These steps were most pronounced when the applied flux on the area O was equal to an integer number of flux quanta ($B_{\perp}O = n\frac{h}{2e}$) and disappeared when the applied flux was equal to a half integer number of flux quanta ($B_{\perp}O = (n + \frac{1}{2})\frac{h}{2e}$). Although the self-inductance can never be ignored, in principle one can understand this behavior if one concludes that the amplitude of the ac voltage is larger when an integral number of flux quanta is applied to the area O. When $B_{\perp}O = n\frac{h}{2e}$, the supercurrents through the individual contacts are oscillating with the same frequency and are in phase so radiation is emitted. When $B_{\perp}O = (n + \frac{1}{2})\frac{h}{2e}$ the supercurrents through the individual contacts are oscillating with the same frequency but are in opposite phase. Since the distance between the two point contacts is very much smaller than the wave length of the radiation, the net emitted radiation is now much smaller and cannot be detected. Therefore steps in the current at constant voltage are most pronounced when $B_{\perp}O = n\frac{h}{2e}$ and absent when $B_{\perp}O = (n + \frac{1}{2})\frac{h}{2e}$, as is shown in Fig. 25 [81].

If the self-inductance of the symmetrical double junction is taken into account ($L \neq 0$), however, it is no longer possible to treat the problem in a simple way and find an analytical solution. The total current through each contact is then equal to

$$i/2 - i_{circ} = i_1 \sin\Delta\phi_a^*(t) - \frac{\hbar}{2e}\frac{\partial\Delta\phi_a^*(t)}{\partial t}/2R_n \qquad (90)$$

$$i/2 + i_{circ} = i_1 \sin\Delta\phi_b^*(t) - \frac{\hbar}{2e}\frac{\partial\Delta\phi_b^*}{\partial t}/2R_n \quad . \qquad (91)$$

By addition and subtraction one obtains again an expression for the total current and the circulating current. Together with Eq. (28) these equations have to be solved numerically. The results obtained from computer calculations for current, flux and dc voltage are shown in Fig 26 [27].

Figure 26 shows (a) the critical current oscillations $i_c(B_{\perp})$; (b) the current-voltage characteristics when $B_{\perp}O = n(h/2e)$ and $B_{\perp}O =$

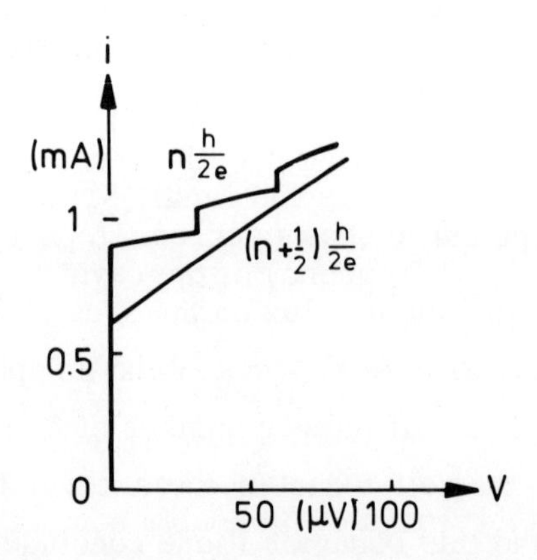

Fig. 25 The current-voltage curve of a small double junction placed in the center of a microwave cavity for the cases $B_\perp O = n\frac{h}{2e}$ and $B_\perp O = (n + \frac{1}{2})\frac{h}{2e}$.

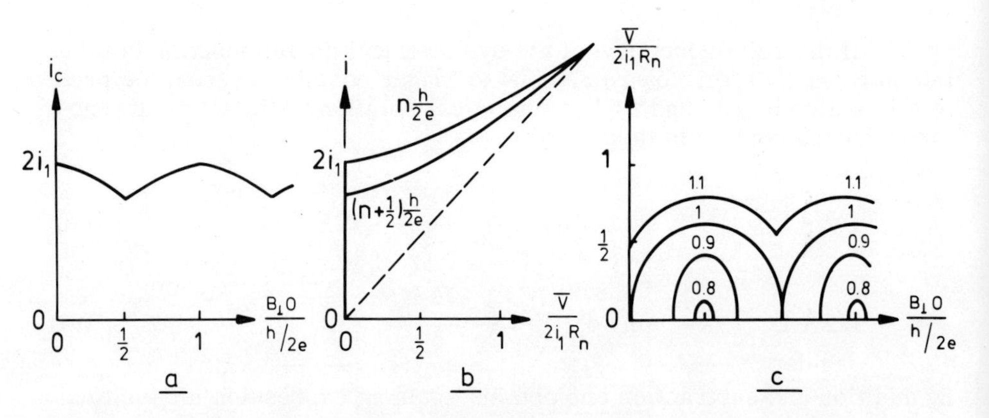

Fig. 26. Description of the i-V-$B_\perp$ dependence of a symmetrical double junction when the self-inductance $L \neq 0$.

$(n + \frac{1}{2}) \frac{h}{2e}$; and (c) the voltage oscillations as a function of the applied magnetic flux $B_{\perp}O$ at several values of a constant applied current.

Finally we discuss the influence of a normal metal in the hole between the two superconductors of a double point contact of the $\langle i \rangle$ vs $\langle V \rangle$ vs $B_{\perp}$ dependence, which improves the voltage sensitivity [71] (see Fig. 27). We previously considered the self-inductance L frequency independent. If however the value of L depends on the frequency of the ac magnetic field in the hole, then an extra V(t) dependence is introduced in the equations. When there is helium or vacuum in the hole between the contacts then L is not a function of the frequency, but just a geometric constant of the sample. The value of L is an increasing function of the area O enclosed by the superconducting blocks and the point contacts. This means that a relatively small magnetic field period $\Delta B_{\perp}$ necessarily implies a small amplitude of the oscillations. What one would like to have, however, is a double point contact with at the same time a small $\Delta B_{\perp}$ and a large amplitude in the dc voltage $(\overline{V})$ oscillations. This can be achieved by putting a normal conductor (copper or platinum for example) in the hole, insulated from the blocks, de Waele, Vergouwen, Matsinger and de Bruyn Ouboter [71]. In this case L is frequency dependent due to the skin effect of the normal metal. In the low frequency limit L is relatively large. Low-frequency magnetic fields penetrate the normal metal completely, consequently the

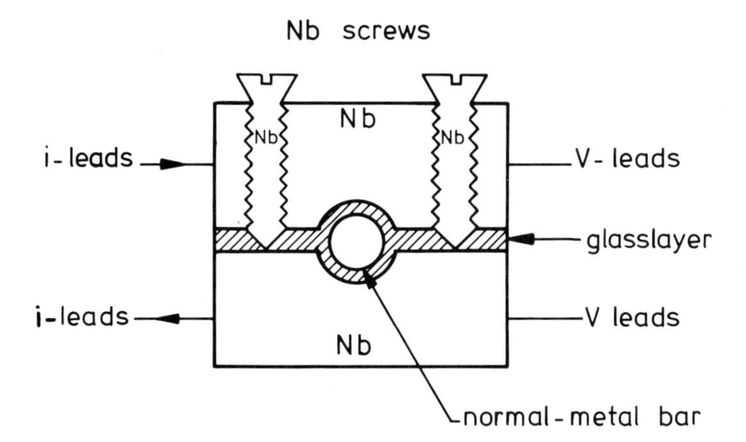

Fig. 27. A double junction with a normal-metal bar in the hole.

period $\Delta B_\perp$ is unchanged. Also the value of the critical current should be unchanged because this is the largest current flowing through the double point contact with V = 0 (and hence f = 0). If, however, $\bar{V}$ is so large that the penetration depth for an ac magnetic field with frequency $f = \frac{2e}{h}\bar{V}$ is much smaller than the dimensions of the normal conductor, the ac magnetic field is completely excluded from the normal conductor. For the high frequency field L is much smaller. From Fig. 10 it can be seen that a smaller self-inductance leads to a larger amplitude in the i_c-oscillations. Also the amplitude of the voltage oscillations is larger. Apart from the period $\Delta B_\perp$ (which remains unchanged), the corrugated $\bar{i}$ vs $\bar{V}$ vs $B_\perp$ dependence for these values of $\bar{V}$ is the same as if the hole was filled with a superconductor instead of a normal metal. In the device, an insulated thin-walled tube of copper or platinum, fitting tightly in the hole, is baked together with the niobium and the glass to form a solid unit. In this way voltage oscillations from double point contacts with large holes can be observed.

The author is very much indebted to Dr. H. van Beelen for valuable discussions while writing this article.

REFERENCES

General References

M. Tinkham, Introduction to Superconductivity, McGraw Hill, New York (1975).

P.G. de Gennes, Superconductivity of Metals and Alloys, Benjamin, New York (1966).

J. Bardeen, J.R. Schrieffer, Progr. in Low Temp. Phys., vol. 3 (C.J. Gorter ed.) North-Holland, Amsterdam (1961).

P.W. Anderson, Progr. in Low Temp. Phys., vol. 5 (1967).

R. de Bruyn Ouboter, A.Th.A.M. de Waele, Progr. in Low Temp. Phys., vol. 6 (1970).

Superconductivity (R.D. Parks, ed.), Dekker, New York (1969), Chap. 8 by J.E. Mercereau and Chap. 9 by B.D. Josephson.

A.H. Silver, J.E. Zimmerman, Applied Superconductivity (V.L. Newhouse, ed.) Vol. I (1975).

O.V. Lounasmaa, Experimental principles and methods below 1 K, Academic Press, New York (1975).

1. F. London, Superfluids, Vol. I, Dover Public, New York (1960).
2. L.N. Cooper, Phys. Rev. 104, 1189 (1956); J. Bardeen, L.N. Cooper, J.R. Schrieffer, Phys. Rev. 108, 1175 (1957).

3. V.L. Ginzburg, L.D. Landau, Zh. Eksperim. i. Teor. Fiz. 12, 1064 (1950).
4. C.J. Gorter, H.B.G. Casimir, Physica 1, 306 (1934), Phys. Z. 35, 963 (1934), Z. techn. Phys. 15, 539 (1934).
5. F. London, Phys. Rev. 74, 462 (1948), F. and H. London, Proc. Roy. Soc. (London) A 149, 71 (1935).
6. M. Meissner, R. Ochsenfeld, Naturwissenschaften 21, 787 (1933).
7. B.S. Deaver, W.M. Fairbank, Phys. Rev. Lett. 7, 43 (1961).
8. M. Doll, M. Näbauer, Phys. Rev. Lett. 7, 51 (1961), Zeitschrift für Physik 169, 526 (1962).
9. N. Byers, C.N. Yang, Phys. Rev. Lett. 7, 49 (1961).
10. C.N. Yang, Rev. Mod. Phys. 34, 694 (1962).
11. Y. Aharonov, D. Bohn, Phys. Rev. 115, 485 (1959).
12. F. Bloch, Phys. Rev. 137, A 787 (1965), Phys. Rev. Lett. 21, 1241 (1968).
13. B.B. Schwartz, L.N. Cooper, Phys. Rev. 137, A 829 (1965).
14. W.A. Little, R.D. Parks, Phys. Rev. Lett. 9, 9 (1963), Phys. Rev. 133, A97 (1964), R.P. Groff, R.D. Parks, Phys. Rev. 176, 567 (1968).
15. B.D. Josephson, Phys. Lett. 1, 251 (1962), Rev. Mod. Phys. 36, 216 (1964), Advances Physics 14, 419 (1965).
16. P.W. Anderson, J.M. Rowell, Phys. Rev. Lett. 10, 230 (1963).
17. J.M. Rowell, Phys. Rev. Lett. 11, 200 (1963).
18. R.C. Jaklevic, J. Lambe, J.E. Mercereau, A.H. Silver, Phys. Rev. 140, A 1628 (1965), Phys. Rev. Lett. 12, 159 (1965), 12, 274 (1975), 14, 887 (1965).
19. R.P. Feynman, R.B. Leighton, M. Sands, The Feynman Lectures on Physics, Vol. 3, chap. 21, Addison Wesley, New York (1965).
20. P.G. de Gennes, Phys. Lett. 5, 22 (1963).
21. A. Baratoff, J.A. Blackburn, B.B. Schwartz, Phys. Rev. Lett. 25, 1096 (1970), J. Low Temp. Phys. 20, 523 (1975).
22. N.F. Pedersen, T.F. Finnegan, D.N. Langenberg, Phys. Rev. B6, 4151 (1972).
23. C.M. Falco, W.H. Parker, S.E. Trullinger, Phys. Rev. Lett. 31, 933 (1973).
24. D.A. Vincent, B.S. Deaver Jr., Phys. Rev. Lett. 32, 212 (1974).
25. J.E. Zimmerman, A.H. Silver, Solid State Commun. 4, 133 (1966), J. Appl. Phys. 39, 2679 (1968), Phys. Rev. Lett. 15, 888 (1965), Phys. Rev. 157, 317 (1967), J. Appl. Phys. 42, 30 (1971).
26. R.P. Giffard, R.A. Webb, J.C. Wheatley, Journ. Low Temp. Phys. 6, 533 (1972).
27. A.Th.A.M. de Waele, R. de Bruyn Ouboter, Physica 41, 225 (1969); 42, 626 (1969).
28. L.D. Jackel, R.A. Buhrman, W.W. Webb, Phys. Rev. 10B, 2782 (1974).
29. C.S. Owen, D.J. Scalapino, Phys. Rev. 164, 538 (1967).
30. T.A. Fulton, R.C. Dynes, Phys. Rev. Lett. 25, 794 (1970); Phys. Rev. 3, 3015 (1971), Solid State Commun. 8, 1353 (1970);

T.A. Fulton, L.N. Dunkelberger, R.C. Dynes, Phys. Rev. 6B, 855 (1972).

31. J.R. Waldram, J.M. Lumley, Revue de Physique Appliquée, 10, 7 (1975).
32. H.H. Zappe, Appl. Phys. Lett. 25, 424 (1974); 27, 432 (1975).
33. P. Guéret, Appl. Phys. Lett. 25, 426 (1974).
34. I.M. Khalatnikov, Introduction to the theory of superfluidity, Benjamin (1965), Chap. 9.
35. S. Putterman, R. de Bruyn Ouboter, Phys. Rev. Lett. 24, 50 (1970).
36. S.J. Putterman, Superfluid Hydrodynamics, North Holland (1974).
37. V. Ginzburg, Zh. Eksperim. i. Teor. Fiz. 14, 177 (1944).
38. N.V. Zavaritskii, J.E.T.P. Lett. 19, 126 (1974).
39. C.M. Pegrum, A.M. Guénault, G.R. Pickett, Low Temp. Phys. LT14, North-Holland (1975), Vol. 2, p. 513; C.M. Falco, Solid State Commun. 19, 623 (1976).
40. H.B.G. Casimir, A. Rademakers, Physica 13, 33 (1947).
41 J.G. Daunt, K. Mendelssohn, Proc. Roy. Soc. (London) A185, 225 (1946).
42. M.L. Yu, J.E. Mercereau, Phys. Rev. Lett. 28, 1117 (1972).
43. J. Clarke, Phys. Rev. Lett. 28, 1363 (1972).
44. M. Tinkham, J. Clarke, Phys. Rev. Lett. 28, 1366 (1972).
45. M. Tinkham, Phys. Rev. B6, 1747 (1972).
46. J.R. Waldram, Proc. Roy. Soc. (London) A345, 231 (1975).
47. S.R. de Groot, P. Mazur, Non-equilibrium thermodynamics, North-Holland (1969), Chap. 13, § 4 - § 6 .
48. T.J. Rieger D.J. Scalapino, J.E. Mercereau, Phys. Rev. Lett. 27, 1787 (1971); Phys. Rev. 6B, 1734 (1972).
49. L.P. Gorkov, G.M. Eliashberg, Sov. Phys. J.E.T.P. 27, 328 (1968), 29, 1298 (1969).
50. M. Cyrot, Repts. Progr. Phys. 36, 103 (1973); J. Low Temp. Phys. 2, 161 (1970).
51. W.J. Skocpol, M.R. Beasley, M. Tinkham, Journ. of Low. Temp. Phys. 16, 145 (1974).
52. S. Shapiro, Phys. Rev. Lett. 11, 80 (1963).
53. M.D. Fiske, Rev. Mod. Phys. 36, 221 (1964).
54. I.K. Yanson, V.M. Svistuno, I.M. Dmitrenko, Sov. Phys. J.E.T.P. 48, 976 (1965).
55. R.E. Eck, D.J. Scalapino, B.N. Taylor, Phys. Rev. Lett. 13, 15 (1964).
56. D.N. Langenberg, W.H. Parker, B.N. Taylor, Phys. Rev. 150, 186 (1966); Phys. Rev. Lett. 18, 287 (1967).
57. W.H. Parker, B.N. Taylor, D.N. Langenberg, Phys. Rev. Lett. 18, 287 (1967); Metrologia 3, 89 (1967).
58. T.F. Finnegan, A. Denenstein, D.N. Langenberg, Phys. Rev. B4, 1487 (1971); Rev. Mod. Phys. 41, 375 (1969).
59. J. Clarke, Phys. Rev. Lett. 21, 1566 (1968).
60. A.H. Dayem, C.C. Grimes, Appl. Phys. Lett. 9, 47 (1966).
61. C.C. Grimes, P.L. Richards, S. Shapiro, J. Appl. Phys. 39,

3905 (1968); Phys. Rev. Lett. 17, 431 (1968); Journ. of Physics 36, 690 (1968).

62. D.B. Sullivan, J.E. Zimmerman, Am. J. of Physics 39, 1504 (1971).
63. J.E. Zimmerman, A.H. Silver, Phys. Lett. 10, 47 (1964); Phys. Rev. Lett. 13, 125 (1964); 15, 888 (1965); Phys. Rev. 141, 367 (1966).
64. J. Clarke, Phil. Mag. 13, 115 (1966); Phys. Rev. Lett. 21, 1566 (1968).
65. R. de Bruyn Ouboter, M.H. Omar, A.J.P.T. Arnold, T. Guinau, K.W. Taconis, Physica 32, 1449 and 2044 (1966); 34, 525 (1967).
66. P.W. Anderson, A.H. Dayem, Phys. Rev. Lett. 13, 195 (1964).
67. H.A. Notarys, J.E. Mercereau, Physica 55, 424 (1971).
68. M. Nisenoff, Revue de Physiqe Appliquée 5, 21 (1970).
69. J.E. Mercereau, Revue de Physique Apliquée 5, 13 (1970).
70. R.A. Buhrman, J.E. Lukens, S.F. Strait, W.W. Webb, J. Appl. Phys. 41, 45 (1971).
71. A.Th.A.M. de Waele, C.P.M. Vergouwen, A.A.J. Matsinger, R. de Bruyn Ouboter, Physica 59, 155 (1972).
72. W.C. Stewart, Appl. Phys. Lett. 12, 277 (1968).
73. D.E. McCumber, J. Appl. Phys. 39, 297 and 2503 and 3113 (1968).
74. J.E. Zimmerman, J.A. Cowen, A.H. Silver, Appl. Phys. Lett. 9, 353 (1966).
75. J. Warman, J.A. Blackburn, Appl. Phys. Lett. 19, 60 (1971).
76. L.G. Aslamazov, A.I. Larkin, Sov. Phys. J.E.T.P. 9, 87 (1969).
77. L.D. Jackel, J.M. Warlaumont, T.D. Clark, J.C. Brown, R.A. Buhrman, and M.T. Levinsen, Appl. Phys. Lett. 28, 353 (1976).
78. L.D. Jackel, W.H. Henkels, J.M. Warlaumont, and R.A. Buhrman, Appl. Phys. Lett. 29, 214 (1976).
79. J.E. Zimmerman, A.H. Silver, Phys. Rev. 141, 367 (1966).
80. R.L. Forgacs, A. Warnick, Rev. Sci, Instr. 38, 214 (1967).
81. R. de Bruyn Ouboter, W.H. Kraan, A.Th.A.M. de Waele, M.H. Omar, Physica 35, 335 (1967).

Chapter 3

SUPERCONDUCTING QUANTUM INTERFERENCE DEVICES FOR LOW FREQUENCY MEASUREMENTS

John Clarke

Department of Physics

University of California

Berkeley, California 94720

and

Materials and Molecular Research Division

Lawrence Berkeley Laboratory

Berkeley, California 94720

I. INTRODUCTION

Quantum interference effects between two Josephson tunnel junctions [1] incorporated in a superconducting ring were first observed by Jaklevic et al. [2] in 1964. These workers showed that the critical current of the double junction was an oscillatory function of the magnetic flux threading the ring, the period being the flux quantum, Φ_o. The implications of this result for instrumentation were quickly realized, and a variety of dc SQUIDS* (Superconducting Quantum Interference Devices) were developed and used. These devices included several designs involving machined pieces of niobium connected with point contact junctions [3-5], and the SLUG [6] (Superconducting Low-Inductance Undulatory Galvanometer), which consisted of a bead of solder frozen onto

*The prefix "dc" indicates that the device is operated with a direct current bias, while the prefix "rf" indicates that the device operates with an rf flux bias. The rf SQUID is mis-named, as no quantum interference takes place.

a niobium wire. In the late 1960's, the rf SQUID [7-9] appeared. The rf SQUID consists of a single Josephson junction on a superconducting ring, and, presumably because only a single junction is required, has become much more widely used than the dc SQUID. Several commercial versions [10-13] of the rf SQUID, complete with sophisticated readout electronics, are available. Close attention has been paid to the optimum coupling of the rf SQUID to the room temperature electronics. As a result of this research, the present rf SQUIDs have a higher sensitivity than the first generation of dc SQUIDs. However, Clarke, Goubau, and Ketchen [14] have recently described a thin-film tunnel-junction dc SQUID that is also ideally coupled to the room temperature electronics. The sensitivity of this dc SQUID is limited by its intrinsic noise and is one-to-two orders of mangitude higher than that of the earlier dc SQUIDs; it compares favorably with that of most rf SQUIDs.

In this article, I outline the principles and operation of both dc and rf SQUIDs, and describe their applications to low frequency measurements. The general outline follows that of an earlier review [15]. Section II very briefly reviews the relevant facts of flux quantization and Josephson tunneling, and mentions the important practical configurations of Josephson junctions. Sections III and IV are concerned with the dc SQUID and the rf SQUID respectively. In each section I have used the same parallel development: Theory, operation, noise theory, fabrication and performance, and future improvements. Section V is devoted to applications of dc and rf SQUIDs. I describe first the principles of the flux transformer, and then discuss in turn magnetometers, gradiometers, susceptometers, and voltmeters, mentioning the principles involved, indicating where improvement in performance is needed, and comparing the SQUID-based devices with alternative instruments. Finally, in Section VI, I mention some of the practical applications in which SQUIDs have been used or in which they have potential use.

II. SUPERCONDUCTIVITY AND THE JOSEPHSON EFFECTS

In a superconductor, some of the free electrons are paired together [16]. These Cooper pairs are in a macroscopic quantum state that can be described by a single wavefunction [17]

$$\psi(\vec{r}, t) = |\psi(\vec{r}, t)| \exp[i\phi(\vec{r}, t)] \quad ,$$

where $|\psi|$ and ϕ are the amplitude and phase of the wavefunction. The existence of this macroscopic quantum state gives rise to several observable phenomena, for example, flux quantization [17, 18] and Josephson tunneling [1].

A. Flux Quantization

The requirement that ψ be single-valued at any point in a closed superconducting ring is expressed by the Bohr-Sommerfeld condition

$$\oint \vec{p}\cdot d\vec{s} = \oint 2m\vec{v}\cdot d\vec{s} + \oint 2e\vec{A}\cdot d\vec{s} = nh \quad . \tag{1}$$

In Eq. (1), $\vec{p} = 2m\vec{v} + 2e\vec{A}$ is the pair canonical momentum, m and e are the electron mass and charge, $\vec{v}$ is the pair-velocity, $\vec{A}$ is the vector potential, $d\vec{s}$ is an element of length, h is Planck's constant, and n is an integer. The term $\vec{v}$ is proportional to the supercurrent flowing around the ring in a penetration depth λ. Provided that the thickness of the ring is much larger than λ, we may choose the path of integration in a region of zero current, so that $\oint \vec{v}\cdot d\vec{s} = 0$. The term $\oint \vec{A}\cdot d\vec{s} = \int \mathrm{curl}\vec{A}\cdot d\vec{S} = \vec{B}\cdot d\vec{S}$ ($\vec{B}$ is the magnetic field and $d\vec{S}$ is a surface element) is just the total magnetic flux Φ in the ring, which in general consists of an applied flux and a screening flux generated by induced supercurrents. Equation (1) then reduces to the condition

$$\Phi = nh/2e = n\Phi_0 \quad . \tag{2}$$

Thus the flux contained in the ring is quantized in units of the flux quantum $\Phi_0 = h/2e \approx 2 \times 10^{-15}$ Wb.

B. The Josephson Equations

The "classical" Josephson [1] tunnel junction consists of two superconducting films separated by a thin insulating barrier. If zero current flows through the junction, the superconductors are coupled by an energy [1, 19, 20]

$$E_c = -I_c\Phi_0/2\pi \quad . \tag{3}$$

Provided that $|E_c| \gg k_BT$, the phases of the two superconductors are locked together, and a time-independent supercurrent can be passed through the barrier up to a maximum value of I_c, the critical current. The difference between the phases of the two superconductors, θ, adjusts to the externally applied current I according to

$$I = I_c \sin\theta \quad . \tag{4}$$

If a current greater than I_c is passed through the junction, a voltage V appears across it, and the supercurrent oscillates at a frequency

$$\nu = (1/2\pi)\, d\theta/dt = 2eV/h = V/\Phi_0 \quad . \tag{5}$$

C. Types of Josephson Junctions and their Current-Voltage Characteristics

There are three main types of junctions that are currently used in SQUIDs. The first is the tunnel junction [20] (Fig. 1(a)), which is fabricated by evaporating or sputtering a strip of superconductor onto an insulating substrate, oxidizing the strip thermally or by a glow discharge in oxygen, and depositing a second strip of superconductor. Early tunnel junctions were not very reliable, and alternative weak-link configurations were consequently developed and used in devices. However, in the past few years, very reproducible and reliable tunnel junctions have been produced. The most useful junctions appear to be Pb - PbOx - Pb [21-24], Nb - NbOx - Pb [25], and Nb - NbOx - Nb [26]. All of these junctions can be stored at room temperature and recycled many times without significan deterioration, and appear to be less prone to damage from electrical transients than some other types of weak link. Their critical currents have very little temperature dependence below 4.2K. Photoresist techniques [27] have been used to produce evaporated lead strips with widths down to about 1 μm.

Josephson tunnel junctions exhibit hysteresis in their I-V characteristic, as shown in Fig. 1(b). For most SQUID applications, it is essential that the I-V characteristic by non-hysteretic. The hysteresis may be removed (Fig. 1(c)) by shunting the junction with a resistance R such that the hysteresis parameter [28]

$$\beta_c = 2\pi I_c R^2 C/\Phi_o \leqslant 1 \quad , \tag{6}$$

where C is the junction capacitance (typically 2 pF for a 10 x 10 μm junction). The shunt may consist either of a small disk of normal metal that underlays the intersection of the two superconductors, or of a diagonal normal strip joining the two superconductors near their intersection [29].

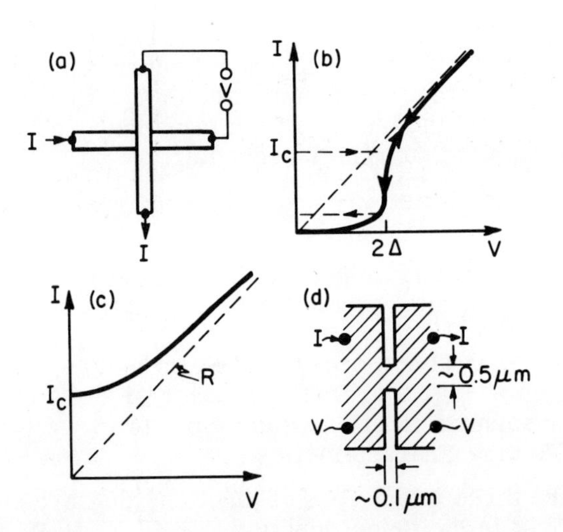

Fig. 1. (a) Josephson tunnel junction; (b) I-V characteristic of Josephson tunnel junction with identical superconductors (Δ is the energy gap); (c) I-V characteristic of shunted Josephson junction ($\beta_c \ll 1$); (d) Anderson-Dayem bridge.

In the latter case, the junction and the strip must be covered with an insulating layer (too thick to permit tunneling) and a superconducting ground plane to reduce the stray inductance to a negligible level. If $\beta_c \ll 1$, the I-V characteristic is described (in the absence of noise) by [28]

$$V = R\left(I^2 - I_c^2\right)^{1/2} . \quad (7)$$

From Eq. (7) we can immediately obtain the dynamic resistance

$$R_D = R/[1 - (I_c/I)^2]^{1/2} . \quad (8)$$

The second kind of commonly used Josephson junction is the Anerson-Dayem [30] bridge shown in Fig. 1(d). In this structure, there is no tunneling barrier, but rather a superconducting "weak link" connecting two superconducting films. The most reliable bridges are probably those fabricated from niobium [31-33], $NbSe_2$ [34], or Nb_3Ge [35]. In one version, the bridge is "weakened" by a thin normal metal underlay or overlay [8, 31].

The third widely used junction is the point contact [36]. Its properties have been reviewed extensively by Zimmerman [37]. The point contact junction consists essentially of a sharpened niobium point pressed against a niobium block. In some versions, the pressure of the contact can be changed by means of a differential screw operated from outside the cryostat. Versions that can be recycled repeatedly without adjustment have also been developed [38].

In point contacts that are clean (non-oxidized), and in bridges whose length is long compared with a coherence length, deviations [39] from the ideal sinusoidal current-phase relation [Eq. (4)] are to be expected. The voltage-frequency relation is always exact. In this article, we shall always assume a sinusoidal current-phase relation. Deviations from this behavior do not affect the principles on which the SQUIDs operate, but may give rise to substantially higher intrinsic noise [40].

III. DC SQUID

A. Theory of the dc SQUID

The dc SQUID consists of two Josephson junctions mounted on a superconducting ring (Fig. 2). When the external flux threading the ring, Φ_e, is changed monotonically the critical current of the two junctions oscillates as a function of Φ_e, with a period Φ_o. At low voltages, the voltage across the junctions at constant bias current is also periodic in Φ_e. Semi-quantitative descriptions of this behavior have been given by De Waele and R. de Bruyn Ouboter [41] and by Tinkham [42]. We will sketch the more recent numerical calculation by Tesche and Clarke [43].

We derive a set of equations describing the time-dependent behavior of the SQUID. The inductance of the ring is L, and the critical

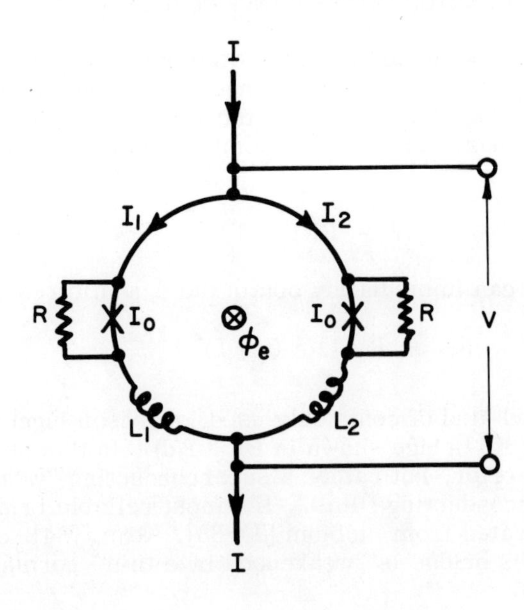

Fig. 2. Configuration of dc SQUID

current and shunt resistance of each junction are I_c and R. The SQUID is biased with a constant current I, and the currents through the two junctions are $I_1(t)$ and $I_2(t)$. Thus

$$I = I_1 + I_2 \quad . \tag{9}$$

We define a circulating current $J(t)$ as

$$J = (I_2 - I_1)/2 \quad . \tag{10}$$

The currents $I_1(t)$ and $I_2(t)$ are related to the voltages $V_1(t)$ and $V_2(t)$ and phase differences $\theta_1(t)$ and $\theta_2(t)$ across the junctions by

$$I_1 = I_c \sin\theta_1 + V_1/R \quad , \tag{11}$$

and

$$I_2 = I_c \sin\theta_2 + V_2/R \quad , \tag{12}$$

where

$$d\theta_1/dt = (2e/\hbar)\, V_1 \quad , \tag{13}$$

and

$$d\theta_2/dt = (2e/\hbar)\, V_2 \quad . \tag{14}$$

The total voltage V developed across the SQUID is

$$V = V_1 + L_1 dI_1/dt + M dI_2/dt \tag{15}$$

$$= V_2 + L_2 dI_2/dt + M dI_1/dt, \tag{16}$$

where L_1 and L_2 are the self-inductances of the two arms of the SQUID, and M is the mutual inductance between the arms.

The phase differences θ_1 and θ_2 are related by

$$\theta_1 - \theta_2 = 2\pi\Phi_T/\Phi_o \quad , \tag{17}$$

where Φ_T is the total flux in the SQUID. The total flux is the sum of the individual fluxes Φ_1 and Φ_2 produced by the currents I_1 and I_2 and the external quasistatic flux Φ_e. (Because the SQUID responses are periodic in Φ_e with period Φ_o, we make the restriction $0 \le \Phi_e \le \Phi_o$.) If we define $\ell_1 \equiv -\Phi_1/I_1$ and $\ell_2 = -\Phi_2/I_2$ we can easily show that $\ell_1 + \ell_2 = L$; we take the symmetric case $\ell_1 = \ell_2 = L/2$. The flux produced by the currents I_1 and I_2 is thus $-LI_1/2 + LI_2/2 = LJ$, and the total flux is just

$$\Phi_T = \Phi_e + LJ. \tag{18}$$

The quantities L, L_1, L_2, ℓ_1, ℓ_2, and M are related in the following way. Suppose that in some time-dependent mode $dI_1/dt \neq 0$ while $dI_2/dt = 0$. The inductive voltage drop around the ring (neglecting any contributions from the junctions) is $V = L_1 dI_1/dt - M dI_1/dt$. The rate of change of flux in the ring yields $V = \ell_1 dI_1/dt$. Hence $\ell_1 = L_1 - M$, and, similarly, $\ell_2 = L_2 - M$. Using these expressions for M and the fact that $dJ/dt = -dI_1/dt = dI_2/dt$ (since I is constant), we can reduce Eqs. (15) and (16) to

$$V = V_1 - (L/2)\, dJ/dt \tag{19}$$

$$= V_2 + (L/2)\, dJ/dt. \tag{20}$$

These equations include the effect of the mutual inductance even though M does not appear explicitly.

We define $\beta = 2LI_c/\Phi_o$. The final set of equations can be derived from Eqs. (9) to (20). Hence, from Eqs. (17) and (18)

$$\frac{J}{I_c} = \frac{\theta_1 - \theta_2}{\pi\beta} - \frac{2\Phi_e}{\beta\Phi_o} \quad ; \tag{21}$$

from Eqs. (13, (17), and (19)

$$V = \frac{\hbar}{4e}\left(\frac{d\theta_1}{dt} + \frac{d\theta_2}{dt}\right) \quad ; \tag{22}$$

and from Eqs. (9) to (14),

$$\frac{d\theta_1}{dt} = \frac{2eR}{\hbar}[(I/2) - J - I_c \sin\theta_1], \tag{23}$$

and

$$\frac{d\theta_2}{dt} = \frac{2eR}{\hbar}[(I/2) + J - I_c \sin\theta_2]. \tag{24}$$

These equations can be solved numerically to find all the characteristics of the dc SQUID (for details of the methods of solution, see Tesche and Clarke [43]). In Fig. 3 we plot the critical current I_m of the SQUID as a function of Φ_e for several values of the parameter $\beta = 2LI_c/\Phi_o$. I_m is periodic in Φ_e with period Φ_o. Notice that the curve is smooth near $\Phi_e = 0$ and Φ_o, and cusped at $\Phi_e = \Phi_o/2$. In the limit of large β, the maximum change in critical current, ΔI_m, approaches Φ_o/L; for $\beta = 1$ (a typical value for practical SQUIDs), ΔI_m is approximately $\Phi_o/2L$. In Fig. 4 we plot I-V characteristics with $\Phi_e = \Phi_o/2$ for $\beta \sim \infty$, 2.0, 1.0, and 0.4. As $\beta \to \infty$, the relative critical current modulation depth $\Delta I_m/2I_m \to 0$; this curve is thus identical with the curves for lower values of β with $\Phi_e = 0$. For $\beta = 1$, as Φ_e is changed from 0 to $\Phi_o/2$, we observe a modification of the I-V characteristic that is considerable at low voltages, but that progressively decreases as the voltage (or bias current) is increased. This decrease can be understood in the following way. The circulating current, J(t), has a maximum value at $\Phi_e = \Phi_o/2$ and oscillates at a frequency that increases as the voltage across the SQUID is increased. When this frequency becomes comparable with R/L, a further increase in the bias current causes the amplitude of J(t) to decrease, and hence the modfication of the I-V characteristic at $\Phi_e = \Phi_o/2$ also decreases.

This effect is also shown in Fig. 5, where we plot the average voltage across the SQUID as a function of Φ_e for several values of bias current and with $\beta = 1$. The decrease in the voltage modulation amplitude ΔV with increasing bias current is clearly demonstrated. Notice also that the cusp in the I_m vs Φ_e curve at $\Phi_e = \Phi_o/2$ is rounded out in the V vs Φ_e curve.

As we shall see later, the important parameter when the SQUID is used as a device is $(\partial V/\partial \Phi_e)_I$. Values of $(\partial V/\partial \Phi_e)_I$ may be deduced from Fig. 5. However, an order-of-magnitude estimate can be obtained as follows. The voltage modulation depth (the change in voltage when Φ_e is changed from $n\Phi_o$ to $(n+1/2)\Phi_o$) for a SQUID with $\beta \approx 1$ is just

$$\Delta V \approx r_D \Delta I_m \approx r_D \Phi_o / 2L \quad . \tag{25}$$

In Eq. (25), r_D ($= R_D/2$) is the dynamic resistance of the two junctions in parallel. If $r_D \approx 1\Omega$ and $L \approx 10^{-9}$ H, we find $\Delta V \approx 1 \mu V$. From Eq. (25) we find

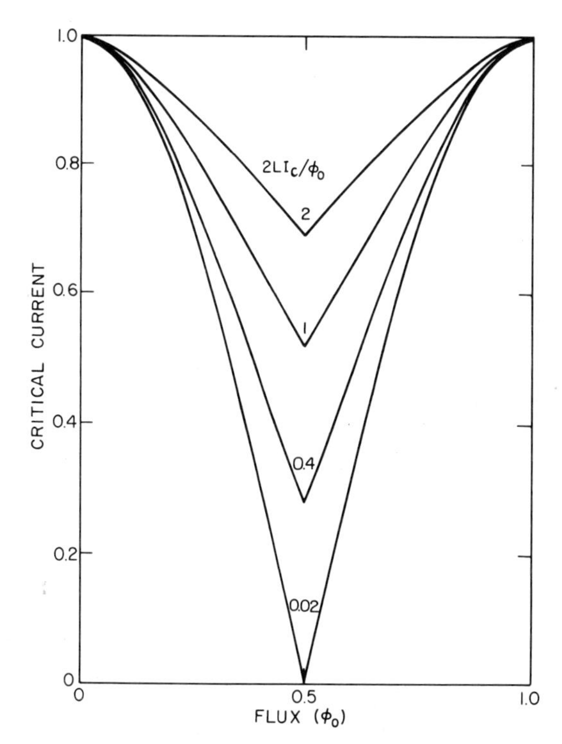

Fig. 3. Reduced critical current vs applied flux for four values of $\beta = 2LI_c/\Phi_0$.

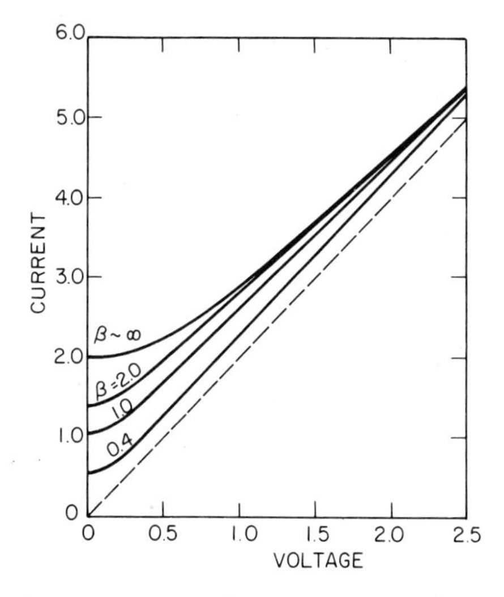

Fig. 4. Current (in units of $2I_c$) vs. voltage (in units of I_cR) for a dc SQUID for four values of $\beta = 2LI_c/\Phi_0$.

$$\left(\frac{\partial V}{\partial \Phi_e}\right)_I \approx \frac{r}{L} \quad , \tag{26}$$

about $2 \mu V \Phi_o^{-1}$. Here, r is the parallel resistance of the two shunts; we have set $r_D \approx r$.

B. Operation of the dc SQUID

To detect changes in the flux threading the dc SQUID, we bias the SQUID at a nonzero voltage with a steady current, I_o, and apply a sinusoidal modulation flux of peak-to-peak amplitude $\sim \Phi_o/2$ and frequency f = 100 kHz. This flux is generated by passing a current in the modulation coil that is mounted inside the SQUID. As illustrated in Fig. 6(a), the ac voltage across the SQUID has a large component at 2f and zero component at f when $\Phi_e = (n + 1/2)\Phi_o$. As Φ_e is increased from $(n + 1/2)\Phi_o$, the amplitude V_f of the ac signal across the SQUID at the fundamental frequency f increases (initially linearly), while the component at 2f decreases. The f component reaches a maximum at $\Phi_e = (n + 3/4)\Phi_o$ (Fig. 6(b)), and becomes zero again when $\Phi_e = (n + 1)\Phi_o$; it reverses phase at $\Phi_e = n\Phi_o$ and $(n + 1/2)\Phi_o$. Figure 6(c) shows the variation of V_f with Φ_e near $(n + 1/2)\Phi_o$. Although the exact value of $(\partial V_f/\partial \Phi_e)_{I_o}$ near $(n + 1/2)\Phi_o$ depends on the detailed shape of the V vs Φ_e curve, a reasonable estimate is $\sim 2\Delta V/\Phi_o$.

The ac voltage across the SQUID is amplified by room temperature electronics, and lock-in detected at frequency f, as shown in Fig. 7. The lock-in detector produces an output that is proportional to the amplitude of the signal across the SQUID at frequency f. The output from the lock-in is integrated. Thus, with the feedback loop open, the quasistatic output from the integrator is periodic in Φ_e (provided that the gain of the broadband amplifier in Fig. 7 is reduced almost to zero: otherwise the lock-in and integrator will saturate). When the feedback switch is closed, the output from the integrator is connected via a resistor R_F to the modulation coil. The feedback system maintains the total flux in the SQUID near either $n\Phi_o$ or $(n + 1/2)\Phi_o$, depending on the sign of the feedback. This configuration is known as the flux-locked SQUID. When the flux applied to the SQUID is changed by $\delta\Phi_e$, a current is fed back into the modulation coil that produces an opposing flux $-\delta\Phi_e$. Thus the SQUID is always operated at a constant flux and serves as a null detector in a feedback circuit. The voltage developed across R_F is proportional to $\delta\Phi_e$.

The ac-modulation and lock-in detection together with the negative feedback minimize certain sources of drift and 1/f noise, for example: changes in the critical current caused by changes in the bath temperature; drifts in the bias current, I_o; drifts in the thermal emf's in the cryostat leads; and 1/f noise in the preamplifier.

A major difficulty in the past has been the satisfactory coupling of the dc SQUID to the room temperature electronics. At 100 kHz, a state-of-the-art low noise FET preamplifier has an optimum noise temperature of about 1 K for a source impedance of about 100 kΩ.

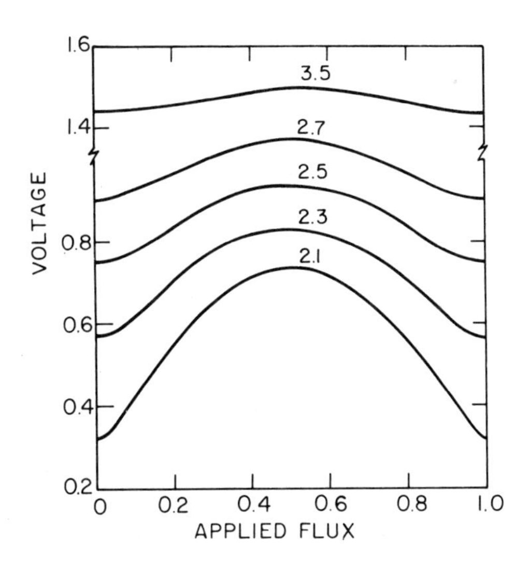

Fig. 5. SQUID voltage (in units of $I_c R$) vs. applied flux (in units of Φ_0) for $I_0/2I_c = 2.1$, 2.3, 2.5, 2.7, and 3.5.

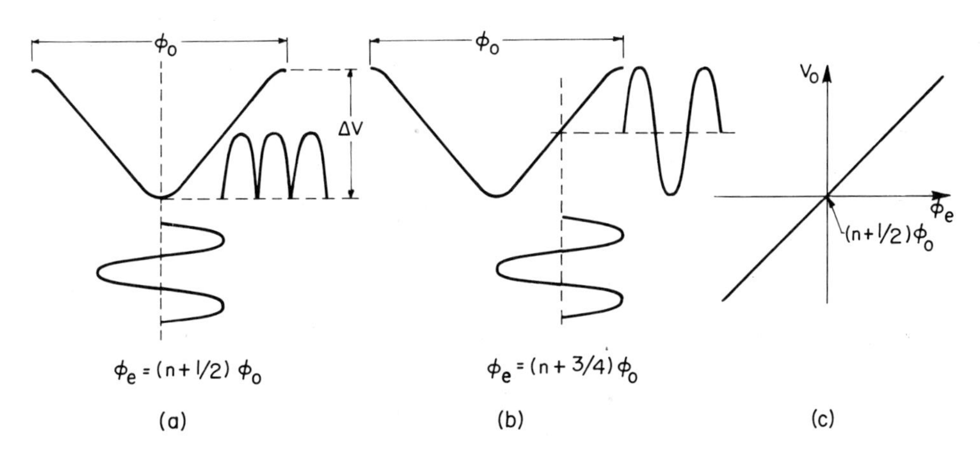

Fig. 6. Voltages across SQUID produced by an ac modulating flux of frequency f. In (a) $\Phi_e = (n+1/2)\,\Phi_0$; in (b) $\Phi_e = (n+3/4)\,\Phi_0$; the amplitude of the output voltage at f as a function of Φ_e is shown in (c) for Φ_e close to $\Phi_0/2$.

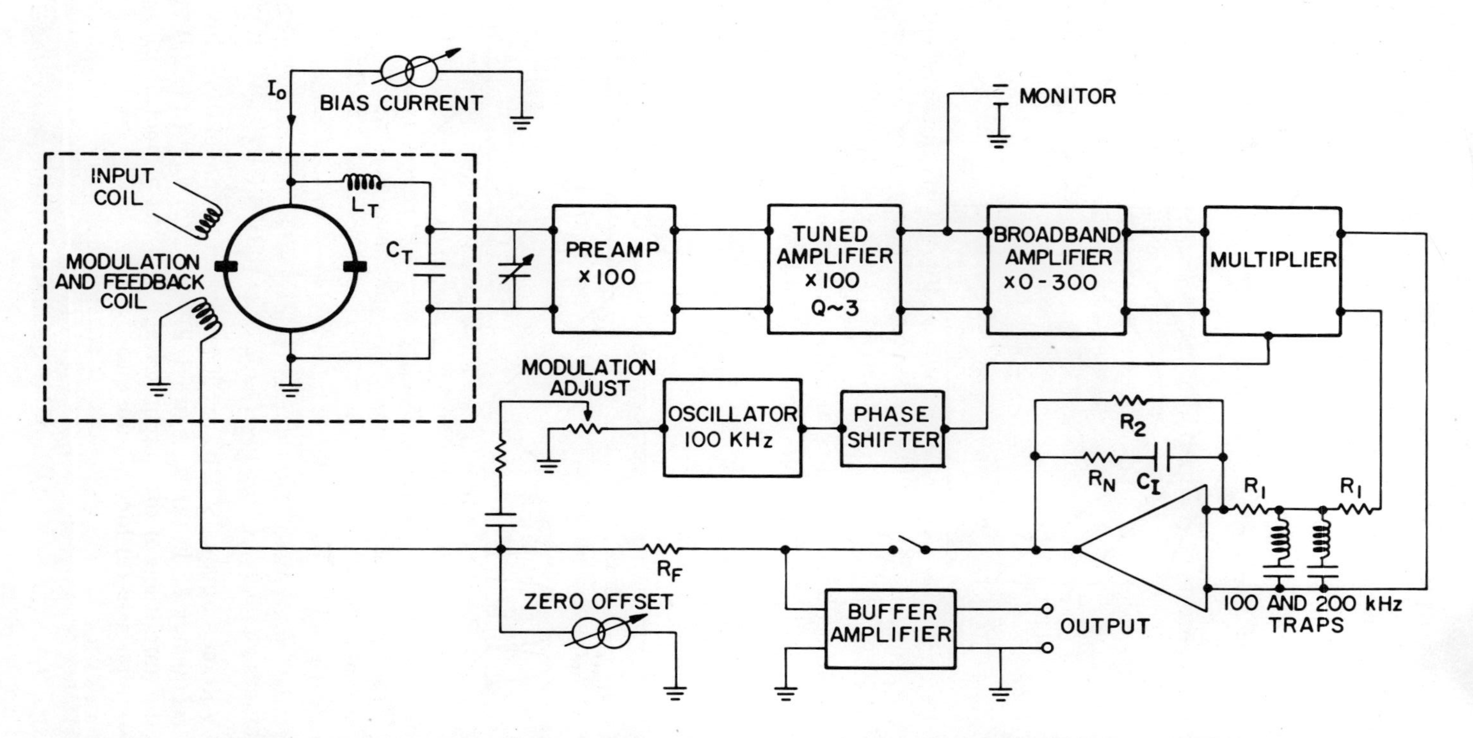

Fig. 7. Schematic of dc SQUID electronics. Components within the dashed box are at liquid helium temperature.

Therefore, it is necessary to enhance the ac impedance of the SQUID ($\sim 1\Omega$) by a factor of about 10^5 to achieve optimum low noise performance. Clarke et al. [14] achieved satisfactory impedance matching by means of a cooled LC circuit resonant at 100 kHz, as shown in Fig. 7. The tank circuit amplifies the signal by $Q \sim 150$, and amplifies the SQUID impedance by $Q^2 \sim 2 \times 10^4$.

A rough estimate of the amplitude of the signal obtained from the tank circuit can be made in the following way. Suppose an ac flux of peak-to-peak value $\Phi_0/2$ is applied to the SQUID with $\Phi_e = (n \pm 1/4)\Phi_0$. A 100 kHz voltage of peak-to-peak amplitude $\sim r_D\Phi_0/2L$ appears across the SQUID. The tank circuit amplifies this signal by an amount $Q = \omega L_T/r_D$, provided that r_D is the dominant resistance in the tank circuit (L_T is the inductance, and $\omega = 2\pi f$). The peak-to-peak signal across the tank circuit is then

$$V_T \approx \Phi_0 \omega L_T / 2L \quad . \tag{27}$$

If $L_T \approx 200\,\mu H$, and $L \approx 1\,nH$, we find $V_T \approx 100\,\mu V$. When the SQUID is operated in the feedback mode, the parameter of interest is $(\partial V_T/\partial\Phi_e)_{I_0}$ near a turning point in the V vs Φ_e curve. Since V_T represents the peak-to-peak voltage for a peak-to-peak flux Φ_0, we have*

$$\left(\frac{\partial V_T}{\partial \Phi_e}\right)_{I_0} \approx \frac{\omega L_T}{L} \quad , \tag{28}$$

about $200\,\mu V \Phi_0^{-1}$ for the parameters gives above. It should be noted that $(\partial V_T/\partial\Phi_e)_{I_0}$ is independent of Q and r_D, provided that the losses in the tank circuit are dominated by dissipation in r_D.

C. Theory of Noise in the dc SQUID

We discuss first the white noise limitations due to intrinsic SQUID noise and preamplifier noise. To properly evaluate the performance of a SQUID, one must specify not only the rms flux noise as a function of frequency, but also the SQUID inductance and the coupling coefficient between the SQUID and input coil. These factors will be discussed at the end of this section.

A detailed analysis of the intrinsic white noise in a dc SQUID has recently been carried out by Tesche and Clarke [43]. They added two independent noise terms to Eqs. (11) and (12), each representing the

---

*If the V vs Φ_e curve is represented as a series of triangles, it can be shown [9, 14] that $(\partial V_T/\partial\Phi_e)_{I_0} = (8/\pi) V_T$. In view of the other approximations involved, we have neglected the factor $(4/\pi)$ in the present analysis.

Johnson noise in one of the resistive shunts, and used a digital computer to calculate the flux noise power spectrum as a function of various parameters. We shall give here an approximate analysis [14] that yields results within a factor of two of the numerical calculation.

Consider first a single shunted junction with $\beta_c = 0$. The I-V characteristic is "rounded" by Johnson noise in the shunt, the degree of rounding increasing as the parameter $\gamma = I_c\Phi_o/2\pi k_B T$ is decreased from ∞ (T = 0) to 0 (I_c = 0). The shape of the noise-rounded I-V characteristic has been calculated by Ivanchenko and Zil'berman [44], Ambegaokar and Halperin [45], and Vystavkin et al. [46]; these calculations are in good agreement with the experiments of Falco et al. [47]. Likharev and Semenov [48] and Vystavkin et al. [46] have calculated the noise power spectrum for a shunted junction. For frequencies much less than the Josephson frequency at the given voltage bias, they find a white voltage noise power spectrum

$$S_V = \left[1 + \frac{1}{2}\left(\frac{I_c}{I_o}\right)^2\right]\frac{4k_B T R_D^2}{R} , \qquad (29)$$

where R_D is the noise-rounded dynamic resistance. An inspection of the noise-rounded I-V characteristics [44-46] for $\gamma \approx 25$ (typical for our SQUIDs) and for $I/I_c \approx 1.3$ (typical operating bias), indicates that R_D is not substantially different from its value in the absence of noise [Eq. (8)]. We shall therefore use the noise-free value of R_D in Eq. (29) to obtain the voltage noise power spectrum.

The Johnson noise in the resistive shunts affects the dc SQUID in two ways: It induces a voltage noise across the SQUID, and it induces a circulating current noise that in turn generates a flux noise in the SQUID. A detailed analysis, however, indicates that the flux noise power is typically an order of magnitude smaller than the equivalent flux noise power due to the voltage noise, and we shall neglect the flux noise contribution.

Since the only dc SQUID on which detailed noise measurements have been made is that of Clarke et al. [14], we shall present a noise analysis in which a tank circuit is used to amplify signals from the SQUID. The analysis is readily adapted to SQUIDs connected directly to a preamplifier, or with transformer coupling. The voltage noise power spectrum referred to the output of the tank circuit is*

---

*This noise estimate is reasonably accurate when the SQUID is biased near a critical current maximum (i.e., $\Phi_e \approx n\Phi_o$). However, it is less accurate when the SQUID is biased near a minimum (i.e., $\Phi_e \approx (n+1/2)\Phi_o$), because the I-V characteristic is significantly distorted from the form of Eq. (7). However, it is thought that Eq. (30) is accurate to within a factor of about 2. A more exact numerical calculation appears in Ref. 43.

$$S_V^{(t)} \approx Q^2\left[1+\frac{1}{2}\left(\frac{I_m}{I_o}\right)^2\right]\frac{4k_BTr_D^2}{R} \approx \left[1+\frac{1}{2}\left(\frac{I_m}{I_o}\right)\right]^2\frac{4k_BT\omega^2L_T^2}{r}. \quad (30)$$

The total voltage noise power spectrum is the sum of $S_V^{(t)}$ and the preamplifier voltage noise power spectra $S_V^{(p)} + S_I^{(p)}Q^4 r_D^2$. $S_V^{(p)}$ is the voltage noise power spectrum when the preamplifier input is shorted, while the second term represents the voltage generated by the preamplifier current noise flowing through a source resistance $Q^2 r_D$. The flux noise power spectrum of the SQUID is found be dividing twice* the total voltage noise power spectrum by $(\partial V_T/\partial \Phi_e)^2_{I_o}$:

$$S_\Phi = 2\left\{\left[1+\frac{1}{2}\left(\frac{I_m}{I_o}\right)^2\right]\frac{4k_BT\omega^2L_T^2}{r} + S_V^{(p)} + S_I^{(p)}\frac{\omega^4L_T^4}{r_D^2}\right\}\Bigg/\left(\frac{\partial V_T}{\partial \Phi_e}\right)^2_{I_o}. \quad (31)$$

Equation (31) indicates how S_Φ can be optimized. For a given SQUID, ac modulation frequency, and preamplifier, the only variable is L_T, which can be varied without affecting the resonant frequency by changing C appropriately. Since $(\partial V_T/\partial \Phi_e)^2_{I_o} \propto \omega^2 L_T^2$, the contribution of the SQUID voltage noise term to S_Φ is independent of L_T, and the problem is reduced to one of choosing L_T to optimize $\left[\left(S_V^{(p)}/\omega^2L_T^2\right) + \left(S_I^{(p)}\omega^2L_T^2/r_D^2\right)\right]$. The optimum value is given by $S_V^{(p)}/S_I^{(p)} = Q^4 r_D^2 = \omega^2 L_T^4/r_D^2$. Since the noise temperature of a state-of-the-art FET preamplifier is about 1 K, the contribution of the preamplifier noise to S_Φ when the SQUID is operated at 4 K is almost negligible.

One very important consideration concerning the noise characterization of SQUIDs will be mentioned here, although it is equally applicable to both dc and rf SQUIDs. In almost all practical applications, the magnetic flux to be measured is coupled into the SQUID be means of a second superconducting coil that we shall refer to as the input coil (see Section V for examples). The relevant noise parameter then involves not only the flux noise of the SQUID, but also how efficiently the input

*The factor of 2 arises because of the effects of the 100 kHz modulation and lock-in detection scheme. Noise is mixed into the bandwidth around 100 kHz both from frequencies near zero, and from frequencies around 200 kHz. This calculation is carried through in detail in Appendix C of Ref. 14.

coil is coupled to the SQUID. Let the input coil of inductance* L_i have a mutual inductance M_i with the SQUID, where $M_i^2 = \alpha^2 L_i L$. Let the minimum detectable current change per $\sqrt{Hz}$ in the input coil be $\delta I_i = S_\Phi^{1/2}/M_i$. We then take as our figure of merit the energy per Hz associated with the current, $\epsilon = L_i(\delta I_i)^2/2$, or

$$\epsilon = S_\Phi L_i/2M_i^2 = S_\Phi/2\alpha^2 L \quad . \tag{32}$$

We see that ϵ represents the minimum energy resolution of the SQUID per Hz, $S_\Phi/2L$, multiplied by $1/\alpha^2$. This figure of merit, or a variation of it, has been used by a number of authors [9, 14, 15, 49-53]. It should be noted, however, that it is strictly valid only in the zero-frequency limit. As has been emphasized by Claassen [53], and as we shall see explicitly in Section V, this figure of merit is appropriate for all SQUID applications at low frequencies, and provides a meaningful way of comparing different SQUIDs. To optimize SQUID sensitivities, one should seek to minimize $S_\Phi/2\alpha^2 L$, rather than S_Φ. In practice, one usually determines ϵ by measuring S_Φ, L_i, and M_i. It is difficult to measure the parameters α and L separately, although $\alpha^2 L$ is of course known once L_i and M_i are determined.

Equation (32) can be used to express the intrinsic SQUID energy resolution in a particularly useful way. If we assume that the preamplifier noise is negligible, by using Eq. (28) with Eq. (31), we find

$$S_\Phi/2\alpha^2 L \approx 4k_B T/(r/L), \tag{33}$$

where we have neglected $(I_m/I_o)^2/2 \le 1/2$. The energy resolution is thus $4k_B T$ divided by the characteristic frequency or "sampling frequency", of the SQUID, r/L.

D. Practical dc SQUIDs: Fabrication and Performance

The earliest practical dc SQUIDs were the point contact version of Zimmerman and Silver [36] (Fig. 8), and the SLUG [6]. The SQUID consisted of a split hollow niobium cylinder, with the two halves rigidly clamped together, but electrically insulated with mylar spacers. Two sharpened niobium screws in one half-cylinder could be adjusted to make point contact junctions with the other half-cylinder. Buhrman, Strait, and Webb [38] used thin sheets of glass expoxied to the niobium

---

*Note that, in general, the inductance of the input coil when it is coupled to the SQUID is lower than when it is free-standing. The reduction in inductance is a result of the ground-planing effect of the SQUID. Throughout this article, L_i refers to the inductance of the coil when it is coupled to the SQUID.

for electrical insulation. The thermal expansion coefficient of the glass was matched to that of the niobium, so that the SQUID could be thermally recycled. The most sophisticated (published) electronics for these early dc SQUIDs was that of Forgacs and Warnick [5], who achieved an rms flux resolution of about $10^{-3}\,\Phi_0\,\mathrm{Hz}^{-1/2}$. This sensitivity seems to be typical of the early dc SQUIDs, and apparently was limited by preamplifier noise rather than intrinsic SQUID noise.

Relatively little development of the dc SQUID took place in the late 1960's and early 1970's, presumably because of the growing interest in the rf SQUID. However, there has recently been a revival of interest in the dc SQUID. Mercereau and coworkers [54, 55] and Richter and Albrect [56] have fabricated thin-film planar dc SQUIDS using Dayem bridges. Mercereau and colleagues used Nb, Nb_3Sn, and NbN thin films that are extremely stable under thermal cycling. Clarke, Goubau and Ketchen [14] developed the thin-film tunnel junction dc SQUID, and since this version appears to have the best performance, I shall discuss it in detail.

The substrate is a fused quartz tube 20 mm long with an outside diameter of 3 mm (Fig. 9). A band of Pb/In alloy (approximately by wt. 10% indium) about 11 mm wide and 3000 Å thick is evaporated around the tube. A 250 μm wide 750 Å thick gold film is then evaporated: this film is the shunt for the tunnel junctions. Next, two 150 μm wide 3000 Å thick niobium films, separated by 1.2 mm, are dc sputtered onto the

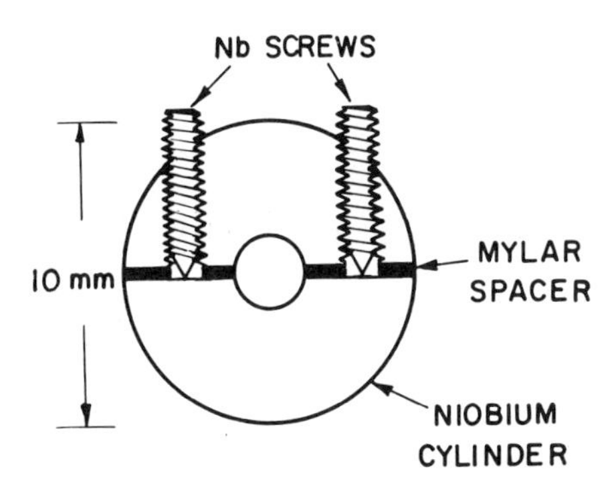

Fig. 8. Point contact dc SQUID

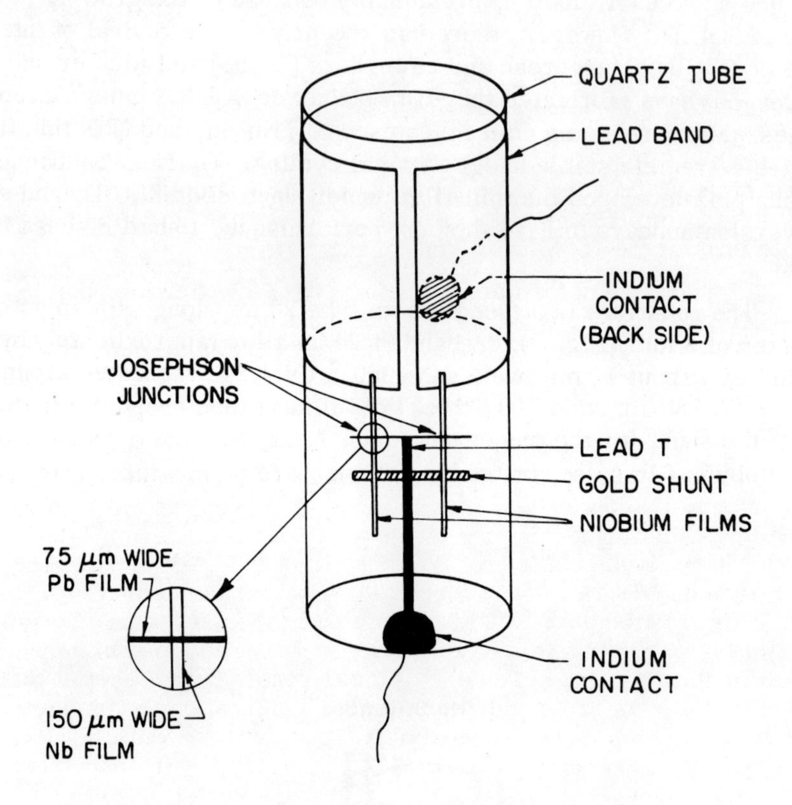

Fig. 9. Thin-film tunnel-junction dc SQUID.

cylinder. (Each niobium film makes a low resistance contact with the gold film, and at low temperatures, a superconducting contact with the Pb/In band.) The niobium is thermally oxidized, and, immediately afterwards, a 3000 Å thick Pb/In tee is deposited. The crossbar of the T overlaps the niobium strips to form two tunnel junctions, each with an area of about 10^{-2} mm^2. The stem of the tee bisects the gold strip between the niobium films to form a shunt for each junctions. Next, the Pb/In band is scribed with a razor blade midway between the niobium strips. Two indium beads are pressed on as contacts, one on the base of the tee, and the other on the Pb/In band on the reverse side of the cylinder. The entire sensor is coated with a thin insulating layer of Duco cement, applied by immersing the sensor in a solution of 1 part Duco cement in 5 parts (by volume) acetone. Finally, a 3000 Å thick Pb/In ground plane (not shown in Fig. 9) is evaporated over the front surface of the SQUID. The ground plane reduces flux leakage through the slit in the Pb/In band and minimizes the inductance of the various metal strips. A 500 Å overlay of silver is deposited on top of the ground plane to protect the Pb/In film from oxidation.

Typical parameters for the sensor are: Capacitance per junction — 200 pF; total critical current per junction — 1 to 5 μA; and parallel shunt resistance — 0.5Ω. The free standing inductance of the cylindrical part of the SQUID is approximately 0.75 nH, while the (parasitic) inductance of the niobium and Pb/In strips is estimated to be about 0.5 nH.

For most applications, it is essential to screen the SQUID from environmental magnetic field fluctuations. Excellent shielding may be obtained by mounting the SQUID inside a cylindrical tube machined from lead, 50/50 tin-lead solder, or niobium, as shown in Fig. 10. The SQUID is mounted on two Delrin rods inserted in support screws, as shown. The ac modulation and feedback coil (see Section III-C), typically two turns of 50 μm diameter Formvar-covered niobium wire of inductance 10 nH, is wound on one of these rods. It is essential that the whole structure be very rigidly mounted to avoid microphonic noise. The cylinder also acts as a ground plane to reduce the inductance of the SQUID cylinder to roughly 0.5 nH. Thus the total SQUID inductance is about 1 nH.

It is usually necessary to couple external signals into the SQUID by means of a superconducting coil coupled as closely as possible to the SQUID. We have made satisfactory coils in the following way. The coil is wound from 75 μm diameter insulated niobium wire on a teflon rod whose diameter is about 50 μm greater than the outer diameter of the SQUID. The coil is coated with Duco cement, and, when the cement is dry, the coil is carefully removed from the teflon rod and mounted on the SQUID. A typical 24-turn single-layer coil had the following parameters: $L_i = 356$ nH, $M_i = 11.5$ nH, and $\alpha \approx 0.6$.

To operate the SQUID, the bias current, I_0, is varied (with the feedback loop open) until a maximum value of V_T is obtained. Typically, $I_0 \approx 1.3\, I_m$. At this bias point, r_D is about 1Ω, and the tank circuit Q

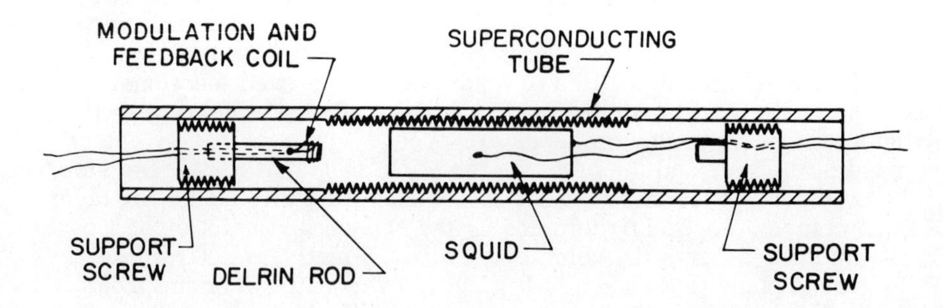

Fig. 10. SQUID mounted in superconducting shield.

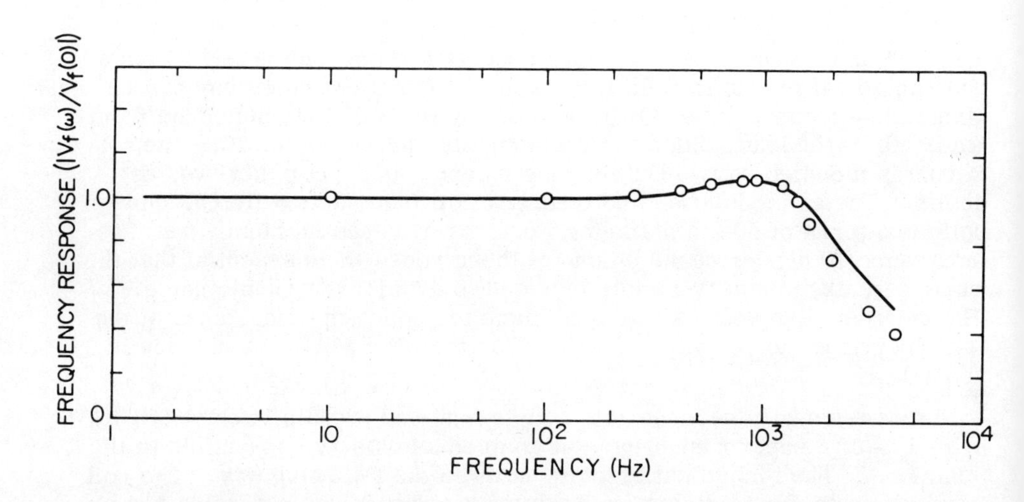

Fig. 11. Typical frequency response for flux-locked dc SQUID: solid line is theoretical, and circles are measured data.

is about 150. The impedance $Q^2 r_D$ presented to the FET is, therefore, about 20 kΩ, somewhat below the value for optimum noise performance. The dynamic range of the flux-locked SQUID in a 1 Hz bandwidth is about $\pm 3 \times 10^6$. A typical frequency response (open circles) for the system is shown in Fig. 11, compared with the theoretical curve (solid line). The response is approximately flat from 0 to 1 kHz.* A typical slewing rate is $2 \times 10^4\ \Phi_0 \mathrm{sec}^{-1}$*. A detailed discussion of the frequency response and slewing rate has been given elsewhere [14].

A typical noise power spectrum for the dc SQUID in a superconducting shield is shown in Fig. 12. The left-hand ordinate is labeled in units of S_Φ, and the right-hand ordinate is labeled in units of $S_\Phi/2\alpha^2 L$ for the 24-turn coil described previously. The power spectrum was taken by digitizing the signal from the output of the flux-locked system, and storing the digitized signal in a PDP-11/20 computer. A fast Fourier transform of this signal was taken, squared, and stored, and the process repeated, typically 30 times, to obtain an averaged power spectrum.

The noise of the SQUID is nearly white between 2×10^{-2} Hz and 200 Hz, with a rms value of about $3.5 \times 10^{-5}\ \Phi_0 \mathrm{Hz}^{-1/2}$. The energy resolution for the 24-turn coil is about $7 \times 10^{-30}\ \mathrm{JHz}^{-1}$. The roll-off above 200 Hz is a result of filtering in the electronics. The rms noise predicted by Eq. (31) with $(\partial V_T/\partial \Phi_e)_{I_0} = 150\ \mu V/\Phi_0$ (measured), $I_0/I_m = 1.3$, $L = 10^{-9}$ H, $L_T = 200\ \mu$H, $T = 4.2$ K, $r = 0.5\Omega$, and $S_V^{(p)} = 2 \times 10^{-18}\ V^2 \mathrm{Hz}^{-1}$ is $3.2 \times 10^{-5}\ \Phi_0 \mathrm{Hz}^{-1/2}$. (The current noise term is negligible because L_T was somewhat below the optimum value.) This calculated value is in remarkably good agreement with the measured value. Given the approximations made in the theory, the excellent agreement must be considered somewhat fortuitous. When the temperature of the SQUID was lowered to 1.8 K, the white noise was reduced to about $2 \times 10^{-5}\ \Phi_0 \mathrm{Hz}^{-1/2}$; the excellent agreement is again probably fortuitous. However, the fact that the flux resolution of the SQUID improves as the temperature is lowered demonstrates that the measured noise is dominated by intrinsic thermal noise in the SQUID.

Below 2×10^{-2} Hz, the power spectrum varies approximately as 1/f, with a mean square value of about 10^{-10} (1 Hz/f) $\Phi_0^2\ \mathrm{Hz}^{-1}$. The origin of this 1/f noise is not firmly established. Clarke and Hawkins [57] measured the 1/f noise in single Josephson junctions, and established that it was generated by equilibrium temperature fluctuations in the junctions. The magnitude of the noise was in good agreement with an appropriately modified version of the theory of Clarke and Voss [58] for 1/f noise in metals. However, the measured 1/f noise power in the SQUID is about two orders of magnitude larger than that expected if the noise originated in thermal fluctuations in the junctions. It is possible that the 1/f noise is produced by the motion of flux pinned in the SQUID or its superconducting shield.

*By replacing the tank circuit with a tuned transformer with a bandwidth of about 20 kHz, we have improved the frequency response to 50 kHz and the slewing rate to $2 \times 10^5\ \Phi_0\ \mathrm{sec}^{-1}$.

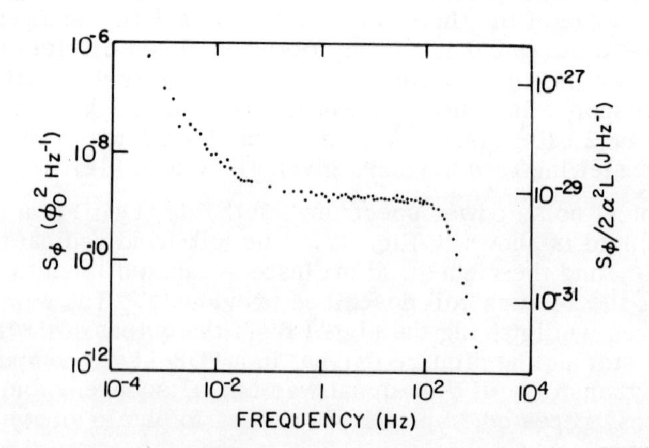

Fig. 12. Typical noise power spectrum for tunnel junction dc SQUID. The right-hand axis specifies the energy resolution with respect to a 24-turn input coil.

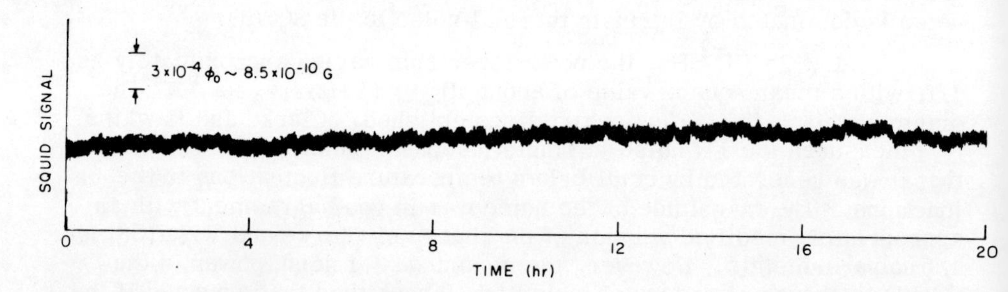

Fig. 13. Long-term drift of a flux-locked SQUID with the temperature of the helium bath regulated at a nominal 4.2 K. The measurement bandwidth is 0 to 0.25 Hz.

The output of the flux-locked SQUID drifted slowly with time. A typical drift observed over a 20 h period is shown in Fig. 13. The average drift is about $2 \times 10^{-5} \Phi_0 h^{-1}$. To achieve this low drift rate, it was necessary to regulate the temperature of the helium both to within ± 50 μK. This regulation was achieved by controlling the pressure of the He^4 vapor in the cryostat. The temperature of the SQUID was measured by a carbon resistor in an ac bridge whose output was used to regulate a valve through which the helium gas was vented. This technique compensated for changes in temperature resulting both from atmospheric pressure fluctuations and from the decrease in hydrostatic pressure of the helium bath as the liquid evaporated.

The drift in the output of the flux-locked SQUID arises from drifts in the temperature of the SQUID. Detailed measurements have been made of this temperature dependence [14]. There are two contributions to the drift. The first contribution is proportional to the change in temperature and to the residual static magnetic field trapped in the superconducting shield around the SQUID. It is likely that the effect arises from the reversible motion of flux lines trapped in the shield as the temperature is changed. The amplitude of the drift produced by a given change in temperature and for a given trapped field depends on the material, being largest for solder, smaller for lead, and smallest for niobium. Niobium is, therefore, the preferred material for the shield. For a niobium shield, the drift was of the order of $1 \Phi_0 K^{-1} G^{-1}$ at 4.2 K.

The second component of temperature-related drift is independent of the trapped magnetic field, and is evident only when the trapped field is small, 10 mG or less [14]. It is thought that this contribution is related to an asymmetry in the SQUID. If the two junctions of the SQUID are not identical, I_0 divides unequally between them, thereby linking flux to the SQUID. If the critical currents change with temperature, the flux generated by I_0 also changes with temperature. The magnitude of the magnetic-field independent component was not greater than $0.1 \Phi_0 K^{-1}$.

Both contributions to the drift can be minimized by regulating the temperature of the helium bath.

E. Future Improvements in the dc SQUID

At frequencies above about 10^{-2} Hz the tunnel-junction dc SQUID is limited by its intrinsic noise. With the present design, no improvements in the noise of the device are possible, except by lowering the temperature. For most applications, it is not practical to operate the SQUID at temperatures other than 4.2 K. From Eq. (31), we see that for the intrinsic noise S_Φ is proportional to L^2/r, and, consequently that $S_\Phi/2\alpha^2 L$ is proportional to L/r.* Thus the figure of merit can be

*It is noteworthy that S_Φ/L is proportional to L/r, the time constant of the SQUID. Under optimum condition, the dc SQUID operates at a Josephson frequency near the frequency r/L; the higher the frequency, the

lowered by decreasing L and/or increasing r. It is not practical to decrease L significantly without also decreasing α: if the cylindrical part of the SQUID is made longer or thinner, the parasitic inductances will become more important. The only way to decrease $S_\Phi/2\alpha^2 L$ is to increase r. Because of the restriction on the hysteresis parameter ($\beta_c = 2\pi I_c R^2 C/\Phi_0 \leqslant 1$), an increase in r must be accompanied by a reduction in either I_c or r. Because of noise-rounding of the I-V characteristic, it is undesirable to reduce I_c much below its present value, and, therefore, C must be reduced. This reduction can be achieved only by decreasing the area of the tunnel junctions. Since, for fixed β_c, $S_\Phi \propto 1/R \propto C^{\frac{1}{2}}$, a reduction in the junction size by four orders of magnitude to 1x1 μm (a size presently attainable by state-of-the-art photoresist techniques) would decrease S_Φ by two orders of magnitude. There is no known reason why such an improvement in resolution should not be obtained in the white noise region. However, it is likely that the reduction in the volume of the junctions would increase the 1/f noise. In the present SQUID, the measured 1/f noise power spectrum is two orders of magnitude greater than that expected from the intrinsic 1/f noise in the junctions, which is inversely proportional to the junction volume. Thus a four-order-of-magnitude reduction in the junction volume is expected to increase the 1/f noise power spectrum of the SQUID by two orders of magnitude.

IV. RF SQUID

A. Theory of the rf SQUID

The operation of the rf SQUID has been described by a number of authors [7-9, 15, 40, 59]. I shall generally follow the description given in my earlier article [15], but include recent ideas of Jackel and Buhrman [40] that are important in the subsequent noise analysis.

The rf SQUID (Fig. 14) consists of a superconducting ring of inductance L (typically 10^{-9} H) containing a single Josephson junction of critical current I_c, shunted by a resistance R and a capacitance C. We assume that the junction obeys the sinusoidal current-phase relation [Eq. (4)], and that $\beta_c \ll 1$, so that the I-V characteristic is non-hysteretic (the cases with a non-sinusoidal current-phase relation and hysteretic I-V characteristics are described by Jackel and Buhrman [40]). The critical current is usually chosen to be about Φ_0/L.

Fluxoid quantization [60] imposes the constraint

$$\theta + 2\pi\Phi/\Phi_0 = 0 \tag{34}$$

---

better is the energy resolution. The upper limit on the Josephson frequency is Δ/h, about 3×10^{11} Hz in lead or niobium, although quasiparticle relaxation processes may set an appreciably lower limit.

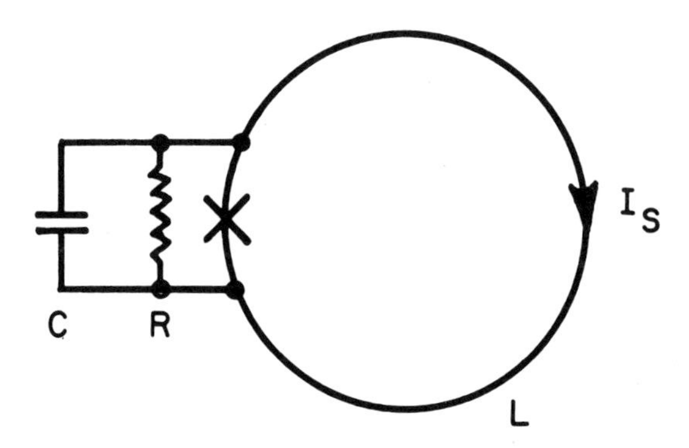

Fig. 14. Configuration of rf SQUID.

on the flux Φ threading the ring and the phase difference θ across the junction. The phase difference θ determines the current I_s flowing around the ring,

$$I_s = -I_c \sin(2\pi\Phi/\Phi_o) \quad . \tag{35}$$

A quasistatic external flux Φ_e will thus give rise to a total flux

$$\Phi = \Phi_e - LI_c \sin(2\pi\Phi/\Phi_o) \tag{36}$$

in the ring. The variation of Φ with Φ_e is sketched in Fig. 15(a) for $LI_c = 1.25\,\Phi_o$. The regions with positive slope are stable, whereas those with negative slope are not. Suppose that Φ_e is now slowly increased from zero (Fig. 15(b)). The total flux Φ will increase less rapidly than Φ_e as the circulating current I_s tends to screen out Φ_e; if the ring were completely superconducting, the screening would be exact, and Φ would remain at zero. When I_s just exceeds I_c, at an applied flux Φ_{ec} and an enclosed flux Φ_c, the junction switches momentarily into a non-zero voltage state, and the SQUID jumps from the $k = 0$ quantum state into the $k = 1$ quantum state. The time for this transition is $\sim L/R$. If Φ_e is increased further, the SQUID will make transitions into the $k = 2, 3 \ldots$ states at $\Phi_e = \Phi_c + \Phi_o, \Phi_c + 2\Phi_o, \ldots$. Suppose Φ_e is now decreased from just above Φ_c . The SQUID will remain in the $k = 1$ state until $\Phi_e = \Phi_o - \Phi_c$, at which point I_s again exceeds the critical current and the SQUID returns to the $k = 0$ state. In the same way, as Φ_e is lowered to below $-\Phi_c$ and then increased again, a second hysteresis loop will be traced out.

This hysteretic behavior occurs if $LI_c > \Phi_0/2\pi$ for a sinusoidal current-phase relation. If $LI_c < \Phi_0/2\pi$, no hysteresis occurs: SQUIDs operated in this limit are in the so-called "inductive mode" [31,61]. In this article we shall be concerned only with the "hysteretic mode" in which most SQUIDs are operated. We also note that in practice, thermodynamic fluctuations cause transitions between quantum states to occur at lower values of Φ_e than those just described. The resultant uncertainty in the value of Φ_e at which transitions occur is the source of intrinsic noise (see Section IV-D).

The energy ΔE dissipated in going around a single hysteresis loop is given by the area of the loop divided by L. By inspection of Fig. 5(b), we find

$$\Delta E = \Phi_0 (2\Phi_{ec} - \Phi_0)(1 - \Phi_c/\Phi_{ec})/L \sim \Phi_0 I_c \tag{37}$$

if $LI_c \sim \Phi_0$ and $\Phi_c/\Phi_{ec} << 1$.

We now consider the rf operation of the SQUID. The SQUID is inductively coupled to the coil of an LC-resonant circuit, as shown in Fig. 16. L_T, C_T, and R_T are the inductance, capacitance, and parallel resistance of the tank circuit, and $\omega/2\pi$ is the resonant frequency, typically a few tens of MHz. The tank circuit is excited at its resonant frequency by an rf current $I_{rf} \sin \omega t$, and the voltage across the tank circuit is amplified by a preamplifier with a high input impedance. Suppose initially that $\Phi_e = 0$. When I_{rf} is very small, the peak flux applied to the ring, $MI_T = QMI_{rf}$, is less than Φ_{ec}, and no dissipation occurs in the SQUID ($Q = R_T/\omega L_T$, $M^2 = K^2 LL_T$, and I_T is the peak current in the tank coil). The tank circuit voltage, V_T, is initially a linear function of I_{rf}, as shown in Fig. 17. As I_{rf} is increased, the peak flux will equal Φ_{ec} when $I_T = \Phi_{ec}/M$ or $I_{rf} = \Phi_{ec}/MQ$, at A in Fig. 17. The corresponding peak voltage across the tank circuit is

$$V_T^{(n)} = \omega L_T \Phi_{ec}/M \quad , \tag{38}$$

where the suffix (n) indicates $\Phi_e = n\Phi_0$, in this case with $n = 0$. At this point, the SQUID makes a transition to <u>either</u> the $k = +1$ <u>or</u> the $k = -1$ state, depending on the direction of the rf flux. Later in the rf cycle, the SQUID returns to the $k = 0$ state. As the SQUID traverses the hysteresis loop, energy ΔE is extracted from the tank circuit. Because of this loss, the peak flux on the next half-cycle does not exceed the critical flux, and no transition occurs. The tank circuit takes many cycles to recover sufficient energy to induce a further transition; this transition may be into either the $k = +1$ or $k = -1$ states. In practice, the

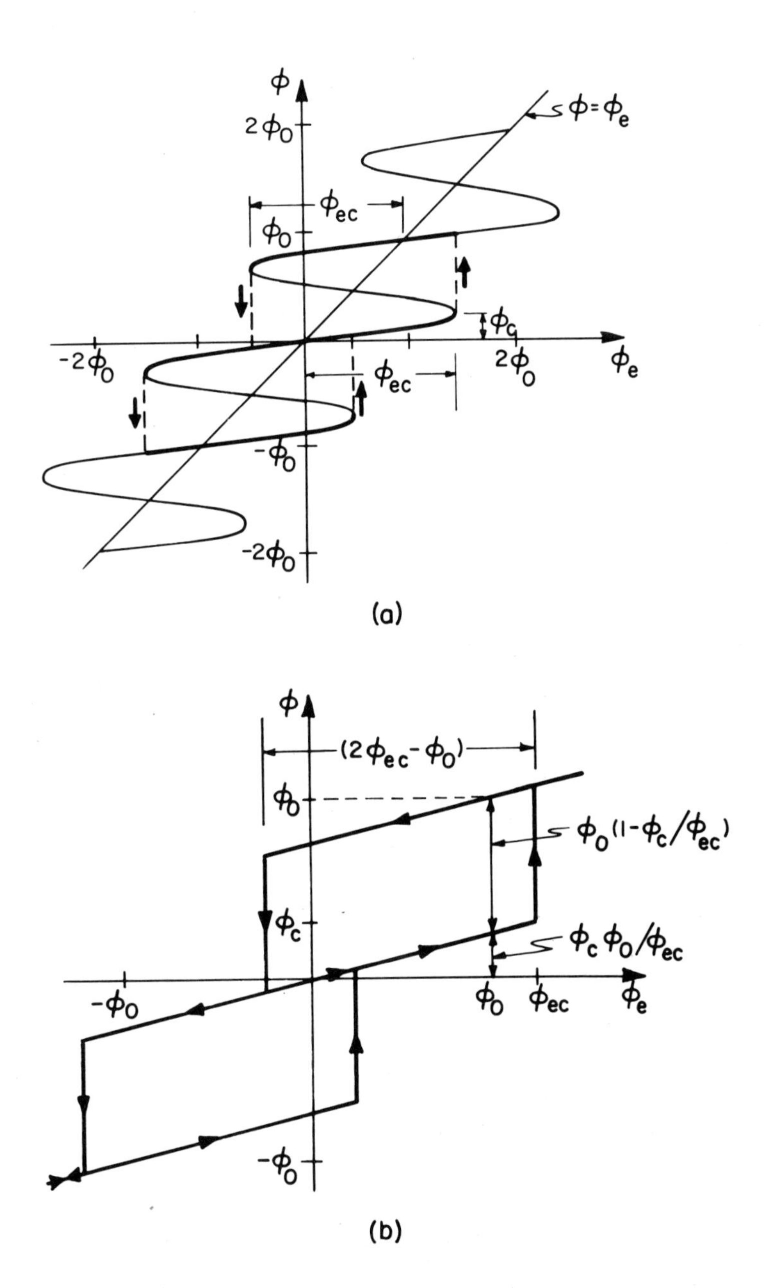

Fig. 15. rf SQUID: (a) plot of enclosed flux (Φ) vs Φ_e for LI_c = $1.25\ \Phi_0$; (b) values of Φ as Φ_e is slowly increased and then decreased.

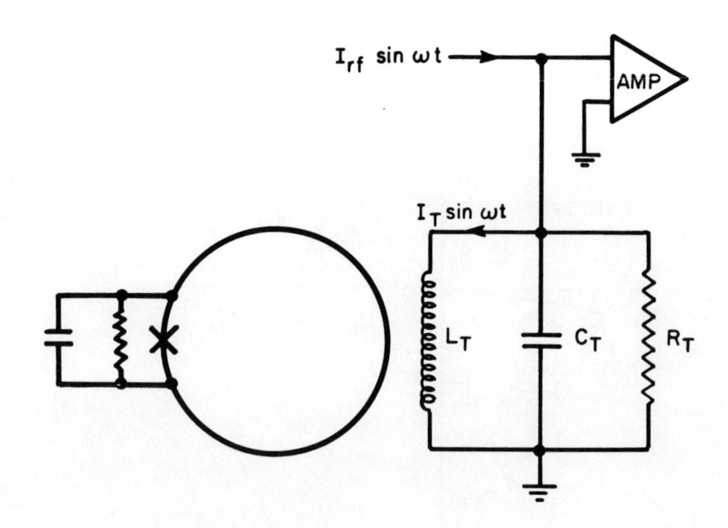

Fig. 16. Tank circuit inductively coupled to rf SQUID.

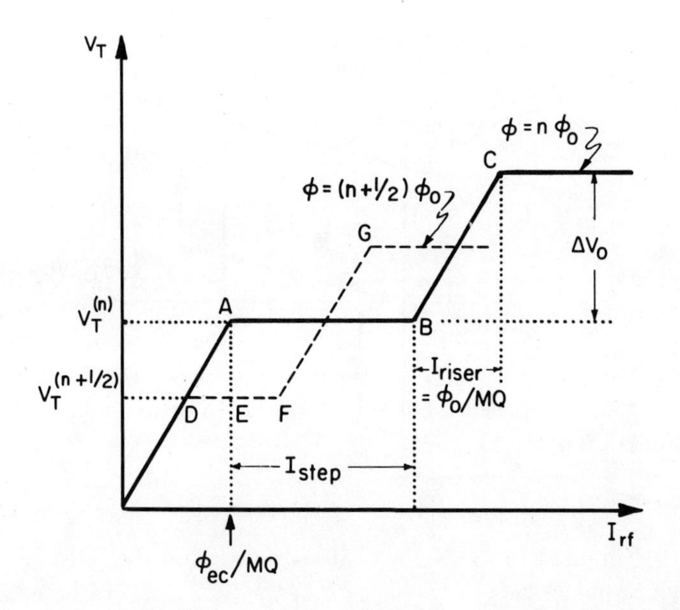

Fig. 17. V_T vs I_{rf} for rf SQUID in absence of thermal fluctuations.

slow recovery of V_T after each pair of transitions is not observed, probably because the fraction loss of tank circuit energy $[\Delta E/(L_T\Phi_{ec}^2/2M^2) \sim 2K^2$ if $LI_c \sim \Phi_0]$ is rather small, and the V_T vs I_{rf} curves are appreciably noise-rounded in practice.

If I_{rf} is now increased, transitions occur at the same values of I_T and V_T, but because energy is supplied at a higher rate, the stored energy builds up more rapidly after each transition, and transitions occur more frequently. As I_{rf} is increased, the "step" AB in Fig. 17 is traced out. At the midpoint of AB, a transition occurs once each cycle on either the positive or negative peak of the rf cycle. At B, a transition is induced on each positive and negative peak. The excess power supplied at B over that at A is just $2\Delta E(\omega/2\pi)$, the power dissipated when two hysteresis loops are traversed on each rf cycle.

The length of the step AB is given by [40]*

$$V_T^{(n)} I_{step}/2 = 2\Delta E(\omega/2\pi) \tag{39}$$

or

$$I_{step} \approx 2I_c M/\pi L_T , \tag{40}$$

where we have used Eqs. (37) and (38), and assumed $\Phi_{ec} \approx \Phi_0$.

A further increase in I_{rf} beyond B produces a "riser" BC (see Fig. 17). At C, transitions from the $k = \pm 1$ to the $k = \pm 2$ states occur. The rf flux applied to the SQUID is Φ_{ec} at B and $(\Phi_{ec} + \Phi_0)$ at C. Thus $I_{riser} = \Phi_0/MQ$. In an analogous way, a series of steps and risers is observed as I_{rf} is further increased. The separation of successive steps is

$$\Delta V_o = Q I_{riser} \omega L_T = \omega L_T \Phi_0/M \quad . \tag{41}$$

Now apply a positive external flux $\Phi_e = \Phi_0/2$ to the SQUID. This flux has the effect of shifting the hysteresis loops of Fig. 14 by $\Phi_0/2$. Thus a transition will occur on the positive peak of the rf cycle at a flux equal to $(\Phi_{ec} - \Phi_0/2)$, whereas on the negative peak, the required flux is $(\Phi_{ec} + \Phi_0/2$. Thus as I_{rf} is increased from zero, the first step will occur at D in Fig. 17 when

---

* Note that $V_t^{(n)}$ and I_{step} are peak rather than rms values, so that the power dissipation is $V_T^{(n)} I_{step}/2$, assuming that the voltage and current are in phase.

$$V_T^{(n+1/2)} = \omega L_T(\Phi_{ec} - \Phi_0/2)/M \quad . \qquad (42)$$

As I_{rf} is increased along DF, only one hysteresis loop is traversed, corresponding to the $k=0$ to $k=1$ transition at $(\Phi_{ec} - \Phi_0/2)$. As I_{rf} is further increased, V_T rises to G, with I_{riser} again equal to Φ_0/MQ. At G, transitions at a peak rf flux - $(\Phi_{ec}+\Phi_0/2)$ begin. Thus a series of steps and risers is obtained for $\Phi_e = \Phi_0/2$, interlocking those obtained for $\Phi_e = 0$.

As Φ_e is increased from 0 to $\Phi_0/2$, the value of V_T at which the first step occurs steadily decreases. For $0 < \Phi_e < \Phi_0/2$ the first step splits into two distinct steps, the lower corresponding to the transition between $k=0$ and $k=1$ at an rf flux $(\Phi_{ec} - \Phi_e)$, and the upper corresponding to the transition between $k=0$ and $k= -1$ at an rf flux $(\Phi_{ec}+\Phi_e)$. For the special case $\Phi_e = \Phi_0/2$, the steps corresponding to the transitions between $k=0$ and $k = -1$ and those between $k=1$ and $k=2$ occur at the same value of V_T. The steps for higher order transitions are similarly split except when $\Phi_e = 0$ or $\Phi_0/2$. For $\Phi_0/2 < \Phi_e \Phi 3\Phi_0/2$, the lowest energy state of the SQUID in the absence of rf flux is the $k=1$ state. Thus, in this range, the lowest order transitions induced by the rf flux are from the $k=1$ to the $k=0$ and 2 states. As Φ_e is increased from $\Phi_0/2$ to Φ_0, V_T increases from $V_T^{(n+1/2)}$ to $V_T^{(n)}$. In an analogous way, as Φ_e is steadily increased, the voltage at which the first step occurs oscillates between $V_T^{(n)}$ and $V_T^{(n+1/2)}$ with period Φ_0. The modulation amplitude $\Delta V_T^{(\Phi_0/2)} = V_T^{(n)} - V_T^{(n+1/2)}$ is given by subtracting Eq. (42) from Eq. (38):

$$\Delta V_T^{(\Phi_0/2)} = \omega L_T \Phi_0/2M \quad . \qquad (43)$$

The incremental change in V_T for a small change $\delta\Phi_e$ $(\Phi_e \neq n\Phi_0/2)$ is

$$\delta V_T = \omega L_T \delta\Phi_e/M \, . \qquad (44)$$

For given values of L_T and L, Eqs. (43) and (44) imply that the sensitivity can be made arbitrarily high by making the coupling coefficient K arbitrarily small. However, K obviously cannot be made so small that the SQUID has a negligible influence on the tank circuit, and we need to find a lower limit on K. In order to operate the SQUID, one must be able to choose a value of I_{rf} that lies on the first step for all values of Φ_e. This requirement, namely that F (in Fig. 17) lie to the right of A, or that DF must exceed DE, sets a lower bound on K. The power dissipation in the SQUID is zero at D and $\Delta E(\omega/2\pi)$ at F. Hence, $I_{rf}^{(F)} - I_{rf}^{(D)}) V_T^{(n+1/2)}/2 = \Delta E(\omega/2\pi)$ (I_{rf} and V_T are peak values rather than rms values). If we note that $I_{rf}^{(E)} - I_{rf}^{(D)} = \Phi_o/2MQ$, and use Eqs. (37) and (42), the criterion DF ≥ DE becomes (with $\Phi_{ec} = \Phi_o$)

$$K^2 Q \geq \pi/4 \tag{45}$$

This result is only approximate because we have assumed in this discussion that the SQUID dissipation is much larger than the dissipation in R_T.

B. Operation of the rf SQUID

A typical (simplified) circuit for the operation of the rf SQUID is shown in Fig. 18. The rf oscillator is adjusted so that the SQUID is biased on the first step of the V_T vs I_{rf} curve for all values of Φ_e. The rf voltage across the tank circuit is amplified with a low-noise preamplifier (usually with an FET input stage) that has a high input impedance. After further amplification the rf signal is demodulated (sometimes with a diode as shown in Fig. 18, although often a more sophisticated demodulation scheme is used), and the demodulated signal is integrated. The smoothed output is periodic in Φ_e. An ac modulation flux, at a frequency of typically 100 kHz, and with a peak-to-peak amplitude $\leqslant \Phi_o/2$, is also applied to the SQUID, just as in the case of the dc SQUID. The signal at the output of the rf integrator, therefore, contains both 100 kHz and 200 kHz components, depending on the value of Φ_e. This signal is lock-in detected at the modulation frequency, integrated, amplified, and fed back as a current in the modulation coil. The rf SQUID is thus operated in a feedback mode in the same manner as the dc SQUID.

C. Noise in the rf SQUID

There are three sources of white noise that we should consider: intrinsic noise, tank circuit noise, and preamplifier noise. A discussion of low frequency noise will be deferred until Section IV-C.

In the previous section, we assumed that all transitions occurred when $\Phi_e = \Phi_{ec}$; in practice, this is not the case. The SQUID has a nonzero probability of making a transition at $\Phi_e < \Phi_{ec}$. Each time Φ_e is increased from zero, the $k = 0$ to $k = 1$ transition occurs at a different value of Φ_e. For a sinusoidal current-phase relation and when $d\Phi_e/dt = \text{const} \ll \Phi_o R/L$, Kurkijärvi [62] showed that the distribution in values of Φ_e at which the transitions occur was

$$\sigma = (3\pi/2\sqrt{2})^{2/3} \sigma_o L I_c (k_B T/\Phi_o I_c)^{2/3} \quad , \tag{46}$$

where T is the SQUID temperature, and $\sigma_o \approx 1$. For $T = 4\,K$, $L = 10^{-9}$ H, and $LI_c = \Phi_o$, $\sigma \approx 0.13\,\Phi_o$. Equation (46) has been experimentally verified [63, 64].

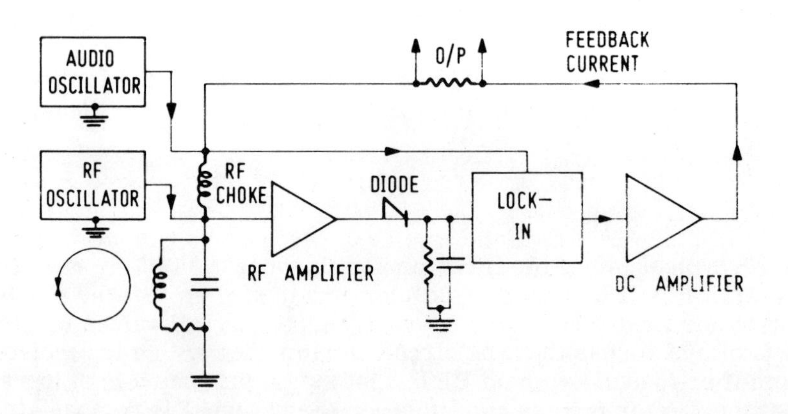

Fig. 18. Simplified schematic of rf SQUID in flux-locked loop.

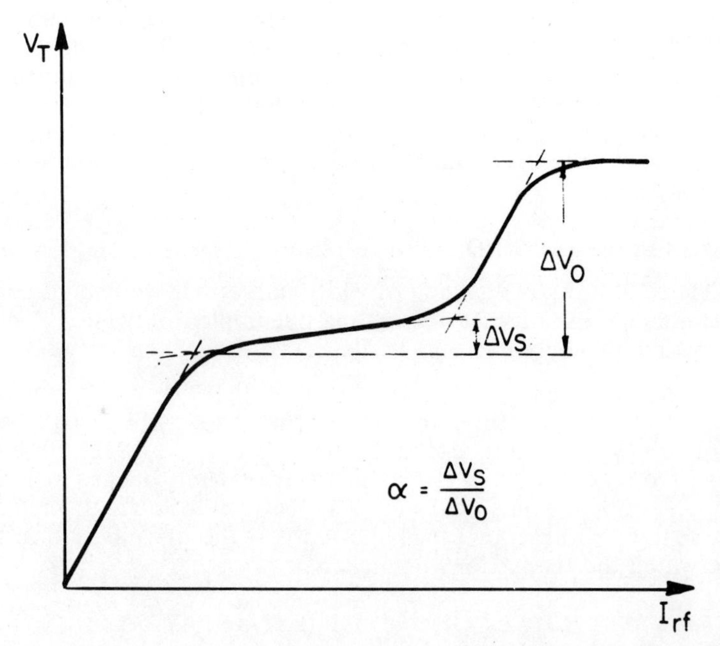

Fig. 19. V_T vs I_{rf} for rf SQUID in presence of thermal fluctuations.

When the SQUID is driven with an rf flux, the fluctuations in the value of flux at which transitions occur have two consequences. First, Kurkijärvi and Webb [65] showed that V_T will fluctuate, giving an equivalent intrinsic flux noise power spectrum

$$S_\Phi^{(i)} \approx 0.7(3\pi/2\sqrt{2})^{4/3}(2\pi/\omega)L^2I_c^{2/3}(k_BT/\Phi_o)^{4/3} \quad . \tag{47}$$

If the current-phase relation is non-sinusoidal, the noise will be higher than that predicted by Eq. (47) [40, 66]. Second, the noise will round the step edges and cause the steps to tilt [65], so that the voltage increases from A to B (see Fig. 19). This tilting arises in the following way [66]. In the absence of thermal fluctuations, the transition from the $k = 0$ to the $k = 1$ state occurs at a precisely defined value of flux $(\Phi_e + \Phi_{rf})$. However, in the presence of thermal fluctuations, the transition has a certain probability of occurring at a lower value of flux. Just to the right of A (Fig. 17), a transition occurs at the peak of the rf flux only once in many rf cycles. The SQUID, therefore, makes many attempts at any given transition, and the probability of its occurring in any one cycle is small. Just to the left of B, a transition must occur at each peak value of the rf flux, with unity probability. To increase the probability of the transition, the peak value of the rf flux increases slightly, so that V_T also increases as I_{rf} is increased from A to B. Following Jackel and Buhrman [40], we introduce a parameter α, the ratio of the voltage rise along a step, ΔV_s, to the separation in voltage of successive steps, ΔV_o (see Fig. 18). Jackel and Buhrman [40] show that α is related to $S_\Phi^{(i)}$ by the relation

$$\alpha^2/(1-\alpha)^2 \approx S_\Phi^{(i)}\omega/\pi\Phi_o^2 \quad . \tag{48}$$

They have verified this relation experimentally, and shown that, apart from small changes in the numerical factor, it remains valid even when the current-phase relationship is non-sinusoidal [40].

The fact that the step is tilted modifies Eq. (44) to

$$\delta V_T' = (1-\alpha)\omega L_T\delta\Phi_e/M \quad . \tag{49}$$

We consider next the tank circuit noise. The tank circuit resistance R_T produces a Johnson noise current in the inductance L_T that adds to I_{rf} [60]. At a given value of I_{rf}, these currents cause the bias point of the SQUID to fluctuate. Because the steps are tilted (dV_T/dI_{rf} is non-zero), the Johnson noise contributes to the SQUID noise. Following Jackel and Buhrman [40], we calculate the equivalent flux noise of this contribution. From Fig. 19, and Eqs. (40) and (41)*, assuming $<I_c \approx \Phi_o$, we find

---

*See footnote to Eqs. (39) and (40).

$$\left.\frac{dV_T}{dI_{rf}}\right|_{step} = \frac{\alpha \Delta V_o}{I_{step}} \approx \frac{\pi \alpha \omega L_T}{2K^2} \quad . \tag{50}$$

We define the effective Q on a step as

$$Q_{eff} = \left.\frac{dV_T}{dI_{rf}}\right|_{step} \Bigg/ \omega L_T \approx \frac{\pi\alpha}{2K^2} \quad , \tag{51}$$

where we have used Eq. (50).

The spectral density of the current noise in the inductance L_T at the resonant frequency of the tank circuit is

$$S_I^{(tc)} = \frac{4k_BT_e}{R_T} = \frac{4k_BT_e}{Q\omega L_T} \quad , \tag{52}$$

where $Q = R_T/\omega L_T$ is the quality factor of the parallel resonant circuit. T_e is the effective temperature of the tank circuit; with a room temperature amplifier, T_e might be 200 K [40]. The corresponding voltage spectral density (from the first equality of Eq. (51)) is

$$S_V^{(tc)} = Q_{eff}^2 \omega^2 L_T^2 S_I^{(tc)} \quad . \tag{53}$$

The equivalent flux noise due to the tank circuit is, using Eqs. (49), (51), and (53), and setting $K^2Q = \pi/4$,

$$S_\Phi^{(tc)} = \frac{M^2 S_V^{(tc)}}{(1-\alpha)^2 \omega^2 L_T^2} \approx \frac{4\pi\alpha^2}{(1-\alpha)^2} \frac{k_BT_eL}{\omega} \quad . \tag{54}$$

The preamplifier also contributes to the total SQUID noise. The spectral density of the preamplifier voltage noise, $S_V^{(p)}$, is equivalent to a flux noise spectral density (from Eq. (49))

$$S_\Phi^{(p)} = \left[\frac{M}{\omega L_T(1-\alpha)}\right]^2 S_V^{(p)} \quad . \tag{55}$$

We define a noise temperature, $T_N^{(p)}$, for the preamplifier through the relation

$$S_V^{(p)} = 4k_B T_N^{(p)} (\partial V_T/\partial I_{rf})_{step} \approx 2\pi\alpha\omega L_T k_B T_N^{(p)}/K^2 \ . \qquad (56)$$

If we take as typical values $[S_V^{(p)}]^{1/2} \approx 2\text{nVHz}^{-1/2}$, $\alpha \approx 0.2$, $\omega/2\pi \approx 30$ MHz, $L_T \approx 5 \times 10^{-7}$ H, and $K^2 \approx 0.2$, we find $T_N \approx 50$ K. From Eqs. (55) and (56) we obtain

$$S_\Phi^{(p)} \approx \frac{2\pi\alpha}{(1-\alpha)^2} \frac{k_B T_N^{(p)} L}{\omega} \ . \qquad (57)$$

The energy resolution is found from Eqs. (48), (54), and (57):

$$\frac{S_\Phi}{2L} \approx \frac{\alpha}{(1-\alpha)^2} \left\{ \underbrace{\frac{\pi\alpha\Phi_0^2}{2\omega L}}_{\text{intrinsic}} + \underbrace{\frac{2\pi\alpha k_B T_e}{\omega}}_{\text{tank circuit}} + \underbrace{\frac{\pi k_B T_N^{(p)}}{\omega}}_{\text{preamplifier}} \right\} \ . \qquad (58)$$

Note that Eq. (58) assumes that $LI_c \approx \Phi_0$. The intrinsic energy resolution is proportional to the energy available per cycle, $\sim \Phi_0^2/L$, divided by the sampling frequency, ω. The second and third terms represent the thermal energies of the tank circuit ($k_B T_e$) and preamplifier ($k_B T_N^{(p)}$) divided by ω. It should be emphasized that these expressions are approximate, and that more detailed estimates of the noise may differ by factors of 2 or 3 [40].

It is instructive to make estimates for the three contributions. If we take as typical values $\alpha \approx 0.2$, $L \approx 10^{-9}$ H, $\omega/2\pi \approx 30$ MHz, $T_e \approx 200$ K, and $T_N^{(p)} \approx 50$ K, we find $S_\Phi^{(i)}/2L \approx 2 \times 10^{-30}\,\text{JHz}^{-1}$, $S_\Phi^{(tc)}/2L \approx 6 \times 10^{-30}\,\text{JHz}^{-1}$, $S_\Phi^{(p)}/2L \approx 4 \times 10^{-30}\,\text{JHz}^{-1}$, and $S_\Phi/2L \approx 12 \times 10^{-30}\,\text{JHz}^{-1}$. The combined rms flux noise is $S_\Phi^{1/2} \approx 8 \times 10^{-5}\,\Phi_0\,\text{Hz}^{-1/2}$. In this example, the tank circuit noise is slightly greater than the preamplifier noise. From Eqs. (55) or (56) we see that $S_\Phi^{(p)} \propto K^2$, whereas Eqs. (51), (53), and the first equality of Eq. (54) yield $S_\Phi^{(tc)} \propto 1/K^2$, for constant Q. Thus the overall resolution could be (very slightly) improved by increasing K so that $S_\Phi^{(p)} = S_\Phi^{(tc)}$ and $K^2 Q > \pi/4$. The intrinsic noise is relatively insignificant, as is usually the case for SQUIDs operated at 30 MHz. Jackel and Buhrman [40] have given a detailed discussion of the optimization of the flux resolution.

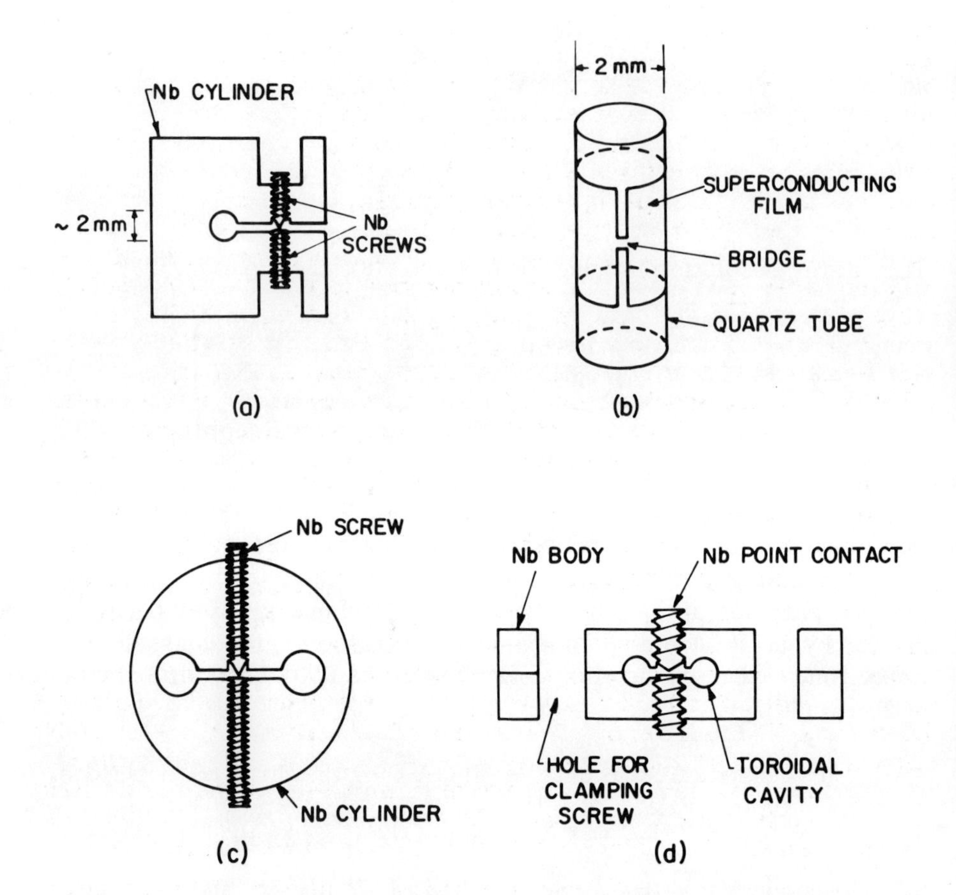

Fig. 20. Selection of rf SQUIDS: (a) point-contact rf SQUID, machined from niobium; (b) thin-film rf SQUID evaporated on quartz tube; (c) two-hole point-contact rf SQUID, machined from niobium; (d) toroidal point-contact rf SQUID, machined from niobium.

D. Practical rf SQUIDs: Fabrication and Performance

A variety of rf SQUID configurations have been used. The earliest versions are shown in Figs. 20(a) and 20(b). The point contact SQUID of Zimmerman et al. [7], Fig. 20(a), was manufactured from a solid cylinder of niobium. The point contact was usually adjusted while it was at liquid helium temperatures to obtain the optimum critical current. A more robust version incorporating a glass spacer in the slot was developed by Buhrman et al. [38]; this SQUID could be recycled without resetting the point contact. The tank circuit coil was rigidly mounted inside the cylinder. Mercereau and co-workers [8] used the thin film rf SQUID shown in Fig. 20(b). A thin film of superconductor (usually a tin alloy or niobium) was evaporated around a quartz or sapphire tube typically 2 mm in diameter and 20 mm long. A single Anderson-Dayem bridge [30] or Notarys bridge [31] was then formed in the superconducting cylinder. Similar devices have been used by Goodman et al. [67] and Opfer et al. [68], and are commercially available from S.C.T. [12] and Develco [10]. The tank coil is placed inside the SQUID or wound around it. A further development of the point contact device is the symmetric rf SQUID [7, 9, 13, 69] shown in Fig. 20(c). Two holes drilled through a niobium cylinder are connected by a slot across which there is an adjustable niobium point contact. The inductance of this SQUID is approximately one-half that of a single-loop SQUID with the same total area. The two-hole SQUID can be recycled successfully without readjusting the niobium screw. A thin-film two-hole SQUID using tunnel junctions has been operated by Ehnholm et al. [70].

A particularly useful point contact configuration is the toroidal SQUID [67, 71] shown in Fig. 20(d). The SQUID is made in two halves that form a toroidal cavity when clamped together. The toroid is connected on its inner surface by a disk-shaped cavity containing a point contact. The tank circuit coil and input coil are placed inside the toroidal cavity. The toroidal SQUID has several advantages: it is extremely rugged; it is self-shielding against external magnetic field fluctuations; its inductance can be rather low, as the length of the cavity is relatively long; and large inductance signal coils can be used. Toroidal rf SQUIDs are commercially available from both S.C.T. [12] and S.H.E. [13].

Yet another very ingenious point contact device, the fractional turn SQUID, has been operated by Zimmerman [72, 73]. This SQUID contains twelve loops in parallel connected across a single point contact. The effective inductance of the SQUID is thereby reduced, since the loop inductances are in parallel, so that the signal available from the SQUID is enhanced. Ehnholm et al. [70] have fabricated a thin-film SQUID with eight loops in parallel across a Nb-NbO_x-Pb tunnel junction.

These devices are most commonly operated at a frequency of about 30 MHz. The inductance, L, of the devices shown in Figs. 20(a)-(c) is 10^{-9} H or less, and their critical current, I_c, is about $\Phi_0/L \sim 2\,\mu A$. In the feedback mode, a dynamic range of 10^6 to 10^7 in a 1 Hz bandwidth, a frequency response that is essentially flat from 0 to several kHz, and a slewing rate of 10^4 to 10^6 $\Phi_0 s^{-1}$ are typical. The flux noise of these devices is usually stated to be white, with an rms value of about 10^{-4} $\Phi_0 Hz^{-1/2}$. This value is in good agreement with the value estimated in Section IV-C.

In the summer of 1975 we measured the noise power spectra for the toroidal SQUIDs available from S.C.T. and S.H.E. using the facilities available at Berkeley [74]. The power spectra obtained, plotted as $S_\Phi/2\alpha^2 L$, are shown in Fig. 21, together with the power spectrum for the dc SQUID. The S.H.E. SQUID had a resistor connected to the input coil that contributed about one-third of the mean square white noise. Since these spectra were taken, the white noise of the S.H.E. SQUID has been reduced to about 5×10^{-29} JHz^{-1} : this value is indicated on the graph. All three SQUIDs exhibit 1/f noise at low frequencies. The value of S_Φ, M_i, L_i and $S_\Phi/2\alpha^2 L$ (in the white noise region) are presented in Table I for five SQUIDs (in the case of the S.H.E. SQUID, the new value of S_Φ has been used).

There have been several attempts to obtain lower noise by working at higher frequencies. The high frequency work was initially motivated by the observation that the signal available from the rf SQUID is proportional to ω, as indicated in Eq. (43). Thus Zimmerman and Frederick [75] and Kamper and Simmonds [76] operated SQUIDs at 300 MHz and 9 GHz respectively, and observed an increase in the available signal that was close to that predicted. They did not report measurements of the noise. It should be realized that an increase in operating frequency does not automatically improve the noise performance since the noise of preamplifiers also tends to increase with frequency. For example, Clark and Jackel [77] operated a SQUID at 450 MHz, and although their rms flux noise was somewhat lower than that of most 30 MHz SQUIDs, the figure of merit $S_\Phi/2\alpha^2 L$ was about 5×10^{-29} JHz^{-1}, comparable with the best 30 MHz SQUIDs (see Table I). However, Pierce et al. [78] operated a thin film cylindrical SQUID at 10 GHz, and obtained a rms flux noise of about $10^{-5} \Phi_0 Hz^{-1/2}$ and $S_\Phi/2\alpha^2 L \approx 2 \times 10^{-30}$ JHz^{-1} at frequencies above a few kHz (see Table I). This is the lowest figure of merit I am aware of. Unfortunately, the noise increased appreciably at lower frequencies. The best flux resolution I know of, 7×10^{-6} $\Phi_0 Hz^{-1/2}$, was achieved by Gaerttner [79] using a 440 MHz SQUID; the parameters required to calculate $S_\Phi/2\alpha^2 L$ do not appear to be available.

Table I. Flux noise power spectrum in the white noise region (S_Φ), mutual inductance with input coil (M_i), inductance of input coil (L_i), and figure of merit ($S_\Phi/2\alpha^2L$) for several SQUIDs.

SQUID	S_Φ (Φ_o^2 Hz^{-1})	M_i (nH)	L_i (nH)	$S_\Phi/2\alpha^2L$ (10^{-30}JHz^{-1})
dc (Clarke et al. [14])	1.2×10^{-9}	11.5	356	7
rf (S.H.E. [13]) 19 MHz	5×10^{-9}	20	2×10^3	50
rf (S.C.T. [12]) 30 MHz	6.6×10^{-10}	3	360	50
rf (Clark and Jackel [77]) 450 MHz	9×10^{-10}	35	3×10^4	50
rf (Pierce et al. [78]) 10 GHz	1×10^{-10}	8	500	2

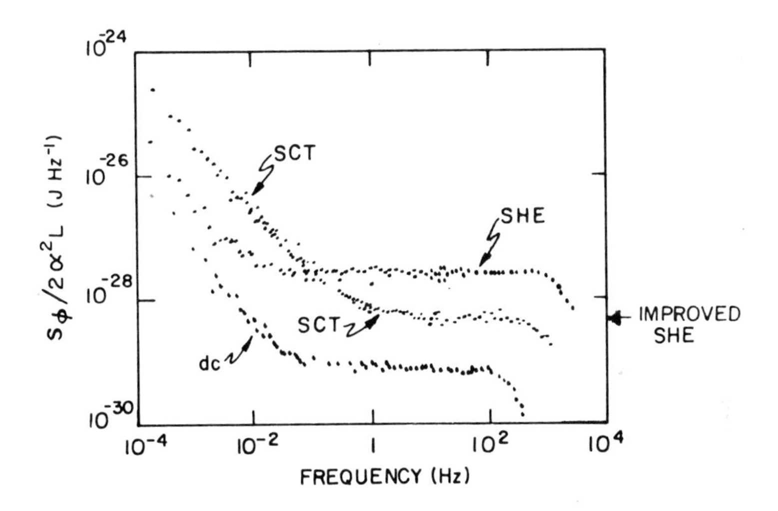

Fig. 21. Noise power spectra (plotted as energy resolution) for tunnel junction dc SQUID and S.C.T. and S.H.E. toroidal rf SQUIDs (Summer 1975). Subsequently, the white noise of the S.H.E. SQUID was improved to the level shown.

E. Future Improvements in the rf SQUID

It appears that SQUIDs operated at about 30 MHz are limited by preamplifier or tank circuit noise rather than by intrinsic noise. The optimization of the performance has been discussed in detail by Jackel and Buhrman [40]. They consider two possible cases: (1) $S_\Phi^{(p)} > S_\Phi^{(tc)}$. From Eqs. (56) and (57) we see that $S_\Phi^{(p)} \propto T_N^{(p)}/\omega \propto 1/Q\omega^2 L_T = 1/\omega R_T$. At a given frequency, one may improve the flux resolution by increasing R_T; (2) $S_\Phi^{(p)} < S_\Phi^{(tc)}$. As noted in Section IV-C, since $S_\Phi^{(p)} \propto K^2$ and $S_\Phi^{(tc)} \propto 1/K^2$ (for constant Q), one can improve the performance by increasing K until $S_\Phi^{(p)} = S_\Phi^{(tc)}$; at the same time, R_T should be made as large as possible. In the best 30 MHz-SQUIDs, the preamplifier and tank circuit noise contributions are optimized, and no significant improvement seems likely.

Another parameter that can be varied is the SQUID inductance, L. If one retains the restriction $LI_c \approx \Phi_0$, and assumes that $\alpha << 1$, one finds that α is proportional to $LI_c^{1/3}$ or to $L^{2/3}$ [Eqs. (47) and (48)]. Thus $S_\Phi^{(i)}/2L \propto L^{1/3}$, $S_\Phi^{(tc)}/2L \propto L^{4/3}$, and $S_\Phi^{(p)}/2L \propto L^0$ (since $\alpha T_N^{(p)}$ is independent of L). If $S_\Phi^{(p)}$ is the dominant noise term, no improvement is gained by reducing L. If $S_\Phi^{(tc)}$ dominates the noise, the performance can be improved by reducing L. However, a similar reduction can be achieved by increasing K, as described earlier. It should be noted that a substantial reduction in L is likely to reduce the coupling coefficient to the input coil, resulting in little overall improvement in the figure of merit. (The toroidal geometry may be an exception, since tight coupling can be achieved even with a relatively low SQUID inductance. Zimmerman's fractional-turn geometry also allows good coupling to a low inductance SQUID [72, 73].)

It is evident from Eq. (58) that all three noise contributions can be reduced by increasing ω, as is demonstrated experimentally (see Section IV-D). The use of higher frequencies appears to be the only way to significantly improve the flux resolution. $S_{\Phi}^{(i)}/2L$ and $S_{\Phi}^{(tc)}/2L$ are both proportional to $1/\omega$, while $S_{\Phi}^{(p)}/2L$ is proportional to $T_N^{(p)}/\omega$ or to $S_V^{(p)}/\omega^2$. It is important to realize that as ω increases, $S_V^{(p)}$ also tends to increase, and that in practice $S_{\Phi}^{(p)}$ usually decreases less rapidly than $1/\omega^2$. For example, $T_N^{(p)}$ might be typically 50 K at 30 MHz and 500 K at 10 GHz. As an example, consider a SQUID with $\omega/2\pi = 10$ GHz, $\alpha = 0.2$ (this may be a low estimate at the higher frequencies), $L = 10^{-9}$ H, $T_e = 200$ K, and $T_N^{(p)} = 500$ K. We find $S_{\Phi}^{(i)}/2L \approx 6 \times 10^{-33}$ JHz^{-1}, $S_{\Phi}^{(tc)}/2L \approx 2 \times 10^{-32}$ JHz^{-1}, and $S_{\Phi}^{(p)}/2L \approx 10^{-31}$ JHz^{-1}. Thus the noise is dominated by preamplifier noise. This result is an order of magnitude smaller than the experimental value of Pierce et al. [78], indicating that their preamplifier had a noise temperature higher than 500 K, and, probably, that α was greater than 0.2. The use of cooled FET preamplifiers [79], which can have lower noise temperatures than room temperature preamplifiers, is a promising development. A cooled preamplifier has the added advantage of reducing T_e appreciably, so that $S_{\Phi}^{(tc)}$ is also reduced.

Jackel and Buhrman [40] have shown that a non-sinusoidal current-phase relation in the weak link will substantially increase the noise over that obtained with a sinusoidal current-phase relation. This result should be borne in mind when one is choosing a weak-link for use in a SQUID.

Buhrman and Jackel [private communication] have found that the ultimate energy resolution of the rf SQUID is on the order of $4 k_B T/\omega_{opt}$ where $\omega_{opt} = R/L$. Thus, the intrinsic sensitivites of the rf and dc SQUIDs with identical values of R/L are comparable. A high resistance weak link (for example, $\sim 100\ \Omega$) with a sinusoidal current-phase relation is necessary to improve the performance of both dc and rf SQUIDs. This requirement tends to favor the use of shunted tunnel junctions of small areas.

Of the various configurations of rf SQUID that have been used, the toroidal geometry has much in its favor. The device is mechanically exceedingly stable, is self-shielding against external field changes, and has a high coupling coefficient to the input coil. It would be of interest to try to incorporate a thin-film junction (for example, a shunted tunnel junction) into such a geometry.

V. SQUIDS AS MAGNETOMETERS, GRADIOMETERS, SUSCEPTOMETERS, AND VOLTMETERS

SQUIDs can be used to measure magnetic fields, magnetic field gradients, magnetic susceptibilities, and voltages. I shall describe each of these applications in turn, indicate possible future improvements, and compare the sensitivity of the SQUID-based measurement with that obtainable by other techniques. I shall begin with a description of the flux transformer, since most of the applications involve this very useful device.

A. Flux Transformer

The flux transformer consists of a superconducting loop of wire, as indicated in Fig. 22(a). The pick-up coil has N_p turns and an inductance L_p, while the input coil has an inductance L_i and a mutual inductance $M_i = \alpha(LL_i)^{1/2}$ to a flux-locked dc or rf SQUID. A magnetic field applied to the pick-up coil generates a persistent supercurrent in the flux transformer that in turn couples a flux into the SQUID. The output of the flux-locked SQUID is then proportional to the applied flux. As we shall see presently, there are a number of variations on this basic principle.

The flux resolution of the transformer is easily calculated. Suppose a magnetic field is applied to the pick-up coil so that a flux $\Delta\Phi$ threads each of the N_p turns. Flux quantization in the flux transformer requires that

$$N_p \Delta\Phi + (L_i + L_p) J = 0 \quad , \tag{59}$$

where J is the induced supercurrent. We have neglected stray inductances. The flux applied to the SQUID is then

$$\Delta\Phi_e = -M_i J = \frac{M_i N_p}{(L_i + L_p)} \Delta\Phi \quad . \tag{60}$$

We have assumed that the inductive coupling between L_i and the feedback/modulation coil is negligible (this is not necessarily true [14, 80]). If the flux noise of the SQUID (including all contributions) has a power spectrum S_Φ, the smallest flux applied to the pick-up loop that can be resolved per $\sqrt{Hz}$ is

$$\delta\Phi = \frac{L_i + L_p}{M_i N_p} S_\Phi^{1/2} = \frac{L_i + L_p}{\alpha(L_i L)^{1/2} N_p} S_\Phi^{1/2} \quad . \tag{61}$$

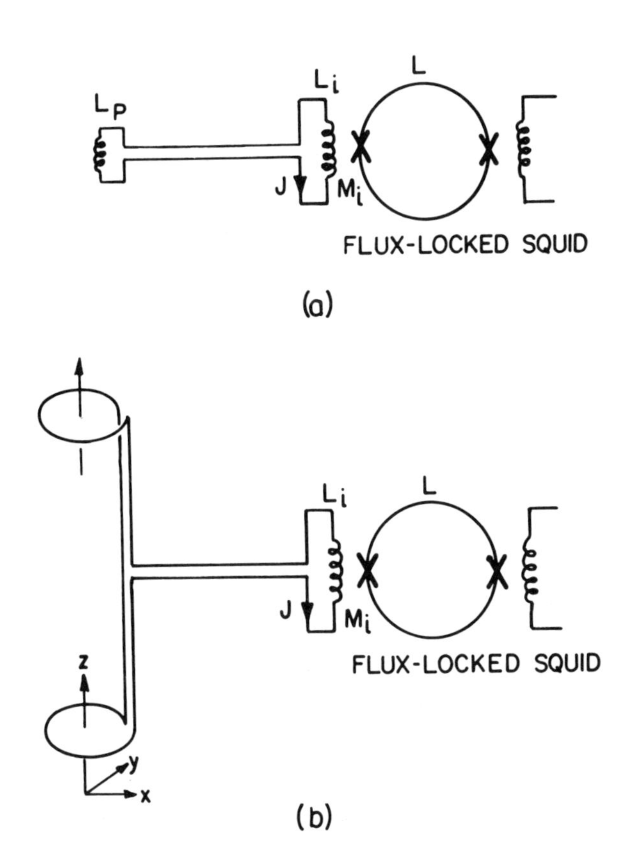

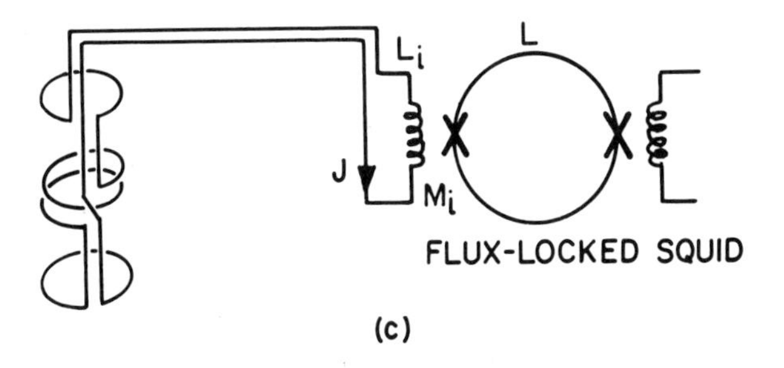

Fig. 22. (a) Flux transformer; (b) first-derivative gradiometer; and (c) second-derivative gradiometer.

For a given pick-up coil (i.e., fixed L_p and N_p), the optimum value of L_i is obtained by differentiating Eq.(61) with respect to L_i. (We assume that α remains constant as L_i is varied.) We find that $\delta\Phi$ has a minimum when $L_i = L_p$ given by

$$\delta\Phi_{min} = \frac{2L_p^{1/2}}{N_p}\left(\frac{S_\Phi}{\alpha^2 L}\right)^{1/2} . \qquad (62)$$

We notice immediately that $\delta\Phi_{min}$ is proportional to $\varepsilon^{1/2}$, where ε is the figure of merit discussed in Section III-C, so that ε is an appropriate figure of merit for the flux transformer/SQUID combination.*

In the configuration just described, the flux applied to the SQUID is cancelled by a flux generated in the modulation/feedback coil. It is sometimes advantageous to couple the feedback flux into the flux transformer rather than into the SQUID. This technique has the advantage of maintaining zero supercurrent in the transformer.

I will next discuss various applications of the flux transformer.

B. Measurement of Magnetic Field

The flux-locked SQUID is of course a sensitive magnetometer for measuring changes in external magnetic field (no highly accurate method of using a SQUID to determine the absolute value of the field has yet been found). For example, the thin-film dc SQUID described in Section III-D has a magnetic field resolution of about 10^{-10} $GHz^{-1/2}$. This is a sufficiently high sensitivity for almost all practical purposes. However, it is sometimes desirable to shield the SQUID from the applied field which, if it is high enough, may cause a deterioration in the performance of the SQUID. This shielding can be achieved by placing the SQUID in a superconducting tube (for example, as shown in Fig. 10), and coupling in the field with a flux transformer. The toroidal SQUID is self-shielding against changes in external magnetic field, and a flux transformer is essential.

The pick-up coil usually has a single turn ($N_p = 1$). The smallest resolvable magnetic field change per $\sqrt{Hz}$ is then found by dividing $\delta\Phi_{min}$ by the area of the loop, A_p, so that

*Eq. (62) is not quite accurate. If a SQUID is coupled to a superconducting circuit of total inductance ℓ with a mutual inductance M_i, the effective SQUID inductance [50, 53] is lowered to a value $L' = (1 - M_i^2/L\ell)$. As a result, the voltage from the SQUID is enhanced, and the flux resolution is improved. Thus the optimization should involve L' rather than L. However, in practice, the correction is not large, and for simplicity I have ignored it.

$$\delta H_{min} = 2 \frac{L_p^{1/2}}{A_p} \left(\frac{S_\Phi}{\alpha^2 L} \right)^{1/2} . \tag{63}$$

Since L_p varies approximately as $A_p^{1/2}$, δH_{min} is approximately proportional to $A_p^{-3/4}$. Thus, in principle, an arbitrarily high magnetic-field sensitivity can be obtained by making the pick-up loop sufficiently large, provided that the condition $L_i = L_p$ can be maintained without decreasing α. As an example, suppose the pick-up loop has a diameter of 60 mm, and an inductance of about 40 nH (40 emu). Then using $S_\Phi/\alpha^2 L \approx 1.4 \times 10^{-22}$ erg$Hz^{-1/2}$, we find $\delta H_{min} \approx 5 \times 10^{-12}$ G$Hz^{-1/2}$. An increase in sensitivity of about one order of magnitude is typical.

Claassen [53] has given a more detailed discussion of coil geometries.

In a practical magnetometer, the flux transformer is almost invariably made of insulated niobium wire, typically 100 μm in diameter. It is vital that the flux transformer be rigidly attached to the SQUID, and that the pick-up loop cannot vibrate in the magnetic field. In the case of the cylindrical SQUIDs, the input coil is usually wound directly on top of the SQUID cylinder. For the dc SQUID, the coupling coefficient is about 0.6; this value appears to be typical. If the magnetometer is operated in an unshieldid environment, the pick-up loop acts as an antenna for radio and television signals, and the resultant interference coupled into the SQUID usually prevents its operation. Shielding against such interference can be obtained by surrounding the cryostat with a Faraday cage made of copper mesh. An alternative method is to place a thin-walled copper tube between the input coil and the SQUID to filter out the rf interference. The filter may introduce additional Johnson noise into the SQUID, however. The leads between the pick-up loop and the signal coil are usually twisted and surrounded by a superconducting shield. The pick-up loop is often wound on a cylinder machined on a quartz block. Three-axis magnetometers are commercially available in which the three pick-up loops are mounted on three orthogonal faces of a precision-ground quartz cube. A superinsulated fiberglass dewar (with no liquid nitrogen jacket) is frequently used. These dewars are relatively non-magnetic, readily portable, and have a boil-off rate of 1 liter or less of liquid helium per day.

These magnetometers have a higher sensitivity than can be used, at least at frequencies above (say) 1 Hz. There are no obvious applications where a sensitivity greater than 10^{-9} or 10^{-10} G$Hz^{-1/2}$ is required. However, the long term drift has sometimes been higher than desirable. It appears that the drift is usually correlated with a drift in the temperature of the helium bath through changes in atmospheric pressure or the steady drop in the hydrostatic head of the liquid. These temperature changes not only cause the SQUID itself to drift, but can also cause changes in the susceptibility of the dewar, which contains paramagnetic impurities. For most applications these difficulties are not too serious, and they can be largely removed by stabilizing the temperature of the helium bath.

The sensitivity of SQUID magnetometers is appreciably higher than that of conventional devices. The fluxgate and proton-precession magnetometers have sensitivities of about 10^{-6} $GHz^{-1/2}$, while the pumped cesium vapor magnetometer has a sensitivity of about 10^{-7} $GHz^{-1/2}$.

C. Measurement of Magnetic Field Gradient

The flux transformer may be readily adapted to measure first or even second spatial derivatives of changes in the ambient magnetic field. A gradiometer for measuring gradients of the form $\partial H_z/\partial z$ is shown schematically in Fig. 22(b). The two pick-up loops are wound so that a uniform magnetic field induces no supercurrent in the flux transformer, while a gradient $\partial H_z/\partial z$ generates a supercurrent that is proportional to the difference in the fluxes threading the two loops. Gradiometers have also been operated in which the loops are in the same plane and measure (for example) $\partial H_z/\partial x$. A straightforward analysis similar to that given in Section V-A shows that the optimum sensitivity for a given pair of coils is obtained when the inductance of the input coil, L_i, is equal to the sum of the inductances of the pick-up loops.

In a practical gradiometer, the two pick-up loops are usually wound on a precision-ground quartz block. The SQUID and the leads coupling the input coil in the gradiometer loops are shielded. It is necessary to take considerable care to ensure that the loops are as closely the same size and as accurately parallel to each other as possible. Despite these precautions, one inevitably finds that not only does the gradiometer respond to a uniform field H_z, but that it is also sensitive to fields along the x- and y-directions. It is therefore necessary to balance the gradiometer. One achieves this balance by adjusting the positions of three small pieces of superconductor along orthogonal axes by means of controls outside the cryostat. The usual procedure is to position the superconductors so as to minimize the response of the SQUID when the cryostat is rotated in the earth's field about orthogonal axes. With care, a balance of better than 1 ppm can be achieved. For two pick-up loops 6 cm in diameter and 20 cm apart, one would expect a sensitivity of better than 10^{-12} $Gcm^{-1}\,Hz^{-1/2}$. In practice, the sensitivity is not as high: values of 10^{-11} $G\,cm^{-1}\,Hz^{-1/2}$ are typical. The coupling of the gradiometer to the SQUID appears to increase the flux noise of the SQUID, for reasons that are not too well understood. The drift encountered in gradiometers is also appreciably higher than that in shielded SQUIDs by themselves. As with the magnetometers, it is believed that the drifts arise from the change in the paramagnetic susceptibility of nearly materials with changes in the temperature of the helium bath. Since the sensitivity of <u>each</u> pick-up loop to magnetic field changes is much higher in gradiometers than in magnetometers (which are not usually operated anywhere near their maximum possible sensitivity), the problem is much more serious in the case of gradiometers. For the most critical low frequency applications, it is quite possible that the liquid helium dewar will have to be made from quartz (which can have a very low level of paramagnetic impurities) to reduce the drift in the gradiometer to an acceptable level.

A further development of the gradiometer is a device to measure the second derivative $\partial^2 H_z/\partial z^2$. Two first derivative gradiometers are wound end-to-end, as indicated in Fig. 22(c). Opfer et al. [68] achieved a balance against changes in a uniform magnetic field of better than 10 ppm, and against changes in magnetic field gradients of better than 1 part in 100. Brenner et al. [81] have also described a second derivative gradiometer.

There is a need for improvement in the sensitivity of gradiometers. The present gradiometers are not able to detect gradient fluctuations in the earth's magnetic field, and are therefore not limited by environmental noise (except when they are operated near machinery producing large field gradients). It is not entirely clear how such improvements can be achieved; certainly, very stringent precautions must be taken to eliminate as much paramagnetic material as possible from the helium dewar, and the whole assembly must be made extremely rigid. It also seems likely that temperature stabilization will also be required to minimize the long term drift.

In the past, all gradiometers have consisted of pickup loops coupled to a SQUID via a coil. Recently, Ketchen et al. [82] operated a thin-film gradiometer in which both the pick-up loops and the SQUID were evaporated onto a single quartz substrate. The device is sensitive to $\partial H_z/\partial x$. This configuration is extremely stable mechanically, and requires balancing only in one dimension. The device has a sensitivity of about 10^{-11} G cm^{-1} Hz$^{-1/2}$. Further development in this direction is likely.

There are no other practical devices that measure magnetic field gradients. Thus one would have to make a gradiometer by subtracting the outputs of, for example, two fluxgate magnetometers. To achieve a resolution of 10^{-11} G cm^{-1} Hz$^{-1/2}$, the magnetometers would have to be $\sim$ 1 km apart! Clearly, one could obtain a high gradient sensitivity by placing two magnetometers far enough apart, but large separations are impracticable for many purposes. As a self-contained gradiometer, the SQUID-based system has no competition.

D. Measurement of Magnetic Susceptibility

An important application of SQUIDs is to the measurement of magnetic susceptibility. The configuration most often used is similar to that of a gradiometer, and is shown in Fig. 23. An astatic pair of niobium coils is connected to the input coil of a flux-locked SQUID. The coils are embedded in an epoxy tube that is in turn epoxied to a niobium tube, the whole assembly being extremely rigid. The sample is placed in one of the coils, as indicated. The niobium tube can be warmed above its superconducting transition temperature and cooled in an axial magnetic

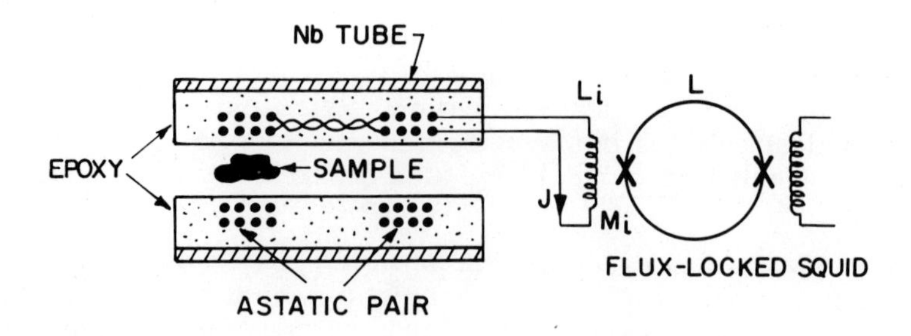

Fig. 23. Configuration used to measure magnetic susceptibility.

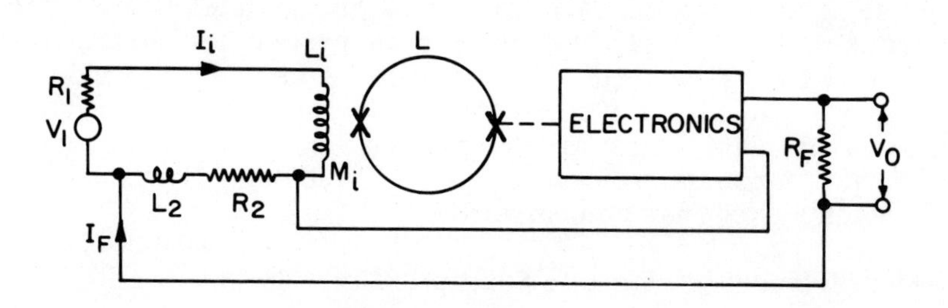

Fig. 24. Configuration of voltmeter.

field so that an exceedingly stable field is trapped (whose value, however, may differ somewhat from the applied field). If the pair of coils were perfectly balanced, the signal applied to the SQUID would be proportional to the applied magnetic field and to the magnetic susceptibility of the sample. In practice, the coils are not perfectly balanced, and it is necessary to subtract the SQUID output in the presence of the sample and sample holder from the SQUID output in the presence of the sample holder alone. (The sample holder may contribute significantly to the total susceptibility.) The temperature dependence of the susceptibility can be determined by measuring the SQUID output as a function of temperature. Again, the temperature dependence of the susceptibility of the sample holder, the epoxy, or even the enamel on the niobium wire [83, 84] may be quite substantial, and it is essential to take measurements with and without the sample.

As an example, we shall quote the results obtained by Giffard et al. [9]. Let Φ be the flux applied to each turn of one of the astatic coils; Φ is thus the out-of-balance flux due to the sample. Each coil has N turns and a self-inductance $L_i/2$ (so that the astatic pair is optimally coupled to the SQUID). The flux coupled to the SQUID is then

$$\Phi_e = M_i N \Phi / 2L_i \quad . \tag{64}$$

The exact calculation of Φ is, in general, very difficult. A relatively simple limiting case [9] is when the sample consists of a small sphere, ellipsoid, or cylinder with diameter equal to height. Then

$$\Phi = 4\pi \chi HV/D \quad , \tag{65}$$

where χ and V are the susceptibility and volume of the sample, H is the trapped field, and D is the average diameter of the coils. Inserting the values [9] $N = 24$, $L_i/M_i = 166$, and $D = 3.6$ mm, we find

$$\frac{\Phi_e}{\Phi_0} = \frac{2\pi M_i N \chi HV}{L_i D \Phi_0} \approx 10^4 \chi \left(\frac{H}{1\,G}\right)\left(\frac{V}{1\,mm^3}\right) . \tag{66}$$

Thus, if the resolution is limited by a SQUID noise of $10^{-4}\ \Phi_0 Hz^{-1/2}$, the sensitivity in a 1 Hz bandwidth is about 10^{-7} emu for a 1 cm^3 sample in a 1 G field. Unfortunately, the resolution is usually limited by vibration and temperature drifts to a lower value.

Improvements in sensitivity have been obtained by using a larger filling factor. Cukauskas et al. [85] achieved a resolution of 10^{-10} emu for a 1 cm^3 sample in a field of 100 G, and Cerdonio and Messana [86] achieved a resolution of 6×10^{-11} emu for 0.5 cm^3 sample in a field of 200 G. A commercially available instrument [12] has a resolution of 2×10^{-11} emu for a 1 cm^3 sample in a field of 1 kG over a temperature range from 4.2 K to 400 K. The sample is rigidly suspended in a thin-walled dewar that passes through the astatic pair, and can be heated above liquid helium temperatures. It might be noted that little increase in sensitivity results from the use of higher magnetic fields; the microphonic and/or background effects are amplified proportionately.

It seems that there is still room for improvement of SQUID-based susceptometers. The main limitations at present appear to be caused by temperature drifts that give rise to changes in the background susceptibility (see Doran and Symko [84]), and vibration of the sample in the astatic pair. It is to be hoped that both of these difficulties can be reduced, although a substantial reduction appears unlikely in the near future.

Other types of instruments have a higher sensitivity provided that one can use a higher magnetic field. Thus Foner [87] has achieved a sensitivity of 10^{-13} emu g^{-1} at 10 kG using a vibrating sample magnetometer. A commercial Faraday balance [88] has a resolution of 10^{-11} emu g^{-1} at 100 kG. These fields may, of course, be too high for some applications; at lower fields the sensitivity will be correspondingly reduced.

E. Measurement of Voltage

SQUIDs have been widely used as voltmeters, usually in the configuration of Fig. 24. The voltage source, V_1, of resistance R_1, is in series with the input coil, L_i, and a known resistance R_2 that has a stray inductance L_2. The current I_i generates a flux in the SQUID, and the electronics feeds a current $I_F = V_o/R_F$ into the resistance so that $I_F R_2 = V_1$ (in the zero frequency limit). Thus $I_i = 0$ on balance, and the voltmeter has a high input impedance. The entire voltmeter circuit and the SQUID are shielded by a superconducting can.

The dynamic behavior of the voltmeter is easily determined. We follow the treatment of Davidson et al. [52]. Let the total inductance of the voltmeter circuit be L_t, so that the time constant $\tau_T = L_t/(R_1 + R_2)$. We assume that the integration time constant in the electronics is small compared with τ_T. At a frequency ω we have

$$V_1 = I_i[R_1 + R_2 + j\omega L_t + g(R_2 + j\omega L_2)] \quad , \qquad (67)$$

where $g = I_F/I_i$. If we combine Eq. (67) with

$$V_o = R_F I_F = g R_F I_i \quad , \qquad (68)$$

and assume $g/(1 + R_1/R_2) \gg 0$, we obtain [52]

$$\frac{V_o}{V_1} = \frac{R_F/R_2}{1 + j\omega[(L_2/R_2) + \tau_T(1 + R_1/R_2)/g]} \quad . \qquad (69)$$

At zero frequency, $V_o/V_1 = R_F/R_2$. Notice that the time constant of the loop, τ_T, is reduced by the loop gain $g(1 + R_1/R_2)$, whereas the time constant $\tau_2 = L_2/R_2$ associated with the feedback resistor is not reduced [52].

Discussions of the noise characterization of the voltmeter have been given by several authors [9, 15, 52, 53]. The voltage noise power spectrum of the voltmeter referred to the source V_1 is just

$$S_V(\omega) = \frac{S_\Phi(R_1 + R_2)^2 (1 + \omega^2\tau_T^2)}{M_i^2} \approx \frac{S_\Phi R_1 \left[1 + \omega^2\tau_T^2\right]}{\alpha^2 L \tau_T} , \quad (70)$$

if $L_t \approx L_i$, and $R_1 \gg R_2$. We then define a noise temperature [15, 49, 52, 53] $T_N(\omega)$ for the voltmeter through the relation $S_v(\omega) = 4k_B T_N(\omega) R_1$:

$$T_N(\omega) \approx \frac{S_\Phi \left[1 + \omega^2\tau_T^2\right]}{4k_B\alpha^2 L\tau_T} . \quad (71)$$

We notice immediately that $T_N(\omega) \propto S_\Phi/2\alpha^2 L$, so that the figure of merit introduced in Section IV-C is also applicable to voltmeters. For low frequencies, $\omega \ll \tau_T^{-1}$, the noise temperature becomes independent of frequency:

$$T_N(0) \approx \frac{S_\Phi}{2\alpha^2 L} \frac{1}{2k_B\tau_T} . \quad (72)$$

For the thin-film dc SQUID described in Section III-D, $S_\Phi / 2\alpha^2 L \approx 7 \times 10^{-30}$ JHz^{-1} in the white noise region, and $T_N(0) \approx 2.5 \times 10^{-7}\, \tau_T^{-1}$ K. Thus the low frequency noise temperature of a given SQUID/coil combination used as a voltmeter ultimately depends only on τ_T. In the case of the dc SQUID, the quoted figure of merit was for an input coil inductance, L_i, of about 360 nH. Thus the noise temperature, $T_N(0)$, is less than 1 K for $R_1 \leqslant 1.4\,\Omega$. We might expect to be able to increase L_i by at least an order of magnitude without significantly reducing α, so that an upper limit on R_1 of at least tens of ohms for a 1 K noise temperature seems perfectly feasible. If necessary, the upper limit on R_1 would be further extended by means of an additional superconducting transformer between the voltmeter circuit and the SQUID, as described by Clarke, et al. [89], and Davidson et al. [52].

For source resistances of 10Ω or less in the liquid He^4 temperature range, the resolution of SQUID voltmeters is limited by Johnson noise in the resistance. Semiconductor voltmeters [90] operated at room temperature (usually with a FET input stage) have an optimum noise temperature of about 1 K for a source impedance of about 1 M. Using a transformer cooled in the helium bath, Prober [91] has been able to couple resistances of a few ohms to a PAR 185 preamplifier and still achieve a noise temperature in the liquid He^4 temperature range.

Thus the SQUID voltmeter and the semiconductor voltmeter complement each other: the SQUID is superior for impedances below 10 Ω, while the semiconductor technology is superior for impedances above 10 Ω. A detailed discussion of the noise characteristics of the two technologies has been given by Davidson et al. [52].

IV. PRACTICAL APPLICATIONS OF SQUID-BASED DEVICES

Broadly speaking, practical applications of SQUID-based devices fall into two classes. Most applications of the magnetometer and gradiometer are "in the field", outside the laboratory, while the applications of the susceptometer and voltmeter tend to be in the laboratory. The use of SQUIDs in a non-laboratory environment has required the development of small, portable, relatively non-magnetic dewars with no nitrogen shielding and with a low boil-off rate of liquid helium. Such dewars, made of fiberglass, are now commercially available, although the paramagnetic susceptibility of the liquid helium container is too high for some applications (especially for gradiometers), and gives rise to drift when the temperature of the helium bath changes. It is hoped that this difficulty will be overcome in the near future with the aid of materials with a very low level of magnetic impurity, such as quartz.

One important use of magnetometers is in geophysics, to measure fluctuations in the earth's magnetic field. These measurements require the magnetometer to be operated at remote sites, far away from man-made magnetic disturbances. For example, W.M. Goubau, T.D. Gamble, H.F. Morrison, E.C. Mosley, and I [92] have used a three-axis SQUID magnetometer in a preliminary magnetotelluric survey in Grass Valley, Nevada. The magnetometer consists of three thin-film dc SQUIDs mounted orthogonally. The magnetotelluric technique enables one to make estimates of the electric conductivity of the earth's surface as a function of depth. Magnetic disturbance in the ionosphere and magnetosphere propagate to the earth's surface. At the earth's surface, one measures simultaneously the fluctuating components of the magnetic field (using a magnetometer) and the electric field (using electrodes buried in the ground). At a given frequency, the ratio of orthogonal components of the electric and magnetic fields is related to the conductivity of the earth's crust averaged over a skin depth at that frequency. Thus, one can estimate the conductivity as a function of depth. These techniques hold promise in surveying for geothermal sources and mineral deposits, which produce anomalies in the electric conductivity.

The short-term fluctuations in the gradient of the earth's magnetic field are probably too small to be detected by a gradiometer. However, it is possible that gradiometers may be useful for detecting relatively slow changes in the magnetization in rocks near the earth's surface. For example, there is evidence [93] that the magnetic field along a fault line changes over periods of a few days prior to an earthquake. These changes are probably due to the piezomagnetism of the rock, that is, the fact that the magnetization of the rock changes when it is stressed.

Techniques such as this might ultimately be used in earthquake prediction.

Both magnetometers and gradiometers have been used to measure magnetocardiograms and magnetoencephalograms. Much of this work has been pioneered by D. Cohen, who has recently reviewed this field [94]. Cohen, Edelsack, and Zimmerman [95] were the first to use a SQUID magnetometer to take magnetocardiograms. The signal strength was typically 5×10^{-7} G, and an elaborate screened room was necessary. Subsequently, gradiometers [67, 96, 97] have been used; because of their discrimination against signals produced by relatively distant sources, no shielding is then necessary. Second derivative gradiometers, with their even higher rejection of signals from distant sources, have been successfully applied to magnetocardiography [68, 81]. SQUIDs have also been used to obtain magnetoencephalograms [94, 98], but in this case some kind of magnetic shielding appears to be essential, as the signals are of the order of 10^{-8} G. Although both magnetocardiology and magnetoencephalography are still at the experimental stage, they offer considerable promise as routine medical techniques in the future.

A further application is to the tracking of magnetic objects. Wynn et al. [99] have used an array of three magnetometers and five gradiometers together with sophisticated processing techniques to track a moving magnetic dipole.

Susceptometers have found application in a variety of applications. One example is the investigation of electronic [100] and nuclear [9, 101, 102] magnetism in solids at ultra-low temperatures and in small magnetic fields. Mercereau and co-workers [103] have measured the susceptibility of minute biochemical samples over a temperature range from 4 to 300 K. Groups at Harvard [104] and Cornell [105] have investigated the susceptibility of small superconducting samples near their transition temperature. Yet another application is to rock magnetometry. A commercial instrument [12] is available with room temperature access that allows the susceptibility of rock samples to be quickly measured in the field.

SQUIDs and SLUGs [6] have been widely used as high resolution voltmeters. All of the measurements have been made inside superconducting shields, and have been restricted to voltages originating in the cryogenic environment. Examples include the measurement of thermoelectric emf's at low temperatures [106], of flux creep in superconductors [107, 108]; and of proximity effects in superconductor-normal metal-superconductor sandwiches [109, 110]. A SLUG was used to compare the Josephson voltage-frequency relation in different superconductors [111] to a precision of 1 part in 10^8, and to measure the quasiparticle potential in non-equilibrium superconductors [112]. Giffard et al. [9] have developed a noise thermometer in which a SQUID was used to measure the Johnson noise in a known resistor. The temperature of the resistor was deduced from the amplitude of the noise in a known bandwidth.

ACKNOWLEDGEMENTS

I am grateful to Professor R.A. Buhrman and Dr. L.D. Jackel for helpful conversations on rf SQUIDs. I should like to thank the members of the Physics Department at the University of Campinas, Brazil, for their hospitality while this article was written. This work was supported in part by the U.S. Energy Research and Development Administration.

REFERENCES

1. B.D. Josephson, Phys. Lett. 1, 251 (1962); Adv. Phys. 14, 419 (1965).
2. R.C. Jaklevic, J. Lambe, A.H. Silver, and J.E. Mercereau, Phys. Rev. Lett. 12, 159 (1964).
3. J.E. Zimmerman and A.H. Silver, Phys. Rev. 141, 367 (1966).
4. M.R. Beasley and W.W. Webb, Proc. Symp. Physics of Superconducting Devices (University of Virginia, Charlottesville, April 28-29, 1967), V. 1.
5. P.L. Forgacs and A. Warnick, Rev. Sci. Instum. 38, 214 (1967).
6. J. Clarke, Phil. Mag. 13, 115 (1966).
7. J.E. Zimmerman, P. Thiene, and J.T. Harding, J. Appl. Phys. 41, 1572 (1970).
8. J.E. Mercereau, Rev. Phys. Appl. 5, 13 (1970); M. Nisenoff, Rev. Phys. Appl. 5, 21 (1970).
9. R.P. Giffard, R.A. Webb, and J.C. Weatley, J. Low Temp. Phys. 6, 533 (1972).
10. Develco, Inc., Mountain View, CA.
11. Instruments for Technology, Helsinki, Finland.
12. S.C.T., Inc., Mountain View, CA.
13. S.H.E. Corp., San Diego, CA.
14. John Clarke, Wolfgang M. Goubau, and Mark B. Ketchen, J. Low Temp. Phys. 25, 99 (1976).
15. John Clarke, IEEE 61, 8 (1973).
16. J. Bardeen, L.N. Cooper, and J.R. Schrieffer, Phys. Rev. 108, 1175 (1957).
17. F. London, Superfluids (Wiley, NY, 1950).
18. B.S. Deaver and W.M. Fairbank, Phys. Rev. Lett. 7, 43 (1961); R. Doll and M. Näbauer, Phys. Rev. Lett. 7, 51 (1961).
19. P.W. Anderson in Lectures on the Many Body Problem, Caianiello, ed. (Academic Press. NY, 1964), Vol. 2, p. 113.
20. P.W. Anderson and J.M. Rowell, Phys. Rev. Lett. 10, 230 (1962).
21. W. Schroen, J. Appl. Phys. 39, 2671 (1968).
22. J.H. Greiner, J. Appl. Phys. 42, 5151 (1972); ibid. 45, 32 (1974).
23. J.H. Greiner, S. Basavaiah, and I. Ames, J. Vac. Sci. and Tech. 11, 81 (1974).
24. S. Basavaiah, J.M. Eldridge, and J. Matisoo, J. Appl. Phys. 45, 457 (1974).
25. J.E. Nordman, J. Appl. Phys. 40, 2111 (1969); J.E. Nordman and W.H. Keller, Phys. Lett. A36, 52 (1971); L.O. Mullen and D.B. Sullivan, J. Appl. Phys. 40, 2115 (1969); R. Graeffe and T. Wiik,

J. Appl. Phys. 42, 2146 (1971); K. Schwidtal, J. Appl. Phys. 43, 202 (1972); P.K. Hansma, J. Appl. Phys. 45, 1472 (1974); S. Owen and J.E. Nordman, IEEE Trans. Magn. MAG-11, 774 (1975).
26. R.F. Broom, R. Jaggi, R.B. Laibowitz, Th.O. Mohr, and W. Walter, Proc. LT14, Helsinki, Finland (1975), p. 172; Gilbert Hawkins and John Clarke, J. Appl. Phys. 47, 1616 (1976).
27. K.E. Drangeid, R.F. Broom, W. Jutzi, Th.O. Mohr, A. Moser and G. Sasso, Intl. Solid-State Circuits Conf. Digest, 68-69 (1971); K. Grebe, I. Ames and A. Ginzberg, J. Vac. Sci. Tech. 458 (1974).
28. W.C. Stewart, Appl. Phys. Lett. 12, 277 (1968); D.E. McCumber, J. Appl. Phys. 39, 3113 (1968).
29. P.K. Hansma, G.I. Rochlin, and J.N. Sweet, Phys. Rev. B4, 3003 (1971).
30. P.W. Anderson and A.H. Dayem, Phys. Rev. Lett. 13, 195 (1964).
31. H.A. Notarys, R.H. Wang, and J.E. Mercereau, Proc. IEEE 61, 79 (1973).
32. R.B. Laibowitz, Appl. Phys. Lett. 23, 407 (1973).
33. P. Crozat, S. Gourrier, D. Bouchon, and R. Adde, Proc. Fourteenth Intern. Conf. on Low Temperature Physics, Otaniemi, Finland, August 14-20, 1975, Matti Krusius and Matti Vuorio, eds. (North-Holland, American Elsevier, 1975), Vol. 4, p. 206.
34. F. Considori, A.A. Fife, R.F. Frindt, and S. Gygax, Appl. Phys. Lett. 18, 233 (1971).
35. M.A. Janocko, J.R. Gavaler, and C.K. Jones, IEEE Trans. Magn. MAG-11, 880 (1975); R.B. Laibowitz, C.C. Tsuei, J.C. Cuomo, J.F. Ziegler, and M. Hatzakis, ibid., p. 883.
36. J.E. Zimmerman and A.H. Silver, Phys. Rev. 141, 367 (1966).
37. J.E. Zimmerman, Proc. 1972 Applied Superconductivity Conf., Annapolis, Maryland (IEEE, NY, 1972), p. 544.
38. R.A. Buhrman, S.F. Strait, and W.W. Webb, J. Appl. Phys. 42, 4527 (1971).
39. A. Baratoff, J.A. Blackburn, and B.B. Schwartz, Phys. Rev. Lett. 25, 1096 (1970).
40. L.D. Jackel and R.A. Buhrman, J. Low Temp. Phys. 19, 201 (1975).
41. A.Th.A.M. De Waele and R. De Bruyn Ouboter, Physica 41, 225 (1969).
42. M. Tinkham Introduction to Superconductivity (McGraw-Hill, 1975), p. 214.
43. Claudia D. Tesche and John Clarke, Proceedings of the Applied Superconductivity Conference, Stanford, August 1976, in press.
44. Y.H. Ivanchenko and L.A. Zil'berman, Zh. Eksperim, I. Teor. Fiz. 55, 2395 (1968) (Sov. Phys. J.E.T.P. 28, 1272 (1969)).
45. V. Ambegaokar and B.I. Halperin, Phys. Rev. Lett. 22, 1364 (1969).
46. A.N. Vystavkin, V.N. Gubankov, L.S. Kuzmin, K.I. Likharev, V.V. Migulin, and V.K. Semenov, Phys. Rev. Appl. 9, 79 (1974).
47. C.M. Falco, W.H. Parker, S.E. Trullinger, and P.K. Hansma, Phys. Rev. B10, 1865 (1974).

48. K.K. Likharev and V.K. Semenov, J.E.T.P. Lett. 15, 442 (1972).
49. V.R. Radhakrishnan and V.L. Newhouse, J. Appl. Phys. 42, 129 (1971).
50. J.E. Zimmerman, J. Appl. Phys. 42, 4483 (1971).
51. J.M. Pierce, J.E. Opfer, and L.H. Rorden, IEEE Trans. Magn. MAG-10, 599 (1974).
52. A. Davidson, R.S. Newbower, and M.R. Beasley, Rev. Sci. Instrum. 45, 838 (1974).
53. J.H. Claassen, J. Appl. Phys. 46, 2268 (1975).
54. S.K. Decker and J.E. Mercereau, Appl. Phys. Lett. 25, 527 (1974).
55. D.W. Palmer, H.A. Notarys, and J.E. Mercereau, Appl. Phys. Lett. 25, 527 (1974).
56. W. Richter and G. Albrecht, Cryogenics 15, 148 (1975).
57. J. Clarke and Gilbert A. Hawkins, IEEE Trans. Magn. MAG-11, 724 (1975); Phys. Rev. B., October 1, 1976.
58. John Clarke and Richard F. Voss, Phys. Rev. Lett. 33, 24 (1974); Richard F. Voss and John Clarke, Phys. Rev. B13, 556 (1976).
59. W.W. Webb, IEEE Trans. Magn. MAG-8, 51 (1972).
60. See C. Kittel, Introduction to Solid State Physics (Wiley, NY, London, Sydney, Toronto, 1971), Edition IV, Appendix J.
61. P.K. Hansma, J. Appl. Phys. 44, 4191 (1973).
62. J. Kurkijärvi, Phys. Rev. B6, 832 (1972).
63. L.D. Jackel, J. Kurkijärvi, J.E. Lukens, and W.W. Webb in the 13th Intern. Conf. Low Temp. Phys., Boulder 1972 (Plenum Press, NY, 1974), Vol. 3, p. 705.
64. L.D. Jackel, W.W. Webb, J.E. Lukens, and S.S. Pei, Phys. Rev. B9, 115 (1974).
65. J. Kurkijärvi and W.W. Webb, Proc. Appl. Superconductivity Conf. Annapolis (IEEE, NY, 1972), p. 581.
66. J. Kurkijärvi, J. Appl. Phys. 44, 3729 (1973).
67. W.L. Goodman, V.W. Hesterman, L.H. Rorden, and W.S. Goree, Proc. IEEE 61, 20 (1973).
68. J.E. Opfer, Y.K. Yeo, J.M. Pierce, and L.H. Rorden, IEEE Trans. Magn. MAG-10, 536 (1974).
69. N.V. Zavaritskii and M.S. Legkostupov, Cryogenics 14, 42 (1974).
70. G.J. Ehnholm, J.K. Soini, and T. Wiik, Proc. of the Fourteenth Intern. Conf. on Low Temperature Physics, Otaniemi, Finland, August 14-20, 1975, Matti Drusius and Matti Vuorio, eds. (North-Holland/American Elsevier, 1975), Vol. 4, p. 234.
71. R. Rifkin and B.S. Deaver, Phys. Rev. B13, 3894 (1976).
72. J.E. Zimmerman, J. Appl. Phys. 42, 4483 (1971).
73. J.E. Zimmerman, Cryogenics 12, 19 (1972).
74. I am grateful to S.C.T. and S.H.E. for their collaboration in these measurements.
75. J.E. Zimmerman and N.V. Frederick, Appl. Phys. Lett. 19, 16 (1971).

76. R.A. Kamper and M.B. Simmonds, Appl. Phys. Lett. 20, 270 (1972).
77. T.D. Clark and L.D. Jackel, Rev. Sci. Instrum. 46, 1249 (1975).
78. J.M. Pierce, J.E. Opfer, and L.H. Rorden, IEEE Trans. Magn. MAG-10, 599 (1974).
79. M.R. Gaerttner, (unpublished — reported at Int. Mag. Conf., Toronto, 1974).
80. R.A. Webb, R.P. Giffard, and J.C. Wheatley, J. Low Temp. Phys. 13, 383 (1973).
81. D. Brenner, S.J. Williamson, and L. Kaufman, Proc. of the Fourteenth Intern. Conf. on Low Temp. Phys., Otaniemi, Finland, August 14-20, 1975, Matti Drusius and Matti Vuorio, eds. (North-Holland/American Elsevier, 1975), Vol. 4, p. 266.
82. Mark B. Ketchen, Wolfgang M. Goubau, John Clarke, and Gordon B. Donaldson, Proc. of the Applied Superconductivity Conf., Stanford, CA, August 17-20, 1976 (to be published).
83. E.C. Hirschkoff, O.G. Symko, and J.C. Wheatley, J. Low Temp. Phys. 4, 111 (1971).
84. J.G. Doran and O.G. Symko, IEEE Trans. Magn. MAG-10, 603 (1974).
85. E.J. Cukauskas, D.A. Vincent, and B.S. Deaver, Jr., Rev. Sci. Instrum. 45, 1 (1974).
86. M. Cerdonio and C. Messana, IEEE Trans. Magn. MAG-11, 728 (1975).
87. For recent improvements in vibrating sample magnetometers and a comprehensive bibliography see S. Foner, Rev. Sci. Instrum. 46, 1425 (1975).
88. Oxford Instruments, England.
89. J. Clarke, W.E. Tennant, and D. Woody, J. Appl. Phys. 42, 3859 (1971).
90. For example, Princeton Applied Research Model 185.
91. D.E. Prober, Rev. Sci. Instrum. 45, 848 (1974).
92. W.M. Goubau, T.D. Gamble, J. Clarke, H.F. Morrison, and E.C. Mosley (unpublished).
93. M. Johnston (private communication).
94. D. Cohen, Physics Today, August 1975, p. 34; IEEE Trans. Magn. MAG-11, 694 (1975).
95. D. Cohen, E.A. Edelsack, and J.E. Zimmerman, Appl. Phys. Lett. 16, 278 (1970).
96. J.E. Zimmerman and N.V. Frederick, Appl. Phys. Lett. 19, 16 (1971).
97. J. Ahopelto, P.J. Karp, T.E. Katila, R. Lukanda, and P. Mäkipää, Proc. of the Fourteenth Intern. Conf. on Low Temp. Phys., Matti Krusius and Matti Vuorio, eds. (North-Holland/American Elsevier, 1975), Vol. 4, p. 202.
98. D. Cohen, Science 175, 664 (1972).
99. W.M. Wynn, C.P. Frahm, P.J. Carroll, R.H. Clark, J. Wellhoner, and M.J. Wynn, IEEE Trans. Magn. MAG-11, 701 (1975).
100. E.C. Hirschkoff, O.G. Symko, and J.C. Wheatley, J. Low Temp.

Phys. 5, 155 (1971).
101. E.C. Hirschkoff, O.G. Symko, L.L. Vant-Hall, and J.C. Wheatley, J. Low Temp. Phys. 2, 653 (1970).
102. J.M. Goodkind and D.L. Stolfa, Rev. Sci. Instrum. 41, 799 (1970).
103. H.E. Hoenig, R.H. Wang, G.R. Rossman, and J.E. Mercereau, in Proceedings of the 1972 Applied Superconductivity Conference (Annapolis, Maryland) (IEEE, NY, 1972), p. 570.
104. J.P. Gollub, M.R. Beasley, R.S. Newbower, and M. Tinkham, Phys. Rev. Lett. 22, 1288 (1969).
105. J.E. Lukens, R.J. Warburton, and W.W. Webb, Phys. Rev. Lett. 25, 1180 (1970).
106. E. Rumbo, Phil. Mag. 19, 689 (1969).
107. M. Wade, Phil. Mag. 20, 1107 (1969).
108. M.R. Beasley, R. Labusch, and W.W. Webb, Phys. Rev. 181, 682 (1969).
109. J. Clarke, Proc. Roy. Soc. A308, 447 (1969).
110. A.B. Pippard, J.G. Shepherd, and D.A. Tindall, Proc. Roy. Soc. A324, 17 (1971).
111. J. Clarke, Phys. Rev. Lett. 21, 1566 (1968).
112. J. Clarke, Phys. Rev. Lett. 28, 1363 (1972); J. Clarke and J.L. Paterson, J. Low Temp. Phys. 15, 491 (1974).

Chapter 4

EQUIVALENT CIRCUITS AND ANALOGS OF THE JOSEPHSON EFFECT

T. A. Fulton

Bell Laboratories, Murray Hill, New Jersey

I. INTRODUCTION

Originally the term Josephson junction referred to the tunnel junction, comprising two bulk superconductors separated by a thin insulating barrier through which current could flow by electron tunneling. Following the original studies by Giaver [1] of the dissipative current flow by quasiparticle tunneling at voltages $V \neq 0$ in this system, Josephson [2] predicted that there could also be supercurrent flow at $V = 0$ and, further, that the supercurrent would continue to flow at $V \neq 0$, but as an alternating current at a frequency proportional to V. Subsequently, the $V = 0$ dc supercurrent was observed by Anderson and Rowell [3] and the ac supercurrent by Shapiro [4]. Since that time similar dc and ac supercurrent effects have been observed in other superconductor systems, notably in small area metal-metal contacts (point contacts) by Zimmerman and Silver [5], in thin-film superconducting constrictions (bridges) by Anderson and Dayem [6], superconductor-normal metal-superconductor thin-film sandwiches (SNS junctions) by Clarke [7], and in bridges weakened by normal metal overlays by Notarys and Mercereau [8]. All these systems are now called Josephson junctions [9-15].

Since these initial discoveries, the nonlinear and active electrical properties of these devices and their ability to respond at high frequencies has held forth great promise for the fabrication of devices of unparalleled sensitivity and speed. Potential applications, particularly in logic and memory elements, magnetometers and microwave detectors, have been vigorously pursued and are now beginning to be realized in practice.

An important practical ingredient in this work is the fact that the equations describing the interplay of voltage and currents in Josephson circuitry are relatively simple. Moreover, they are similar in form to those encountered in certain mechanical systems, notably the ordinary simple pendulum [9]. The ability to understand and indeed to visualize the behavior of Josephson junctions in terms of the analogous systems has greatly facilitated the understanding of their behavior.

In this chapter we review the electrical behavior of Josephson junctions, emphasizing the interaction of the junction with passive circuit elements and relying heavily on pictorial descriptions based on the analogs. We will avoid, as far as possible, analytic and numerical treatment of the equations. Because the range of even the qualitative behavior is large, a comprehensive treatment is not attempted. Instead we will discuss representative cases whose behavior, we hope, will provide a general insight. In this discussion we will draw consciously or unconsciously on the contributions of many different workers. The references are very far from complete, and we wish to apologize at the beginning to those whose contributions are not referenced explicitly.

Section II surveys the "small" junction limit, touching on topics including I-V curves, interaction with rf currents, inductive shunts and SQUIDs. Section III discusses the "large" junction limit and especially vortex properties.

II. SMALL JUNCTIONS

A. Model of the Supercurrent Flow

The term "small junction" refers to point contacts, bridges, and also to tunnel junctions and SNS junctions if the latter are sufficiently "small" according to a criterion set forth in Section III. The standard equivalent circuit model used to describe the dc and ac supercurrent behavior in all these systems assumes that the active region of the device, e.g., the constriction of a bridge or the barrier of a tunnel junction, acts as though it contains a nonlinear inductive element, called the supercurrent element. Through this element flows a time-dependent current $I_S(t) = I_c \sin \phi(t)$. Here I_c is the maximum supercurrent the generator can provide, and $\phi(t)$, called the phase, is related to the voltage $V(t)$ across the active region by

$$d\phi/dt = 2\pi V(t)/\Phi_o$$

or equivalently by

$$\phi(t) = (2\pi/\Phi_o) \int_o^t V(u)\, du$$

(where t = 0 is some time at which $I_S(t) = 0$, and $dI_S/dt > 0$ for $V > 0$). The quantity Φ_o is the flux quantum, $\Phi_o = h/2e = 2.068 \times 10^{-15}$ wb.

It is often useful to regard this supercurrent element as a voltage-controlled current generator which delivers a current $-I_S(t)$ to other circuit elements, as indicated in Fig. 1(a). These other elements are lumped or distributed passive impedance elements, external current generators, and/or other junctions. The passive elements may be externally attached or they may be part of the junction structure, such as the capacitance or the quasiparticle tunneling conductance of a tunnel junction. The interaction of this nonlinear supercurrent generator with these elements is the subject of all subsequent discussion in this chapter.

B. Voltage-Biased Model

In the lowest approximation we suppose that V can be controlled and modulated in any desired way [2]. As sketched in Fig. 1(b), if V = 0, a dc supercurrent $I_c \sin \phi$ will be generated, where ϕ is presumed to adjust itself to fit any applied bias current. If $V \neq 0$, the generator current will be sinusoidal in time with a frequency $f = V/\Phi_o$, where $(1/\Phi_o) \approx 484$ GHz/mV. (Since typical voltages occurring in the junctions are fractions of millivolts, high-frequency applications are an obvious possibility.)

As Fig. 1(c) depicts, however, in treating the supercurrent element as a generator one must keep in mind that the voltage V(t) that controls $I_S(t)$ is that which appears across the output terminals through which $I_S(t)$ flows, as if the input and the output were shorted together in an ordinary voltage-controlled oscillator. If the current generated by the junction produces a non-negligible voltage by its interaction with associated impedances, then this voltage acts back on the junction to cause the current to be non-sinusoidal in time. In many practical situations such feedback effects alter or dominate the behavior and V is not really under external control. The voltage-biased picture is generally most accurate in the high-frequency limit where the dc component of V becomes large compared to the feedback voltages.

C. Stewart-McCumber Model

1. Circuit equations

The basic model which treats feedback effects is that shown in Fig. 2(a) and first discussed by Stewart [16] and by McCumber [17]. Here the supercurrent element acting as a current generator is shunted by a passive resistor R and capacitor C. A constant current bias I is applied to the parallel combination.

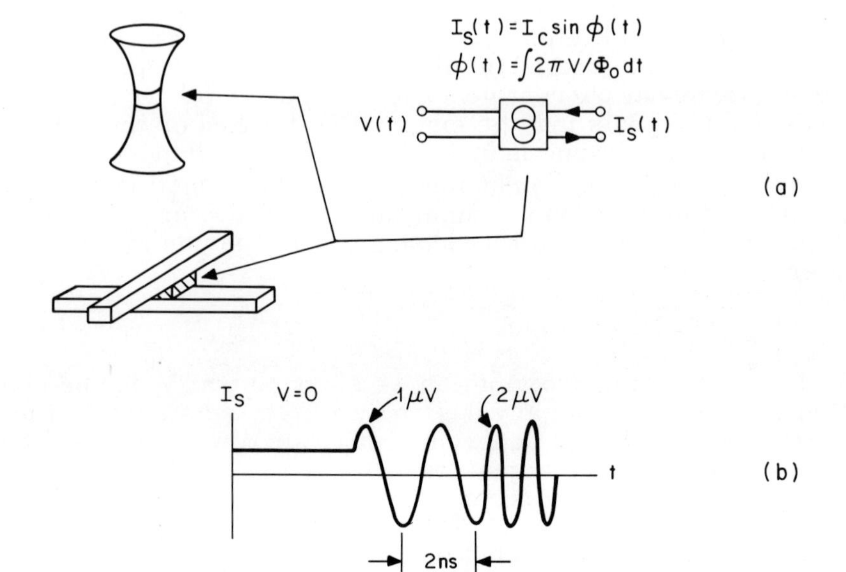

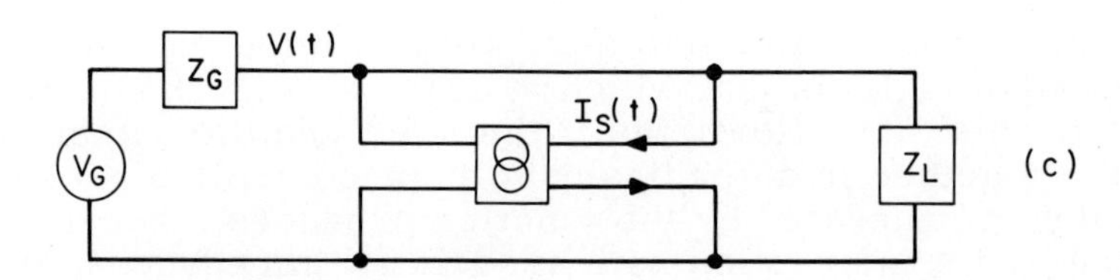

Fig. 1 (a): Equivalent voltage-controlled generator, a model for supercurrent behavior of the active region of a Josephson junction.
(b): Supercurrent waveform for three constant voltages.
(c): Equivalent voltage-controlled generator connected to load impedance Z_L and an external generator of voltage V_G and impedance Z_G. Note that the input and output voltages are equal.

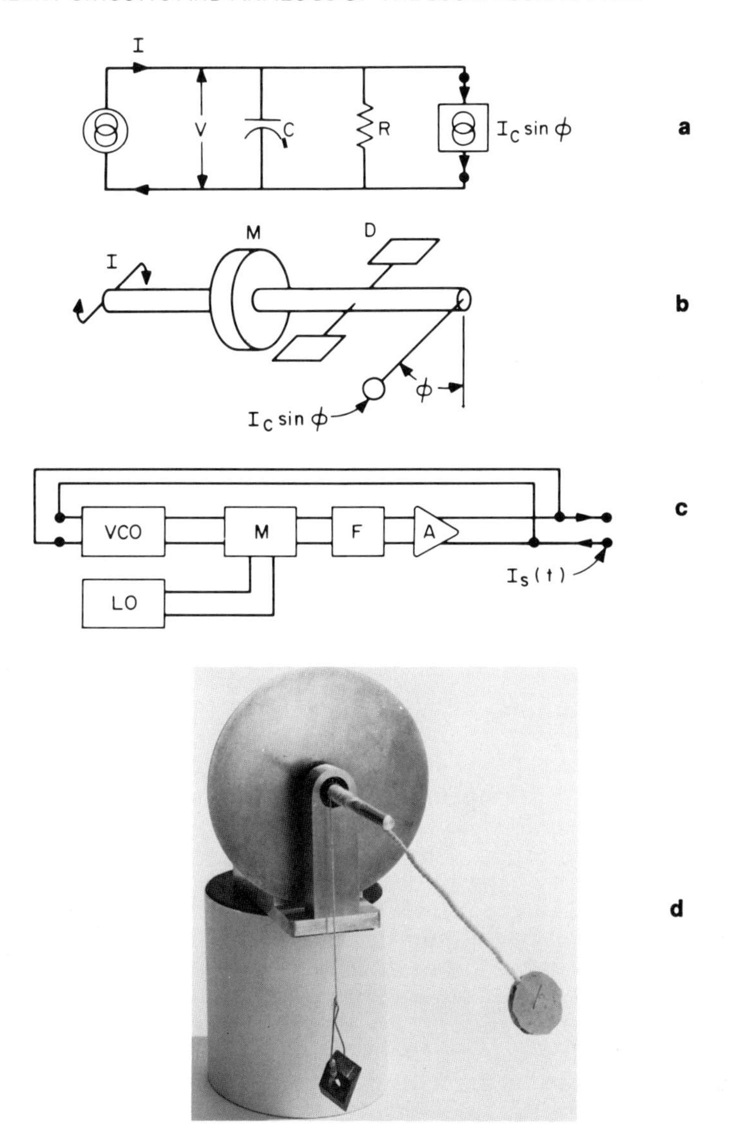

Fig. 2 (a): Stewart-McCumber circuit.
(b): Pendulum analog of Stewart-McCumber circuit.
(c): Electronic analog (after Ref. 22) of supercurrent element: VCO - voltage controlled oscillator at frequency $f_o + kV$; LO - local oscillator at frequency f_o; M - mixer; F - low-pass filter; A - high output impedance amplifier. Load impedances C, R, etc., and external current sources are placed in parallel across terminals marked $I_S(t)$.
(d): Photograph of pendulum analog.

The equation of current continuity is

$$I = I_c \sin\phi + V/R + C\, dV/dt\,. \qquad (1)$$

Using the relation $d\phi/dt = 2\pi V/\Phi_o$, we may express (1) in terms of ϕ as

$$I = I_c \sin\phi + D\, d\phi/dt + M\, d^2\phi/dt^2 \qquad (2)$$

where $D = \Phi_o/2\pi R$ and $M = C\Phi_o/2\pi$, or in a dimensionless form as

$$(I/I_c) = \sin\phi + (\omega_J^{-2}\tau^{-1}) d\phi/dt + \omega_J^{-2}\, d^2\phi/dt^2 \qquad (3)$$

where $\tau = RC$, and ω_J, the Josephson plasma frequency, is

$$\omega_J = \left(\frac{2\pi I_c}{\Phi_o C}\right)^{\frac{1}{2}}\,.$$

This equation, given the initial conditions on ϕ and $d\phi/dt$, controls the behavior of ϕ and hence the voltages and currents for all subsequent time.

2. Mechanical analogs

Consider the familiar simple pendulum with moment arm I_c, moment of inertia M, damping coefficient D, phase angle ϕ measured from vertically downward, and subjected to an external torque I. The pendulum is sketched in Fig. 2(b) with M provided by a flywheel and D by a paddle-wheel arrangement. A photograph of an assembled model of Fig. 2(b) is shown in Fig. 2(d). Evidently the Newton's law equation governing the time evolution of ϕ has exactly the same form as Eq. (2) [9]. The analogies between the torques and the currents and between the angular velocity and $d\phi/dt$ are apparent. The use of the same symbols is intended to emphasize the correspondence. There is a further correspondence between the kinetic energy and $\frac{1}{2}CV^2$, and between the gravitational potential energy and what is called the Josephson coupling energy, $-(\Phi_o I_c/2\pi)\cos\phi$. Similarly the rate of work done by external torque, $I\, d\phi/dt$, corresponds to IV. Table I gives a list of corresponding quantities in this and other analogies to be discussed.

The correspondence between the mechanical motion of the simple pendulum, and the electrical behavior of the junction makes much of the junction behavior self-evident, although even a pendulum can surprise one at times. In all subsequent discussions we will stress this analogy, often mixing mechanical and electrical terms together. While the current-torque and $d\phi/dt$-angular velocity correspondences are straightforward, the voltage V has no obvious familiar analog and must be connected to the angular velocity via $2\pi V/\Phi_o = d\phi/dt$. In this regard it is useful to

Table 1: A list of major corresponding quantities for Josephson circuits and their pendulum analogs as described in Section II. Very similar correspondences between torque and current densities and between the energy densities of the two systems occur in Section III.

Josephson	Common Symbol	Mechanical
Critical Current	I_c	Pendulum Moment Arm
Phase Difference	ϕ	Phase Angle Pendulum
Voltage x $2\pi/\Phi_o$	$d\phi/dt$	Angular Velocity
Applied Current	I	Applied Torque
$\Phi_o/(2\pi R)$	D	Damping Coefficient
$C\Phi_o/(2\pi R)$	M	Moment of Inertia
$\Phi_o/(2\pi L)$	K	Torsion-Bar Spring Constant
Current in Inductor	I_L	Torque in Torsion Bar
Josephson Coupling Energy, $-(\Phi_o I_c/(2\pi))\cos\phi$		Gravitational Potential Energy, $-I_c \cos\phi$
Electrostatic energy, $\frac{1}{2}CV^2$		Kinetic Energy, $\frac{1}{2}M(d\phi/dt)^2$
Magnetic Energy, $\frac{1}{2}L\,I_L^2$		Torsional Energy, $\frac{1}{2}I_L^2/K$
Power Input from Applied Current, IV		Power Input from Applied Torque, $I\,d\phi/dt$

remember that V = IR corresponds to the pendulum moving with its terminal velocity $d\phi/dt = I/D$.

Another mechanical analog [18] which is sometimes useful is that of the damped motion of a particle subjected to a linear plus cosinusoidal potential, giving a Newton's law of the same form as Eq. (2). The behavior here, though completely analogous to the pendulum system, has some pictorial advantages. It is easily fabricated by taking a hollow cylinder, fastening a weight to one rim and allowing it to roll along an inclined plane.

In common with other workers [19, 20], we have constructed and made use of both mechanical models for several years. Much of the discussion in subsequent portions of this section is based upon experience with their behavior. Where unsubstantiated and unfamiliar statements concerning the junction electrical properties appear in this discussion, these properties will usually correspond to more or less obvious mechanical behavior.

Several workers [21-24] have designed and studied electrical elements acting as current generators with a sinusoidal dependence of the current on the time-integrated voltage across the input terminals, thereby simulating a Josephson supercurrent element. For example, one such circuit [22] sketched in Fig. 2(c) employs an oscillator with a voltage-controlled frequency f obeying $f = f_0 + kV$, whose signal is mixed with a reference signal at frequency f_0 to provide a difference-frequency voltage signal driving a current amplifier. Such circuits are very useful in making analog computers for simulating the behavior of Josephson circuits and complement the visual, intuitive quality of the mechanical analogs. A number of the figures in this chapter are taken from oscilloscope displays of such a circuit.

3. I-⟨V⟩ curves

The simplest experiment one can perform on a Josephson junction is to measure the low-frequency current-voltage characteristic [16, 17], or I-⟨V⟩ curve where the bracket notation ⟨V⟩ indicates the time average or dc component. For the pendulum analog, initially at rest with I = 0 and $\phi = 0$, the analogous procedure for measuring the I-⟨V⟩ curve is to increase I slowly and note the corresponding average frequency of rotation. If I is increased slowly, the pendulum angle will adjust slowly so as to balance the applied and gravitational torques, $I = I_c \sin\phi$. Correspondingly, the junction remains at $\langle V\rangle = 0$ and carries a dc supercurrent. As I approaches I_c, ϕ approaches $\pi/2$. For still larger I, the pendulum will overbalance and begin to rotate, so that ⟨V⟩ becomes nonzero. The nature of the rotation and the nature of the I-⟨V⟩ curve depend upon the value of $\omega_J\tau$. The limiting regimes in which $\omega_J\tau \ll 1$ and $\omega_J\tau \gg 1$, the overdamped and underdamped regimes, respectively, are of special interest. In the overdamped regime inertial effects can be neglected and the time scale is set by $(\omega_J^2 \tau)^{-1} = \Phi_o/2\pi RI_c$, whereas in the underdamped regime inertial effects dominate and ω_J and τ are the time scales of interest.

In describing the I-$\langle V \rangle$ curves it will be useful to keep in mind two general relations resulting from Eq. (1). (We denote time averages by brackets.) In steady state there will be an average current balance given by

$$\langle I \rangle = \langle I_c \sin \phi \rangle + \langle V \rangle / R \tag{4}$$

so any departure from an ohmic I-$\langle V \rangle$ curve involves a nonzero average supercurrent, $\langle I_c \sin \phi \rangle$. Further, there is an average power balance of

$$\langle VI \rangle = \langle V^2 \rangle / R . \tag{5}$$

Decomposing I and V into steady and alternating components

$$I = \langle I \rangle + I_{ac} \quad \text{and} \quad V = \langle V \rangle + V_{ac} ,$$

then

$$\langle V \rangle \langle I \rangle + \langle V_{ac} I_{ac} \rangle = \langle V \rangle^2 / R + \langle V_{ac}^2 \rangle / R . \tag{6}$$

If I is fixed, as it often is, then $\langle I_c \sin \phi \rangle = \langle V_{ac}^2 \rangle / R$, and the departure of the I-$\langle V \rangle$ curve from ohmic behavior also indicates that the voltage has an alternating component.

Overdamped case: In the overdamped regime the pendulum obeys the equation

$$D\, d\phi/dt = I - I_c \sin \phi \tag{7}$$

where the velocity is proportional to the difference between the external and the gravitational torques. Figure 3 shows the quantities $\sin \phi(t)$, $V(t)$, and $\cos \phi(t)$, as predicted by this equation for the cases (a) $I = 4.0\, I_c$ and (b) $I = 1.05\, I_c$. Each point in the plot of $\cos \phi(t)$, the solid circles located at coordinates $(t, -\cos \phi)$ is connected by a straight line to a point at the coordinates $(t - \alpha \sin \phi, 0)$, where α is 60% of the time intervals. The intention is to give the appearance of the pendulum at these times as seen from the front and slightly out of the plane of the motion. Thus for $I \gtrsim I_c$ (see Fig. 4(b)) the pendulum will move very slowly in the region of $\phi = \pi/2$ and relatively swiftly in the region of $\phi = 3\pi/2$. The angular velocity vs time then has the form of a series of widely separated pulses of width $D/2I_c = \Phi_o/\pi RI_c$, with ϕ advancing by $\sim 2\pi$ in each pulse, and the time-average velocity $\langle d\phi/dt \rangle$ is then much smaller than the "terminal

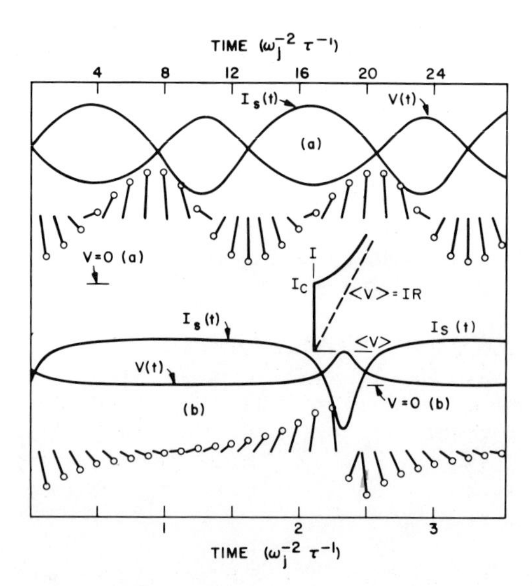

Fig. 3 $I_S(t) = I_C \sin\phi(t)$ and V(t) for an overdamped junction. Also shown is a representation of a pendulum performing analogous motion as described in the text. (a): $I = 1.05\ I_C$; (b) $I = 4.0\ I_C$. The I-$\langle V\rangle$ curve is shown in the inset.

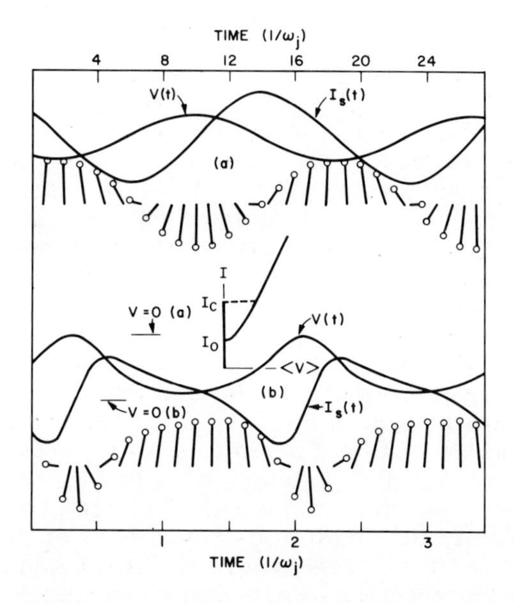

Fig. 4 $I_S(t) = I_C \sin\phi(t)$ and V(t) for an underdamped junction having $\omega_J\tau = 3$. Also shown is the representation of the analogous motion of a pendulum. (a): $I = 1.01\ I_C$; (b): $I = 0.425\ I_C \gtrsim I_O$. The I-$\langle V\rangle$ curve is shown in the inset.

velocity" I/D, and correspondingly the time-average $\langle V \rangle << IR$ for $I/I_c \sim 1$. Because $\phi \sim \pi/2$ for most of each cycle, one has $\langle I_c \sin\phi \rangle \approx 1$, as required by the current balance relation, Eq. (4). For $I >> I_c$, as in Fig. 4(a), the applied torque is nearly independent of ϕ and terminal velocity $d\phi/dt = I/D$ or $V = IR$ is approached.

In a more analytic vein [17, 25], Eq. (7) can be solved by simple integration, giving in dimensionless form

$$(I/I_c \sin\phi)(I/I_c - \sin\sqrt{I^2 - I_c^2}\, t/D) = (I/I_c)^2 - 1 \, .$$

Averaging this result over one period gives an I-$\langle V \rangle$ curve of the form $\langle V \rangle = R\sqrt{I^2 - I_c^2}$ for $I > I_c$ and $\langle V \rangle = 0$ for $I < I_c$, as shown in the inset of Fig. 3. This sort of I-$\langle V \rangle$ curve is frequently observed for point-contact junctions [26] and, more arguably, in thin-film bridges [27]. Both of these devices appear to correspond to the overdamped limit, as one might expect from the small capacitance of such structures.

Underdamped regime: Here inertial forces dominate. The pendulum begins to rotate for $I > I_c$ and accelerates over a time $M/D = \tau$ to nearly terminal velocity of $\langle V \rangle = IR$. Figure 4(a) shows an example of the steady-state rotational behavior of $\sin\phi(t)$, $V(t)$ and $\cos\phi(t)$ (in a pendulum motif) as calculated from Eq. (7) for $\omega_J\tau = 3$ at $I = 1.01\, I_c$. At this velocity the ratio of the kinetic energy $\frac{1}{2}M\dot{\phi}^2$ or $\frac{1}{2}CV^2$ to the maximum gravitational potential energy $I_c\pi$ or $\Phi_0 I_c/\pi$ is $\sim(\omega_J\tau)^2$, so that V is nearly constant and $\langle \sin\phi \rangle$ is correspondingly small. Suppose then that from this steady rotating condition I is reduced slowly on the time scale of $\tau = RC$. The pendulum will continue to rotate at $\langle V \rangle \approx IR$ for I substantially less than I_c. Thus, the I-$\langle V \rangle$ curve for an underdamped junction is hysteretic in this range, with the value of $\langle V \rangle$ depending on the previous history. One must reduce I to the level where $\langle V \rangle \sim \Phi_0 \omega_J/2\pi$ before the kinetic and the gravitational potential energy become comparable. In this regime an increasing velocity modulation occurs. In the limit where the kinetic plus potential energy barely exceeds the maximum potential energy, the pendulum rotates very slowly past $\phi = \pi$ in each cycle and $d\phi/dt$ and V go to a pulse-like form as is shown in Fig. 4(b) for $I = 0.425\, I_c$. Note that in this regime the pendulum spends the bulk of its time in an inverted position.

Because V_{ac} is appreciable in this regime the power balance, Eq. (6), requires that $\langle I_c \sin\phi \rangle = \langle V_{ac}^2 \rangle / R$ become non-zero and the I-$\langle V \rangle$ curve as shown in the inset of Fig. 4 must bend at low $\langle V \rangle$. At very small $\langle V \rangle$, where the voltage pulses are separated by more than their width, the energy loss due to damping in a cycle approaches a constant value and $\langle V \rangle$ goes to zero at $I = I_0 = 4I_c/\pi\omega_J\tau$ for $\omega_0\tau >> 1$ [16]. If I is reduced below this level the pendulum will fail to pass through the inverted position and comes to rest after oscillating for a time $\sim\tau$. One would then need to increase I to $I > I_c$ in order to cause $\langle V \rangle$ to become non-zero again. Underdamped behavior is usually exhibited by tunnel junctions [28] for which $\omega_J\tau$ typically is $\sim$10 to 10^3. The resistance of tunnel junctions is actually nonlinear with voltage, but no serious modification [29, 30] of the previous arguments results from this, except that the I-$\langle V \rangle$ curve for $\langle V \rangle \gtrsim \Phi_o \omega_J/2\pi$ follows the nonlinear resistance. In addition, since the operation of the pendulum is so delicate at low $\langle V \rangle$, any slight noise perturbation will tend to interrupt the motion and cause the junction to return to $\langle V \rangle = 0$. Stable operation of a small tunnel junction for $\langle V \rangle \lesssim \Phi_o \omega_J 2\pi$ is seldom observed, and the junction returns discontinuously to $\langle V \rangle = 0$ in this range.

Finally, between the two limiting regimes of $\omega_J\tau$, the character of the pendulum motion and the junction I-$\langle V \rangle$ curve changes in a straightforward manner. The point at which the I-$\langle V \rangle$ curve changes from an hysteretic to a single-valued behavior is numerically computed [31] to be $\omega_J\tau = 1.193$. Experimental cases of junctions lying in this intermediate regime have been fabricated by adding a shunt resistance to a tunnel junction [30, 32].

4. Plasma oscillations

From Eq. (2) small oscillations in ϕ about a fixed value lead to $dI_s/dt = (I_c \cos\phi)\, d\phi/dt = (\cos\phi)(2\pi I_c/\Phi_o)V$. Thus the supercurrent responds to V as though the junction were an inductor [33], and the quantity $\Phi_o/2\pi I_c$ is called the Josephson inductance, L_J. The junction and the parallel capacitance then behave as a resonant tank circuit for small amplitude oscillations and, for $\omega_J\tau >> 1$, have a resonant frequency of $\omega_J\sqrt{\cos\phi} = \omega_J(1 - I^2/I_c^2)^{\frac{1}{4}}$. These oscillations, called plasma oscillations, are the analog of the familiar small oscillations of a pendulum. Note that the decrease in the frequency to zero at $I = I_c$ is an indication that the system becomes statically unstable at this bias.

For oscillations of large amplitude the anharmonicity of the potential energy reduces the frequency of the oscillations and makes them decidedly non-uniform in time. Figure 5 shows some examples. Note that at I = 0 the oscillation possesses only odd harmonics of the fundamental frequency, whereas at $I \neq 0$ all harmonics occur. This anharmonic oscillation has been observed and employed in a variety of resonance and parametric generation experiments [34-37].

5. Punchthrough

As an example of effects occurring when I is varied non-adiabatically, consider an underdamped junction in an initial $I > 0$ and $\langle V \rangle > 0$ steady state. If I is suddenly changed to $I = I' < -I_o < 0$

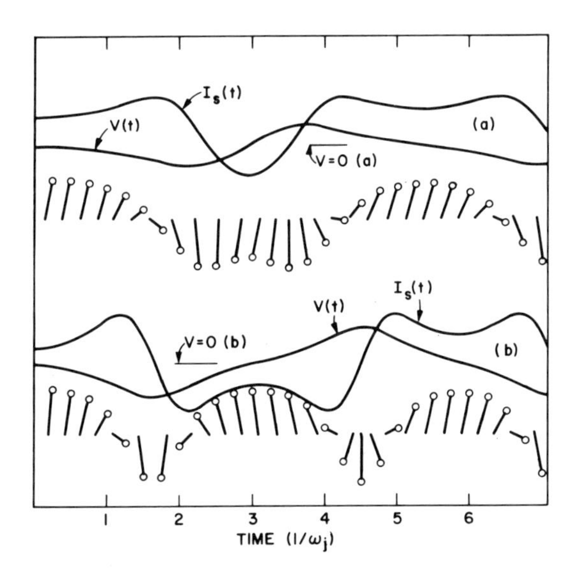

Fig. 5 $I_S(t) = I_C \sin \phi(t)$ and V(t) for plasma oscillations in an underdamped junction with $\omega_J\tau = 100$. (a): $I = 0.5\ I_C$; (b): $I = 0$. Also shown is a representation of the analogous motion of a pendulum.

in a time $\ll$ RC, then as it happens the voltage may end up either at $\langle V\rangle = 0$ or at $\langle V\rangle \neq 0$. To understand this, consider an analogous situation in which a particle initially coasting downhill in the periodic plus linear potential described previously suddenly finds itself coasting uphill, as in Fig. 6. Gravity and damping bring it to a halt eventually and, for $\omega_J\tau \gg 1$, this always occurs at a point A near the top of a potential maximum M. The particle falls back into the minimum and ascends the next downhill M + 1. If by chance point A is at the peak of M, the particle will have enough energy to pass over M + 1 and, since $|I'| > |I_o|$, continue on with $\langle V\rangle < 0$, whereas if the particle just barely surmounts M + 1 going uphill, damping losses will prevent it from escaping the minimum and it will settle to $\langle V\rangle = 0$. The probability of being trapped in the minimum evidently depends on the residual kinetic energy at M + 1 going uphill, which is determined essentially randomly by the details of the previous history. A rough calculation gives a trapping probability of $(|I'| - |I_o|)/(|I'| + |I_o|)$.

This phenomenon, called punchthrough [38], is observed in the I-$\langle V\rangle$ characteristic of tunnel junctions for sweep rates of I which exceed some critical rate, dependent mainly on RC and I_o. It is especially common for junctions of low critical current density which have correspondingly large values of $\omega_J\tau$.

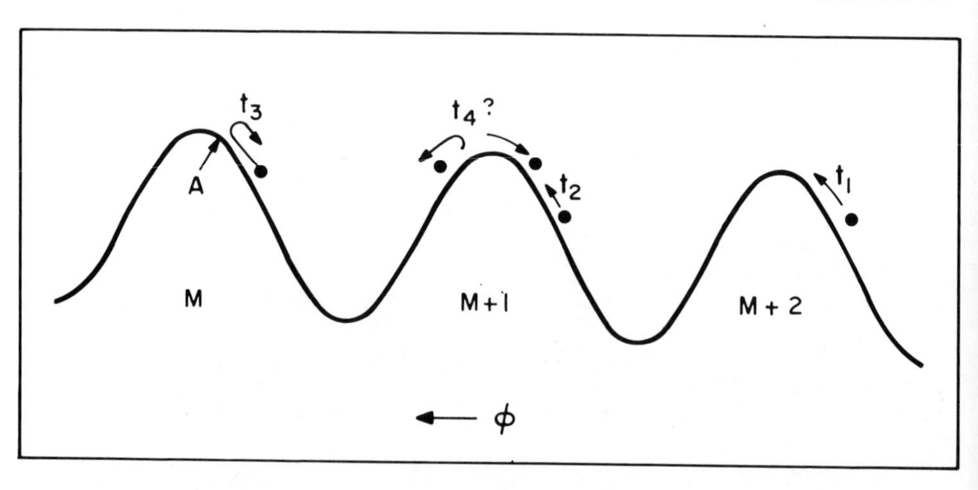

Fig. 6 Schematic representation of the particle-in-a periodic-potential analog executing punchthrough. The particle is shown moving uphill at successive times t_1 and t_2, coming to a halt at t_3 and either escaping downhill or being trapped in a well at t_4.

6. Interaction with rf currents

Probably the best-known aspect of the alternating supercurren in Josephson junctions is the effect of application of a high-frequenc or "rf" current at a frequency f near the junction frequency $\langle V \rangle / \Phi_o$. This causes the ac supercurrent to lock into phase-coherence with the applied currents [2, 11] such that $\langle V \rangle = f\Phi_o$ over a range of applied current I, giving an exactly constant voltage step in the I-$\langle V \rangle$ curve. Besides being responsible for extraordinarily precise measurements [39] of $\Phi_o = h/2e$, this interaction is involved in a number of detection schemes for rf radiation [40, 41]. It is also closely related to the behavior shown by the junction in interaction with passive resonant circuits. We shall attempt to describe some of the details of the interaction in underdamped and then overdamped junctions.

For tunnel junctions the behavior seems well-described [2, 11] by an extension of the voltage-biased model. It is assumed that the junction is subject to a controlled constant plus alternating voltage $V = V_{dc} + V_{rf} \cos \omega t$. By a straightforward computation of $\phi(t)$ and $\sin \phi(t)$, one obtains a non-zero value of $\langle I_c \sin \phi \rangle$ only if $V_{dc} = n\Phi_o \omega/2\pi$, where n is an integer. Specifically, for n = 1 one obtains $\langle \sin \phi \rangle = J_1(V_{rf}/V_{dc}) \sin \phi_o$, where J_1 is a Bessel function and ϕ_o is the position of the pendulum at t = 0. It is assumed that ϕ_o adjusts to conform to any applied current, and we shall see how this comes about. Note that $\langle I_c \sin \phi \rangle$ can actually be negative, in which case the rf source is providing power to the current source via the junction, according to Eq. (6).

To see the phase-locking in pictorial terms, consider an underdamped pendulum subjected to a torque $I = I_{dc} + I_{rf} \cos \omega t$. Suppose initially that $I_{dc}R \approx \omega\Phi_0/2\pi$, where $\omega >> \omega_J$, so that the pendulum is rotating almost uniformly in time in approximate synchronism with the "rf" torque. The latter acts on the moment of inertia M and causes a modulation in the pendulum velocity $d\phi/dt$, speeding the rotation in half of the cycle and retarding it in the other half. For simplicity we will assume that this velocity modulation is much larger than that induced by the gravitational torque, which requires that $I_{rf} >> I_c$, but that the net velocity modulation is small compared to the average rotation frequency.

If the phase of the modulation of $d\phi/dt$ is such that the pendulum spends more time in the region of $\sin \phi > 0$, then $\langle I_c \sin \phi \rangle$ is positive and acts to slow the average frequency of rotation $\langle d\phi/dt \rangle$. This causes the position of the minimum of $d\phi/dt$ to shift to smaller values of $\sin \phi$. The condition $\langle I_c \sin \phi \rangle > 0$ remains and $\langle d\phi/dt \rangle$ continues to decrease until the position of minimum $d\phi/dt$ has moved to the region of $\phi = 0$. By this time, however, $\langle d\phi/dt \rangle$ is substantially less than the rf frequency and the relative phases will continue to drift apart so that $\langle I_c \sin \phi \rangle$ becomes negative, acting to increase $\langle d\phi/dt \rangle$ again. Evidently the equilibrium condition comes for the minimum in $d\phi/dt$ occurring around $\phi = 0$, and underdamped small oscillations can occur about this value. This tendency of small deviations in relative phases to develop an effective force opposing this deviation is a necessary and familiar feature of phase-locked systems. Figure 7(a) shows an example of this phase-locking procedure for a junction having $\omega_J\tau = 10$. Shown is the phase difference $\phi(t) - \omega t$ between the supercurrent $I_c \sin \phi(t)$ and the rf current. In this calculation an artificial displacement from the equilibrium phase-locked condition is assumed at $t = 0$. The relative phases show a small short-term oscillation in each cycle of the rf, but also the larger amplitude long-term plasma-like oscillation mentioned above, which eventually is damped and allows the equilibrium phase relation to be established.

If a small increase in I_{dc} occurs, the pendulum begins to rotate faster, and the velocity minimum moves toward values of $\sin \phi > 0$, leading to $\langle I_c \sin \phi \rangle > 0$. This tends to slow the pendulum again, counteracting the change in I_{dc} and maintaining the phase-lock. The maximum I_{dc}, which still allows phase-locking, comes where the minimum of $d\phi/dt$ occurs at $\phi \approx \pi/2$. Figure 8 shows photographs of the oscilloscope displays of the electronic analog for $\sin \phi(t)$, $V(t)$, and $I_{rf} \cos \omega t$ in the equilibrium phase-locked condition. Figure 8(a) corresponds to the case in which I is set such that the unperturbed frequency is equal to the

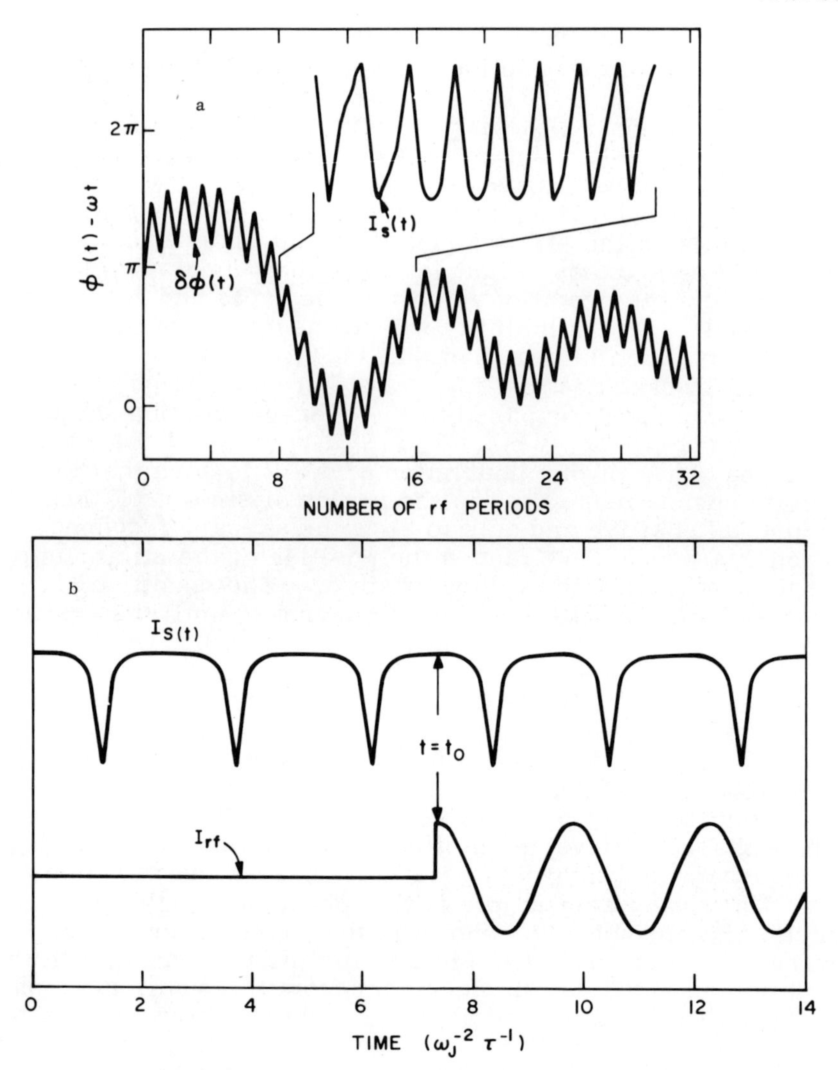

Fig. 7 (a): Behavior of an underdamped junction of $\omega_J\tau = 10$ under applied rf current, $I_{rf} \cos \omega t$, for $I = 0.5\, I_c$, $I_{rf} = 15\, I_c$, and $\omega = 2\pi RI/\Phi_0$. The lower curve plots the difference $[\phi(t) - \omega t]$ beginning at $\phi = \pi$ at $t = 0$, and shows the locking-on process. The upper curve shows $I_s(t) = I_c \sin \phi(t)$ during a portion of this time.
(b): Behavior of an overdamped junction under applied rf current. For $t < t_o$, $I_s = I_c \sin \phi(t)$ is shown for $I = 1.05\, I_c$ and $I_{rf} = 0$. At $t = t_o$, rf current is applied with amplitude $I_{rf} = 0.05\, I_c$ with frequency equal to that of the free-running junction and with the maxima of $I_s(t)$ and the rf current initially coincident.

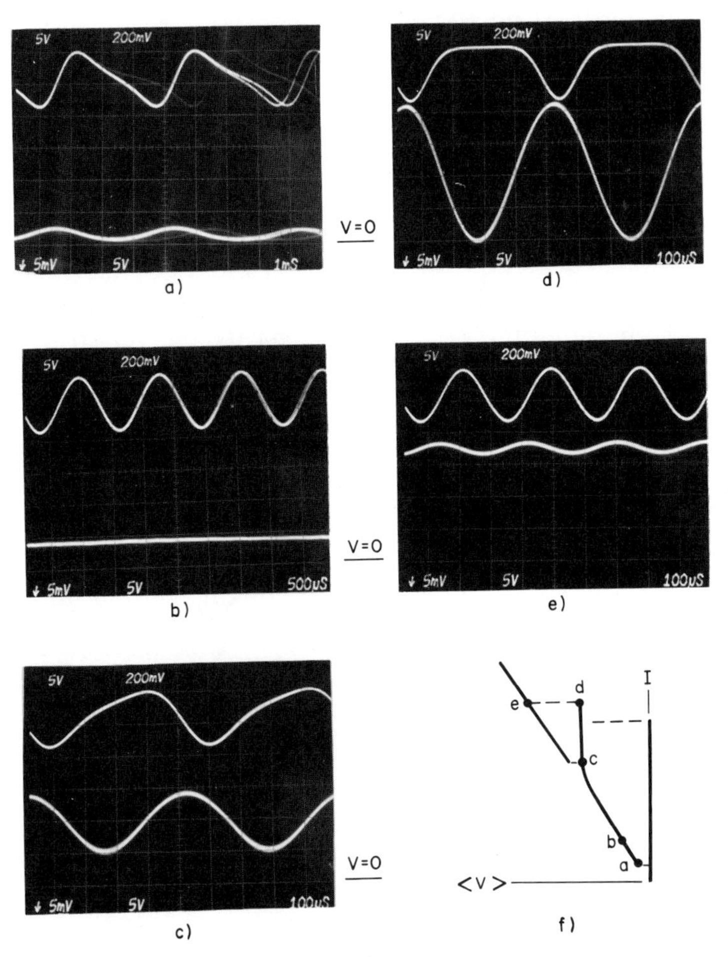

Fig. 8 Oscilloscope traces taken for an electronic analog of a junction phase-locked to an rf current source. Each photograph shows, top to bottom, $I_s(t) = I_c \sin \phi(t)$, $V(t)$, and $I_{rf} \cos \omega t$. (a) and (b): An underdamped junction having $I_c = 2\ \mu A$, $R = 3.35\ k\Omega$, $C = 2.1\ \mu F$, $\Phi_o = 5 \times 10^{-6}$ V-sec. (The effective value of Φ_o is the voltage/frequency ratio of the VCO), $I_{rf} = 75\ \mu A$ and $\omega = 1400$ Hz. This gives $\omega_J \tau = 7.7$. (c): The I-⟨V⟩ curve corresponding to (a) and (b) with the points at which (a) and (b) are recorded indicated. (d) and (e): An overdamped junction with $I_c = 2\ \mu A$, $R = 3.35\ k\Omega$, $C = 0$, $\Phi_o = 5 \times 10^{-6}$ V-sec, $I_{rf} = 0.8\ \mu A$ and $\omega =$ 1400 Hz. (f): The I-⟨V⟩ curve corresponding to (d) and (e) (solid line) and the I-⟨V⟩ curve for the $I_{rf} = 0$ (dashed line). The points at which (d) and (e) are recorded are indicated.

rf frequency, so that no power is drawn from the rf, whereas Fig. 8(b) shows these quantities for the maximum value of I which allows phase-locking to be maintained. The I-$\langle V \rangle$ curve produced by the analog under these conditions is shown in Fig. 8(c), with the points corresponding to (a) and (b) indicated.

From the power-balance considerations of Eq. (6), the dc and ac powers should be approximately equal for small damping, i.e., $\langle VI \rangle = \langle V_{ac}^2 \rangle / R \approx 0$, or $\langle V \rangle I_{dc} = - \langle V_{ac} I_{ac} \rangle$. Thus, in order to have $I_{dc} \neq 0$ on an rf-induced step, there must be some component to V_{ac} and the velocity modulation other than that provided by the capacitance, which is out of phase with the applied current. This in-phase component is provided by $I_c \sin \phi$ which, to the extent that it is approximately sinusoidal in time, generates a voltage $I_c/\omega C$ which has a maximum at $\phi = 0$. This is the position of the minimum of the applied torque when I_{dc} takes on its maximum value. From the power balance condition we obtain the maximum value for I_{dc} as $I_{dc} = \frac{1}{2}(I_c/\omega C)(I_{rf}/\langle V \rangle) = \frac{1}{2} I_c (I_{rf}/\omega C)(1/\langle V \rangle)$, in agreement with the small-amplitude prediction of the voltage-biased model. In principle the voltage modulation caused by $I_c \sin \phi$ should probably be included in the voltage-biased model from the first since it causes a comparable modulation to that of I_{rf} for $I_c \gtrsim I_{rf}$. However, its contribution to $\langle I_c \sin \phi \rangle$ is zero to first order, so that the voltage-biased model gives the right answer, even when the $I_c \sin \phi$ term is ignored.

As a second example, consider the overdamped small junction [40, 41]. Here, also, rf currents induce constant-$\langle V \rangle$ steps in the I-$\langle V \rangle$ curve by a similar phase-locking mechanism, but the details differ. From Eq. (7), $d\phi/dt = (I - I_c \sin \phi)/D$, so that the pendulum moves slowly through the region of $\sin \phi > 0$ and faster through the region of $\sin \phi < 0$. For emphasis, suppose that $I_{dc} \gtrsim I_c$, so that the bulk of the time required for a single rotation is spent near $\phi = \pi/2$, with the rest of the rotation occurring in a short pulse. If a weak rf torque is applied at the same frequency as this motion, and if the relative phases of the torque and the motion are such that the rf torque reaches its maximum during the time the pendulum is moving very slowly near $\phi = \pi/2$, the time spent in this region, and therefore the time required to complete a cycle, is substantially reduced. As a result, the pendulum reaches $\phi = \pi/2$ on the next cycle before the torque quite reaches its maximum. The period of the motion is again shortened, but by not as much. In this fashion the pendulum phase continues to shift ahead of that of the rf torque until the time at which the pendulum reaches $\phi = \pi/2$ occurs at a time when the rf torque is approximately zero (and increasing). A further shift in the two phases would result in the pendulum running slower than the rf frequency, thus correcting the offset. The situation resembles that of the underdamped case except that the phase of the rf torque and $\phi(t)$ are offset by $\pi/2$ and that small excursions from the equilibrium position are restored by an overdamped motion. An example of sin $\phi(t)$ with and without an applied rf current at the unperturbed frequency is shown in Fig. 7(b). The attainment of equilibrium is quite rapid in this example, within one or two cycles.

If a small decrease in I_{dc} now occurs it causes the period of the rotation to decrease and the $\phi = \pi/2$ point to shift to a later time at which the rf torque is positive again. This brings the pendulum back up to speed so that $\langle V \rangle$ maintains the value of $\Phi_o \omega/2\pi$. It also decreases the value of $\langle I_c \sin \phi \rangle$, since the pendulum spends less time near $\phi = \pi/2$, so that the current balance is obeyed. The maximum decrease in I_{dc} that can be balanced out in this way occurs when the $\phi = \pi/2$ point is aligned with the maximum rf torque, so that the motion is speeded as much as possible. Figures 8(d) and 8(e) show the quantities sin $\phi(t)$, V(t) and $I_{rf} \cos \omega t$ for the equilibrium phase-locked condition in this mode as taken from oscilloscope photos of the electronic analog simulation. In Fig. 8(d), I is set such that the unperturbed frequency equals ω, whereas Fig. 8(e) corresponds to the minimum I which retains the phase locking. The corresponding I-$\langle V \rangle$ curve produced by the analog is shown in Fig. 8(f) with the points corresponding to those of Figs. 8(d) and 8(e) indicated.

A further slight decrease in I_{dc} below the value corresponding to Fig. 8(e) causes the phases to become unlocked. However, the drift between the two phases is relatively slow whenever the $\phi = \pi/2$ point is in approximate alignment with the maximum rf torque, and becomes relatively rapid when the alignment is π away, so that for the bulk of the time the phase relation between the rf torque and the pendulum motion is essentially the same as at point e in Fig. 8(f). Note that for the power balance, V_{ac} is exactly opposite in phase with the rf current at e in Fig. 8(f), and is predominently opposite in phase in the region of slightly smaller I_{dc} below the step. The junction acts as a negative resistance [42, 43] supplying power to the rf source under this bias. By contrast the underdamped junction possesses similar negative-resistance properties when I_{dc} is set to the maximum value on that current step.

D. Inductively-Connected External Elements

1. Circuit and mechanical analogs

If a small junction is connected to an external passive element, Z, as in Fig. 9(a), account must be taken of the inductance, L, of the connection [19, 44]. If in Fig. 9(a) we define the phases $\phi = 2\pi\int(V/\Phi_o)dt$ and $\theta = 2\pi\int(V_z/\Phi_o)dt$, where V and V_z are the voltages across the junction and the element Z, then we may write the current I_L flowing in L towards the junction as $I_L = K(\theta - \phi)$, where $K = \Phi_o/2\pi L$. An analogous mechanical system is pictured in Fig. 9(b), which shows the inevitable pendulum attached via a torsional spring (hereinafter called a T-bar) to an element simulating Z, e.g., a paddle-wheel for a resistance. A photograph of an assembled model is shown in Fig. 9(c). Denoting the pendulum angle by ϕ and the total twist of the T-bar as $(\theta - \phi)$, the torque acting on the pendulum from the T-bar is $K(\theta - \phi)$, where K is the T-bar "stiffness". To the extent that the relationship between I_L and V_Z can be simulated mechanically, which is straightforward for simple parallel circuits, the equations of motion for ϕ and θ in the junction and the mechanical analog are the same. In addition to the previous analogies we now have that I_L corresponds to the T-bar torque and that V_Z is proportional to the angular velocity of the Z element. Further, the torsional energy corresponds to the magnetic energy $LI_L^2/2$, and the magnetic flux linking L in units of Φ_o corresponds to the twist in the T-bar in units of 2π.

2. Resistive Shunts

Consider the case of Z = r, a resistor. In addition to ω_J and $\tau = RC$, the new parameters LI_c/Φ_o and L/r also play a role in determining the behavior. For $LI_c/\Phi_o \ll 1/2\pi$, or $L \ll L_J$, the two resistive elements are nearly rigidly connected and act as a single equivalent resistor. Such low-inductance shunts have been employed in practice to convert the underdamped tunnel junction into a critically-damped or overdamped junction [20, 32].

For $LI_c/\Phi_o \gg 1/2\pi$, the T-bar is very soft, and motion of the pendulum through angles of 2π or so causes only slight changes in the torque exerted by the T-bar. An overdamped pendulum will then have nearly the same behavior as if driven from a constant current source, except that additional dc current $\langle V \rangle/r$ is drawn by the shunt.

For an underdamped junction having $LI_c/\Phi_o \gg 1/2\pi$, there is the possibility of a relaxation oscillation [45] driven by the hysteretic I-$\langle V \rangle$ curve if $L/r \gg (LC)^{\frac{1}{2}}$. Suppose that I starts at I = 0 and is increased slowly to $I \gtrsim I_c$ where the pendulum begins to rotate. For $L/r \gg (LC)^{\frac{1}{2}}$ the end of the T-bar is almost clamped in place and as the pendulum accelerates the T-bar will unwind, with I_L decreasing from I_c. The inertia of the pendulum will keep it rotating until the T-bar becomes wound up in the opposite sense to almost $I_L = -I_c$. This amounts to a half cycle of an LC oscillation. When the pendulum comes to a stop it may be captured at $\langle V \rangle = 0$ (depending on punchthrough) or it may execute one or more additional half cycles of oscillation. However, sooner or later it will cease to rotate and will return to the $\langle V \rangle = 0$ state until the torsion bar is rewound by I, which requires a time

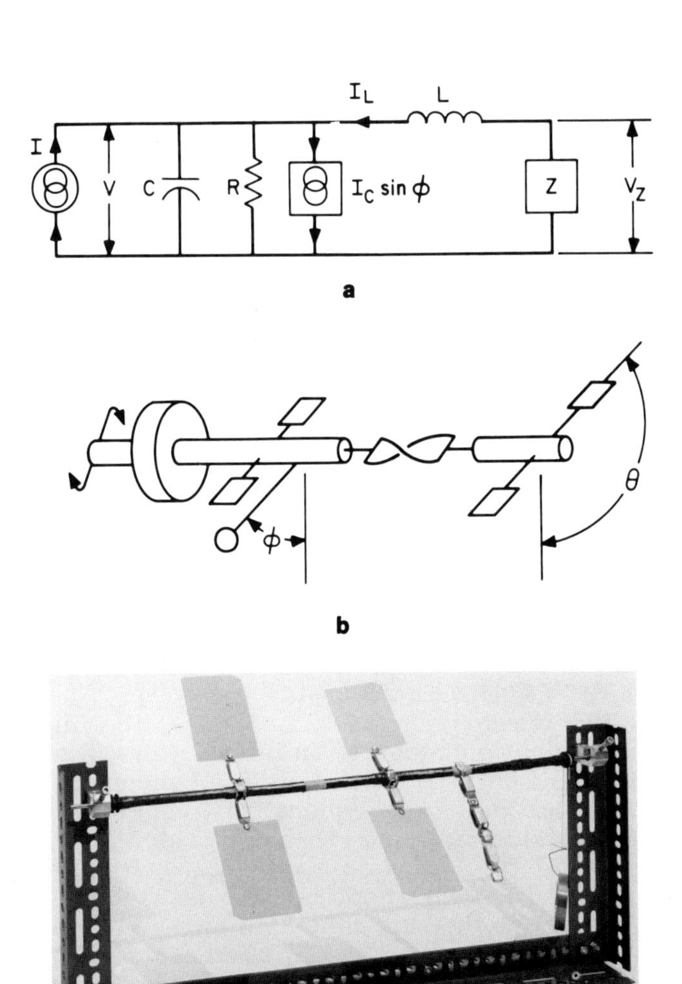

Fig. 9 (a): Stewart-McCumber circuit shunted by element Z through an inductance.
(b): Pendulum analog of circuit in (a) for Z a resistor. The twisted ribbon in the center represents a torsional spring corresponding to L.
(c): Photograph of mechanical analog of (b) constructed using battery clips for the pendulum and rubber tubing for the torsional spring.

$\sim L/r$. In principle the pendulum might be able to rotate smoothly at the very low $\langle V \rangle = Ir$ instead of engaging in the oscillation, but the motion is so delicate at low $\langle V \rangle$ that any noise perturbation will interrupt it and return the system to $\langle V \rangle = 0$ again to begin the oscillation anew. Figure 10 shows $\sin \phi(t)$ and $V(t)$ for a typical oscillation. Since the junction is carrying a current $\sim I_c$ during the long-lasting $\langle V \rangle = 0$ portion of the cycle, the I-$\langle V \rangle$ curve shows a substantial excess current of roughly the appearance shown in the inset of Fig. 10. Such oscillations are easily induced in tunnel junctions by placing a small resistance in parallel with the leads [45].

3. Capacitive shunts

If Z is a capacitor, C_z, there is the prospect of LC oscillations interacting with the alternating supercurrents. We can only discuss a few of the many possible varieties. We will restrict ourselves mainly to underdamped junctions operating at $\langle V \rangle \neq 0$ and shall suppose that C and C_z are comparable.

For $LI_c/\Phi_o << 1/2\pi$, the flywheels representing C and C_z are nearly rigidly attached and the low-frequency behavior (including the plasma oscillations) is that of a simple pendulum with an effective capacitance $C + C_z$. The interesting region occurs at the resonance of L and $C_{eff} = (1/C + 1/C_z)^{-1}$. At this resonance which occurs at a frequency $\omega_{res} = (L\ C_{eff})^{-\frac{1}{2}}$, the impedance of the C-L-C_z combination becomes infinite so that $I_c \sin \phi(t)$ cannot be sinusoidal as it is for the simple pendulum, and nonlinear effects can occur. In a general way we know from the current and power balance Eqs. (4) to (6), which also apply to this circuit, that the relation $\langle V \rangle \langle I_c \sin \phi \rangle = \langle V_{ac}{}^2 \rangle / R$ still applies. Consequently, in the neighborhood of the resonance, since there must be an increase in the V_{ac} causing additional loss, there also will be an increase in $\langle I_c \sin \phi \rangle$. This will show up as a current step in the I-$\langle V \rangle$ curve near $\langle V \rangle = \Phi_o\ \omega_{res}/2\pi$.

In more detail, in the pendulum model let the applied torque I be such that $\omega_J << \langle d\phi/dt \rangle << \omega_{res}$. The pendulum rotates smoothly as a result of the inertia of the two flywheels with a small alternating component in $d\phi/dt$ or V induced by the gravitational torque. Now if one looks at the impedance of the C-L-C_z circuit versus ω, Z decreases faster than $1/\omega$ and goes to zero at $\omega = (LC_z)^{-\frac{1}{2}}$. This is an anti-resonant frequency lying below ω_{res}. As I is increased and $\langle V \rangle$ approaches this antiresonance, V_{ac} goes to zero and the value of $\langle I_c \sin \phi \rangle$ also becomes zero. (Of course, for an underdamped junction $\langle I_c \sin \phi \rangle$ is not large anyway. This effect is more marked for moderately damped or overdamped junctions, as we shall see.) Above the antiresonance the impedance increases with the opposite sign (behaves inductively) and at $\omega = \omega_{res}$ goes to ∞. Through this region of $\langle V \rangle$ the pendulum exhibits the non-intuitive property

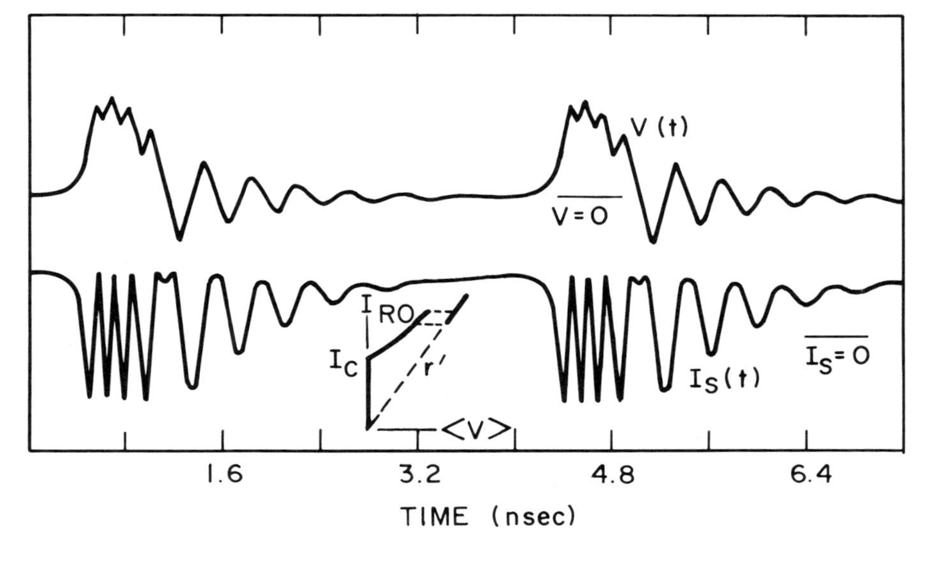

Fig. 10 Circuit of Fig. 9a, with $Z = r \ll R$, undergoing relaxation oscillations. Shown are $I_s(t) = I_c \sin \phi(t)$ and $V(t)$ for $I_c = 5\ \mu A$; $C = 50$ pF; $R = 5\ \Omega$; $r = 0.5\ \Omega$ (giving $\omega_J \tau = 4.3$); $L = 600$ pH; and $I = 7.5\ \mu A$. A schematic I-$\langle V \rangle$ curve is shown in the inset. The region marked RO corresponds to relaxation oscillations, while at larger $\langle V \rangle$ the rotation in nearly uniform.

of speeding up near $\phi = \pi$ (at the peak of its motion) and slowing down near $\phi = 0$. This motion, which is easily observed in a mechanical analog, becomes increasingly rapid as $\langle V \rangle$ approaches the resonance. This increase of V_{ac} causes the ohmic losses to increase, leads to an increase in $\langle I_c \sin \phi \rangle$ and hence the value of I required to maintain the value of $\langle V \rangle$, i.e., a current step occurs at $\langle V \rangle = \Phi_o \omega_{res}/2\pi$.

As V_{ac} increases, the supercurrent becomes less sinusoidal in time. Intuitively one might guess that the maximum velocity modulation that can be self-induced in this fashion would be on the order of $V_{ac} \sim \langle V \rangle$, so that the pendulum comes almost to a stop in each cycle. This would seem the most effective motion in producing a maximum in the value of $\langle \sin \phi \rangle$ as it permits $\sin \phi$ to be non-zero over an extended portion of the cycle. In any case some limit must be reached for V_{ac} since too large a value would induce $\sin \phi$ to alternate in sign several times in each cycle and depress the value of $\langle \sin \phi \rangle$. This behavior is familiar from the oscillatory dependence on V_{ac} of the amplitude of the rf-induced steps that occur in the voltage-biased model. In the mechanical model, and also the electronic analog, this limiting does indeed seem to occur, giving a maximum in $\langle I_c \sin \phi \rangle$ at $\langle V \rangle \approx \Phi_o \omega_{res}/2\pi$. A further increase in I causes the pendulum to break free of the resonance and accelerate again to a higher voltage of $\langle V \rangle \approx IR$. From the power relation $\langle V \rangle \langle I_c \sin \phi \rangle = \langle V_{ac}^2 \rangle / R$, if $V_{ac} \sim \langle V \rangle$ then $\langle I_c \sin \phi \rangle$ will be $\sim \langle V \rangle / R$, which may be small compared to I_c.

A more analytic but still approximate analysis of this behavior may be made as follows. The non-sinusoidal current $I_s(t)$ produced by the supercurrent element generator can be written as a sum of the fundamental and higher harmonics. Since the impedance of the C-L-C_z-R shunting elements for high Q is large at and very near the fundamental frequency, but small near the harmonics, the voltage V_{ac} produced by $I_s(t)$ will be nearly sinusoidal and at the fundamental frequency. Self-consistently, the form of $I_s(t)$ will be that produced by the combination of $\langle V \rangle$ and the sinusoidal V_{ac} as in the voltage-biased model. The fundamental component of I_s will have amplitude $I_f = I_c[J_o(V_{ac}/\langle V \rangle) - \cos 2\phi_o J_2(V_{ac}/\langle V \rangle)]$ and the dc component will have amplitude $\langle I_c \sin \phi \rangle = I_c J_1(V_{ac}/\langle V \rangle)$, where J_0, J_1, and J_2 are Bessel functions of the first kind and $\tan \phi_o = R/X$, where X is the reactance of the C-L-C_z combination. To complete the self-consistency, the amplitude of V_{ac} is $I_f (1/R^2 + 1/X^2)^{-\frac{1}{2}}$. Rewriting, we obtain $V_{ac}/\langle V \rangle = (RI_c/\langle V \rangle) [J_o(V_{ac}/\langle V \rangle) - \cos 2\phi_o J_2(V_{ac}/\langle V \rangle)] (\sin \phi_o)$. Then as one passes through the resonance with increasing I, thereby increasing ϕ_o from 0 to π, the value of $V_{ac}/\langle V \rangle$ increases. However, $V_{ac}/\langle V \rangle$ never gets larger than that in the region of the first positive lobe of $J_1(V_{ac}/\langle V \rangle)$ because of the term $J_o - \cos 2\phi_o J_2$, which tends to zero in the vicinity of the first zero of J_1. At the resonance $X \to \infty$, $\phi = \pi/2$ and $(V_{ac}/\langle V\rangle)^2 = (2RI_c/\langle V\rangle)\, J_1(V_{ac}/\langle V\rangle)$ by the usual Bessel identities. This determines the value of $V_{ac}/\langle V\rangle$, which in turn determines $\langle I_c \sin\phi\rangle$ self-consistently for any value of $\langle V\rangle/RI_c$. Note that this expression also obeys the power balance equation. This gives a maximum value for $\langle I_c \sin\phi\rangle = 0.58\, I_c$ at the value of $V_{ac}/\langle V\rangle = 1.84$ and $\langle V\rangle/RI_c = 0.344$, in fair agreement with our crude estimate of $V_{ac}/\langle V\rangle \sim 1$ and $\langle I_c \sin\phi\rangle \sim \langle V\rangle/R$. The value of $\langle I_c \sin\phi\rangle$ decreases rather slowly for $\langle V\rangle/RI_c$ above and below this value.

In Figs. 11(a) to (e) we show a series of photographs of $\sin\phi(t)$ and $V(t)$ taken from an oscilloscope display of the electronic analog interacting with such a resonant circuit. Figure 11(a) shows the low-$\langle V\rangle$ waveforms, 11(b) the antiresonant condition, 11(c) the inductive response of V to I_s just below the resonant frequency, 11(d) the resistive relation and increased response of V to I_s at the peak of the resonance and 11(e) the capacitive response at higher frequency. The I-$\langle V\rangle$ curve corresponding and the respective points at which the photos were taken appear in Fig. 11(f). Note the sinusoidal form of V(t) near the resonance condition. Figure 12(a) shows $\sin\phi(t)$ and V(t) for a somewhat lower frequency resonance, and the corresponding I-$\langle V\rangle$ curve is in Fig. 12(c).

If the currents in the C-L-C_z circuit feel some frequency-dependent loss mechanism which is not included in the dc resistance and is approximately equivalent to a parallel resistance R_{eff}, then a larger loss, $\langle V_{ac}^2\rangle/R_{eff}$, can occur, leading to a current step of larger amplitude than for R alone. If R_{eff} is only significant near the resonance, one is able to use impedance calculations by simply replacing R by R_{eff}.

For LC resonances of lower frequency approaching ω_J, one has that $LI_c/\Phi_0 \gtrsim 1/2\pi$. Since the supercurrent is non-sinusoidal with time at these voltages, even without the resonant circuit, an impedance analysis is complex. From experience with the analog it appears that at large currents, where $\langle V\rangle \gg \Phi_0 \omega_{res}/2\pi$, the pendulum rotates smoothly. As the current decreases and $\langle V\rangle$ approaches $\Phi_0\, \omega_J/2\pi$, the pendulum without the resonance begins to move in a pulse-like form, spending considerable time in an inverted position. At this level, as described earlier, the excess energy which keeps the pendulum rotating is small. What usually happens is that (because the T-bar is relatively soft) the pendulum tends to fall over from its inverted position, thereby exciting an LC oscillation and borrowing some of the kinetic energy for use in winding the T-bar at just the moment when all available kinetic energy should be devoted to keeping the pendulum rotating through the inverted position. This usually interrupts the rotation after a very few cycles and causes a return to the

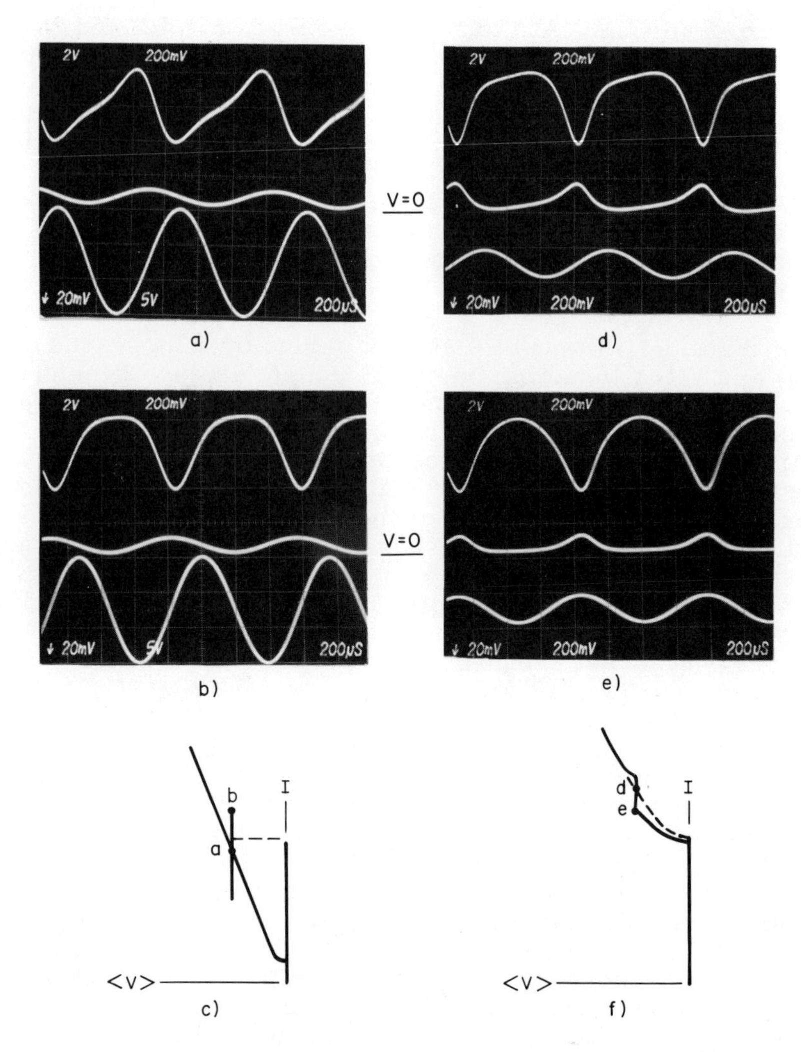

Fig. 11 Oscilloscope traces (a) to (e) taken for an electronic analog of a junction in the configuration of Fig. 9a with $Z = C_Z$. In each photograph the upper trace is $I_c \sin \phi(t)$ and the lower $V(t)$. The parameters are $I_c = 4\ \mu A$, $R = 3.35\ k\Omega$, $C = 0.34\ \mu F$, $L = 20$ mH, $C_Z = 2.1\ \mu F$, and $\Phi_o = 5 \times 10^{-6}$ V-sec, giving $L_J = 200$ mH. The conditions for (a)-(e) are described in the text. (f): The I-⟨V⟩ curve and the points corresponding to (a) to (e). For the low-⟨V⟩ picture, a noise and 60 Hz pickup cause considerable variation in the Josephson frequency resulting in the multiple images in the photograph.

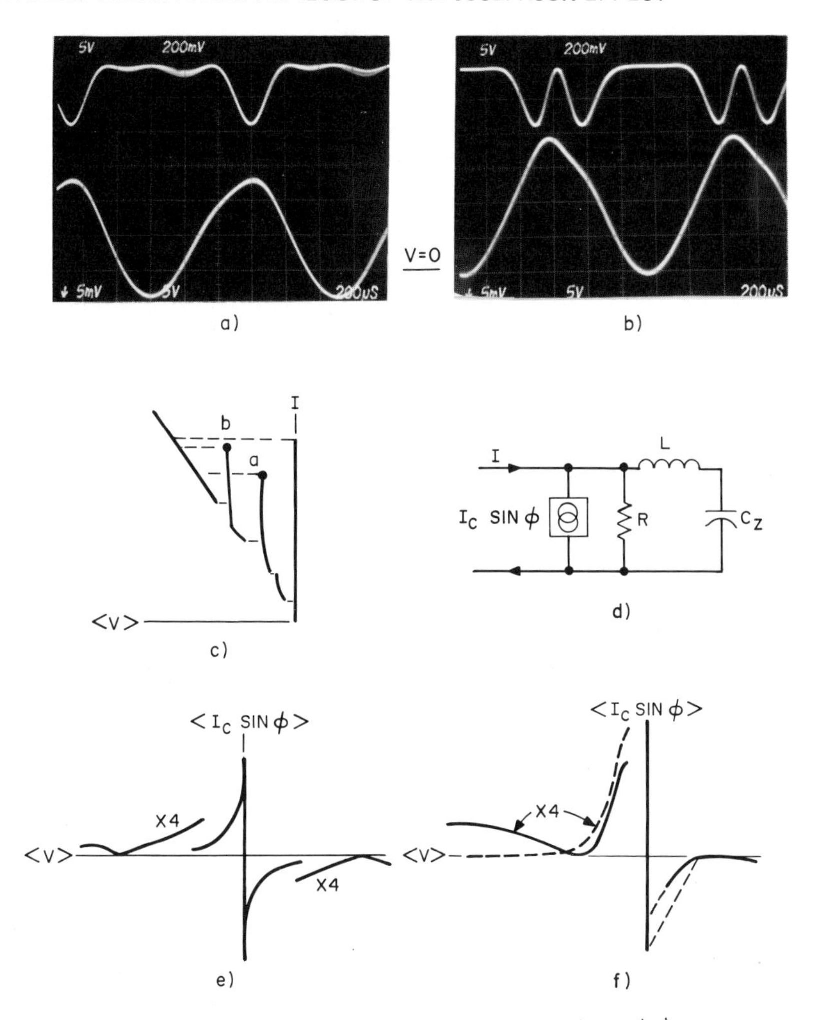

Fig. 12 Oscilloscope traces (a) to (c) and the I-⟨V⟩ curve taken for the electronic analog of the same circuit as Fig. 11, except L = 100 mH. Resonant interactions occur at ⟨V⟩ corresponding to both (a) the L-C_{eff} resonance frequency, and (b) twice that frequency as described in the text. The I-⟨V⟩ curve in (c) shows the points corresponding to (a) and (b). (d): A Stewart-McCumber circuit of an overdamped junction with an L-C_Z shunt. (e) and (f): ⟨I_c sin φ⟩ vs ⟨V⟩ curves produced by the electronic analog in the configuration of (d) for the cases (e): I_c = 2 μA, R = 790 Ω, L = 100 mH, and C_Z = 8.4 nF, and (f): I_c = 2 μA, R = 3.35 kΩ, L = 100 mH (solid curve) or L = 0 (dashed curve), and C_Z = 0.1 μF. ⟨I_c sin φ⟩ vs ⟨V⟩ is shown rather than I vs ⟨V⟩ to bring out the resonant structure.

$\langle V \rangle = 0$ state.

Occasionally a more complex interaction between the LC oscillation and the supercurrent will allow a relatively stable low-frequency rotation. For the analog the most stable such situation seems to be one in which $LI_c/\Phi_o \sim 1$; the pendulum advances by making two complete rotations and then halting momentarily while the C_z flywheel accelerates, unwinds the T-bar and winds it up in the opposite direction. This causes the pendulum to flip twice again, unwind the T-bar, and rewind it in the original sense. This motion, an example of which is shown in Fig. 12(b) for the electronic analog, seems to be most robust, i.e., not easily interrupted, if the damping is reasonably large and the applied current is correspondingly large. More complex oscillations involving three or more flips of the pendulum can occur, but are rather sensitive to the particular range of bias and circuit parameters.

As another example consider the circuit in Fig. 12(d), an overdamped junction ($C = 0$) shunted by an L-C_z combination [27]. It is useful to think of this model in terms of the supercurrent element generator feeding $I_s(t)$ into an L-C_z-R resonant circuit. Complicated behavior ensues for the case $L \gtrsim L_J$. We will restrict ourselves to $L < L_J$ so that the resonant frequency $\omega_{res} = (LC_z)^{-\frac{1}{2}}$ occurs at a frequency large compared to $(2\pi I_c/\Phi_o C_z)^{\frac{1}{2}}$ which is where the plasma frequency would be in the absence of L. Note that ω_{res} may lie above or below the characteristic frequency $2\pi RI_c/\Phi_o$, depending on the Q of the L-C_z-R circuit.

If the resonance is high-Q, $\omega_{res} L >> R$, so that $\omega_{res} >> 2\pi RI_c/\Phi_o$, then the impedance of the L-C_z arm is large at all non-resonant frequencies and $I_s(t)$ flows primarily through R. Just at the resonance the impedance of the L-C_z arm goes to zero and takes all the current. The supercurrent $I_s(t)$ becomes sinusoidal in time, V_{ac} goes to zero and the I-$\langle V \rangle$ curve is indented at the value of $\langle V \rangle = IR$, as shown in the $\langle I_c \sin \phi \rangle$ - $\langle V \rangle$ curve taken from the electronic analog in Fig. 12(e). Above and below the resonance the I-$\langle V \rangle$ curve retains its overdamped form (except when the harmonics of V(t) coincide with ω_{res} and are shorted out). It is noteworthy that, because the supercurrent element acts as a current generator, the behavior at resonance is tame despite the high Q, with only a maximum current of I_c flowing.

If the circuit Q is low, $\omega_{res} L << R$, so that $\omega_{res} \leqslant 2\pi RI_c/\Phi_o$, then the L-C_z arm passes considerable current both near and below ω_{res}. Depending on the value of $(2\pi I_c/\Phi_o C_z)^{\frac{1}{2}} RC_z$, the I-$\langle V \rangle$ curve at low $\langle V \rangle$ may have either a damped or an underdamped form, since the inductive impedance is so small that the junction essentially sees only the capacitance. At ω_{res} the impedance goes to zero and the I-$\langle V \rangle$ curve again touches the ohmic value at this point. At higher frequencies the inductive impedance gradually increases and the I-⟨V⟩ curve approaches the overdamped form at voltages ⟨V⟩ ~ $\Phi_o R/2\pi L$, as shown in Fig. 12(f).

Finally, suppose that Z is a parallel resonant circuit consisting of an inductance L_Z and a capacitance C_Z with perhaps some damping. If L >> L_Z, the resonant circuit is weakly coupled to the junction and not all its current flows through the junction, a situation more nearly related to experiment than the previous cases. If the Q is high enough, there can be a fairly high level of excitation in the resonant circuit before the voltage modulation at the junction approaches ⟨V⟩. Consequently, $I_L(t)$ is nearly sinusoidal and the behavior resembles that caused by an external current source. As in that case, a current step will occur in the I-⟨V⟩ curve at the resonant frequency. Since the L_Z-C_Z circuit is excited by the junction, only that portion of the step which has a negative rf resistance can occur. An underdamped junction shows a step with positive excess current and an overdamped junction shows a step with negative excess current. Both types are observed experimentally [46, 47].

4. AC SQUID

If Z in Fig. 9 is a pure inductance, then one has a so-called ac SQUID circuit used in magnetometry (see the article by J. Clarke in this volume for discussions and references). Figures 13(a), (b) and (c) show three equivalent versions of this circuit. In Figs. 13(a) and (b), bias current is applied at different positions, and in Fig. 13(c) a source of external magnetic flux Φ_x linking the loop is substituted. All these biasing agents produce the same behavior if we equate Φ_x in (c) with L'I in (b) and (L' + L")I in (a). The subsequent discussion will be in terms of the flux- riven circuit of (c).

The mechanical analog [19] of the circuit in Fig. 13(c) is the usual pendulum attached to a T-bar of stiffness corresponding to L = L' + L". The far end of the T-bar is attached to a fixed plate whose angular position θ_x is under external control. The usual analogs apply along with a correspondence between θ_x and $2\pi\Phi_x/\Phi_o$. Both the circuit and the analog have the new feature that a non-zero Φ_x can cause persistent currents to flow in the inductance around the loop.

The behavior depends strongly on LI_c/Φ_o, which corresponds to the T-bar twist (in units of 2π) required to balance the pendulum in a horizontal position. For $LI_c/\Phi_o \ll 1/2\pi$ the T-bar is very stiff so that ϕ follows θ_x and the supercurrent and I_L are periodic functions of Φ_x. As the T-bar is made softer, at $LI_c/\Phi_o = 1/2\pi$, the pendulum is no longer stable at all angles, but tends to fall over sideways from an inverted position. Since it can

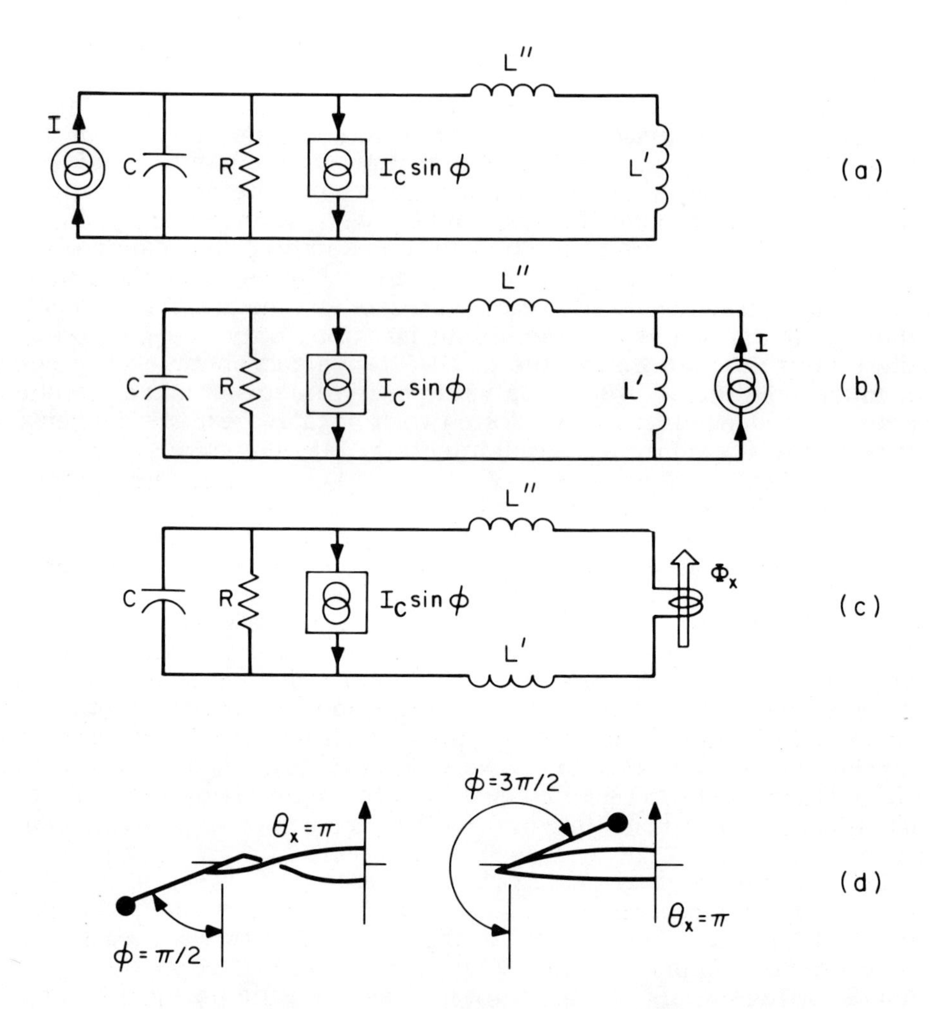

Fig. 13 Three equivalent versions (a) to (c) of the circuit of Fig. 9a when Z is an inductance. Note that (c) is an rf SQUID circuit. (d): A pendulum analog of the circuit of (c) for $\Phi = \frac{1}{2}\Phi_o$ ($\theta_x = \pi$) and $LI_c/\Phi_o = 0.25$.

fall in either direction, there is more than one possible static configuration for given Φ_x, differing in the value of I_L. The two possibilities for $\theta_x = \pi$ and $LI_c/\Phi_o = 0.25$ are shown in Fig. 13(d). This multi-valued behavior becomes more and more prominant for $LI_c/\Phi_o >> 1$, where a number of full twists in the T-bar can be balanced by the weight of the pendulum in a stable configuration.

As an example of the operation of an ac SQUID, suppose we are dealing with an overdamped junction and, say, $LI_c/\Phi_o = 3.75$. Beginning with $\theta_x = 0$ and $I_L = 0$, let θ_x be increased by four full turns (or $\Phi_x = 4\Phi_o$), at which point the pendulum reaches the horizontal and $I_L = I_c$. The pendulum then overbalances and executes a single not-quite 2π rotation, reducing the T-bar torque I_L by $\sim \Phi_o/L$ and coming to rest again. The energy lost to the damping in this flip is of order $\Phi_o I_c$. If θ_x is increased by another 2π the pendulum flips again, and so on. This energy dissipation is exploited in the ac SQUID magnetometers as a means of monitoring Φ_x. In oversimplified form, for this example one procedure would be to apply a magnetic flux to the SQUID consisting of the dc component Φ_{dc} and an alternating component Φ_{ac} of amplitude $4.2\ \Phi_o$, large enough to cause $I_L \gtrsim I_c$ in one or both polarities. The ac frequency is usually low enough so that a pendulum flip occurs in a time much shorter than the rf period. Suppose that $\Phi_{dc} = 0$. Then, as indicated in Fig. 14(a), which shows I_L vs $\Phi_x = \Phi_{dc} + \Phi_{ac}$, the pendulum will flip four times in each cycle of Φ_{ac}, consuming an energy/cycle of $4\Phi_o I_c$. If $\Phi_{dc} = 0.4\ \Phi_o$, on the other hand, only two flips/cycle occur (Fig. 14(b)) since the minimum flux is not quite sufficient to cause I_L to reach $-I_c$ and the energy consumption decreases to $2\Phi_o I_c$/cycle. In actual circuits this energy consumption is made to dominate the ac losses, giving a Q dependent on Φ_{dc}. In practice, the perturbation of the ac circuit by the flipping of the SQUID is large enough that the assumption of a constant ac flux amplitude is invalid, but the above description is still roughly correct.

It is also possible to make SQUIDs with $LI_c/\Phi_o < 1/2\pi$ for which ϕ follows θ_x smoothly. SQUIDs in this regime rely on reactive impedance variations of the ac circuit induced by the $\cos\phi$ dependence of the inductance presented by the junction [48, 49].

E. The dc SQUID

Another widely used circuit [50] in magnetometry is the dc SQUID which consists of two junctions connected in parallel by an inductive shunt (Fig. 15(a)). Its mechanical analog [19, 51], by the usual arguments, consists of two pendulums connected by the familiar T-bar (Fig. 15(b)). All previous analogies hold here in subscripted form, e.g., I_{c1} and I_{c2} are the critical currents of the two junctions or the moment arms of the two pendulums respectively. In addition, an external magnetic flux Φ_x is coupled

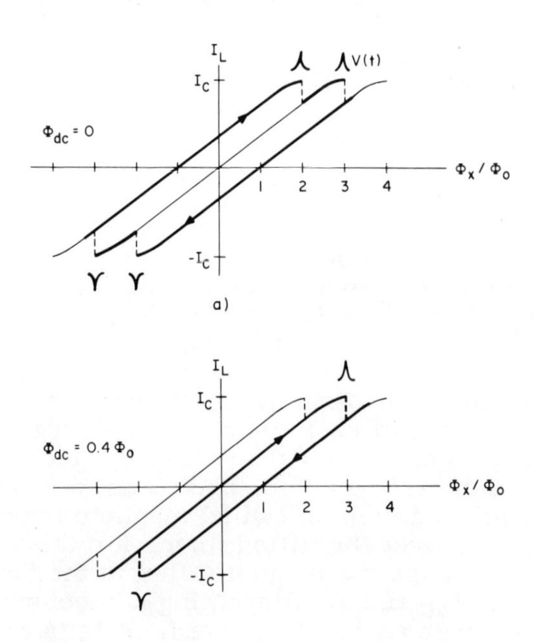

Fig. 14 rf SQUID operation. Shown is the current in the inductance, I_L vs $\Phi_x/\Phi_o = (\Phi_{dc} + \Phi_{ac})/\Phi_o$ for $LI_c/\Phi_o = 3.75$. The heavy line indicates the path followed for Φ_{ac} alternating from $+4.2\,\Phi_o$ to $-4.2\,\Phi_o$ with (a): $\Phi_{dc} = 0$, and (b): $\Phi_{dc} = 0.4\,\Phi_o$. A voltage pulse of area $\approx \Phi_o$ occurs each time that $I_L = \pm I_c$ in this cycle, as indicated schematically.

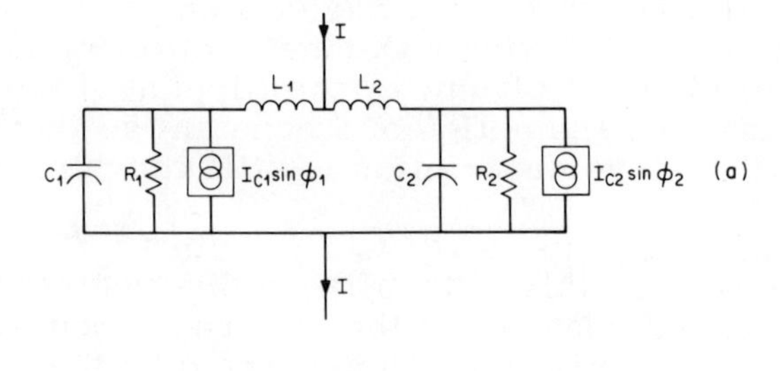

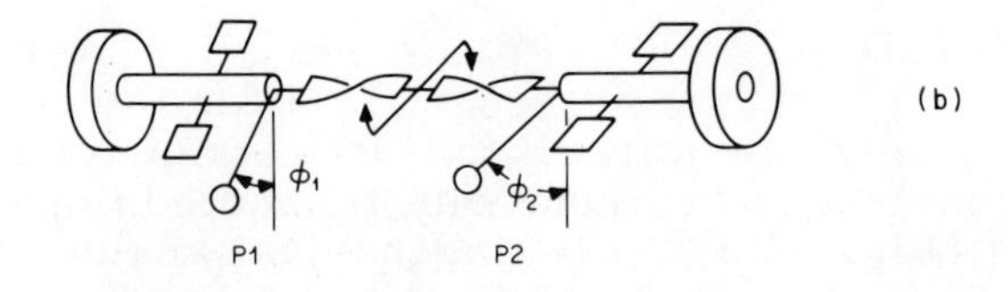

Fig. 15 (a): A dc SQUID comprising the Stewart-McCumber junctions connected in parallel through an inductance.
(b): Pendulum analog of a dc SQUID.

into the junction loop. In the mechanical analog the effect of Φ_x is simulated by attaching the pendulums to the T-bar with an angular offset θ_x such that in the absence of gravity they would adopt angles ϕ_1 and ϕ_2 related by $(\phi_2 - \phi_1) = 2\pi\Phi_x/\Phi_o = \theta_x$. Note that here, as in the ac SQUID, one has persistent loop current $I_L = (I_{s1} - I_{s2})/2$ induced in response to Φ_x. Also, in Fig. 15(a) the division of the total inductance L into L_1 and L_2 in the separate arms is required to allow for the fact that the bias current I may cause a magnetic flux, $I(L_1 - L_2)/2$, to link the circuit. In the analog, the T-bar stiffness is inversely proportional to $L = L_1 + L_2$, and the bias torque I is applied at a point spaced between the pendulums in the proportion of L_1/L_2.

Here again the behavior depends strongly on LI_{c1}/Φ_o, where we assume that $I_{c1} \leq I_{c2}$. Obviously, for $LI_{c1}/\Phi_o << 1/2\pi$ the pendulums are rigidly bound together at an angle θ_x and act as a single pendulum for static behavior, and for rotation frequencies below any L-C resonance frequencies. The critical torque I_c required to begin rotation depends on Φ_x according to $I_c = (I_{c1}^2 + I_{c2}^2 + 2I_{c1}I_{c2} \cos \theta_x)^{\frac{1}{2}}$, so that both I_c and the shape of the I-$\langle V\rangle$ curve vary periodically with increasing external magnetic field; this property leads to the use of dc SQUIDS in magnetometry (see the articles by Clarke and by de Bruyn Ouboter in this volume).

As LI_{c1}/Φ_o increases, the T-bar becomes flexible and the pendulums will sag. Here, as in the ac SQUID, this introduces the possibility of multiple static configurations in the $\langle V\rangle = 0$ state. For example, even for the small value of $LI_{c1}/\Phi_o = 0.05$, in the case $I_{c1} = I_{c2}$, $I = 0$ and $\theta_x = \pi$, the pendulum static configuration could be either of those in Fig. 16(a), which differ in having the opposite sign for the circulating current I_L and the self-magnetic flux LI_L. In another example, let $I_{c1} = I_{c2}$, $I = 0$, $\theta_x = 0$ and $LI_{c1}/\Phi_o = 1.053$. Then any of the three static configurations shown in Fig. 16(b), with $I_L = 0, \pm 0.707\ I_{c1}$ are possible. Such multi-valued behavior is a common feature of dc SQUIDs. Note that any of the states in Figs. 16(a) or 16(b) could be smoothly transformed into the others by appropriately increasing or decreasing the value of θ_x through 2π or 4π.

In the limit $LI_{c1}/\Phi_o >> 1/2\pi$ the pendulums are loosely connected. Here, as in the similar limit for the ac SQUID, the system has many possible static ($\langle V\rangle = 0$) configurations which involve a nearly integral number of full twists in the T-bar being held in place by the weight of the two pendulums.

The critical current I_c of a dc SQUID having $LI_{c1}/\Phi_o > 1/2\pi$ will also vary periodically with Φ_x as will the shape of the I-$\langle V\rangle$ curve, and such SQUIDs are widely used in magnetometry. The

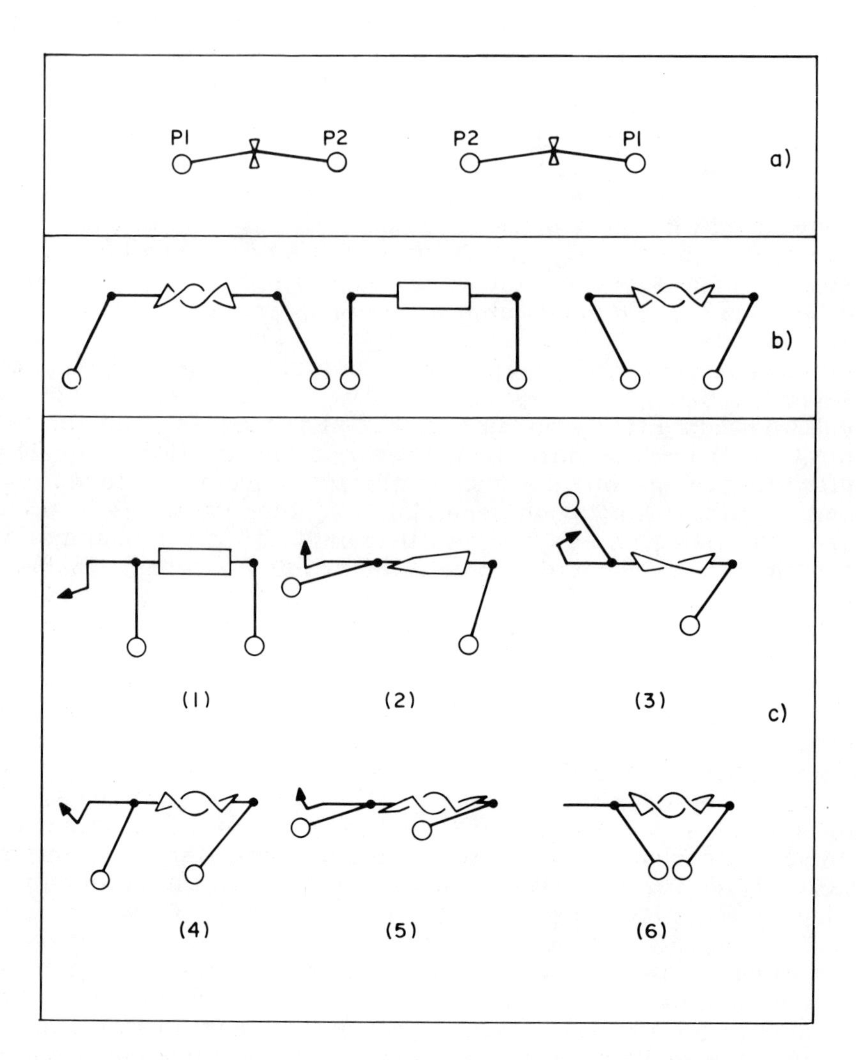

Fig. 16 (a): End view of a dc SQUID pendulum analog having $LI_c/\Phi_o = 0.05$ and $\theta_x = \pi$, showing the two possible static states with $I_L \approx \pm I_c$.

(b): Oblique front view of dc SQUID analog having $LI_c/\Phi_o = 1.053$ and $\theta_x = 0$, showing three possible static states with $I_L = 0, \pm 0.707\ I_c$.

(c): Behavior of a dc SQUID analog under increasing torque applied at the left end. The sequence is described in the text.

multivalued behavior can have its effect on I_c, especially for underdamped junctions. As an example, consider the behavior shown in Fig. 16(c) for the case $L_1 = 0$, $L_2 = L$, $I_{c1} = I_{c2}$ and $LI_{c1}/\Phi_o = 1.0$. Initially, for $I = 0$ and $\theta_x = 0$, both pendulums P1 and P2 can hang straight down (Fig. 16(c), (1)). As I is increased to $I \approx 1.25\ I_{c1}$, P1 tilts towards $\phi = \pi/2$ and P2 tilts through a smaller angle, $\phi_2 \approx 0.25$ (Fig. 16(c), (2)). For I just a bit larger, P1 overbalances and begins to rotate (Fig. 16(c), (3)). The pendulum swings through most of a full circle, winding up the T-bar and transferring some of the applied torque to P2. If the system is underdamped, P1 acquires momentum in this process and the additional inertial torque applied to P2 will, in this example, cause it to overbalance and also begin rotation. The combination of P1 and P2 will continue to rotate and accelerate to terminal velocity. However, if the system is overdamped, P1 does not acquire momentum in flipping and the two pendulums will come to rest in the configuration shown in Fig. 16(c), (4). The applied torque I must be increased further, to the level of $I = I_{c1} + I_{c2}$ in this example (Fig. 16(c), (5)) before rotation begins. Still a third possibility is that initially at $I = 0$ the pendulums do not hang down, but rather there is an initial twist in the T-bar as in Fig. 16(c), (6). Increasing I to $1.25\ I_{c1}$ from this initial configuration recreates the situation of Fig. 16(c), (4), without the necessity for flipping P1. Rotation will not begin in this situation until (Fig. 16(c), (5)) $I = I_{c1} + I_{c2}$ for either underdamped or overdamped junctions. The upshot is that underdamped systems may display multiple critical currents which reflect different initial states for the loop current I_L and the junctions. This is a common observation in tunnel junction dc SQUIDs. In contrast, overdamped systems have only one value of critical current although dynamic changes in configuration may occur at smaller I_c. This behavior is characteristic of point-contact and bridge dc SQUIDs.

In an underdamped dc SQUID there is the possibility of LC resonances which can cause effects very much like those of single junctions in interaction with resonance circuits. The most obvious and easily simulated situation is that of the small oscillations in the $\langle V \rangle = 0$ state which involve a mix of coupled plasma and LC oscillations whose character depends on I and θ_x.

As an example of $V \neq 0$ behavior, for $LI_{c1}/\Phi_o << 1/2\pi$ and identical junctions, the LC resonance is again at a frequency $\omega_{res} = (1/LC_1 + 1/LC_2)^{\frac{1}{2}} >> \omega_J$. For $\theta_x = 0$ the two junctions rotate in synchronism to first order and the torques acting on the T-bar are in phase, giving $I_L = 0$, so that excitation of the resonance is not favorable, and does not appear to occur in the mechanical analog. In contrast, if $\theta_x = \pi$ a strong alternating torque $I_L = I_{c1}$ occurs to first order and the resonance is easily excited, as is observed in the analog. Here, as for the single junction, the velocity modulation under an applied current seems to be limited to the equivalent of $V_{ac} \sim \langle V \rangle$ such that each of the two pendulums alternately approximately stops in each cycle. The net effect is to produce an increase in $\langle I_{c1} \sin \phi_1 \rangle + \langle I_{c2} \sin \phi_2 \rangle$, giving rise to a current step in the I-$\langle V \rangle$ curve at $\langle V \rangle \approx \Phi_o \omega_{res}/2\pi$, whose amplitude varies periodically with ϑ_x. The same sort of power-balance considerations and impedance arguments as previously can be used to estimate the amplitude of the step with similar results. Figure 17 shows $\sin \phi_1(t)$, $V_1(t)$, $\sin \phi_2(t)$, and $V_2(t)$ for a numerical calculation of such a resonance.

At larger values of LI_{c1}/Φ_o the LC frequencies again approach ω_J and effects similar to those occurring in the single-junction-resonator combination occur. At high $\langle d\phi/dt \rangle$ the rotation is smooth, but as lower values $\langle d\phi/dt \rangle$ are reached, where each pendulum spends a considerable amount of time in an inverted position, there is a tendency for one pendulum to fall over forward and the other backward in the excitation of an LC oscillation. This decreases the kinetic energy and interrupts the motion, returning the system to $\langle V \rangle = 0$. This tendency is especially violent if $\vartheta_x \neq 0$, so that the pendulums are not simultaneously inverted and therefore exert alternating torque on the T-bar in each cycle. There is, however, always the possibility that low-frequency rotation can occur in synchronism with the main rotation. Experience with the analog suggests that the situation in which this is most common is that numerically simulated in Fig. 18. In this example the junctions are identical and $LI_{c1}/\Phi_o = 1.45$. The sequence is that first P2 is fixed and P1 advances its phase with two full rotations, the first rotation unwinding the T-bar and the second winding it up in the opposite direction, reversing I_L. Then P2 goes through a corresponding motion while P1 remains more or less fixed. Here, as in the similar behavior of Fig. 12(b), the motion seems to be more robust if some damping is present and the bias current is fairly large. Note that the phase advances by 4π in approximately the period of an LC oscillation which involves only one of the capacitors, the other being fixed at V = 0. The oscillator motion is so perturbed by the supercurrent, however, that the frequency of this motion is only very roughly $(LC_1)^{-\frac{1}{2}}$ and the step in the I-$\langle V \rangle$ curve will occur only roughly at $V = 2\Phi_o(LC_1)^{-\frac{1}{2}}/2\pi$.

III. LARGE JUNCTIONS

A. Two-Dimensional Systems

1. Circuit models

A large junction is one for which the Josephson supercurrent

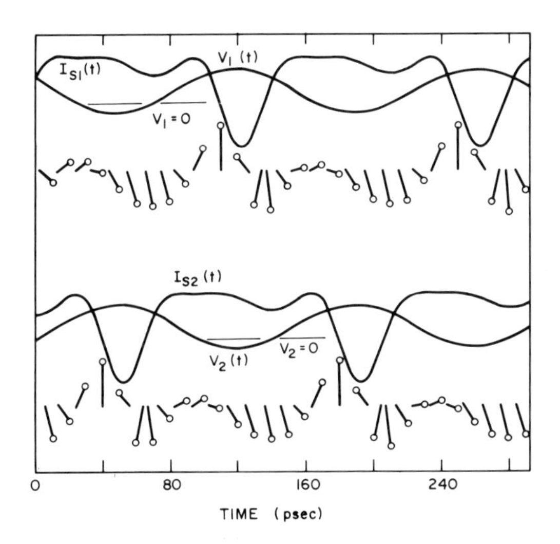

Fig. 17 Behavior of a dc SQUID (circuit of Fig. 15a) when the LC resonance is highly excited. Parameters are $I_{c1} = I_{c2} = 5\ \mu A$, $C_1 = C_2 = 50$ pF, $R_1 = R_2 = 5\Omega$, $L_1 = L_2 = 10$ pH, $\theta_x = \pi$, and $I = 12.5\ \mu A$ near the peak of the current step corresponding to the resonance. Shown are $I_c(t) = I_c \sin \phi(t)$ and $V(t)$ for junctions 1 and 2.

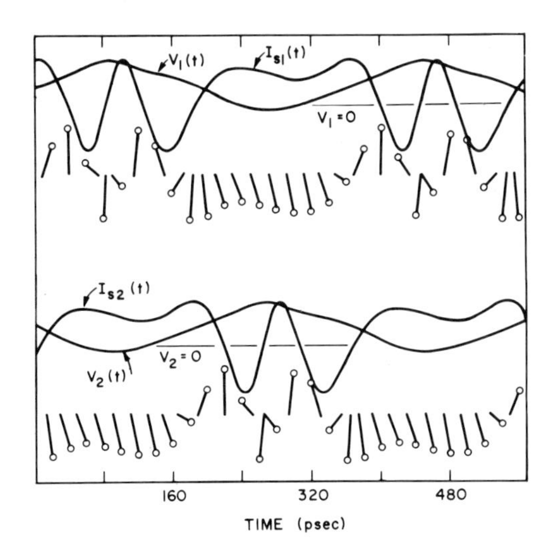

Fig. 18 Behavior of a dc SQUID in a condition for which the LC resonance is excited in a double pulse mode described in the text. Parameters are $I_{c1} = I_{c2} = 5\ \mu A$, $C_1 = C_2 = 50$ pF, $R_1 = R_2 = 5\ \Omega$, $L_1 = L_2 = 60$ pH, $\theta_x = 0$, and $I = 7.5\ \mu A$ near the peak of the current step corresponding to this mode. Shown are $I_s(t) = I_c \sin \phi(t)$ and $V(t)$ for junctions 1 and 2.

is not uniform throughout the area of the junction, a phenomenon induced by the presence of a magnetic field. Because the effects produced by this are especially important in tunnel junctions, the subsequent discussion will be directed towards this system. The usual model [2, 9-11] is to consider the active region of the junction as a surface of small thickness d lying in, say, the x-y plane and sandwiched between two bulk superconducting electrodes. The direction normal to this surface is denoted by a unit vector $\hat{z}$. Each element of this surface acts as though it contains a current generator of the same sort as for the small junction, so that at any point P current flows between the electrodes with a density $j_s(P, t) = j_c(P) \sin \phi(P, t)$, where $j_c(P)$ is the maximum supercurrent density at P and $\phi(P, t)$, the phase at P is related to the voltage V(P) by $d\phi(P, t)/dt = 2\pi V(P, t)/\Phi_o$. These currents are fed by surface currents flowing on the electrodes.

The spatially non-uniform flow of $j_s(P, t)$ comes from any non-uniformity in $j_c(P)$, but more importantly from non-uniformity in ϕ caused by a dependence of ϕ on magnetic field [2, 3, 9], which comes about from the Faraday voltage induced by application of the field. Consider a line integral of the electric field around the contour C as in Fig. 19(a), which shows a cross section of the junction. The contour crosses the barrier region at points P1 and P2 and stays well inside the electrodes elsewhere at a depth much greater than λ, the superconducting penetration depth. Consequently the only significant contribution to the integral comes in crossing the barrier, and we have $V(P2) - V(P1) = d\Phi/dt$, where Φ is the magnetic flux linking C and pointed into the figure. In time-integrated form this becomes $\phi(P2) - \phi(P1) = 2\pi\Phi/\Phi_o$ just as for the dc SQUID. Since the electrodes are bulk superconductors, the magnetic flux lines lie in and parallel to the barrier and extend with exponentially decreasing strength a distance λ into the electrodes on either side. The field $\vec{B}(x, y)$ in the barrier is accompanied by surface currents/unit length $\vec{k}(x, y)$ on the upper electrode and $-\vec{k}(x, y)$ on the lower electrode, where $\vec{k} = -\hat{z} \times \vec{B}/\mu_o$. In differential form the relation between ϕ and $\vec{B}$ is

$$\nabla\phi = \hat{z} \times \vec{B}\,(2\pi(2\lambda + d)/\Phi_o) = -(2\pi/2\lambda + d)\mu_o/\Phi_o)\vec{k}$$

Taking the two-dimensional divergence of $\vec{k}$ gives

$$-\nabla \cdot \vec{k} = (2\pi(2\lambda + d)\mu_o/\Phi_o)^{-1}\,\nabla^2\phi = j_c \sin\phi + \sigma V + (\epsilon_o \varkappa/d)\,dV/dt + I_{edge} \tag{8}$$

$$= j_c \sin\phi + (\sigma\Phi_o/2\pi)\,d\phi/dt + (\epsilon_o \varkappa \Phi_o/2\pi d)d^2\phi/dt^2 + I_{edge}$$

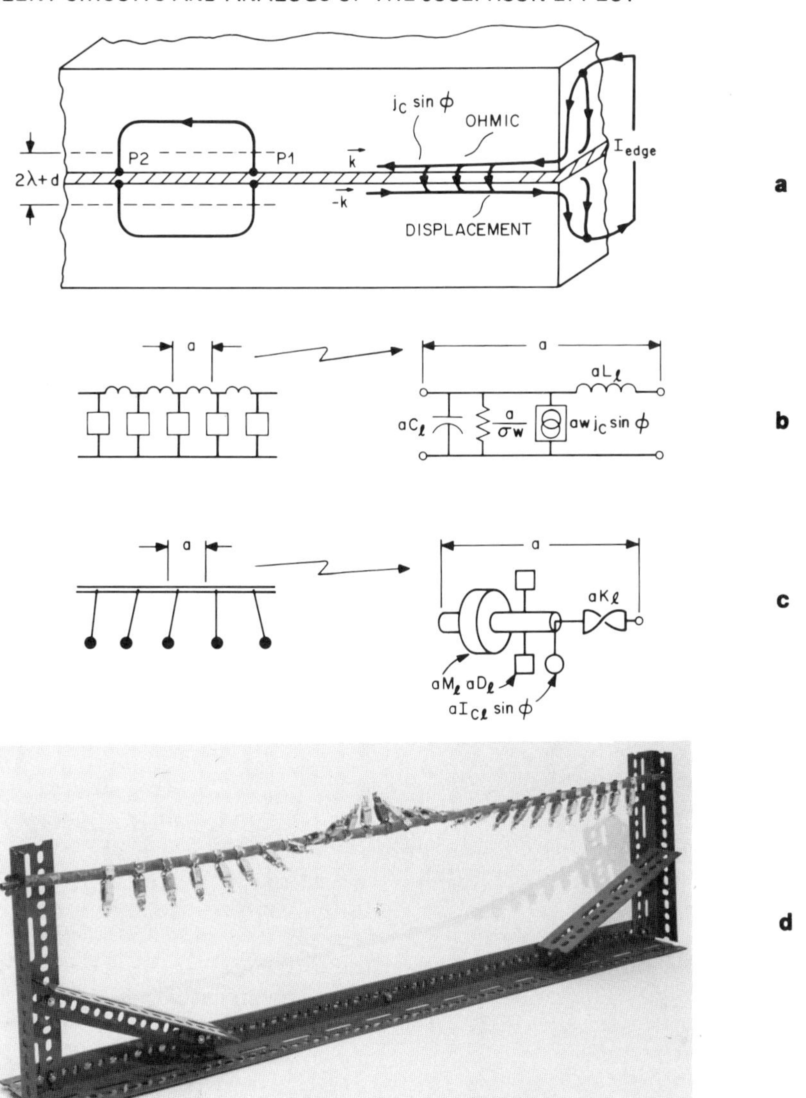

Fig. 19 (a): Cross-section of a tunnel junction. The barrier is shown cross-hatched. On the left is the contour C involved in determining the dependence of ϕ on B. On the right, the surface current density $\vec{k}$ flowing on the electrodes is schematically shown as being fed by external current sources I_{edge} and drained by the three current densities through the barrier.

(b): The equivalent circuit of a one-dimensional junction.

(c): The pendulum-torsion bar analog of the circuit in (b).

(d): Photograph of analog of long junction.

where σ and $\epsilon_o \varkappa / d$ are the conductivity and the capacitance density. As indicated schematically in Fig. 19(a), the value of $-\nabla \cdot \vec{k}$ within the junction is equal to the sum of the current densities through the barrier, namely those of the supercurrent, the dissipative current (taken as ohmic), and the displacement current. The term I_{edge} is a delta-function source term at the edge of the junction which takes account of the current flowing into the junction across the edges and gives effectively the boundary conditions on Eq. (8). Note that I_{edge} can originate either from external current sources or from applied magnetic fields or both. While the behavior predicted by Eq. (8) does not explicitly depend on the origin of the currents feeding the junction, there is an implicit dependence from the constraint that the net current supplied by a magnetic field is zero.

For a junction of uniform or near-uniform properties the behavior of the junction as determined by Eq. (8) depends especially upon the parameters $\omega_J = (2\pi j_c d / \epsilon_o \varkappa \Phi_o)^{\frac{1}{2}}$ and $\lambda_J = (\Phi_o / (2\pi \mu_o (2\lambda + d) j_c))^{\frac{1}{2}}$. As will emerge in more detail later, ω_J is a small oscillations frequency corresponding to the Josephson plasma frequency in small junctions, and λ_J is a screening length over which the magnetic field caused by the Josephson supercurrents can cause appreciable changes in ϕ.

2. Two-dimensional mechanical analog

A mechanical system obeying an equation of the same form as Eq. (8) can be constructed in principle [53, 54] by taking an elastic sheet with finite mass density and under constant tension lying in the x-y plane. The sheet is allowed to undergo damped vertical motion of small displacement z under the influence of vertical forces at the edges proportional to I_{edge} and with a vertical force density throughout its area proportional to $j_c \sin(2\pi z / z_o)$. Newton's law for any element of the sheet then takes the same form as Eq. (8). The periodic force tends to cause the sheet to try to lie in any of a set of planes parallel to the x-y plane and spaced by z_o, while the $\nabla^2 \phi$ term tends to cause the sheet to lie as flat as possible consistent with I_{edge}. Conceptually the periodic force might be provided by attaching the sheet to a two-dimensional array of pendulums.

While construction of an analog does not seem very practical, one can at least imagine qualitatively how it would behave. From such consideration we believe that while there are some phenomena peculiar to the two-dimensional junction [53-55], most of the important properties are the same as occur in a simpler and more easily constructed one-dimensional version to which we shall subsequently restrict ourselves.

B. One-Dimensional Junctions

1. Circuit equations

A reasonable approximation to a one-dimensional junction [56-58] is a junction having a strip-line structure with, say, the y-dimension of constant width $w \ll \lambda_J$, while the x-dimension is unrestricted. If the y-dependence of all quantities in Eq. (8) is neglected, then Eq. (8) becomes

$$K_\ell \, d^2\phi/dx^2 = I_{c\ell} \sin\phi + D_\ell \, d\phi/dt + M_\ell \, d^2\phi/dt^2 + I_{x\ell} . \quad (9)$$

Here $I_{c\ell}$ is the critical current/unit length, $I_{x\ell}$ is the current/unit length applied to the junction, $K_\ell = \Phi_o w/2\pi\mu_o(2\lambda+d) = \Phi_o/2\pi L_\ell$, where L_ℓ is the inductance/unit length, $M_\ell = \epsilon_o \varkappa w\Phi_o/2\pi d = \Phi_o C_\ell/2\pi$, where C_ℓ is the capacitance/unit length, and $D_\ell = \sigma w\Phi_o/2\pi$, where σw is the conductance/unit length. K_ℓ is independent at x but the other quantities may vary with it. Note that $K_\ell d\phi/dx$ is the surface current I_L flowing in the upper electrode, so that Eq. (9) is an equation of current density conservation similar to Eq. (2) for the small junction. For completeness we note that the magnetic flux Φ threading the junction in the y-direction obeys the relation $d\Phi/dx = L_\ell I_L = (\Phi_o/2\pi)\, d\phi/dx$.

In Fig. 19(b) is sketched a lumped equivalent circuit for this structure which comprises a set of many junctions in parallel separated by small inductances such that the LI_c/Φ_o ratio for each element is $\ll 1$.

In the limit $I_{c\ell}(x) = 0$ this simply becomes the wave equation for a damped stripline with current sources $I_{x\ell}$, with the current I_L and voltage V being obtained from $d\phi/dx$ and $d\phi/dt$ respectively. In another limit, if $I_{c\ell}(x) \neq 0$ at just two small, separated regions, this system becomes a dc SQUID.

2. Mechanical analog

The standard mechanical analog [59, 60] of a one-dimensional junction is a sort of continuous limit of a dc SQUID, namely an array of simple pendulums spaced attached to a T-bar as in Fig. 19(c), with the equivalent LI_c/Φ_o of each segment being $\ll 1$. For this analog one can write Newton's law at any point, balancing the gravitational torque density $I_{c\ell} \sin\phi$, the applied torque density $I_{x\ell}$, and the gradient in the T-bar torque $K_\ell d^2\phi/dx^2$ against the damping and inertial torque densities $D_\ell d\phi/dt$ and $M_\ell d^2\phi/dt^2$. This gives an equation exactly the same as Eq. (9). Here again there is an analogy between the torques and currents and between angular velocity and ϕ (or V) just as for the small junction case. Similarly, the T-bar torque is analogous to I_L and the twist in the T-bar is related to the magnetic flux Φ threading the junction with a 2π twist occurring between $x = x_1$ and $x = x_2$ corresponding to a flux Φ_o threading the junction in this region. Further analogies between the torsional, kinetic and gravitational energies and the magnetic, electric and Josephson coupling energies occur as previously.

If the quantities $I_{c\ell}$, D_ℓ and M_ℓ are independent of position, a condition we will subsequently refer to as a "uniform" array or junction, one could cause all pendulums to be set swinging in phase, with no torque in the T-bar. Then the motion would be like that of a single pendulum. If we denote $\omega_J = (I_{c\ell}/M_\ell)^{\frac{1}{2}}$ and $\tau = M_\ell / D_\ell$, then the motion will be called overdamped or underdamped according as $\omega_J \tau \lessgtr 1$. We will be dealing usually with underdamped cases, which are more characteristic of tunnel junctions.

In common with other workers [59-63] we have constructed and used extensively such a pendulum analog, using a rubber tube strung on a stretched steel wire for the T-bar and "crocodile" battery clips for pendulums. Much of the subsequent discussion and most of the unsupported assertions therein are based on our observations of this analog. A photograph of it is shown in Fig. 19 (d).

3. Small oscillations and displacements

Consider a uniform, underdamped array in the quiescent $\phi(x) = 0$ condition. Any small oscillations or disturbances from this configuration can be expressed in terms of normal mode solutions $\exp(i(kx - \omega t))$ resulting from approximating $\sin\phi$ by ϕ in Eq. (9). One has $\omega^2 = \omega_J^2 + \bar{c}^2 k^2$, where $\bar{c} = (M_\ell K_\ell)^{-\frac{1}{2}}$ is the velocity of propagation of vibrational waves in the pendulum array neglecting gravity, or of electromagnetic waves in the stripline neglecting Josephson effects. For $\omega \geq \omega_J$ these are traveling waves called plasma oscillations [10, 11], which at high frequencies $\omega \gg \omega_J$ are essentially the same as stripline modes with phase velocity of $\bar{c}$, while at the cutoff frequency $\omega = \omega_J$ the phase velocity becomes infinite and the motion becomes uniform. For junctions with finite length, standing-wave versions of this model have solutions with $k = n\pi/\ell$ with $n = 0, \pm 1, \pm 2, \cdots$. If $\ell \ll \lambda_J$ these modes all go over into the usual stripline modes with frequencies $n\,\bar{c}/2\ell$, $n = \pm 1, \pm 2, \cdots$ plus the plasma mode at ω_J. Figures 20(a) and (b) show two such oscillations in a pendulum array for a half-wavelength (a) and for two wavelengths (b) within the array. Note the boundary condition $d\phi/dx = 0$ at the ends.

For $\omega < \omega_J$ there are only solutions with imaginary k, which can be excited only at the ends and extend into the array with exponentially decreasing amplitude. At $\omega = 0$ one has a static disturbance [56] resulting from, say, a small constant torque applied to the end of the junction at $x = 0$ in which $\phi(x) = \phi_o e^{-x/\lambda_J}$.

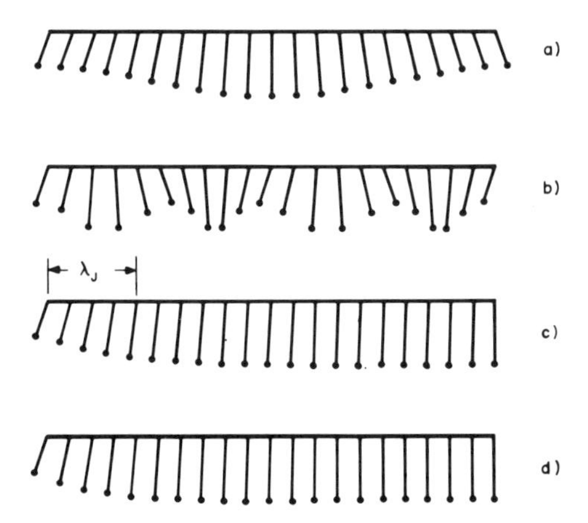

Fig. 20: Oblique front view of small (plasma) oscillations of a pendulum array model of a one-dimensional junction for (a): a half wavelength, and (b): two wavelengths fitting in the junction. Displacements caused by small oscillatory torques at frequencies (c): $\omega = 0.8\ \omega_J$, and (d): $\omega = 0$, applied at the left end.

Figures 20(c) and (d) show two such displacements, one (c) caused by an oscillating torque at frequency $\omega = 0.8\ \omega_J$ and a resultant decay length $\lambda = 1.666\ \lambda_J$, and the other (d) caused by a static torque.

4. Magnetic diffraction and fiske modes for $\ell \ll \lambda_J$

If the junction length ℓ is smaller than λ_J the T-bar is relatively stiff and the behavior is very like that of the small junction described earlier. Even here, however, $\phi(x)$ can vary with position under the influence of an applied magnetic field. For the pendulum array the effect of a transverse applied magnetic field is simulated by application of equal and opposite torques on the two ends of the array, corresponding to the magnetic-field-induced surface currents $I = \pm\ wB/\mu_0$ entering the junction. This torque pair causes the T-bar to be wound up uniformly with a pitch $d\phi/dx \propto B$. There is negligible additional twist caused by the gravitational torque since $\ell < \lambda_J$. In a static configuration this gives $\phi(x) = [2\pi B(2\lambda + d)/\Phi_0]x + \phi_0$, where ϕ_0 is such that the net gravitational torque $\int I_{c\ell} \sin \phi(x)_{dx}$ balances any applied torque $\int I_{x\ell}\, dx$. The array thus behaves as a single pendulum with a moment arm which depends on B in an oscillatory fashion, giving the familiar diffraction-pattern form [54] for $I_c(B)$, whose exact form depends on the detailed shape of $I_{c\ell}(x)$.

Figures 21(a) to (d) show schematic front and side views of the pendulum array for values of applied torque at the two ends which cause a variation in $\phi(x)$ from one end of the array to the other of 0, π, 1.8 π and 2.2 π, respectively. The configuration is for zero applied torque. Note that the array turns upside down between (c) and (d).

While the array behaves as a single pendulum at $\langle V \rangle = 0$ and small rotational frequencies, there is always the possibility of interacting with the resonant modes of vibration of the line which occur at frequencies $f = n\bar{c}/2\ell$, $n = \pm 1, \pm 2, \cdots$, where $f > \omega_J/2\pi$. Just as was discussed earlier concerning the interaction of small junctions and dc SQUIDs with LC resonances, one would expect that the alternating torque will tend to excite the vibrational resonance if it occurs at the resonant frequency and provided that the spacial distributions of the gravitational torque and the voltage of the resonant mode are not orthogonal. Thus, for example, the fundamental mode would be expected to be most strongly excited if the magnetic field $B = \Phi_o/2\ell(2\lambda + d)$, giving a half-twist from one end to the other. The ac voltages resulting from this excitation will again lead to an increase in $\langle \sin \phi \rangle$ averaged over the junction, in the vicinity of $\langle V \rangle = \Phi_o \bar{c}/2\ell$, causing a current step in the I-$\langle V \rangle$ curve. Such magnetic-field dependent steps, sometimes called Fiske modes [65-67], are a familiar sight in tunnel junctions.

Figures 21(e) to (h) show schematically four configurations, separated by $\pi/2$ in time phase, of a pendulum array rotating synchronously with the fundamental oscillation, i.e., on a Fiske mode step. Note how the time-average supercurrent is positive as a result of the difference between steps (f) and (h).

A perturbation treatment [68, 69] of this interaction can be made based on the assumption of small $V_{ac}(x) = V(x) - \langle V \rangle$. However, the excitation of the mode is often so large as to preclude such treatment. In this case experience with the analog suggests that a self-limiting behavior will occur for $V_{ac} \sim \langle V \rangle$ for reasons similar to those described previously, giving a maximum amplitude for the step of $\sim \langle V \rangle / R_{eff}$. A more analytical treatment [70] gives a similar result. In tunnel junctions R_{eff} is generally large compared to the quasiparticle tunneling resistance and is usually thought to result from losses in the surface impedance of the electrodes.

5. Junctions having $\ell >> \lambda_J$ - vortices and critical currents

We will now be interested in the supercurrent flow at $\langle V \rangle = 0$ for a junction extending from $x = 0$ to $x = \ell >> \lambda_J$ [56]. This is described by

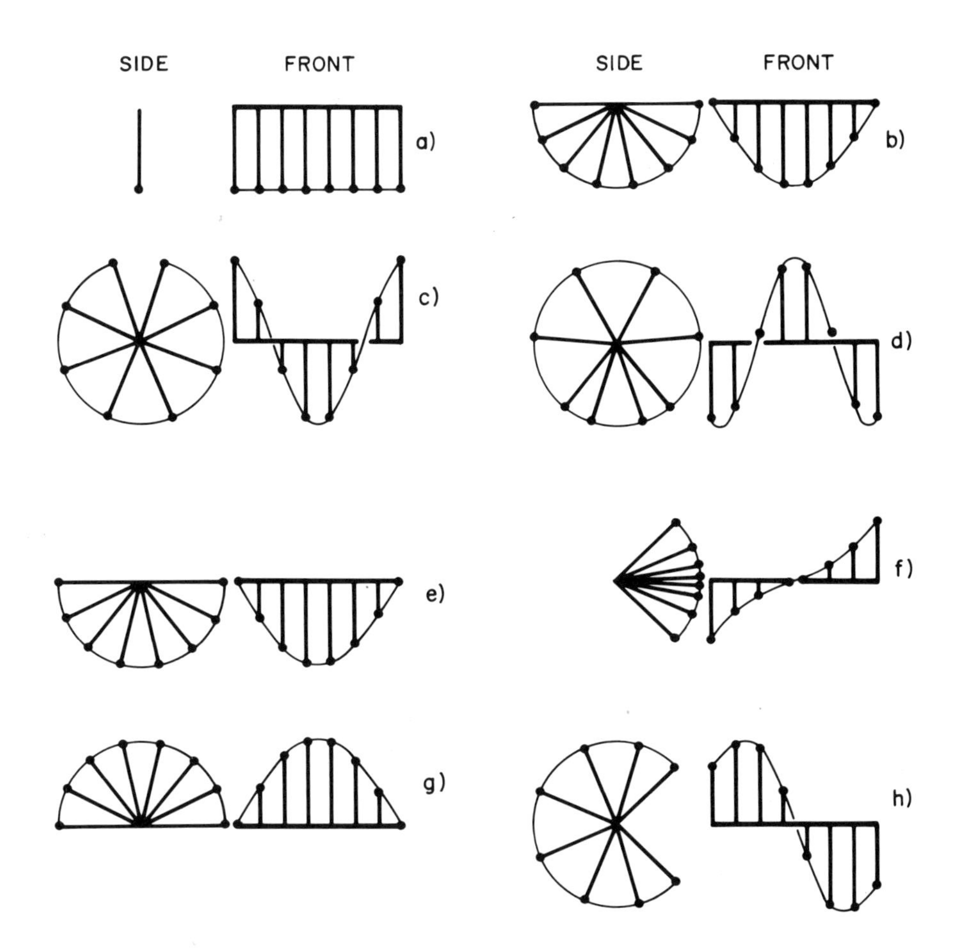

Fig. 21 Schematic front and side views (a) to (d) of a short pendulum array subject to increasing equal and opposite torques at the two ends (external magnetic field). $\Phi(x)$ varies linearly between the two ends in (a) to (d) by 0, π, 1.8 π, and 2.2 π, respectively. Schematic front and side views (e) to (h) of a short array displaying the behavior expected on a Fiske step. The array is vibrating in its lowest mode of oscillation and is rotating at the same frequency. Successive configurations are spaced by one-quarter of the period with (f) and (h) occurring at the two voltage nodes. The equivalent of 0.5 Φ_o of static flux links the array.

$$K_\ell \, d^2\phi/dx^2 = I_{c\ell} \sin\phi + I_{x\ell} . \tag{10}$$

Some insight into Eq. (10) is gained by nothing that if ϕ is replaced by $\phi' + \pi$, then Eq. (10) is the same as

$$M \, d^2\phi'/dt^2 = -I_c \sin\phi' + I_x \tag{11}$$

which describes the motion of an undamped pendulum subject to an external torque. Caution is required, however, because the familiar small-oscillation solutions of Eq. (11) are only metastable in the junction, e.g., $\phi' = 0$ for the pendulum corresponds to the upside-down configuration $\phi = \pi$ for the junction which is an energy saddle point.

It is best to describe at the outset a special solution of Eq. (10) for a uniform junction of infinite length, the "vortex" solution [9], which plays a central role in the behavior. This solution comprises a complete 2π twist in the array such that $\phi \to 0$ at $x \to -\infty$, and $\phi \to \pm 2\pi$ at $x \to +\infty$. Intuition suggests correctly that this twist will mostly occur in a limited region over a few λ_J in length, because this provides the best compromise between minimizing the torsional energy and gravitational energy. In a junction, a vortex comprises a localized current loop generating a magnetic flux of Φ_o linking the junction and providing the 2π increase of in phase.

In the analogy of Eq. (11), the corresponding situation is that of a pendulum which starts in a nearly inverted position at $t \to -\infty$, falls over and swings through 2π, coming to rest at $t \to +\infty$ in an inverted position again. The analog of conservation of kinetic and potential energy gives $\frac{1}{2}(d\phi/dx)^2 = (1 - \cos\phi)/\lambda_J{}^2$ for this motion, which can be integrated to provide the functional form of a vortex positioned about $x = x_o$, $\phi(x) = \pm 4 \tan^{-1} \exp((x - x_o)/\lambda_J)$. We will use the convention of calling solutions with $d\phi/dx > 0$ a vortex and those with $d\phi/dx < 0$ an antivortex. Figures 22(a) and (b) show a front view and a top view (or $-\cos\phi(x)$ and $-\sin\phi(x)$) of a pendulum array containing a vortex, Fig. 22(c) shows $d\phi/dx$, and Fig. 22(d) shows $\phi(x)$ for the vortex.

Although this solution applies for an infinite junction, it is apparent that a 2π twist in a long but finite pendulum array will look much the same. The only important difference is that the twist can escape from the ends of the array. One could prevent this escape by applying small external torques in appropriate places or by placing the twist in a region where $I_{c\ell}(x)$ has a local minimum [71], as indicated in Fig. 22(d) where the end pendulums are especially massive.

To proceed now with the discussion of the static supercurrent flow, let us denote the net external applied torque applied

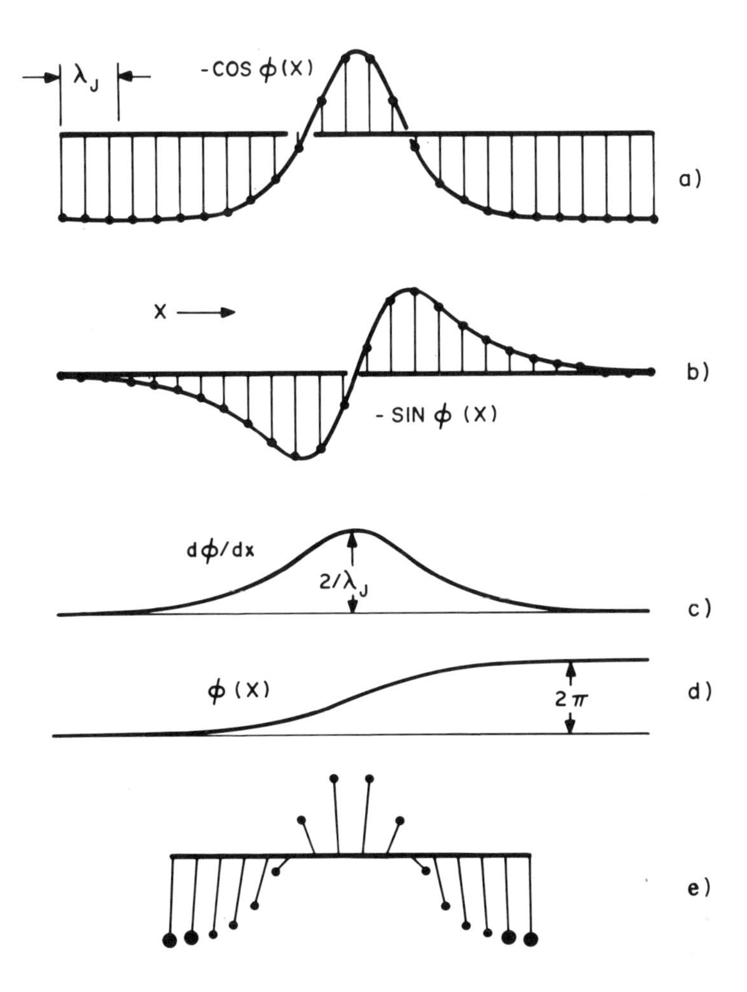

Fig. 22 Various views (a) to (d) of a segment of an infinitely long pendulum array containing a vortex. (a): Front view ($-\cos\phi(x)$), (b): Top view ($-\sin\phi(x)$), (c): $d\phi/dx$, (d): $\phi(x)$. (e): A vortex trapped in an array of finite length by the excessive weight of the end pendulums.

to the array as $I_x = \int I_{x\ell}dx$. The maximum I_x that can be applied to a uniform array without setting it into rotation, which is the analog of the critical current I_c, depends upon the form of $I_{x\ell}$. If $I_{x\ell}$ is constant, all pendulums will be displaced uniformly and the value of I_c is the maximum possible (if $I_{c\ell}$ is uniform). If $I_{x\ell}$ is applied only at one end of the array, the situation is more interesting. The end pendulums rotate to $\phi \neq 0$, but the flexibility of the T-bar permits the interior pendulums to sag back towards $\phi = 0$ over a length $\sim \lambda_J$, the exponential decay mentioned in the previous section. The supercurrent flow and the associated magnetic field are thus confined to the end of the junction. If the torque is further increased there will be some maximum value $I_x = I_{c1}$ at which the array begins to rotate. Since the applied torque fixes the value of $d\phi/dx$ at the end of the array, I_{c1} corresponds in the language of Eq. (11) to the maximum $d\phi'/dt$ that a simple pendulum can have at $t = 0$ and still come to rest in an inverted position at $t \to \infty$. This velocity is just that occurring at the bottom of a full swing, or in the middle of a vortex. The configuration at $I_x = I_{c1}$ is then that of a half-vortex extending into the junction from the end and I_{c1} is given by the value of $d\phi/dx = 2/\lambda_J$ at the center of a vortex, or $I_{c1} = 2\lambda_J I_{c\ell}$. Figures 23(a) to (d) show four configurations in this process with increasing I_x. Extending the argument a little further, the application of a torque at a point in the center of the array will produce a value of $I_c = 4\lambda_J I_{c\ell}$ with a configuration for $\phi(x)$ consisting of a half-vortex facing a half-antivortex on this site.

Once the critical torque is exceeded, the array will start to rotate. For an underdamped array the transient behavior is complex for non-uniform I_x with various oscillations, but generally we expect that the array will come to terminal velocity $\langle V \rangle = I_c R$ with uniform rotation and some parabolic variation in $\phi(x)$ due to the finite I_L. The behavior at lower velocities $\langle V \rangle \sim \Phi_o \omega_J/2\pi$ or for an overdamped array at any speed is more complex, and we defer discussion of such cases.

6. Magnetic field behavior for $\ell >> \lambda_J$

The static behavior of a long junction in an applied transverse magnetic field B is closely related to its behavior under an applied current [57], since B generates surface currents on the electrodes which produce equal and opposite bias currents $\pm Bw/\mu_o$ at the two ends. Beginning with $B = 0$ and $\phi(x) = 0$, and increasing B slowly, the junction will confine the current flow and magnetic field penetration to within λ_J of the ends of the junction so long as the currents are less than I_{c1}. The corresponding situation in the pendulum array is the application of equal and opposite torques on the two ends. The maximum torque that can be screened out at $B = B_{c1}$ comes when a half-vortex is formed at each end of the junction as in Fig. 23(e). If the torque

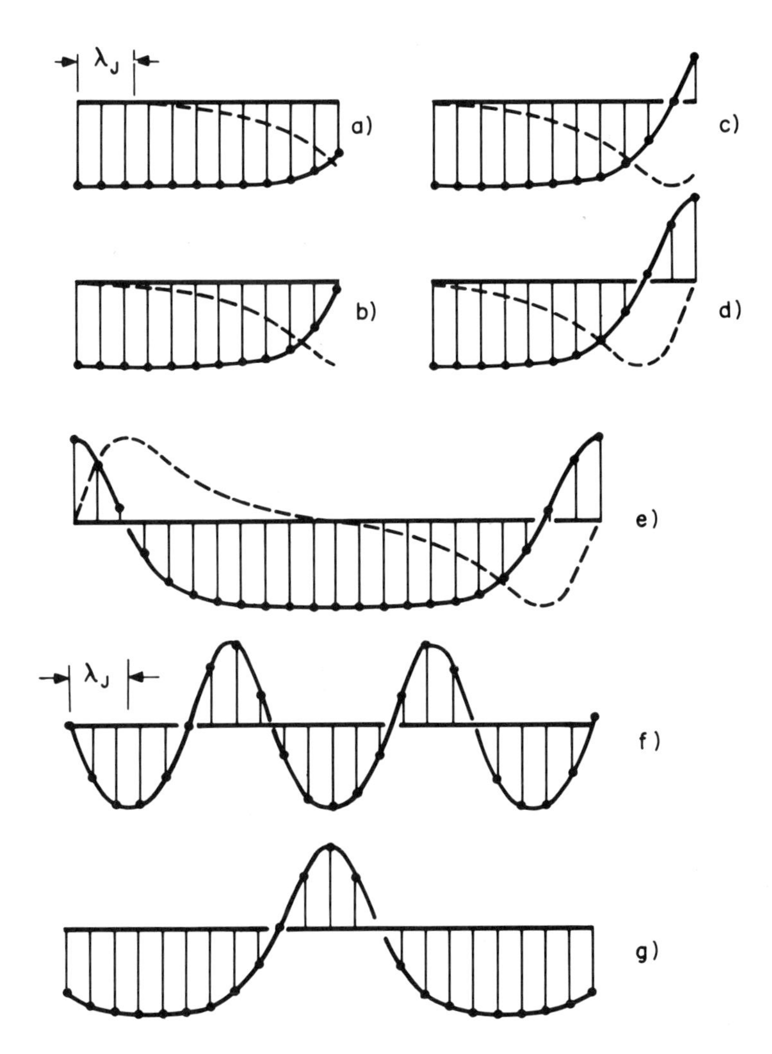

Fig. 23 Front view ($-\cos\phi(x)$) of a long array subject to applied torque at the right end of (a): $I = 0.425\ I_{c1}$, (b): $I = 0.648\ I_{c1}$, (c): $I = 0.887\ I_{c1}$, and (d): $I = I_{c1}$. The dashed line is a top view ($-\sin\phi(x)$). (e): Similar views of an array subjected to equal and opposite torques $I = \pm I_{c1}$ at the two ends corresponding to a magnetic field B_{c1}. (f): Front view of array of (e) with $B = 1.05\ B_{c1}$ in a multivortex configuration. (g): Array of (e) with $B = 0.5\ B_{c1}$ with a single trapped vortex.

is increased further one or both ends will begin to rotate and several 2π twists will enter the array. A typical configuration is shown in Fig. 23(f).

The gravitational torque will cause each 2π twist to occur as a localized object rather like a distorted version of the isolated vortex, and one may think of this configuration as an array of vortices. The vortices act as though they possess a short-range mutual repulsive force due to the increased magnetic field energy caused by their overlap, and are held together by the torques at the end of the array. The spacing between the vortices is determined by the repulsion between them. The magnetic field in the interior of the junction is comparable to B_{c1} though with considerable modulation from the vortex structure.

If B is increased further the vortices pack themselves more closely and new ones are formed at the end. The modulation of $\phi(x)$ decreases and in the limit that the flux linking the junction is $\gtrsim \Phi_o$ in a length $\lesssim \lambda_J$, $\phi(x)$ becomes approximately linear. If B is decreased from B_{c1}, the vortex-vortex repulsion spreads the array apart and vortices are expelled from the end of the junction. As the vortices get farther apart than a few λ_J their interaction becomes exponentially weak. For $\ell \gg \lambda_J$, one or more vortices can remain trapped in the interior for $B \ll B_{c1}$, as in the configuration of Fig. 23(g). One way of describing this configuration is to say that the small B-induced bias currents create an exponential variation in $\phi(x)$ near the ends, which is equivalent to the tail of a vortex, and the mutual repulsion between these truncated vortices and the vortices in the central part of the junction prevents the latter from escaping. If the central region of the junction possesses a local minimum in $I_{x\ell}$, the vortex can remain in the junction even at $B = 0$. A negative B, forming truncated antivortices at the ends of the junction which attract the trapped vortex, may be used to release it from this region.

7. Vortex motion

In a uniform pendulum array of infinite length, a full twist or vortex can be moved from place to place with no cost in energy [59, 60, 72]. Application of a weak torque in the vicinity of a vortex causes the pendulums to rotate in the direction of the applied torque and the twist moves along the array away from the torque. Since rotation occurs when the torque is applied, work is done on the array and in an underdamped array this shows up as kinetic energy of the pendulums in the neighborhood of the twist. Remarkably, this kinetic energy is transferred to neighboring pendulums in just such a way as to cause the twist to move uniformly down the array rather as if it were a localized particle. Furthermore, in interactions with applied torques, other vortices and irregularities in the array the vortex tends to keep its localized nature. The effect of the interaction is basically to speed, slow, stop or reflect the vortex in its motion, enhancing the resemblance to the motion of a particle. While experience with the pendulum array tends to make this behavior seem quite natural and even trivial after a while, the mathematics of the behavior is actually quite subtle and complex [73], particularly as regards interaction between moving vortices and the stability of the motion in non-uniform and damped lines. We will of course ignore these complexities and will simply adopt the attitude that the vortex does indeed behave as a kind of particle in motion acted on by various forces in many situations.

The following are some characteristic features of this particle-like behavior. As previously noted, a gradient in $I_{c\ell}(x)$ causes a force tending to move either a vortex or an antivortex towards the minimum in $I_{c\ell}$, as in Fig. 24(a). Application of a small torque I_x (small compared to I_c) to the line provides a localized force pushing a vortex to the right and an antivortex to the left, as in Fig. 24(b). Damping forces act as a drag on the vortex motion, reducing its velocity, and the motion may either be underdamped if $\omega_J\tau >> 1$, or overdamped if $\omega_J\tau << 1$. The short-range repulsive interactions between static vortices persist between moving vortices, and one can have billiard ball style collisions in underdamped systems in which two moving vortices approach each other, collide and bounce away with kinetic energy conservation. Collision between a vortex and an antivortex involves short-range attractive interactions since the total energy is reduced by bringing the two into superposition. In an overdamped system the two will be annihilated in such a collision, whereas in an underdamped system the collision has the startling result that the two pass right through each other and continue on with their relative kinetic energies intact. Such a collision is shown at four different times in Figs. 24(c) to (f). Shown are $-\cos\phi(x)$ and $V(x)$.

Analytically the form of $\phi(x,t)$ for a vortex moving with velocity v on an infinite, undamped line is $4\tan^{-1}\exp((x-vt)/\lambda_J(1-v^2/\bar{c}^2)^{\frac{1}{2}})$, and the moving vortex is accompanied by a voltage $V = (\Phi_o/2\pi)d\phi/dt$ and a magnetic field threading the junction $B = \mu_o I_L/w = (\Phi_o/2\pi(2\lambda+d))d\phi/dx$. The movement of a vortex past a point generates a time-integrated voltage Φ_o sufficient to advance the phase by 2π at that point. As $v \rightarrow \bar{c}$, its maximum value, the width becomes smaller and the ratio of the total energy to the energy at rest $(4/\pi)\Phi_o\lambda_J I_{c\ell}$ becomes larger by the factor $(1-v^2/\bar{c}^2)^{-\frac{1}{2}}$ An increasingly small part of the energy is formed by the Josephson coupling energy and the moving vortex becomes more and more like an ordinary voltage pulse on the strip transmission line.

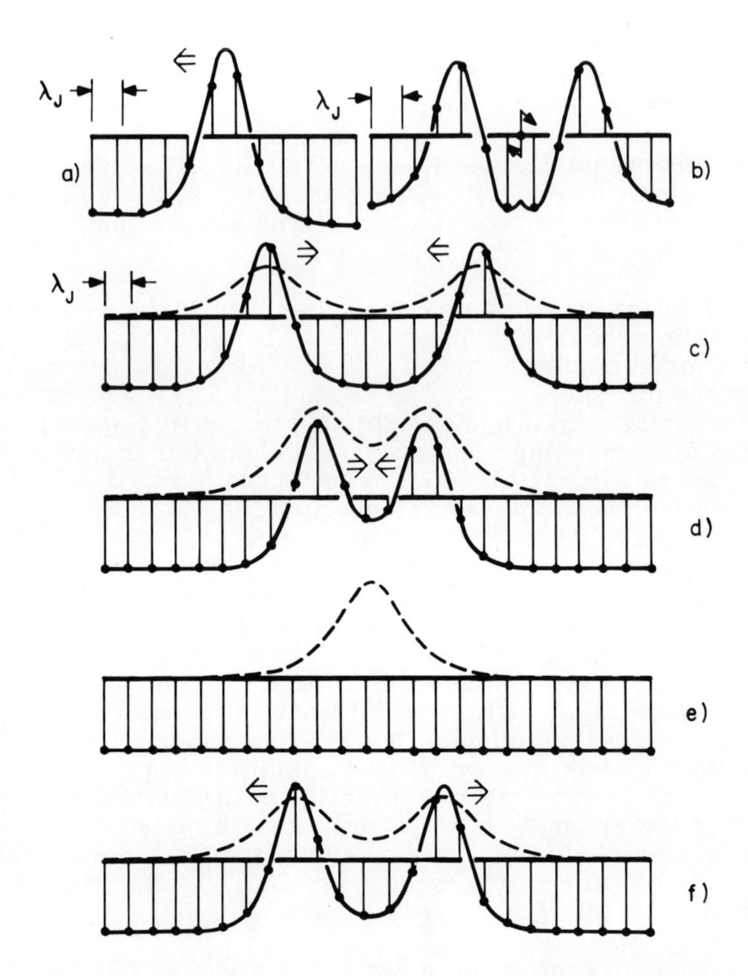

Fig. 24 Front view ($-\cos\phi(x)$) of arrays containing vortices responding to various forces. (a): The gradient in $I_{c\ell}(x)$ causes an effective force pushing the vortex to the left. (b): A vortex and antivortex being pushed in opposite directions by a local applied torque. (c) to (f): Collision of a moving vortex and antivortex in an underdamped array at four successive moments. The dashed line is $V(x)$.

Although vortex motion is a smooth process in a uniform junction, any sharp irregularities in, e.g., $I_{c\ell}(x)$ can cause kinetic energy to be lost in the generation of plasma oscillations. Both this energy and that lost to the drag forces has to be made up by the power supplied from the external current sources. The energy supplied to a vortex in passing a point source of I_x can be obtained from integrating VI_x with time and is $\Phi_o I_x$.

8. Resonant vortex propagation

We now describe a variety of vortex motions frequently encountered in tunnel junctions [74, 75]. Consider the behavior of a moving vortex as it approaches the end of a uniform, underdamped array of length $\ell \gg \lambda_J$, as shown in the sequence of Figs. 25(a) to (e). Since the end of the junction is effectively a region of reduced $I_{c\ell}$, the vortex speeds up as it approaches the end , then passes out of the end, rotating the end pendulums by 2π and disappears. However, the energy released by its destruction, plus any kinetic energy it possessed, is given to the kinetic energy of rotation of the pendulums at the end of the junction. These continue to rotate in the same sense after the vortex disappears and smoothly create an antivortex (with the same kinetic energy as the original vortex) which propagates back towards the other end of the array. Remarkably, the process is devoid of loss in the zero-damping limit, with no energy given to plasma oscillations. At high energies the vortex is effectively a voltage pulse and the process becomes the familiar one of the reflection of a voltage pulse from the end of an open-circuited transmission line. In fact, since the boundary condition at the end is $d\phi/dx = 0$, the process is the same as the collision of a vortex and an antivortex on an infinite line (as in Figs. 24(c) to (f), with the end of the junction at the symmetry point. Thus the force felt by a vortex on nearing the end of the junction is the same as would be provided by a mirror-image antivortex.

When the antivortex reaches the other end of the junction a similar reflection occurs, creating a vortex proceeding in the original direction. If damping is present the frictional losses will eventually interrupt this process, but it may be kept running by application of an external torque I_x anywhere along the line. For $\omega_J \tau \gg 1$, the vortex velocity will approach $\bar{c}$ for all but the smallest I_x, and applications of larger I_x simply cause the vortex to decrease in size, increasing the amplitude of the accompanying voltage pulse and the ohmic losses. In steady-state the phase advances by 4π in every cycle, 2π for each vortex and antivortex passage at a frequency of $\bar{c}/2\ell$, giving $\langle V \rangle = \Phi_o \bar{c}/\ell$, which is twice the voltage of the first Fiske mode. If $I_{x\ell}$ is distorted so that it is small in the center and large near the ends, this process goes over into that depicted for the dc SQUID in Fig. 18. Alternately, if $\ell \leqslant \lambda_J$, the voltage and current patterns go over into

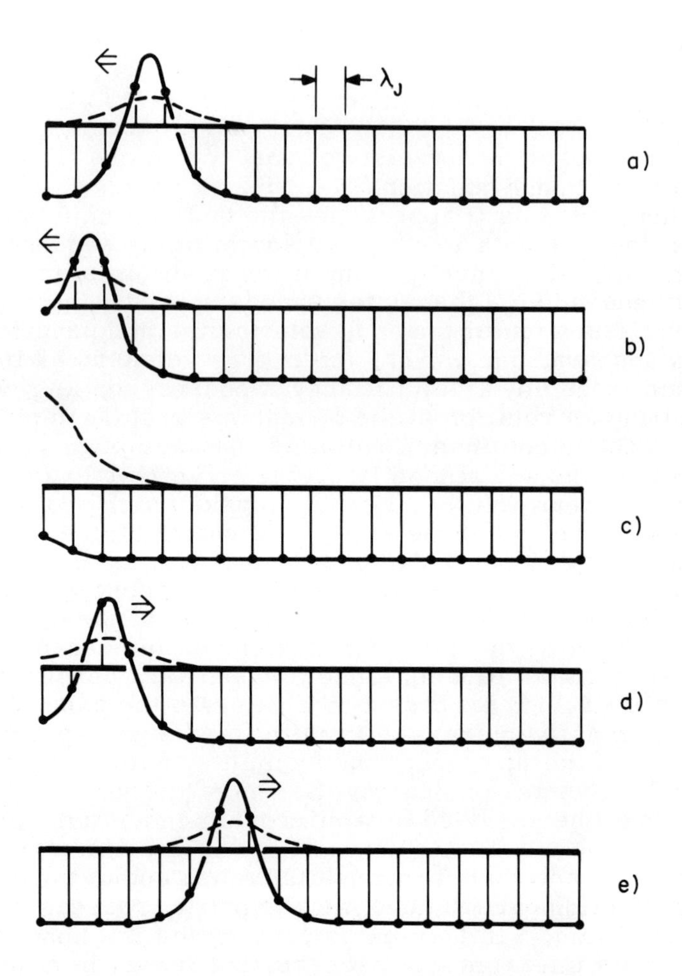

Fig. 25 Resonant propagation of a single vortex in an array showing five successive configurations (a) to (e). A reflection from the end occurs in (b) to (d). Solid line - front view ($-\cos\phi(x)$). Dashed line - $V(x)$.

those of the first stripline resonance.

This process has been observed to occur in tunnel junctions [74, 75] having $\ell \gg \lambda_J$, where it creates a current step at $\langle V \rangle = \Phi_o \bar{c}/\ell$ at zero applied magnetic field. Steps are also seen at low integer multiples of this voltage, which are interpreted as more than one vortex simultaneously engaged in such motion. Figure 26 shows two vortices in this mode. The motion is very stable in both junctions and the mechanical analogs, and persists to levels of I_x which are an appreciable fraction of I_c.

9. Finite $\langle V \rangle$ behavior for $\ell \gg \lambda_J$

Consider an underdamped long junction biased at a current I_x such that $\langle V \rangle \gg \omega_J \Phi_o/2\pi$. In this case the I_x-$\langle V \rangle$ curve will be approximately ohmic. As the bias is reduced to where $\langle V \rangle \sim \Phi_o \omega_J$, this corresponds to a pendulum array rotating at relatively low frequency, and spending a good deal of its time in an inverted position. This configuration is evidently prone to collapse, transferring some of its kinetic energy to the T-bar torsion and interrupting the motion and returning to $\langle V \rangle = 0$. Such a collapse is sketched in Fig. 27. The collapse, however, will necessarily involve creation of 2π twists in the array, and there is the prospect that as they are pushed towards the end of the array by the applied torque a resonant propagation mode of the sort just described may be initiated. In that case, the junction will be able to continue running at a relatively low $\langle V \rangle = (n\bar{c}/2\ell)\Phi_o \leqslant \omega_J \Phi_o/2\pi$. This is in practice how these modes are excited in tunnel junctions. From the fact that the oscillations period of a pendulum described in Eq. (11) is always $> 2\pi/\omega_J$, we can deduce by analogy that such collapse behavior requires that $\ell \geq 2\pi\lambda_J$.

The behavior of an overdamped junction at finite voltage is relatively dull [60]. If the current bias is applied, say at one end, and I_x just exceeds I_{c1}, the half-vortex will sluggishly develop into a full vortex which will be pushed into the junction by I_x. Another vortex will then form, and another, each successive one pushing its predecessor along the junction until they ooze from the far end. At larger I_x the vortex spacing becomes small compared to λ_J and $\phi(x)$ takes on a smooth parabolic form reflecting a linear decrease in I_L along the junction.

10. Vortex oscillations

As our final category, we discuss the role played by vortices in $\langle V \rangle = 0$ oscillations [55, 58, 72]. It can be shown that an infinite, undamped line with a single vortex has only extended, plasma-like oscillation modes with no localized modes about the vortex [72]. However, a vortex bound by any sort of irregularity

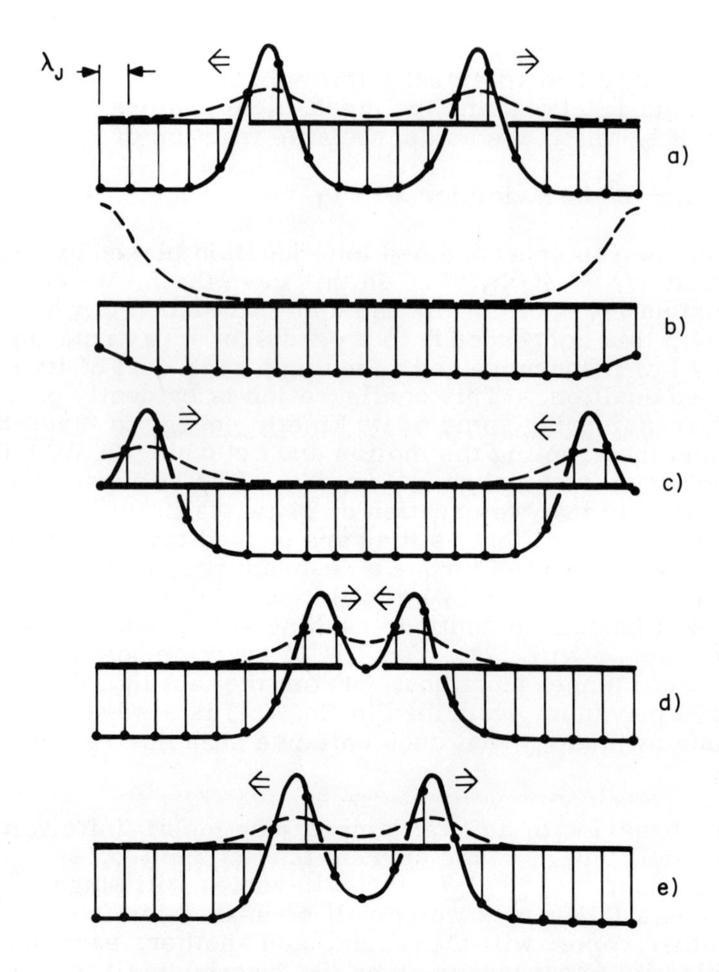

Fig. 26 (a): Resonant propagation of the vortices in an array showing five successive configurations. Reflection from the ends occurs in (b) and (c), while a collision in the center occurs in (d) and (e). Solid line - front view (- cos ϕ(x)). Dashed line - V(x).

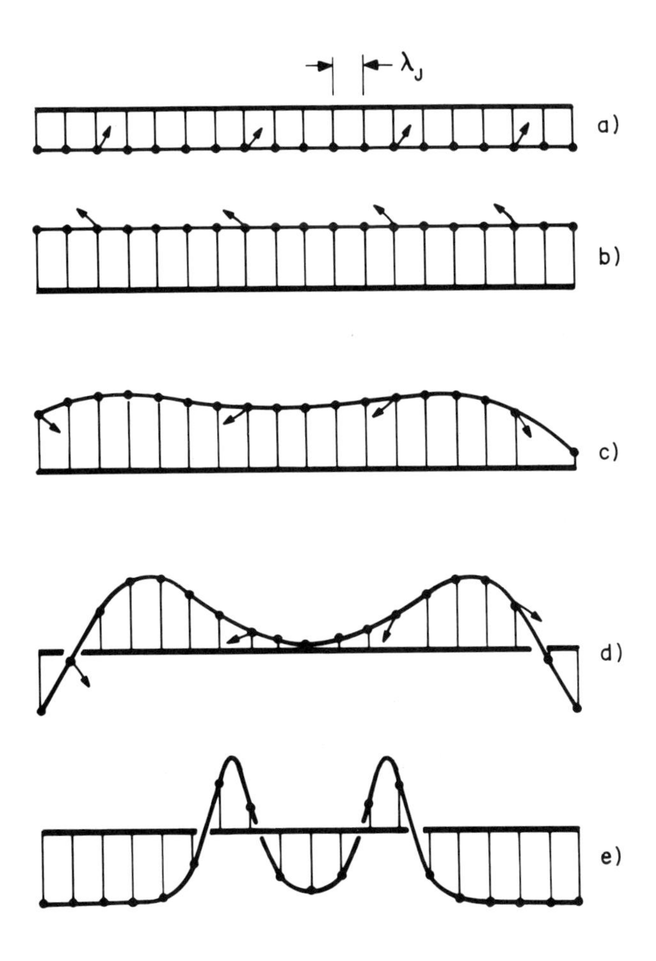

Fig. 27 Return to V = 0 of a long pendulum array. Front view ($-\cos \phi(x)$) at five successive times. In (a) and (b) the array is rotating uniformly with low energy. In (c) the hesitation near $\phi(x) = \pi$ allows a fluctuation in $\phi(x)$ to develop and grow, causing a collapse. In (d) the central part of the array falls forward and the ends backwards. In (e) the collapse has resulted in a vortex-antivortex pair.

such as a local minimum in $I_{c\ell}(x)$ may oscillate back and forth about this site as in Figs. 28(a) and (b). Unless the perturbation is so large as to cause a severe distortion of the vortex, the frequency of the oscillation will be well below ω_J. Another such situation occurs in a junction of $\ell >> \lambda_J$ containing a vortex trapped by a weak applied magnetic field B, with the field-induced currents at the two ends causing two opposed forces creating a potential well. If several vortices are trapped in the well, their oscillations within it are modified by their mutual repulsion. As the strength of the trapping field is increased to $B \gtrsim B_{c1}$, these oscillations go over into the stripline modes appropriate for a linear $\phi(x)$.

Another interesting oscillation is a bound state version of the by-now familiar vortex-antivortex collision. If the kinetic energy of these two is less than that required for them to escape from each other, they will oscillate back and forth through each other, as in Figs. 28(c) and (d). The frequency of the oscillation is low if the energy is high, since the vortices nearly escape from each other, and increases to ω_J at lower energy as the oscillation goes over into a long wavelength plasma oscillation. Since the intersection of a vortex with the end of a junction is the same as occurs between a vortex and an antivortex, this oscillation should also occur as a localized mode at the end of the junction. Once excited, the mode should lie below ω_J and possess high Q, but this has not yet been noticed in tunnel junctions.

Finally, a more complex version of the vortex-antivortex collision can occur in an undamped, infinite junction in which an array of equally spaced vortices propagates through a similar array of oppositely-moving antivortices. There is a corresponding bound-state version of this process which would involve similar arrays of vortices and antivortices flopping back and forth through each other as in Figs. 28(e) and (f). The low-energy limit of this motion is the $k = 0$ plasma oscillation. If the sequence of vortex and antivortex in Fig. 28(e) were changed to vortex, antivortex, antivortex, vortex, etc., a similar oscillation can occur in which the sign of the voltage pulses changes from one site to the next, and whose low-frequency limit is a $k \neq 0$ plasma mode.

IV. CONCLUSIONS

It seems appropriate finally to mention by name only some topics which we chose not to discuss here. These include non-sinusoidal current-phase relations, noise properties, frequency-dependent I_c's, the related cos ϕ damping term and properties of series arrays. Any and all of these topics are of current or past interest and can be treated in a more or less adequate fashion in

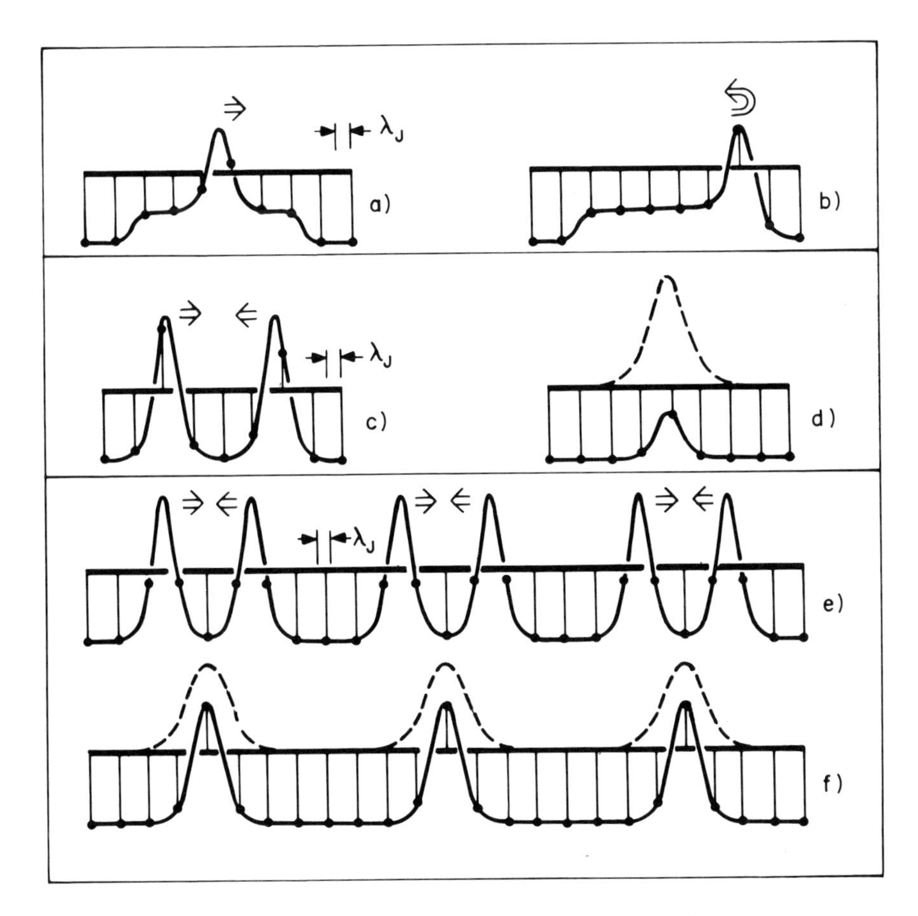

Fig. 28 Various modes of oscillation in a long array. Solid line - front view (- cos $\phi(x)$). Dashed line - $V(x)$. (a) and (b): Oscillation of a vortex trapped in a local minimum of $I_{c\ell}(x)$. (c) and (d): Bound state oscillation of a vortex-antivortex pair shown at (c) maximum separation and (d) ≈ one-quarter period later. (e) and (f): Array of bound vortex-antivortex pairs undergoing oscillations shown at (e) maximum separation and (f) ≈ one-quarter period later.

equivalent circuit language, but their complexity has prevented us from attempting their treatment here. Some of these topics are covered in other articles in this text.

We hope that the discussion presented here will indicate to the reader the very considerable insight into the behavior of Josephson junctions which can be gained from equivalent circuits and mechanical analogs. It seems fair to suggest that if the simple pendulum did not exist, then, apart from other perhaps serious side effects, the study of Josephson junctions would be made considerably more difficult.

We wish to thank our colleagues at Bell Laboratories who over the years have helped us to try to understand the properties of Josephson junctions. These include J. M. Rowell, R. C. Dynes, D. E. McCumber, P. W. Anderson, L. N. Dunkleberger, A. Contaldo, A. F. Hebard, J. H. Magerlein, G. J. Dolan and R. H. Eick among others. Special thanks go to R. C. Fulton and M. H. T. Fulton for their considerable patience during the writing of this manuscript.

REFERENCES

1. I. Giaver, Phys. Rev. Lett. 5, 147 (1960).
2. B. D. Josephson, Phys. Lett. 1, 251 (1962).
3. P. W. Anderson and J. M. Rowell, Phys. Rev. Lett. 10, 230 (1963).
4. S. Shapiro, Phys. Rev. Lett. 11, 80 (1963).
5. J. E. Zimmerman and A. H. Silver, Phys. Rev. 141, 367 (1966).
6. P. W. Anderson and A. H. Dayem, Phys. Rev. Lett. 13, 195 (1964).
7. J. Clarke, Phil. Mag. 13, 115 (1966).
8. H. A. Notarys and J. E. Mercereau, Physica 55, 424 (1969).
9. P. W. Anderson in Lectures on the Many-Body Problem (Ravello, 1963), Vol. 2, E. Caianello, Ed., New York: Academic Press, 1964, p. 115.
10. B. D. Josephson, Rev. Mod. Phys. 36, 216 (1964).
11. B. D. Josephson, Adv. Phys. 14, 419 (1965).
12. P. W. Anderson, in Progress in Low Temp. Phys., Vol. 5, C. J. Gorter, Ed., Amsterdam: North Holland, 1967, p. 1.
13. B. D. Josephson, in Superconductivity, R. D. Parks, Ed., New York: Marcel Dekker, 1969, p. 423.
14. J. E. Mercereau in Superconductivity, R. D. Parks, Ed., New York: Marcel Dekker, 1969, p. 393.
15. I. O. Kulik and I. K. Yanson, Josephson Effect in High-Frequency Tunnel Structures, Nauka, Moskva, 1970: L. Solymar, Superconductive Tunneling and Applications, Wiley-Interscience, New York, 1972.

16. W. C. Stewart, Appl. Phys. Lett. 12, 277 (1968).
17. D. E. McCumber, J. Appl. Phys. 39, 3113 (1968).
18. V. Ambegaokar and B. I. Halperin, Phys. Rev. Lett. 22, 1364 (1969).
19. D. B. Sullivan and J. E. Zimmerman, Am. J. Phys. 39, 1504 (1971).
20. P. K. Hansma and G. I. Rochlin, J. Appl. Phys. 43, 4721 (1972); G. I. Rochlin and P. K. Hansma, Am. J. Phys. 41, 878 (1973).
21. C. A. Hamilton, Rev. Sci. Instr. 43, 445 (1972).
22. C. K. Bak and N. F. Pedersen, Appl. Phys. Lett. 22, 149 (1973).
23. N. R. Werthamer and S. Shapiro, Phys. Rev. 164, 523 (1967).
24. Y. Taur, J. H. Claassen, and P. L. Richards, Rev. de Phys. Appl. 9, 263 (1974).
25. L. G. Aslamazov and A. I. Larkin, JETP Lett. 9, 87 (1969).
26. See J. E. Zimmerman, Proc. Annapolis Appl. Sup. Conf. IEEE Pub. No. 72CH0682-5-TABSC, 1972, p. 544, for a review and reference.
27. e.g., P. E. Gregors-Hansen, E. Hendricks, M. T. Levinson and G. Fog Pedersen, Proc. Annapolis Appl. Sup. Conf. IEEE Pub. No. 72CH0682-5-TABSC, 1972, p. 597.
28. See J. Matisoo, Proc. Annapolis Appl. Sup. Conf. IEEE Pub. No. 72CH0682-5-TABSC, 1972, p. 555, for a review and references.
29. W. C. Scott, Appl. Phys. Lett. 17, 166 (1970).
30. W. C. Stewart, J. Appl. Phys. 46, 1505 (1976).
31. M. Urabe, as quoted in Non-Linear Differential Equations, G. Sansone and R. Conti (Pergamon, New York, 1964), pp. 278 and 302.
32. J. Clarke, W. M. Goubau and M. B. Ketchen, IEEE Trans. Mag. 11, 724 (1975).
33. A. H. Silver and J. E. Zimmerman, Phys. Rev. 157, 317 (1967).
34. A. J. Dahm, A. Denestein, T. F. Finnegan, D. N. Langenberg and D. J. Scalapino, Phys. Rev. Lett. 20, 859 (1968).
35. N. F. Pedersen, T. F. Finnegan, and D. N. Langenberg, Phys. Rev. B6, 4151 (1972).
36. N. F. Pedersen, M. R. Samuelsen and K. Saermark, J. Appl. Phys. 44, 5120 (1973).
37. J. Mygind, N. F. Pedersen and O. H. Sorensen, Appl. Phys. Lett. 29, 31 (1976).
38. T. A. Fulton and R. C. Dynes, Sol. St. Comm. 9, 1069 (1971).
39. W. H. Parker, B. N. Taylor and D. N. Langenberg, Phys. Rev. Lett. 18, 287 (1967).
40. The proceedings of a recent conference on this subject is contained in Revue de Physique Appliquée 9, 1-312 (1974). See also Chapter VI by R. Adde and G. Vernet in present volume.

41. P. L. Richards, R. Auracher and T. Van Duzer, Proc. IEEE 61, 36 (1973).
42. F. Auracher and T. Van Duzer, Proc. Annapolis Appl. Sup. Conf. IEEE Pub. No. 72CH0682-5-TABSC, 1972 , p. 603.
43. H. Kantor and F. L. Vernon, Jr., J. Appl. Phys. 43, 3174 (1972).
44. E. H. Shin and B. B. Schwartz, Phys. Rev. 152, 207 (1966).
45. F. L. Vernon, Jr. and R. J. Pedersen, J. Appl. Phys. 39, 2661 (1968).
46. S. Shapiro, private communication.
47. A. H. Dayem and C. C. Grimes, Appl. Phys. Lett. 9, 47 (1966).
48. P. K. Hansma, J. Appl. Phys. 44, 4191 (1973).
49. R. Rifkin, D. A. Vincent, B. S. Deaver, Jr., and P. K. Hansma, J. Appl. Phys. 47, 26545 (1976).
50. R. C. Jaklevic, J. Lambe, A. H. Silver and J. E. Mercereau, Phys. Rev. Lett. 12, 274 (1974); A. Th. A. M. de Waele and R. de Bruyn Ouboter, Physica 41, 225 (1969).
51. T. A. Fulton, L. N. Dunkleberger and R. C. Dynes, Phys. Rev. B6, 855 (1972).
52. J. E. Mercereau, Proc. U.S. / Japan Seminar on Low Temp. Physics (1967); A. M. Goldaman, private communication; T. A. Fulton and L. N. Dunkleberger, J. Appl. Phys. 45, 2283 (1974).
53. K. Nakajima, Y. Onodera, T. Nakamura and R. Sato, J. Appl. Phys. 45, 4095 (1974).
54. T. A. Fulton, IEEE Trans. Mag. 11, 749 (1975).
55. I. O. Kulik, JETP 24, 1307 (1967).
56. R. A. Ferrell and R. E. Prange, Phys. Rev. Lett. 10, 479 (1963).
57. C. S. Owen and D. J. Scalapino, Phys. Rev. 164, 538 (1967).
58. A. E. Gorbonosov and I. O. Kulik, JETP 33, 374 (1971).
59. A. C. Scott, Am. J. Phys. 37, 52 (1969).
60. J. R. Waldram, A. B. Pippard and J. Clarke, Phil. Trans. Roy. Soc. London, Ser. A, 268, 265 (1970).
61. K. Nakajima, T. Yamashita and Y. Onodera, J. Appl. Phys. 45, 3141 (1974).
62. K. Nakajima, Y. Sawada and Y. Onodera, J. Appl. Phys. 46, 5272 (1975).
63. T. Yamashita, L. Rinderer, K. Nakajima and Y. Onodera, J. of Low Temp. Phys. 17, 191 (1974).
64. J. M. Rowell, Phys. Rev. Lett. 11, 200 (1963).
65. A. B. Pippard, discussion remark in Rev. Mod. Phys. 36 225 (1964).
66. M. D. Fiske, Rev. Mod. Phys. 36, 221 (1964).
67. D. N. Langenberg, D. J. Scalapino, and B. N. Taylor, Proc. IEEE 54, 560 (1966).
68. R. E. Eck, D. J. Scalapino and B. N. Taylor, Phys. Rev. Lett. 13, 15 (1964).

69. I. O. Kulik, JETP Lett. 2, 84 (1965).
70. I. O. Kulik, Sov. Phys. Tech. Phys. 12, 111 (1967).
71. T. A. Fulton, R. C. Dynes and P. W. Anderson, Proc. IEEE 61, 28 (1973).
72. P. Lebwohl and M. J. Stephen, Phys. Rev. 163, 376 (1967).
73. R. M. B. Fogel, S. E. Trullinger, A. R. Bishop and J. A. Krumhansl, Phys. Rev. Lett. 36, 1411 (1976); A. C. Scott, F. Y. F. Chu and D. W. McLaughlin, Proc. IEEE 61, 1443 (1973).
74. J. T. Chen, T. F. Finnegan and D. N. Langenberg, Physica 55, 413 (1971).
75. T. A. Fulton and R. C. Dynes, Sol. St. Comm. 12, 57 (1973).

Chapter 5

SUPERCONDUCTING DEVICES FOR METROLOGY AND STANDARDS

Robert A. Kamper

National Bureau of Standards

Institute for Basic Standards

Boulder Colorado, USA

I. INTRODUCTION

This topic falls naturally into five almost independent parts:

a) Voltage standards,
b) Current comparators and measurements of ratios of current and voltage,
c) Measurements of rf power and attenuation,
d) Noise thermometry, and
e) Measurements of frequency.

These are discussed in Sections II to VI respectively. General familiarity with the properties of superconductors and the Josephson effect is assumed; these topics are discussed in detail in other parts of these proceedings.

The basic goal of metrology is to refer all measurements to a common, self-consistent set of units. The system that has been adopted for all serious scientific work is the Systéme International [1] (SI), based on the meter, kilogram, second, ampere, kelvin, mole, and for the time being, the candela. The function of standards laboratories is to realize these base units, and the other units derived from them, in a practical form that can be made available as a reference for measurements made out in the field. Some of these units cannot be realized from their definitions with sufficient accuracy for all practical needs. This makes it necessary to define "legal" standards that are consistent with the most accurate available realization of the corresponding SI units, but defined with greater precision.

The Josephson volt is such a "legal" standard. In contrast, the noise thermometers I shall describe provide an independent realization of the Kelvin scale of temperature. The other superconducting devices I shall describe are intended for use in the practical measuring process, to refer measured quantities to the basic standards.

II. VOLTAGE STANDARDS

A. The SI Volt

The electrical units may be derived from the basic mechanical units (mass, length and time) if the velocity of light is known and a value is assigned to the magnetic permeability μ_0 of a vacuum. The most accurate practical realization [2] uses a measurement of the force between two current-carrying conductors to determine the ampere, and a comparison of the AC impedance of a resistor and a capacitor of calculated capacitance to determine the ohm. The volt may then be derived using Ohm's law. Because the direct realization of the ampere is a tedious and expensive process, it is used only for the establishment and occasional checking of the other electrical quantities.

Standards of voltage and resistance are kept by all national standards laboratories for daily use, and regular international intercomparisons of these standards are organized by the Bureau International des Poids et Mesures (BIPM) to maintain the consistency of measurements made throughout the world. One benefit of this arrangement is that the results of determinations of the absolute units made in different places at different times may be combined to form a composite result of greater accuracy [2] than the individual results it represents.

The determination of the ampere is the step that limits the absolute accuracy of the electrical units. The best measurements that have been made used either a current balance [3, 4] or a Pellat electrodynamometer [5, 6]. Both these instruments are based on the same principle. A set of moving coils mounted on a balance beam is inductively coupled to a set of fixed coils connected in series with them. When a current I flows through the coils the balance beam experiences a torque $I^2 dM/d\theta$, where M is the mutual inductance of the fixed and moving coils and θ is the angular displacement of the beam. This torque is measured by restoring the balance of the beam with weights, and M is calculated from the geometry of the coils. The acceleration of gravity must also be determined at the site of the measurement to convert the weighings into absolute measurements of force. The current I flows also through a standard resistor, and the voltage across it is compared with the emf of a set of standard cells, which are used to preserve the absolute value of voltage so obtained. Combining the best recent measurements, the uncertainty in the determination of the ampere is ± 4 parts per million (standard deviation) [2].

The best determinations of the ohm have all used the "cross capacitor," which takes advantage of a theorem in electrostatics discovered by Thompson and Lampard [7] to facilitate the calculation of capacitance of an arrangement of conducting rods from its linear dimensions. The practical form of this device has a rather low capacitance (less than 10^{-12}F), so careful work with ratio transformers and impedance bridges is required to compare its capacitance with the resistance of a standard 1Ω resistor. The most recent results of a measurement of this kind were those of Cutkosky [8], who reported an uncertainty in the establishment of the ohm of ±0.06 parts per million (95% confidence). There has been some difficulty in transferring the ohm to other laboratories at this level of precision, but the uncertainty is less than one part per million.

Combining the uncertainties in the ampere and the ohm, we see that the SI Volt can be realized with an uncertainty of ±4 parts per million (standard deviation).

B. Standard Cells and the Defined Volt

Even before the advent of the Josephson effect, it was possible to compare voltages in distant laboratories with a precision of a few parts in 10^7, and international intercomparisons of voltage showed that the leading standards laboratories in the world maintained voltage standards that were stable to a few parts per million over periods of several years. All this was done with carefully tended sets of Weston cells.

The proper construction and maintenance of standard cells has been described in full detail in a monograph by Hamer [9]. The U.S. standard of voltage, which was similar to those maintained by other national laboratories, consisted of 44 Weston cells immersed in an oil bath to provide an environment with constant temperature and minimum mechanical disturbance. The mean of the emf of the set of 44 cells was taken as the reference voltage. The cells were kept under constant surveillance, and if the emf of one of them were to drift significantly away from the mean the cell would be replaced. In addition, the gyromagnetic ratio of the proton was measured periodically and referred to the standards of voltage and resistance. Its constancy was regarded as good evidence of the stability of both these standards. Intercomparisons of voltage standards between distant laboratories was accomplished with sets of four cells in temperature-controlled traveling containers [10]. By these means voltage standards throughout the world were kept consistent with each other more precisely than the absolute volt could be determined. The voltage standards maintained at NBS and at BIPM drifted apart at a rate generally less than one part per million per year. However, the desirability of a voltage standard based on some reproducible physical process that would not drift to any significant extent at all was clearly recognized.

C. The Josephson Effect and e/h

Among the earliest manifestations of the Josephson effect to be observed in the laboratory were the constant voltage steps induced in the direct current-voltage characteristic of a Josephson junction by microwave radiation [11]. It was quickly recognized that these steps are a result of the oscillation of the current through a Josephson junction under the influence of an applied voltage. The current oscillates at a frequency

$$f = 2eV/h \qquad (1)$$

where V is the electrochemical potential difference across the junction, e is the electron charge, and h is Planck's constant. If the DC power supply has a reasonably high impedance, the oscillation through the junction phase-locks to the applied microwave signal (or to its harmonics generated by the non-linear response of the junction), holding the voltage constant over a range of current that may be as wide as several mA, and is generally significantly smaller than the critical current of the junction. This creates a series of constant voltage steps at integer multiples of the voltage V of Eq. (1), where f is now taken to be the frequency of the applied microwave radiation. If $f \approx 10$ GHz (X-band), then V is approximately $20\,\mu V$. It is possible to generate well defined steps extending to a voltage of over 10 mV, and relatively easy to exceed 1 mV, as discussed in Section II F.

Inspection of Eq. (1) shows that an accurate measurement of the separation of the Josephson steps generated by a microwave signal of known frequency would yield a direct measurement of the fundamental constant e/h, and this stimulated the first thorough experimental and theoretical exploration of this phenomenon, pioneered by the group at the University of Pennsylvania. The major experimental problems were: the optimization of the steps, in both flatness and number, and the measurement of a voltage of the order of a millivolt at a temperature of a few kelvin with the full available accuracy of a warm voltage standard. The outstanding theoretical problem was whether the appealing simplicity of Eq. 1 is exact or merely a first-order approximation. These problems are common to both the measurement of e/h and the establishment of voltage standards, and work is still continuing on their solution. I shall discuss them in detail later.

By 1966 the group at the University of Pennsylvania had measured e/h with an accuracy of a few parts per million [12]. This was sufficient to enable them to derive a new value for the fine structure constant independent of the assumptions of quantum electrodynamics, which enabled them to discover some minor errors in that theory and adjust the values of all the fundamental constants into a new, self-consistent set [2]. They also recognized that the accuracy of their measurement was limited by the transfer of the volt to their laboratory from the national standard, and raised the possibility of turning their experiment around to use the Josephson effect, with a defined value of e/h, to maintain the standard of voltage [13]. By 1970 the measurement had been repeated at NBS (USA) [14], NPL (England) [15], PTB (West Germany) [16], and NML (Australia) [17]. After a careful intercomparison of the respective national voltage standards, the results of all these measurements were found to be in agreement [10] within the estimated experimental errors. In July 1972 the NBS volt was redefined [18] by adopting the value $4.835\,934\,20 \times 10^{14}$ Hz/V for 2e/h. In October 1972 the Comité Consultatif d'Electricité (CCE) recommended adoption of the value

$$2e/h = 4.835\,944 \times 10^{14} \text{ Hz/V} \tag{2}$$

to define the international volt. Those laboratories that possessed the necessary equipment adopted Eq. (2) as the working definition [19-22]. Standard cells are still used as transfer standards for daily calibration and transfer of the volt to users out in the field.

D. Practical Josephson Voltage Standards

The essential elements of a Josephson voltage standard are: a stable microwave signal generator with means to measure its frequency; the Josephson junction for frequency to voltage conversion; and a voltage divider and null detector capable of comparing a few millivolts in liquid helium with the emf of a set of standard cells in the warm laboratory. Of these three, the first is straightforward, the second has been the subject of considerable research, and the third was a challenge that has stimulated some ingenious developments that I will discuss in Section III. The most detailed exposition of a complete Josephson volt system to be found in the literature is the classic paper by Finnegan, Denenstein and Langenberg [23]. There have been improvements since it was written, which I will discuss, but it still represents a perfectly adequate standard.

E. The Microwave Signal Source

The requirements for the signal generator are: the frequency must be stable and measurable to one part in 10^8, and the power level must be stable to 1%. The power level required depends upon the matching of the Josephson junction to the transmission line. It is usual to use a klystron oscillator stabilized by a commercial stabilizer that uses a quartz crystal oscillator for reference frequency. An oil bath is usually used to stabilize the temperature of the klystron. This reduces the fluctuations in power level to an acceptable amount. Frequencies from 9 GHz to 70 GHz have been used. For the measurement of frequency, sufficiently accurate counters are available commercially. Their internal quartz crystal reference oscillators are calibrated against broadcast standard frequency signals.

A Josephson voltage standard makes only modest demands on the art of frequency metrology, that can be satisfied with commercially available equipment.

F. The Josephson Junction

For practical purposes a Josephson junction may be functionally represented by a "pure" supercurrent $I_c \sin\Phi$ (where Φ is the quantum mechanical phase difference) shunted by a displacement current associated with the capacitance C of the junction and the quasi-particle current. The quasi-particle current is a combination of leakage and tunneling currents. In a point contact junction it may be approximated linearly by a constant conductance G. In a tunnel junction between evaporated films its behavior is more complicated because of the influence of the energy gap. However, practical Josephson volt standards usually operate at a voltage of several millivolts, which is greater than the energy gaps of the lead and tin films that are commonly used. The parameters I_c, C and G are the characteristics of the junction that are more or less under control, and there has been some effort to find their optimum values.

The simplest theory assumes a microwave source of voltage V_μ at angular frequency ω_μ, with impedance much lower than that of the junction. The shunt capacitance and conductance would then have no influence, and the amplitude of the nth step would be

$$I_n = I_c J_n \ (2eV_\mu/\hbar\omega_\mu) \tag{3}$$

where J_n is the nth order Bessel function of the first kind. In practice it is usually found that the higher-order steps are smaller, relative to the zero-order step, than this formula would predict. Russer [24], among others, found that the opposite extreme assumption, that the impedance of the microwave source is very high compared to that of the junction, gives a much better theoretical description of what is usually observed. He used an analog computer to investigate the influence of varying the values of G and I_c, ignoring the displacement current through C. He found that the behavior of the junction depends on the value of the parameter

$$\xi = n\omega_\mu G/2eI_c . \tag{4}$$

When $\xi = 1$ the amplitudes of the high-order steps approximately follow Eq.(3). For smaller values of ξ they become smaller relative to the zero-order step, and their maxima and minima occur at closer intervals of microwave power level, just as is commonly observed.

The conspicuous feature of the steps that this theory does not account for is the high-voltage limit. The fullest discussion of this in the literature is given by McDonald et al [25]. The observation to be explained is that the steps fade out above a limiting voltage, of the order of a few millivolts, that varies widely from one junction to another. The frequency dependence of the Josephson effect itself has been studied by Werthamer [26]. It cannot account for the low limit that is usually found, which is probably set by shot noise. The argument of McDonald et al is that when a current I flows through the junction there is a shot noise current with rms value i given by

$$i = eIB/\pi \tag{5}$$

where B is an effective bandwidth that is probably set by the plasma frequency of the junction (of the order of 10^{10} to 10^{12} Hz). The effect of noise currents on the steps has been studied by Kose and Sullivan [27], who found that the amplitude of a step is reduced by approximately an amount i. Therefore if i exceeds I_n the nth step disappears. This sets a limit which appears to correspond approximately with what is observed. Both the super-current and the quasi-particle current contribute to the shot noise, but the displacement current does not. Detailed calculations of the variation of this high-voltage limit with the adjustable parameters of the junction have not been reported, but the optimum value of ξ [Eq.(4)] is probably small ($\ll 1$) so Eq.(3) does not describe the optimum situation.

The importance of the capacitance of the junction appears to be greatest when we consider the coupling of microwave energy from its source to the junction. A large value of C causes an intolerable impedance mismatch. This can be avoided either by keeping C small by using a point contact, or by choosing the linear dimensions of the junction to make it resonate at the applied microwave frequency. This enables tunnel junctions between thin evaporated films to be used. These tunnel junctions will support electro-magnetic waves with a very low phase velocity, because the effective spacing for the electric field is the thickness of the dielectric barrier ($\sim 10^{-9}$m) while the effective spacing for the magnetic field is extended by the penetration depth into the superconductors ($\sim 10^{-7}$m). A typical junction with dimensions of the order of 1 mm in the plane of the films resonates at a frequency of the order of 10^{10} Hz.

These theoretical considerations have given some guidance to the development of acceptable Josephson junctions for voltage standards. However, success was achieved mainly by a great deal of empirical trial and error. Both point contacts and evaporated film tunnel junctions are in use today.

The point contacts are usually niobium with one electrode chemically etched or electropolished to a fine point that touches a flat, clean niobium surface. The only parameters susceptible of adjustment are the contact pressure and the degree to which manipulation of the point breaks through the inevitable oxide layer on the surface of the niobium. Adjustment is accomplished by various combinations of steady hands, levers, and differential screws, and success depends upon the operator developing the necessary skill. There has been some progress in developing permanent pre-set point contacts. The critical current is usually set at a few mA.

Point contacts can perform almost as well as the best evaporated film junctions, and have the advantage of easier impedance matching to the microwave source. They are used in connection with the maintenance of the volt at PTB (West Germany) [19], NML (Australia) [20] and NPL (England) [21].

The preferred material for evaporated films has been lead, with a dielectric barrier of lead oxide. The junctions used at NBS [18] are formed at the perpendicular intersection of two lead strips, 1 mm and 0.3 mm wide respectively, and 150 nm thick. The oxide barrier is formed, after vacuum deposition of the first film, by exposure to oxygen at 1/2 atmosphere pressure and 40° C temperature before restoration of the vacuum and deposition of the second film. Junctions are chosen with normal-state resistance of about 0.1 Ω and critical current of 9 mA. Two junctions are operated in series to create usable steps at 10 mV. "Usable" is taken to mean greater than 50 μA in amplitude, to accommodate drift and fluctuations in the current supply. This performance can also be attained with a single junction with the overlapping films in-line instead of perpendicular to one another.

The substrate holding the junctions is placed in the H-plane of a length of regular rectangular X-band waveguide. The microwave source is tuned to the resonant frequency of the junctions (which must be a matched pair), and impedance matching is assisted by adjusting a movable short that terminates the guide. However, impedance matching is still a severe problem and microwave power levels as high as 500 mW are occasionally required. Recently Finnegan et al [28] have been experimenting with a promising arrangement of striplines, both for impedance matching the junction to the microwave source and for filtering the microwave power out of the DC circuits, where it might interfere with the null detectors. They report the observation of steps of amplitude 100 μA at 10 mV with a single in-line tunnel junction drawing 75 mW from the microwave source through a stripline structure.

G. Shielding, Filtering, and Tempering

Some special precautions must be taken because of the low signal levels at which all Josephson devices operate.

The Josephson junction must be shielded from external (radio) interference. Interfering currents greater than about 10^{-6} A cannot be tolerated. Since the long leads associated with cryogenic equipment function as efficient antennas, it is usual to operate a Josephson volt standard in an RF shielded room. These are available commercially, and usually have a double layer of wire mesh giving attenuation over 100 dB in the broadcast frequency ranges, which takes care of the prime nuisance from local radio and television transmitters. Interference at very low frequencies from electric motors, etc., is usually screened out by enclosing the cryostat in a nest of two or three cans made of sheet metal with high magnetic permeability. It is also helpful to have lead shields around individual parts of the system that operate in liquid helium.

The DC leads to the null detector must be filtered to avoid coupling out an excessive amount of the microwave power that drives the Josephson junction. SQUID null detectors cannot tolerate microwave interference at power levels over about 10^{-10}W, and sometimes 0.5 W must be used to generate the high-order steps. Fortunately, the required 100 dB of attenuation is not too difficult to attain at a single microwave frequency. The SQUID is usually enclosed in a conducting can with low-pass filters on all DC leads entering it.

Precautions must be taken to stabilize the thermoelectric emf on the DC measuring leads that pass out of the liquid helium bath. This emf is caused by variations in composition in nominally uniform wire. It is typically of the order of 10^{-6}V for strain-free copper wire, but a kink in a region of large temperature gradient can generate an emf of several tens of microvolts. This emf can be averaged out by reversing the polarity of the leads for successive measurements, but it must also be stabilized against variations in the position of the thermal gradient. This is done by encasing the leads in a combination of copper and stainless steel tubes.

H. Theoretical Uncertainty

In this section I shall review briefly the theoretical and experimental evidence that establishes the Josephson frequency-voltage relationship [Eq.(1)] to be sufficiently accurate and universal for the definition of a voltage standard.

The state of knowledge in 1971 was summarized very well by Finnegan *et al* [23]. There had been theoretical discussion of possible corrections to Eq. 1 to take account of collective many-particle interactions in metals and a perturbation due to the non-equilibrium interaction of the Josephson junction with the radiation field. Both effects would be of the order of one part in 10^8 or smaller, and the existence of both has been refuted by counter-arguments. The situation is simplified considerably by the fact that the measurements that are made really do measure electrochemical potentials, by balancing them in circuits and finding null conditions in which no current flows. Thus we are on firm ground if we take eV in Eq. (1) to be a single quantity representing the electrochemical potential of electrons exchanged between the Josephson junction and the chemical cells used for reference. There is then no need to worry about internal fields in the various metals connected in the circuit, since these must be compensated by contact potentials.

Many experiments referred to by Finnegan *et al* [23] have investigated the influence of different materials, type of junction, temperature, magnetic field, step number, frequency, and power. All have been shown to affect the frequency-to-voltage conversion factor by less than the experimental uncertainty of one or two parts per million. Their own measurements showed that the influence of temperature, magnetic field, and step number is less than a few parts in 10^8. Two experiments of somewhat higher precision deserve special mention. Clarke [29] found that junctions with electrodes of lead, tin, and indium gave results differing by less than one part in 10^8, and Bracken and Hamilton [30] found junctions with lead and tin electrodes to give equal results within five parts in 10^9.

There is little doubt that Eq. (1) can be taken to be exact to at least one part in 10^8 for practical purposes.

I. Present Activities

The national voltage standards of the U.S., West Germany, and Australia now depend upon the Josephson effect. England, Canada, Japan, and BIPM (in Paris) possess the necessary equipment and use it to monitor the variations in the emf of sets of standard cells in preparation for adopting a Josephson volt. The national standards laboratories in all these countries are capable of comparing the emf of standard cells with the Josephson steps with accuracy of the order of one part in 10^7. Current activities are directed towards improvements such as better matching of the Josephson junctions to the microwave source and all-cryogenic voltage dividers (see Section III). At NBS a portable Josephson transfer standard is being developed. It is designed for simplicity and convenience rather than the highest accuracy. One part per million is considered adequate for this purpose. There is not much effort to increase the accuracy of national standards, because not much need can be seen at present for a super-precise volt that is related to the basic SI units by measurements of lower accuracy.

III. CURRENT COMPARATORS AND RATIO MEASUREMENTS

Measuring any electrical quantity that does not happen to coincide in value with one of the primary standards requires the measurement of a dimensionless ratio of voltages or currents, with the full accuracy required of the original measurement. The advent of the Josephson volt stimulated several improvements and additions in the technique of ratio measurements, among them being the superconducting current comparator. Before describing the new superconducting devices, I shall set them into context by reviewing briefly some of the principles previously used to make the most accurate measurements. I shall include only techniques capable of accuracy better than one part per million.

A. Resistive Networks

If the ratio of two resistors can be established very accurately, then an equal ratio of voltage can obviously be set up by connecting them in series to a common current supply. However, it does require ingenuity to apply this simple principle with errors less than one part per million. A voltage divider ratio is usually set up by trimming various combinations of resistors to have equal resistance before connecting them in the network. It is difficult to do this with true four-terminal resistors, so special precautions must be taken to stabilize and compensate for the resistance of the connections. All resistors vary a small amount with changes in temperature, so calibrating and using resistors at different current levels is a possible source of error. Leakage currents must be kept very low. Thermoelectric emf must be cancelled by always taking the average of measurements with reversed current, but does not contribute error provided that thermal gradients are sufficiently stable during measurements. This often calls for long settling times while the system reaches a steady state. This last point applies also to conventional galvanometers when they are used for null detectors. Thermal noise ultimately limits the precision with which measurements can be made. In the following paragraphs I shall describe two methods that have been used to establish very accurate ratios of resistors, namely cascaded-interchange and series-parallel exchange. Both of these have been discussed in detail by Finnegan and Denenstein [31, 32], in connection with their application to the Josephson voltage standard.

In its simplest form, the cascaded-interchange technique [31] requires two identical "volt boxes", each consisting of a string of $n + 1$ resistors in the nominal ratio $1:1:2:4:8 \ldots 2^{n-1}$, connected in series to a stable current supply (Fig. 1). First, the smallest resistors in both boxes are connected into a configuration similar to a Wheatstone bridge. The resistors are trimmed until the bridge is balanced and remains so when two resistors are interchanged. The four resistors are then of equal value, and the process is repeated to trim the next larger resistor in each box to be equal to the sum of the two smallest ones, and so on until all n resistors are trimmed to be accurately in simple proportion. The potential drop across one of the smallest resistors in the box is then equal to

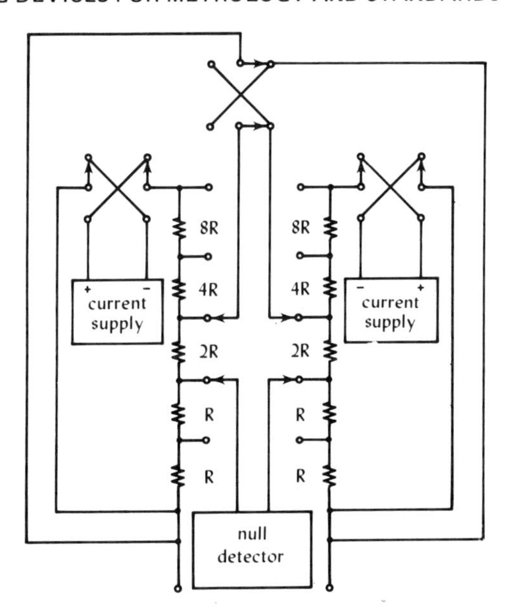

Fig. 1. 16:1 Cascaded-interchange voltage comparator, in second cycle of calibration.

2^{-n} of the potential drop across the whole box, as accurately as the trimming of the resistors allows. Variations of this technique can generate integer ratios other than those in the simple binary sequence.

The great advantage of the cascaded-interchange technique is that the resistors can be calibrated at the same level of current as they carry during the measurement. The resistance of the interconnections of the Wheatstone bridge circuits are a source of error. However, the two volt boxes are driven by independent current supplies, and this balances the arrangement, so current flows through the interconnections only when the balance is not perfect. Thus the errors due to the resistance of the interconnection are reduced to the degree that the power supplies can be balanced. This effectively reduces them to second-order errors.

The series-parallel exchange technique (Fig. 2) uses 2n nominally equal resistors. It depends on the principle that the ratio of the resistance of a set of n resistors connected in series to that of the same set in parallel is n^2, to first order in the differences between the individual resistors (these appear in a second-order correction). If a divider is constructed out of two sets of n resistors, one in series and one in parallel, then the average resistance ratio of the two permutations that differ by the interchange of the two sets of resistors is n^2.

The advantage of this technique is the elimination of the first-order effect of differences between the individual resistors. Thus if an accuracy of one part in 10^8 is desired, it is sufficient to trim the resistors to be equal to one part in 10^4. The main disadvantage is that the resistors

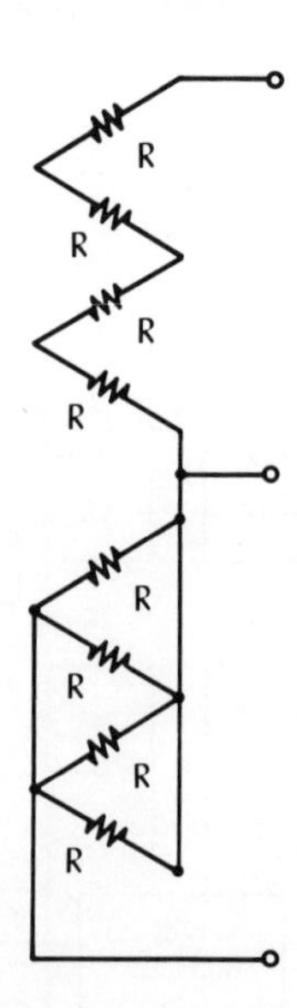

Fig. 2. Primitive form of series-parallel voltage divider. Switches and fan resistors are not shown.

carry different currents in the series and parallel configurations, so care must be taken to avoid errors due to variations in self-heating. Also, the resistors connected in series cannot avoid being used in a two-terminal configuration, so great care must be taken with the distribution of current in the connections. The connections themselves are usually made in a tetrahedral shape, and auxiliary "fan" resistors are used to distribute the current properly in the parallel resistors.

B. Inductive Devices

The great advantage of inductive ratio devices is that it is quite easy to wind inductors that are very precisely equal to each other, thereby eliminating the need to trim the values of components designed for stability. The heart of one of these inductors is a toroidal core of very high relative permeability (greater than 10^5). Wires for several sets of windings are twisted together into a cable, which is wound onto the toroidal core. This is sufficient to produce a set of coupled inductors that are equal to one another to a fraction of a part per million. There are two types of practical ratio devices that use this principle: the inductive voltage divider and the current comparator.

In the inductive voltage divider, or ratio transformer, the sets of windings are all connected in series to make a tapped auto-transformer [33, 34] (Fig. 3). When the ends of the chain of windings are connected to an audio frequency power source, equal voltages appear at the pairs of taps in phase with one another. Because of the impedance converting

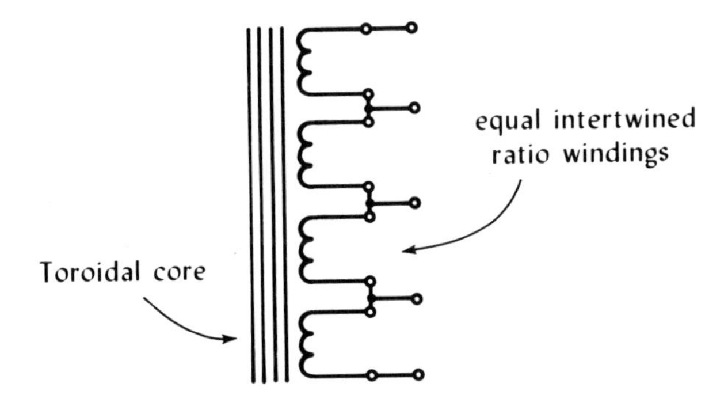

Fig. 3. Basic ratio transformer.

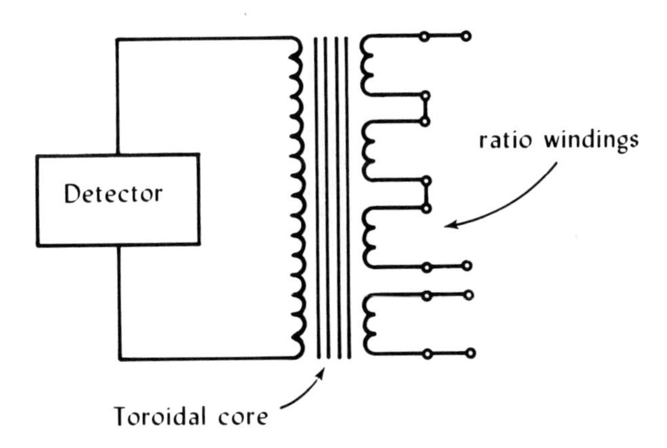

Fig. 4. Basic AC current comparator.

property of a transformer, a series of ratio transformers may be connected in cascade to subdivide a reference voltage over many decades without serious problems with loading. At frequencies above about 1 kHz a combination of core loss, leakage inductance, and inter-turn capacitance begins to degrade the accuracy of division. These effects may be reduced by a more complex design with guarded windings [35].

The current comparator has been developed into a fine instrument by Kusters [36]. It consists basically of a set of equal ratio windings, as described above, and some means of sensing the magnetic flux in the core. The instrument is used by passing currents through two different sets of windings so that their contributions to the flux in the core cancel, indicating an exact balance of ampere-turns. The ratio of currents is then the reciprocal of the ratio of turns.

The detection of the magnetic flux in the toroidal core is straightforward when alternating currents are being compared. All that is required is an extra detection winding. In order to avoid errors due to unbalance in the leakage inductance and inter-turn capacitance, this detection winding is wound close to the core and is protected by both magnetic and electrostatic shields from local interaction with near parts of the ratio windings (Fig. 4). For comparing direct currents, modulation is used. The core is split into two identical toroids with identical modulation windings connected in series opposition (Fig. 5). These are driven at an audio frequency to sufficient amplitude to saturate the magnetization

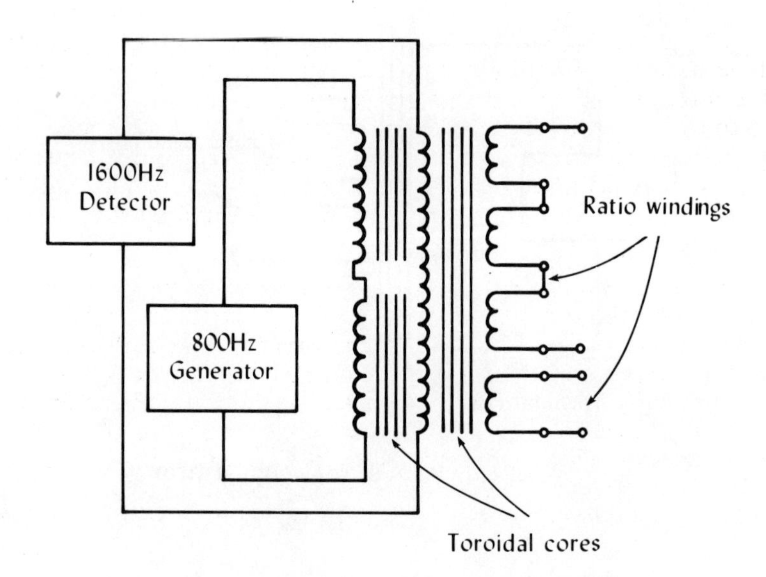

Fig. 5. Basic DC current comparator.

of the half cores at the peaks of every cycle. The two half cores share common detection and ratio windings. When the currents in the ratio windings are balanced, there is no signal at the detection winding. When an unbalance leaves a resultant magnetic flux in the cores, they are driven by the modulation around different parts of their hysteresis loops, and a signal at double the modulation frequency appears at the detection winding. Kusters and his associates have developed this principle into a very versatile and very precise measuring system [37].

The advent of the Josephson volt challenged the performance of these systems with a demand for ratio measurements accurate to a few parts in 10^8 to be made at very low voltage levels and in the face of the thermal emf associated with conductors running from room temperature into liquid helium. This stimulated the development of several devices that took advantage of the properties of superconductors for improvements in performance.

C. Cold Resistive Dividers

The technique and advantages of operating the voltage divider and null detector of a Josephson voltage standard in the same liquid helium bath as the Josephson junction were first explored and described comprehensively by Sullivan [38], who demonstrated a 100:1 series-parallel divider as an example.

The first and most obvious advantage is the elimination of resistance from the contacts, connections, and switches, and the elimination with it of the need for elaborate networks to compensate for this unwanted resistance. Also, placing all the components that carry signals at low levels, together with their interconnections, in a bath of superfluid helium virtually eliminates the nuisance of thermoelectric emf, except in the comparison (at the level of approximately one volt) of the emf at the high side of the voltage divider with the standard cell. Leakage conductances also tend to be lower at liquid helium temperature, but still need careful attention in the parts of the circuit outside the cryostat. A properly chosen SQUID null detector enables measurements to be made quickly (no thermal delays in settling) with a precision limited by thermal noise in the cold resistors.

The components of a cold voltage divider requiring the most development were the resistors themselves. First, it was necessary to find an alloy with resistivity insensitive to changes in temperature at the operating temperature of the Josephson junction (usually 2K). Evanohm and other alloys used for stable resistors at room temperature do not perform very well in this condition. What is needed is an alloy with a high resistivity caused by impurity and defect scattering, but with a tolerably small Kondo effect. This requires the elimination of magnetic impurities and is aided by using metals in which the Kondo effect is intrinsically weak.

Sullivan chose a Silicon bronze [38], consisting of 96% copper, 3% silicon, and 1% zinc. It was prepared from components with a nominal purity of 99.999% to avoid magnetic impurities. In the cold drawn (unannealed) state it has a resistivity of 18.3 $\mu\Omega$-cm at 2K, which varies by approximately -2 ppm per K at that temperature. Other workers have found other bronzes that are almost as good as this one. Gallop and Petley [21] use a commercial phosphor bronze consisting nominally of 95% copper, 5% tin, and 0.2% phosphorus. This has a temperature coefficient of resistance (TCR) of -4 ppm/K at 4.2 K, which is attributed mainly to the Kondo effect associated with a trace (80 ppm) of iron in the alloy. Warnecke and Kose [39] measured the properties of an alloy consisting of 94% copper and 6% germanium, made from 99.999% pure components. Its TCR is -2.5 ppm/K at 4.2K and -6 ppm/K at 2K. More recently, the same group [40] found a commercial aluminum alloy consisting nominally of 95% aluminum and 5% magnesium. Although it has about 3000 ppm of magnetic impurities, its TCR is less than 0.2 ppm/K between 2K and 4K, because aluminum has a much weaker Kondo effect than copper. This alloy has a residual resistivity of 3.6 $\mu\Omega$ cm, somewhat lower than that of the bronzes mentioned above.

The terminations of these resistors call for careful attention to avoid a variation of resistance with current due to perturbation of the superconducting/normal interface. Sullivan found it sufficient to tin the ends of the silicon bronze resistance wires with soft solder (which is a superconductor), making sure that the tinned layer had clean edges. Harvey and Collins [41] improved the termination by silver soldering copper cups to the ends of the resistance wires. These were then filled with soft solder to increase and define the area of superconducting/normal interface.

These wire-wound resistors can be trimmed by abrading away small amounts of metal (and then replacing the insulation with thin varnish). This must be done when they are warm, so several iterations may be necessary to achieve the desired ratio in the liquid helium bath. It is common experience to find that the resistance measured in the liquid helium bath changes by a few parts in 10^5 on cycling the temperature of a resistor up to room temperature and back. The cause has not yet been traced, and it is an operational nuisance. Sullivan's divider was provided with leads to check the values of all resistors in their operating condition. These leads were broken by a connector that could be disconnected during the actual measurements to avoid the introduction of warm leakage paths into the measuring circuit.

The stability of these cold resistors against changes in current is generally good, because of the efficiency of heat transfer into liquid helium. Harvey and Collins [41] measured the load coefficient of a 10 Ω silicon bronze resistor, and found it to be -1.3×10^{-5}/W at a temperature of 2K. Variations in the temperature of the helium bath from other causes may be cancelled by a servo-controlled heater. Attaining a stability of 10^{-3} K is quite straightforward.

At the levels of current encountered in a Josephson voltage standard, superconducting switches do not require particularly careful design. A wiping contact is best,preferably with one electrode of a soft metal such as an alloy of lead. Currents of several amperes may be switched with no detectable contact resistance.

The null detector described by Sullivan uses a multi-loop RF-biased SQUID devised by Zimmerman [42]. This device is particularly convenient for this purpose because of its sensitivity (the noise level is about 10^{-13} AHz$^{-1/2}$), and physical shape. It is a cylinder approximately 1.5 cm diameter on which a coupling coil carrying the current to be detected may be wound directly. Sullivan used a 600-turn coil to get a noise level of 5×10^{-12} AHz$^{-1/2}$. The SQUID must be housed in a superconducting can with microwave filters on the incoming leads to isolate it from external noise and interference. The settling time of the whole system is a fraction of a second, i.e., it is not limited by thermal settling.

Cold resistive dividers are used in new Josephson voltage standards being developed at NBS [43], NML [44], PTB [40] and NPL [21]. Some of these are series-parallel networks, while others are simple chains of resistors calibrated by ratio transformers or current comparators. Recently, Kose (private communication) reported a very compact resistive divider in the form of a serpentine strip of thin film on an insulating substrate. It will be interesting to learn further details of this device.

D. Superconducting Inductive Current Comparators

The principle of the superconducting current comparator was devised by Harvey [45], and realized with various topological variations at several laboratories. In its simplest form it consists of a bundle of insulated wires encased in a superconducting tube. Because of the Meissner effect, a current flowing along any of the wires in the bundle induces an equal and opposite current on the inside wall of the tube. This current must return along the outside of the tube, and at a reasonable distance from the ends the distribution of current on the outside is independent, to a very high degree of approximation, of the position of the wire that is the original source. Thus if equal and opposite currents flow through two wires in the bundle, their contributions to the magnetic field outside the superconducting tube exactly cancel, and this cancellation may be readily detected with a flux detector such as a SQUID. Currents in simple integer ratios may be balanced in the same way by connecting groups of wires in the bundle in series, with some precaution to shield the SQUID detector from the magnetic influence of the interconnections. With a good design and some care in eliminating pinholes in the superconducting shields, simple integer ratios of currents may be determined with systematic error less than one part in 10^9, and a noise level below 10^{-9} ampere-turns.

In Harvey's realization of the principle [45], the superconducting tube containing the bundle of wires for the ratio windings is wound on a toroidal mumetal core to form the primary of a 50:1 transformer. The secondary is a single toroidal turn made of lead, that encloses the device completely and acts also as the outer shielding can. It is inductively coupled to the simple SQUID detector. The ends of the primary windings are taken outside the toroid through a long superconducting tube. Figure 6 shows the arrangement of a refined version of this device [46].

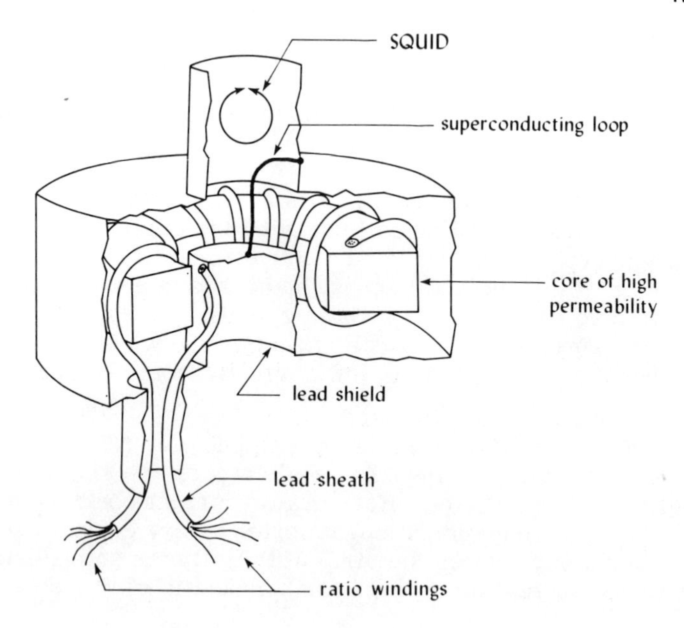

Fig. 6. Harvey superconducting current comparator.

Sullivan and Dziuba [47] simplified the design of the superconducting current comparator by adapting it to Zimmerman's multi-loop SQUID [42]. They simply wound the superconducting tube containing the ratio windings directly onto the outside of the SQUID, thereby eliminating the toroidal transformer. It was still necessary to shield the whole device in a superconducting can, because variations in the external magnetic field would cause errors in the measurement of current ratios if the SQUID were exposed to them.

Harvey's basic arrangement of ratio windings in a common superconducting tube is somewhat cumbersome to use when high ratios are desired, because all interconnections of the basic sets of windings (which must have equal numbers of turns) are made outside the shields. Sullivan and Dziuba [47] evolved a more convenient design, which uses but a single turn of superconducting shielding containing ratio windings with arbitrary numbers of turns (Fig. 7). The form of the shield was described by its inventors as resembling a snake swallowing its own tail. The ends of the windings are taken out through long nested superconducting side tubes, one for each layer of shield that is penetrated. The most developed form of this device [48] has a binary set of windings with 14 x (1, 1, 1, 2, 4, 8, 16, 32, 32, 64) turns respectively. This enables any simple integer ratio of current up to 160:1 to be measured.

Another current comparator developed independently by Grohmann et al. [49, 50], is topologically similar to Harvey's but turned inside out so as to resemble Kuster's ac current comparator (see Section IIIB). It is

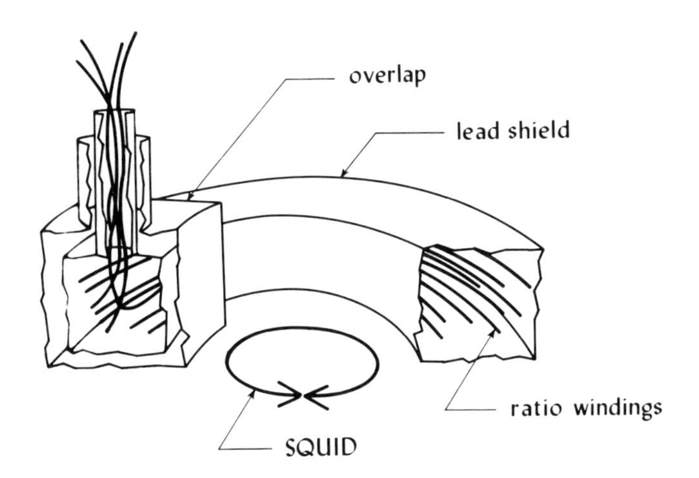

Fig. 7. Sullivan and Dziuba superconducting current comparator.

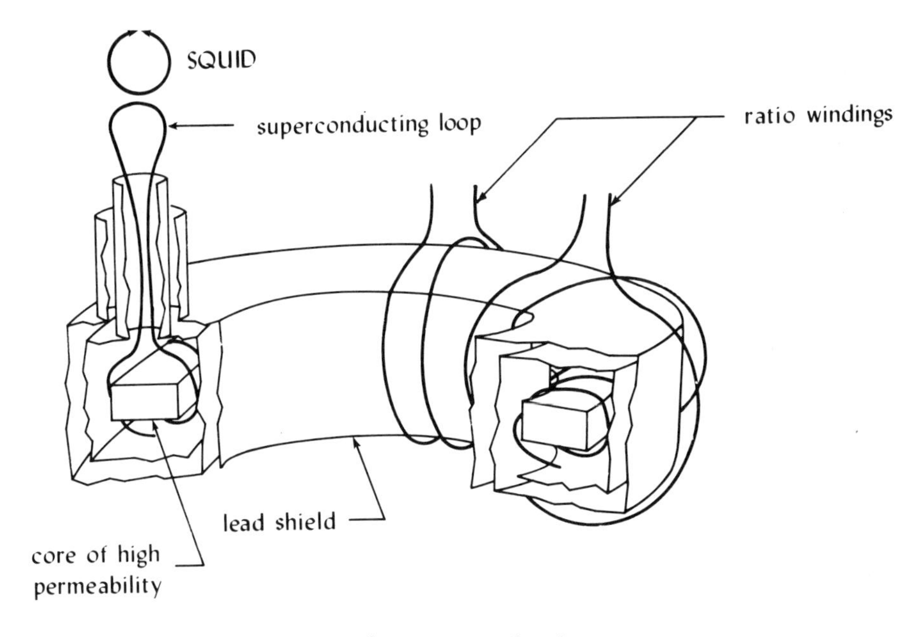

Fig. 8. Grohmann _et al._ superconducting current comparator.

shown in Fig. 8. The superconducting tube of Harvey's device has been rolled back on itself to become a hollow toroid that is not completely closed, but is left open through some labyrinthine folds that permit the entry of magnetic flux only if it has the right toroidal symmetry. The ratio windings are then simply wound on the outside of the shield, and the SQUID null detector is inductively coupled to a secondary winding inside it.

All these variants of the superconducting current comparator perform similarly. The choice between them is largely a matter of taste.

At the time of writing, there are working voltage standard systems under test at NML [44] and NBS [51], both of which depend upon superconducting current comparators to calibrate the voltage dividers. Compared to their predecessors, they are simple and compact systems.

IV. MEASUREMENTS OF RF POWER AND ATTENUATION

A. Some General Remarks on RF and Microwave Measurements

I will confine this discussion to transmission line systems, comprising various active or passive components connected together by uniform transmission lines. These last could be coaxial lines, strip lines, or singly connected waveguides. Until quite recently, it was almost universal practice to choose transmission lines that would support only a single mode of propagation at the operating frequency. The objectives of the measurements to be made are to determine the phase and amplitude of the electromagnetic waves travelling in either direction along the transmission lines, and the effect on these quantities of interaction with the components at the ends of the lines. I shall not discuss measurements of frequency in this section.

In general, RF measurements do not attempt to attain the degree of accuracy that is possible for DC measurements. The basic reason for this is that connectors and bends in the transmission lines introduce small (of the order of 1%) reflections that can vary during normal use of a system, so with very few exceptions practical systems are designed to tolerate errors of comparable magnitude. Measurements with uncertainty less than 0.1% are rarely called for.

At this level of accuracy, it is comparatively straightforward to make an independent realization of the ohm, in the form of a uniform coaxial transmission line whose characteristic impedance Z_0 may be calculated from the radii of the inner and outer conductors and the permittivity of the dielectric (usually air) between them. The impedance Z at the port of a component connected to this line is referred to Z_0 by measuring the complex reflection coefficient Γ, and using

$$Z = Z_0 (1+\Gamma)/(1-\Gamma) \quad . \tag{6}$$

When the characteristic impedance of the line is known, the amplitudes of electromagnetic waves can be referred to the SI units by measuring the power absorbed in an approximately matched (i.e., reflectionless) termination of a transmission line. The device commonly used is a bolometer, with which the heating effect of the absorbed wave is compared with that of a direct (or low frequency) current flowing through the same resistive element, thereby converting the measurement to one of DC (or low frequency) current and voltage. Corrections must be made for: residual reflections; absorption of some RF power in parts of the bolometer other than the intended resistive element; and differences between the spatial distributions of the heating effects of the DC and RF currents. The last two are evaluated by calorimetry.

For some purposes it is sufficient to measure the relative amplitudes of two waves, and this can be done somewhat more accurately than the measurement of absolute power. The reference standard for this is usually a calculable attenuator. Various principles are used for these devices. One is the waveguide beyond cutoff, in which the wave is propagated between a launching coil and a receiving coil through a section of circular waveguide whose cutoff frequency is above the frequency of the wave. The attenuation then depends exponentially on the distance between the two coils. If it is expressed in decibels, this is then converted to a linear dependence that can be calculated from the dimensions of the waveguide. This device calls for fine machining to attain the desired accuracy, and is used mainly for frequencies below 1 GHz.

Another principle is applied in the rotary vane attenuator, in which one of the two degenerate modes of propagation in a section of circular waveguide is suppressed by a resistive vane in the plane of the electric field. Rotating this section about its axis couples a calculable proportion of a wave of fixed polarization into the "dead" mode, where it dies also.

The measurement of phase angle requires more trouble and expense than almost any other quantity. The phase angle between the signal of interest and a coherent reference signal must be measured. This can be done by allowing the two signals to combine in an adjustable bridge and seeking a balance at which they cancel. A more elaborate way that is commonly used in automatic systems is to use mixers and a local oscillator to convert both signals down to an intermediate frequency that is low enough to permit the use of phase-sensitive detectors to compare phases directly. The most recent and sophisticated method employs the mathematics of scattering theory, recognizing that measurements of the amplitude of the signals at four properly chosen points in a transmission line network supplied with two coherent signals contain enough information to determine the relative phase and amplitude of the two signals. A simple calibration technique and a small computer to perform the necessary matrix manipulations comprise the remainder of a practical measuring system. This is known as a 6-port system [52, 53].

The principles I have sketched here are applied to practical measurements in a very wide variety of systems covering a wide range of accuracy, versatility, convenience, sophistication, and expense. The classic texts describing the beginning of this field are the MIT Radiation Laboratory

Series [54] and Ginzton's book[55]. Most of the more recent advances can usually be found in the proceedings of the biennial Conference on Precision Electromagnetic Measurements [56].

B. The SQUID as an RF Measuring Device

The RF-biased SQUID is a magnetic flux detector with a periodic response to flux that can follow extremely fast variations. This is converted to a periodic response to current by inductively coupling the SQUID to the circuit in which the current to be measured flows. The basic arrangement is shown in Fig. 9. The period of the response of the SQUID to current in the input line is Φ_0/M, where M is the mutual inductance between the input circuit and the SQUID, and Φ_0 (the flux quantum) is the period of the response of the SQUID to magnetic flux. Because of the Meissner effect, it is possible to design superconducting circuits in which the distribution of current (and hence M) is insensitive to frequency in the range from zero to several GHz. Thus the SQUID can serve two functions in RF metrology [57, 58]: (1) its regular periodic response to current can be used to determine ratios of amplitudes of RF signals, and hence to measure attenuation or to function as a detector with amplitude markers for a system such as a 6-port; (2) the frequency-independence of the basic SQUID's response to current can be used to compare RF and direct currents directly, and hence to relate the amplitudes of RF signals to the SI units by comparing magnetic fluxes instead of comparing heating effects as a bolometer does.

The advantages a SQUID has to offer in these functions appear to be: (1) it is a much cheaper device than a variable attenuator of the quality that would be used as a primary standard; (2) it is sensitive to much lower power levels than any bolometer; (3) it can be used to make measurements over a wide range of frequencies.

SQUIDs for RF metrology differ from those used for magnetometers and other low frequency sensors in two respects: they must be provided with properly designed inductively coupled RF input circuits (Fig. 9); and it is desirable that the pump frequency be considerably higher than the frequency of the RF signal to be measured. In the following section I shall describe two SQUIDs developed at NBS: first, a broad band SQUID pumped at 9 GHz with which the early demonstrations of the feasibility of RF measurements were made [59]; and then the most recent design that Frederick, Sullivan, and Adair [60] have evolved for a system to measure attenuation at 30 MHz. Several other designs have been published elsewhere [61], including one that is pumped at 90 GHz [62].

The basic principle of using a SQUID to make RF measurements is illustrated in Figs. 9 and 10. The "microwave readout circuit" (Fig. 9) is designed to display the variation of microwave reflection coefficient of the SQUID as the magnetic flux coupled into it by the current I in the "input line" varies. Figure 10a shows an oscilloscope display of the amplitude of the reflected microwave signal as a slowly varying direct current is applied to the "input line." The only remarkable feature of this compared with SQUIDs operating at lower pump frequencies is that the periodic

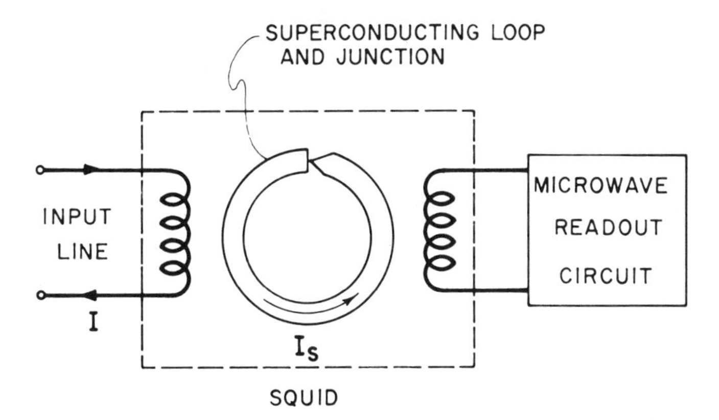

Fig. 9. The elements of a SQUID system for RF measurements.

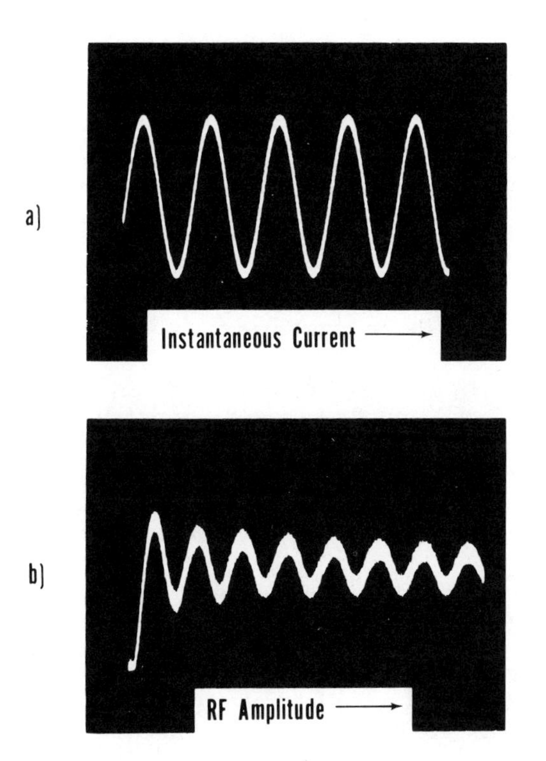

Fig. 10. Oscilloscope displays of the microwave signal reflected from the SQUID of Fig. 9:
(a) the response to slow variations of current in the "input line";
(b) the averaged response to a slowly amplitude modulated RF current in the "input line".

response to magnetic flux has degenerated into an almost pure sinusoidal form. This is partly a consequence of the high pump frequency (9 GHz in this instance). It is also deliberately cultivated in the design of the SQUID by the choice of the characteristics of the Josephson junction and its coupling to the microwave readout circuit. Presented with an RF current at the input line, the SQUID and its readout circuit records an average over a segment of Fig. 10a. The width of the segment is determined by the amplitude of the RF current, and its location can be moved by simultaneously applying a DC bias. As the amplitude of the RF current varies, this averaged response reflects the basic periodicity of the SQUID. This is shown in Fig. 10b, which is an oscilloscope display obtained by applying a slow amplitude modulation to an RF current at 65 MHz applied to the SQUID. A simple analysis shows, in the approximation that the basic response to current shown in Fig. 10a is a sine function, that the averaged response to RF current shown in Fig. 10b is a zero-order Bessel function J_o. The familiar shape of this function can be seen in Fig. 10b. We have developed a simple system to use a lock-in detector to locate the zeros of the approximate Bessel function describing the response to RF current shown in Fig. 10b. The values of the argument of J_o at these zeros have been computed and published in tables of Bessel functions [63]. They form the natural scale on which this system of RF measurements is based. We have been able to locate the positions of the first 900 points on this scale, covering a dynamic range of over 60 dB.

The basic form of the "microwave readout circuit" (Fig. 9) is shown in Fig. 11. The microwave circuit in the upper part of the picture is essentially a simple reflectometer, and the displays shown in Fig. 10 were obtained by connecting an oscilloscope to the "signal" input of the lock-in detector. The function of the lock-in detector can be understood by making the assumption that the microwave power P reflected from the SQUID has the form:

$$P = P_o + P_1 \cos (2\pi \ I/I_o) \tag{7}$$

where I_o is the current required to change the magnetic flux in the SQUID by one quantum. The average reflected microwave power $\overline{P}$ would then vary with the amplitude I_1 of an RF current $I_1 \cos (2\pi ft)$ as follows:

$$\overline{P} = P_o + P_1 J_o (2\pi I_1/I_o) \tag{2}$$

If a steady bias is applied to the SQUID to change the magnetic flux linking it by $\Phi_o/2$, the effect will be to interchange the positions of the peaks and valleys shown in Fig. 10a. The reflected microwave power P would now have the form

$$P = P_o + P_1 \cos (2\pi I/I_o + \pi), \tag{9}$$

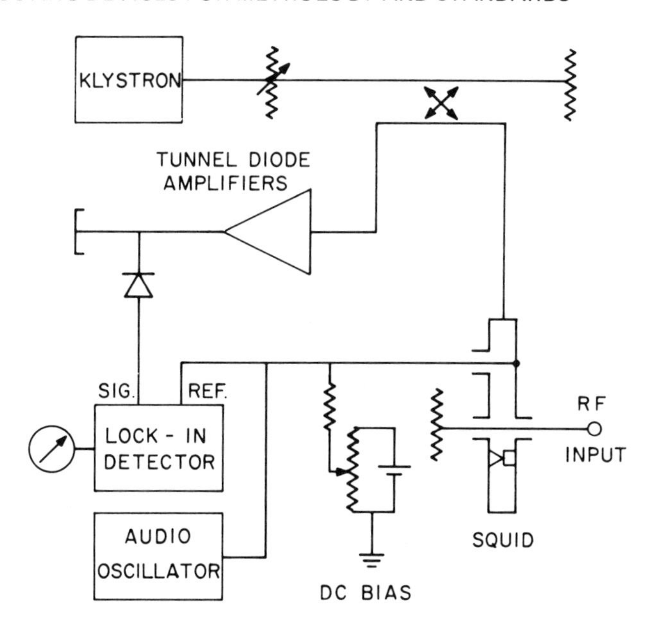

Fig. 11. The elements of the microwave readout circuit of Fig. 9.

and the average reflected microwave power would vary with the amplitude I_1 of an applied RF current as

$$\overline{P} = P_o - P_1 J_o (2\pi I_1/I_o), \quad (10)$$

i.e., the response of the SQUID shown in Fig. 10b is inverted. This inversion occurs at every half cycle of the modulation, and the lock-in detector responds only to that part of the reflected signal that is affected. This enables us to eliminate the unwanted offset P_o (which carries no information) and look to the detected output of the lock-in detector to locate the zeros of the Bessel function $J_o (2\pi I_1/I_o)$ on which our measurement system is based.

C. Practical SQUIDs for RF Metrology

In this section two SQUIDs are described: a broadband SQUID pumped at 9 GHz, with which the initial demonstrations of feasibility of this technique for RF measurements were made [57, 58]; and a SQUID with a permanently set Josephson junction, pumped at 1 GHz, developed recently by Sullivan [60] for a practical system to measure attenuation at 30 MHz.

The basic form of the broadband SQUID is sketched in Fig. 12. It consists of a short section of rectangular X-band waveguide, reduced in the E-direction to a height of 0.25 mm in order to give a characteristic impedance of 10Ω (at the midpoint). It is lined with superconducting metal (babbitt). The Josephson junction closing the loop is placed on the center line a quarter wavelength from the short circuit termination which forms the loop itself.

The current to be measured is coupled inductively to the active part of the SQUID. In the version shown in Fig. 12 this is accomplished by passing the center conductor of the coaxial input line through the device parallel to the E-field. This mode of coupling has the merits of a very broad bandwidth (the lowest resonance is at the cutoff frequency of the waveguide, approximately 6.6 GHz), and a very small perturbation of the input line. It is equivalent to a series inductance of about 10^{-10} H. Even without capacitive compensation this causes a smaller reflection than most RF connectors. The sensitivity of this version of the SQUID to current in the input line is only moderate. Depending upon the position of the input line relative to the Josephson junction, the current I_0 required to couple one flux quantum into various SQUIDs we have tested varies from 16 μA (line 2 mm from point) to 100 μA (line 1 cm from point, towards the side of the waveguide). The sensitivity could be reduced arbitrarily by moving the input line further away.

We often provide our SQUIDs with a pair of input lines. One is carefully terminated by a matched load and carries the RF current to be measured. The other is usually terminated by a simple short and carries DC bias and modulation at an audio frequency (usually 1 kHz). Although the input lines are drawn on the center line of the SQUID for clarity in Figs. 12 and 13, we usually place them toward the sides. Because of the high impedance of the 50 Ω coaxial lines compared with the 10Ω waveguide, the loading of the microwave system is small.

In order to increase the sensitivity of the SQUID to current in the input line, we employ the toroidal coupling coil shown in Fig. 13. This acts as a transformer with a multi-turn primary and a single-turn secondary. With a 50-turn coil we have observed I_0 down to 1 μA, or 24 dB less than the minimum we have observed with the broadband arrangement of Fig. 12. The price one pays is reduction of the bandwidth. This same 50-turn coil has an inductance of 1.5×10^{-7} H and a self-resonant frequency of 200 MHz. For maximum sensitivity, we connect one end of the coupling coil to ground and use a tuned matching network. The objective is to dissipate maximum RF power to be measured in the Josephson junction rather than a passive load outside the SQUID.

The microwave readout circuit (Fig. 9) is connected with WR 90 waveguide to the port shown on the left of Figs. 12 and 13 by a tapered waveguide section (10 cm long). The point contact Josephson junction is adjusted from outside the cryostat by means of a push-rod with a micrometer drive. A hook on the end of this rod engages a small lever on the end of one of the two niobium screws that comprise the junction. When the junction is adjusted, the hook is backed off a short distance to prevent it transmitting mechanical shocks.

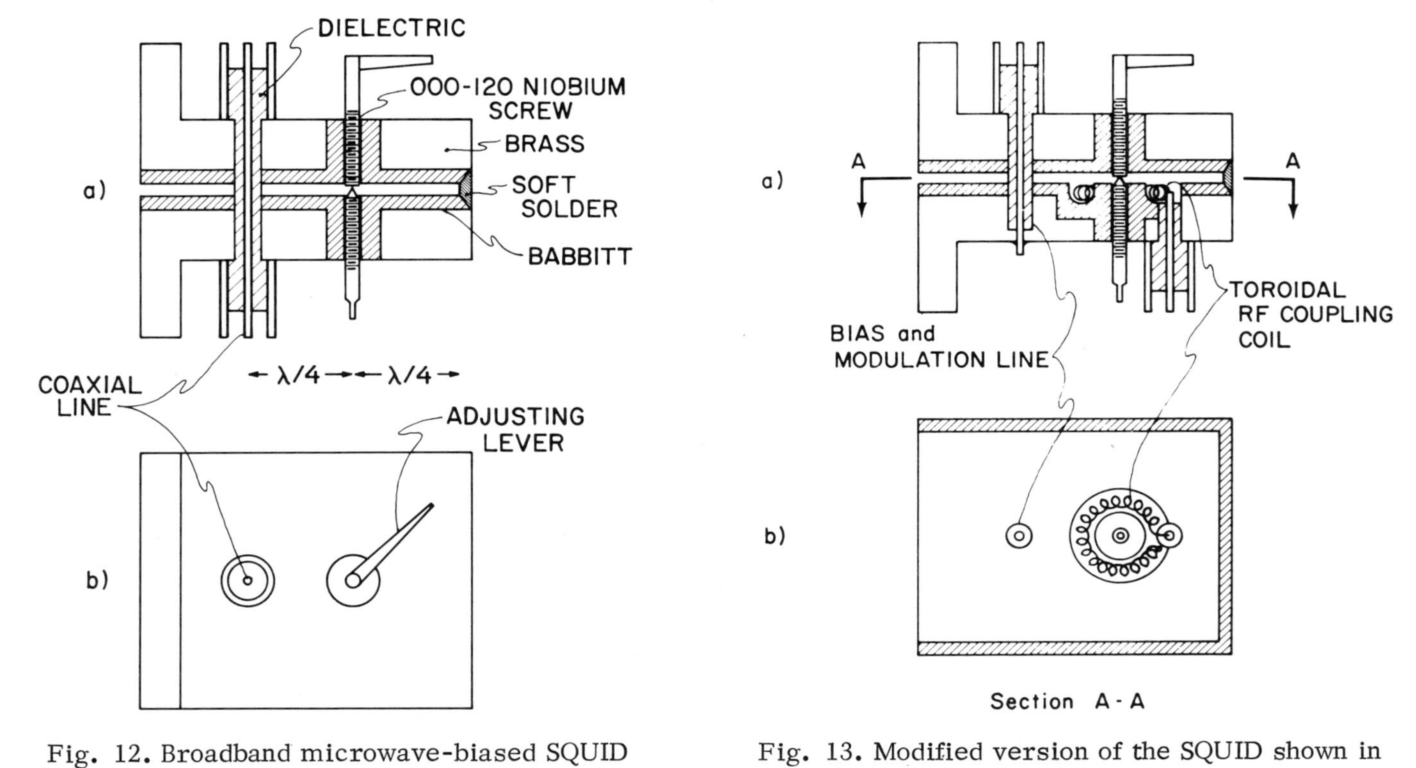

Fig. 12. Broadband microwave-biased SQUID connected to the microwave readout circuit via a taper section of waveguide connected to the flange at the left. The "coaxial line" corresponds to the "input line" of Fig. 9. Some dimensions have been enlarged for clarity.

Fig. 13. Modified version of the SQUID shown in Fig. 12. The toroidal input coupling circuit gives it greater sensitivity. A separate input line for bias and modulation is shown.

Although this SQUID performs well enough for demonstration purposes, the adjustable point contact is an intolerable nuisance in a routine measurement system. Also, for a practical system, the pump frequency must be chosen carefully. As a general rule, a higher pump frequency has the advantage of giving a stronger microwave readout signal, which can be exploited if an amplifier with low enough noise figure is available. At 9 GHz we used a tunnel diode amplifier, but at 1 GHz, transistor amplifiers are available at a much lower price, and the next SQUID I shall describe was designed to use one of these [60].

This SQUID is shown in Fig. 14. It is built around a permanently set point contact Josephson junction that is mounted in a replaceable cartridge, shown in Fig. 15. The body of the cartridge is made by fusing together two niobium blocks separated by an insulating layer of borosilicate glass. The contacts are niobium screws, one pointed and one flat, that are oxidized by brief exposure to a flame. After adjustment, the screws are locked in position with nuts and sealed with a minute quantity of epoxy resin right at the point of contact. The cartridge is held in the SQUID body by set screws. The microwave readout circuit is coupled to the SQUID by a simple lumped L-C network, and the signal to be measured (at 30 MHz) is coupled in by a toroidal coil surrounding the cartridge. This SQUID is used to measure attenuation at 30 MHz as described in Section III C.

Having a preset Josephson junction in the SQUID presents the embarassing responsibility of deciding what the setting should be. The operating characteristics of a SQUID can be described well enough by a model

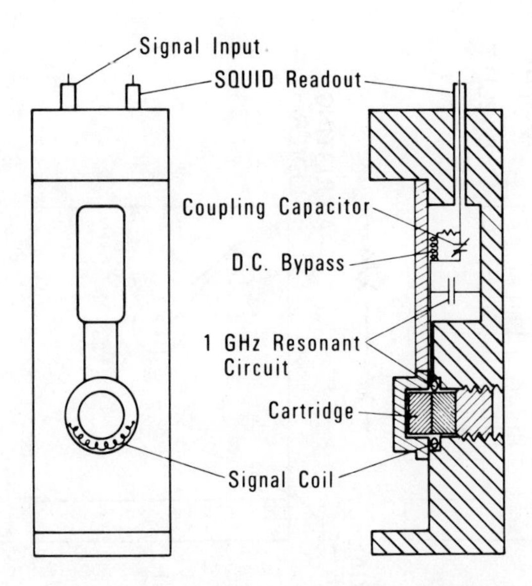

Fig. 14. Microwave-biased SQUID using a cartridge-mounted junction, coupled to the microwave readout circuit (operating at 1 GHz) via a lumped-element resonant circuit shown symbolically.

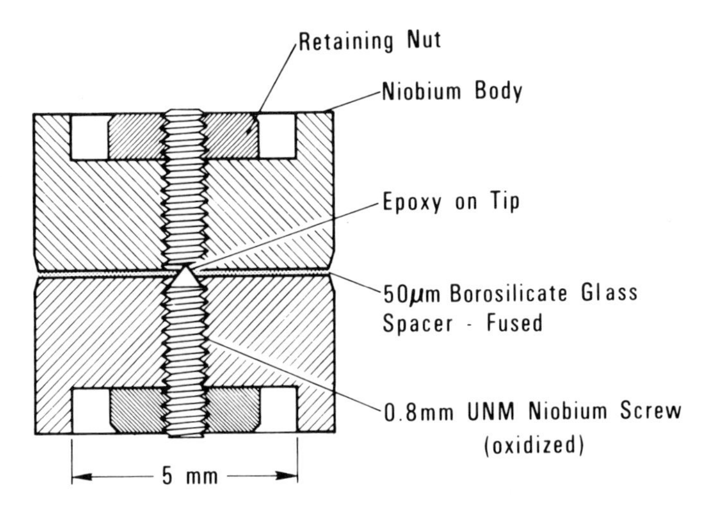

Fig. 15. Cartridge-mounted Josephson junction, used with the SQUID shown in Fig. 14.

consisting of a pure inductance L connected in parallel with an ideal Josephson junction with critical current I_c that is shunted by a quasi-particle current flowing through resistance R. The displacement current is usually not important. The objective for a SQUID for RF measurement is to make the basic response to magnetic flux shown in Fig. 10a as nearly sinusoidal as possible while preserving a reasonably high signal-to-noise ratio. The operation of a SQUID is a complex, non-linear process [64], but two general guidelines can be stated.

(1) The SQUID should operate in a non-hysteretic mode [64]. This requires that $LI_c < \Phi_0$.
(2) The SQUID should be under-damped, i.e., its natural decay time should be longer than the period of the pump. This requires that $\omega L > R$, where ω is the angular frequency of the pump.

The optimum setting found by Sullivan for his preset SQUID was with $I_c \simeq 2\mu A$ and $R \simeq 1\Omega$.

D. The Measurement of Attenuation

A system to use a SQUID to measure attenuation at 30 MHz [58] is shown in Fig. 16. A signal at 30 MHz is supplied to the measuring port of the SQUID via the attenuator under test. As the attenuator is set to different values (or inserted and removed) the SQUID is used to measure the variation in received signal level. The SQUID and its standard readout circuit (Fig. 11) can be seen in the lower part of the diagram. The

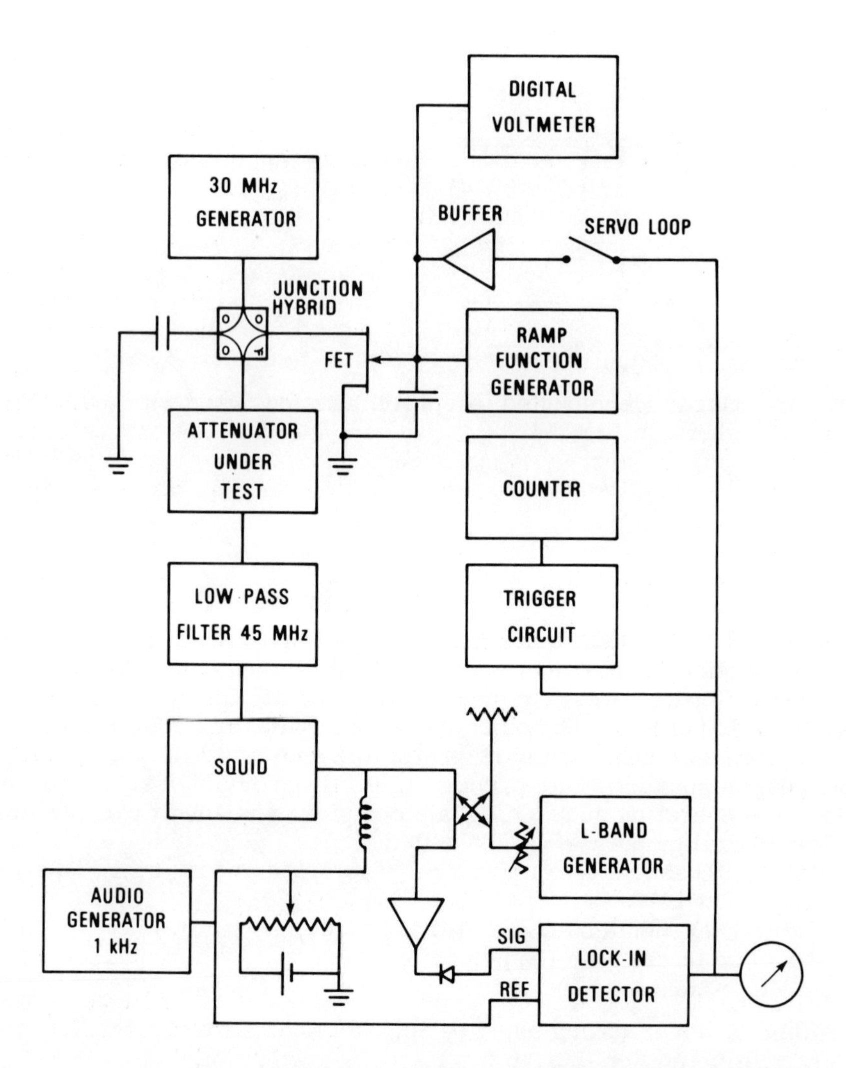

Fig. 16. System for measuring attenuation at 30 MHz using a SQUID. This has provision for automatic counting of RF flux quanta and interpolation between them.

basic scale for the measurement is defined by the nulls of the approximate function $J_0(2\pi I_1/I_0)$ discussed in Section B. The remainder of the system shown in Fig. 16 is used to determine which are the nearest nulls to the level I_1 to be measured, and to interpolate between them.

The hybrid junction functions as a voltage-controlled attenuator with low insertion loss and wide dynamic range (60 dB). It responds to a voltage applied to the field-effect transistor that terminates one of its side ports. If this voltage is supplied from a ramp function generator, the signal at 30 MHz supplied to the system can be raised from a very low level to its preset working level slowly enough for nulls in the function $J_0(2\pi I_1/I_0)$ to be counted electronically as they are passed. A servo-loop can then be closed to bring the (30 MHz) signal level at the SQUID exactly to the nearest null. A measurement of the voltage required to do this supplies sufficient information for interpolation after the attenuation vs. voltage characteristic of the combination of hybrid junction and field-effect transistor has been calibrated.

An early version of this system, using a SQUID with an adjustable point contact and no refinements such as the interpolation circuit described above, was tested with a variable transfer standard attenuator [58]. It was found to be capable of agreement with the NBS Calibration Service to ± 0.002 dB (rms) over a dynamic range of 60 dB, a result that was recently repeated at NPL [65]. This agreement was better than the estimated systematic error of the NBS system, and served as a valuable check on its performance. However, because of the variability of adjustable point contacts, and interference between various components of the system that carried signals at very low levels, the SQUID system could not be relied upon to perform so well every time. Considerable effort was required to develop the SQUID with a preset junction in a cartridge and a reliable system. The result of this effort has been reported by Frederick Sullivan and Adair [60], who describe the system in its final form.

E. The Measurement of Power

In order to measure RF power, the current I_0 at the RF measuring port of the SQUID required to induce one quantum of magnetic flux [Eq. (8)] must be measured. If the frequency of the signal to be measured is well below the resonant frequency of the input circuit, this may be done with DC [Eq. (7)]. The SQUID is then calibrated to measure the amplitude I_1 of the RF current in its input circuit. If this is terminated with a matched load Z_0, then the power incident on the SQUID is $1/2\ I_1^2\ Z_0$.

In practice, this method can be used to measure power levels down to about 10^{-10} W, using a system almost identical with the one shown in Fig. 16. Care must be taken to account for all the attenuation and reflections in the transmission line connecting the point where the measurement of power is desired and the port of the SQUID. This is common to all systems for measuring power, but is complicated by the properties of the lines connecting the warm laboratory to the cold SQUID in the cryostat. These must be designed with a tolerably low thermal conductance, and this usually results in a significant attenuation that depends on both temperature and frequency. The thermal gradients must therefore be stabilized against variations in the level of liquid helium by judicious use of copper cladding, and the attenuation must be measured under operating conditions. We have tested the feasibility of measuring power by this method [58], and shown that measurements with errors of the order of ± 0.1 dB can be made at frequencies from 0 to 1 GHz, using the broadband SQUID described in Section IV C.

Measuring power at lower levels requires increasing the sensitivity of the SQUID to RF signals by adopting a resonant input circuit. This precludes direct use of the simple technique of measuring I_0 with DC. A technique being developed by Sullivan (private communication) is to use two SQUIDs in series on the same RF line. The first has a broadband coupling to the input line with only moderate sensitivity to current. The second terminates the line with a resonant coupling circuit with the greatest possible sensitivity. The first SQUID is then calibrated with DC and used to calibrate the second at the working frequency. Sullivan has designed a system based on this principle that is intended to measure RF power at levels down to 10^{-15}W.

F. Systematic Errors

In this section I shall discuss two sources of systematic error in measurements of RF current by the techniques I have described. These are variation with frequency of the geometrical factor (mutual inductance M) relating current in the RF circuit to magnetic flux linking the SQUID, and distortion of the basic scale defined by the nulls of the function $J_0(2\pi I_1/I_0)$. I will dismiss the first of these topics with a few estimates of orders of magnitude: it is not a serious source of error, it affects measurement of absolute power but not attenuation, and it is well understood by microwave circuit designers. Since the second topic is unique to the new measuring system, I will analyze it in some detail.

The two geometrical effects to consider are: the variation with frequency of the depth of penetration of current into the conducting walls of the SQUID, with the consequent variation in the distribution of current; and the variation with frequency of the distribution of electromagnetic fields in the hollow spaces of the SQUID because of the formation of standing waves.

The penetration depth for current into the surfaces of superconducting lead and niobium (the two materials we most commonly use for SQUIDs) is about 4×10^{-8} m. This quantity varies slightly with frequency. Waldram [66] found that the penetration depth in tin changes by 8% between DC and 9.4 GHz. One would expect lead and niobium to be less sensitive to frequency, but a change of 8% would amount to 3×10^{-9} m. The smallest linear dimension of any SQUID is usually greater than 2.5×10^{-4} m, so the effective linear dimensions of the superconducting walls of the SQUID might change by 10 parts per million between DC and X-band. We can neglect this for present purposes.

The presence of standing waves will cause a significant perturbation of the distribution of electromagnetic fields inside a hollow conductor when the wavelength is comparable to the linear dimensions of the conductor. For smaller conductors, the perturbation will be of the order of 1 - cos (ℓ/λ) where ℓ is the longest linear dimension and λ is the wavelength. Integrating this over the length of one of the simple broadband SQUIDs (Fig. 12), yields an error of one part per thousand at 1 GHz. SQUIDs coupled to the RF line by coils (see Figs. 13 and 14) are affected by standing waves at much lower frequencies, because they are slow-wave structures. However, our mode of using them does not rely on their calibration being independent of frequency.

Let us now consider distortion of the basic scale. The microwave SQUID, together with its microwave readout system, converts variations in magnetic flux Φ (induced by current I in the coaxial input line) to variations in an output voltage V which depends periodically on Φ. The functional form is very nearly sinusoidal:

$$
\begin{aligned}
V &= V_o + V_1 \cos(2\pi I/I_o) \\
&= V_o + V_1 \cos(2\pi \Phi/\Phi_o) \qquad (11)
\end{aligned}
$$

where I_o is the increment of current required to change Φ by one flux quantum Φ_o. Faced with current alternating at a radio frequency f_{RF}, we assume that the recording system would register an average value $\overline{V}$ of the output voltage V. In the approximation that the basic response to current [Eq. (11)] is sinusoidal,

$$
\overline{V} = V_o + V_1 J_o(X_o) \qquad (12)
$$

where X_o is defined by

$$
I = \frac{X_o I_o}{2\pi} \sin(2\pi f_{RF} t) .
$$

With some technique to locate the zeros of $J_o(X_o)$, the system is then capable of measuring ratios of RF current and hence power. Equations (11) and (12) represent such a close approximation to the observed performance of the system that we can regard them as "normal". Any deviation is an "error" to be analyzed and corrected.

We observe that the system always appears to respond symmetrically to increasing and decreasing magnetic flux. We may therefore represent the basic response with a Fourier cosine series

$$
V = V_0 + \sum_n V_n \cos(nX + n\delta) \qquad (13)
$$

where $X = 2\pi I_{RF}/I_o = X_o \sin(2\pi f_{RF} t)$. It represents the RF current in units of the flux quantum. The quantity δ is defined in a similar way to include contributions to the input current (such as DC bias and low frequency modulation) which vary with time much more slowly than the RF current (slowly enough for the recorder to follow). Expanding the cosines in Eq. (13).

$$V = V_o + \sum_n \left[V_n \cos(nX)\cos(n\delta) - V_n \sin(nX)\sin(n\delta) \right], \quad (14)$$

and integrating over a complete RF cycle:

$$\bar{V} = V_o + \sum_n \Big[\left(V_n \cos(n\delta) / 2\pi \right) \int_0^{2\pi} \cos(nX_o \sin y)\, dy - \left(V_n \sin(n\delta)/2\pi \right) \int_0^{2\pi} \sin(nX_o \sin y)\, dy \Big] \quad (15)$$

where $y = 2\pi f_{RF} t$. The second term in the square brackets vanishes on integration, and the first term defines the Bessel function J_o. Hence

$$\bar{V} = V_o + \sum_n 2V_n \cos(n\delta)\, J_o(nX_o) \quad . \quad (16)$$

We now examine δ in more detail, including both DC bias and low frequency modulation,

$$\begin{aligned} \delta &= a + b\sin(2\pi f_M t) \\ &= a + b \sin z \end{aligned} \quad (17)$$

where $a = 2\pi I_{DC}/I_o$; $b = 2\pi I_M/I_o$, I_{DC} is the DC bias and I_M is the depth of modulation at frequency f_M. The modulated signal $V_M \sin z$ at the output of the microwave system may then be used with a lock-in detector to locate the zeros of the Bessel function part of the response expressed in Eq. 16. The quantity V_M is the first Fourier coefficient of $\overline{V}$:

$$V_M = \frac{1}{\pi}\int_{-\pi}^{\pi} \overline{V} \sin z \, dz. \tag{18}$$

Using Eq. (16),

$$V_M = \frac{1}{\pi}\sum_n V_n J_o(nX_o)\int_{-\pi}^{\pi} \sin z \cos(na + nb \sin z)\, dz. \tag{19}$$

Expanding the cosines

$$V_M = \frac{1}{\pi}\sum_n V_n J_o(nX_o)\int_{-\pi}^{\pi}\Big[\cos(na)\cos(nb \sin z)\sin z$$

$$- \sin(na)\sin(nb \sin z)\sin z\Big]dz. \tag{20}$$

The first term in the square brackets vanishes on integration, leaving

$$V_M = -\frac{1}{\pi}\sum_n V_n J_o(nX_o)\sin(na)\int_{-\pi}^{\pi}\sin(nb \sin z)\sin z\, dz, \tag{21}$$

or

$$V_M = \sum_n 2V_n J_o(nX_o)\sin(na) J_1(nb). \tag{22}$$

The quantity V_M in Eq. (22) is the signal that would be recorded by a lock-in detector following the microwave crystal detector, with the reference at the modulation frequency. The approximate form of this signal, given in Eq. (12), corresponds to the leading term of Eq. (22) with $n = 1$. The other terms distort this signal by adding Bessel functions of multiple argument.

Let us consider the simplest case, where only second harmonic distortion of the basic response occurs ($n = 1, 2$). Then

$$V_M = 2V_1 \sin(a) J_1(b) J_o(X_o) + 2V_2 \sin(2a) J_1(2b) J_o(2X_o). \tag{23}$$

The presence of the second term shifts the zeros from where they would be with the first term alone. In order to find the amount of this shift, let us consider the function

$$V_M/2V_1 \sin(a) J_1(B) = J_o(X_o) + \alpha J_o(2X_o)$$

where (24)

$$\alpha = \left[V_2 \sin(2a) J_1(2b)\right] / \left[V_1 \sin(a) J_1(b)\right].$$

Neglecting second derivatives, a simple analysis is shows that the zeros of the function in Eq. (24) occur at

$$X_o = j_o + \Delta X$$

where

$$\Delta X = \alpha J_o(2j_o) / \left[J_1(j_o) + 2\alpha J_1(2j_o)\right], \tag{25}$$

and

$$J_o(j_o) = 0,$$

i.e., j_o is the argument of the Bessel function at the unperturbed zero. If $2\alpha << 1$, we can neglect the second term in the square brackets and

$$\Delta X \simeq \alpha J_o(2j_o) / J_1(j_o). \tag{26}$$

The quantity $J_o(2j_o)/J_1(j_o)$ is approximately equal to $0.5 \times (-1)^n$, where n is the order number of the zero of J_o under consideration. Thus, a second harmonic distortion of the basic response of the SQUID to current has the effect of displacing odd and even zeros from those of J_o by approximately equal amounts in opposite directions. Inspection of Eq. (24) shows that this effect (and indeed the effect of all even-order Fourier components) may be eliminated by adjusting the DC bias so that $a = \pi/2$. This also happens to be the adjustment that optimizes the desired signal!

To analyze the effect of third harmonic distortion, let us assume that the response function of the SQUID system takes the form

$$V \sim J_o(X) + \alpha J_o(3X). \tag{27}$$

Then, to first order, the arguments of the zeros are displaced by an amount

$$\Delta X = \alpha J_o(3j_o)/J_1(j_o), \tag{28}$$

where

$$J_o(3j_o)/J_1(j_o) \simeq +0.58.$$

Thus, a perturbation of this type displaces all the zeros by the same amount ΔX with respect to current. This displacement has an arbitrary sign, depending on the coefficient V_3[Eq. (13)]. It may be eliminated by adjusting the modulation depth so that $J_1(3b) = 0$.

RF leakage also causes a displacement of all the zeros by an equal amount in a common, arbitrary direction, since signals travelling by different paths recombine with arbitrary phase.

When measurements and errors are expressed in decibels, a displacement ΔX in a zero causes a measurement error of $20 \log_{10}(1 + \Delta X/j_o)$. Thus these errors all affect the first few zeros most strongly, and have a diminishing effect on those of higher order.

One final source of error to consider is external RF noise. Let us assume that an external RF signal of amplitude X_1 is interfering with the signal of amplitude X_o which is being measured. The approximate form of the resulting signal X is then

$$X = X_o + X_1 \cos \omega t \tag{29}$$

where ω is the angular beat frequency. The functional form of the response of the SQUID system is then

$$V \sim \omega \int_0^{2\pi/\omega} J_o(X_o + X_1 \cos \omega t)\, dt. \tag{30}$$

Using the integral form of J_o, and inverting the order of integration, we find

$$V \sim \frac{1}{\pi} \int_0^{\pi} J_o(X_1 \sin \Phi) \cos(X_o \sin \Phi)\, d\Phi. \tag{31}$$

Making the approximation

$$J_o(X) \simeq 1 - \frac{1}{4} X^2,$$

we find

$$V \sim J_o(X_o) \frac{X_1^2}{8} \left[J_o(X_o) - J_2(X_o) \right]. \tag{32}$$

At the zeros of J_o,

$$J_o(X_o) = J_o(j_o) = 0$$

and

$$V \sim X_1^{\ 2} J_2(j_o) / 8 .$$

Because

$$J_2(j_o) = 2 J_1(j_o) / j_o,$$

we find that

$$V \sim \frac{X_1^{\ 2}}{4j_o} J_1(j_o). \tag{33}$$

Hence, to first order, the displacement ΔX of the zeros of the response by this perturbation is

$$\Delta X = -V / \frac{dV}{dX_o} \simeq X_1^{\ 2} / 4j_o. \tag{34}$$

Hence, the displacement in dB is 20 $\log_{10}(1 + X_1^{\ 2} / 4j_o^{\ 2})$. Note that this error is always positive. Paradoxically, in the presence of RF interference, more power (less attenuation) is needed to set on the nulls. The error affects the first null strongly, and the others very little.

These sources of error appear to cover the deviations of a "well behaved" system. More pathological behavior can be caused by: interference between microwave, RF, and modulation circuits; strong line resonances; or a SQUID with excessively large I_c or small shut conductance. These can be eliminated only at their sources by careful design of the system [60].

V. THERMOMETRY

The central problem in thermometry is the realization of the Kelvin scale of temperature. Here we consider only the range below 10 K, where the contributions of superconducting devices have been most significant. These contributions include absolute noise thermometry, the definition of fixed points by the transition temperatures of superconductors, and a thermometer based on the temperature dependence of the energy gap of a superconductor.

A. The Kelvin Scale Below 1K

The general topic of thermometry at very low temperature was recently reviewed comprehensively by Hudson et al. [67]. Only a few general remarks are made here, and the reader is referred to this review for the details.

A practical thermometer must be based on some measurable property of a material or system that varies in a predictable way with temperature. A primary thermometer has a law of variation that is known from basic principles, so it may be used to define the Kelvin scale. A secondary thermometer is a device that must be calibrated against a primary thermometer, but is usually more convenient to use for everyday measurements. The distinction between primary and secondary thermometers is a little blurred, because primary thermometers all depend on a knowledge of some fundamental constant such as Boltzmann's constant or the gas constant, and these depend on measurements of temperature, albeit at the fixed points that define the Kelvin degree. However, in the range below 1K the distinction is a real one, because all these fundamental constants are determined by measurements made at higher temperatures, where accurate thermometry is an easier art because of the higher thermal conductivity of materials and the closer approximation to linearity of the equations of state of gases.

The most common primary thermometers for very low temperatures are based on the temperature dependence of the polarization or alignment of paramagnetic systems. The measurable quantities are the magnetic susceptibilities of paramagnetic salts, or of the nuclear spins of metals, and the anisotropy of γ-rays emitted by radioactive decay of nuclei aligned in the strong internal fields of ferromagnetic metals. These are all localized systems that obey Boltzmann statistics. This would lead to a simple temperature dependence (Curie's Law) for the magnetic susceptibility of a system of non-interacting spins, but in paramagnetic salts the interaction (either dipolar or exchange-coupled) between spins causes nonlinear effects that depend on the shape of the specimen and introduce uncertainty into measurements of very low temperatures (below a few mK). Nuclear spin susceptibility measurements also suffer from this effect, but it is neglibly small at temperatures above 10^{-6} K. However, nuclear spins also have the disadvantage of generally poor thermal contact with the outside world. Metals are the most favorable hosts in this respect, but even there the operating magnetic field strength is restricted by the need to minimize the magnetic heat capacity of the spins in order to keep the time required to reach thermal equilibrium reasonably short. This makes the measurement of the nuclear magnetic susceptibility with the accuracy required for good thermometry a challenging problem. The measurement of nuclear alignment by γ - ray anisotropy is free of this problem, but has difficulties of its own connected with the heating of the thermometer by its own radioactivity.

In summary, several thermometers have been developed with which many good measurements have been made, but they are all prone to sufficient uncertainty to make an independent check, by a thermometer working on an entirely different principle, very desirable. This role can be filled by noise thermometry, which can be extended to very low temperatures with the aid of SQUID sensors. These sensors then provide a completely independent realization of the Kelvin scale that appears to be largely in agreement with measurements made by other techniques.

B. Noise Thermometry with SQUID Sensors

Two different noise thermometers using SQUIDS have been developed, one at NBS [68-70] and one by Wheatley and his colleagues at the University of California at San Diego [71,72]. I shall discuss the first in some detail, since many of the principles and sources of error are common to both, and then I shall describe the second and discuss the significant differences between the two.

The principle of noise thermometry with the Josephson effect is illustrated in its simplest form in Fig. 17. The Josephson junction is connected in a structure of very small inductance ($<10^{-9}$ H) to a shunt of small resistance R($\sim 10^{-5}\,\Omega$) to form a resistive SQUID. A direct current I of the order of 10^{-6}A through this shunt will maintain a bias voltage of the order of 10^{-11}V across the junction, driving it to oscillate at a frequency of the order of 5 kHz. The frequency f of oscillation is related to the bias voltage V across the junction by

$$hf = 2eV$$

or

$$f = V/\Phi_0 \; . \tag{35}$$

The voltage V across the junction is

$$V = IR + \text{noise} \tag{36}$$

and the fluctuations in V due to noise will be converted into fluctuations in the frequency f. If the shunt resistance R is small compared with the resistance of the junction and the impedance of the current supply, its thermal noise will dominate the fluctuations in the frequency of oscillation. These fluctuations will then contain sufficient information to determine the absolute temperature of the shunt resistor if R is determined by measuring the bias current and the average frequency of oscillation.

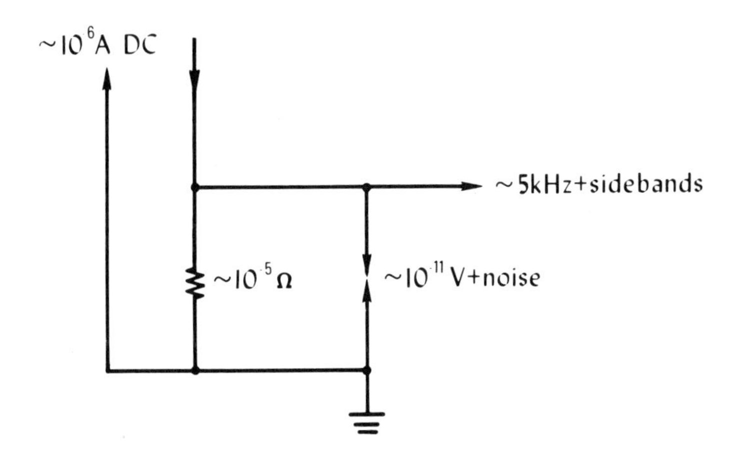

Fig. 17. The principle of noise thermometry by frequency modulation of an oscillating Josephson junction.

The essential components of a noise thermometer are a resistor to generate thermal noise and a preamplifier, with noise temperature less than the temperature to be measured, and with sufficient gain to generate a signal detectable by a receiver at ambient temperature. Here the Josephson junction fulfills the function of the preamplifier. A simple analysis [68] shows that the components of the noise spectrum which have the greatest effect on the RF power spectrum of the Josephson oscillation are all at low frequency (<1 kHz at 4 K, less at lower temperatures). They create a narrow spectrum of sidebands close to the center frequency IR/Φ_o. Thus the Josephson junction amplifies the noise signal by parametric up-conversion. The performance of a Josephson junction in this mode is such that the equivalent of an ordinary radio frequency communications receiver should suffice for noise measurements.

The actual measurement of noise voltage can be done by either spectrum analysis or frequency counting.

Spectrum analysis requires a narrow band receiver to explore the long-term average of the spectrum of sidebands near the Josephson oscillation frequency. The quantity to be determined is the linewidth Δf, which is the frequency interval between the two points at which the spectral density is one half of its peak value at the unperturbed frequency IR/Φ_o. Assuming that the only significant source of broadening is thermal noise from the shunt resistor R, the mean-square fluctuation in voltage $\overline{V_n^2}$ across the junction is given by

$$\overline{V}_n^{\,2} = \int 4kTRdf \tag{37}$$

where k is Boltzmann's constant, T is the absolute temperature, and df is an interval of frequency. Application of standard FM theory to the corresponding fluctuation in frequency yields a formula for the observed linewidth Δf,

$$\Delta f = 4\pi kTR/\Phi_o^2$$

or

$$\Delta f = (4.03 \times 10^7)\ RT. \tag{38}$$

For a practical system we preferred the frequency counting method. A signal derived from the Josephson oscillation is amplified and used to drive a frequency counter, which repeatedly counts the number of cycles in a fixed gate time τ. Fluctuations in the frequency then appear directly as fluctuations in the count. A convenient measure of these fluctuations is the variance, or mean-square deviation σ^2. If only white noise is generated by the resistance R, then

$$\sigma^2 = \overline{(f - \overline{f})^2} = 2kTR/\tau\Phi_o^2$$

or

$$\sigma^2 = (6.4 \times 10^6)\ RT/\tau\ . \tag{39}$$

This approach to noise measurement has the advantages that it is easy to automate the recording of data and to discriminate against spurious sources of noise with anomalous power spectra, such as flicker noise.

Equations (38) and (39) both relate the absolute temperature to measured quantities by simple expressions containing fundamental constants. This is exactly what is required of a primary thermometer, but we must consider the possible errors and limitations.

First, let us discuss the validity of the Nyquist formula, Eq. (37), on which Eqs. (38) and (39) are based. This formula is based on Boltzmann statistics, whereas the photons involved in the generation of noise should properly be described with Bose statistics. The difference is only significant at very low temperatures, such that

$$hf \geqslant kT$$

or

$$f/T \geqslant 20\ \text{GHz/K}\ . \tag{40}$$

The components of the noise spectrum which have the major effect on the Josephephson oscillation are those at frequencies of the order of the line-width Δf [Eq. (38)] or less [68]. A comparison of Eq. (38) with the limit given in Eq. (40) shows that Boltzmann statistics, and hence the Nyquist formula, should be an excellent approximation at any temperature.

Next we consider the current supply to maintain the constant bias voltage on the shunt resistor R. In its simplest form this would consist of a battery with an emf of a few volts in series with a resistor R_0, of the order of $10^3\ \Omega$, at ambient temperature T_0. Neglecting the battery, the effective noise temperature T_n of the combination of R and R_0 in parallel is

$$T_n = (RT_0 + R_0 T)/(R + R_0). \tag{41}$$

The contribution to T_n from the resistor R_0 at ambient temperature is $RT_0/(R + R_0)$. If $R = 10^{-5}\ \Omega$ and $R_0 = 10^3\ \Omega$, then this amounts to 3×10^{-6} K, which is clearly negligible. The question of noise contributed by the battery has been investigated by Knott [73]. He tested several different dry cells and found that the noise they generated in the frequency range of interest was comparable with thermal noise on a resistor of $10^3\ \Omega$ at ambient temperature. Thus the combination of battery and resistor would be a satisfactory power supply for noise thermometry down into the micro-kelvin range.

The ultimate low temperature limit to the Josephson noise thermometer will most likely be set by the noise temperature of the junction itself. This has been analyzed by several authors, most completely and recently by Stephen [74]. I will summarize his conclusions briefly.

A perfect junction transmits current by tunneling of Cooper pairs and by tunneling of quasi-particles. The respective contributions of these two processes to the total current depend upon voltage and temperature. The significant differences so far as noise is concerned are in the quantities I/V, $\partial I/\partial V$, and the double charge per particle carried by Cooper pairs. At high temperatures, when $kT \gg eV$, both processes generate thermal noise appropriate to the temperature and some suitably defined effective resistance [Eq. (37)]. At very low temperature ($kT \ll eV$) they generate only shot noise. This shot noise is equivalent to thermal noise at an effective temperature of T_{eff}, where

$$T_{eff} \simeq eV/k$$

or

$$T_{eff} \simeq hf/2k \tag{42}$$

where f is the frequency of Josephson oscillation. Our circuit for noise thermometry in which the junction oscillates at a frequency of a few kHz will be described later. This would generate shot noise at a level corresponding to a noise temperature of about 10^{-7} K. Furthermore, in the circuit used for thermometry, the shunt resistor damps the noise emf generated by the junction. The effective noise temperature of the combination is given by an expression of the same form as Eq. (41), with a suitably defined effective resistance for the junction. Inspection of Eq. (41) shows that it would be possible to measure temperatures below T_{eff} with a very small shunt resistance.

In addition to the two tunneling processes, real junctions can transmit some current by ohmic leakage through the barrier. This would presumably generate thermal noise. It can be considered as part of the shunt conductance with no further complication of the problem.

Thus the fundamental limitations to noise thermometry with the Josephson effect appear to permit its use in the temperature range from about 10^{-6} K up to the critical temperatures of conveniently available superconductors, near 10 K. However, a real system is subject to noise from sources other than those considered by Stephen. There are external sources such as radio stations, atmospheric and galactic noise, etc., which can be attenuated with a screened enclosure. Both mechanical vibration of nearby superconducting parts and thermally driven eddy currents in nearby normal conducting parts can inductively couple additional noise into the junction. Vacuum pumps and gauges, digital electronic equipment, and other thermometers with AC readout should be considered as sources of noise. If the temperature is unstable, fluctuating thermoelectric emf can also contribute to the noise. The lowest observable noise temperature depends upon one's degree of success in careful elimination of these spurious effects.

So far I have discussed limitations which are common to measurements by both spectrum analysis and frequency counting. I will now discuss some points which are peculiar to frequency counting.

First, consider random error. A simple analysis shows that the rms scatter ΔT of the value T of a noise temperature measured by the frequency counting method is

$$\Delta T = T\sqrt{2/n} \tag{43}$$

where n is the number of measurements of frequency used to compute the variance of the fluctuations. This is a purely random error arising from the statistics of measuring a fluctuating quantity.

A frequency counter also introduces a systematic error, which arises because it counts only complete cycles. Thus the measured frequency is always an integral multiple of τ^{-1}, where τ is the gate time. If a counter is set to measure a signal with an arbitrary frequency f, then its reading will switch randomly between the two integral multiples of τ^{-1} bracketing f. Let us denote these quantities f_0 and $(f_0 + \tau^{-1})$. Assuming that there is no phase correlation between the signal and the time base of the counter, then a fraction $1 - (f - f_0)\tau$ of the readings will be f_0 and a fraction $(f - f_0)\tau$ will be $(f_0 + \tau^{-1})$. Thus the mean value of a large number of readings is just equal to f. However, this also introduces an extra fluctuation in the measured frequency. The variance σ_f^2 of this extra fluctuation is

$$\sigma_f^2 = (f - f_o)\, \tau^{-1} - (f - f_o)^2 . \tag{44}$$

In practice, f will fluctuate also, so we should average σ_f^2 over all values of $(f - f_o)$ between zero and τ^{-1}. The effect is to introduce a shift δT in the measured temperature, where

$$\delta T = \Phi_o^2 / (24kRT) \quad . \tag{45}$$

Equation (45) differs by a factor 2 from the corresponding expression in our previous publications on this topic [69]. I thank Robert Soulen and James Filliben for pointing out the error in the previous work.

It is a trivial matter to correct measured temperatures for this effect, but the extra fluctuation increases the random scatter of the measurements and will ultimately limit the resolution of temperature when δT exceeds T. Thus, the minimum resolvable temperature T_{min}, taking n readings with a counter of gate time τ, is

$$T_{min} = \frac{\Phi_o^2 \sqrt{2/n}}{24kR\tau}$$

or

$$T_{min} = 1.7 \times 10^{-8} \, (R\tau\sqrt{n})^{-1} . \tag{46}$$

Thus a fairly long gate time is desirable for measuring very low temperatures. With $R = 10^{-5}\ \Omega$ and $\tau = 1$ second, millikelvin resolution could be obtained in a run of a few minutes.

An alternative way to use a counter is to measure the period of the signal rather than the frequency. The counter then counts the number of complete cycles of its internal oscillator which occupy the same time interval as a specified number of cycles of the signal. If the frequency f_c of the internal oscillator (usually 1 MHz) exceeds that of the signal f_o, then the variance σ_c^2 of the fluctuations introduced by counting only complete cycles would be reduced by a factor f_o^2/f_c^2 for a given counting time. The same improvement could be obtained by counting a harmonic of the signal frequency. However, both these tactics amount to measuring phase fluctuations of a fraction of a cycle of the Josephson oscillation. The benefit of doing this is limited by amplifier noise, which is added to the signal and causes an uncertainty (of a fraction of a cycle) in the time of each zero crossing. Thus it is possible to obtain a resolution of temperature better than that specified in Eq. (46), but only when an uncommonly good signal-to-noise ratio is available at the detector (the maximum power available from the fundamental oscillation of a point-contact Josephson junction at 30 MHz is of the order of 10^{-13} W with a loop inductance of 10^{-9} H).

In analyzing white noise, care must be exercised to discriminate against other sources of noise, such as flicker noise, which have an anomalously large part of their spectral distribution at very low frequencies (the power spectrum for flicker noise is proportional to f^{-1}). These will always interfere with measurements requiring long periods of time. Methods of analyzing and handling flicker noise have been developed by Barnes [75] and Allan [76]. Following them, we arranged a computer program to calculate the variance σ^2 from the formula

$$\sigma^2 = \frac{1}{2}\overline{(f_i - f_j)^2} \tag{47}$$

where f_i and f_j are two successive measured frequencies. The advantage of Eq. (47) over Eq. (39) is that Eq. (47) is only sensitive to fluctuations with period comparable to the interval between successive measurements or less, whereas Eq. (39) is sensitive to fluctuations with period up to the total length of time required to take all the data. It is also helpful to use as short a gate time τ as possible, since this increases the sensitivity of the variance σ^2 to white noise. The dependence of σ^2 on τ for noise sources with with other commonly found power spectra is weaker, the one major exception being the additive noise imposed on an FM signal by the RF amplifier. This would generate a variance σ^2 proportional to τ^{-2}, and impose a lower limit on usable values of τ. As a check on the frequency spectrum of the observed fluctuations, our computer program also calculates σ^2 from Eq. (47) taking measurements with double and quadrupole intervals. That is, as well as taking $j = i + 1$, it takes $j = i + 2$ and $j = i + 4$. If we are observing pure white noise, this leaves the value of σ^2 unchanged apart from random scatter. If there is a component with a different power spectrum then these three values of σ^2 will differ significantly.

The first experimental test of this principle for thermometry was a verification of the linewidth predicted by Eq. (38) by Silver et al. [77]. Honesty was assured by ignorance of the exact numerical coefficient, and the agreement of the measured linewidth with Eq. 38 was within the experimental uncertainty of about 5%. Subsequently, a system was developed for a more rigorous test of thermometry using the frequency counting technique [70]. The SQUID that was used (see Fig. 18) has a toroidal geometry to minimize its sensitivity to external noise and interference. It has a point contact junction between a niobium screw and a niobium block that is connected to the main (niobium) body through a silicon-bronze resistor. Both RF readout and DC bias are connected through a single coaxial line. The readout circuit is shown in Fig. 19. For the early tests the readings of the frequency counter were punched on paper tape and processed by a separate computer.

This sytem was refined by Soulen, with the addition of a dedicated mini-computer and other improvements. He and Marshak then undertook

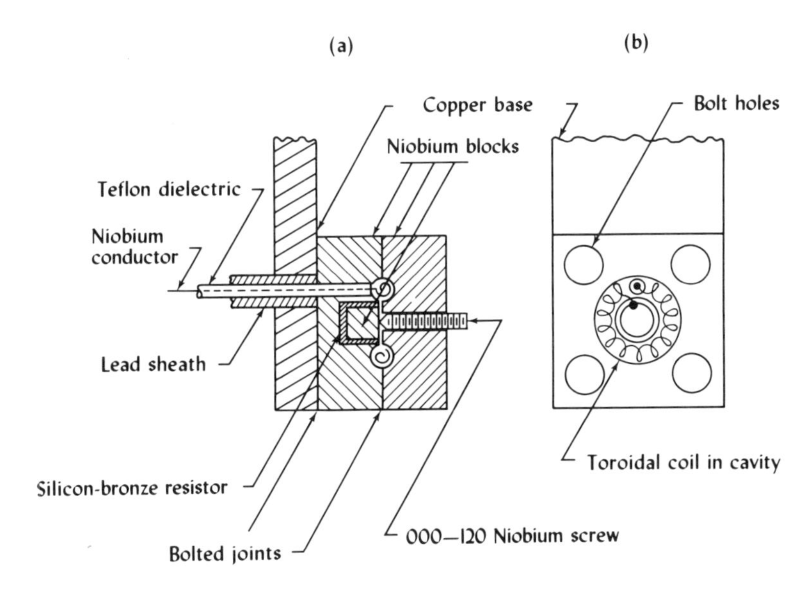

Fig. 18. Details of a resistive SQUID designed for noise thermometry.

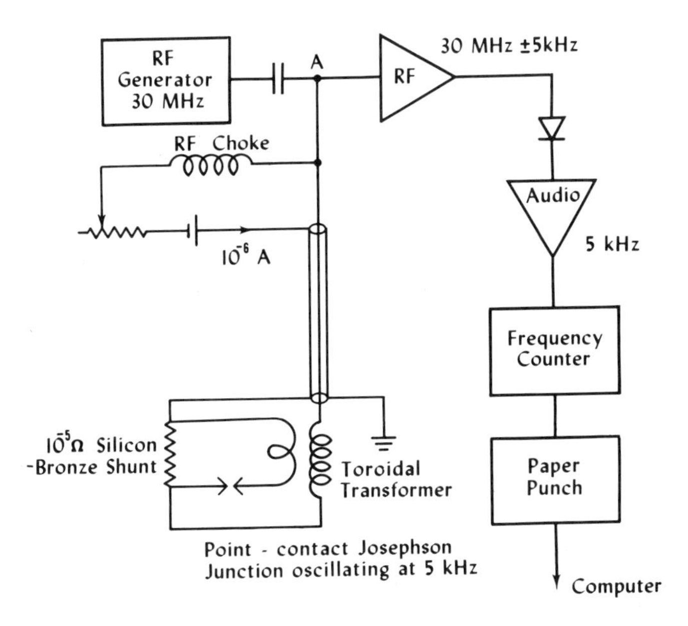

Fig. 19. System for noise thermometry using a resistive, oscillating SQUID. The paper punch was later eliminated by dedicating a small computer to the system.

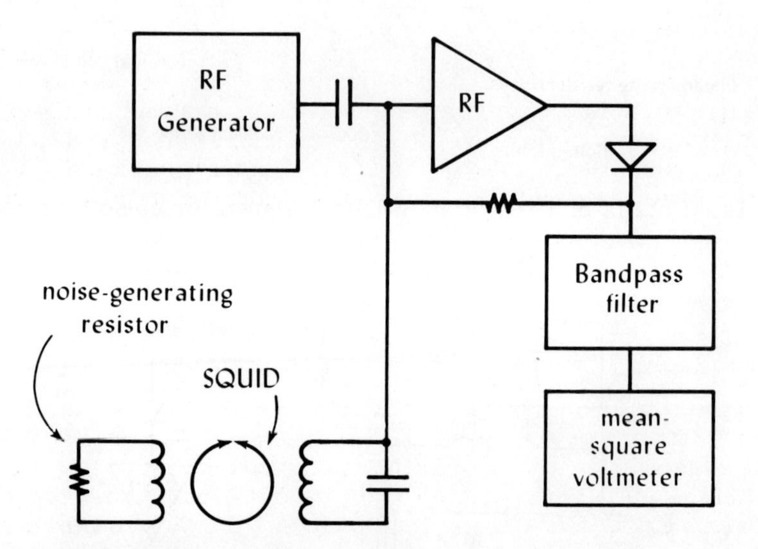

Fig. 20. System for noise thermometry using a flux-locked SQUID magnetometer inductively coupled to the noise-generating circuit.

a very careful comparison of the temperature scale defined by this thermometer and that defined by the anisotropy of γ - radiation from ^{60}Co. After discovery of the error in Eq. (45) they found that the two theromometers agree to within 1% in the range from 12 mK to 35 mK [78]. Some of their earlier publications report a discrepancy of approximately 1 mK due to this error, for which I take responsibility.

Let us now turn to Wheatley's noise thermometer [71, 72], shown schematically in Fig. 20. In this arrangement the SQUID is used as a regular flux-locked magnetometer to detect the fluctuations in current in a coil connected in parallel with the resistor generating the noise to be measured. The output of the magnetometer is passed through a bandpass filter (to define the bandwidth that appears in Nyquist's formula) to an integrating voltmeter that records the mean square voltage. The fundamental difference between this and the NBS noise thermometer is that the resistor is coupled to the SQUID via a transformer, so the requirements for the design of a low-noise SQUID do not constrain the choice of resistor. Also, the effective bandwidth of this system is greater than that imposed by the effects of frequency modulation on the NBS design. The result is that greater precision can be achieved in a given integrating time t. The available precision of all noise thermometers is described by a formula derived by Rice [79]

$$\Delta T/T = (t\ \delta f)^{-1/2} \quad , \tag{48}$$

where ΔT is the rms uncertainty in the measured noise temperature, t is the integrating time, and δf is the bandwidth of the system.

There is no doubt that these advantages make Wheatley's sytem a superior thermometer for everyday use. However, for absolute measurements it is necessary to know: (a) the effective bandwidth of the filter; (b) the gain of the SQUID magnetometer; (c) the value of the resistor; and (d) the transformer ratio of the coupling to the SQUID. These can be calculated or measured independently with a total uncertainty [71] of ±3%. Alternatively, this thermometer could be calibrated at one fixed point and used as a secondary thermometer. In this mode it was used for some interesting measurements of the melting curve of ^{3}He [80]. For the special purpose of realizing the absolute Kelvin scale, the NBS design has the advantage of fewer uncertainties.

C. Magnetic Thermometry with SQUIDs

The advantage that the SQUID magnetometer brings to magnetic thermometry is its great sensitivity, which radically extends the useful range [67], particularly of nuclear magnetic thermometers. With a SQUID magnetometer the nuclear magnetic susceptibility of copper may be measured directly at temperatures up to 1K (the susceptibility is inversely proportional to temperature). This does not create a good primary thermometer, however, because other contributions to the total magnetization measured by the SQUID can be eliminated only by calibration at two fixed points. The SQUID also brings an advantage to magnetic thermometry using paramagnetic salts, because it allows a drastic reduction in the size and hence the heat capacity of the salt specimen.

D. Superconducting Fixed Points

The primary thermometers used to establish the Kelvin scale are often slow and inconvenient to use, so practical thermometry is usually done with handier secondary thermometers that must be calibrated before use. The transfer of calibration from primary to secondary thermometers is usually done by means of fixed points defined by changes of state in various well characterized substances. In the temperature range below 10 K, the transition temperatures T_c of pure superconducting metals have proven to be the most suitable for this purpose [67].

Great care must be taken with the metallurgical preparation of specimens. The metals must be pure, strain-free, and with large grain size in order to show a sharp transition at a reproducible value of T_c. Also, when making measurements of temperature the ambient magnetic field must be either eliminated or measured. The transition temperatures of lead, indium, aluminum, zinc, and cadmium have been established for this purpose [81]. Specimens of these metals with certified transition temperatures are available as Standard Reference Materials from NBS [82], mounted in a device with which calibrations of secondary thermometers may be reproduced with a precision of ± 0.001K. The use of other superconductors to extend the range of this device is under investigation at NBS.

VI. MEASUREMENTS OF FREQUENCY

Frequency can be measured much more accurately than any other quantity. The second is defined to ten significant figures with reference to the frequency of the hyperfine structure interval in the ground state of ^{133}Cs, and the best national frequency standards are designed to realize this definition with an accuracy of about one part in 10^{13}. The statistical description of the uncertainty of measurements of time and frequency is also in a more advanced state than has been reached in the analysis of the measurements of other quantities. The entire field has been surveyed very thoroughly by Blair et al. [83]. The contributions of superconducting devices have been in the development of very stable oscillators and in the extension of accurate measurement of frequency into the sub-millimeter wave region of the spectrum. I will review these briefly after a discussion of the generally accepted language for specifying the performance of stable oscillators.

A. The Stability of Oscillators

The driving force behind the development of stable oscillators has been the need to measure time. This has encouraged a long-term view of the question of fluctuations and stability, which led to the realization that no natural processes are truly stationary, in the sense of being describable by variables that fluctuate about a time-invariant mean value. Another way of stating this is to assert that all observed noise spectra diverge at very low frequencies. Consequently the description of an oscillator by a stationary measure such as the frequency spectrum of its output has only limited usefulness that is confined to short-term activities such as interferometry. A generally useful measure of performance must be capable of specifying random, long-term, time-dependence.

The measure that is commonly used is the Allan variance [76, 83]. This is defined with reference to a set of N measurements of the frequency in question, each measurement occupying an interval of time τ and being separated from the next by an interval T. The Allan variance is the average value of the mean square deviation of the measured frequencies in the set divided by their mean. The question of confidence in the measured Allan variance and its relationship to the number of repetitions of the measurements making up the set has received attention [84]. The Allan variance is a very useful measure of performance if it is expressed as a function of the parameters N, T, and τ. Otherwise it contains but limited information.

For many practical purposes the particular Allan variance with N = 2, T = τ contains sufficient information. This is a function $\sigma_y^2(\tau)$ of the single parameter τ given by

$$\sigma_y^2(\tau) = 2\,\overline{(f_{k+1} - f_k)^2 / (f_{k+1} + f_k)^2} \tag{48}$$

where f_k is one of a set of measurements of frequency made in successive intervals of time τ with no dead time in between. This definition is obviously related to the description of practical measurements of frequency with a counter.

Using the square root, $\sigma_y(\tau)$, of the Allan variance as an index of performance, Fig. 21 shows a comparison of the best published examples of all the types of oscillator that might be considered for use as frequency standards. It is taken from an unpublished CCIR report, with extra material added. This plot shows clearly that great spectral purity (short-term stability) does not necessarily imply the best long-term stability. Of course, Fig. 21 does not tell the whole story. For example, the hydrogen maser appears to be capable of greater long-term stability than the cesium beam-controlled oscillator, but the latter is preferred for a frequency standard because, with proper precautions, systems of this type built and operated independently of each other can be expected to generate the same frequency within one part in 10^{13}. Also, the τ-axis of Fig. 21 spans only ~12 days. All the curves shown will eventually turn upwards if they are continued long enough in time.

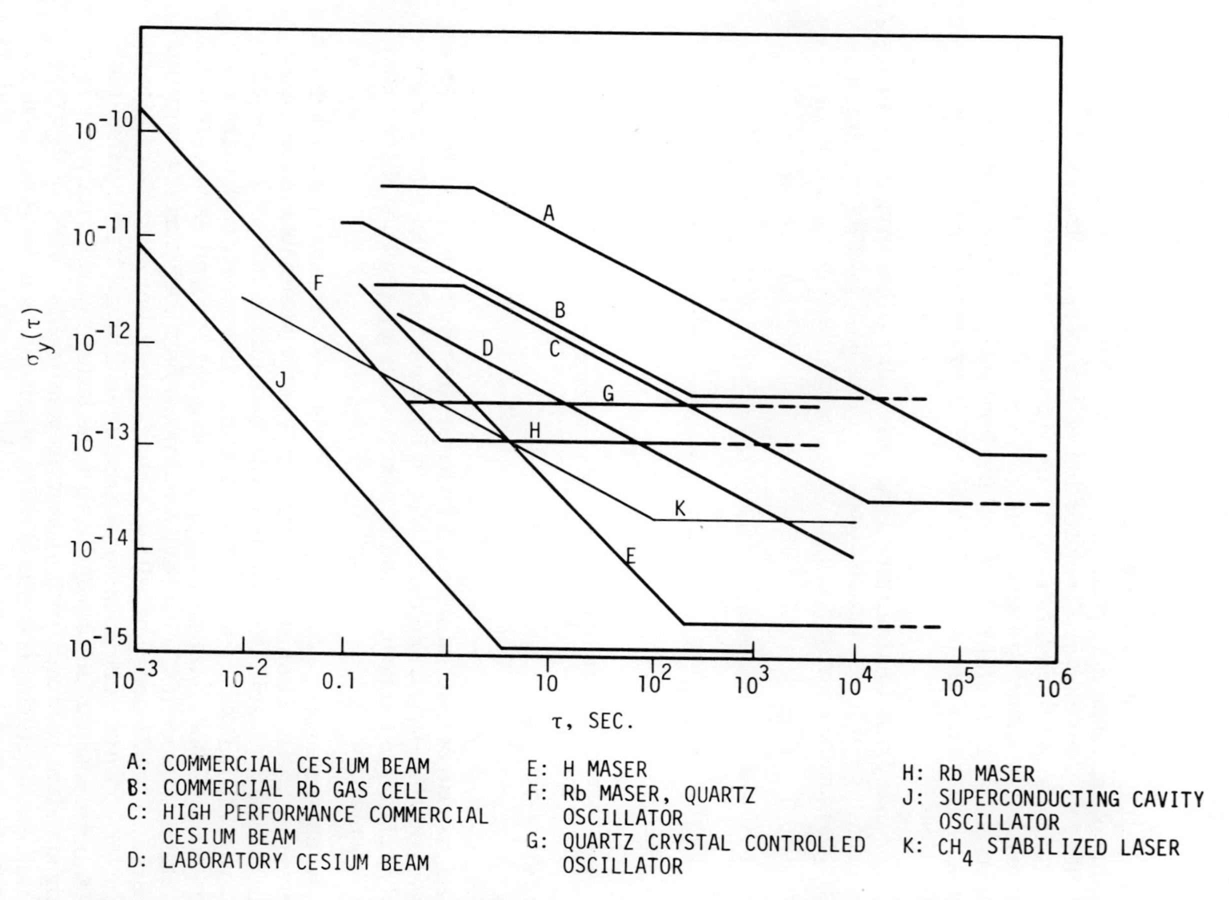

Fig. 21. Plot of square root of Allan variance $\sigma_y(\tau)$ versus measurement time interval τ for the best published examples of various types of very stable oscillators.

With these words of caution, Fig. 21 does show the remarkable performance of a microwave oscillator using a superconducting cavity resonator. In the range of integrating times less than 1000 seconds it is unsurpassed in stability. It cannot become an independently realizable primary frequency standard to rival the cesium beam, but practical uses will probably be found for its unique characteristics.

B. Oscillators with Superconducting Cavity Resonators

The obvious attraction of a superconducting cavity for use as the resonator of an oscillator is the possibility of using its very high Q to suppress the effects of noise and instability in the other components. This has proven to be a reasonable expectation after careful attention was given to stabilizing the resonant frequency of the superconducting cavity and to the design of the driving circuit. The performance shown in Fig. 21 was obtained [85] with a cavity resonator with a Q of 10^{11}. It is interesting to note that the next best oscillator for short-term stability (which is affected most by the Q of the resonator) uses a quartz crystal resonator with a Q of only a few million, and its performance only falls short by a factor 10.

A second advantage of the superconducting cavity resonator is that its natural place is in a microwave oscillator, so if the ultimate purpose is some measurement at a microwave or higher frequency, the degradation of the signal introduced by frequency multipliers is at least partly avoided.

The early work on oscillators with superconducting resonators [86, 87] coincided with the beginning of a large effort to develop superconducting cavities for linear accelerators, mainly in the universities of Stanford and Karlsruhe and at Siemens. This effort led to the emergence of niobium as the best material, and the development of the technology required to make and use cavities with Q up to 10^{11}. Excellent reviews have been published by Turneaure [88], Pierce [89], and Stein [85]. Stein's review concentrates on the design and performance of oscillators, and covers most of what I will say here in greater detail.

At high frequencies electric fields can penetrate into superconductors, so the current induced by a microwave field consists of both quasi-particles and Cooper pairs. The loss associated with the quasi-particle current sets the ultimate limit to the attainable Q of a cavity resonator. Since this scales with the quasi-particle density, it is necessary to use a superconductor with a high transition temperature operating at a very low temperature. To get a Q of 10^{11} required a niobium cavity operating at 1.25 K [90]. Very careful preparation of the internal surface of the cavity is also required, either by firing in ultra high vacuum and chemical polishing or by anodizing [89]. Cavities with "natural" surfaces usually have values of Q below 10^6, the extra loss being due to hysteresis in the distribution of surface currents.

In addition to using a cavity with very high Q, mechanical vibration must be kept to a very low level in order to attain the short-term stability shown in Fig. 21.

For long-term stability of the resonant frequency of the cavity, it is necessary to stabilize the factors that influence its linear dimensions and the penetration depth of the superconducting walls. The most important of these are: the temperature; the microwave power level; the ambient magnetic field; and the orientation of the cavity in the gravitational field.

The driving circuit also needs careful attention. Performance of the system can be degraded by phase noise in the amplifier that drives the oscillation, and by instability of the phase delay in the transmission lines connecting the various components. This latter effect is a particularly severe problem with the lines that pass in and out of the cryostat. It can be solved either by using components that can operate in liquid helium for the whole system, so that lines can be kept short and stable, or by encoding the essential phase information in such a way that its transmission is insensitive to the phase delay of the lines.

The oscillator developed by Stein and Turneaure [91] (whose performance is represented in Fig. 21) used the latter technique. It consists of a voltage-controlled Gunn diode oscillator (operating at ambient temperature) with its frequency stabilized by a modified Pound stabilizer using the superconducting cavity as a reference. Part of the output of the oscillator (at 8.6 GHz) is phase modulated at 1 MHz and reflected from the port of the superconducting cavity. This converts the phase modulation to an amplitude modulation with a phase that depends on the sign of the offset of the center frequency from the resonant frequency of the cavity. This amplitude modulated signal is then rectified and its phase is compared with that of the original frequency modulation to derive an error voltage that is applied to control the frequency of the Gunn diode. One extra advantage of this system is that its power output does not pass through the cavity and hence is not limited by breakdown of the resonator by the high fields that are a consequence of the large value of Q.

No other oscillator exists at present that can match the performance of this one. Stein (private communication) is working on another oscillator that is driven parametrically by a varactor diode that operates in liquid helium. The superconducting cavity will be a part of the primary oscillating circuit. This will limit the available power, but is expected to improve the spectral purity of the signal. This oscillator has not been tested yet.

An interesting application of one of these super-stable oscillators was reported recently by Turneaure and Stein [92]. They compared the frequencies of an ensemble of superconducting cavity oscillators with those of an ensemble of cesium-beam oscillators over a 12-day period in order to check the constancy of the Fine Structure Constant α, which is directly connected to the frequency of the cesium hyperfine transition, but not to the resonant frequency of the superconducting cavity. They concluded that the rate of change of α is less than 4×10^{-12} per year. Other applications are possible, for example to long baseline interferometry, but they have not been realized yet.

C. Far Infrared Frequency Synthesis

The function of a Josephson junction in frequency synthesis is as a harmonic mixer, to generate various combinations of harmonics and their sum and difference frequencies from the frequencies of a pair of applied signals. The rival devices are the Schottky barrier diode at lower frequencies and the tungsten whisker diode in the sub-millimeter range. In the range in which a Schottky barrier diode can operate (at frequencies below about 100 GHz) the Josephson junction has no great advantage and there is no good reason to use it. In the sub-millimeter range the Josephson junction does have a clear advantage in its ability to convert a large portion of the power in an incident signal into harmonics of very high order.

Accurate frequency metrology in the far infrared region of the spectrum has received so much attention because of the development of extremely stable lasers [93], some of them intrinsically stable and others stabilized by a saturated absorption cell. Of these, the one that has received the most attention is the helium-neon laser stabilized by saturated absorption in methane. Its frequency has been measured [93] by establishing a chain of frequency markers (a klystron at 74 GHz, and HCN, H_2O, and CO_2 lasers at higher frequencies) by harmonic mixing to compare its frequency indirectly with that of the Cs beam frequency standard at 10 GHz. Tungsten whisker diodes were used throughout. The Allan variance has also been measured [94], and is plotted against integrating time in Fig. 21. It can be seen that the methane-stabilized laser is a possible rival to the cesium beam as a frequency standard. It is also several orders of magnitude more stable and reproducible than the krypton lamp that serves for the present definition of the meter, and there has been talk of using it for a unified standard of length and time. One of the conditions required for this to be taken seriously is that frequency measurements must be referable to the methane-stabilized laser with something less than the experimental tour de force that is required now.

Other frequency markers have been established in the far infrared. The carbon dioxide laser has a copious spectrum of rotation-vibration transitions, many of which have been measured with the aid of a Josephson junction mixer [95]. These can be used to synthesize frequencies close to an unknown frequency anywhere in the band from 0.025 to 100 THz [96], thereby opening up a new field of very accurate spectroscopy that may eventually cover the whole infrared range.

The performance of the Josephson junction as a harmonic mixer was demonstrated by McDonald et al [97], who reported the 400th order harmonic mixing of signals from a klystron at 9.5 GHz and a water-vapor laser at 3.8 THz to generate a measurable IF signal at 9 GHz. Also, Blaney and Knight [98] have reported 825th order mixing between a signal at 1.08 GHz derived from a quartz crystal oscillator and a signal at 891 GHz from a HCN laser, generating an IF signal at 30 MHz. For this purpose point-contact junctions are used, with one electrode being a finely etched niobium wire.

A question of great interest, to which only a partial answer is available at present, is the upper limit to the frequency at which a Josephson junction can operate. The current theory of the Josephson effect would place this limit near 10 THz for presently known superconductors [99]. However, the measurement of the rotational structure of the CO_2 spectrum [95] required mixing signals at frequencies near 30 THz. McDonald et al. [25] have published an interesting discussion of the significance of this observation. They concluded that the Josephson effect was probably the dominant mechanism, but that other non-linear effects could have been contributing. It appears, then, that Josephson junctions are unlikely to be useful for frequency synthesis to significantly higher frequency than a few tens of THz. However, if a need arises for more activity in measuring frequencies in the far infrared, then the use of Josephson junctions could simplify the process by eliminating the need for multiple reference frequencies below 10 THz. All other harmonic mixers are only usable at order numbers less than about 10.

D. Recent Work

The most active group in applying superconductivity to frequency metrology at present is at NBS. The intention is to start with a superconducting cavity oscillator at about 10 GHz and to multiply its frequency with a Josephson junction in a single step to fix the frequency of an optically pumped methyl alcohol laser at 4.25 THz. Tungsten whisker diodes will then be used to complete the chain for a redetermination of the frequency of the methane-stabilized helium-neon laser. The hope is that various improvements, including the elimination of intermediate steps below 4.25 THz, will enable the measurement of frequency to be made with at least an order of magnitude more accuracy than heretofore.

REFERENCES

1. C.H. Page and P. Vigoureux, editors, "The International System of Units (SI)", NBS Special Publication 330 (1974).
2. B.N. Taylor, W.H. Parker, and D.N. Langenberg, "The Fundamental Constants and Quantum Electrodynamics", Academic Press, New York (1969).
3. P. Vigoureux, Metrologia 1, 3 (1965).
4. R.L. Driscoll and R.D. Cutkosky, J. Res. NBS 60, 297 (1958).
5. R.L. Driscoll, J. Res. NBS 60, 287 (1958).
6. R.L. Driscoll and P.T. Olsen, report to CCE of the CIPM (1968).
7. A.M. Thompson and D.G. Lampard, Nature 177, 888 (1956).
8. R.D. Cutkosky, IEE Trans. I and M, IM-23, 305 (1974).
9. W.J. Hamer, "Standard Cells, their Construction, Maintenance, and Characteristics", NBS Monograph 84 (1965).
10. W.G. Eicke and B.N. Taylor, IEEE Trans. I and M, IM-21, 316 (1972).
11. S. Shapiro, Phys. Rev. Letters 11, 80 (1963).
12. D.N. Langenberg, W.H. Parker, and B.N. Taylor, Phys. Rev. 150, 186 (1966).
13. B.N. Taylor, W.H. Parker, D.N. Langenberg, and A. Denenstein, Metrologia 3, 89 (1967).

14. T.F. Finnegan, T.J. Witt, B.F. Field, and J. Toots, "Atomic Masses and Fundamental Constants," edited by J.H. Sanders and A.H. Wapstra, Plenum Press, New York, p. 403 (1972).
15. B.W. Petley and K. Morris, Metrologia 6, 46 (1970).
16. V. Kose, F. Melchert, H. Fack, and H.J. Schrader, PTB Mitteilungen 81, 8, (1971).
17. I.K. Harvey, J.C. MacFarlane, and R.B. Frenkel, Phys. Rev. Letters 25, 853 (1970).
18. B.F. Field, T.F. Finnegan, and J. Toots, Metrologia 9, 155 (1973).
19. V. Kose et al, IEEE Trans. I and M, IM-23, 271 (1974).
20. I.K. Harvey, J.C. MacFarlane, and R.B. Frenkel, Metrologia 8, 114 (1972).
21. J.C. Gallop and B.W. Petley, IEEE Trans. I and M, IM-21, 310 (1972).
22. G.H. Wood, A.F. Dunn, and L.A. Nadon, IEEE Trans. I and M, IM-23, 275 (1974).
23. T.F. Finnegan, A. Denenstein, and D.N. Langenberg, Phys. Rev. B4, 1487 (1971).
24. P. Russer, J. Appl. Phys. 43, 2008 (1972).
25. D.G. McDonald, F.R. Petersen, J.D. Cupp, B.L. Danielson, and E.G. Johnson, Appl. Phys. Letters 24, 335 (1974).
26. N.R. Werthamer, Phys. Rev. 141, 255 (1966).
27. V.E. Kose and D.B. Sullivan, J. Appl. Phys. 41, 169 (1970).
28. T.F. Finnegan, J. Wilson, and J. Toots, IEEE Trans. Mag. MAG-11, 821 (1975).
29. J. Clarke, Phys. Rev. Letters 21, 1566 (1968).
30. T.D. Bracken and W.O. Hamilton, Phys. Rev. B6, 2603 (1972).
31. A. Denenstein and T.F. Finnegan, Ref. Sci. Instr. 45, 735 (1974).
32. T.F. Finnegan and A. Denenstein, Rev. Sci. Instr. 44, 944 (1973).
33. J.J. Hill and A.P. Miller, J. IEE 109B, 157 (1962).
34. T.L. Zapf, ISA Trans. 2, 195 (1963).
35. D.N. Homan and T.L. Zapf, ISA Trans. 9, 201 (1970).
36. N.L. Kusters and W.J.M. Moore, IEEE Trans. I and M, IM-13, 107 (1964).
37. N.L. Kusters and M.P. MacMartin, IEE Trans. I and M, IM-24, 331 (1975).
38. D.B. Sullivan, Rev. Sci. Instr. 43, 499 (1972).
39. P. Warnecke and V. Kose, Rev. Sci. Instr. 46, 1108 (1975).
40. V. Kose, B. Fuhrmann, P. Warnecke, and F. Melchert, Proc. AMCO-5 (to be published).
41. I.K. Harvey and H.C. Collins, Rev. Sci. Instr. 44, 1700 (1973).
42. J.E. Zimmerman, J. Appl. Phys. 42, 4483 (1971).
43. R.F. Dziuba, B.F. Field, and T.F. Finnegan, IEEE Trans. I and M, IM-23, 264 (1974).
44. I.K. Harvey, Metrologia (to be published).
45. I.K. Harvey, Rev. Sci. Instr. 43, 1626 (1972).
46. I.K. Harvey, J. Phys. E. Sci. Instr., 6, 812 (1973).
47. D.B. Sullivan and R.F. Dziuba, Rev. Sci. Instr. 45, 517 (1974).
48. D.B. Sullivan and R.F. Dziuba, IEEE Trans. I and M, IM-23, 256 (1974).

49. K. Grohmann, H.D. Hahlbohm, H. Lübbig, and H. Ramin, PTB Mitteilungen 83, 313 (1973).
50. K. Grohmann, H.D. Hahlbohm, H. Lübbig, and H. Ramin, IEEE Trans. I and M, IM-23, 261 (1974).
51. B.F. Field, private communication.
52. G.F. Engen and C.A. Hoer, IEEE Trans. MTT, IM-21, 470 (1972).
53. C.A. Hoer and K.C. Roe, IEEE Trans. MTT, MTT-23, 978 (1975).
54. MIT Radiation Laboratory Series, edited by L.N. Ridenour, McGraw-Hill, New York.
55. E.L. Ginzton, "Microwave Measurements", McGraw-Hill, New York (1957).
56. IEEE Trans. I and M: the proceedings of CPEM usually appear as special issues in December of even-numbered years.
57. R.A. Kamper, M.B. Simmonds, R.T. Adair, and C.A. Hoer, Proc. IEEE 61, 121 (1973).
58. R.A. Kamper, M.B. Simmonds, C.A. Hoer, and R.T. Adair, NBS Tech. Note 643 (1973); and NBS Tech. Note 661 (1974).
59. R.A. Kamper and M.B. Simmonds, Appl. Phys. Letters 20, 270, (1972).
60. N.V. Frederick, D.B. Sullivan, and R.T. Adair, 1976 Applied Superconductivity Conference, to be published in IEEE Trans. Mag.
61. R.A. Kamper, IEEE Trans. Mag. MAG-11, 141 (1975).
62. H. Kanter and F.L. Vernon, 1976 Applied Superconductivity Conference, to be published in IEEE Trans. Mag.
63. British Association Mathematical Tables, Vol. VI (1950).
64. A.H. Silver and J.E. Zimmerman, "Applications of Superconductivity", vol. 1, edited by V.L. Newhouse, p.1, Academic Press, New York (1975).
65. B.W. Petley, K. Morris, R.W. Yell, and R.N. Clarke, Electronics Letters 12, 237 (1976).
66. J.R. Waldram, Adv. Phys. 13, 1 (1964).
67. R.P. Hudson, H. Marshak, R.J. Soulen, and D.B. Utton, J. Low Temp. Phys. 20, 1 (1975).
68. R.A. Kamper, Proc. Symposium on the Physics of Superconducting Devices, Charlottesville, Virginia, P.M-1 (1967).
69. R.A. Kamper and J.E. Zimmerman, J. Appl. Phys. 42, 132 (1971).
70. R.A. Kamper, J.D. Siegwarth, R. Radebaugh, and J.E. Zimmerman, Proc. IEEE 59, 1368 (1971).
71. R.A. Webb, R.P. Giffard, and J.C. Wheatley, J. Low Temp Phys. 13, 383 (1973).
72. J.C. Wheatley and R.A. Webb, Science, 182, 531 (1969).
73. K.F. Knott, Electronics Letters 1, 132 (1965).
74. M.J. Stephen, Phys. Rev., 182, 531 (1969).
75. J.A. Barnes, Proc. IEEE, 54, 207 (1966).
76. D.W. Allan, Proc. IEEE, 54, 221 (1966).
77. A.H. Silver, J.E. Zimmerman, and R.A. Kamper, Appl. Phys. Letters, 11, 209 (1967).
78. R.J. Soulen and H. Marshak, Proc. LT 14, Vol. 4, edited by M. Krusius and M. Vuorio, North Holland (1976), p. 60.

79. S.O. Rice, Bell System Tech. J., 23, 282 (1944).
80. R.T. Johnson, D.N. Paulson, C.B. Pierce, and J.C. Wheatley, Phys. Rev. Letters, 30, 207 (1973).
81. R.J. Soulen, Cryogenics, 14, 250 (1974).
82. J.F. Schooley, R.J. Soulen, and G.A. Evans, NBS Special Publication 260-44 (1972).
83. B.E. Blair, editor, "Time and Frequency: Theory and Fundamentals", NBS Monograph 140 (1974).
84. P. Lesage and C. Audoin, IEEE Trans. I and M, IM-22, 157 (1973).
85. S.R. Stein, Proc. 29th Symp. on Freq. Control (obtainable from Electronic Industries Association) (1975).
86. J.L. Stone and W.H. Hartwig, SWIEEECO Record (obtainable from IEEE) p. 9-3-1 (1967)
87. F. Biquard, P. Grivet, and A. Septier, Electronics Letters 4, 143 (1968).
88. J.P. Turneaure, Proc. 1972 Applied Superconductivity Conference, IEEE Publication 72 CHO682 - 5 - TABSC, p. 621 (1972).
89. J.M. Pierce, "Methods of Experimental Physics", vol. 2, edited by R.V. Coleman, Academic Press, New York, p. 541 (1974).
90. J.P. Turneaure and Nguyen Tuong Viet, Appl. Phys. Letters, 16, 333 (1970).
91. S.R. Stein and J.P. Turneaure, Proc. 27th Symp. on Freq. Control (1973).
92. J.P. Turneaure and S.R. Stein, Proc. AMCO-5 (to be published).
93. K.M. Evenson and F.R. Petersen, "Topics in Applied Physics", vol. 2, edited by H. Walter, Springer-Verlag, Berlin, p. 349 (1976).
94. C. Borde and J.L. Hall, Proc. of the Laser Spectroscopy Conference, Vail, edited by R.G. Brewer and A. Mooradian, Plenum Press, New York, p. 125 (1974).
95. F.R. Petersen, D.G. McDonald, J.D. Cupp, and B.L. Danielson, Phys. Rev. Letters, 31, 573 (1973).
96. F.R. Petersen, K.M. Evenson, D.A. Jennings, J.S. Wells, K. Goto, and J.J. Jimenez, IEEE J. Quantum Electronics QE-11, 838 (1975).
97. D.G. McDonald, A.S. Risley, J.D. Cupp, K.M. Evenson, and J.R. Ashley, Appl. Phys. Letters, 20, 296 (1972).
98. T.G. Blaney and D.J.E. Knight, J. Phys. D: Appl. Phys., 7, 1882 (1974).
99. P.L. Richards, F. Auracher, and T. Van Duzer, Proc. IEEE, 61, 36 (1973).

Chapter 6

HIGH FREQUENCY PROPERTIES AND APPLICATIONS OF JOSEPHSON JUNCTIONS FROM MICROWAVES TO FAR-INFRARED

R. Adde and G. Vernet

Institut d'Electronique Fondamentale, Université Paris-Sud

91405 - Orsay, France

I. GENERAL PROPERTIES OF JOSEPHSON JUNCTIONS FOR HIGH FREQUENCY APPLICATIONS

A. High Frequency Fundamental Properties of the Ideal Josephson Junction

A domain where the Josephson effect may be promising is the very high frequency region (submillimeter and far infrared). The fundamental reason is the very high sensitivity of the Josephson junction (JJ) to any applied electric signal. The second of the two familiar Josephson equations for an ideal junction [1]

$$I(t) = I_c \sin\phi \qquad (1)$$

and

$$d\phi/dt = (2e/\hbar)\,V(t) \qquad (2)$$

shows that a small variation of the applied voltage to a junction leads to a very strong variation of the phase difference between the two electrodes ($2e/h = 484\,GHz/mV$). This very high frequency modulation coefficient is a characteristic of the Josephson junction and is responsible for its high nonlinearity. It is exploited in high frequency generation, detection, mixing or parametric amplification. Other characteristics of Josephson junctions are high cut-off frequencies above which the nonlinearities get worse for different reasons, and low level intrinsic noise.

B. The Parallel Impedance of Real Josephson Junctions

The Josephson current described by Eqs. (1) and (2) consists of a transfer of electron pairs without energy loss from one side of the weak link to the other. In real junctions there is a parallel component of lossy single electron current, and a component of displacement current related to the junction capacitance. A corresponding equivalent circuit (Fig. 1) may be used to represent these three current components and analyze many properties of Josephson devices [2-4]. A good agreement is obtained with most experimental results in detection and mixing applications.

The parallel impedance of a JJ plays a fundamental role in the behavior of the real junction both in the dc and ac regimes. The nature of this impedance depends on the type of JJ. At not too high frequencies (microwave region) the point contact parallel impedance is essentially a constant resistance which takes into account the current flow due to an imperfect dielectric and (or) metallic shorts. The equivalent circuit of a point contact (still the preferred high frequency Josephson device despite its mechanical instability) is therefore the familiar Resistive Shunted Junction (RSJ) model. The capacitive impedance of oxide tunnel SIS junctions is preponderant at microwaves except for very small junctions. At the higher frequencies it is necessary to include the parasitic capacitance of a point contact and the strongly voltage-dependent resistance $R_{qp}(V)$ which is related to the quasiparticle tunneling current appearing at the gap voltage. Microbridges have a very small capacitance and may often be treated with the RSJ model.

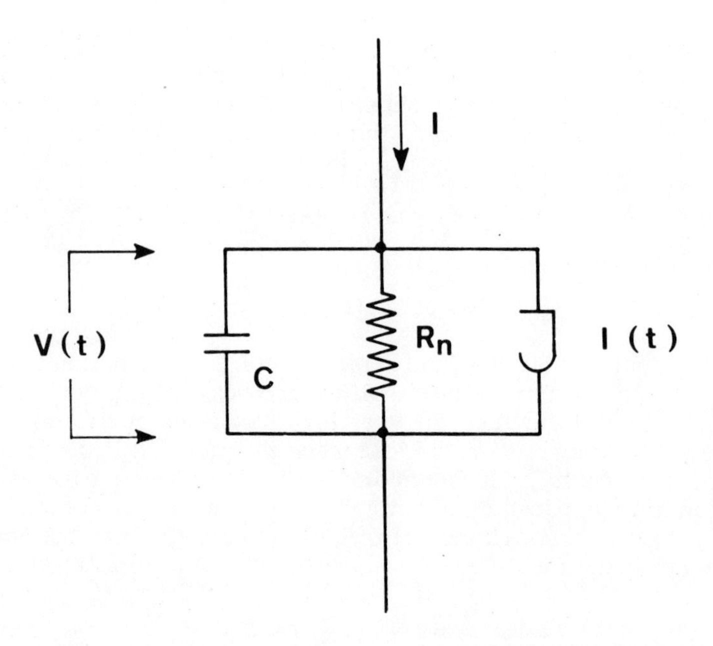

Fig. 1. Circuit model of a Josephson junction.

A characteristic of Josephson junctions is their low impedance both at dc and ac (~0.1 to 10 Ω). In particular it is low compared to the usual rf impedances (free-space impedance = 377 Ω). This means that JJs are usually current biased. Therefore, the voltage at the junction is driven by the parallel impedance and the alternative Josephson current flowing through this impedance modulates the junction voltage. This feedback (a resistive feedback for point contacts) produces a strong harmonic distortion of the Josephson oscillation (modulated or not by an external radiation) and is responsible for its spectrum rich in harmonic components [4]. It is this phenomenon which is at the basis of the operation of a point contact as a mixer or harmonic mixer with an external local oscillator (see Section IF).

This feedback mechanism is responsible for the shape of the (VI) characteristic (later referred to as VIC) of Josephson junctions. An ideal junction, i.e., a junction whose properties are described by the two fundamental Josephson equations, can only be voltage biased since $\overline{I_j(t)} = \overline{I_c \sin \phi} = 0$ for $V \neq 0$. The corresponding VI characteristic is shown in Fig. 2a. For point contacts, the parallel resistive impedance creating a distortion of the Josephson oscillation, a dc component $\overline{I_J(t)} \neq 0$ appears for $V \neq 0$. The measured dc current flowing through the junction is the sum of this component and the current carried by the shunt. The corresponding VI characteristic is shown in Figs. 2b and 3. It can be seen that the harmonic distortion decreases at large bias values. If there were no harmonic distortion, the VIC at $V \neq 0$ would be the line $V = R_N I$.

The influence of a large capacitance may be understood in the following way. By shorting out the harmonics of the fundamental Josephson oscillation, the capacitance suppresses any possible current bias of the junction because the solution $\overline{V(t)} \neq 0$ does not exist, and the VIC switches directly from the zero voltage current to the ohmic characteristic (Fig. 2c). This is the situation corresponding to large area point contacts and to oxide tunnel junctions. When the latter are current biased, the VIC switches from $I = I_c (V = 0)$ to the quasiparticle current tunneling characteristic (Fig. 2d).

C. Limiting Factors of Josephson Junctions at High Frequencies

The limiting factors of JJ's have several origins: (a) the physical mechanism itself responsible for the Josephson effect, (b) the geometrical structure of the junction related to its coupling to the radiation at the different frequencies involved, (c) fluctuation phenomena (noise), and (d) thermal effects. These different aspects are often intimately interrelated.

1. Frequency limitation related to the physical mechanism

The Josephson effect is due to a weak coupling between two superconducting electrodes I and II. The current in the junction is a function $I(\phi)$ of the quantum phase difference $\phi = \phi_I - \phi_{II}$. The particular

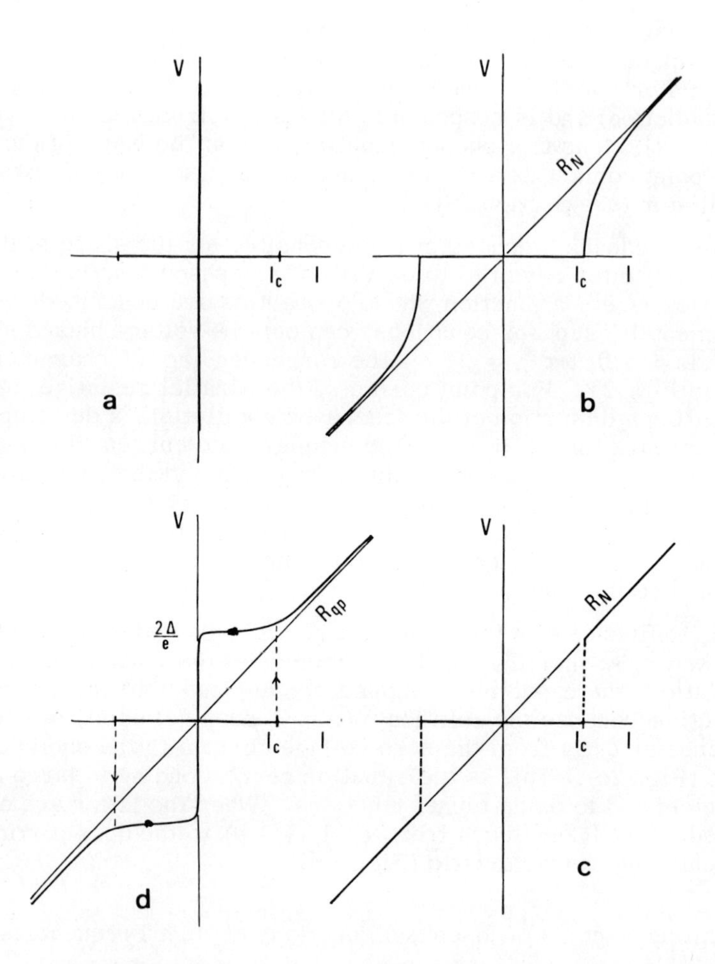

Fig. 2. VI characteristic of a Josephson junction: (a) voltage biased ideal junction; (b) current-biased real junction obeying the RSJ model; (c) current-biased real junction with strong capacitive effect; (d) tunnel SIS junction (the dotted line corresponds to switching in a current-biased junction).

relation $I(t) = I_c \sin\phi$ is a good approximation for many real junctions at not too high frequencies. This relation gives a proper description of the behavior of many types of junctions, independent of the physical mechanism responsible for the observed Josephson effect. On the contrary, at very high frequencies, limitations appear which are inherent to the physical processes in the particular junction.

a. Tunnel junction

The coupling in tunnel junctions is realized with a very thin oxide barrier ($\simeq 20$ Å). The alternating Josephson effect is interpreted in this case as a coherent tunneling current of Cooper pairs through the barrier with emission (or absorption) of photons. This process is nondissipative. Its limitation at high frequencies is related to the nature of the tunnel effect and therefore to the energy gap 2Δ of the superconducting electrodes. A more complete description of SIS junctions from microscopic theory leads to the following expression of the Josephson current [1]

$$I_J(V,T) = I_p(V,T)\sin\phi + I_{p-qp}(V,T)\cos\phi + I_{qp} + CdV(t)/dt \ . \quad (3)$$

The first term in $I_J(V,T)$ is the phase dependent Josephson pair tunneling current and is preponderant at bias voltages $V_o < 2\Delta/e$ for small capacitance junctions. The third term I_{qp} is the dissipative quasiparticle tunneling current and the last is the displacement current related to the junction capacitance. The second term is referred to as the $\cos\phi$ term or quasiparticle - pair interference term. It can be regarded as a "modulation" of the normal current due to coherence effects on the quasiparticle distribution [1, 5]. The voltage dependence of the components of $I_J(V,T)$ were first calculated at 0 K by Werthamer [6], but were given graphically at finite temperature recently [7]. The important result is the existence of a singularity in the $\sin\phi$ component, first predicted by Riedel [8], associated with a discontinuity in the $\cos\phi$ and quasiparticle components when the applied voltage is such that $eV_o = 2\Delta(T)$ as shown in Fig. 4. The amplitude of the normal and $\cos\phi$ terms becomes important only at $V_o \gtrsim 2\Delta/e$. The possible consequences of the $\cos\phi$ term have received attention only recently and are not yet well appreciated. Its experimental manifestations are probably rather subtle.

The high frequency properties of Josephson devices are therefore determined essentially by the $\sin\phi$ component amplitude which peaks at $hf_J = 2eV_o = 4\Delta$ and then decreases slowly (Fig. 4). The observation of the Riedel singularity in SIS junctions has been observed indirectly by the measurement of induced steps amplitudes, but there is a strong damping at the maximum [9, 10]. The frequency $f_G = 4\Delta/h$ may be considered as characterizing possible performances of SIS junctions at high frequencies (Table 1). Far above the gap at zero temperature the dependence of the Josephson current [6] with the voltage V_o is $I_j(V_o)/I_j(0) \simeq 2\Delta/eV_o (2\Delta/eV_o << 1)$. For example, in the case of

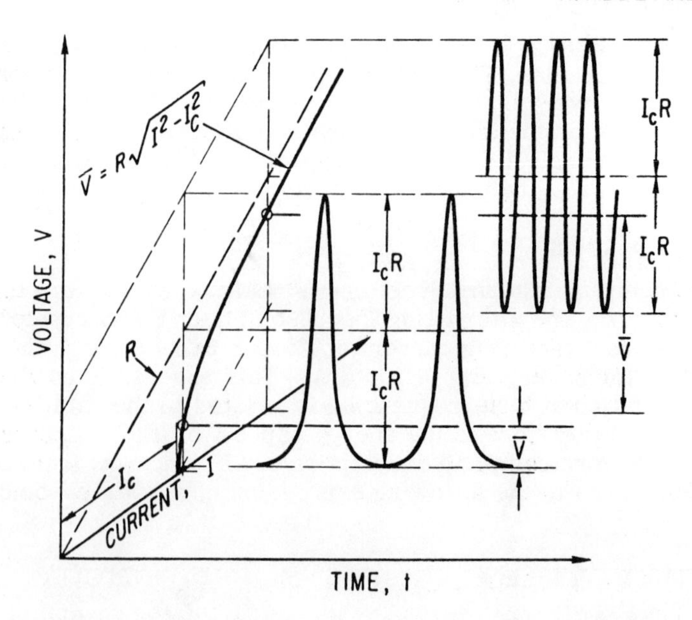

Fig. 3. Time evolution of the voltage in a current-biased junction obeying the RSJ model (Fig. 2b) showing the strong harmonic distortion related to the resistive feedback (taken from Ref. [44]). Note that in this figure $R = R_N$.

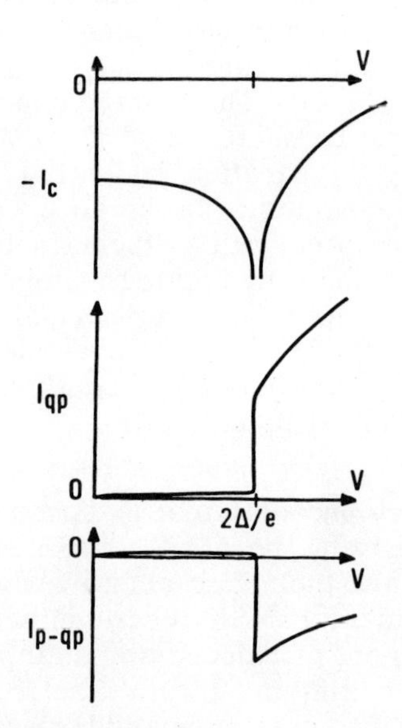

Fig. 4. Amplitudes of the tunnel current components. The current is written in the form $I = -I_c \sin\phi + I_{qp} + I_{p-qp} \cos\phi$ (Taken from Ref. [48]).

Table I. Characteristics of superconducting materials for high-frequency Josephson junctions.

Material	T_c (K)	2Δ (mV)	$f_G = 4\Delta/h$ (GHz)	$f_c = \frac{\pi}{4} f_G$ (GHz)	f_{Debye} (GHz)
Sn(T=0 K)	3.72	1.15	560	440	3400
Nb (T=4 K)	9.25	2.8	1350	1060	5850
NbN	16	4.8	2300	1800	
Nb_3Sn	18	5.4	2700	2100	
Nb_3Ge	21	6.4	3000	2400	

(For NbN, Nb_3Sn, Nb_3Ge, the energy gap is calculated assuming $2\Delta = 3.5\ kT_c$).

Niobium at 4 K, $f_G \simeq 1.35$ THz and at $f_J = 10\ f_G$, $I_J(f_J)/I_J(0)$ is still $\simeq 0.1$. These numbers indicate that the Josephson effect may still give an appreciable response well above the gap voltage and that high energy gap materials (Nb_3Sn) should be preferred.

A steeper decline of the Josephson current amplitude is predicted [11] at frequencies above the limiting frequency of the phonon spectrum in the metal (Debye frequency, see Table I).

b. Point contacts

Although point contacts are the most frequently used Josephson devices in high frequency applications, the physical processes leading to the observed Josephson effect still receive different interpretations. The situation is complicated since many parameters may influence the device operation (preparation technique, nature of the electrode surfaces, shape of the contact, etc.) and an analysis of the microphysical structure of the contact is very difficult. However, recent reports of the observation of the Riedel peak in superconducting point contacts [12, 13] prove that contacts with good high frequency detection performances have some physical properties which make them look similar to small oxide barrier tunnel junctions. The main difference between these two types of junctions is that the normal current component in a point contact is described by a resistance R_N in the RSJ model. It corresponds to a time constant τ connected with the current redistribution process in the superconductor between the normal and superconducting components (see Section IF2) where

$$\tau = f_c^{-1} = \left[\left(\frac{2e}{h}\right) R_N I_c\right]^{-1} \quad . \tag{4}$$

It can be shown that the junction sensitivity to external radiation of frequency f is often proportional to $(f_c/f)^2$. Therefore good high frequency performances of point contacts imply large values of the product $R_N I_c$.

The microscopic theory of SIS junctions allows us to calculate this product where the resistance in the case of an oxide tunnel junction is the resistance R_{qp} related to the quasiparticle current which becomes dominant when $eV \geqslant 2\Delta$. The result is

$$R_{qp} I_c = \pi\Delta/2e \quad . \tag{5}$$

When it is applied to point contacts one obtains

$$f_c = (\pi/4) f_G \quad \text{with} \quad f_G = 4\Delta/h \quad ,$$

a result which may be considered as a theoretical limit. Numerical values are listed in Table I. In practice the observed values of $R_N I_c$ with point contacts are often much smaller than the above limit (sometimes an order of magnitude). The reproducible realization of Josephson point contacts with good high frequency performances (large value of $R_N I_c$) requires a better understanding of the physical processes in these junctions, particularly the nature of the normal current component and the influence of the chemical preparation technique.

c. Microbridges.

The situation is complex for microbridges due to the variety of structures, the diversity of experimental results and the different mechanisms which have been proposed (variation of the order parameter in time and space, instantaneous 2n phase slip, vortex motion...). Important frequency limitation may be related to a corresponding characteristic relaxation time larger than the time constant related to the $R_N I_c$ product. Gap effects have also been observed in microbridges. However, thermal effects are found to be the dominant source of high frequency limitation for microbridges made from uniform thickness films. These effects are strongly reduced with variable thickness structures (Sec.I-C 3). The first theoretical studies of the time dependent Josephson effect in microbridges have been based on simple time-dependent generalizations of the Ginzburg-Landau theory which have a restricted range of validity. More elaborate analyses are actually developed [14] and may help to predict quantitatively the high frequency limitations of microbridges.

2. Geometrical structure and coupling

The geometrical structure of the junction has two important aspects.

a. Geometrical capacitance of the junction

The capacitance is an important factor mostly for SIS junctions. The corresponding cut-off frequency is the plasma frequency [1]

$$f_{pl} = \frac{1}{2\pi}\left(\frac{2e\,I_c}{\hbar c}\right)^{1/2} . \qquad (6)$$

For SIS junctions with standard dimensions ($\simeq 100\,\mu$), the capacitance is relatively large ($C \simeq 10 - 100$ pF) due to the very thin oxide layer ($\simeq 20$ Å) and the value of $f_{pl} < f_c$ determines the junction limitations at high frequencies. It must be noted that very small SIS junctions ($1 \times 1\,\mu$) are now realized [24]. In the case of point contacts, the capacitance may also be very low ($C \simeq 0.01 - 1$ pF) if the contact area is reduced. For example, with $I_c = 0.1$ mA and $C = 10^{-2}$ pF, $f_p = 880$ GHz. However, very light pressure contacts are necessary, which means mechanical problems and a tendency to be microphonic.

In microbridges, the geometrical capacitance has generally a negligible influence.

b. Coupling to an incident electromagnetic radiation

Tunnel junctions have been considered less promising for high frequency applications due to their relatively low impedance which gives a strong impedance mismatch at the junction edges. Their closed structure makes radiation coupling difficult.

Point contacts are still the favorite device for high frequency applications due to the small capacitance which may be achieved ($\simeq 10^{-2}$ pF) and their open structure which allows antenna techniques to be used in the submillimeter range.

The application of integrated circuit technology, as for metal-oxide-metal junctions, may solve the coupling problems for tunnel junctions and microbridges if the hf performances of the latter become of sufficient interest.

As discussed later, the transfer of signal power from the radiation field to the junction is optimized in most cases if the device is fed from a source of comparable resistance and the effects of the device reactance are minimized. Whereas in the longer wavelength part of the 1 mm - 0.1 mm range, waveguide structures are probably the best choice, antenna systems are to be preferred at the shorter wavelengths. To realize the potential applications of Josephson detectors in radio-astronomy, providing a coupling system with an effective area and an angle of view appropriate for coupling to a telescope optics is necessary.

3. Thermal effects

Thermal effects produce an increase of the junction temperature which reduces the critical current, and the performance may be strongly degraded. Operation at a temperature far enough below T_c is necessary to realize an efficient cooling of the active part of the junction.

In SIS junctions, thermal effects are generally small due to the relatively low current densities (and the comparatively large area) which allow an efficient cooling on substrates with a high thermal conductivity (sapphire).

Thermal effects are also generally negligible in point contacts despite the normal current component; here the three-dimensional structure allows an efficient thermal coupling with the bath. However, these effects may become important in the infrared,and at 10 μa thermal rather than a Josephson response cannot yet be excluded [15] for detectors in this region.

In microbridges thermal effects are important in structures with uniform thickness. There are two main reasons: (1) the physical process is dissipative; and (2) the volume in which the process takes place is very small, and the corresponding area in thermal contact with the substrate is much smaller than for tunnel junctions. The heating can be sufficiently large that the phase-slip centers develop in normal spreading hot spots [16]. The formation of such hot spots is the dominant cause of the hysteresis observed in the VIC at low temperatures, and imposes a high-voltage limit on the ac Josephson effect in these devices. Variable thickness bridges again provide a solution to prevent heating due to their three dimensional structure.

4. Noise

The very high frequency sensitivity to any electrical signal ($2e/h = 484$ MHz/μV) makes the Josephson junction attractive in very high frequency applications. However, it may be expected that any electrical noise (internal or external) will play an important role on the device behavior.

The most direct example concerning the influence of fluctuations in a tunnel junction is the Josephson oscillation linewidth [17] given by

$$\Delta f_o \simeq \left(\frac{2e}{\hbar}\right)^2 \frac{P_v}{2} \quad , \tag{7}$$

where P_v is the spectral density of the voltage fluctuations defined by

$$V^2_{rms} = \int_B P_v \, d\omega \tag{8}$$

where B is the noise bandwidth.

The origin of the Josephson oscillation linewidth is easy to understand. The low frequency fluctuations of the dc current flowing through the junction are converted into voltage fluctuations at the junction impedance. These voltage fluctuations are responsible for the phase fluctuations of the Josephson oscillation which lead to the linewidths observed experimentally.

For example, for a point contact with a critical current $I_c \simeq 0.1$ mA, Δf_o may be several 100 MHz. Here lies an important source of difficulty in using the JJ as an oscillator. This allows us to understand why the Josephson oscillator or the Josephson oscillator-mixer is seldom used in practical applications. In mixing experiments, the Josephson junction (and more precisely point contacts) are generally used as any other conventional mixer, i.e., with an external local oscillator. In this mode of operation it may be shown (see IF3) that there are frequency converted signals which are independent of the Josephson oscillation and of its phase fluctuations.

D. The Main Detection Mechanisms

Josephson junctions may be used as detecting elements following different schemes.

1. Wide band detection

a. Bolometer

A bolometer is a thermal detector, i.e., it yields an output in direct proportion to the temperature rise ΔT caused by absorption of a small signal radiation. J. Clarke et al. [18] have taken advantage of the temperature dependence of the critical current in a Josephson junction for the thermometric element of a bolometer. The temperature is measured using the strong critical current variation versus T of a normal metal tunnel junction (S-N-S), e.g., a sandwich Pb/Cu-Al/Pb. The critical current is given by

$$I_c(T) = I_c(0) \exp(-T/T_o)^{1/2} \tag{9}$$

with $T_o \simeq 0.1$ K.

If the junction is biased at a constant current slightly larger than I_c, the voltage variation with critical current may be appoximated by the differential resistance R_D at the bias point. A chopper on the incident beam converts the change of I_c to an ac output voltage.

b. Square-law video detection

Josephson detectors have a square-law response to small ac signals as conventional diodes, i.e., the output voltage is proportional to the input power. One makes use of the deformation of the VIC under application of an amplitude modulated signal [19]. A given VIC corresponds to each signal level and if the junction is biased at a constant current slightly larger than I_c, the output voltage follows the signal modulation (Fig. 5). We shall see later that the critical current varies with the amplitude of the rf signal current as

$$I_c(I_s) = I_c(I_s = 0)\,[\,1 - \gamma I_s^2\,] \quad . \tag{10}$$

If R_D is the dynamic resistance of the VIC at the operating point, the output voltage at the modulation frequency is

$$S = -\gamma\, I_c I_s^2 R_D \quad . \tag{11}$$

2. Narrow band detection (linear)

Two heterodyne modes of operation may be used as the local oscillator, either internal or external.

a. Internal local oscillator

The external signal to be detected is mixed with the fundamental Josephson oscillation ($\omega_0 = [2e/\hbar]\,V_0$) or one of its harmonics. The dc voltage at the junction has to be perfectly determined: $V_0 = \frac{\hbar}{2e}(\omega_s \pm \omega_{if})$. This mode of operation presents the great advantage of simplicity. However, it is difficult to realize due to the great sensitivity

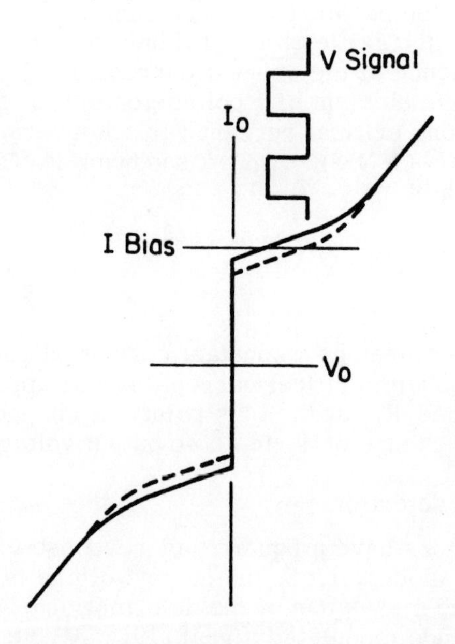

Fig. 5. Video square law detection with a current biased junction.

of the Josephson oscillation to fluctuations (D4). The linewidth of the frequency converted signal is equal to the Josephson linewidth Δf_o if the signal to be detected is monochromatic. Practically, this means that the if chain must have a wide bandwidth, and consequently a large if (see Sec.II-D).

b. External local oscillator

The external local oscillator (LO) is the more common mode of operation because it is easier to operate. The frequency converted signal does not depend explicitly on the Josephson frequency; its linewidth is determined by the linewidths of the signal and of the local oscillator. The junction may be biased to any point of the VIC. The junction behavior is similar to a classic resistive mixer although the nonlinearity here is related to the phase modulation.

With a low level signal compared to the LO ($I_S \ll I_L$) and a small if frequency ($\omega_{if} \ll \omega_L$), the current induced in the junction may be written

$$I_L \sin \omega_L t + I_S \sin \omega_S t \simeq (I_L + I_S \cos \omega_{if} t) \sin[\omega_L t + (I_S/I_L) \sin \omega_{if} t] \quad .$$

Neglecting the phase variation of amplitude I_S/I_L, the net effect is that of the LO amplitude modulated at the if. Since the amplitude of the induced step by the LO is a function of LO current amplitude, the characteristic is modulated at the if, and for a current biased device, a voltage at the if proportional to R_D is observed. Practically, the junction is biased at a point of the VIC where R_D is large, i.e., near the critical current between two successive induced steps induced by the LO.

E. The Josephson Junction and Parametric Amplification

The Josephson junction may be considered as a nonlinear inductance. For example, with an ideal junction, we may write from Eqs. (1) and (2)

$$\frac{dI_J}{dt} = L_J^{-1}(\phi)\, V(t) \quad , \tag{12}$$

with

$$L_J(\phi) = \frac{L_o}{\cos\phi} \tag{13}$$

and

$$L_o^{-1} = \frac{2e}{\hbar} I_c \quad . \tag{14}$$

In a real junction, the plasma resonance frequency given by

$$\omega_{pl} = (L_o C)^{-1/2}$$

shows when the capacitance begins to play the leading role. We consider first junctions where both the capacitance and the resistance may be disregarded. $L_J(\phi)$ is a variable reactance function of the applied voltage and parametric effects are possible. The situation may be compared to varactors where the capacitance is modulated by the voltage.

For example, the frequency conversion process seen above may be considered as a 3 photon parametric effect (with one idler frequency)

$$\omega_p = |\omega_s + \omega_i| \quad .$$

Since an ideal Josephson junction is nondissipative, the Manley-Rowe relations generalized to Josephson junctions [20,21] permit it the prediction of the ultimate conversion loss given by

$$\frac{P_{if}}{P_S} = \frac{\omega_i}{\omega_S} \quad . \tag{15}$$

Therefore, as any other parametric device with a nonlinear reactance, a JJ in this mode of operation presents only a practical interest if $\omega_i > \omega_S$, a situation where frequency conversion with gain is possible (up-converter).

A real Josephson junction differs from varactors in two ways, because it is a highly nonlinear active element able to convert dc to ac energy, and furthermore, it is a dissipative element.

The first fundamental difference is that the internal Josephson oscillation may be used as the pump. Then the nonlinear inductance is modulated at the Josephson frequency

$$L_J^{-1}(\omega_o) = \frac{2eI_c}{\hbar} \cos \omega_o t \quad . \tag{16}$$

A JJ may consequently be used as a parametric element following different regimes: (a) at zero voltage (V = 0) with an external pump. This situation corresponds to varactor parametric amplifiers [26]; (b) with an internal pump: $\omega_p = \omega_o = \frac{2eV_o}{\hbar}$; and (c) with an external pump and $V_o \neq 0$. The Josephson oscillation frequency is arbitrary, but the nonlinear element receives energy from the dc power supply. The behavior is different if the Josephson oscillation is synchronized on the pump frequency or not.

From the Manley-Rowe relations, generalized to the case where a dc power supply delivers energy to the junction which is converted in ac energy [20, 21], the conversion losses corresponding to cases (b) and

(c) may be calculated.

The second fundamental difference is that the real JJ is dissipative (point-contact or microbridge) due to the normal shunt resistance, with the consequence that a very complicated spectrum (generation of many harmonics of the Josephson oscillation or of an applied radiation) is induced by the resistive feedback. Then energy exchange may happen between these many components, and the Manley-Rowe relations cannot be applied to determine the conversion losses for two reasons: (a) the losses of the parametric device must be small; and (b) simple relations are obtained only if frequency components other than the signal, pump and the one or two idlers are open or short circuited. The first condition corresponds to $\omega L_J(\phi) << R_N$, and it can be shown that this relation means that $\omega \leqslant \omega_c$ (see IC1, IC2). This limits the application of this mode of operation at high frequencies. The second condition is practically very difficult to realize at high frequencies and the difficulty is increased by the low impedance of the JJ. Consequently, at very high frequencies, the JJ will be used mostly as a resistive mixer (with internal or external local oscillator). In this situation it must be recalled that Eq. (15), deduced from the Manley-Rowe relations, is not valid. The analysis of the RSJ equivalent circuit (Section II D) shows that the conversion efficiency decreases as $\Omega^{-2} = (\omega_c/\omega)^2$ when $\Omega \gg 1$.

On the contrary, in the microwave region the above two conditions may be fulfilled using resonant circuits and parametric amplification with self-pumped or externally pumped Josephson junctions has been realized.

F. The Real JJ Analyzed with the RSJ Model

For Josephson junctions of small area (small capacitance) the RSJ model forms the basis of most theoretical calculations of junction performances. Two electrical models may be used to analyze the hf JJ junction behavior: the voltage source model [19], where the device is considered to be driven by sources of vanishingly small impedances; and the current source model, which is a better representation of the real situation due to the low device impedance.

Only the latter model permits a complete quantitative description of the junction properties. However, the equations describing this model have no analytical solutions when rf sources are considered and must be solved by numerical computations. On the contrary, the voltage source model gives a simple mathematical treatment since the phase variations (second Josephson equation) are simply proportional to the applied voltage at the junction. This model allows us to understand many fundamental properties of JJ. The two models will be presented here for the case where a single external signal is applied to the junction.

1. Voltage source model [19]

The applied voltage at the junction irradiated at ω_1 is

$$V(t) = V_o + V_1 \sin\omega_1 t \quad . \tag{17}$$

Using Eq. (2), we obtain

$$I_j(t) = I_c \sin\phi = I_c \sin \int_o^t (2e/\hbar) V(t')\,dt' \quad , \tag{18}$$

$$I_J(t) = I_c \sin\left[\omega_o t + \frac{2eV_1}{\hbar\,\omega_1} \sin\omega_1 t + \phi_o\right] \quad , \tag{19}$$

and

$$I_J(t) = I_c \sum_{k=-\infty}^{+\infty} J_k(2\alpha_1) \sin(\omega_o t + k\omega_1 t + \phi_o) \quad . \tag{20}$$

The initial phase of the external radiation is chosen to be zero and $J_k(2\alpha_1)$ is the Bessel function of the first-kind of order k, $2\alpha_1 = 2eV_1/\hbar\omega_1$, and $\omega_o = 2eV_o/\hbar$.

This model predicts two situations: (a) a non synchronized junction where $\omega_o \neq -k\omega_1$. This situation corresponds to the Josephson oscillator mixer. The intermediate frequency $\omega_i = |\omega_o - n\omega_1|$, with $(n = -k)$ and has an amplitude $I_c J_k(2\alpha_1)$; and (b) a synchronized junction where $\omega_o = -k\omega_1$. The phase of the Josephson oscillation is locked on the phase of the applied signal (if the latter has an intensity higher than the noise level). This gives a dc current component in the Josephson current spectrum with amplitude $I_c J_k(2\alpha_1)$ which corresponds to an energy transfer to the junction. The VIC in this model has current spikes at the voltages $V_o = n(\hbar/2e)\omega_1$ $(n = 1, 2, \ldots)$. This model allows us to predict that the JJ sensitivity decreases at high frequencies since $J_n(2\alpha_1) \neq 0$ only for $2\alpha_1 = (2eV_1/\hbar\omega_1) > n$.

2. The current source model [22, 23]

The response of a current driven junction fed by a dc source I_o and an ac source $I_1 \sin\omega_1 t$ is given by

$$I_o + I_1 \sin\omega_1 t = I_c \sin\phi + \frac{V(t)}{R_N} + C\,\frac{dV(t)}{dt} \quad . \tag{21}$$

This equation is valid at $f \leq f_G$ (see I C 1) and can be written in reduced variables as

$$i_o + i_1 \sin\omega_1 t = \sin\phi + \frac{\dot\phi}{\omega_c} + \frac{\ddot\phi}{\omega_p} \quad , \tag{22}$$

with $i_o = \frac{I_o}{I_c}$, $\omega_c = \frac{2e}{\hbar} R_N I_c$, $\tau = R_N C$, and

$$\omega_{pl} = \left(\frac{\omega_c}{\tau}\right)^{1/2} = \left(\frac{2e\,I_c}{\hbar\,c}\right)^{1/2} . \tag{23}$$

As seen above ω_c is a cut-off frequency related to the damping introduced by the normal current component in the junction, and ω_{pl} (plasma frequency) characterizes the high-frequency limitation of the junction related to its capacitance. Physically, ω_{pl} is the resonant frequency $(L_o C)^{-1/2}$ determined by the capacitance C and the inductance $L_o = (\hbar/2eI_c)$ of the junction.

The situation which has been most studied up to now is the RSJ model (C = 0) valid for point contacts.

Without external radiation, Eqs. (21) and (22) give

$$I_o = I_c \sin\phi + \frac{V(t)}{R_N} \quad , \quad i_o = \sin\phi + \frac{\dot{\phi}}{\omega_c} \tag{24}$$

which have an analytical solution. The resulting voltage at the junction [4] is a periodic wave of amplitude $R_N I_c$ (Fig. 3) which contains harmonics of the fundamental frequency ω_o given by

$$V(t) = V_o \left(1 + \sum_{m=1}^{\infty} A_m \cos m\,\omega_o t\right) , \tag{25}$$

where $V_o = \overline{V(t)} = R_N (I_o^2 - I_c^2)^{1/2}$,

$$\text{and} \quad A_m = 2\left\{\left[1 - \sqrt{1 - \frac{1}{i_o^2}}\right] i_o\right\}^m \quad \text{where } i_o = I_o/I_c . \tag{26}$$

At high frequency, when $\omega/\omega_c \gg 1$, V(t) is practically sinusoidal and Eq. (25) becomes

$$V(t) = V_o + R_N I_c \sin\omega_o t . \tag{27}$$

With an external radiation this equation has been solved numerically on a computer when the junction is synchronized on the applied signal. The important result is the presence of constant voltage steps in the VIC instead of the current spikes predicted by the voltage model. The variation of the step heights as a function of the applied signal intensity differs markedly from the Bessel function dependence previously obtained in the voltage source model. However, at very high frequencies ($\Omega = \omega/\omega_c \gtrsim 1$), since the resistive path of the junction carries most of the rf current, the voltage source model is again a good approximation and the step heights approach the Bessel dependence (Fig. 6). This result may be understood in the following way. When $\Omega \gtrsim 1$ the term in $\sin\phi$ of Eqs. (21) and (22) becomes negligible, and the voltage at the junction can be written as

$$V(t) = R_N I_o + R_N I_1 \sin\omega_1 t$$
$$= V_o + V_1 \sin\omega_1 t$$

as in the voltage source model.

3. An important example: the Josephson heterodyne mixer with an external oscillator

If the simple voltage biased model of IF 1 is applied to the situation where a Josephson junction is irradiated simultaneously at two frequencies ω_1 and ω_2, the generalization of Eq. (20) gives the frequencies of the output current components

$$\omega' = (\omega_o + \ell\omega_1 + m\omega_2) \quad (28)$$

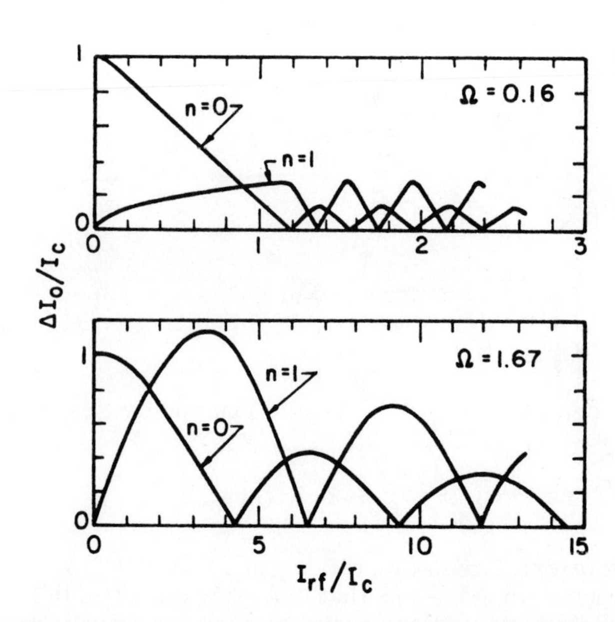

Fig. 6. Dependence of the (half) height of the n = 0 step and the (full) height of the n = 1 step on the rf current for two values of the normalized frequency Ω (taken from Ref. [42]).

where ℓ and m are integers. The above relation involves only combinations of ω_0 with ω_1 and ω_2, and consequently does not give beat signals independent of the Josephson oscillation. However, it has been found in the experiments [19] that beats are obtained which do not shift in frequency as V_0 is changed. Moreover, the linewidths of the beat signals are also independent of the Josephson oscillation linewidth. Therefore, experiments show that ω_0 is not involved directly in these beat signals which may be written as

$$\omega' = \ell\omega_1 + m\omega_2 \quad .$$

These results have been confirmed by analog computer simulation [25] of the complete equation

$$I_0 + I_1 \sin\omega_1 t + I_2 \sin\omega_2 t = I_c \sin\phi + \frac{V(t)}{R_N} \quad . \tag{29}$$

The existence of the components independent of ω_0 may be understood qualitatively using the voltage source model as follows. The resistive feedback mechanism leads to a distorted periodic voltage at the junction. In a first order approximation we take into account the two first terms of the expansion of Eq. (25), and can write the total voltage at the irradiated junction in a phenomenological way as

$$V(t) = V_0 + A_0 \cos\omega_0 t + A_1 \cos\omega_1 t + A_2 \cos\omega_2 t \quad . \tag{30}$$

Consequently, the Josephson current in the junction may be expanded into a Fourier-Bessel series

$$I_j(t) = I_c \sin\frac{2e}{\hbar}\int_0^t V(t')\,dt'$$

$$= I_c \sin\left(\phi_0 + \omega_0 t + \sum_{i=0}^{2} 2\alpha_i \sin\omega_i t\right), \text{ where } 2\alpha_i = \frac{2eA_i}{\hbar\omega_i}$$

$$= I_c \sum_{k=-\infty}^{+\infty}\sum_{k=-\infty}^{+\infty}\sum_{k=-\infty}^{+\infty} J_k(\alpha_0)J_\ell(\alpha_1)J_m(\alpha_2) \times$$

$$\sin[(\omega_0 + k\omega_0 + \ell\omega_1 + m\omega_2)t + \phi_0] \quad . \tag{31}$$

The result is a complex spectrum which may be separated in two groups of components: (a) components dependent of the Josephson oscillation frequency or its harmonics for all values of $k \neq -1$. They are of interest when using the Josephson oscillation or one of its har-

monics as a local oscillator (they are the components found in I F 1).
(b) Components independent of the Josephson oscillation for the value $k = -1$. Among these components one may recognize in Eq. (31) a dc component ($k = -1$, $\ell = m = 0$) which corresponds in the present phenomenological description to the dc pair current flow calculated by Aslamazov and Larkin, current components at ω_1 ($k = -1$, $\ell = 1$, $m = 0$), ω_2 ($k = -1$, $\ell = 0$, $m = 1$) and their harmonics and the mixing current components ($\ell\omega_1 + m\omega_2$). The latter correspond to the case where a junction is operated as a mixer with an external local oscillator (ω_1) used either at the fundamental ($\ell = 1$) or as an harmonic mixer ($\ell > 1$).

G. Noise

1. Physical origin of fluctuations in Josephson junctions

The analysis of fluctuations in Josephson junctions has been performed for the two structures: oxide barrier tunnel (SIS) junctions, and superconductor-constriction-superconductor (ScS) junction obeying the Aslamazov and Larkin model [4].

a. SIS tunnel junctions

The theory of electrodynamic fluctuations in tunnel junctions has been developed mainly by Scalapino, Stephens, and Rogovin [26-29]. For dc biased junctions above the critical current (which is the usual situation in detector applications) two principal sources of noise are to be considered related respectively to the quasiparticle and pair current fluctuations. In a well designed device, noise from the dc biased source is generally negligible. If the bias voltage contains both a dc voltage, V_0, and a modulating noise voltage, $V_n(t)$, then

$$I_J(t) = I_c \sin\left[\omega_o t + (2e/\hbar) \int_o^t V_n(t')dt'\right] \quad , \tag{32}$$

with $\omega_o = 2eV_o/\hbar$. We have seen when studying the voltage source model of a JJ (see I F 1) that the Josephson harmonic current components of order n, which have an appreciable amplitude, are such that $2eV/\hbar\omega > n$ where V is the alternative voltage amplitude at frequency ω. Assuming a white noise spectrum, it is therefore the fluctuations of frequency $\omega \ll \omega_o$ which will be dominant in determining the noise effects in JJ. In this way the high frequency spectrum of the current [Eq. (32)] arises from, and reflects, the low-frequency current fluctuations.

The mean square fluctuation current is related to the current power spectrum $P_I(\omega)$ by $\langle i(\omega)^2 \rangle = \int_0^\infty P_I(\omega)\, d\omega$. If we know $P_I(\omega)$ then the noise calculation for the device is straightforward.

Scalapino [26] has calculated the quasiparticle current noise spectrum in terms of the current response $I_{qp}(V_o)$ of the junction taking into account first-order tunneling processes which, for bias voltages V_o large compared with $\hbar\omega/e$, reduces to

$$P_{I_{qp}}(\omega) = \frac{e}{\pi} I_{qp}(V_o)\coth\frac{eV_o}{2kT} \quad . \tag{33}$$

Equation (33) is of the shot-noise type with the coth factor arising from the sum of the contributions by forward and backward qp electron-flow crossing the junction tunneling barrier.

Stephens [28] has calculated the contribution from fluctuations of the pair current which arise from the coupling of the coherently tunneling Cooper pairs to random fluctuations due to the background radiation (e.g., blackbody radiation field). He considered the specific geometry of an SIS sandwich structure which constitutes a cavity in which the coherently tunneling pairs excite an electromagnetic field at ω_o .

The strong interaction between the electromagnetic field and the Josephson current is responsible for a dc pair current component $I_p(V_o)$ which is illustrated for example by the self induced steps [35] in the VIC. (In the case of point contacts the dc pair current (see IB) is related to the interaction with the normal component through the resistance R_N, i.e., to the resistive feedback). Therefore fluctuations of the electromagnetic field (blackbody radiation) create in turn fluctuations in the dc pair current $I_p(V_o)$. This corresponds to the fact that when a photon is absorbed (or emitted) in the cavity a pair must tunnel for energy conservation. It is these corresponding low frequency fluctuations which give phase diffusion (through the capacitance C of the junction) and the broadening of the Josephson oscillation. The related power spectrum $P_{I_p}(\omega)$ at voltages $V_o \gg \hbar\omega/e > 0$ is given by

$$P_{I_p}(\omega) = \frac{2e}{\pi} I_p(V_o)\coth\frac{eV_o}{kT} \quad . \tag{34}$$

The overall factor of 2 reflects the fact that the current noise is due to coherent tunneling of Cooper pairs.

An analogous expression [29] may be obtained for the current power spectrum of the interference pair-quasiparticle current $I_{qp\text{-}pair}$ where

$$P_{I_{qp\text{-}pair}}(\omega) = \frac{2e}{\pi} I_{qp\text{-}pair}(V_o)\coth\frac{eV_o}{2kT} \quad . \tag{35}$$

The influence of $I_{qp\text{-}pair}$ on the behavior of high frequency Josephson devices is not yet well known, and we will drop this latter term in the following, but investigations in this direction seem necessary.

Therefore the total current-power spectrum of a tunnel SIS junction is for the quasiparticle and pair contributions is

$$P_I(\omega) = \frac{e}{\pi}\left[I_{qp}(V_o)\coth\frac{eV_o}{2kT} + 2\,I_p(V_o)\coth\frac{eV_o}{kT}\right] \qquad (36)$$

for ($\hbar\omega \ll eV_o$).

Two limiting cases should be considered for practical applications:

(a) At low dc bias such that $eV_o \ll kT$ ($kT/e = 360\,\mu V$ at 4 K),

$$P_I(\omega) = \frac{2kT}{\pi}\,\frac{I_p + I_{qp}}{V_o} = \frac{2kT}{\pi R_o}, \qquad (37)$$

with $R_o = V_o/I_o$, $I_o = I_p + I_{qp}$. This is a Johnson noise type expression (thermal noise), related to the junction resistance R_o (and not to its normal resistance R_N).

(b) At T = 0 or at very high frequencies, when $V_o \gg kT/e$,

$$P_I(\omega) = \frac{e}{\pi}\,(I_{qp} + 2\,I_p) \qquad . \qquad (38)$$

This corresponds to diode shot noise related to the electric charge flow (here the pairs and quasiparticles) through the junction barrier. Often, at very high frequencies, the dc pair current is much smaller than the quasiparticle current, Eq. (38) reduces to

$$P_I(\omega) \simeq \frac{e\,I_{qp}}{\pi} = \frac{e\,I_o}{\pi} \qquad . \qquad (38\text{ bis})$$

Experimentally, only the case $eV_o \ll kT$ has been investigated [32]. It was found that the fluctuation current is dominated by pair and quasiparticle shot noise. The measured linewidths were found generally in excess of that calculated with Eq. (36). This result was attributed to excessive radio-frequency power in the junction.

b. S.C.S. Junctions

In S.C.S. junctions (superconductor-constriction-superconductor) of the Aslamazov and Larkin (A.L) model [4], the structure is a weak link created between two superconducting electrodes, in the form of a short section of effective dimensions less than the coherence length ξ. It is shown that these junctions obey the RSJ model. Microbridges which may be realized with dimensions $< \xi$ are well-known realizations corresponding to this model.

Since the junction has dimensions $< \xi$, the fluctuations of the order parameter in the bridge area are very small. Therefore, the only source of fluctuations which is considered is the noise of the normal resistance R_N of the junction. The junction current-power spectrum is [30, 31]

$$P_I(\omega) = \frac{1}{\pi} \frac{\hbar\omega}{R_N} \coth \frac{\hbar\omega}{2kT} \quad . \tag{39}$$

Except at very low temperatures, $\hbar\omega \ll kT$ and $P_I(\omega)$ reduces to

$$P_I(\omega) = \frac{2}{\pi} \frac{kT}{R_N} \quad . \tag{40}$$

No experimental data on the current-power spectrum of the Josephson current in microbridges exists which obeys this model.

c. Comparison between the two models

At high frequencies such that $eV_0 \gg kT$ one obtains

$$\frac{P_I(\omega)_{SIS}}{P_I(\omega)_{SCS}} = \frac{eV_0}{2kT} \quad . \tag{41}$$

Therefore, at $eV_0 \simeq 2kT$, the two models give similar results. It is only in experiments made at $eV_0 \ll kT$ or $eV_0 \gg kT$ that precise quantitative indications for the validity of the two models may be expected.

d. Point contacts

The experimental situation is more complex for point contacts. Some results may be interpreted with the noise model which is valid for tunnel junctions [33, 34] while others [36] correspond to lower noise levels and have been interpreted with the SCS model involving only thermal noise. This is a complicated problem since the result is related to the nature of the physical mechanism giving rise to the observed Josephson phenomena in point contacts.

If tunneling is the preponderant mechanism in this type of junction, as is probably the case in junctions where structures are observed at the gap voltage in the IV characteristic or (and) when the Riedel singularity is observed [12, 13], one may expect that the noise level will be higher than in the SCS junction model, and better interpreted with the SIS junction model. As before, since it is the low-frequency fluctuations which are of importance, the current-power spectrum in a point contact is related to the dc current carried through the junction which may be written $I_0 = I_M + I_{qp} + I_p$. I_M is the dc current through the normal resistance associated with the metallic path, I_{qp} is the current corresponding to quasiparticle tunneling through the tunnel paths in the junction, and I_p represents the dc pair current carried in the contact which results from the harmonic distortion of the Josephson alternating current produced by the resistive feedback (see I B). This latter term has a different physical origin than the pair current component in SIS junctions which results from the strong interaction of the Josephson current with the electromagnetic field in the junction cavity. However, it may be expected that a similar current power spectrum may be associated with this term.

The current power spectrum in a point contact may be written

$$P_I(\omega) = \frac{2}{\pi}\frac{kT}{R_M} + \frac{2eI_p}{\pi}\coth\frac{eV_o}{kT} + \frac{eI_{qp}}{\pi}\coth\frac{eV_o}{2kT} \quad . \qquad (42)$$

If $eV_o \ll kT$, $\quad P_I(\omega) \simeq \frac{2kT}{\pi}\left(\frac{1}{R_M} + \frac{I_p + I_{qp}}{V_o}\right)$;

if $eV_o \gg kT$, $\quad P_I(\omega) \simeq \frac{1}{\pi}\left(\frac{2kT}{R_M} + I\right)$.

Therefore, the noise level in point contacts may be found following their microphysical structure between the two limits given by Eq. (42) and $P_I(\omega) = \frac{2}{\pi}\frac{kT}{R_N}$ corresponding to the SCS model. Careful investigations at high frequencies, where point contacts find most of their applications, are still necessary.

e. Microbridges

The theoretical lower limit for noise in very small bridges (dimensions $< \xi$) is given by the SCS model. However, recent results in proximity effect bridges [110] are better described by existing theories of electron tunneling. A better knowledge of the physical processes in microbridges is necessary to interpret the noise effects.

2. Josephson junction response in the presence of fluctuations

We consider here the consequences of the junction current-power spectrum discussed in IG 1. The junction current fluctuations are converted into voltage fluctuations via the junction impedance. Because only the low frequency fluctuation current components are important, they give rise to a modulation of the VIC (see also II) and the junction dynamic resistance R_D monitors the voltage fluctuations amplitude

$$P_V(\omega) = R_D^2 P_I(\omega) \quad .$$

Under the assumption of white noise, the rms noise voltage in a bandwidth B is

$$V_{rms}^2 = \int_B P_V(\omega)\, d\omega = P_V B = R_D^2 B P_I(\omega) \quad .$$

The effects of current fluctuations on the linewidth of the radiation emitted by a Josephson junction are first considered. Then we analyze the different manifestations of noise respectively in the junction modes of operation where the Josephson oscillation is: (a) directly involved as in the oscillator-mixer; (b) indirectly involved as in the heterodyne mixer with external local oscillator.

a. Calculation of the Josephson linewidth

i. Voltage source model

When a Josephson junction is biased at a voltage V_o, and if there is also a noise voltage $V_n(t)$, the ac pair current

$$I_J(t) = I_c \sin\left[\omega_o t + \frac{2e}{\hbar}\int^t V_n(t')dt'\right]$$

is frequency modulated by $V_n(t)$. The high-frequency spectrum of the Josephson pair current arises from the low-frequency voltage fluctuations defined above by the voltage power spectrum $P_V(\omega)$. The linewidth $\Delta\omega_o$ may be obtained by a method classic in electronics [37]. The spectrum of fluctuations may be considered as white noise up to frequencies of order $(CR_D)^{-1}$ where C is the junction capacitance. Since the voltage fluctuations of interest are of low frequency compared to $(CR_D)^{-1}$, the fraction of the total voltage fluctuation contributing to the linewidth is small, so that the resulting lineshape is Lorentzian with

$$\Delta\omega_o = \pi\left(\frac{2e}{\hbar}\right)^2 R_D^2 P_I(\omega) \quad . \tag{44}$$

ii. Current source model

The resistive feedback which occurs when a junction is current-biased gives a different relation between $\Delta\omega_o$ and $P_I(\omega)$ which has been calculated by Likharev and Semenov [31] in the small-noise limit of no noise-rounding of the VIC. The result is

$$\Delta\omega_o = \pi\left[\frac{2e}{\hbar}\right]^2 R_D^2\left[1 + \frac{I_c^2}{2I^2}\right] P_I(\omega) \quad . \tag{45}$$

For small currents $(I \simeq I_c)$,

$$\Delta\omega_o \simeq \frac{3\pi}{2}\left[\frac{2e}{\hbar}\right]^2 R_D^2 P_I(\omega) \quad , \tag{46}$$

and for large currents $(I \gg I_c)$,

$$\Delta\omega_o \simeq \pi \left[\frac{2e}{\hbar}\right]^2 R_D^2 P_I(\omega) \quad . \tag{47}$$

iii. Comparison between the two models

Equation (47) shows that at high frequencies, since $eV_o \gg kT$ and practically $I \gg I_c$, the voltage model [Eq. (44)] and the current model [Eq. (47)] give the same linewidths as a function of the current-power spectrum $P_I(\omega)$. This is not surprising since at high frequencies the RSJ model junction is practically voltage-biased. At low frequencies, when $eV_o/kT \ll 1$ and $I \simeq I_c$, the two models give a linewidth ratio of 3/2.

b. Junction response dependent of the Josephson oscillation (oscillator-mixer)

The junction response is affected in two ways: (a) In Section I F 3 we have analyzed the Josephson junction in the heterodyne mode and found that the spectrum of a junction irradiated at ω_1 and ω_2 contains all the components

$$\omega' = (1+k)\omega_o + \ell\omega_1 + m\omega_2 \quad .$$

All the components $k \neq -1$ depend explicitly of the Josephson oscillation. Therefore, all these components are affected by noise following the same mechanism which gives the Josephson oscillation linewidth $\Delta\omega_o$, and their linewidths are related to $\Delta\omega_o$. Thus, if the applied signals at ω_1 and ω_2 have narrow linewidths (e.g., with a microwave generator) the linewidth of the mixing components ($k \neq -1$) will be determined by $\Delta\omega_o$ as shown experimentally [34, 38]. This may produce a severe loss in the frequency converted signal if the bandwidth B of the detection window is smaller than $\Delta\omega_o$. In this situation the mixer sensitivity decreases and its noise temperature increases (see I H). (b) The junction-current fluctuations create a rms noise voltage given by $V_{rms}^2 = R_D^2 B P_I(\omega)$ which allows us to define an equivalent noise temperature at the mixer output (see I H).

c. Junction response independent of the Josephson oscillation (external local oscillator mixer)

In Section I F 3 a qualitative analysis has shown that the mixing components ($k = -1$) are independent of the Josephson oscillation frequency and

$$\omega' = \ell\omega_1 + m\omega_2 \quad .$$

It is shown by experiment [19, 23] and analogue computer simulation [25] that the linewidth of these components is independent of the Josephson oscillation linewidth $\Delta\omega_o$. This result may be understood qualitatively when fluctuations are taken into account because only the low frequency ones are of importance, and the Josephson junctions have a very high speed response, so that the voltage and current in the junction are synchronous. The phenomenological analysis of Section I F 3 may be applied in a time scale of the order of the low-frequency fluctuations. Therefore, Fourier analysis may still be used showing that the components $\omega' = \ell\omega_1 + m\omega_2$ are completely independent of the Josephson oscillation, and particularly that their linewidth is independent of $\Delta\omega_0$. The consequences of noise in the heterodyne mode with external local oscillator are only related to the rms noise voltage $V^2_{rms} = R^2_D BP_I(\omega)$ at the junction and to the associated equivalent noise temperature at the mixer output.

H. Noise Temperature, Minimum Detectable Temperature, NEP [39, 41]

1. Noise temperatures

a. Noise from a thermal source

The thermal noise power available from a resistor at temperature T(K) in a bandwidth B is

$$P = \frac{h\nu B}{\exp\left(\frac{h\nu}{kT}\right) - 1} \simeq kTB \quad \text{if} \quad h\nu \ll kT \quad . \tag{48}$$

We assume in the following that this low-frequency limit is applicable; this means that at $\lambda = 1$ mm, $T \geqslant 30$ K, and for $\lambda = 0.1$ mm $T \geqslant 300$ K.

If the resistor is replaced by a lossless matched antenna of radiation resistance R whose field of view is filled with a blackbody source at temperature T_A, the noise power will be kT_AB. The antenna temperature T_A is determined by the temperature of the emitting region which the antenna sees through its directional pattern. The use of temperature to express noise powers is particularly convenient when the noise powers are small and when the frequencies are in the normal frequency region ($h\nu \ll kT$) where noise from thermal origin overcomes noise having its origin in the quantized nature of electromagnetic radiation.

b. Noise temperature of a receiver

In a receiver, noise is generated in the antenna (T_A), but extra noise is also associated with the detector or mixer element, amplifiers, etc. The receiver noise contribution is characterized by an effective input noise temperature T_e at the antenna terminals, and the noise contribution of the overall system by an operating noise temperature T_S. The power required of an input signal to make the output signal to noise ration S/N = 1 is kT_SB. The total system noise power at the antenna terminals is

$$P_n = k(T_A + T_e)B = kT_S B \quad , \tag{49}$$

and the noise power available at the receiver output terminals is $G_o kT_S B$ where G_o is the receiver system available gain.

c. Effective input noise temperature of a heterodyne receiver

i. Mixer conversion efficiency

The maximum conversion efficiency G of a mixer is defined by

$$G = \frac{\text{IF power available from the mixer}}{\text{Power available from the signal source}} \quad . \tag{50}$$

This definition assumes lossless, perfectly matched, coupling systems at the input and output of the mixer. In practice, only a fraction C_{hf} of the available signal is coupled to the mixer, and a fraction C_{if} of the available IF power is coupled to the IF amplifier. Therefore, the total conversion gain η_T may be defined such that

$$P_{if} = \eta_T P_S = C_{hf} C_{if} G P_S = \eta C_{if} P_S \tag{51}$$

where η, conversion gain taking into account the input coupling efficiency C_{hf}, will be the important parameter to discuss mixer performances, P_S is the total signal power available from the antenna, and P_{if} is the if power effectively coupled in the if amplifer. Then

$$\eta_T = C_{if}\eta = C_{hf} C_{if} G \quad . \tag{52}$$

C_{if} and C_{hf} may be written as a function of the reflection coupling coefficients Γ_{if} and Γ_{hf} (assuming that the reactive components of the impedances are compensated) so that

$$C_{if} = 1 - \left[\Gamma_{if}\right]^2 = 1 - \left[\frac{R_{\ell f} - R_{if}}{R_{\ell f} + R_{if}}\right]^2 = \frac{4 R_{\ell f} R_{if}}{(R_{\ell f} + R_{if})^2} \quad \text{and}$$

$$C_{hf} = 1 - \left[\Gamma_{hf}\right]^2 = \frac{4 R_{hf} R_S}{(R_{hf} + R_S)^2} \quad . \tag{53}$$

Here, R_{hf} and $R_{\ell f}$ are the impedances of the mixer (assumed to be real) relative to the input (high frequency) and to the output (low frequency), R_S is the impedance of the high frequency source, and R_{if} is the input impedance of the if amplifier following the mixer. The discussion of these parameters in Josephson junctions is given in Section II2.

ii. Receiver noise temperature

Making the reasonable assumption that the noise generated in the mixing element has a flat spectrum within the if bandwidth B, one can define a mixer output noise temperature T_N ($\geqslant$ physical temperature T of the device) such that the effective noise power from the mixer at the if amplifier input is $kT_N B C_{if}$.

The if amplifier noise is defined by a noise temperature T_{if} (or a noise factor $F = 1 + \frac{T_{if}}{T_o}$, $T_o = 290$ K) so that the effective noise input power to the amplifier from this source is $kT_{if}B$.

The receiver noise power P_R (single-sideband receiver) referred to the input of the if amplifier under these conditions is

$$P_R = kB\,(T_N C_{if} + T_{if}) = kB\eta_T T_e \quad , \tag{54}$$

and

$$T_e = \frac{T_N C_{if} + T_{if}}{\eta_T} = \frac{T_N}{\eta} + \frac{T_{if}}{\eta_T} = T_M + \frac{T_{if}}{\eta_T} \quad . \tag{55}$$

Here, T_M is the mixer noise temperature referred to its input $= (T_N/\eta)$. The total system noise temperature is

$$T_S = T_A + T_M + (T_{if}/\eta_T) \quad . \tag{56}$$

Much of the analysis of JJ devices as mixers will be concerned with estimates of C_{hf}, C_{if}, G which allow calculation of η and T_e. G depends on the junction parameters and its mode of operation. T_e is proportional to the hf conversion losses through G and C_{hf}. Matching the junction is therefore very important since it has a low impedance. If low values of T_M are achieved, it is necessary to have a low-noise if amplifier (cooled parametric amplifier) and a good coupling efficiency C_{if} at the if.

2. System sensitivity

a. Minimum detectable power of a receiver

The ultimate consideration in a low-noise receiver is the minimum detectable signal power. The system sensitivity may be defined as the available signal power at the antenna terminals when the output signal-to-noise ratio is unity. The minimum detectable power for a linear receiver without quadratic postdetection is $P_{min} = RT_S B$ where T_S is the total system noise temperature and B its bandwidth.

For a linear receiver (if bandwidth B) with quadratic postdetection (bandwidth B′), the if chain is assumed to have enough gain so that all the noise sources to be considered are in the predetection section of bandwidth B. The quadratic detector has a broad input bandwidth and an ouput low pass filter with time constant τ ($B' = 1/2\tau$). The output noise components come from the noise components within the predetection bandwidth beating with each other, which produce low-frequency noise components in the postdetection bandwidth. For signals small enough to ignore beating between signal and noise, it may be shown [39] that the minimum detectable signal is given by

$$P_{min} = kT_S (2BB')^{1/2} \quad . \tag{57}$$

b. Minimum detectable temperature T_{min}

For a receiver of predetection bandwidth B, and postdetection bandwidth B', one may define a differential temperature, ΔT_{min}, which is the minimum detectable temperature variation of a signal to be detected (with S/N = 1). This is related to P_{min} by

$$P_{min} = k\Delta T_{min} B \quad . \tag{58}$$

Therefore, from Eqs. (57) and (58)

$$\Delta T_{min} = T_S \sqrt{\frac{2B'}{B}} = \frac{T_S}{\sqrt{B\tau}} \quad . \tag{59}$$

In a more general way, depending on the type of receiver (total power, Dicke receiver, etc....), ΔT_{min} may be written [39]

$$\Delta T_{min} = K \frac{T_S}{\sqrt{B\tau}}$$

where $K \sim 1$, depending on the nature of the system. It must be noticed that the predetection bandwidth B is not exactly the 3 dB bandwidth of the if chain. If B limits the bandwidth, as is usually the case, $B = c\Delta\nu_{if}$ with $c \sim 1$ to 3 [39].

c. Noise-equivalent-power

A receiver sensitivity is often expressed by its noise-equivalent-power (NEP) which is the minimum detectable power (S/N = 1) generally referred to a 1 Hz postdetection bandwidth (B' = 1 Hz). Thus

$$NEP = P_{min}(B' = 1\ Hz) = kT_S\sqrt{2B} \quad . \tag{60}$$

The NEP and ΔT_{min} are related to each other by

$$\Delta T_{min} = \frac{NEP\sqrt{B'}}{kB} = \frac{NEP}{kB\sqrt{2\tau}} \quad . \tag{61}$$

When a square law receiver is considered (video receiver) instead of a heterodyne receiver, B is the input bandwidth of the receiver.

I. Coupling and Impendance Matching [39-42]

1. General remarks

The purpose of coupling and impedance matching is to optimize the energy transfer from the electromagnetic field to be detected to the junction (e.g., with a telescope optic) and from the junction to the post amplifier, so that the receiver noise temperature (antenna + mixer + if amplifier) is minimum [See Eq. (56)]. Two fundamental parameters characterize the coupling between the antenna and the junction (Fig. 7): (a) The effective area which is defined as the ratio $A = W/P$ where P is the power per unit area carried by the wave, and W is the available power at the antenna terminals; (b) The input impedance (or antenna impedance), $Z_A = R_A + jX_A$ at the antenna terminals. The resistive component of the input impedance is generally closely connected to the radiation resistance.

The maximum energy transfer to the junction will be realized if the antenna effective area is of the order of the focussed spot size, and if the junction impedance $Z_{hf} = R_{hf} + jX_{hf}$ is the complex conjugate of the antenna impedance.

When the coupling parameter C_{hf} in Eq. (53) is calculated, the above two aspects must be taken into account. As we shall see later, this problem is the most important, and also the most difficult to solve.

On the other hand, to maintain a minimum noise contribution of the if amplifier, it has to be matched to the junction ($C_{if} = 1$, $Z_{\ell f} = Z_{if}^*$). With a Josephson mixer, impedance matching and optimum coupling with the antenna are problems which are more difficult to solve than with conventional mixers for the two following reasons: (a) The Josephson junction impedance is relatively low, and, moreover, it is frequency dependent; (b) The junction impedance R_{hf} which gives the best sensitivity does not correspond always to the impedance matching condition $R_{hf} = R_S$.

2. Impedance matching

a. The Josephson junction impedance in the RSJ model

This impedance can only be determined with a computer in the general case. When the applied signal to the junction has a low level, a perturbation calculation has been used [44]. The real part $R = \langle R(t) \rangle$ of the junction impedance may then be written

$$R = R_D - \frac{\omega_s^2}{\omega_s^2 - \omega_o^2}(R_D - R_N) \tag{62}$$

where $R_D = \Delta V/\Delta I$ is the dynamic resistance of the VIC at the bias point. There are two interesting limit situations to consider:

high frequencies, $\omega_s \gg \omega_o$, for which $R \to R_N$, and

low frequencies, $\omega_s \ll \omega_o$, for which $R \to R_D$.

In both cases the reactive part, X, of the junction impedance, $Z = R + jX$, becomes zero ($X_{hf} = 0$, $X_{\ell f} = 0$). When $\omega_s = \omega_o$, the junction impedance cannot be determined from Eq. (62).

In the general case, the results are obtained on a computer [73]. They show that when $\omega_s \simeq \omega_o$ the impedance variations are complicated. In particular, it must be noted that if $\omega_s \lesssim \omega_o$, R may have negative values, and if $\omega_s \gtrsim \omega_o$, R does not differ much from R_N. The reactive part X cancels unless the junction is phase-locked on the external signal (junction biased on an induced step, $\omega_s = n\omega_o$).

However, in practical situations R_{hf} may be taken as $\simeq R_N$. In mixers with external LO, the junction is usually biased between two successive induced steps by the LO signal, and generally $\omega_s \gg \omega_o$. In mixers using the internal Josephson oscillation, the junction is biased such that

$$\omega_o = \omega_s \pm \omega_{if} \quad .$$

It has been shown (see Section II D 2) that this device may only be used with relatively large if and in the detection of low level signals. Here again one obtains $R_{hf} \simeq R_N$.

On the other hand, the device output impedance is $R_{\ell f} = R_D$ (for $\omega_{if} \ll \omega_o$). In mixers with external LO, R_D is generally larger than R_N (up to $\simeq 10\ R_N$) because the junction is biased between two steps. With the oscillator-mixer, practically $R_{\ell f} = R_D \simeq R_N$ if the VIC is not distorted by the low level applied signal.

Finally, in any case, the reactive part of the junction impedance may be neglected.

b. Intermediate frequency matching

Matching to the if requires satisfying the condition: $C_{if} = 1$ ($R_{\ell f} = R_{if}$). For an external LO, we must realize $R_{\ell f} = R_D = R_{if}$. The if amplifier impedance R_{if} is $\simeq 50\ \Omega$. Since R_D may be several times larger than R_N, which is of the order of a few ohms, matching to the if is a simple task. For an internal LO the matching condition is $R_{\ell f} \simeq R_N = R_{if}$. The R_N and R_{if} values differ strongly and matching is more difficult to perform. It is necessary to use impedance transformers or amplifiers with low input impedance, as is the case with some cooled parametric amplifiers.

c. High frequency matching

As already mentioned, the condition $R_{hf} = R_N = R_S$ ($C_{hf} = 1$) does not correspond to the best sensitivity of the Josephson mixer. If the junction is tightly coupled to the hf source in a large bandwidth around the signal frequency, when the source resistance R_S is in parallel with the normal junction resistance R_N, the characteristic frequency ω_c [Eq. (4)] becomes

$$\omega_c' = \frac{2e}{\hbar} R_{eff} I_c \quad \text{with} \quad R_{eff} = \frac{R_N R_S}{R_N + R_S} \quad . \tag{63}$$

The sensitivity of the Josephson mixer is directly related to ω_c (generally proportional to ω_c^2 as will be seen in Section II D), it may be easily understood that $R_{hf} = R_S$ is not the optimum condition. Figure 8 represents the variation of the optimum value of the ratio R_S/R_N as a function of $\Omega = \omega_S/\omega_c$ for the mixer with external LO [42]. It shows that $(R_S/R_N)_{opt}$ is much smaller than 1 for small Ω. On the other hand, at high frequencies ($\Omega \simeq 1$), the condition $R_S = R_N$ may be considered as a good approximation. It may be expected that this result is also valid for the Josephson mixer with internal LO.

3. Signal input coupling

The choice of a coupling system to the incident radiation is a function of several criteria: (a) The antenna effective area (or effective aperture) A_e which may be calculated knowing the antenna pattern is $A_e \simeq \lambda^2/\Omega_A$, where λ is the wavelength, Ω_A is the beam solid angle (rad^2) of the antenna = $(1/g)$ and g is the antenna gain; (b) The antenna losses; and (c) The antenna impedance. Consequently, a high sensitivity implies an effective area of the same order as the focussed spot size, low antenna losses and an antenna impedance such that R_A/R_N is optimized as a function of Ω (Section II 3 C).

a. Frequencies $\leqslant$ 300 GHz: Horn-fed waveguides

Up to frequencies of the order of 300 GHz ($\lambda \geqslant 1$ mm), waveguide mounts seem well adapted to Josephson detectors or mixers associated to a horn fed waveguide. For this range the effective area of a horn is large ($\simeq$ half the physical area), the surface resistance losses are still low with careful design (high conductivity material, surface smoothness), and waveguide technique allows matching of the reactive part of impedances using an adjustable stub and plunger system. However, the efficiency of the tuning stub appears difficult to obtain at higher frequencies. The bandwidth is also rather restricted (a few percent). A reduced height waveguide may help to obtain a reasonable match of the junction resistance.

b. Frequencies $\geqslant$ 300 GHz: Antenna systems

At frequencies higher than 300 to 500 GHz, the antenna technology

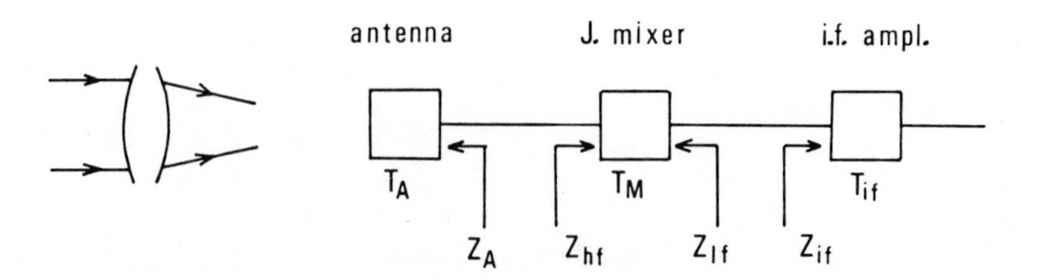

Fig. 7. Impedances and noise temperatures in a receiver.

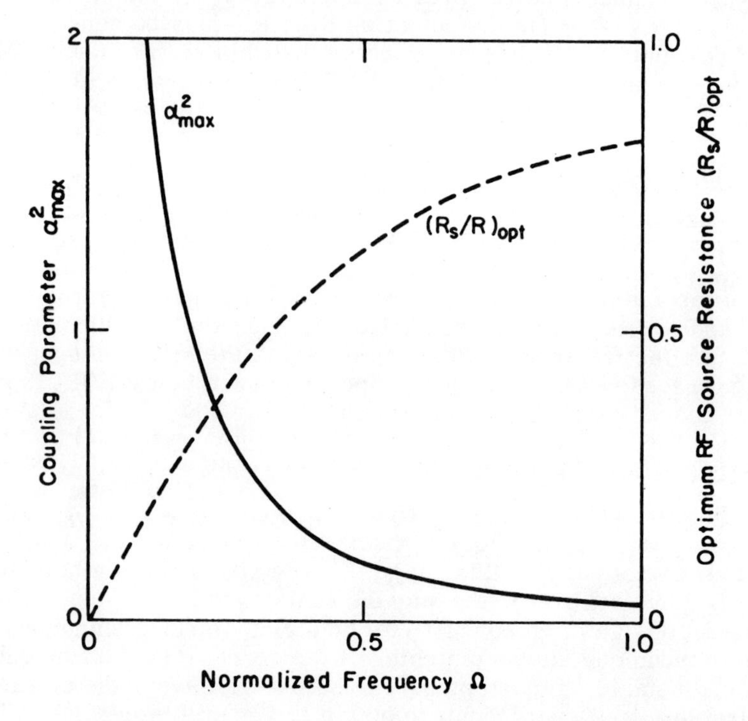

Fig. 8. Maximum value of the coupling parameter α and optimum rf source impedance R_S computed from the RSJ model as a function of the normalized frequency Ω (taken from Ref. [42]). Note that in this figure $R = R_N$.

seems to be preferable and the junction is integrated in the antenna structure. Ohmic losses are negligible if the transverse dimensions are not too small relative to the length. However, simple dipole antennas have a rather small effective area. For a short dipole, $A_e \simeq 3\lambda^2/8$. Therefore, it is necessary in practice to have secondary reflectors to increase A_e (up to a factor of ten [43]). Small f number optics decrease the focussed spot size, but the plane-wave assumption of antenna calculations is no longer a good approximation at the focus. Another solution is to use long-wire ($\ell \gg \lambda$) or more complex antennas which have a more directive radiation pattern (higher gain g) and a larger effective area ($\gtrsim \lambda^2$). The impedance of long-wire antennas is rather large ($\gtrsim 100\,\Omega$) which makes the impedance matching difficult. Moreover, these antennas have a narrow bandwidth ($\simeq$ 5-10%) which may be a drawback in some applications.

The type of antenna to be used and its realization depends on the type of Josephson junction which is considered:

i. Planar structures: SIS junctions and microbridges

Here the application of integrated circuit technology must be considered. At frequencies of a few GHz microstripline techniques have been applied either for single junction or series arrays [45, 102]. Another approach is the integration of junctions with deposited resonant antenna structures. Successful experiments have been realized with MOM junctions (metal-oxide-metal) [46]. It should be noted that these latter junctions have a much larger impedance ($\simeq 100\,\Omega$) than Josephson junctions.

ii. Point contacts

In a dipolar antenna structure, the contact electrodes constitute a dipolar antenna which may be short or large relative to $\lambda/2$. Short antennas have relatively low values of R_A, but their reactance X_A is important. Their gain is rather low and their directivity is broad. Half-wave resonant dipoles have large values of R_A ($\simeq 75\,\Omega$) which are not well suited to the impedance of superconducting contacts. Long-wire antennas have a larger effective area ($A_e \simeq \ell\lambda/2\pi$), but the radiation pattern is more complicated and the radiation must be carefully focussed along the direction of the major antenna lobe. The corresponding angle of incidence θ_m between the lobe and the whisker axis is [43,47] (Fig. 9)

$$\theta_m = \cos^{-1}(1 - 0.371/\ell\lambda^{-1}) \quad , \tag{64}$$

which shows that the greater the length ℓ of the whisker and the radiation frequency, the smaller is θ_m. For example, to have $\theta_m = 30^\circ$, $\ell = 0.93$ mm at $\lambda = 337\,\mu$, and $\ell = 31\,\mu$ at $\lambda = 10\,\mu$. These antennas have a large impedance ($\gtrsim 100\,\Omega$) and their reactance is probably of the same order of magnitude.

In a conical antenna (Fig. 9) the electrode transverse dimensions may be important [38]. The radiation pattern of such a structure is always

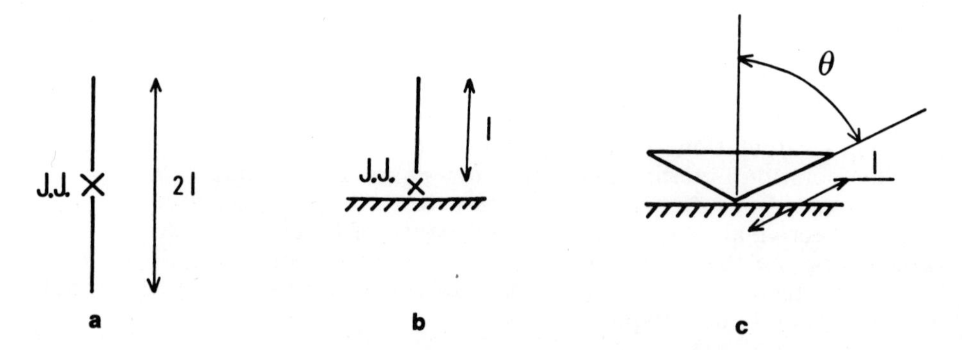

Fig. 9. Antenna structures with a Josephson point contact: (a), (b) dipole antenna; (c) conical antenna.

similar to that of a short dipole antenna, whatever the transverse dimensions, if the cone angle 2θ is large ($2\theta > 60°$). The effective area of such an antenna is also of the same order of magnitude ($\simeq \lambda^2/8$). A considerable advantage is that, if the transverse dimensions are large ($\ell \geq 2\lambda$), the input impedance is very near the characteristic impedance (TEM mode) [43]

$$K(\text{Ohms}) = 60 \log_e \cot(\theta/2) \tag{65}$$

The impedance is a decreasing function of the cone angle 2θ and goes to zero for $\theta = 90°$. Therefore, the conical structure makes it possible to obtain the value R_A/R_N corresponding to the optimum sensitivity. Another advantage of this antenna is its broadband properties related to the small variation of impedance with frequency. The practical difficulty is to fabricate junctions with a small contact area (small junction capacitance) with large cone angles.

II. ANALYSIS AND PERFORMANCES OF HIGH FREQUENCY JOSEPHSON DEVICES

The studies which have concerned high-frequency Josephson devices up to now and are analyzed in the following sections have been treated theoretically using the RSJ model. Consequently all the complications involved at frequencies of the order of the superconducting gap or larger are neglected, e.g., variation of the amplitude of the Josephson and quasiparticle current components, influence of the pair-quasiparticle interference current component, and photon assisted tunneling effects when an external LO is used. It should also be noted that for the noise properties most often the only source of fluctuation which is considered is the thermal noise from the normal resistance R_N.

A. Generation of Radiation

The principal problems involved with the Josephson junction used as a voltage-tunable source of high frequency radiation are the small amount of radiation power, and the wide linewidth of the emitted radiation.

The Josephson generator converts dc energy from the bias source to ac energy. At high frequency, using the voltage-source model, the alternating voltage is

$$V(t) = R_N I_c \sin \omega_o t \quad .$$

The antenna impedance is in parallel with the junction impedance and R_N must be replaced by

$$R_{eff} = \frac{R_N R_S}{R_N + R_S} \quad .$$

At high frequency $R_N \simeq R_S$ is a good approximation (Section I-I2 c) R_{eff} = (1/2) R_N. Therefore the maximum radiation power is

$$P_{max} = (1/8)\, R_N\, I_c^2 \quad .$$

Theoretically $R_N I_c \simeq \frac{\pi \Delta}{2e}$. However, in practical devices $R_N I_c$ is often 0.3 to 0.5 of its optimal value, and one finds that $P_{max} \sim 10^{-8}$ W. At lower frequencies (centimeter or millimeter region) P_{max} is even lower.

Experimentally, up to 10^{-11} W of emitted radiation has been obtained at 9 GHz with tunnel junctions when the sandwich structure acts in a resonant mode [49, 50]. A better coupling of a JJ to the outside was obtained with point contacts in a coaxial cavity. Emitted powers at 9 GHz of 10^{-10} W [51] and 10^{-11} W [52] were observed in this way. It is, of course, of little practical interest to produce such small amounts of microwave power with a cryogenic device. However, the radiation emission in the 9 GHz range permitted investigation of fluctuation phenomena in SIS tunnel junctions [32] and in point contacts [34] by means of radiation linewidth measurements.

Extension of the frequency limit to the submillimeter range is of interest. Continuously tunable millimeter and submillimeter radiation has been generated with a Nb point contact JJ in a superconducting transmission line [53]. The transmission line is centered in an oversized multimode brass cavity. Radiation was detected for all bias $\leqslant$ 2mV ($\nu \leqslant$ 1 THz). The maximum power absorbed with an external detector was $\simeq 2 \times 10^{-10}$ W and the power emerging from the cavity was estimated to be $\simeq 10^{-9}$ W, a number which compares favorably to the maximum calculated power (a few 10^{-8} watts). The possibility of performing low temperature spectrometry in the same frequency range with this device has also been demonstrated [54].

The potential applications of large arrays of Josephson junctions as microwave generators have been investigated. Tilley [55] has predicted theoretically that m identical junctions connected in series can interact with a cavity mode so that the number of coherently emitted photons is proportional to m^2. The observation of such a superradiant state with two tunnel junctions connected in series has been performed successfully [56]. However, the extension to a large array is difficult.

Microwave generation (X-band) has been observed recently in microbridges [57,58], which represents the most direct verification of the AC Josephson effect in these devices. The results are in close analogy with observations on point contacts, but the detected power level is lower (total integrated emitted power $\simeq 10^{-12}$ W).

B. Bolometer

1. Bolometer characteristics

Bolometers generally have a slow response compared to quantum detectors (photon or Josephson effect detectors) and diode detectors (semiconductor or metal). Their advantages in the far-infrared (FIR) - millimeter range include broadband response, the possibility of absolute calibration, and designs which are separable into two parts (absorption of radiation and sensing of resultant temperature rise). The development of nearly ideal single-photon noise-limited detectors in the visible, and heterodyne detectors at microwave frequencies, has largely limited the use of bolometers to the infra-red where cryogenic bolometers [59-62] are the most sensitive broadband detectors available today. Semiconductor bolometers are widely used, but in applications of low level background blackbody radiation (far-infrared, cooled or space environments) superconducting bolometers allow better performances.

A bolometer consists of a radiation absorbing element and a resistance thermometer with total heat capacity C (Joules/K) which are attached to a thermal bath through a thermal conductance G(Watt/K). The heat equation for a thermal detector in its simplest form is $C\frac{d}{dt}(\Delta T) + G(\Delta T) = We^{i\omega t}$, which has the steady state solution

$$\Delta T = \frac{W}{G} \frac{1}{(1 + \omega^2 \tau^2)^{1/2}} , \tag{66}$$

where W is the absorbed peak radiation power, ΔT the temperature rise, and τ the bolometer time constant = C/G. It may be seen from Eq. (66) that the ideal bolometer must have the lowest possible heat capacity for fast response, obtained by minimizing the detector volume, and the thermal conductance compatible with the required τ and an acceptable ΔT. The responsivity must be high so that the noise is limited by the thermal fluctuations as shown below.

The various contributions to low temperature bolometer noise are independent, so that their squares can be added to give a value of the system NEP [65] referred to a 1 Hz bandwidth

$$(NEP)^2 = \frac{4kTR}{R_V^2} + 4kT^2G + 8e\sigma kT_B^5 A\Omega + \frac{V_N^2}{R_V^2} \quad . \tag{67}$$

Each term is written as an equivalent optical noise power entering the detector. In the first term 4kTR is the squared Johnson noise voltage associated with the bolometer resistance R. The voltage responsivity ($R_V = dV/dP_{signal}$) converts this term to an optical power. The second term, found only in thermal detectors, is the thermal or phonon noise from statistical fluctuations in the bolometer temperature. Its amplitude is proportional to the thermal conductance G between the bolometer and heat sink. The third term represents the background black body fluctuations for a detector of emissivity σ, area A and throughput $A\Omega$. In the final term, V_N is the noise voltage corresponding to noise sources such as the current noise in the bolometer element, amplifier noise, etc.

The first and last term in Eq. (67) can be made negligible with a detector of high responsivity R_V, and the third term may be strongly reduced when a low background temperature is present, or by reducing the bandwidth with cooled filters. Therefore, the ultimate sensitivity is limited (in the FIR) by the thermal fluctuations contribution

$$NEP\ (1\ Hz) = (4kT^2G)^{1/2} \quad . \tag{68}$$

2. SNS and superconducting transition edge bolometers

In the Josephson junction bolometer [18] the radiation is absorbed in a thin Bi film deposited on a sapphire substrate (Fig. 10). The temperature is measured using the critical current variation of an SNS junction given by

$$I_c = I_o \exp(-T/T_o)^{1/2} \quad ,$$

and

$$\frac{dI_c}{dT} = -\frac{I_c}{2\sqrt{TT_o}} \quad , \tag{69}$$

with $T_o \simeq 0.1$ K.

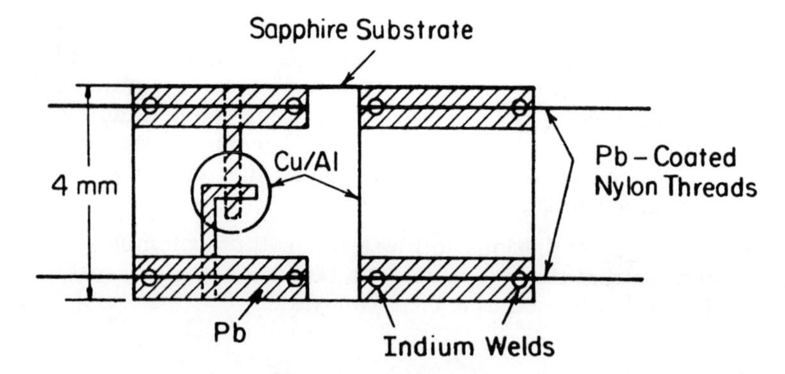

Fig. 10. SNS Bolometer. The SNS junction is evaporated on the left side. A Cu/Al alloy heater is evaporated on the right side to make NEP measurements (Ref. [18]).

The minimum detectable amount of absorbed radiation power is obtained from Eq. (66), $NEP = G\Delta T_m$, where ΔT_m is the minimum detectable temperature change, which can be written as a function of the rms noise voltage V_N as

$$\Delta T_m = \frac{\partial T}{\partial I_c}\,\Delta I = \frac{2\sqrt{TT_o}}{I_c}\;\frac{V_N}{R_D} \quad .$$

If the noise voltage generated in the SNS is only due to Johnson noise in the shunt resistance,

$$V_N = R_D\left(\frac{4kT}{R}\right)^{1/2} \tag{70}$$

and

$$NEP = \frac{4GT}{I_c}\left(\frac{kT_o}{R}\right)^{1/2} \quad . \tag{71}$$

The numerical parameters of an SNS bolometer [18] are shown in Table II, so that a comparison between the thermal limit and calculated and measured results can be seen.

This bolometer has a very high detectivity: $D^* = \sqrt{A}/NEP \simeq 10^{14}$ cm $\sqrt{Hz}/W$, which is the largest reported for a He^4 bolometer. On

Table II. Comparison of Bolometer performances

The detectivity D^x is the reciprocal of the NEP normalized to a detector area of 1 cm^2 and a 1 Hz electrical bandwidth. An NEP of 2×10^{-15} $W/\sqrt{Hz}$ at a chopping frequency of 10 Hz has been obtained recently with the transition edge bolometer.

SNS Bolometer [18]	Superconducting transition edge [65] Bolometer	Ge Bolometer [66]
$T = 1.5$ K	$T = 1.5$ K	$T = 4$ K
$R_D \simeq 3 \times 10^{-6}\ \Omega$		
$I_c = 2 \times 10^{-5}$ A		
$G = 10^{-8}$ W/K	$G = 5 \times 10^{-8}$ W/K	$G = 2 \times 10^{-4}$ W/K
$C = 10^{-9}$ J/K	$C = 6.5 \times 10^{-9}$ J/K	
$NEP = 5 \times 10^{-5}\ W/\sqrt{Hz}$	$NEP_{(meas.)} = 10^{-14}\ W/\sqrt{Hz}$	$NEP_{(meas.)} = 10^{-13}\ W/\sqrt{Hz}$
$NEP_{(calc)} = 2 \times 10^{-15}\ W/\sqrt{Hz}$	NEP (th. fluct. limit) $2.5 \times 10^{-15}\ W/\sqrt{Hz}$	
$\tau = 0.1$ sec	$\tau = 0.13$ sec	$\tau = 0.01$ sec
NEP (thermal fluctuation limit) $\simeq 1 \times 10^{-15}\ W/\sqrt{Hz}$	$D^* = 4 \times 10^{13}$ cm $Hz^{1/2}W^{-1}$	$D^* = 1.4 \times 10^{12}$ cm $Hz^{1/2}\ W^{-1}$

the other hand, its very low impedance ($R_D \simeq 3 \times 10^{-6}\ \Omega$) makes it necessary to use a superconducting quantum interference amplifier (SQUID) to measure the voltage output. The temperature dependence of $I_c = f(T)$ in proximity effect bridges can be similar to that in SNS junctions, which also depend on the proximity effect for their operation. A series array of such bridges could be used to increase the impedance, but because of the very large number (10^5 to 10^6) required to raise the impedance to $\simeq 1\ \Omega$, the construction problem would be formidable.

Previous superconducting bolometers suffered from appreciable low frequency noise. Clarke and Hsiang [64] have shown recently that this noise can be reduced by a strong thermal coupling of the film to its substrate. Using this result Clarke et al. [65] have realized an Al superconducting transition edge bolometer on a sapphire substrate which is comparable in sensitivity (Table II) to the SNS bolometer, but considerably easier to construct and operate, particularly because conventional room temperature FET electronics may be used.

3. Comparison of devices

The well developed technology of heavily doped semiconducting bolometers (Si or Ge) allows one to approach optimum performance in far infrared experiments which are limited by fluctuations in the room temperature blackbody background radiation. An NEP = 10^{-13} W/$\sqrt{\text{Hz}}$ [66] with $\tau = 10^{-2}$ sec has been obtained at 4 K (Table II). In applications with low level background blackbody radiation, superconducting bolometers allow superior performances. The superconducting edge bolometer and the SNS bolometer are comparable in performance but the former is easier to construct and operate.

C. Video Detection

1. Junction quadratic response

a. Qualitative analysis

A Josephson junction (generally a point contact) used as a video square law detector is biased at a current I_o slightly larger than I_c. If it is irradiated with a very small rf signal of amplitude I_S, the zero voltage current (n = 0 step) is depressed quadratically for any rf frequency ω_S as shown in Fig. 6 (Section I F 2). The zero voltage current $I_c(I_S)$ may then be written as given by Eq. (10),

$$I_c(I_S) \simeq I_c\,(I_S = 0)\left[1 - \gamma I_S^2\right] \quad , \tag{10}$$

and the output voltage

$$V_{rf} = R_D\,\delta I = -\gamma R_D I_c I_S^2 \quad . \tag{11}$$

At high frequencies ($\Omega = \omega/\omega_c \gtrsim 1$), the voltage bias model may be used (Section I F) and the height of the zero voltage step is given by Eq. (20)

so that

$$\frac{I_c(I_S)}{I_c} = J_o\left(\frac{2eV_S}{\hbar\,\omega_S}\right) \simeq 1 - \frac{e^2V_S^2}{\hbar^2\omega_S^2} = 1 - \frac{e^2R_N^2I_S^2}{\hbar^2\,\omega_S^2} \quad . \tag{72}$$

From Eqs. (10), (11) and (72), the amplitude of the output voltage is

$$V_{rf} \simeq \frac{R_D\,I_S^2}{4\,I_c\,\Omega^2} \quad . \tag{73}$$

It will be seen in the next section that this expression is also valid for frequencies $\Omega < 1$. The important result is the Ω^{-2} frequency response.

2. Voltage response in the general case

A more precise analysis has been performed by Kanter and Vernon [44] and also by Likharev and Semenov [67]. For a current driven JJ obeying the RSJ model, the dc voltage response and impedance to external high frequency currents is calculated with a second-order perturbation method based on the unperturbed solution for the time evolution of the voltage as given by Aslamazov and Larkin [4]. In the absence of fluctuations, the amplitude of the detected voltage is [44]

$$V_{rf} = \frac{1}{4}\frac{I_S^2}{I_c}\;\frac{R_N}{\left(\frac{I_o^2}{I_c^2} - 1\right)^{3/2}}\;\frac{\omega_o^2}{\omega_S^2 - \omega_o^2} = \frac{I_S^2}{4}\;\frac{R_D}{I_o}\;\frac{\omega_o^2}{\omega_S^2 - \omega_o^2} \quad . \tag{74}$$

The response is proportional to the signal power and expressed in terms of measurable parameters of the VIC. There are two regions of wideband detection corresponding to the internal self-governed Josephson frequency larger than or smaller than the signal frequency, separated by resonant detection when $\omega_S = \omega_o$.

Region i) $\omega_o \gg \omega$: From $V_o = R_N(I_o^2 - I_c^2)^{1/2}$ we obtain

$$\left(\frac{I_o^2}{I_c^2} - 1\right)^{-3/2} = \frac{I_c}{R}\;\frac{d^2V_o}{dI_o^2} \quad , \tag{75}$$

where d^2V_o/dI_o^2 is the second derivative of the VIC at this bias point, and in the limit $\omega_o \gg \omega_S$ Eq. (74) assumes the simpler form

$$V_{rf} \simeq \frac{1}{4}\,I_S^2\;\frac{d^2\,V_o}{dI_o^2} \quad . \tag{76}$$

The detection is "classical", i.e., proportional to the curvature of the VIC. It is not of practical importance with JJ because the response is much lower than for $\omega_o << \omega_S$.

Region ii) $\omega_o << \omega_S$: this corresponds to biasing at a current slightly larger than I_c. Equation (74) gives

$$V_{rf} = \frac{I_S^2}{4} \frac{R_D}{I_o} \frac{1}{\Omega^2} \tag{77}$$

which is almost the same as Eq. (73) obtained using the voltage bias model, but is valid for any value of Ω. In this regime, the characteristic time to reach the steady-state Josephson oscillation is larger than the signal period, which explains the decrease of the response with increasing frequency.

Region iii) $\omega_o \simeq \omega$: Equation (74) predicts singular response for $(\omega_S - \omega_o) \to 0$ because fluctuations have been omitted in deriving this expression. Since the Josephson oscillation usually has an appreciable linewidth, this singularity is partially washed out when noise is taken into account. A typical response obtained at 90 GHz is shown in Fig. 11. It clearly shows the high sensitivity obtained near zero voltage at $\omega_o << \omega_S$, and the dependence of V_2 with the different parameters of Eq. (77).

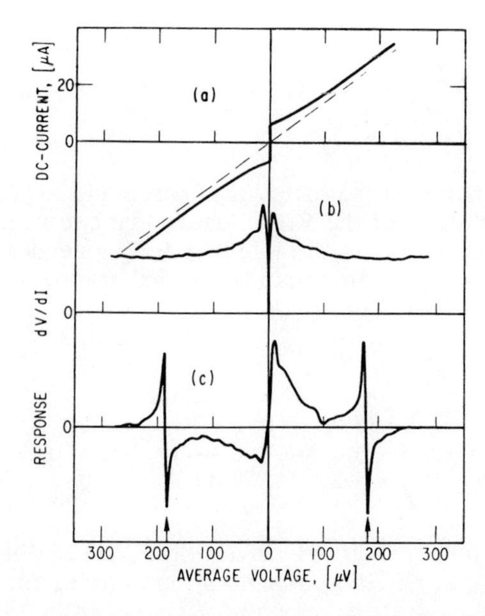

Fig. 11. Video detection. (a) Measured VI characteristics, (b) dV/dI, and (c) the voltage response V_{rf} at 90 GH_z. The arrows indicate the bias equivalent to the applied frequency (after Ref. [44]).

If P is the applied signal power and the coupling efficiency is C_{hf}, since the junction impedance when $\omega_o \ll \omega$ is $\simeq R_N$ (see Section I-12), the signal power in the circuit is $C_{hf}P = \frac{1}{2} I_S^2 R_N$ from which we obtain the voltage response

$$R_V = \frac{V_{rf}}{P} = \frac{C_{hf}}{2} \frac{R_D}{R_N} \frac{1}{I_o} \frac{1}{\Omega^2} \tag{78}$$

3. Noise equivalent power

We now consider the situation where the noise consists mainly of the intrinsic noise of the junction and the amplifier noise may be neglected. The relevant figure of merit of the detector now is its NEP. The NEP is obtained by equating the signal voltage $V_{rf} = R_V \times$ NEP with the mean square root value of the output noise voltage across the junction (Section I G 2)

$$V_{rms}^2 = R_D^2 B' P_I(\omega) \quad ,$$

where B' is the detector postbandwidth and $P_I(\omega)$ is the current power spectrum which reduces to $\frac{2}{\pi} \frac{kT}{R_N}$ [see Eq. (42)] if only thermal noise from the junction normal resistance is considered, and to $\frac{2}{\pi} \frac{kT}{R_o}$ with $R_o = \frac{V_o}{I_o}$ [see Eq. (37)] if the junction current fluctuations are preponderant. Then

$$\mathrm{NEP}\ (B' = 1\ \mathrm{Hz}) = \frac{V_{rms}(B' = 1\ \mathrm{Hz})}{R_V} = \frac{2}{C_{hf}} R_N I_o \Omega^2 \sqrt{P_I(\omega)} \quad . \tag{79}$$

To reach high sensitivities, it is necessary [see Eq. (79)] to have the best coupling efficiency. This implies a rather large impedance R_N (a few $10^2\ \Omega$) in the far infrared if points contacts of the long wire antenna type are selected (effective area $\simeq \lambda^2$). However, I_c becomes small in that case compared with the noise current, and the voltage response R_V may suffer so that the system becomes limited by the amplifier noise (R_V is proportional to R_D which is decreased by the rounding of the $n = 0$ step). Since the junction is biased at a low voltage, $R_N I_o$ does not differ much from $R_N I_c$ and $(R_N I_o \Omega^2)$ is proportional to $(R_N I_c)^{-1}$. Junctions with a large $R_N I_c$ product are therefore required.

4. Discussion of experimental results and comparison with other video detectors

In Table III the data calculated and measured by Kanter and Vernon at 90 GHz are compared. The device is mounted in a waveguide structure at this frequency and a reasonable matching may be expected

Table III. Video detector performance

Josephson junction [44]	Schottky diode [70]	Super-Schottky diode [70]
$F = 90$ GHz	$F = 10$ GHz	$F = 10$ GHz
$I_c = 10^{-5}$ A	$T = 300$ K	$S = 6000\ V^{-1}$
$R_N = 40\ \Omega$	$S = 40\ V^{-1}$	$(NEP)_{th} = 2.4 \times 10^{-16} W/\sqrt{Hz}$
$I_o = 1.2 \times 10^{-5}$ A	$R = 100\ \Omega$	$T = 4$ K
$V_o \simeq 10^{-5}$ V	$(NEP)_{th} = 4.5 \times 10^{-13} W/\sqrt{Hz}$	$R_i = 300$ A/W
$T = 4$ K		$(NEP)_{meas} = 8 \times 10^{-15} W/\sqrt{Hz}$
$(NEP)_{meas} = 5 \times 10^{-15} W/\sqrt{Hz}$	$T \leqslant 77$ K	
$(NEP)_{calc} = 10^{-16} W/\sqrt{Hz}$	$S = 276\ V^{-1}$	$T = 1$ K
(thermal noise from R_N)	$(NEP)_{th} = 2.5 \times 10^{-14} W/\sqrt{Hz}$	$R_i = 1100$ A/W
$(NEP)_{calc} = 8 \times 10^{-16} W/\sqrt{Hz}$		$(NEP)_{meas} = 2 \times 10^{-15} W/\sqrt{Hz}$
(junction current shot noise)		

using a low impedance structure (reduced height waveguide) and a tuning short plunger. The calculated NEP corresponding to the limit of thermal noise of R_N seems to be much too low compared to the measured data. The data seem to be much better fitted if the device noise is assumed to be dominated by the shot noise associated to the junction current.

It should be noted that the NEP of the video detector increases with Ω^2. Comparing its performances with the bolometer performances in Table II, it may be seen that at wavelengths shorter than 1-2 mm, the bolometer has a better sensitivity. However, the video detector may be preferred for its high speed compared to the bolometer. A JJ has been used as a detector for astronomy applications by Ulrich [68] and in the far infrared by Tolner et al. [69]. For 300 K radiation ($\lambda \simeq 0.1$ mm) it has been reported [69] an NEP of 2.8×10^{-14} $W/\sqrt{Hz}$. A comparison with theoretical estimates is difficult for this latter case due to the uncertainties concerning the coupling coefficient value.

Finally, it should be noted that series arrays of JJ (e.g., microbridges) have been suggested by Richards [42] to help match at infrared frequencies and for detecting multimode sources.

Table III also compares the performance of the Schottky barrier diode and the Super Schottky barrier diode used as video detectors. The latter which is a superconductor-semiconductor barrier diode has been developed recently by McColl, Millea and Silver [70]. The parameter $S = q/k(T + T_0)$ is a measure of the nonlinearity of the diode. T_0, which is related to the semiconductor gap energy ($\simeq 1$ eV) for Schottky diodes and to the superconducting gap ($\simeq 1$ meV) for super-Schottky diodes, is much smaller in the latter case. Therefore, extremely high values of S are obtained with super-Schottky diodes. Table III indicates that if the cooled Schottky barrier diode does not allow very high performances in video detection, the super-Schottky diode appears promising. Experimentation in the submillimeter range is necessary for a useful comparison with Josephson junctions.

D. Heterodyne Detection

Here we examine the sensitivity of the Josephson heterodyne mixer with either an external or internal local oscillator, and then compare the predicted or obtained performances with the corresponding results for other available heterodyne mixers. Because the actual far-infrared mixers are generally not photon shot noise limited, their sensitivity is well characterized by the mixer noise temperature referred to its input

$$T_M = T_N/\eta , \quad \text{with } \eta = C_{hf} G = \eta_T / C_{if} \quad \text{[Section IHIc, Eqs. (50), (52) and (55)]} .$$

Because the coupling coefficient C_{hf} does not depend a priori on the mixer mode of operation, the difference between the external LO and internal LO will occur in the value of the conversion efficiency G. The coupling to the external radiation and hf impedance matching has been considered in Section I-I. As for video detection, heterodyne detection experiments have been mostly restricted to point contacts. It must be noted that in the following all the complications related to the energy gap (Riedel peak, pair-quasi particle interference current, photon assisted tunneling) are not considered.

1. External local oscillator

We have seen in Section ID2 that in this mode the signal to be detected at ω_S is mixed directly in the junction with the signal from the local oscillator at ω_L. The fact that the frequency converted signal does not depend explicitly on the Josephson frequency is related to the resistive feedback created by the normal shunt resistance of point contacts (Section IF3). In particular, a fundamental result is that the linewidth of the frequency converted signal is independent of the Josephson linewidth (Section IG3) and generally much smaller than the bandwidth of the intermediate frequency amplifier. Therefore, the relations given in Section IH1c are valid in the present situation and the problem is to determine the conversion efficiency η.

a. Calculation of the conversion efficiency η

Such a determination in the most general case is a very complicated problem which can be solved only on a computer [71-73]. However, in practical situations one is mainly interested in the detection of very low level signals with small intermediate frequencies. The high-frequency current delivered by the signal and LO sources (Fig. 12) may then be

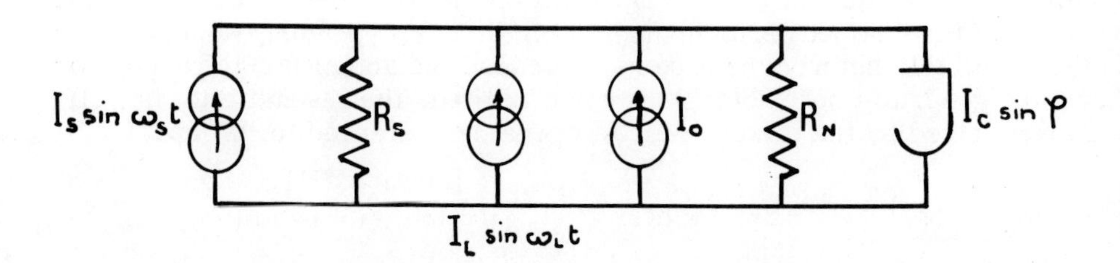

Fig. 12. Circuit model of the calculations for the Josephson mixer with external LO.

written (Section I D 2):

$$I = I_L \sin\omega_L t + I_S \sin\omega_S t \simeq (I_L + I_S \cos\omega_{if} t) \sin\omega_L t \tag{81}$$

with $\omega_{if} \ll \omega_S$, ω_L and $I_S/I_L \ll 1$.

With a current-biased junction, the result is a modulation of the VIC at the intermediate frequency which gives rise to an if voltage with amplitude

$$V_{if} = R_D I_{if} \tag{82}$$

with $R_D = \left(\frac{\partial V_o}{\partial I_o}\right)_{I_L}$ and $I_{if} = I_S \left(\frac{\partial I_o}{\partial I_L}\right)$.

As the signal to be detected has a very low level and the junction is current biased at a point midway between two successive steps induced by the LO, the hf impedance is practically R_N (Section I-I 2 a). Consequently the mismatch at the high frequency may be taken into account writing Eq. (82) as

$$I_{if} = I_S \frac{R_S}{R_S + R_N} \mu \tag{83}$$

$$\text{with} \quad \mu = \left(\frac{\partial I_o}{\partial I_e}\right) \tag{84}$$

where I_e is the high frequency current effectively coupled into the junction $\left(I_e = I_L \frac{R_S}{R_S + R_N}\right)$.

The power coupled at the intermediate frequency is

$$P_{if} = (P_{if})_{av.} C_{if} , \tag{85}$$

where $(P_{if})_{av.}$ is the available power at the if from the mixer with

$$(P_{if})_{av} = V_{if}^2 / 8 R_d$$

because the low frequency impedance of the junction is R_d (Section I-I 2 a).

From Eqs. (82) to (85) and the definitions in Section I-I 1 c, the total conversion loss may be determined [75] as

$$\eta_T = \frac{P_{if}}{(P_S)_{av}} = C_{if}\,C_{hf}\,\frac{\mu^2}{4}\,\frac{R_d}{R_N} \tag{86}$$

where the available power from the signal source is

$$(P_S)_{av.} = R_S I_S^2/8 \quad .$$

Therefore, $$\eta = \frac{\eta_T}{C_{if}} = C_{hf}\,\frac{\mu^2}{4}\,\frac{R_d}{R_N} \quad . \tag{87}$$

This result is also presented in the literature [74] in the form

$$\eta_T = C_{if}\,\alpha^2\,\frac{R_d}{R_N} \quad , \tag{88}$$

with $$\alpha^2 = \left(\frac{\partial I_o}{\partial I_L}\right)^2 \frac{R_N}{R_S} = \left[\frac{\partial I_o}{\partial\left(\frac{8P_L}{R_N}\right)^{1/2}}\right]^2 \tag{89}$$

and $P_L = R_S I_L^2/8$.

Equation (88) is more convenient to describe the properties of Josephson mixers in practical situations because the expression of η_T in Eq. (86) is a function of Ω through μ^2 and of the ratio R_S/R_N through C_{hf} and μ^2 (Ω is also a function of R_S/R_N as seen in Section II2c). This is a reason why α^2 is determined directly on a computer [76]. In this way Fig. 8 shows the value of R_S/R_N optimizing α^2 and the corresponding value $(\alpha^2)_{max}$ as a function of Ω.

It is interesting to consider the following two limit cases:

(i) Very high frequencies ($\Omega \gtrsim 1$)

The calculations are much simplified and Eq. (86) may be used. The ratio R_S/R_N to obtain an optimum sensitivity is near 1 (Fig. 8). On the other hand, the variation of the dc current with the hf signal follows approximately a Bessel function dependence in this frequency range (Section IF2). Therefore, if we assume that the junction is biased near the first step induced by the LO

$$\mu = \frac{\partial I_o}{\partial I_e} \simeq \frac{\partial}{\partial I_e}\left[I_c J_1(2x)\right] \simeq \frac{\partial}{\partial I_e}(x\,I_c) \quad , \tag{90}$$

where here $x = \dfrac{e R_N I_e}{\hbar \omega_L} = \dfrac{I_o}{I_e} \quad \dfrac{1}{2\Omega}$.

Then $$\mu \simeq \frac{1}{2\Omega} \tag{91}$$

and $$\eta \simeq \frac{C_{hf}}{16\Omega^2} \frac{R_d}{R_N} \quad . \tag{92}$$

A very similar result has been obtained numerically on a computer [71, 72, 79] assuming $C_{hf} = 1$,

$$\left(\alpha^2\right)_{max} = \frac{1}{12\Omega^2} \quad \text{and} \quad \eta \simeq \frac{1}{12\Omega^2} \frac{R_d}{R_N} \quad , \tag{93}$$

where α^2 corresponds to the asymptotic value of Fig. 8 for $\Omega > 1$. These results must be considered with care since the RSJ model is not expected to be valid at $\Omega > 1$.

(ii) Low frequencies ($\Omega \ll 1$)

The results computed numerically [71] are here:

$$(\alpha^2)_{max} \simeq \frac{1}{4\Omega} \quad \text{with} \quad \frac{R_S}{R_N} = 1.67$$

and $$\eta \simeq \frac{1}{4\Omega} \frac{R_d}{R_N} \quad . \tag{94}$$

b. Noise temperature

i. Preliminary remarks

As we have already pointed out, the Josephson junction is such a highly non-linear element that it is particularly sensitive to fluctuations and especially to internal fluctuations. This has the following consequences.

(i) A decrease of the dynamic resistance R_D at the operating point. This effect has been calculated without applied radiation by Ambegaokar and Halperin [77] and their results are confirmed by numerical calculations by Auracher [72]. When an external radiation is applied, the consequence of noise in the VIC is a rounding of the induced steps as shown by Stephens [28] and investigated experimentally by Henkels and Webb [78]. The physical origin of this rounding is related to a partial unlocking of the Josephson oscillation by the applied radiation which results from the fluctuations. In practical situations R_d is an easily measurable parameter. From this decrease of R_d, a decrease of the conversion factor η results as shown by Eqs. (86-88).

(ii) The current fluctuations in the junction (Section IG 1), characterized by $I^2_{rms} = \langle i^2(\omega)\rangle = \int P_I(\omega)\, d\omega$, which are situated outside the if bandwidth B are through complex frequency conversion processes converted inside this bandwidth. The corresponding increase of the noise level at the if may be described with a noise parameter (see Eq. (76)]

$$\beta^2 = I^2_{rms}/i^2_{rms} \quad , \tag{95}$$

which is the ratio of the output to the input noise current. In general the determination of β^2 is complicated. Analog computer calculation results [74] show that without a local oscillator $\beta^2 \gtrsim 1$ except for $\Omega \ll 1$. With an applied LO, a similar behavior is obtained, but β^2 presents a minimum between successive steps.

Besides noise mixing in the junction, which is described with the parameter β, the origin and the intensity of the fluctuation components defined by $P_I(\omega)$ must be considered. This problem has already been discussed in Section IG 1. At the present time the existing models do not explain satisfactorily all the experimental results concerning point contacts. Some are well interpreted using the current power spectrum derived for tunnel SIS junctions [33, 34] whereas others are interpreted using the mechanism of noise mixing in the junction, and assuming that the only source of noise is the thermal noise associated with R_N and is described by

$$P_I(\omega) = \frac{2}{\pi} \frac{kT}{R_N} \quad .$$

In the following we present the calculation of the mixer noise temperature considering the more favorable case where only the noise contribution of R_N is considered.

ii. Mixer noise temperature

From the low frequency (if) point of view, the junction may be described with the circuit of Fig. 13. The available noise power in the if bandwidth B is

$$P_N = \beta^2 i^2_{rms} R_D = \beta^2 R_d P_I(\omega) B \qquad \text{or,} \tag{96}$$

$$P_N = \beta^2 \frac{2}{\pi} kT \frac{R_d}{R_N} B \quad , \tag{97}$$

where T is the ambient temperature of the junction. It corresponds to a noise power available from the resistance R_D at a temperature referred as T_N. Then

$$P_N = \frac{2}{\pi} kT_N B \tag{98}$$

with

$$T_N = \beta^2 \frac{R_d}{R_N} T \quad . \tag{99}$$

Consequently, the minimum noise temperature referred to the mixer input is from Eq. (88)

$$T_M = \frac{T_N}{\eta} = \frac{\beta^2}{\alpha^2} T \quad , \tag{100}$$

a result which is independent of R_d.

If $\Omega < 1$, β^2 and α^2 have similar variations and T_M is not very sensitive to Ω [71].

If $\Omega \geqslant 1$, $\beta^2 \simeq 1$ and from Eq. (93) assuming $R_d \simeq R_N$

$$T_M \simeq \frac{T}{\alpha^2} = 12\Omega^2 T \tag{101}$$

[or $16\,\Omega^2 T$ if one considers Eq. (92)].

In a real mixer the losses in the high frequency circuit have to be taken into account and the minimum mixer noise temperature is

$$T_M \simeq \frac{12\Omega^2 T}{C_{hf}} \quad , \tag{102}$$

where in C_{hf} the respective losses due to the impedance mismatch and to the coupling to the external radiation must be included (Section II).

The contribution to the complete receiver input effective temperature T_e due to the if amplifier may be estimated simply. In the particular case in which we are mostly interested here ($\Omega \geqslant 1$) this contribution may be written [Eq. (55)]

$$\frac{T_{if}}{\eta_T} = \frac{T_{if}}{\eta C_{if}} = T_{if} 12\Omega^2 \frac{R_N}{R_d} \quad . \tag{103}$$

This contribution will be small if

$$T_{if} \frac{12\Omega^2}{C_{if}} \frac{R_N}{Rd} < 12\Omega^2 T \quad , \text{ i.e.,}$$

$$T_{if} < \frac{Rd}{R_N} C_{if} T \quad . \tag{104}$$

For example, assuming $R_d/R_N \simeq 5$, $C_{if} \simeq 1$ and $T = 4.2\,K$, Eq. (104) gives $T_{if} < 20\,K$.

This condition is difficult to fulfill even with cooled parametric amplifiers. However, as we have seen above, the Josephson mixer noise temperature generally will be higher than given by Eq. (102).

2. Internal local oscillator

The signal to be detected is mixed directly with the internal Josephson oscillation (Section I D 2). An essential difference with the external LO mode is that the junction response [82] is affected by noise in two ways (Section I G 2 b): the noise level at the mixer output characterized by T_N, and the linewidth of the frequency converted signals which is related to the Josephson linewidth Δf_o [38]. There will be supplementary conversion losses if the condition $B > \Delta f_o$ is not satisfied.

a. Conversion efficiency

The conversion gain is simply determined with the voltage source model (Section I F 1). The result obtained in this way is therefore valid a priori at high frequencies such that $\Omega \gtrsim 1$ (Section I F 2). However, a recent calculation by Vystavkin [75] shows that the result appears to be valid for any value of Ω.

The Josephson current component at the intermediate frequency $\omega_{if} = |\omega_o - \omega_S|$ has an amplitude (Section I F 1)

$$I_{if} = I_c J_1\left(\frac{2e\,V_S}{\hbar\omega_S}\right) \quad , \tag{105}$$

where V_S is the voltage at frequency ω_S across the junction. The if mixing occurs with the n = 1 harmonic component induced by the applied signal. This one is assumed to have a low level and I_{if} may be written

$$I_{if} \simeq I_c \frac{eV_S}{\hbar\,\omega} \simeq \frac{I_S}{2\Omega} \tag{106}$$

Because the junction impedance at high frequencies is $\simeq R_N$ (Section I-I 2 a), $V_S \simeq R_N I_S$.

If the source signal has an impedance R_S (Fig. 14), in the same way as in Section I-I D 1 a, Eq. (106) must be replaced by

$$I_{if} = \frac{I_S}{2\Omega} \frac{R_S}{R_S + R_N} \tag{107}$$

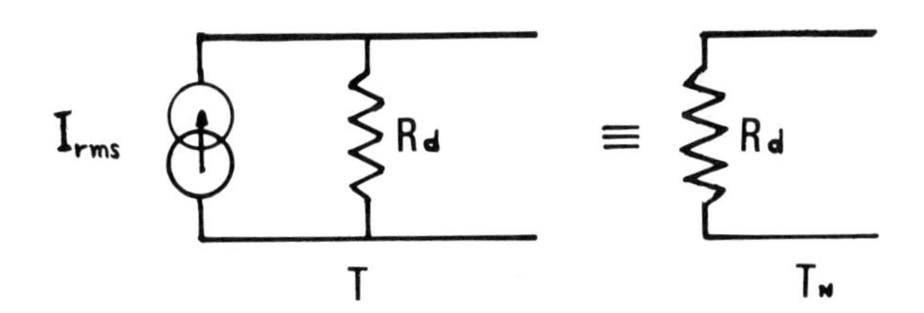

Fig. 13. Equivalent output noise circuit of a JJ.

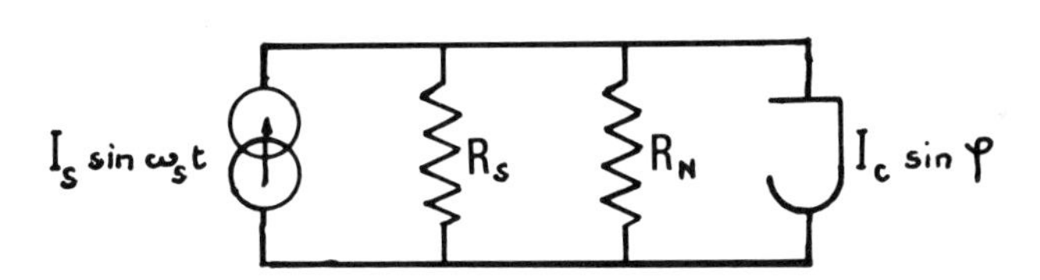

Fig. 14. Circuit model of the calculations for the Josephson self-oscillator mixer.

and

$$\eta_T = \frac{P_{if}}{(P_S)_{av}} = C_{if} C_{hf} \frac{1}{16\Omega^2} \frac{R_d}{R_N} \quad . \tag{108}$$

The signal level to be detected has a very low level and there is no visible induced step in the VIC as is the case with the external LO. If we consider high frequency signals with $\Omega \geqslant 1$, the bias voltage is important, and $R_d \simeq R_N$. Finally

$$\eta_T \simeq C_{if} C_{hf} \frac{1}{16\Omega^2} \quad . \tag{109}$$

If $C_{hf} \simeq 1$ (Section I-I 2 c), Ω must be replaced by $\Omega_{eff} = \omega_S/\omega_c'$.

Comparing Eqs. (92) and (109), it may be seen that within the factor R_d/R_N, the conversion gains of the Josephson mixers with internal and external LO have the same value at high frequencies.

We have disregarded phase-locking effects between the applied signal and the Josephson oscillation which may appear if the applied signal does not have a sufficiently low level when the intermediate frequency is small [80].

b. Influence of the Josephson linewidth

If the if bandwidth B is not larger than the linewidth Δf_o of the if frequency converted signal, only the fraction $(B/\Delta f_o)$ of the available energy will be amplified in the if chain (Section I G 2 b). Equation (109) must be replaced by

$$\eta_T = C_{if} C_{hf} \frac{1}{16\Omega^2} \frac{B}{\Delta f_o} \tag{110}$$

which is valid only if $\Delta f_o > B$. As an example, with $R_N = 5\Omega$ and if we assume that R_N is the only source of noise for simplification, Eq. (40) gives

$$\Delta f_o = \frac{\Delta \omega_o}{2\pi} \simeq 1 \text{ GHz} \quad .$$

This corresponds to a selectivity $(\Delta f_o/f_S) \simeq 10^{-3}$ at 1 THz. In practice, linewidths two or three times broader are observed at this frequency which are better explained assuming junction current shot noise [81]. Therefore, the Josephson oscillator-mixer implies that wide band if amplifiers are used, which means that large if are selected, e.g., $f_i \sim$ 1-5 GHz with $B \sim 0.1 - 1$ GHz. The practical consequences will be considered in the following section.

The linewidth Δf_o may be reduced if contacts of lower resistance are used ($\Delta f_o \sim R_d^2$). However, this makes the matching both at the hf and at the if worse (Section I-D) and a compromise is necessary.

A potential way to narrow the Josephson linewidths consists of reducing the low frequency internal fluctuations with a resistive shunt R_{sh} such that $R_{sh} \ll R_d \simeq R_N$. In this case, the current power spectrum of the junction becomes $P_V(\omega) = R_{sh}^2 P_I(\omega)$, and the Josephson linewidth

$$\Delta\omega_o = \pi \left(\frac{2e}{\hbar}\right)^2 P_I(\omega) R_{sh}^2 \quad , \tag{111}$$

with $R_{sh} = 0.1\,\Omega$ we obtain $\Delta f_o \simeq 20$ MHz and $\Delta f_o/f_o \simeq 2 \times 10^{-5}$ at 1 THz. These last theoretical predictions have not yet been tested experimentally. The realization of a shunt in parallel at the junction is difficult. The shunt resistance has an associated inductance L_{sh}. The cut-off frequency ω_{sh} must be large enough so that the low frequency noise components generated in the junction are shorted out (Section IG2a). The noise bandwidth is $\leqslant \Delta f_o$. On the other hand, the if signal must not be affected by the shunt which means $\omega_{sh} \ll \omega_i$. This problem can only be solved if large if are used ($f_i \sim$ 1-5 GHz).

c. Noise temperature

Equations (109) and (110) allow an estimate to be made of the theoretical minimum noise temperature of the oscillator-mixer. If we assume perfect matching of the hf and if channels ($C_{if} = 1$, $C_{hf} \simeq 1$) and $\Delta f_o \leq B$, at $\Omega \geqslant 1$,

$$\eta_T \simeq 1/16\Omega^2 \quad , \tag{112}$$

and

$$T_M \simeq T_N\, 16\Omega^2 \quad . \tag{113}$$

The discussion concerning T_N given in the case of the external LO is also valid here. In the most favorable case (thermal R_N noise) and at high frequencies $\Omega \geqslant 1$, we have $\beta^2 \simeq 1$,

$$T_N \simeq T \qquad (R_d \simeq R_N) \quad ,$$

and

$$T_M \simeq 16\Omega^2 T \quad .$$

If matching at the hf is not performed or the hf coupling is not optimized and if B is too narrow,

$$T_M = \frac{16\Omega^2 T}{C_{hf}} \frac{\Delta f_o}{B} \quad .$$

The contribution of the if noise to the overall receiver noise may be treated in the same way as in Section II D 1 b.

3. Discussion of the results and comparative performances of other mixers.

a. Experimental results

An interest in hf Josephson devices arises from the need for sensitive heterodyne receivers in the submillimeter range. However, most of the actual quantitative results concern the region < 100 GHz due to the difficulties of making accurate measurements at high frequencies. These results are related mainly to the heterodyne mode with external LO. They are summarized in Table IV and concern only point contact junctions.

The most systematic experiments have been performed at 36 GHz by Taur, Claassen, and Richards [71, 76, 83]. Niobium and vanadium junctions with non-hysteretic VIC have been tested which agree with the predictions of the RSJ model, but they have low RI_c products between one and two tenths of the theoretical value. Resonant impedance matching with a choke plunger and a tuning stub is used. Figure 15 shows typical measured values of η_T; the maximum of η_T occurs between steps when R_D is maximum. The optimum values of η_T range up to 4 and are 80-100 per cent of the value predicted by Eq. (88) using the experimental R_D. The values of the noise parameter β^2[Eq. (95)] are up to a factor of two greater than the values corresponding to the thermal noise from R_N. The lowest value of T_M = 54 K obtained from Eq. (100) is very indicative of the performances of the JJ mixer. The very low level of

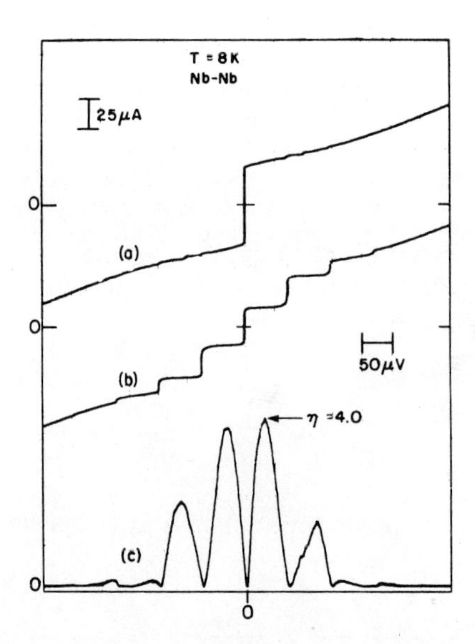

Fig. 15. Josephson mixer with external LO at 36 GHz. Junction VI characteristic: (a) without; (b) with 36 GHz LO; (c) conversion efficiency for a Nb point contact mixer with conversion gain (After Ref. [42]).

Table IV. Performance of high frequencies mixers

Junction	References	Signal Frequency (GHz)	Material	Temperature (K)	IF (B) (MHz)	$\Omega = \frac{\omega}{\omega_c}$	η	$T_{M(exp)}$ (K)	NEP (W/Hz)	P_{LO} Watt
Josephson mixer	[71,76]	36	Nb	4	50 (20)	0.36	0.47	140	7.5×10^{-22}	10^{-9}
	[83,84]		V	1.4		0.3	1.35	54		
J. harmonic mixer	[74]	36	Nb	7	50 (20)	0.8	0.5	400		
J. mixer	[86]	891	Nb	4	0.01-1	1	10^{-3}(0.2)(1)	7×10^5(7000)(1)	10^{-17}	10^{-7}
J Osc – mixer	[88]	206±5%(2)	Nb	4	280(400)		6×10^{-2}	100-600		
Cooled Schottky	[94]	33	GaAs	18	1400 (60)		0.38	200	2.8×10^{-21}(3)	10^{-3}
Cooled Schottky	[94]	85	GaAs	18	1400 (60)		0.42	280	3.9×10^{-21}(3)	
Cooled Schottky	[95]	115	GaAs	77	4500		0.39	300	4.1×10^{-21}(3)	
Schottky	[93]	230	GaAs	300				5000		
Schottky	[96]	340	GaAs	300	2300				10^{-15}	50×10^{-3}
Schottky	[106]	891	GaAs	300					10^{-15}	50×10^{-3}
Super Schottky	[97]	9	Pb-GaAs	1.1	1400		0.125	13	1.8×10^{-22}	3×10^{-7}
InSb mixer	[98]	230	InSb	4	0.02-2					

Notes: In Column 5 the if are given with the bandwidth in parentheses.
(1) The data in parentheses are the intrinsic conversion loss and mixer noise temperature assuming an input coupling equal to 1.
(2) The input signal is wideband.
(3) Calculated from the relation $NEP = kT_M$.

local oscillator power is also remarkable and makes efficient harmonic mixing possible. A conversion efficiency of 0.5 has been obtained for a 4th order harmonic mixing experiment $\omega_s/2\pi = 36$ GHz.

Mixing at 95 GHz has been investigated by Kanter [84]. However, the conversion efficiencies obtained are much less than predicted by the RSJ model. At 140 GHz, preliminary measurements by Richards, Claassen and Taur [85] indicate agreement with their results at 36 GHz.

At 891 GHz Blaney [86] has investigated fundamental mixing using two HCN lasers. The highest overall conversion efficiency is low ($\simeq 10^{-3}$), but subsequent experiments [87] have indirectly shown that the input coupling efficiency is only $\simeq 0.01$, so that the intrinsic conversion efficiency is $\simeq 0.1$, and compares favorably with the theoretical estimate $\simeq 0.2$. The system noise in these experiments is dominated by amplifier noise since $T_{if} = 500\,K > T R_D/R_N$ [see Eq. (104)]. However, the NEP obtained indicates clearly that Josephson junctions should be useful mixers in this frequency range.

The Josephson oscillator-mixer has been used for wide band millimeter waves (206 GHz ± 5%). The obtained conversion efficiency $\eta_T = 6 \times 10^{-2}$, is far from the optimum calculated value (0.16). However, the estimated mixer temperature T_M which ranges between 100 and 600 K indicates its potential value when extrapolated in the THz region.

High-order harmonic mixing of microwave local oscillators with infrared laser signals has been studied successfully with Josephson junctions. Harmonic orders of $\simeq$ 100 [89, 90], 400 [91] and 1000 [92] have been realized. Frequency comparison is the main purpose of such experiments. The especially high nonlinearity of JJ compared to Schottky diodes of MIM diodes allows conversion of an unusually large part of the microwave incident power into very high order harmonics. This makes the JJ an outstanding device to measure laser frequencies in the THz region in a single step with a 9 GHz reference oscillator. Impedance matching and optimum coupling is less severe in these experiments than in low noise level receiver operation because the hf power available is less limited ($\sim$ 1-100 mW), which explains why JJ devices are already in active development for frequency synthesis in metrology. The main problem encountered concerns the requirements on the spectral purity of the microwave source. The noise power in the sidebands of its frequency spectrum increases approximately as n^2, where n is the harmonic order relative to the signal power. Therefore, the rf source noise may become dominant in the frequency fluctuations of the mixing signal in an harmonic mixer. Microwave solid state oscillators or klystrons stabilized on a high Q superconducting cavity should allow this difficulty to be overcome up to a few THz.

The potential properties of series arrays for heterodyne mixers have been discussed by Richards [42]. The conversion efficiency and the noise will be similar to that in a single junction with the same R_D as in the array. The expected benefit is that higher total impedance may be reached in the latter case and help matching to the hf and if channels.

Since optimum mixer operation is rather sensitive to the junction parameters, it is important that the junctions have quite identical properties, which is difficult to realize. Another actual limitation is that microbridges which are to be used in series arrays have not yet given high performance in the submillimeter domain where the very small area tunnel SIS junctions are actually under development [10, 24] to allow good high frequency sensitivity. It would be interesting to investigate the mixing properties of such single junctions and also of arrays.

b. Respective interests of the external LO and internal LO Josephson mixers.

Since the theoretical noise temperatures are of the same order of magnitude, especially when $\Omega \gtrsim 1$, this criterion is not of importance in the choice between the two modes of operation. The problems related to linewidths of the frequency conversion signals on one hand, the local oscillator on the other hand will be the main factors.

The main advantage of the oscillator-mixer is its great simplicity in principle since no external LO is necessary. This point is mostly important at submillimeter and far infrared frequencies where the available sources are CO_2 pumped molecular lasers which are not very simple to operate and have only discrete (if many) oscillation frequencies. This aspect does not exist in the millimeter range since the JJ mixer with external LO only needs small amounts of LO power. On the other hand, a drawback of the oscillator-mixer is its wide linewidth which may limit its use to applications needing only moderate frequency selectivity such as in radioastronomy or infrared spectroscopy. The fact that wide band if amplifier chains must be used ($B > \Delta f_0$) to keep the conversion losses to the lowest possible level leads to an increase in sensitivity since $\Delta T_{min} \sim B^{-1/2}$ [Section I H 2 b, Eq. (59)].

To sum up, practically it appears difficult to define the best mode of operation at high frequencies ($\gtrsim 300$ GHz) because there is no detailed data available (conversion gain, noise temperature), either for the oscillator-mixer or for the harmonic mixer with external LO in this frequency range.

Actually the JJ mixer with external LO seems successful in the millimeter range, and in experiments where frequency selectivity is required. Since the LO powers are relatively low, harmonic mixing may also be expected to be performed at sufficiently weak levels so that the LO noise does not increase the mixer noise temperature too much. The oscillator-mixer may become a competing device in the future in the applications where selectivity is required, if the shunting of the junction is well resolved.

c. Comparative performances with non JJ mixers

The Schottky barrier diode [93-96], either at room temperature or cooled is the most widespread hf mixer. More recently the Super-Schottky diode mixer [97] has been developed. The high nonlinearity is related to the low energy gap of the superconductor. The hot electron bolometer mixer may also be considered [98]. MIM (metal-insulator-metal) diodes find use mainly as mixers or harmonic mixers at frequencies from a few THz up to the visible due to their very high cut-off frequency ($\sim 10^3$ THz). Typical data concerning non Josephson mixers are given in Table IV for comparison.

Schottky barrier diodes [93] have a frequency response independent of frequency up to their cut-off frequency $\omega_c = (R_{sp}C_o)^{-1}$ related to the junction capacitance C_o and its spreading resistance R_{sp}. The former is the capacitance of the space charge limited region, and effectively shunts the nonlinear diode resistance R_j. The spreading resistance R_{sp} consists mainly of the resistance of the undepleted part of the epitaxial layer below the junction, because the degenerately doped substrate and the ohmic contacts have low resistance. R_{sp} is in series with the parallel combination of C_o and R_j. R_{sp} and C_o have the following parasitic effects: C_o allows current to bypass R_j while R_{sp} is a source of dissipation for the signal if and LO, giving conversion losses, heat production and consequently diode excess noise. In general, a high ω_c correlates well with improved high frequency operation. In order to increase ω_c, a reduction in diode area may be used ($\omega_c \sim r^{-m}$, $0<m<1$, where r is the diode radius), but has its limitations. An excessive decrease of the diode area gives unacceptable values of R_{sp} and excessive ohmic heating of the junction degrading the noise figure. It has also been shown recently [99] that the large current densities reached in small area junctions do not allow full utilization of their nonlinearity, and give excess intrinsic conversion losses. Cut-off frequencies of up to $\simeq 2$ THz have been obtained [93]. Table IV also clearly indicates that a large amount of LO power is necessary with Schottky diodes, which is an important disadvantage at high frequencies.

Super Schottky diode mixers have only been investigated at 9 GHz. The small superconducting gap ($\simeq 1$ meV) may introduce a frequency limitation here. The extremely low level of LO power is to be noted. The influence of the small superconducting gap on the frequency limitation has not yet been investigated.

The data presented in Table IV allow the following remarks:

(i) At frequencies ≤ 150 GHz, there is not doubt that Schottky barrier diodes are to be preferred unless ultimate sensitivities are required. The super-Schottky diode represents actually at microwaves the major advance in low noise frequency conversion with $T_M = 13$ K at 10 GHz. The extension of comparable performances to higher frequencies remains to be accomplished. JJ point contacts also have good performance but their sensitive mechanical adjustment make them less convenient. The development of small tunnel Josephson junctions or microbridges with good high frequency performance is possible.

(ii) In the 150-300 GHz range, data are not available concerning cooled Schottky diodes. The results obtained with the Josephson oscillator-mixer may be improved with a better conversion ratio. With an InSb mixer a total receiver noise temperature T_{DSB} = 300 K has been obtained, but the bandwidth is restricted to 2 MHz because of the slow time response of the device.

(iii) Between 300 GHz and 1 THz, the Josephson junction appears to be very competitive (low noise temperature, low LO power) even with a poor input coupling (NEP $\simeq 10^{-17}$ W/Hz). The main improvements of the device concern the input coupling and the impedance matching. The Schottky diodes and MIM diodes require a prohibitive amount of LO power (50 mW) and have lower sensitivity.

(iv) From 1 to 10 THz the discussion is wide opened. It is reasonable to expect that JJ with high cut-off frequencies ω_c (large $R_N I_c$) may have an important role in both modes (external and internal LO).

(v) Above 10 THz the MIM diodes are ahead due to their very high cut-off frequency.

(vi) In the particular application of high order harmonic mixing, Schottky diodes are more practical (no cryogenic cooling) at frequencies below 1 THz when the harmonic order is not too high ($\leq$ 30). Point contacts JJ are under development in several laboratories when high order harmonic is required ($\simeq$400) at not too high frequencies (a few THz). MIM diodes which allow only small harmonic orders ($\leqslant$ 10) are the best harmonic mixers approaching the visible range.

E. Parametric Amplification

We have found in Section I E that the high nonlinearity of Josephson junctions may be exploited in parametric effects. A majority of parametric devices are related to two basic types of parametric amplifiers: (i) The up-converter (upper sideband upconverter) where the output frequency is equal to the sum of the input and pump frequencies. The pump frequency is large compared to the signal frequency in order to have high conversion gain and this device is not of practical use at high frequencies. (ii) The negative resistance parametric amplifier which becomes of interest at the higher frequencies where the upconverter is no more used.

If the ideal Josephson junction may be considered as a pure reactance, the real JJ (point contact, microbridge) which is often well described by the RSJ model is a lossy device of rather low impedance generating high order mixing components. Therefore the main problem for operation in a parametric mode is to suppress as much as possible the unwanted components which dissipate energy at the combination frequencies. They represent active idlers which determine the impedance of the junction at the corresponding frequencies. The suppression of undesired frequency components is still relatively easy at microwave frequencies (centimeter and millimeter) using resonant cavities or striplines. Its extension to submillimeter waves seems less probable.

Two lines of development have been pursued with JJ as for heterodyne mixers which are based either on self-pumped or on externally-pumped Josephson parametric amplifiers. In each case the modes of operation which have been preferred have only a small number of important idlers in order to fulfill the above requirement more easily.

1. Parametric amplification with self-pumped JJ

The reactance variation is provided by the internal oscillations due to the average junction potential. Russer and Thomson [20,21] have shown that the energy exchange between channels obeys the Manley-Rowe relations modified in the following way. The power injected at the pump frequency is replaced by the dc power corresponding to the Josephson self-oscillation. Several modes of operation have been considered.

a. Negative resistance amplifier with single idler

This mode corresponds to the three photon parametric amplifier defined by $\omega_p = \omega_s + \omega_i$ which is the mode generally employed in conventional parametric amplifiers with varactor diodes. An example of this mode of operation is represented in Fig. 16. It is essential to keep unavoidable junction losses small compared to the dissipation of signal and idler currents in the external circuits. This is realized because: (i) a resistive-SQUID permits use of a voltage-biased JJ in the microwave range, i.e. eliminates the unwanted harmonics induced by resistive feedback, (ii) coupling the junction to a microwave cavity can be done ($\omega_M \simeq 9$ GHz). The frequencies ω_i and ω_s are chosen slightly off resonance ($\omega_i \simeq \omega_s \simeq \omega_M$) with

$$\omega_p = \omega_o \simeq 2\omega_s \quad .$$

Energy exchanges are only possible between ω_s, ω_i and ω_o. An 11 db gain has been obtained. Similar results have already been reported at 30 MHz.

b. Two idler negative resistance amplifier

The experimental arrangement is the same as with one idler, but $\omega_p = \omega_o \leqslant \omega_s$. One makes use here of the negative resistance of the junction when it is biased just below the first step [Section IB1, Eq. (62)]. Then $\omega_1 = \omega_o - \omega_s << \omega_o$. It may be shown from the RSJ model that there are mostly two idlers involved in the parametric process at the frequencies ω_1 and $\omega_2 = 2\omega_o - \omega_s \simeq \omega_s$. Here again a 12 dB gain has been obtained at $\omega_s = 9$ GHz. The noise level in the two idler mode is consistently higher than for the single idler. This result happens often when a low-frequency idler participates in the process. Noise from ω_1 is up-converted with gain into the signal circuit. The corresponding contribution to the noise temperature may be approximately written [107] $T = (\omega_s/\omega_1)T_1$, where T_1 is the idler termination temperature. Because of the extra noise, this amplifier is less attractive than the one-idler amplifier which is pumped at $\omega_o \simeq 2\omega_s$.

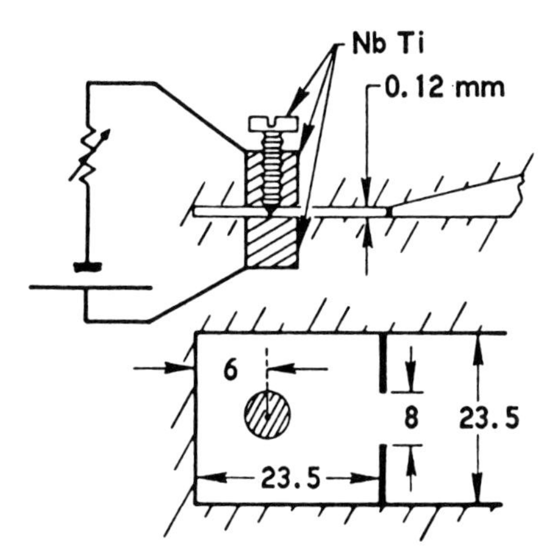

Fig. 16. Parametric amplification with self-pumped Josephson junction. Cavity with point contact for single and two idler experiments, and bias circuit (resistive SQUID) (After Ref. [100]).

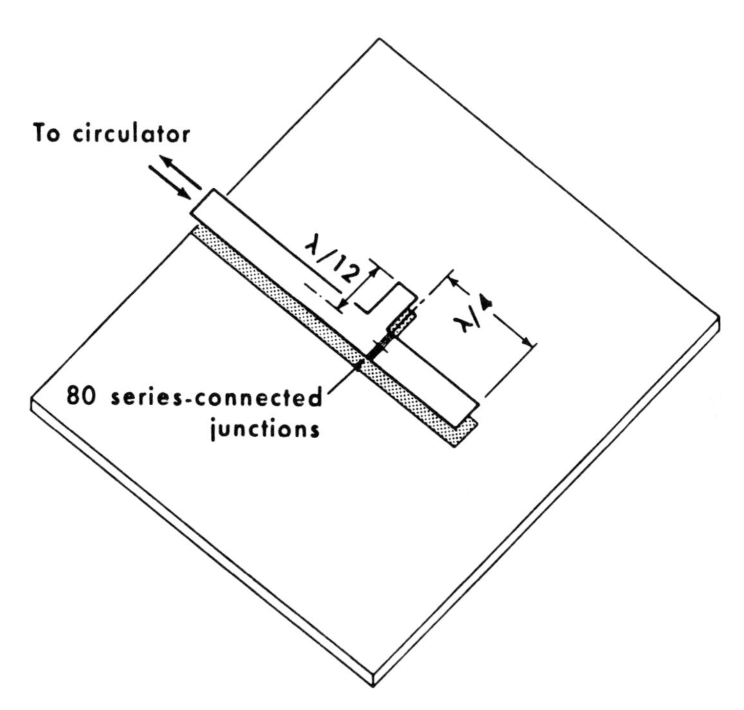

Fig. 17. Parametric amplification with unbiased Josephson junctions-Microstrip circuitry (After Ref. [102]).

c. Parametric up converter

The device of Fig. 16, slightly modified, is now used as a parametric up-converted. The Josephson oscillation at 9 GHz produces parametric upconversion of the power into sidebands at 9 ± 0.115 GHz in a 115 MHz circuit placed inside the cavity. A conversion gain has been obtained which is a factor of 3 lower than the theoretical value $G = \omega_0/\omega_1$ (Section I E). The amplifier noise temperature referred to the input at 115 MHz, has been deduced from the internal noise conversion process and is found to be $\leqslant$ 15 K.

2. Externally pumped JJ parametric amplifier

Chiao, Feldman and Parrish have developed the externally pumped four photon parametric amplifier which has been operated at 9 GHz [102] and 33 GHz [103] and efforts are undertaken to extend the device at 100 GHz. The junction has no dc bias and this fact gives it remarkable properties which may be deduced from symmetry considerations; the unbiased JJ is symmetric to a physical inversion. Then the non-linear Josephson inductance (Section I E) developed in power series of the current contains only even harmonics. In other words, if the junction is pumped at ω_p, the non-linear inductance is modulated only at the even harmonics, the first of which is $2\omega_p$. A similar effect has been observed in superconducting films [108].

If a signal is applied at ω_s, an idler appears at $\omega_i = 2\omega_p - \omega_s$ and signal and idler frequencies are symmetric about ω_p. This is a four photon process in which two pump photons decay in the system to create a signal photon and an idler photon. When ω_s and ω_i are closed sidebands of ω_p, the mode of operation is said to be doubly degenerate. The same transmission line may be used for all three components, ω_s, ω_p, ω_i which simplifies the design. For the same symmetry reason, only even photon processes are possible. Therefore all odd-numbered photon processes are not allowed which eliminates all their mixing components. A detailed theoretical treatment has been performed, and such important quantities as the gain, which is a function of pump power, the bandwidth have been computed.

A device has been fabricated with a series array of Sn micro-bridges (80 and 160) integrated in a microstrip transmission line (Fig. 17) which has an open circuit at a distance $\lambda_g/4$ from the junction array, so that the junction effectively terminates the transmission line and an open-circuited line at $\lambda_g/12$ is placed at right angles to short out the third harmonic frequencies. The system is operated as a reflection amplifier. Only the overall performance of the array may be measured. Due to inhomogeneities in their properties, only a fraction of the micro-bridges participate in the amplification process. Table V shows the performance obtained at 10 GHz and 33 GHz which is promising. Its wide bandwidth and low pump power level are particularly interesting.

Table V. Performance of series array of Sn microbridges

	f = 10 GHz	f = 33 GHz
Temperature	2.9 K	2.9 K
$R_N I_c$	50 μV = 0.06 $(R_N I_c)_{th}$	
Gain (db)	12	15
Bandwidth (GHz)	1	3.4
Noise Temperature (K)		20 ± 10
Pump Power (W)	10^{-7}	5×10^{-8}

3. Discussion

The development of Josephson parametric amplifiers is relatively recent compared to Josephson mixers, which explains why fewer results are available. For the first time, a series array of microbridges has been operated in a device. As a counterpart, the parameters of the individual junctions cannot be measured; this does not allow easy comparison of the experimental and theoretical results which is very useful in the first steps. The low noise temperature obtained at 33 GHz with the unbiased JJ parametric amplifier compares very favorably with the results which may be expected with conventional parametric amplifiers for radioastronomy applications. With a 200 GHz pump frequency, the minimum noise temperature for a 300 K amplifier at 100 GHz is 300 K. At 46 GHz, Edrich [105] has obtained $T_N \simeq$ 60 K with a 20 K cooled preamp. The JJ parametric amplifier presents considerable advantages in this frequency range (bandwidth, LO power). However, its very low level saturation may be inconvenient. The relative interests of the self-pumped pumped (one idler) and externally pumped (two idler) amplifiers cannot yet be defined.

F. Conclusions

The main difference between a JJ and other junctions (varactor, Schottky) is that it is an active highly nonlinear element (the Super-Schottky is passive, but is also highly nonlinear). The fundamental reason is the voltage-phase Josephson relationship which has consequences both from the theoretical and application points of view.

The analysis of the equivalent circuit (even the simple RSJ model) is complicated and very often requires numerical computing methods. The usual methods of analysis of parametric devices are generally not valid except in some particular situations. However, one may use the simple voltage source model to understand certain basic phenomena. Similarly the analysis of noise phenomena is difficult. This is the reason why in Section I we have often used a qualitative description of the main phenomena.

The ac Josephson effect has a wide range of applications. In Section II we have selected the hf modes of operation which appear to be the most important potentially, or are the most developed. We have omitted some special detection or parametric effects which are described in several recent reviews [22, 23, 42].

At $\Omega = (\omega/\omega_c) \geqslant 1$, the different operating modes have a common criterion: the sensitivity decreases as Ω^2. Efforts should therefore be directed toward realizing reproducible junctions with large characteristic frequencies (proportional to $R_N I_c$) both by improving the fabrication techniques and developing high T_c materials. This will help simultaneously to solve the impedance problem which is a weakness of JJ. This defect is intrinsic to the Josephson effect because $(R_N I_c)_{opt} = \pi\Delta/2e$. The critical current cannot be reduced below a certain limit because it must compete with the fluctuations. High T_c materials (i.e., high Δ) will increase the maximum R_N falues obtainable.

In each type of application there is a strong competition between the JJ and other He temperature devices (mostly semi-conductors). The Super-Schottky appears to have very high performance and it is difficult to predict which detectors will be preferred in the future. The Super-Schottky development may be helped by the strong semiconductor technology, but it also needs a well developed superconducting thin-film technology, which may also benefit the Josephson junction technology.

The domain where the Josephson junction may have the most significative advantages is as a submillimeter and far infrared mixer, either harmonic mixer with external LO or oscillator-mixer depending on the application. At millimeter frequencies, Josephson parametric amplifiers with their broadband may be useful when associated with a far infrared Josephson mixer. In this frequency range, series arrays may permit elimination of the impedance mismatch. The design of all superconducting low noise receivers is also considered [109].

Finally, the fundamental problem to solve concerning weak links (point contacts, microbridges) is related to the nature of the physical processes responsible for the "Josephson type effects" and to their theoretical treatment. This is important for analysis of the high frequency properties of weak links. Moreover,the ultimate performance of a device depends on its internal noise which should be determined by the physical processes in the junction.

REFERENCES

1. B.D. Josephson, Phys. Lett. 1, 251 (1962); Adv. Physics 14, 419 (1965).
2. W.C. Stewart, Appl. Phys. Lett. 12, 277 (1968).
3. D.E. McCumber, J. Appl. Phys. 39, 3113 (1968).
4. L.G. Aslamazov and A.I. Larkin, Sov. Phys. JETP Lett. 9, 87 (1969); Zh. ETF Pis. Rad. 9, 150 (1968).
5. D.N. Langenberg, Rev. Phys. Appl. 9, 35 (1974).
6. N.R. Werthamer, Phys. Rev. 147, 255 (1966).
7. R.E. Harris, Phys. Rev. B10, 84 (1974).
8. E. Riedel, Z. Naturforsh. A19, 1634 (1964).
9. C.A. Hamilton, Phys. Rev. B5, 1912 (1972).
10. S.A. Buckner, J. Low Temp. Phys., 22, No. 5/6 (1976).
11. L.L. Malinovskii, Sov. Phys. JETP Lett. 20, 250 (1974).
12. G. Vernet and R. Adde, Appl. Phys. Lett. 28, 599 (1976).
13. Thomé and Couder, Int. Conf. Low Temp. Physics, LT14, Otaniemi, 1975.
14. A. Baratoff and L. Kramer, IC Squid Int. Conf. Berlin, 1976.
15. D.G. McDonald, F.R. Petersen, J.D. Cupp, B.L. Danielson and E.G. Johnson, Appl. Phys. Lett. 24, 335 (1974).
16. W.J. Skocpol, M.R. Beasley and M. Tinkham, J. Appl. Phys. 45, 4054 (1974).
17. D. Rogovin and D.J. Scalapino, Annals of Physics 86, 1 (1974).
18. J. Clarke, G.I. Hoffer and P.L. Richards, Revue de Phys. Appl. 9, 69 (1974).
19. C.C. Grimes and S. Shapiro, Phys. Rev., 169, 397 (1968).
20. P. Russer, Proc. IEEE 59, 282 (1971).
21. E.D. Thompson, IEEE ED20, 680 (1973).
22. P.L. Richards, F. Auracher and T. Van Duzer, Proc. IEEE 61, 36 (1973).
23. Proc. of the Int. Conf. on Detection and Emission of Electromagnetic Radiation with Josephson Junctions, Rev. Phys. Appl. 9, 1-312 (1974).
24. J.T.C. Yeh and D.N. Langenberg IEEE Mag 11, 683 (1975).
25. C.A. Hamilton, J. Appl. Phys. 44, 2371 (1973).
26. D.J. Scalapino, University of California, Report No. AD 66 1848, 1967 (unpublished).
27. D. Rogovin and Scalapino, Physica 55, 399 (1971).
28. M.J. Stephens, Phys. Rev. Lett. 21, 1629 (1968); Phys. Rev. 182, 531 (1969) and 186, 393 (1969).
29. D. Rogovin and D.J. Scalapino, Annals of Physics 86, 1 (1974).
30. A.I. Larkin and Yu. N. Ovchinnikov, J.E.T.P. 26, 1219 (1968).
31. K.K. Likharev and V.K. Semenov, J.E.T.P. Lett. 16, 442 (1972).
32. A.J. Dahm, A. Denenstein, D.N. Langenberg, W.H. Parker, D. Rogovin and D.J. Scalapino, Phys. Rev. Lett. 22, 1416 (1969).
33. H. Kanter and F.L. Vernon, Phys. Rev. Lett. 25, 588 (1970).
34. G. Vernet and R. Adde, Appl. Phys. Lett. 19, 195 (1971).
35. D.N. Langenberg, D.J. Scalapino, B.N. Taylor and R.E. Eck, Phys. Rev. Lett. 15, 294 (1965).

36. J.H. Claassen, Y. Taur and P.L. Richards, Appl. Phys. Lett. 25, 759 (1974).
37. J.L. Stewart, Proc. IRE 14, 1539 (1954).
38. G. Vernet and R. Adde, J. Appl. Phys. 45, 2678 (1974).
39. J.D. Kraus, Radioastronomy, McGraw-Hill (1966).
40. L.A. Blackwell and K.L. Kotzebue, Semiconductor Parametric Amplifiers, Prentice Hall, 1961.
41. R.R. Arams, Infrared to Millimeter Wavelengths Detectors (Artech House, 1973).
42. P.L. Richards, Semiconductors and Semimetals, Vol. 12, Infrared Detectors, Willardson and Beer, Eds.
43. H. Jasik, Editor, Antenna Engineering Handbook (McGraw Hill, New York, 1961).
44. H. Kanter and F.L. Vernon, J. Appl. Phys. 43, 3174 (1972).
45. T.F. Finnegan, J. Wilson and J. Toots, IEEE Trans. Mag-11, 821 (1975).
46. J.G. Small, G.M. Elchinger, A. Javan, A. Sanchez, F.J. Backner and D.L. Smythe, Appl. Phys. Lett. 24, 275 (1975).
47. L.M. Matarrese and K.M. Evenson, Appl. Phys. Lett. 17, 8 (1970).
48. D.N. Langenberg, Rev. Phys. Appl. 9, 35 (1974).
49. D.N. Langenberg, D.J. Scalapino and B.W. Taylor, Proc. IEEE 54, 460 (1966).
50. I.M. Dmitrenko and I.K. Yanson, Intern. Conf. on Low Temp. Phys. LT10, S32, p. 228.
51. A.H. Dayem and C.C. Grimes, Appl. Phys. Lett. 9, 47 (1966).
52. I. Ya. Krasnopolin and M.S. Khaikin, JETP Lett. 6, 129 (1967).
53. R.K. Esley and A.J. Sievers, Applied Superconductivity Conference, IEEE Pub. No. 72CHO 682-S-TABSC, p. 716 (1972).
54. R.K. Esley and A.J. Sievers, Rev. Phys. Appl. 9, 295 (1974).
55. D.R. Tilley, Phys. Lett. 33, 205 (1970).
56. T.F. Finnegan, J. Wilson and J. Toots, Rev. Phys. Appl. 9, 199 (1974).
57. V.N. Gubankov, V.P. Koshelets and G.A. Ovsyannikov, JETP Lett. 21, 226 (1975).
58. N.F. Pedersen, O.H. Sorensen, J. Mygind and P.E. Lindelof, M.T. Levinsen, T.D. Clark, Appl Phys. Lett. 28, 562 (1976).
59. D.H. Martin and D. Bloor, Cryogenics 1, 159 (1961).
60. J.M. Andrews and M.W.P. Strandberg, Proc. IEEE 54, 523 (1966).
61. C.L. Bertin and K. Kose, J. Appl. Phys. 39, 2561 (1968).
62. R.M. Katz and K. Rose, Proc. IEEE 61, 55 (1973).
63. F.J. Low and A.R. Hoffman, Appl. Opt. 2, 649 (1963).
64. J. Clarke and T.Y. Hsiang, IEEE Mag 11, 845 (1975).
65. J. Clarke, G.I. Hoffer, P.L. Richards, and N-H Yeh, Int. Conf. Low Temp. Physics LT 14, Otaniemi, Finland, p. 226 (1975).
66. N. Coron, G. Dambier, J. Leblanc, J-P. Moalic, Rev. Sci. Inst. 46, 492 (1975).
67. K.K. Likharev and V.K. Semenov, Radio Eng. and Elect. Physics 18, 1734 (1973).

68. B.T. Ulrich, Rev. Phys. Appl. 9, 111 (1974).
69. H. Tolner, C.D. Andriesse and H.H.A. Schaeffer, Infrared Physics 16, 213 (1976).
70. M. McColl, M.F. Millea and A.H. Silver, Appl. Phys. Lett. 23, 263 (1973).
71. Y. Taur, Ph.D. Dissertation, Univ. Cal., Berkeley, Apr. 1974 (unpublished).
72. F. Auracher, Ph.D. Dissertation, Univ. Cal. Berkeley, March 1973 (unpublished).
73. F. Auracher and T. Van Duzer, Rev. Phys. Appl. 9, 263 (1974).
74. Y. Taur, J.H. Claassen and P.L. Richards, Rev. Phys. Appl. 9, 263 (1974).
75. A.N. Vystavkin, V.N. Gubankov, L.S. Kuzmiu, K.K. Likharev, V.V. Migulin, V.K. Semenov, Rev. Phys. Appl. 9, 79 (1974).
76. Y. Taur, J.H. Claassen and P.L. Richards, IEEE MTT-22 , 1005 (1974).
77. V. Ambegaokar and B.I. Halperin, Phys. Rev. Lett. 22, 1364 (1969).
78. W.H. Henkels and W.W.Webb, Phys. Rev. Lett. 26, 1164 (1971).
79. C.A. Hamilton, Rev. Sci. Instr. 43, 445 (1972).
80. G. Vernet and R. Adde, IEEE Mag 11 , 825 (1975); Rev. Phys. Appl. 10, 165 (1975).
81. G. Vernet (unpublished).
82. G. Vernet, These de Doctorat d'Etat, Université Paris-Sud, Orsay, 1976 (unpublished).
83. J.H. Claassen, J.H. Taur and P.L. Richards, Appl. Phys. Lett. 25, 759 (1974).
84. H. Kanter, Rev. Phys. Appl. 9, 363 (1974).
85. P.L. Richards, J.H. Claassen and Y. Taur, Int. Conf. LTT14, Otaniemi, Finland, p. 238, Aug. 1975.
86. T.G. Blaney, Rev. Phys. Appl. 9, 279 (1974).
87. T.G. Blaney, Report SI no. 89/0373, Feb. 1976 (unpublished).
88. R.S. Avakjan, A.N. Vystavkin, V.N. Gubankov, V.V. Migulin and V.D. Shtykov, IEEE MAG 11, 838 (1975).
89. D.G. McDonald, A.S. Risley, J.D. Cupp and K.M. Evenson, Appl. Phys. Lett. 18, 162 (1971).
90. J-M. Lourtioz, R. Adde, G. Vernet and J-C. Henaux, Rev. Phys. Appl. (in press).
91. D.G. McDonald, A.S. Risley, J.D. Cupp, K.M. Evenson, and J.R. Ashley, Appl. Phys. Lett. 20, 296 (1972).
92. T.G. Blaney and D.J.E. Knight, J. Phys. D7, 1882 (1974).
93. G.T. Wrixon, IEEE MTT-22, 1159 (1974).
94. S. Weinreb and A.R. Kerr, IEEE SC-8, 58 (1973).
95. A.R. Kerr, IEEE MTT-23, 781 (1975).
96. H.R. Fetterman, B.J. Clifton, P.E. Tannenwald, C.D. Parker, and H. Penfield, IEEE MTT-22, 1013 (1974).
97. M. McColl, R.J. Pedersen, M.F. Bottjer, M.F. Millea, A.H. Silver, and F.L. Vernon, Appl. Phys. Lett. 28, 159 (1976).
98. T.G. Phillips and K.B. Jefferts, IEEE MTT-22, 1290 (1974).

99. M. McColl (to be published).
100. H. Kanter, IEEE Mag 11, 789 (1975).
101. H. Kanter, J. Appl. Phys. 46, 4018 (1975).
102. M.J. Feldman, P.T. Parrish, and R.Y. Chiao, J. Appl. Phys. 46, 4031 (1975).
103. R.Y. Chiao and P.T. Parrish, J.Appl. Phys. 47, 2639 (1976).
104. H. Kanter and A.H. Silver, Appl. Phys. Lett. 19, 515 (1971).
105. J. Edrich, IEEE MTT-22, 581 (1974).
106. W. Reinert, Space Sci. Review (Netherlands) 17, 703 (1975).
107. P. Penfield and R. Rafuse, Varactor Applications, MIT Press, 1962.
108. A.S. Clorfeine, Appl. Phys. Lett. 4, 131 (1966).
109. A.H. Silver, IEEE Mag 11, 794 (1975).
110. S.K. Decker and J. Mercereau, Appl. Phys. Lett. 27, 466 (1975).

Chapter 7

FABRICATION OF JOSEPHSON JUNCTIONS†

B. T. Ulrich

Department of Electrical Engineering and Computer Sciences and the Electronics Research Laboratory, University of California, Berkeley, California 94720 and Department of Physics, University of Nijmegen Netherlands

T. Van Duzer

Department of Electrical Engineering and Computer Sciences and the Electronics Research Laboratory, University of California, Berkeley, California 94720

I. INTRODUCTION

This paper discusses the materials and microcircuit-fabrication techniques that can be applied to Josephson junctions. All practical applications of Josephson junctions require reproducibility in fabrication and physical ruggedness to withstand the stresses imposed by thermally cycling a structure containing materials with different expansion coefficients. In addition, the various applications require different electrical characteristics. For example, junctions for use in SQUID magnetometers and high-frequency electromagnetic detectors and mixers must have non-hysteretic I-V characteristics, while those for most switching circuits should have hysteretic characteristics. In most applications, the ratio of critical current to capacitance (I_c/C) of the

---

† Research sponsored by the U.S. Army Research Office, Grant DAAG29-76-G-0191.

junction should be as large as possible. The choice of materials plays a key role in achieving these ends. Also, achievement of the optimum junction properties requires using sophisticated microfabrication techniques such as photo- and electron-lithography advanced sputtering techniques, and silicon doping and etching.

The first Josephson junction was reported in 1963 by Anderson and Rowell [1]. It consisted of a thin film of tin on an insulating substrate, which was subsequently oxidized to form a tunneling barrier, and a thin-film cross strip of lead to form the counter electrode. Since that time, a large number of experiments have been done with similarly constructed oxide-barrier junctions. Most have been done in a research environment where the lack of cyclability usually inherent in the choice of the convenient low-melting-temperature materials has been dealt with by keeping the junctions in liquid nitrogen in the periods between liquid-helium-temperature experiments. In other work, cyclability has been achieved by employing stable, high-melting-temperature materials. The research on Josephson tunnel junctions has been carried out mostly by persons without access to facilities for making very small thin-film structures so the junctions have had too much capacitance for high-frequency applications.

The need for low capacitance has led to work on point contacts which can be easily made to have low capacitance, but which must be carefully adjusted. In spite of the difficulty of keeping the point contacts adjusted, they have been the preferred device for many experiments because of the ease of fabrication and low capacitance. All of the best results in the use of Josephson junctions for detection and mixing have been achieved with point contacts.

The microbridge, a two-dimensional, thin-film simulation of the point contact has received a considerable amount of attention over the past decade. Some clever, inexpensive techniques have been developed to fabricate the weak link of the microbridge, which is a thin-film constriction which must have submicron dimensions. Present-day constrictions are being made by a variety of means of a simple nature, but the use of electron lithography is gaining favor. These devices share with the point contact the advantage of low capacitance and some success in high-frequency experiments has been obtained recently. They usually have a lower resistance than is desirable for impedance matching and there has been work on series arrays designed to improve the matching.

Another structure having both electrodes in the same plane with a weak connecting link employs a thin-film strip of constant width, but with a short section of weakened superconductivity. The weak region may be made in a variety of ways such as by overlaying a thin cross strip of normal metal to suppress the superconductivity by the proximity effect, by weakening the link by etching it to a very thin layer, or by using a different material such as a normal metal, a weak superconductor, a semimetal, or a semiconductor as the connection. These structures offer the advantage of low capacitance but, except in the case of the semimetal and semiconductor, their resistance is low. Work is continuing on these structures.

Sandwich-type structures using evaporated films of metal or semiconductor have appeared in the literature, mainly in the period between 1968 and 1972. The use of a semiconductor is suggested by the possibility of thereby having a weak tunneling barrier. The attempts to make these devices have usually required patching the pinholes that are almost always present in very thin evaporated films of semiconductors by oxidizing the base superconducting film after deposition of the semiconductor. So far, no evaporated semiconductor junctions have been made with high values of the ratio I_c/C.

A new technique has been developed to make sandwich-type junctions in which the barrier is a single-crystal layer of silicon and was reported in 1974. It is made by etching a hole almost all the way through a silicon wafer and, by a special technique, leaving a thin, uniform, unetched membrane as the floor of the hole. This technique seems to eliminate the need to fill pinholes and it should allow the use of superconductors with high transition temperatures, which will improve the I_c/C ratio.

A number of potentially important application proposals for Josephson junctions have been verified with uncyclable tunnel junctions or with point contacts that require adjustment. The present thrust is to show that cyclable, optimized junctions can be made using microcircuit fabrication techniques so that these devices can be incorporated in practical systems.

II. FABRICATION TECHNOLOGY

In this section we will examine masking techniques for controlling the patterns of various materials deposited or etched to form Josephson junction devices and circuits and we will discuss some of the means for preparation of the films to be masked. It will be assumed that the tasks to be achieved are deposition of thin films and formation of patterns in the films. More details on most of the techniques in this section may be found in Ref. 2.

A. Evaporation Masks

The simplest way to define a pattern for a thin-film device is to cut the desired pattern in a thin metal mask, place the mask in nearly direct contact with the substrate, and evaporate the metal or other material through the mask. The limit on the usefulness of this technique comes in the dimensional accuracy and complexity of a mask that a machine shop can fabricate at reasonable cost. It is obviously also excluded in situations with complete 360° rings of thin film, like the letter "O". Typically, the smallest feature that can be machined in a high-quality machine shop is about 0.2 mm; somewhat smaller features could be achieved in an instrument shop. Simple shapes like straight lines and circular holes of those dimensions can be readily machined. The cost rises rapidly with increasing complexity.

B. Photolithography

For dimensions smaller than about 0.2 mm and larger than about 2 μm, pattern control should be done using photolithography. We will discuss the currently established procedures for contact photolithography wherein a glass slide containing a pattern of opaque regions is placed in contact with the photoresist-coated substrate to control the exposure of the resist. There is another kind of photolithography ("projection lithography") in which the image is projected through a lens system onto the resist-coated substrate; this method is little used to date, but it has certain advantages (one of which is the avoidance of abrasive destruction of the pattern as occurs in contact lithography). We will outline the photolithographic procedures that are used routinely in the fabrication of integrated circuits. We will also discuss the limits of feature size that can be achieved for Josephson junction fabrication.

The ability to achieve feature sizes on the order of several micrometers is based on photographic reduction of masks made of much larger dimensions. Typically, two stages of reduction are used. The pattern is first cut in the red, semi-opaque layer of a double-layered plastic called Rubylith. The other layer is transparent and serves to support the patterned opaque layer. The cutting can be done by hand but computer-controlled cutting tables are also available. Rubylith patterns are usually made with dimensions of about 80 x 80 cm^2. These are first reduced to about 5 x 5 cm^2 so that a linear reduction of 16 is involved. This is variable, typically in the range 10 to 30. The reduced pattern is reproduced on a light-sensitive emulsion in a glass slide. After development, the pattern of the Rubylith then appears in reduced form and, with clear and opaque regions interchanged, on the 5 x 5 cm^2 slide. Another reduction of a linear factor of 10 (which could be variable,but which is usually fixed) is then used to reduce and transfer the image onto another 5 x 5 cm^2 slide. With the factor of 10 reduction, 100 images can be made on the slide. This number can be larger if the image at the intermediate stage is smaller than the allowed 5 x 5 cm^2. The images are projected onto the light-sensitive emulsion on the final slide in a "step-and-repeat" apparatus which (sometimes automatically) exposes the 100 or so images. After development, the glass slide has 100 or so duplicated images, the linear dimensions of which have been reduced by a factor of about 160 from the original. Thus, a 5 μm feature in the final mask was 0.8 mm in the Rubylith mask. To make the original pattern on the Rubylith requires care, but it can be done readily by hand, for simple circuits.

The mask is then used to control the exposure of a layer of photoresist on the substrate where the device is to be fabricated. There are two general types of resist, positive and negative. The positive resist has the property that the regions which have been exposed to light will be removed in an appropriate developer. The negative resist is removed by its corresponding developer in the regions where it is not exposed. There are strategic reasons for using one or another, depending on a variety of factors including feature size, available etchants for the materials being patterned, relative protective and adhesive properties of the resists, etc..

As a simple example, let us consider the preparation of a 10 μm wide strip of a superconductor which is to be one electrode of a Josephson junction. A Rubylith pattern is cut in which the opaque layer is removed in a strip 1.6 mm wide. With two stages of reduction one obtains a mask which is opaque except for a 10 μm wide strip (or many duplicated strips if the step-and-repeat feature is used).

Suppose first that a positive resist is to be used. The steps are as follows. The substrate is cleaned and coated with photoresist. The coating is done on a "spinner" which spins the substrate about an axis perpendicular to its surface at some thousands of rpm as a drop of liquid resist is dropped on the substrate to cover its entire surface. The result is a solidified, mostly flat, light-sensitive, plastic film typically 0.5 - 2.0 μm thick. The result is step (a) in Fig. 1. The mask is placed in contact as shown in Fig. 1(b) and the resist is exposed with a mercury arc lamp. After development, the sample has no resist in the 10 μm wide strip, as seen in Fig. 1(c). A film of the superconductor is then deposited over the whole substrate as seen in Fig. 1(d). The final step is to place the sample in a solvent of the original resist material. (Note that the molecular structure of the resist is changed where exposed; the unexposed resist is not dissolved in the developer.) The solvent works its way under the metal film to dissolve the resist and remove the metal on top of it, leaving the desired strip as in Fig. 1(e). It is sometimes necessary to

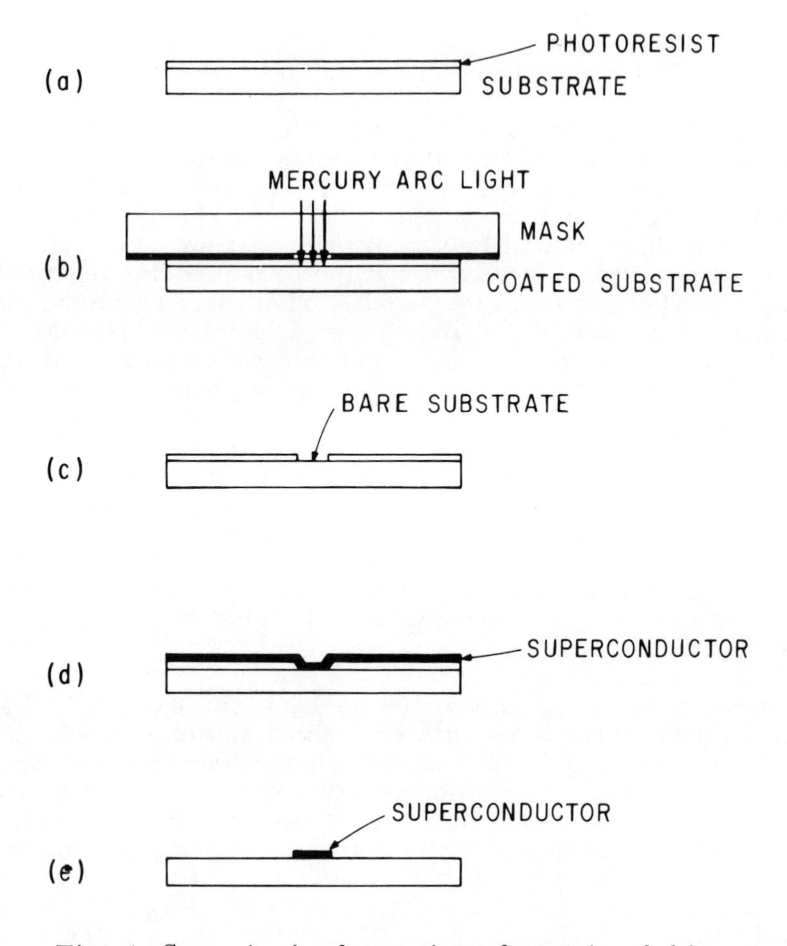

Fig. 1 Steps in the formation of a strip of thin film using a positive photoresist.

use ultrasonic vibration to achieve the so-called "lift-off" of the metal not in contact with the substrate. As can be appreciated from Fig. 1(d), the lift-off of the undesired portion of the film will remove that on the substrate too if the contact between them is stronger than the adhesion of the film strip to the substrate. This becomes an increasingly difficult problem as the strip width is made smaller. There is usually some irregularity of the edges due to tearing.

Consider instead the use of a negative resist process in which we use the same mask for exposure. In this case, one begins by coating the entire substrate with the superconductor and a covering layer of resist as shown in Fig. 2(a). The subsequent exposure changes the negative resist in such a way as to make it insoluble in the developer so that, after the development, one has the situation shown in Fig. 2(c). The sample is then subjected to an etchant for the superconductor with the result shown in Fig. 2(d). The remaining resist that protected a strip of the metal film is then removed with the appropriate solvent, leaving the desired film strip as in Fig. 2(e). One of the problems with this procedure is that the resist sometimes does not adhere well to the metal and the etchant cuts in under the protective resist strip. Later we will discuss sputter etching, which is done in a gaseous environment, and ion milling, which involves a directed beam of ions, to etch the metal film; these usually give better resolution than chemical etching.

The current integrated-circuit industry standard for feature size is 5 μm. This choice is dictated by the economic need to achieve a certain limit on defects in complex circuits. Careful photolithography can lead to 2 μm feature sizes with a reasonable yield on single devices. There have been some successful exposures of 0.4 μm achieved by having the pattern on a flexible mask which conforms under pressure to the irregularities of the surface of the resist-coated substrate. Sometimes tricks such as controlled over-etching can be used to produce lines narrower than would be expected otherwise.

C. Electron Lithography

When it is desired to produce circuits and devices with feature sizes in the range of 0.1 μm to 2 μm, one should consider the use of electron lithography [3]. There is a great deal of similarity between photolithography and electron lithography, but the exposure is done in an electron microscope for the latter. Either positive or negative resists may be used, and the criteria for the choice between them are similar to the photoresist case. One very important difference is that the electron beam must be instructed to follow a pattern which will lead to exposure of the

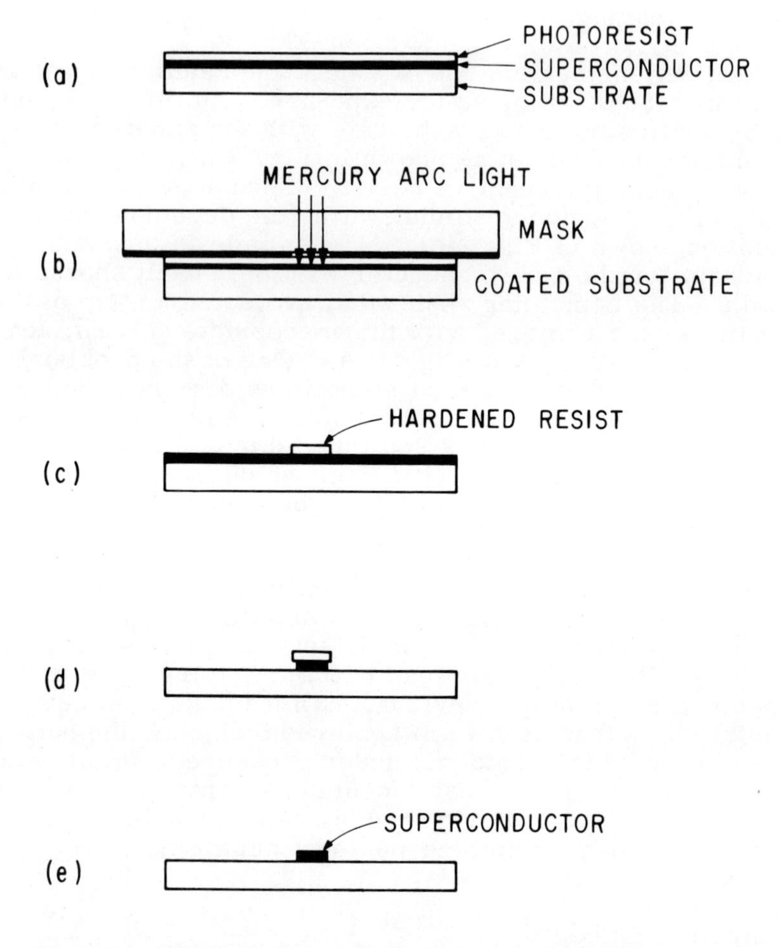

Fig. 2 Steps in the formation of a strip of thin film using a negative photoresist.

desired areas. This usually is done by interfacing the electron microscope deflection system to a computer which can be given the pattern instructions. A minicomputer is adequate for device work.

Figure 3 shows a metal pattern with the shape of a constricted thin film (Dayem bridge) which was made using a computer-controlled electron microscope. In this case a program was written to generate the pattern seen. Other programs exist in which patterns can be generated as combinations of rectangles so complex circuits can be programmed. For some simple patterns, like single lines or sets of single lines, the internal raster generator in the microscope can be used and no computer is required. However, if one were setting up a fabrication facility, the computer control should be included. Another feature of electron lithography is that the area over which high-resolution exposure can be made by deflecting the beam is limited by the aberrations which occur with large deflection. Typically, the maximum linear dimension for submicrometer features is about 0.3 mm. Means for accurately registering successive steps in a multilayer device fabrication are available.

One important difference in the lift-off process using electron resist from that involved in photolithography should be mentioned. The nature of the exposure process with electrons is such that the spreading of the electron beam in passing through the resist and backscattering from the substrate leads to an undercut profile in the resist layer after development, as seen in Fig. 4(a). Upon deposition of the metal film, there is no connection between the part on the substrate and that on the resist if the resist is thicker than the metal, as shown in Fig. 4(b). Thus, when the unwanted metal is lifted off using the solvent of the unexposed resist, even very narrow lines can be left on the substrate (Fig. 4(c)). The smallest line that has been made in this way was 0.06 μm wide.

Electron-beam exposure can have an advantage over contact photolithography even for feature size of 5 μm and larger when the exposure is to be made on a surface with height variations comparable to or greater than the desired feature sizes, in which case the contact photolithographic method cannot be used because of diffraction.

D. Thin-Film Deposition and Ion Etching

The most common way to make thin films is by vacuum deposition. The substrate is located in a vacuum system, typically at a pressure of 10^{-7}-10^{-5} Torr, and it is sometimes cooled or heated to achieve certain special ends. The material

Fig. 3 Constricted-thin-film Josephson junction as an example of the use of electron lithography. The narrowest part is about 0.2 μm wide.

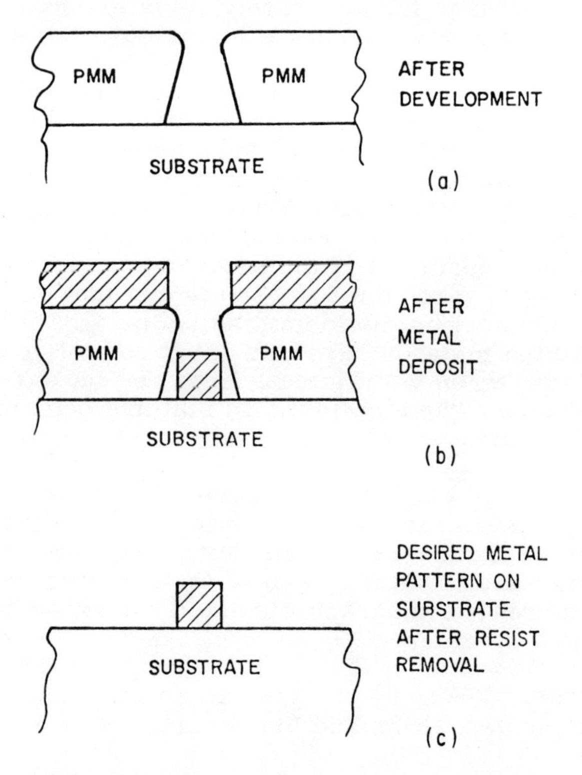

Fig. 4 Steps in the use of electron lithography with a positive photoresist to form a strip of thin film.

to be deposited is either melted and evaporated in a current-heated boat of a refractory metal or by an electron gun. Usually, higher quality films can be obtained using the latter. A shutter is normally provided between the source and the substrate in order to allow evaporation from the contaminated surface of the metal onto the shutter rather than onto the substrate. Sometimes a cooled liquid-nitrogen "chimney" surrounds the deposition activity because the cold surface condenses out contaminants and improves the vacuum.

Typical film thicknesses are in the range of 0.05 μm to 2.0 μm. Films much thicker than 2.0 μm suffer from a tendency to peel off. Films much thinner than 0.05 μm have a tendency to be discontinuous unless care is taken to avoid it. In this regard it is worth pointing out that when the deposited atoms reach the substrate they can move about the surface and they tend to form clusters which subsequently grow together to form a continuous film. If a very thin film, or a film more free of island structure is desired, it is advantageous to assure that the substrate is cooled by making good thermal contact to the cooled chimney.

An alternative, and often preferable, way to obtain thin films is to sputter atoms and groups of atoms off the surface of a target made of the material desired in the film [4]. This is done in a vacuum system containing typically 10^{-2} - 10^{-1} Torr of an inert gas such as argon. The gas is ionized and accelerated by a discharge caused by the application of rf or dc voltages applied to the electrodes. The sputtered target material crosses the gap to the substrate. In the most common and oldest kind of sputtering system, the substrate is also exposed to the ionized gas and is thereby inadvertently heated, typically to a few hundred degrees Celsius unless deliberate efforts are made to make good thermal contact to a water-cooled baseplate. This can be done by placing a small amount of a material like gallium, which is in the liquid phase near room temperature, between the substrate and the baseplate.

In the last several years, another kind of sputtering system has come into use. There is a magnetic field set up in the neighborhood of the sputtering target and the dc voltage is applied between the target and a nearby electrode as shown in Fig. 5. The ionized gas is confined to the region of the target and the ion and electron motions are like those in a magnetron. The result is a much higher rate of sputtering and an absence of substantial heating of the substrate.

In addition to using a sputtering system to deposit material onto a substrate, one may also reverse the process (in the type without magnetron motion) and place the substrate on the target location to do dry etching. Etching in this manner has a distinct

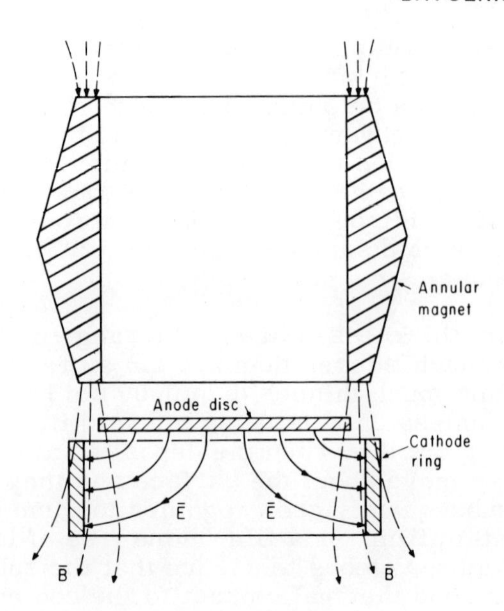

Fig. 5 Schematic representation of a magnetron-type sputtering source.

advantage over chemical etching in some cases in that sharply defined regions can be selectively etched by using an overlying resist or other kind of film as a mask. We will later see examples of structures that have been made this way (Section IV). There are situations in which chemical etching has an advantage in that all but one material of a structure may be impervious to the etchant; sputter etching is material-dependent but not so strongly as chemical etching.

Better ion etching can be achieved with a system that generates a directed beam of ions in an ion gun, neutralizes the beam after accelerating it, and directs the neutralized beam at the sample to be etched. It is a broad beam, so that like sputter etching it depends on masking. However, it has the advantage of directed ion velocities and achieves sharper edges.

III. SANDWICH-TYPE JUNCTIONS

One of the common and important configurations of Josephson junctions is the sandwich form in which two superconductive films are separated by about 20 - 50 Å of oxide, hundreds of angstroms of semiconductor, or up to a micrometer of normal metal. We will concentrate on the oxide- and semiconductor-barrier junctions, because that with a normal-metal barrier has extremely low resistance and its practical application is less likely than the others.

A. Oxide-Barrier Junctions

The first Josephson junctions were of the oxide-barrier type. Because of the large ratio of the capacitance to single-particle tunneling conductance, these junctions have large hysteresis. This property is valuable for some switching applications, but not for the usual mode of operation as detection and mixing elements or in SQUIDs where nonhysteretic I-V characteristics are needed. (They can be made nonhysteretic by resistive shunting at a price in I_cR value.) As we will see, thermal cyclability has been a problem with many oxide-barrier junctions but cyclable types have been developed.

The superconducting elements in order of decreasing transition temperatures are Nb (9.2 K), Tc (8.2 K), Pb (7.2 K), La(β) (6.1 K), V (5.3 K), Ta (4.5 K), Hg (4.2 K), Sn (3.7 K), In (3.4 K), Tl (2.4 K), Re (1.7 K), Pa (1.4 K), Th (1.4 K), Al (1.2 K), Ga (1.1 K), etc.. Those with transition temperatures above 4.2 K, the transition metals Nb, V and Ta and the rather soft metal Pb, find the most use. The most used of the next-lower-T_c group between 4.2 K and about 1.8 K, the temperature achievable by pumping with a small mechanical pump on helium vapor, are Sn and In, which are both soft metals. Aluminum is a convenient metal for lower-temperature test systems.

Superconducting Nb, V and Ta can be vacuum-deposited only if done in a high vacuum system and/or certain other techniques are employed. These materials are good getters for oxygen and their transition temperatures are reduced by absorbed oxygen. For example, the T_c of Nb is reduced about 5% for 0.5% oxygen content [5]. The oxygen is gettered prior to evaporation of the Nb onto the substrate, and a heated substrate is used. Generally speaking, superconductive Nb, V and Ta films are more easily fabricated using sputtering systems as discussed below.

The most common elements used in Josephson junction fabrication are Pb, Sn and In. These are very easy to vacuum-deposit with transition temperatures close to the bulk values. The difficulty with these materials is that they form hillocks and whiskers upon temperature cycling between room and cryogenic temperatures. As the temperature of the sample is lowered, the metal contracts at a greater rate than the insulating substrate and the film is stretched inelastically. When the sample is returned to room temperature, the film is then compressed. Hillocks and whiskers form to relieve the resulting stress. Whiskers can be as long as several micrometers and hillocks are typically 0.2 - 0.5 μm high. Whiskers grow only under certain circumstances;

hillocks are unavoidable in these soft elemental films. Even small hillocks are large enough to puncture the thin oxides used in tunnel junctions, so junctions made with these materials are usually kept continuously at cryogenic temperatures. The junctions can be stored in liquid nitrogen between experiments in liquid helium so that the stress-relief mechanism apparently requires raising the temperature above 77 K. (It has been reported [6] that some amelioration of the cycling problem is obtained by coating the completed junctions with a photoresist layer of a few tenths micrometer thickness and by making the temperature transitions as rapid as possible. In this way, Pb-PbO-Pb junctions have been cycled between 4 K and 293 K eight times with no significant change in the critical current, but they cannot be stored for more than one day at room temperature.) For practical applications where many temperature cycles must be anticipated, the junctions made of pure soft metals are not acceptable.

Alloys of soft metals have been employed to achieve more cyclability of Josephson junctions. These have been of both binary and ternary types. It has been shown [7] that hillock formation in Pb films can be reduced by adding small quantities of In. Cyclable junctions have also been produced using In and In-Sn (50:50) alloying [8]. The usual procedure is to deposit the In or In-Sn with Pb on top and to allow interdiffusion at room, or slightly elevated, temperature levels.

Indium apparently plays a twofold role in addition to aiding in the prevention of hillock formation. It is usually deposited first and serves to improve bonding of the film to the substrate. More importantly, substantial decrease of the tunneling barrier is obtained by virtue of the presence of indium [9]. In the range 0-20 percent indium in a Pb-In alloy, the oxide is a mixture of PbO and In_2O_3. For a junction with a Pb-In base electrode and a 32 Å barrier thickness, the tunneling current density is a strong function of In content, as seen in Fig. 6, and can be higher than 10^4 A/cm^2.

The use of other alloys for lowering the barrier height has also been reported [9]. These include Te-Pb, Sn-Pb, Bi-Pb, and combinations of In, Te, Sn and Bi with other superconductors, though no experimental results have been given.

There is reason to believe that the most effective element for the prevention of hillock and whisker formation in lead junctions is gold [10]. The stress relaxation phenomenon which leads to the growth of protrusions is apparently related to grain-boundary migration and diffusion. Measurements on bulk lead have shown that the best alloying elements for inhibiting grain-boundary motion are ones with atomic sizes much different from

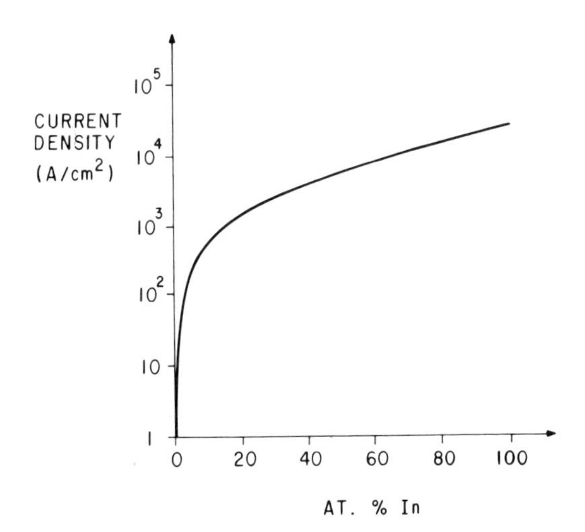

Fig. 6 Dependence of the critical current density on indium content in a Pb-In alloy oxide-barrier Josephson junction. The oxide thickness is 32 Å. (After reference 9.)

that of lead atoms and also with large valence difference. The valence difference seems to play the dominant role [11], so gold and silver (valence of 1 compared with Pb valence of 4) should be good choices. Experiments have shown that gold is more effective in inhibiting hillock growth than is silver [10]. This suggests that one might advantageously use gold to inhibit hillock formation and indium mainly for bonding to the substrate and lowering of the tunneling barrier.

The use of a ternary alloy of lead, gold and indium has been reported [12]. The alloy used in the base electrode was formed by depositing 500 Å of In, 100 Å of Au, and 3500 Å of Pb in succession, and by subsequently heating to aid interdiffusion. After formation of the oxide barrier, a counter electrode of Pb with a small amount of added Au to suppress the above-mentioned stress-relaxation effect was formed. It was made by depositing 3000 Å of Pb, 50 Å of Au, and 2000 Å of Pb in succession with subsequent annealing to homogenize the alloy. It should be pointed out that the layered alloys are best made without exposure to air between deposition of the layers, because a small layer of oxide can inhibit interdiffusion.

Another way to form reliable junctions with transition temperatures well above 4.2 K is to use niobium (or, less commonly,

Ta or V) for the base electrode. As discussed above, the formation of the niobium film can be done using electron-beam evaporation if special precautions are taken. The partial pressure of oxygen in the vacuum system must be kept at a very low level either by having a special very-high-vacuum system or by using the gettering property of niobium on large metal surfaces by continuously depositing it during the deposition of the film on the substrate. It is usually necessary to also raise the substrate temperature by hundreds of degrees Celsius. The more common procedure is to sputter deposit the niobium in either a dc [13] or rf [14] diode system or to use getter sputtering [15, 16]. Probably the method which gives good results most easily employs the magnetron-type unit described briefly above in Section IID [17, 18], in which the rate of deposition is fast enough that inappreciable amounts of oxygen are absorbed. Attempts to use niobium also for the counter electrode have generally failed due to superconducting shorts in the oxide, but some success has been achieved [18]. More commonly, the counter electrode is made of Pb or a Pb alloy and good stable junctions are obtained [17, 19]. The advantage of using a Nb counter electrode is that it is not subject to corrosion (a severe problem for Pb). Alloying the Pb in the top electrode with a small amount of In improves its corrosion resistance [17]. Junctions with either Pb or alloys thereof should be coated with a thin layer of photoresist to prevent moisture from reaching the Pb.

A variety of methods of forming the oxide barrier have been described in the literature. The principal problem experienced with Pb is the formation of porous oxides that contain pinholes which lead to short circuits upon deposition of the top electrode. Oxidation in room environment with the sample either heated or not appears to be the least controllable for Pb, possibly because of climatic variations, but it is often used in experimental work because of its simplicity and has been used for Nb with rather reproducible results [17, 18]. Oxides can also be formed in an oxygen flow with heated or unheated samples. A more reliable method is to form the oxide in the vacuum system immediately after deposition of the base electrode so the environment can be more completely controlled. This is done by introduction of oxygen and by heating of the sample for a period of hours. A similar procedure employs a glow discharge in the vacuum system to form the oxide [20].

A recently developed method involves a competition between growth of an oxide and its sputter removal [21]. The sample is placed in a sputtering system, perhaps with the junction tunneling region defined by a resist mask, and the surface is first cleaned in a low-power discharge of an inert gas such as argon. Then with the argon replaced by oxygen or a combination of argon and oxygen, the oxide begins to form and sputtering of the oxide also proceeds. The rate of sputtering is independent of the oxide thickness. At a thickness which depends on the gas pressures and discharge power, an equilibrium is reached. Thus one can control the oxide thickness accurately by setting the sputtering parameters.

If lithographic steps are used between the depositions of the base and counter electrodes, some method of cleaning the area to be oxidized should be employed. The last mentioned method achieves this cleaning, but one could also clean with a glow discharge and then oxidize thermally inside the vacuum system.

B. Evaporated Semiconductor Barrier Junctions

Another approach to lowering the barrier of a sandwich-type Josephson junction is to replace the oxide with a deposited pad of a semiconductor. The semiconductor materials used have included CdS, CdSe, C, Ge, Te, PbTe, InSb, GaAs, and ZnS. Electrode materials have included Pb, Sn, Tl, Pb_xIn_{1-x}, Pb_xSn_{1-x}, Nb, Nb_xRe_{1-x}, and Al [22]. In no case has a current density been obtained comparable with that achievable with oxide-barrier junctions. In almost all cases, it has been necessary to patch the pinholes through the semiconductor by forming a heavy oxide on the base electrode in the holes. These junctions have some desirable properties such as the suppression of junction resonances by the semiconductor losses and the achievement of parameters equivalent to shunted oxide-barrier junctions for use in certain switching circuits. Generally, the properties of a thin deposited film of a semiconductor are very difficult to control and this will probably limit their application. Some of the best junctions with deposited semiconductor barriers have been formed with Pb electrodes and Te barriers [23]. These can be made with either hysteretic or nonhysteretic I-V characteristics depending on the Te thickness and have been shown to have uniform, though low, current density.

C. Single-Crystal Silicon-Membrane Junctions

A recently developed fabrication technique forms a barrier for a Josephson junction by etching a silicon wafer in a localized region [24]. The etching is initiated through a mask on one side of a standard wafer and is stopped at a predetermined distance from the opposite face by a layer of boron doping. The result is a single-crystal membrane of uniform thickness which can subsequently be coated with a superconductor having the desired electrode pattern, an example of which is shown in Fig. 7. This structure has the advantage over the deposited-semiconductor

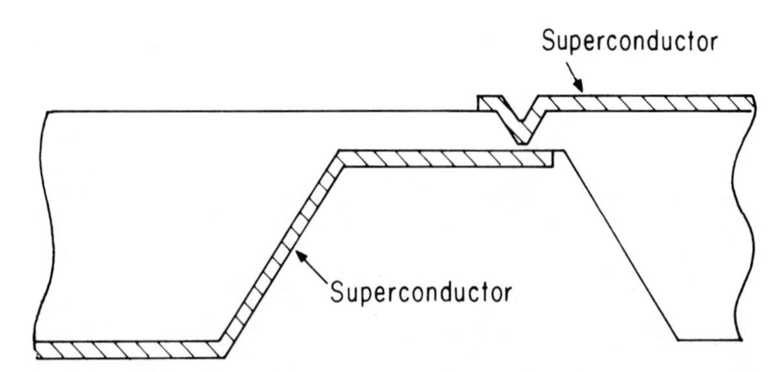

Fig. 7 A Josephson junction employing a thin silicon membrane as the tunneling barrier with a possible electrode configuration having low stray capacitance.

junction that the membrane can be processed after fabrication and before electrode deposition to adjust its properties.

The fabrication procedure is shown in Fig. 8. In step (a) a SiO_2 layer is grown on one side of a (100) oriented wafer. In step (b) a boron diffusion is made in the opposite side. The later etching will stop at the plane of boron density $N_A = 7 \times 10^{19}$ cm^{-3}; the diffusion is designed so that density exists at the depth desired for the membrane thickness. The process is calibrated to account for layers of Si-B and boron glass that must be removed. In step (c), a window (preferably rectangular and aligned with the crystal axes for smoothness of the resulting etch pit) is opened on the top side and the diffused layer is protected with a sputtered SiO_2 layer after removal of a film of boron glass and the Si-B phase layer. Step (d) shows the result of subjecting the wafer to a heated etchant, ethylene-diamine-water-pyrocatechol. This so-called preferential etchant has the properties that it etches (100) planes many times faster than (111) planes, etches SiO_2 only very slowly, and will not etch silicon where the boron concentration exceeds 7×10^{19} cm^{-3}. Membranes as thin as 400 Å and as thick as about 4 μm have been made this way.

The doping profile in the membrane as fabricated is shown in Fig. 9(a), and the corresponding energy-band profile with electrodes is shown in Fig. 9(b). It is seen that Schottky barriers form at the silicon surfaces and the region between is degenerate.

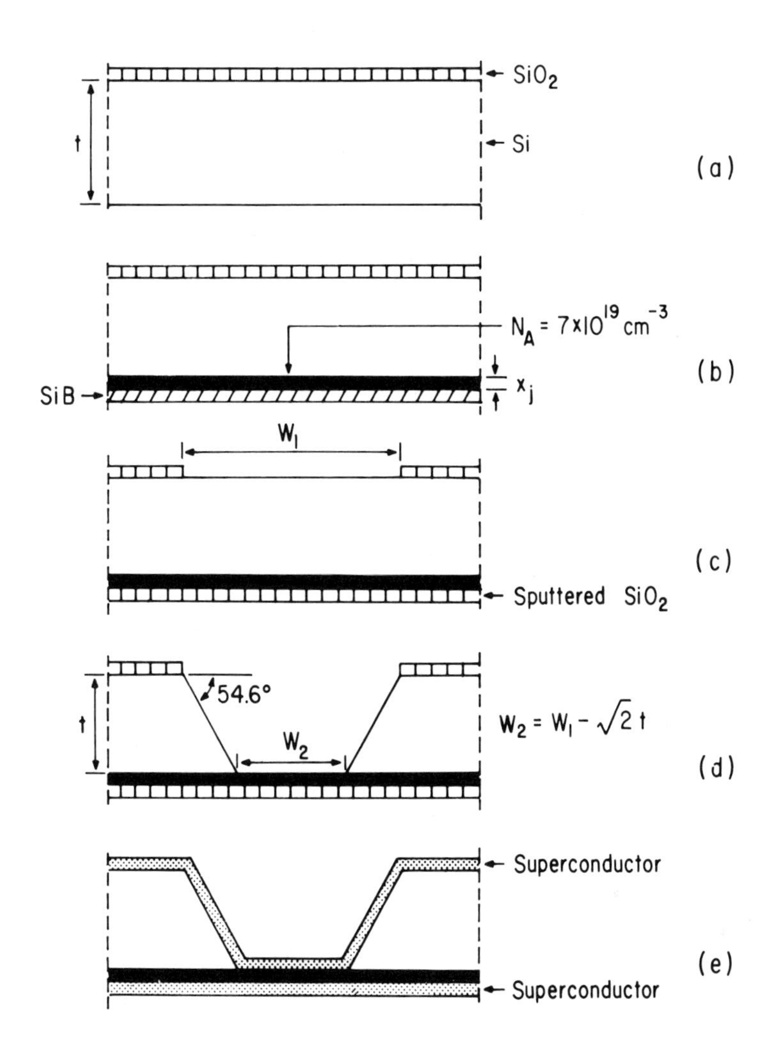

Fig. 8 Steps in the fabrication of a Josephson junction in which the barrier is an integral part of a silicon wafer.

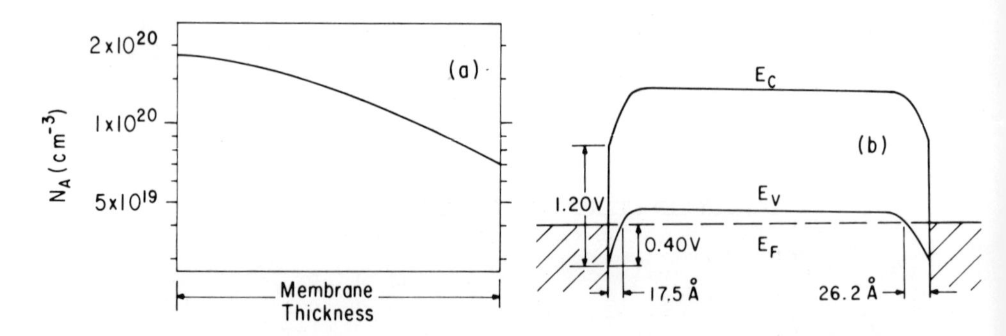

Fig. 9 (a) Typical distribution of doping in a crystalline silicon membrane after formation by the procedure shown in Fig. 8. (b) Energy-band profile in the silicon membrane corresponding to the doping shown in (a). Schottky barriers form at the interfaces between the silicon and the electrodes.

The process described above is limited with respect to the minimum achievable dimensions in the plane of the junction; a process has been developed to overcome this limitation [25]. As seen in Fig. 8(d), the dimension w_2 of the membrane depends on the wafer thickness, a quantity which can only be known to an accuracy of several micrometers. Thus, if membranes with w_2 on the order of one micrometer are desired, as is the case for some applications, the process in Fig. 8 must be modified. This can be done by first using the process of Fig. 8 to make a membrane of about 2 μm thickness which can be measured to an accuracy of about 100-200 Å. The heavy boron doping is out-diffused and a second etching procedure like that in Fig. 10 is followed to make a second etch pit. It is estimated that its dimension (w_2' in Fig. 10(c)) can be controlled to better than 0.1 μm. This work is still at an early stage and the double-etching procedure has only been used for $w_2' \geq 7$ μm to date.

Josephson junctions are made by removing the remaining SiO_2 films on the Si membrane and depositing superconductive films as in Fig. 8(e). A 600 Å-thick membrane with an area of 50 μm^2 was uniformly doped to 1.2×10^{20} cm^{-3} and, when coated with Pb electrodes, had a critical supercurrent density of 1.1×10^4 A/cm^2, which is comparable with the best oxide barrier junctions. The resistance was 0.55 Ω, so that the I_cR product exceeds that theoretically obtainable in oxide-barrier junctions; the implications of that fact have not yet been determined. The resistance can easily be raised to the order of 20 Ω, which is desirable for matching to rf systems.

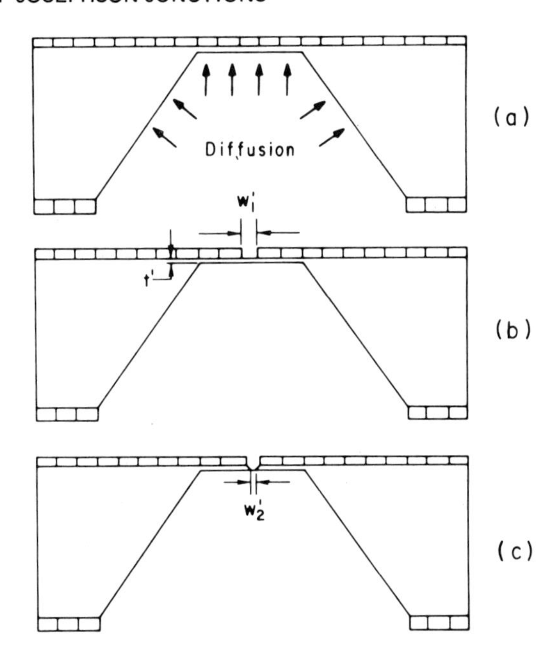

Fig. 10 A double-etching procedure for formation of barrier membranes of submicrometer dimensions. After outdiffusing the dopant from the 0.2 - 0.4 μm-thick membrane (a), a new shallow diffusion is made (b), after which, (c), the etching procedure of Fig. 8 is employed.

The flexibility of being able to tailor the energy-band profile has not yet been exploited, nor have electrode materials other than Pb and In been tested. The dependence of the I-V characteristic on applied rf currents has not yet been determined.

IV. JUNCTIONS WITH COPLANAR ELECTRODES

An important class of junctions contains those with both electrodes in the same plane. Weak coupling between the electrodes may be achieved in a number of ways, and a representative group of methods will be discussed. Generally, the weak coupling is done in a way which leads to lower capacitance than in the sandwich structure. The size of the coupling region is larger than the typical thickness of the oxide layer in a tunnel junction and, consequently, although sizes of the order of the superconducting coherence length must be achieved, the coplanar structures are less prone to problems caused by hillock formation or whisker growth during thermal cycling. In most cases of the coplanar configuration, the region that couples the two electrodes is directly accessible from above the substrate, and can be examined after the completion of the junction, whereas, in the sandwich structure, the coupling region is hidden below the top electrode of the junction. An important practical distinction between coplanar junctions and the sandwich configuration arises because the current density at the weak link between the electrodes in the coplanar configuration is generally much higher than is the case in the sandwich structure. In many of the coplanar structures, this higher current density, which approaches the critical current density of the superconductor, is the mechanism by which weak coupling is achieved between the two electrodes. As a consequence of the high current density in the weak coupling region where the voltage drop across the junction occurs, the power density is higher in most coplanar configurations than in sandwich structures. The resultant heating in many of the coplanar structures has been recognized as an important practical limitation in operating the junctions at voltages corresponding to high frequencies. Because of the importance of this limitation, we will comment on the relative ability of various coplanar structures to remove the heat generated in the weak coupling region, and we will emphasize structures in which the heating effects can be minimized.

The coplanar junctions can be divided roughly into two categories: (1) Structures of constant width in which the current density is approximately uniform so they can be considered one-dimensional. In these structures, the weakening of the coupling between the two electrodes (now often called "banks") is achieved by weakening the superconductivity (lowering the transition temperature) of the region between the two electrodes, or by coupling between the electrodes through a semimetal or through a semiconductor. (2) Coplanar junctions in the second important category are those with a higher current density in the weak coupling region as there is appreciable constriction of the cross section. In both categories, the weak connection can be made in a number of different ways. We will survey representative ways of making junctions in each category. The fabrication techniques include the lithographic techniques mentioned earlier for sandwich structures, as well as new techniques specific to some of the coplanar structures: ion implantation to weaken or strengthen a superconducting region, anodization to decrease film thickness, and various highly developed arts of scratching films of soft superconductors to leave only a small bridge of superconductor connecting the two electrodes.

A. Variable-Composition Junctions

Starting with the first category of coplanar junctions, where the current density is constant, we discuss variable composition type of junctions. Within this type, we include all junctions in which the transition temperature of the weakened section between the electrodes is reduced by changing its composition. In one technique, layers of superconducting materials which are thinner than, or on the order of, the coherence length in thickness are superposed to give a film with a transition temperature intermediate between the transition temperatures of the two films via the proximity effect. Then the higher T_c material is removed in a narrow region across the electrode strip leaving a section of low T_c. In another technique, atoms of a second material are injected directly into a superconductor film to vary its composition. In this technique of "ion implantation", the transition temperature of the superconducting layer can be either increased outside the junction or decreased inside.

An example of a proximity-effect junction [26] is shown in Fig. 11. In this example, a thin film of Ta about 200 Å thick is vapor deposited on a sapphire substrate. Without breaking the vacuum, to avoid oxidation of the Ta, a layer of Nb about 100 Å thick is vapor-deposited over the Ta. The net result is to form a superconducting film with a transition temperature intermediate between that of Ta and Nb. By photolithography, now, all of the film is protected by photoresist except for a transverse slit of length on the order of a micron along the film. The superconducting transition temperature of this transverse region is now

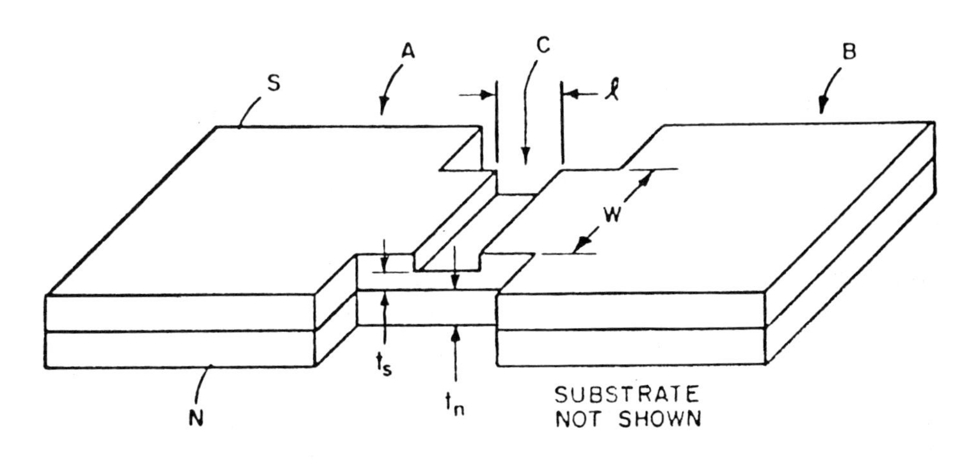

Fig. 11 Proximity-effect junction using a two-layered structure of Nb on Ta. (After reference 26.)

depressed by anodizing the exposed region of Nb, forming a layer of niobium oxide above a now-thinner layer of Nb metal. The anodization is done in an electrolyte which is a weak acid, such as boric acid. For typical dimensions of the slit region of 1 μm long by 20 μm wide across a double-layered film of this width, a 10 msec pulse at about 20 volts has been used successfully for the anodization. The short pulse of relatively high voltage prevents undercutting beneath the protective resist.

Another technique which can be used to weaken the superconductivity of a region is ion-beam etching. The advantage of this technique is that it can be used for combinations of superconductors which cannot be thinned by anodization, as well as for those which can be anodized. The range of such materials is larger, and includes materials such as Nb, Ta, W, Ti, Zr, Al, Hf, Re, Rh, Ir, Pd, and Pt as well as some of their alloys and compounds. Josephson-like effects have been observed over a wide range of parameters for these proximity-weakened structures of Nb and Ta, with lengths of the weakened region ranging from 0.3 μm to 5 μm, widths from 0.5 μm to 300 μm, and normal resistances ranging from 10^{-2} ohm to 50 ohms. The junctions fabricated of hard superconductors are structurally more stable than those of soft materials, although the latter are more easily made. The most important example of the latter type was formed by weakening the superconductivity of an Sn film by a transverse line of Au which was deposited beforehand on a sapphire substrate. The Au is evaporated first to minimize the possibility of an oxide layer between the two films, which would lower the strength of the proximity effect.

The effective composition of a superconducting region also may be changed by implanting ions into an already formed film [27, 29]. The implanted ions may raise the transition temperature of the film of a material such as Mo. A film of Mo, typically 800 Å thick, is first grown by sputtering onto a sapphire substrate at room temperature. The bulk transition temperature of Mo is 0.92 K, and the films thus grown are not superconducting at temperatures which can be reached by usual cryostats in which helium is evaporated by a modest vacuum pump ($\approx$ 1.4 K). However, by injecting either S^{32} or N^{14}, the transition temperature can be raised. A dose of 4×10^{16} N^+ ions/cm^2 implanted with 45 keV energy brings T_c up to about 1.5 K. Such an initial dose was given to an entire Mo film which had been formed already into a strip 12 μm wide by standard photolithographic techniques. Then the strip was masked by photoresist as shown in Fig. 12 to protect a 1 μm length of the strip while 10^{16} S^+ ions/cm^2 were implanted with an energy of 110 keV in the unmasked region. This second implantation further raised the transition temperature of the unmasked region, and the masked region remained as the weakly superconducting connection between

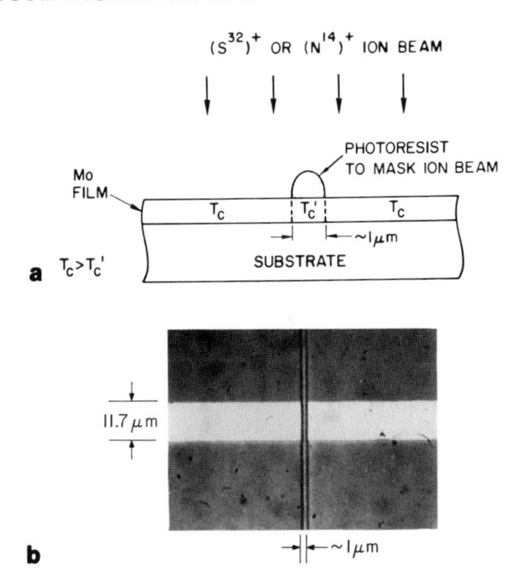

Fig. 12 (a) Photoresist mask for ion implantation to raise the transition temperature outside the junction region. (b) Top view of the masked Mo film.

these two electrodes.

We now consider coplanar structures which are intermediate between those which have one-dimensional current flow, and those which are dominated by a current concentration. These are the semimetal barrier and the semiconductor barrier coplanar junctions. In both cases, a narrow gap is made in a superconducting strip. For example, in the coplanar semimetal barrier junction in which a Pb film is first evaporated on a glass substrate [28], a narrow gap in the Pb film may be cut by a diamond stylus. Ethanol is used to reduce the friction between the diamond and the film. Any shorts that remain across the gap are burned off by a 10-volt pulse. Before evaporation of a 3000 Å thick Bi film to join the two Pb electrodes, the Pb film must be sputter-cleaned by a glow discharge to remove its oxide layer. Arcing across the gap during the sputter cleaning is prevented by shorting the gap at one edge by low temperature solder. During these processes (the evaporation of the Pb film, sputter cleaning, and evaporation of the Bi film) the substrate is kept at liquid-nitrogen temperature. This procedure is especially important during evaporation of Bi, because Pb melts at a lower temperature than the evaporation temperature of Bi at a pressure of 10^{-6} Torr. In addition, the Bi must be evaporated at a rate less than 10 Å/sec. The main advantage of these semimetal junctions, when compared to a superconductor-normal-superconductor sandwich junction is their higher normal resistance (on the order of an ohm). The Pb-Bi-Pb junctions must be stored at liquid nitrogen temperature to prevent alloying of the materials which takes place at room temperature. In principle, an NbN-Bi-NbN junction would not suffer from this difficulty, but NbN is mechanically tenacious and chemically stable, so that it is difficult to form a narrow gap mechanically or chemically after lithography. (A possible approach to the problem of forming a small gap in NbN, not yet tried, to the authors' knowledge, is ion beam machining or sputter etching.)

B. Semiconductor Bridge

Another way of providing the weak coupling between two coplanar superconducting electrodes is through a highly doped substrate [29]. The structure shown in Fig. 13 was fabricated by depositing a 30 μm-wide Pb strip on the cleaned, heavily doped surface of a silicon wafer and by subsequently cutting a crosswise gap in the Pb strip. The gap was cut by making a resist mask by electron lithographic means and by sputter etching the Pb. This is very recent work and little data is available. Junctions with gaps of widths 0.1 - 0.4 μm showed supercurrent. Both hysteretic and nonhysteretic I-V characteristics have been seen. Junction resistances are in the range 0.2 - 1.0 Ω and critical currents are typically in the range of 1 - 10 mA.

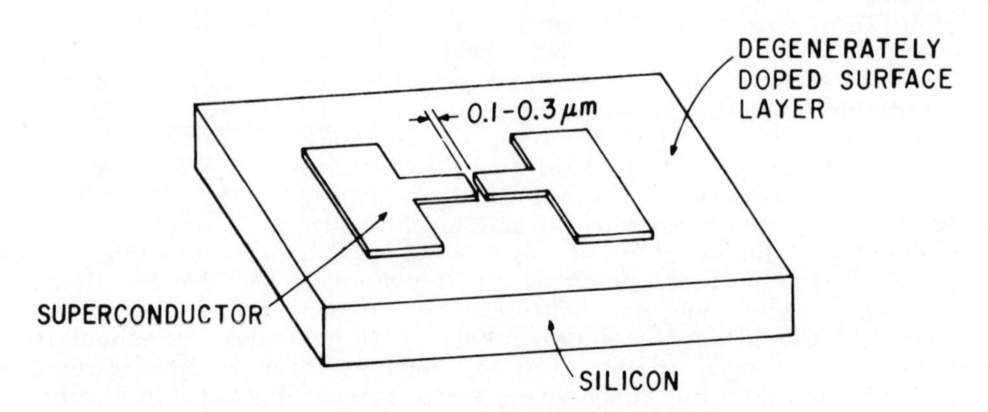

Fig. 13 Coplanar lead-on-silicon Josephson junction.

C. Microbridges

Microbridges can be divided into three broad categories: (1) those which are formed by lithographic techniques in superconducting film of constant thickness; (2) those which are formed by a combination of lithographic techniques and other techniques so that the bridge region is thinner than the electrodes (or "banks"); and (3) those which are formed by a mechanical technique, such as by scratching, and which consequently are variable in thickness. Because of heating problems, recent results indicate that the variable-thickness types are most useful when high voltages must be applied across the bridge, as in the case of microwave detection, but the other types can be used in quantum-interference devices, such as magnetometers. As with junctions that depend on variation in composition of the material, microbridges generally must be operated at temperatures close to T_c for the Josephson effects to be observable. In the case of microbridges, the lowest temperature at which the junction can be operated is usually determined by heating effects. As the temperature is decreased, the critical current increases, and thus, for a given voltage across the junction, a greater amount of power is dissipated. A careful discussion of this effect is given by Octavio, et al. [30], who used the voltages at which energy gap structure appeared in variable thickness microbridges to determine the local temperature in the bridge region. This recent work is consistent with earlier studies that showed that hysteresis in the I-V curves of long narrow superconducting strips could be interpreted in terms of heating at local regions along the strip [31].

We begin our discussion of microbridges with constant thickness bridges. If the relevant parameter that determines the maximum size of the microbridge is the coherence length in the superconductor, then the bridge sizes that can be obtained by photolithographic techniques (about 0.5 μm, at best) are not small enough except for superconductors with long coherence lengths and/or in a narrow temperature region near the critical temperature where the coherence length diverges. Consequently, electron lithography techniques generally must be used. It is convenient to have the electron beam controlled by a computer, although a flying spot scanner also can be used to control the exposure, or an aperture can be placed in the electron beam. To the authors' knowledge, the minimum bridge size that has been obtained by electron lithographic techniques is shown in Fig. 14, in which the bridge region in the Pb film 1000 Å thick has length 0.18 μm and width 0.1 μm [32]. The production of such bridges requires a rather large investment in a high quality scanning electron microscope and the associated exposure pattern control. Fortunately, another technique is available to produce microbridges by mechanical means at lower total cost.

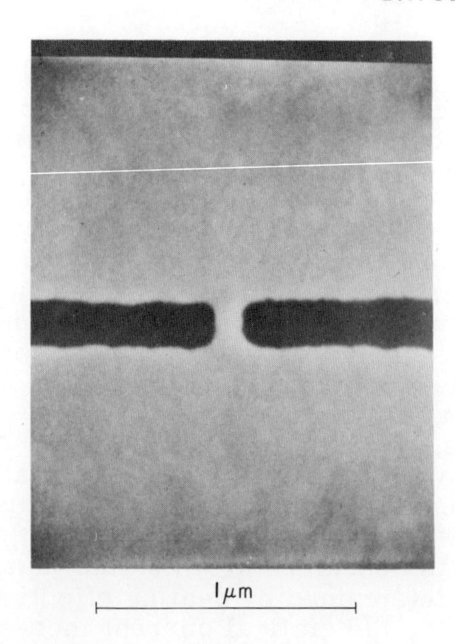

Fig. 14 Microbridge formed by electron lithography. (Courtesy of R. Laibowitz.)

In the basic technique of double-scratching, invented by Levinsen [33], a clean glass slide was scratched with a razor blade to make a submicron line depression in the glass. Then after evaporation of a thin film of Sn, a second scratch was made perpendicular to the first. The second scratch cut through the Sn film everywhere except at the depression left by the first scratch, where a bridge of Sn film remained, joining the two electrodes. By this technique it was possible to make bridges smaller than 0.5 μm x 0.5 μm. In the original technique, the razor blades were held by hand, and only one type of Danish razor blade, a type which is no longer produced, was sharp enough to produce small microbridges. Modern improvements of this basic technique use a diamond stylus for scratching, and a machine to move the stylus [34]. Recently it has been shown that the machine to move the stylus need not be large and complicated, but can be simply a stage for translating laser mirrors in two dimensions [35]. The stage is moved by an oil hydraulic piston, with the rate of motion during scratching controlled by a needle valve in the oil line. A bidirectional valve allows rapid return in the other direction. An automatic controller can be used to produce arrays of weak links. Although the movement stage can be simple, it is crucial that it be placed on a massive vibration isolated table, of the type used for laser experiments, to achieve small bridge sizes. An electron micrograph from an angle 50° to the normal of a microbridge made in this manner is shown in Fig. 15 [35]. The Sn film is 1000 Å thick. The same tool, a

Fig. 15 Microbridge formed by double-scratch method. Bridge width is about 0.2 μm. (Courtesy of R. Chiao.)

diamond with a 90° angle and tip radius of 0.5 μm, was used for both the first and second scratch on the fused silica substrate. The angle between the tool and the substrate was 5°, and the force of the tool was 1/24 gm-weight.

An important improvement in the basic microbridge can be made by making the banks on both sides of the bridge thicker to improve heat removal from the bridge region [36]. Such bridges, with banks up to 3 μm thick, can be produced by a modification of the double scratch technique [30]. The first scratch in a sapphire substrate is made as before. Special cleaning techniques are required to make the thick Sn film adhere to the substrate [37]. A layer of tin oxide 50 to 100 Å thick is formed on the substrate after cleaning, by evaporating a layer of tin and allowing it to oxidize in a glow discharge for 2 to 4 hours. Then the thick film is evaporated in a vacuum of 2×10^{-6} torr at a rate of 90 to 120 Å/sec. [30]. The second cut is done by gently pressing the long edge of a diamond knife through the thick Sn layer (as opposed to dragging the diamond stylus across the film) as shown in Fig. 16. By this technique, it has been possible to produce 0.1 μm bridges. Presumably to avoid burnout, the series combination of a 0.1 ohm current sensing resistor and the bridge is shunted by a 0.1 ohm resistor. Steps produced by X-band microwaves were observed up to a bias voltage of 3.7 mV in a junction with 0.33 ohm normal resistance, I_c of 4.2 mA, and

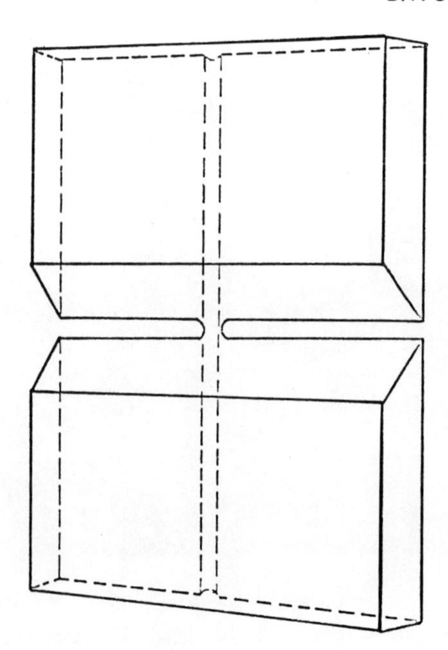

Fig. 16 Microbridge formed as described in the text. (Courtesy of M. Octavio.)

1.7 μm bank thickness, indicating that heating was not a problem up to this voltage.

V. POINT CONTACTS

Point-contact junctions have always been useful for observation of the Josephson effects because they are easier to construct than any of the above types of junctions. They are presently in use in two of the commercially available quantum-interference magnetometers. In addition, all of the best results reported to date in which a Josephson junction is used for high-frequency experiments have been obtained with point-contact junctions.

The characteristics of point-contact junctions used for magnetometers differ from those used in high-frequency experiments in that a higher junction capacitance is permissible in a magnetometer. In this case it has been possible to design point contact junctions which are mechanically rugged and can be cycled from room temperature to helium temperature with only minimal changes in their characteristics. The basic principle in the design of such point-contact structures is to use materials which have the same total mechanical expansion between helium temperatures and room temperatures, such as by making the entire structure out of Nb. This construction is possible in the rf biased SQUID where it is not necessary to apply a dc bias on the junction. When it is desired to place a dc bias on the junction, two sides of the junction must be separated by an insulating material, such as by a glass which has the same overall expansion as niobium [38].

Although some of the thermally cyclable point-contact junctions have been designed for high-frequency applications, generally, the best results have been obtained with junctions which can be adjusted by a differential screw mechanism while they are at liquid helium temperature. For these applications, it is important that the shunt capacitance be as low as possible, and consequently the total diameter of the point should be 1 μm or less. With point contacts, such as those used for astronomical observations at 1 mm wavelength [39] it was found that a Nb point on an Sn flat surface gave a mechanically more stable junction than a Nb-on-Nb junction. The reason for this is that Sn remains ductile at helium temperatures, whereas Nb becomes much less ductile than at room temperature. Such a junction could be cycled between helium temperature and 77 K repeatedly over periods of several days with less than 10 percent change in the junction's critical current.

The technique used to form the point for point-contact junctions varies considerably depending on the application. Mechanical sharpening with an abrasive works well enough for junctions used in magnetometers, and also has been used for junctions in some high frequency experiments. The junction used for astronomical applications was formed by chemical etching in a solution of 50 percent concentrated HF solution and 50 percent concentrated HNO_3. A 125 μm diameter Nb wire was lowered slowly into this solution by a micrometer drive. The position of the wire above the surface of the etchant was observed by looking at the reflection of the wire in the surface of the liquid under illumination by a microscope lamp. When the wire touched the liquid, the surface tension caused the liquid to rise up the wire. As the wire diameter decreased during etching, there was less total surface energy to lift the level of the liquid, and the etchant dropped lower. Finally, the liquid dropped off the wire, leaving a cone-shaped point with an included angle of about 50°. Such an angle is convenient in forming junctions, because a sharper point tends to curl up when pressed against a flat surface.

For applications at high frequencies, it is important that the characteristic voltage of the junction, $V_c = I_cR_n$ of the junction be as high as possible. For Nb-Nb point-contact junctions, a technique has been developed to remove the niobium oxide layer just prior to forming the contact [40]. After formation of the point, it and the opposed flat surface are placed in a chamber with helium gas flowing through it, and a drop of HF is placed on the point. The HF is allowed to evaporate and the point contact system is kept in flowing helium gas until it is cooled to liquid helium temperature. Such junctions have values of V_c near the theoretical limit.

REFERENCES

1. P. W. Anderson and J. M. Rowell, Phys. Rev. Lett. 10, 230 (1963).
2. L. V. Gregor, Proc. IEEE 59, 1390 (1971).
3. R. F. Herzog, J. E. Greeneich, T. E. Everhart, and T. Van Duzer, IEEE Trans. Electron Devices ED-19, 635 (1972); F. Auracher and T. Van Duzer, Rev. Phys. Appl. 9, 233 (1974).
4. L. Maissel, Chapter 4 of Handbook of Thin-Film Technology, McGraw-Hill, New York, 1970.
5. C. A. Neugebauer and R. A. Ekvall, J. Appl. Phys. 35, 547 (1964).
6. T. F. Finnegan, private communication.
7. A. Emmanuel, G. B. Donaldson, W. T. Band, and E. Dew-Hughes, IEEE Trans. Magn. MAG-11, 763 (1975).
8. W. Anacker, K. R. Grebe, J. H. Greiner, S. K. Lahiri, K. C. Park, and H. H. Zappe, U.S. Patent No. 3,733,526, May 15, 1973, "Lead Alloy Josephson Junctions".
9. J. M. Eldridge and J. Matisoo, U.S. Patent No. 3,816,173, June 11, 1974, "Fabrication of Variable Current Density Josephson Junctions".
10. C. Y. Fu and T. Van Duzer, Bulletin Amer. Phys. Soc., Series II, 21, 339 (1976).
11. J. W. Rutter and K. T. Aust, Trans. Metallur. Soc. AIME 218, 682 (1960).
12. I. Ames, U.S. Patent No. 3,852,795, December 3, 1974, "Josephson Tunneling Circuits with Superconducting Contacts". S. K. Lahiri, J. Vac. Sci. Technol. 13, 148 (1976).
13. J. Sosniak and G. W. Hull, Jr., J. Appl. Phys. 38, 4390 (1967).
14. S. Owen and J. E. Nordman, IEEE Trans. Magn. MAG-11, 774 (1975).
15. H. C. Theuerer and J. J. Hauser, J. Appl. Phys. 35, 554 (1964).
16. J. E. Nordman, J. Appl. Phys. 40, 2111 (1969).
17. P. K. Hansma, J. Appl. Phys. 45, 1472 (1974).
18. G. Hawkins and J. Clarke, J. Appl. Phys. 47, 1616 (1976).
19. K. Schwidtal and R. D. Finnegan, Proceedings of 1972 Applied Superconductivity Conference, Annapolis, Md., May 1-3, p.562, IEEE Pub. No. 72CHO682-5-TABSC.
20. J. L. Miles and P. H. Smith, J. Electrochem. Soc. 110, 1240 (1963).

21. J. H. Greiner, J. Appl. Phys. 45, 32 (1974).
22. P. Cardinne, J. Nordman and M. Renard, Rev. Phys. Appl. 9, 167 (1974).
23. J. Seto and T. Van Duzer, Appl. Phys. Lett. 19, 488 (1971).
24. C. L. Huang and T. Van Duzer, IEEE Trans. Mag. MAG-11, 766 (1975); Appl. Phys. Lett. 25, 753 (1974).
25. C. L. Huang and T. Van Duzer, IEEE Trans. Electr. Dev. ED-23, 579 (1976).
26. H. W. Notarys, J. E. Mercereau, J. Appl. Phys. 44, 1821 (1973), and U.S. Patent 3,798,511, March 19, 1974.
27. E. P. Harris, IEEE Trans. on Magnetics MAG-11, 785 (1975); E. P. Harris, J. Vac. Sci. Tecnol. 12, 1383 (1975).
28. H. Ohta, M. J. Feldman, P. T. Parrish and R. Y. Chiao, Rev. Phys. Appl. 9, 187 (1974).
29. M. Schyfter, J. Maah-Sango, N. Raley, R. Ruby, B. T. Ulrich and T. Van Duzer, Proceedings of 1976 Applied Superconductivity Conference - to be published in IEEE Trans. Magnetics.
30. M. Octavio, W. J. Skocpol and M. Tinkham, Proceedings of 1976 Applied Superconductivity Conference - to be published in IEEE Trans. Magnetics.
31. W. J. Skocpol, M. R. Beasley and M. Tinkham, J. Appl. Phys. 45, 4054 (1974).
32. R. Laibowitz, private communication.
33. P. E. Gregers-Hansen, E. Hendricks, M. T. Levinsen and G. R. Pickett, Phys. Rev. Lett. 31, 524 (1973).
34. J. E. Mooij, C. A. Gorter and J. E. Noordam, Rev. Phys. Appl. 9, 173 (1974).
35. R. Y. Chiao, M. T. Levinsen, P. T. Parrish, D. W. Deterson, R. P. Plambeck, private communication and to be published.
36. T. M. Klapwijk and J. E. Mooij, IEEE Trans. on Magnetics MAG-11, 858 (1975).
37. R. Y. Chiao, M. J. Feldman, H. Ohta and P. T. Parrish, Rev. Phys. Appl. 9, 183 (1974).
38. R. A. Buhrman, J. E. Lukens, S. F. Strait and W. W. Webb, J. Appl. Phys. 42, 4527 (1971); H. Tolner, C. D. Andriesse and H. H. A. Schaeffer, Infrared Physics 16, 213 (1976); V. A. Kulikov, N. N. Kurdjumov, G. F. Leshchenko, L. V. Matveets, V. V. Migulin and E. S. Soldatov, Rev. Phys. Appl. 9, 293 (1974).
39. B. T. Ulrich, Rev. Phys. Appl. 9, 111 (1974).
40. D. G. McDonald, private communication.

Chapter 8

BIOMAGNETISM*

S. J. Williamson,† L. Kaufman,‡ and D. Brenner†

New York University, New York, N.Y. 10003

I. INTRODUCTION

Advances in the technology of the superconducting quantum interference device (SQUID) during the past decade provide a magnetic field detector of high sensitivity and have stimulated efforts to use this device to study the weak magnetic fields associated with biological activity. This is the area of research that has become known as "biomagnetism". It is not to be confused with the study of effects produced by the application of a magnetic field to biological systems, an area known as "magnetobiology".

Contemporary interest in biomagnetism has focussed principally on magnetic fields produced by sources within humans. The wide spectrum of phenomena which have been observed includes: the magnetocardiogram (MCG) from heart activity; fetal magnetocardiogram (FMCG) from heart activity in a fetus; magnetomyogram (MMG) from muscle action; magneto-oculogram (MOG) produced by currents associated with the eye; magnetoencephalogram (MEG) from spontaneous activity of neurons within the brain; and visually evoked fields (VEF) from visually evoked brain activity. Each of these effects has an electrical analog which had been observed previously in voltage measurements.

---

* The preparation of this manuscript was supported by the National Science Foundation (Grant BNS75-13162), New York State Health Research Council (Grant C106219) and Office of Naval Research (Contract 00014-76-C-0568).

† Department of Physics

‡ Department of Psychology

This is an appropriate time to review past achievements, assess the capabilities of present technology, and infer which directions future progress may take. This review deals only with fields of biological origin and does not include the discovery and subsequent important studies by D. Cohen [1] of fields produced by magnetic contaminants of the body such as these in the lung and stomach. Nor will we discuss measurements of the magnetic susceptibility of biological systems, such as the studies of the heart as applied to plethysmography by J. Wikswo, Jr., J. E. Opfer, and W. M. Fairbank [2]. Several previous reviews have been written by D. Cohen [3, 4, 5] that include some aspects of biomagnetism not covered here.

We shall use MKS units to express field strengths and express the magnetic induction B in units of the Tesla (1 T = 10^4 gauss). Biomagnetic fields are so weak that they are often expressed in units of nanotesla (1 nT = 10^{-9} T), picotesla (1 pT = 10^{-12} T) or femtotesla (1 fT = 10^{-15} T). For reference we have listed in Table 1 the typical magnitudes for biomagnetic fields and their corresponding electrical analogs. By comparison, the most sensitive fluxgate magnetometers are available with an rms noise level of 20 pT $Hz^{-\frac{1}{2}}$, which implies an rms noise of 20 pT for a 1 Hz bandwidth or 200 pT for a 100 Hz bandwidth, and this is not sufficient to detect directly any of these biomagnetic fields. Induction coils with ferrite cores for enhanced sensitivity have a frequency dependent sensitivity which is limited by Johnson noise in the wire. Noise levels as low as 300 fT $Hz^{-\frac{1}{2}}$ at 10 Hz have been achieved, and this is sufficient to detect the MCG and MEG with signal averaging. But by far the record is held by SQUID systems with a noise level of $\sim$ 10 fT $Hz^{-\frac{1}{2}}$. This affords a factor of 2 x 10^3 greater sensitivity than the fluxgate magnetometer.

Electrical measurements are generally - but not always! - easier to perform than magnetic measurements, because use of a SQUID presently requires liquid helium as a coolant to keep the SQUID superconducting and there is a more severe noise problem from background ac magnetic fields. Nevertheless a number of advantages from magnetic measurements may be realized: (1) it is contactless and therefore can be used to monitor dc effects which in voltage measurements may be masked by contributions from contact and skin potentials; (2) magnetic fields depend on currents, which are generally larger in the interior of a human or animal than at the dermis, and therefore may provide a direct indication of activity in the interior without complications imposed by weaker currents in the intervening medium; (3) the different weighting of source contributions as provided by a magnetic field detection geometry when compared with a voltage detection geometry may provide additional information about the sources [6, 7]; and (4) magnetic fields are produced by a different set of

TABLE 1: Typical biomagnetic field amplitudes for normal humans and the corresponding electrical voltages obtained with skin electrodes. The numbers are very approximate owing to large inter-subject variability.

Effect	Magnetic Measurement	Magnetic Induction Amplitude (pT)	Electric Measurement	Voltage Amplitude (μV)
cardiogram	MCG	50	ECG	2,000
fetal cardiogram	FMCG	5	FECG	20
myogram	MMG	2	EMG	1
oculogram	MOG	10	EOG	300
encephelogram	MEG	1	EEG	50
visually evoked response	VEF	0.2	VEP	10

generators than electric potentials and the two may be partially independent, depending on constraints [8, 9, 10]. Some of these advantages have been proven through measurements, whereas others remain untested.

Before we consider the many interesting biomagnetic phenomena, we shall briefly summarize an ongoing discussion as to what magnetic field measurements can determine about the nature of electromotive sources that voltage measurements cannot. Then we shall examine instrumental capabilities and strategies for contending with magnetic background noise. We shall conclude with a few observations as to what the future promises.

II. FORWARD AND INVERSE PROBLEMS

A question of fundamental importance is whether from a knowledge of magnetic field patterns the distribution of generating current sources can be deduced. If so, aspects of the electrophysiology of biological systems could be studied in detail. This is known as the "inverse problem", since one starts with the effect (magnetic field) and works back to the cause (electrical current). Unfortunately, as will be explained shortly the inverse problem does not have a unique solution. This is true even if data from voltage measurements supplement the magnetic data. The "forward problem" by contrast does have a unique solution, and the magnetic field everywhere in space can be deduced in principle from the known current distribution by evaluating a volume integral as prescribed by the Biot-Savart law [11]. Numerical procedures for solving the forward problem have been discussed by Horacek and Ritsema [12].

The non-uniqueness of the solution to the inverse problem follows from elementary considerations of electromagnetism which we shall briefly summarize because they provide useful insight into the nature of field generators and establish some essential terminology. Our discussion will focus on the MCG and ECG as an example, but the results have more general validity. The starting point is the observation that relatively low frequency effects give rise to the MCG and ECG, and therefore as argued by D. Geselowitz [13] and R. Plonsey [14] the quasistatic limit may be assumed for Maxwell's equations. Thus all terms containing time derivatives are taken to be zero. This is equivalent to neglecting capacitive and inductive effects [15]. An additional consequence is that at each instant of time the charge distribution is taken to be constant, and therefore conservation of charge implies that the current density $\vec{J}(\vec{r})$ everywhere satisfies $\nabla \cdot \vec{J}(\vec{r}) = 0$. This condition, which can be called the conservation of current, states that the net current into a given region of space is zero. The condition permits currents everywhere in the biological medium to be expressed in terms of a primary current source density $\vec{J}^{i}(\vec{r})$ which is imposed by biological activity. In the case of the heart, $\vec{J}^{i}(\vec{r})$ describes the transmembrane current of sodium and potassium ions set up by biological pumps. As a passive response to this primary current source (or the equivalent electromotive force established across the membrane) a current also flows in the surrounding, conducting medium. This responding current is called the "volume current". Assuming for simplicity an unbounded, homogeneous medium with linear electrical response, the total current density at any location can be written as $\vec{J}(\vec{r}) = \vec{J}^{i}(\vec{r}) + \sigma\vec{E}(\vec{r})$, where σ is the conductivity of the medium and $\vec{E}(\vec{r})$ is the electric field associated with the volume current. Since the total current is conserved, the electric potential at each point, which is associated with $\vec{E}(\vec{r})$, can be expressed in terms of $\vec{J}^{i}(\vec{r})$. The expression gives the potential $V(\vec{r})$ as a volume integral of $\nabla \cdot \vec{J}^{i}(\vec{r})$ [11]. For an unbounded medium $\vec{J}^{i}(\vec{r})$ can be viewed as the primary current source from which the potential $V(r)$, total current density $\vec{J}(\vec{r})$, and magnetic induction $\vec{B}(\vec{r})$ can be obtained uniquely by appropriate volume integrations.

That $V(\vec{r})$ and $\vec{B}(\vec{r})$ are related to different aspects of the spatial configuration of $\vec{J}^{i}(\vec{r})$ is shown by first noting that a vector field such as $\vec{J}^{i}(\vec{r})$ that is continuous and vanishes at infinity can be represented by the sum of two vector fields, one that is irrotational (zero curl) and one that is solenoidal (zero divergence).

This is known as Helmholtz's theorem. Consider $\vec{J}^i$ as the sum of a "flow" source $\vec{J}^i_F$ and a "vortex" source $\vec{J}^i_V$ which have these respective properties. With the assumption that the medium is homogeneous and unbounded, F. Grynszpan [8] noted that as $V(\vec{r})$ is expressible in terms of the divergence of the primary current source, it depends only on the spatial distribution of $\vec{J}^i_F(\vec{r})$; and $\vec{B}(\vec{r})$ which is expressed in terms of the curl of the primary current source can depend only on the distribution of $\vec{J}^i_V(\vec{r})$. Thus the ECG and MCG measured within such a medium yield information about contributions to $\vec{J}^i$ that are electromagnetically independent [9].

It should not be inferred from this conclusion that the ECG and MCG provide distinctly different information. The reason, as emphasized by S. Rush [10], is that additional physiological constraints may cause $\vec{J}^i_F$ and $\vec{J}^i_V$ to be at least partially interdependent.

A more realistic model for the human torso must include inhomogeneities which represent the various organs as well as the boundary of the torso. Only domain-wise homogeneous models have been considered to date in the general theory, in which σ is assumed uniform within each domain that taken together represent the torso. The analysis by D. B. Giselowitz [16, 17] shows that alterations in the electric potential $V(\vec{r})$ and magnetic induction $\vec{B}(\vec{r})$ caused by discontinuities in σ at the boundaries between domains can be described in terms of a surface integral involving the value of the electric potential everywhere on the boundaries. Equivalently, these contributions can be described by a volume integral over the torso by introducing the notion of a secondary current source $\vec{K}^i(\vec{r})$ distributed along the boundaries. Since $\vec{K}^i(\vec{r})$ is expressible in terms of $V(\vec{r})$, it can depend only on $\vec{J}^i_F(\vec{r})$.

The secondary current source $\vec{K}^i$ can also be represented by the sum of a flow contribution $\vec{K}^i_F$ and a vortex contribution $\vec{K}^i_V$. Since $V(\vec{r})$ depends on a volume integral involving $\nabla \cdot [\vec{J}^i(\vec{r}) + \vec{K}^i(\vec{r})]$, the ECG on the torso depends only on the flow contribution to the primary current source $\vec{J}^i_F$. As $\vec{B}(\vec{r})$ is given by an integral involving $\nabla \times [\vec{J}^i(\vec{r}) + \vec{K}^i(\vec{r})]$, it depends on the vortex contributions to the primary and secondary current sources; but since the latter is determined by $\vec{J}^i_F(\vec{r})$ the MCG depends on <u>both</u> $\vec{J}^i_V$ and $\vec{J}^i_F$.

The non-uniqueness of the solution to the inverse problem [18] has been known for a long time and can be indicated quite simply by an argument recently given by J. P. Wikswo, Jr. [19]. The vector $\vec{J}^i$ at a particular point is specified <u>a priori</u> by three independent variables; therefore, not all of the total of six components of $\vec{J}^i_F$ and $\vec{J}^i_V$ can be independent. Indeed, the three conditions expressed by $\nabla \times \vec{J}^i_F = 0$ imply there is only one independent parameter for the flow current; that is, if the component of $\vec{J}^i_F$ along a particular direction is known at all points in space the other two components can be deduced from this condition. Also, $\nabla \cdot \vec{J}^i_V = 0$ imposes a constraint that leaves only two independent parameters for the vortex current. The total number of independent parameters for $\vec{J}^i_F$ and $\vec{J}^i_V$ is consistently the same as for $\vec{J}^i$. In seeking a solution for the inverse problem, suppose for the sake of argument that $\vec{J}^i_F(\vec{r})$ is known everywhere within the torso (this complete knowledge cannot be gained even from ECG measurements everywhere on the torso, because there is no unique solution to the electrostatic problem [18]). This gives one of the three independent parameters which describe $\vec{J}^i(\vec{r})$. But because $\nabla \cdot \vec{B} = 0$ in the source-free region outside the torso, there can be only one additional independent magnetic field parameter determined by the MCG. This must represent some combination of the two independent parameters of $\vec{J}^i_V$, when the contribution of $\vec{K}^i_V$ is subtracted. That leaves one other parameter unestablished in $\vec{J}^i$. Thus, at most only two of the three independent parameters of the primary current source can be determined from an analysis of ECG and MCG data. The inverse problem whether or not inhomogeneities are present has no unique solution. This fundamental limitation indicates that data must be compared with predictions of current models using the Biot-Savart law in order to make contact between sources and observed fields. Source modeling will play an important role in data analysis if determining the electrical structure of the source is a goal of a biomagnetic research program.

III. SQUID MEASUREMENT TECHNIQUES

The earliest biomagnetic studies were conducted with multiturn coils which by Faraday induction provide an output voltage proportional to the rate of change of the total magnetic flux linking the coil. [6, 20, 21]. With a ferrite core to enhance sensitivity, this technique permits detection of the MCG and MEG with signal averaging; and with proper shielding to reduce pickup from fluctuations in the ambient field, the sensitivity is limited by thermal noise in the induction coil [22, 23]. Devices based on the SQUID provide two advantages: much lower system noise (at least below $\sim 10^2$ Hz), and equal sensitivity to dc as well as ac fields. The frequency-independent sensitivity derives from the fact that the SQUID system responds to magnetic flux and not the rate of change of flux. The principles of rf bias and flux-locked loop operation which are used in SQUID systems to enhance sensitivity and ensure linear response are described in the literature and will not be discussed here [24, 25, 26]. The reader is also referred to a paper by L. D. Jackel and R. A. Buhrman [27] for the most recent analysis of noise contributions which limit the magnetic field sensitivity of SQUID systems.

Operated in fiberglass dewars with minimal rf shielding and superinsulation in order to minimize internal field noise from Johnson currents, commercial SQUID systems are presently available with a flat noise power spectrum from less than $\sim 10^{-1}$ Hz to $\sim 10^{4}$ Hz, and the equivalent rms field noise is ~ 10 fT Hz$^{-\frac{1}{2}}$ [28]. For most biomagnetic applications, the major experimental problem is not the detection system sensitivity, but interference from fluctuating magnetic fields in the experimental area. Typical ambient fields include the dc field of the earth which is $\sim 50\ \mu$T. In addition, geomagnetic field fluctuations amount to ~ 1 pT Hz$^{-\frac{1}{2}}$ at 10 Hz and have a frequency dependence which is approximately f^{-m}, with m ranging from 0.9 to 1.4 depending upon solar activity [29]. Local field sources in a laboratory such as elevators, mechanical pumps, and ventilation conduits contribute ac fields of 0.1 to 1 μT; and fields of comparable magnitude are found at the power line frequency and harmonics. The steel structure in a building distorts these field patterns, so that it is not uncommon to find low frequency variations in field with an amplitude as large as $\sim 1\ \mu$T and variations in field gradient of $\sim 10\ \mu$T/m. The difficulty in detecting biomagnetic fields as weak as those listed in Table 1 in the presence of this large background noise is evident. Four approaches have been taken to minimize the effect of background noise: signal averaging of repetitive signals, band-pass filtering to isolate the desired signal, using spatial discrimination afforded by the geometry of the field detector, and shielding of the subject and detector. The first two techniques are standard procedures which are not peculiar to magnetic detection and alone are not sufficient to insure success with SQUID systems. Of the last two techniques, spatial discrimination is the simplest and for many applications eliminates the need for shielding. The principle of spatial discrimination is advantageous because the background field is produced by sources that are distant from the detector and therefore the background field is relatively more uniform in space than the biomagnetic field produced by a source close to the detector. Thus if the detector is constructed with a geometry which is insensitive to uniform field, but responds to the gradient or higher spatial derivative of the field, its sensitivity to nearby sources is enhanced relative to distant sources. This advantage has been appreciated for more than a decade and was exploited when induction coils were first used in biomagnetic research [6, 7].

For SQUID applications either the SQUID itself can be designed with appropriate geometry to detect the signal directly [30], or if instead a superconducting flux transporter couples the magnetic signal to the SQUID, the detection coil in the primary of the transporter can be given the appropriate configuration [31]. The flux transporter is a closed superconducting loop of wire or ribbon with a primary coil connected in series with a secondary coil (also called the "signal coil") which is inductively coupled to the SQUID. Owing to magnetic flux conservation in the loop, when a field is applied in the region of the primary a current flows around the loop with appropriate magnitude and direction so as to maintain the total flux within the loop invariant. As the current flows through the signal coil it couples flux into the SQUID, and is sensed by the electronic systems which monitor the state of the SQUID. Thus the electronic output signal is proportional to the signal coil current, which in turn is proportional to the original net flux applied to the primary. One advantage of the flux transporter lies in the versatility of the primary, which can be wound in different configurations to achieve various degrees of spatial discrimination. If the primary is to function as a detection coil with a single loop, as illustrated in Fig. 1A, it provides no spatial discrimination because it is equally sensitive to uniform and nonuniform fields. The system is called a "SQUID magnetometer", and the output signal is proportional to B_z averaged over the area of the loop.

The configuration consisting of two identical coils wound in opposition as in Fig. 1B is insensitive to uniform field. It is called a "gradiometer", because the lowest order spatial variation in field that couples a non-zero net flux to the coils is the gradient dB_z/dz. Its sensitivity to a field gradient is proportional to the distance between coils, or "baseline", and thus a shorter baseline reduces the signal produced by changes in the local field

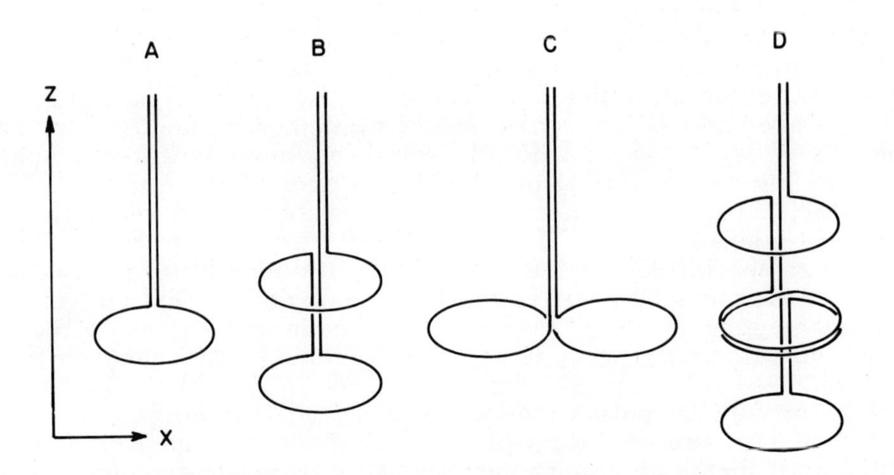

Fig. 1 Detection coil configurations include A: magnetometer, B: gradiometer, C: off-diagonal gradiometer, and D: second derivative gradiometer. The plane of each loop is perpendicular to the z-axis.

gradient. But sensitivity to a biomagnetic field is not reduced if the baseline remains sufficiently long that the field couples strongly only to one coil, usually the lowest coil, called the "pickup" coil. In this case the detector functions as a difference magnetometer, with the signal from the pickup coil reduced only by the amount coupled to the second coil. The configuration in Fig. 1C is called an "off-diagonal gradiometer" because it detects the off-diagonal gradient dB_z/dx of a nearly uniform field. The application of gradiometer techniques to SQUID systems was first reported by J. E. Zimmerman [30, 31], and an alternative approach which takes the difference signal between two magnetometers was described by A. Rosen, G. T. Inouye, A. L. Morse, and D. L. Judge [32].

For further reduction of background noise a second-derivative gradiometer illustrated in Fig. 1D has proved useful because it is insensitive to uniform fields and uniform field gradients. It also reduces vibration-induced noise, since it is less sensitive to flux changes when tipped in a uniform gradient or moved laterally in a non-uniform gradient [33]. It can be considered as two gradiometers in opposition; and the lowest order spatial variation to which it responds is d^2B_z/dz^2. As for the gradiometer, the second derivative gradiometer will respond to a biomagnetic field if the baseline between the end pickup coil and middle coil is sufficiently long that the source has negligible coupling to the middle coil. More quantitative discussions of the spatial discrimination or equivalent sensitivity patterns for gradiometers have appeared in the literature [7, 34, 35] and the extension of this analysis to second derivative gradiometers is straightforward [33].

Although a major advantage of reduction in background noise can be derived from use of a detection coil in gradiometer or second derivative gradiometer configuration, a penalty of decreased sensitivity is also exacted if the intrinsic noise of the SQUID system limits sensitivity. This comes from the fact that the flux Φ_S coupled into the SQUID by the signal coil decreases with increasing total inductance of the flux transporter for a given flux Φ_p produced by the biomagnetic field in the pickup coil. The reason is that the magnetic field energy sensed by the pickup coil is shared by all of the components of the flux transporter. Simple flux division arguments [30] give the relation

$$\Phi_S = \left(\frac{M_S}{L_D + L_S}\right)\Phi_P , \tag{1}$$

where M_S is the mutual inductance between the signal coil and SQUID, and the inductance of the flux transporter $L_D + L_S$ consists of contributions from the detection coil L_D and signal coil L_S. In practice, the dependence of Eq. (1) on L_D is seen by noting that

$M_S = k(L_S L_{SQ})^{\frac{1}{2}}$, where k is the coupling coefficient of mutual inductance, L_{SQ} is the SQUID inductance, and $L_S = L_D$ for optimal transfer of flux from the detection coil to SQUID. The last condition is only approximately valid if an optimal signal to noise ratio is desired [30]. Then Eq. (1) can be rewritten as

$$\Phi_S = \frac{k}{2}\left(\frac{L_{SQ}}{L_D}\right)^{\frac{1}{2}} \Phi_P \ . \tag{2}$$

Therefore the signal coupled into the SQUID varies as $L_D^{-\frac{1}{2}}$. For a given spatial resolution as determined by the dimensions of the pickup coil, L_D is larger for multi-coil configurations as compared with a magnetometer detection coil. Thus the sensitivity of a gradiometer is lower by a factor of $\sqrt{2}$ than the sensitivity of a magnetometer; and the sensitivity of a second derivative gradiometer is lower by a factor of $\sqrt{6}$ for the configuration shown in Fig. 1C or a factor of $\sqrt{4}$ if the center turns are separated so they have negligible mutual inductance.

Recently J. E. Zimmerman [36] has introduced an asymmetri configuration for gradiometers and second-derivative gradiometers that virtually eliminates the loss of sensitivity from flux division and promises to be an integral part of future advances in instrument ation. The SQUID itself takes the place of the pickup coil, and for a second derivative gradiometer the additional two coils mounted some distance above it are of low inductance and are coupled inductively by a third coil to the SQUID. Additional sensitivity is achieved as well through use of a fractional-turn SQUID [30].

Even the most careful construction does not insure perfect balance in a gradiometer, and to compensate for residual field imbalance it is advantageous to provide an adjustment mechanism. This can be a set of small superconducting tabs placed near one coil of the gradiometer [25], or a set of separate small gradiometers oriented orthogonally, with adjustable superconducting shields, connected in series with the detection coil in the flux transporter [33, 37]. Experience indicates that a field balance of ~10 ppm or better is desirable [38]. Figure 2 gives an example of a SQUID gradiometer which operates at the Helsinki University of Technology [39]. The baseline is 10 cm and coil diameter 3.9 cm. When operating in a wooden cabin in a suburban environment the typical noise during daytime averages to 250-300 fT $Hz^{-\frac{1}{2}}$ for the bandwidth 0.05-50 Hz.

The circuit diagram for a SQUID second-derivative gradiometer used at New York University is shown in Fig. 3. The flux transporter is wound from niobium wire on a phenolic form, [37, 38] and the trim coils provide a field balance of ~ 3 ppm. The cryogenic portions of the circuit are mounted within the dewar shown at the top of Fig. 4, and the system operates in a normal laboratory

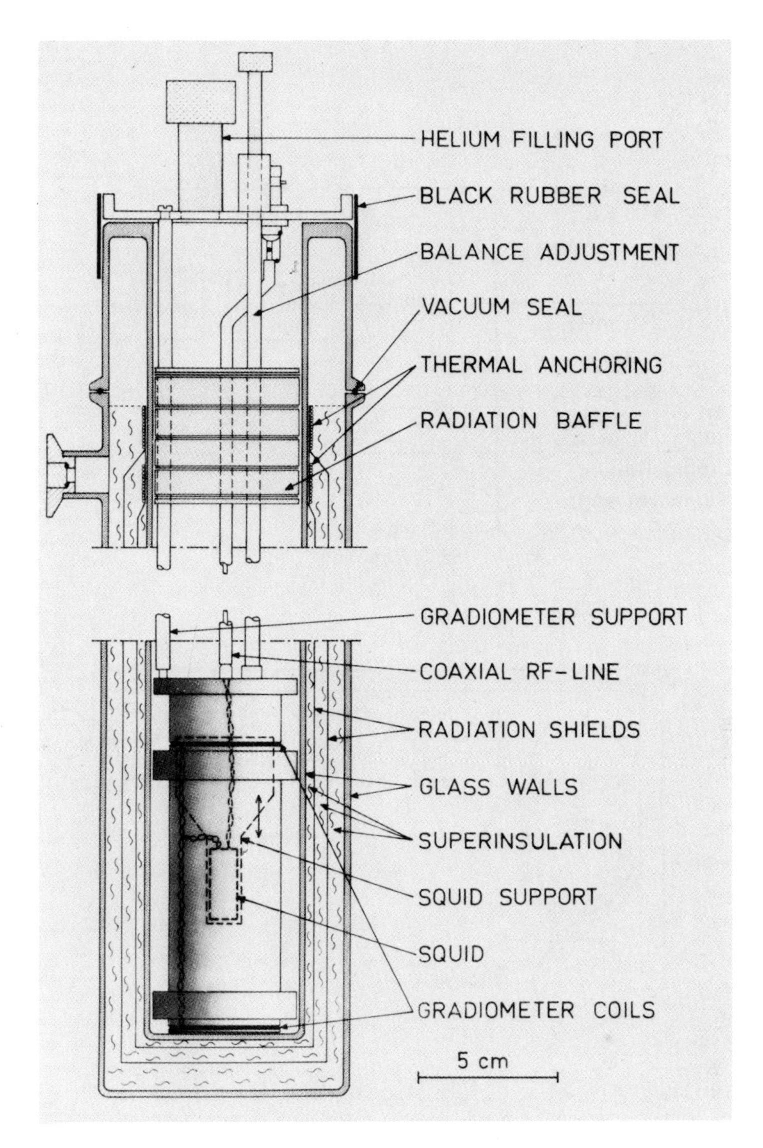

Fig. 2 SQUID with a gradiometer detection coil mounted in a cryostat containing liquid helium. For convenience the SQUID is mounted in the center of the detection coil. The balance adjustment rod moves a superconducting object close to one coil to render the gradiometer insensitive to uniform field. From M. Saarinen, P. J. Karp, T. E. Katila, and P. Siltanen, Cardiovascular Res. 8, 820 (1974).

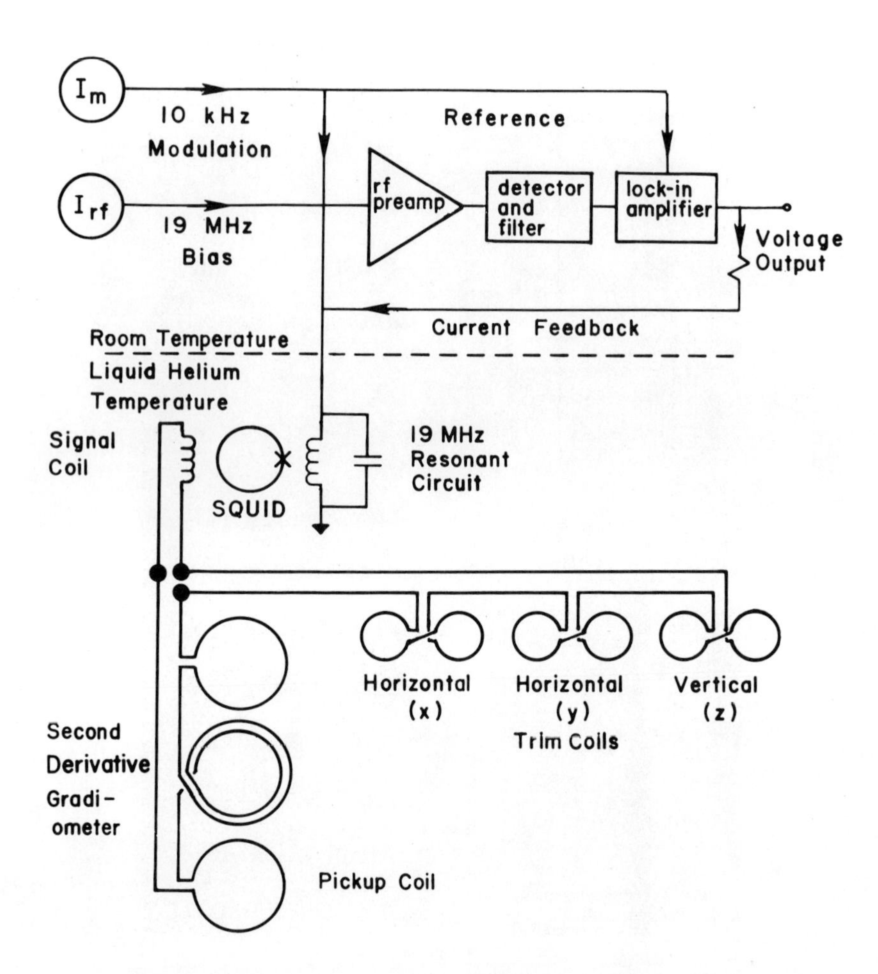

Fig. 3 Diagram of a conventional SQUID electronics system with a superconducting flux transporter consisting of a series connection of a detection coil wound as a second derivative gradiometer, orthogonal trim coils, and signal coil coupling to the SQUID. Superconducting sleeves (not shown) are used to adjust the net area exposed in each pair of trim coils to achieve field balance in the primary of the flux transporter.

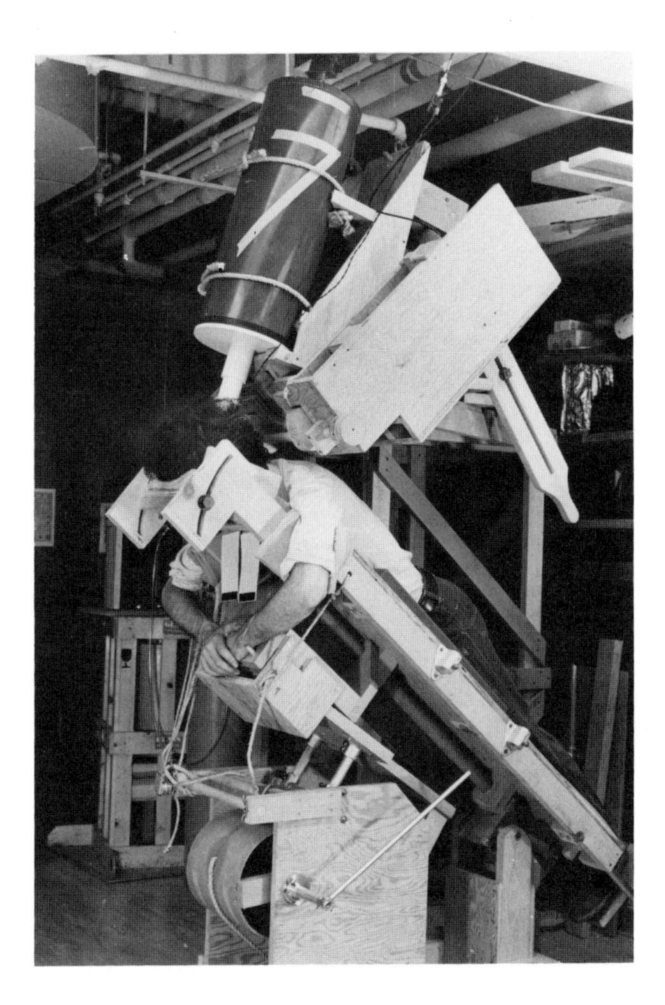

Fig. 4 Arrangement at New York University for recording magnetic fields produced by the brain without recourse to magnetic shielding. The cryostat with SQUID and second derivative gradiometer is positioned with the pickup coil near the scalp and can be rotated between the vertical and 45° to the vertical. The inclined subject support permits versatile adjustment of position while the subject views an oscilloscope screen (not shown).

setting as shown, with no magnetic shielding. The noise spectrum during daytime in this urban environment with the subject absent is illustrated in Fig. 5 for a second-derivative gradiometer having a baseline of 1.9 cm and diameter of 2.2 cm [40]. At high frequency above 400 Hz the level represents the SQUID system intrinsic noise of 340 fT $Hz^{-\frac{1}{2}}$ referred to the pickup coil or $3.8 \times 10^{-4}\ \Phi_o\ Hz^{-\frac{1}{2}}$ flux noise at the SQUID, where Φ_o is the flux quantum. Below 10 Hz the dramatic increase appears associated with geomagnetic fluctuations because it has the characteristic features of decreasing by a factor of ~2 at night and exhibits a broad peak at about 8 Hz. However, the nearly f^{-2} dependence is stronger than expected for this source. The sharp increase is not observed outdoors and evidently represents the effect of gradients introduced by field perturbations from the steel structure of the building.

Shielding is the fourth technique for reducing background noise, and careful design and construction of large shielded enclosures has provided the lowest noise levels achieved to date at low frequencies. A particularly simple method to reduce the ac component of the field is use of an electrically conducting enclosure with sufficiently thick walls that eddy currents induced in the walls by the changing field shield the interior. An effective eddy current shielded enclosure has been designed by J. E. Zimmerman [36] at the National Bureau of Standards in Boulder, Colorado. Fabricated from seam-welded aluminum plate ~ 4 cm thick, this N.B.S. shielded enclosure has inner dimensions of 1.8 x 1.8 x 2.4 m and provides an attenuation of 30 dB at 10 Hz. Some additional magnetic shielding is contributed by an outer enclosure of 0.3 cm thick soft iron that reduces the dc field to a few hundred nanotesla and provides unknown ac attenuation Figure 6 shows the noise level which is achieved in different bandwidths for a fractional turn SQUID having a gradiometer configuration [30]. The effective baseline is 3.5 cm and diameter 1.3 cm. Aside from excess noise in the 2 to 5 Hz record and a periodic artifact of unknown origin in the ½ to 1 Hz record the noise level of 30 fT $Hz^{-\frac{1}{2}}$ is due solely to the SQUID system. It is remarkable that eddy current shielding performs so well at frequencies as low as ½ Hz.

Magnetic shielding using materials of high permeability has been the traditional method to reduce dc and ac fields, and a number of groups have enclosures of this type. For optimal performance the enclosures should be degaussed, magnetically "shaken", and perhaps supplemented by field-sensing feedback circuits which set up fields to oppose the background field. Considerable care must also be devoted to reduce mechanical vibrations in the shielding. Using all these techniques, the best performance to date has been achieved in a shielded room designed by D. Cohen at the Francis Bitter National Magnet Laboratory at Massachusetts Institute of Technology [41, 42]. Figure 7 illustrates the 26-side room which

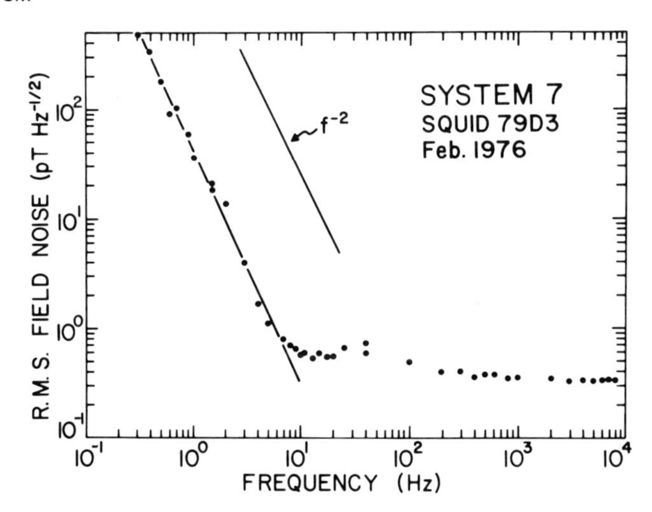

Fig. 5 Noise spectrum of a SQUID with second derivative gradiometer produced by ambient field gradients and intrinsic SQUID noise without magnetic shielding at New York University. The rms noise is expressed as the equivalent field noise at the pickup coil. From D. Brenner, L. Kaufman, and S. J. Williamson, IEEE Trans. Magn., in press.

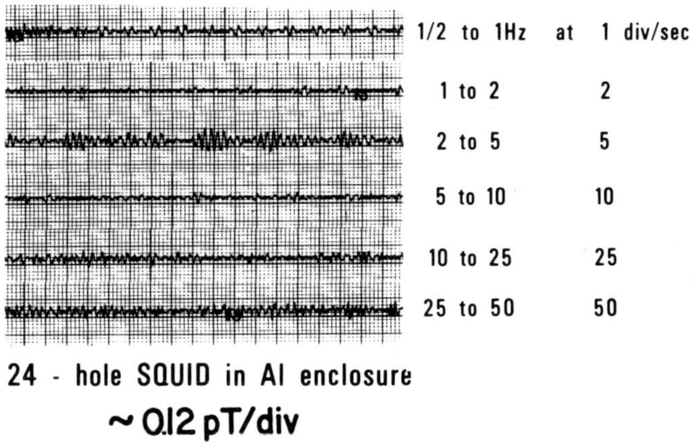

1/2 to 1Hz at 1 div/sec
1 to 2 2
2 to 5 5
5 to 10 10
10 to 25 25
25 to 50 50

24 - hole SQUID in Al enclosure
~ 0.12 pT/div

Fig. 6 Noise level of a fractional turn SQUID gradiometer operated in the National Bureau of Standards eddy current shielded enclosure. Each trace was recorded on a strip chart recorder with the chart speed indicated at the right. The field calibration is expressed as the equivalent field at the effective pickup coil of the device. From J. E. Zimmerman, J. Appl. Phys., in press.

Fig. 7 Magnetically shielded room at the Francis Bitter National Magnet Laboratory at Massachusetts Institute of Technology with doors not yet installed. Electrical connections at the corners join current loops used for demagnetizing, cancellation of the earth's field and magnetic shaking. Subsequently, a pneumatic suspension system was installed to minimize vibration. The interior is sufficiently large to accommodate four or five standing people. From D. Cohen, E. A. Edelsack, and J. E. Zimmerman, Appl. Phys. Letters 16, 278 (1970).

has the shape of a rhombicubocatahedron with outside diameter of 3.8 m and inside diameter of 2.8 m. Three layers of high permeability material provide magnetic shielding; and two layers of aluminum, eddy current shielding. The dc field in the interior of the M.I.T. shielded room can be made as low as 5 nT, with typical gradients of 1 nT/m. The noise detected by a SQUID magnetometer of 2.8 cm coil diameter in the bandwidth 1 Hz - 5 kHz is 8 fT $Hz^{-\frac{1}{2}}$ and represents not the noise within the room, but intrinsic SQUID system noise. It is equivalent to 8 x 10^{-5} Φ_0 $Hz^{-\frac{1}{2}}$ flux noise in the SQUID. Background noise which appears below 1 Hz can also be reduced by operating two magnetometers in a gradiometer configuration. For practical purposes the intrinsic SQUID system noise of 11 fT $Hz^{-\frac{1}{2}}$ then determines the observed noise level all the way to dc. This represents the state-of-the-art in achieving a low noise background within a large working space.

To summarize this section we note that the principal advantage in eddy current or magnetic shields is for low frequency applications. At frequencies above ~ 10 Hz, even in magnetically noisy environments, second-derivative gradiometers without shielding compete favorably. Commercial SQUID second-derivative gradiometers are now available with a noise level of 40 fT $Hz^{-\frac{1}{2}}$ down to 1 Hz when operated outdoors and 10 Hz when in a laboratory [28]. The asymmetric second-derivative gradiometer introduced by J. E. Zimmerman [36] should see this noise reduced still further. In preliminary tests, D. Brenner, L. Kaufman, and S. J. Williamson have found that the excess noise at low frequencies can be reduced by at least a factor of 2 by a buckout technique whereby the ambient field was sensed by a fluxgate magnetometer and the output was appropriately scaled and subtracted from the SQUID electronics output. With these ongoing efforts there is every reason to believe that a wide variety of biomagnetic effects will become accessible to study in normal laboratory settings, without need for shielded enclosures.

IV. MAGNETOCARDIOGRAM

Early efforts to detect biomagnetic fields were directed toward those associated with heart activity, since the electrocardiogram (ECG) was known as one of the strongest bioelectric signals, and by analogy the corresponding magnetocardiogram (MCG) measured near the chest was expected to be comparatively strong. The origin of the ECG is the electrical potential difference established across the heart muscle, or myocardium. When a muscle fiber is stimulated by an action potential, the transmembrane voltage of each cell rapidly changes in a time span of ~ 2 ms from -80 mV (interior relative to exterior potential) to ~20 mV. This is associated with the movement of sodium and potassium ions through the membrane in response to biological pumps, whose action is incompletely understood. Subsequently this polarization charge flows back through the membrane, and during this depolarization the cell contracts, reaching a maximum tension in about 150 ms, and returning to its resting length in an additional 50 ms. The sequential action of different regions of the myocardium as it pumps blood to the lung and body gives rise to a polarization charge distribution which moves in step with the cardiac cycle.

A typical ECG waveform and the standard nomenclature are shown in Fig. 8. The P wave is observed during depolarization of the atria, which begin to contract during the latter portion of the wave and force blood into the ventricles. The QRS complex is associated with the subsequent ventricular depolarization and contraction, which increases blood pressure during the S wave sufficiently to close the atrio-ventricular valves and force blood into the aorta and pulmonary arteries. Ventricular contraction continues during the ST segment, until the muscles commence to relax, and repolarization occurs during the T wave. Toward the end of the T wave ventricular blood pressure has decreased to the point where the aortic and pulmonic valves close and the atrial valves open to re-admit additional blood, and the cycle begins again. Thousands of articles have been written on the electrophysiology of the heart and the ECG, and references to a number of review articles and books are given in a review article by R. McFee and G. M. Baule [43].

Interest in studying the magnetic analog of the ECG is stimulated by the potential clinical applications and the possibility that the MCG will provide new information about electrophysiological features of the heart. Clinical applications of the ECG are based on an empirical approach whereby an association is established in a large number of cases between abnormal features of the waveform and pathologic condition of the heart. The utility of the ECG does not rely on theoretical knowledge of electrical characteristics of the heart, although such models have been the subject of considerable research since W. Einthoven's in 1913 [44].

The first observation of the magnetocardiogram was reported by G. Baule and R. McFee [6] in 1963, nearly 90 years after the first human electrocardiogram was recorded [45]. This achievement marks the birth of experimental biomagnetism. Detection of the human cardiac signal was based on the principle of Faraday induction whereby the time-varying magnetic field from the heart induced a voltage in a coil of 2×10^6 turns with a shaped ferrite core positioned over the chest of the subject. To reduce noise the induction coil had a gradiometer configuration as illustrated in Fig. 1C, the measurements were carried out in an open field away from laboratory disturbances, and the detected voltage

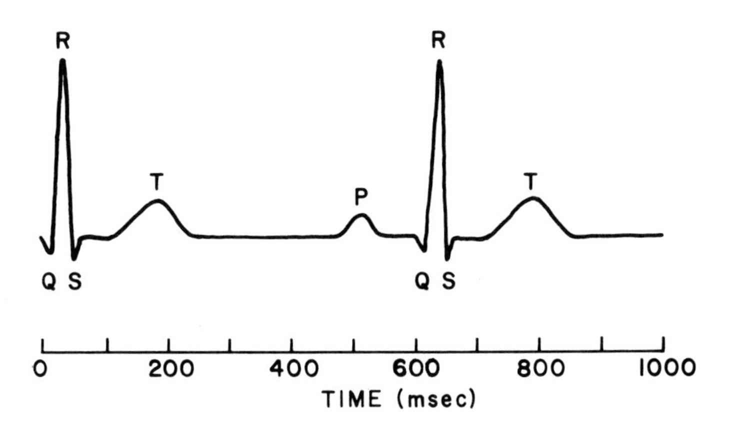

Fig. 8. Electrocardiogram with characteristic features indicated by standard notation. After A. C. Burton, Physiology and Biophysics of the Circulation, 2nd Ed. (Year Book Medical Publishers, Chicago, 1972).

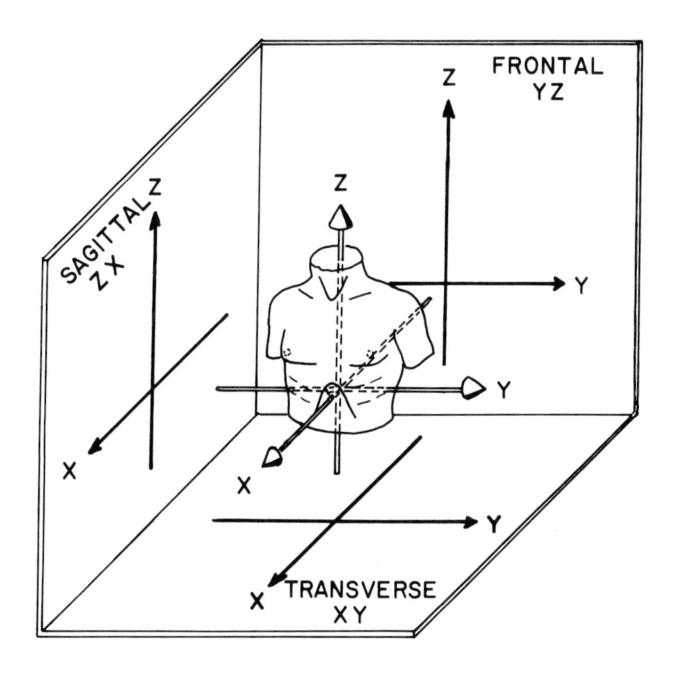

Fig. 9 An orthogonal coordinate system oriented relative to the torso. The x-axis passes through the xiphoid (bottom edge of the sternum).

was signal-averaged by using the ECG as a reference trigger. A voltage of ~ 0.3 μV corresponding to a peak field of ~ 50 pT for the QRS complex was observed. The noise was sufficiently high that the P wave could not be resolved. It has been brought to the present authors' attention by D. Cohen that in the same year R. A. Stratbucker, C. M. Hyde, and S. E. Wixson [46] reported observing the time-varying field from an isolated heart of a guinea pig, but in this case the finding has not yet been confirmed by other investigators.

The Baule and McFee results on human subjects were reproduced by Y. D. Safonov, V. M. Provotorov, V. M. Lube, and L. I. Yakinenkov [20], who also attempted to reduce background noise by use of a magnetically shielded enclosure. Improved magnetic shielding techniques and use of a smaller induction coil with a ferrite rod core enabled D. Cohen [21] to detect the MCG with improved spatial resolution and without averaging during periods of low background noise. The form of the signal near the heart was similar to the ECG, but noise often obscured the P and T waves. The first determination of the three components of the magnetic field vector through the QRS complex for various locations around the torso was reported by D. Cohen and L. Chandler [47], also using an induction coil. These data provided the first extensive test for electrophysiologic heart models which had been proposed by G. M. Baule and R. McFee to explain the MCG.

The introduction of greatly improved magnetic shielding and a SQUID magnetometer by D. Cohen, E. A. Edelsack, and J. E. Zimmerman [42] in 1970 made it possible to obtain the MCG without signal averaging and with a quality which was comparable to that of the ECG. This success encouraged three other groups to develop SQUID gradiometer systems which allow MCG measurements without the need for magnetic shielding in magnetically quiet environments [31, 32, 39, 48].

Although the sensitivity of SQUID systems is amply demonstrated by the high quality MCG's that have been reported, records of nearly comparable quality, but lower spatial resolution, can also be obtained from induction coils with proper magnetic shields and cores, accompanied by signal averaging of the induced voltage for ~ 1 min. This has been achieved in a hospital setting by D. Matelin [49], who used a gradiometer with 1.5×10^4 turns in each coil to map across the chest the MEG components parallel to the torso. Subsequent studies on a large number of patients by B. Denis and D. Matelin [50] have demonstrated the practicality of this simple measurement technique.

Figure 9 illustrates a conventional coordinate system which is oriented relative to the torso. The yz plane is also known as the frontal plane; zx plane, the sagittal plane; and xy plane, the transverse plane. The earliest MCG systematic field mapping that included quality data for P and T waves was reported by D. Cohen and D. McCaughan [51] from measurements in the MIT shielded room with a SQUID magnetometer. An example of records taken at locations across the chest on a 5 x 5 cm grid is given in Fig. 10. The same general features as exhibited by the ECG are present, but the polarity of P, QRS, and T features depend on location, as do their relative strength. Studies show that there is also considerable inter-subject variability in magnitude with the QRS complex ranging from 20 pT to 300 pT between one subject and another. Some tentative correlations between magnitude and subject characteristics have been reported by A. Rosen and G. T. Inouye [52]. Records such as those in Fig. 10 as they stand may have clinical value, and a number of studies have reported data in this general form for both normal and abnormal patients [39, 51, 53, 55, 56]. Vector representations of the observed field with two or more components included have also been reported in map form [47, 49, 52].

There is interest in finding more condensed modes for presenting data. This is motivated by desire to decrease the effort in taking data, to establish a format that conveys a more intuitive appreciation for the actual heart action, and to make closer contact with electrophysiological models for the heart. One operationally simple approach is to assume that the observed fields originate from a magnetic dipole which is located at the center of the heart and whose magnitude and direction vary through the cardiac cycle. This dipole is called the "magnetic heart vector", and several methods have been described to determine its parameters from field data [19, 57]. It is the analog of the "electric heart vector" which represents the three voltages in orthogonal directions obtained from the Frank ECG electrode lead system [58]. The magnetic heart vector does not necessarily correspond to the magnetic dipole model for the heart unless it can be established that the observed field is due solely to primary current sources ($\vec{J}^i$) in the myocardium and not to the responding volume currents in the surrounding medium. The Frank ECG lead system was empirically devised on the basis of an electric dipole model for the heart, and provides an electric heart vector which is parallel to this dipole by compensating to first order for the torso boundary which perturbs the volume currents.

Examples of the magnetic heart vector and electric heart vector deduced from the MCG and ECG taken on the same subject is given in Fig. 11 [59, 60]. The MCG data were obtained from a "unipositional lead", which will be discussed later [35]. The upper traces show the computer calculated variation through the cardiac cycle of vector components in the x, y, and z directions. Aside from polarity differences the waveforms are similar for this normal subject. The lower diagrams

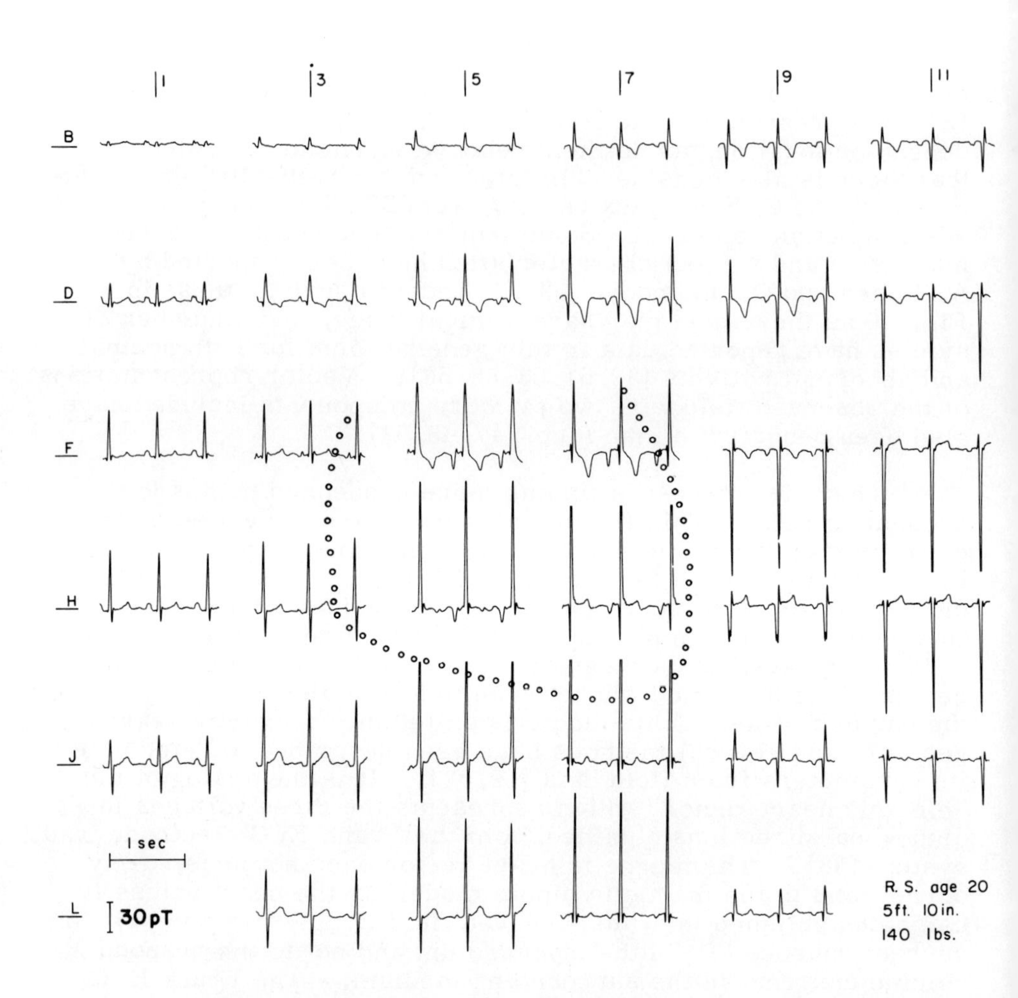

Fig. 10 MCG of a normal male for the component of field normal to the chest at positions on a 5 cm x 5 cm grid, with position H5 at the xiphoid. The outline of the heart was taken from a roentgenogram of the subject. The bandwidth is 0.03-40 Hz. From D. Cohen and D. McCaughan, Am. J. Cardiology 29, 678 (1972).

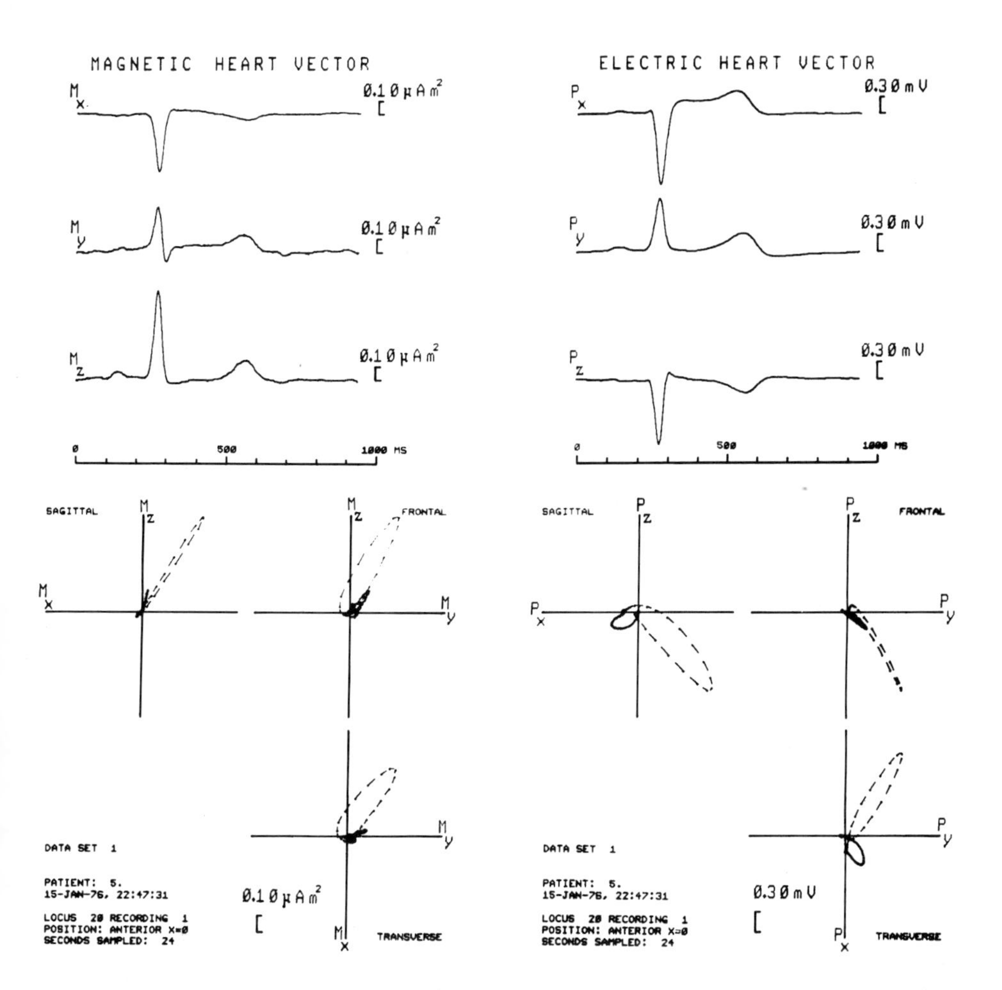

Fig. 11 Examples of computer processed data of the MCG and MEG from a normal subject provided by J. P. Wikswo, Jr., of Stanford University. Components of the magnetic heart vector (equivalent magnetic dipole M) and electric heart vector (equivalent electric dipole P) vary with time as shown at the top. The dashed curves at the bottom are the trajectories of the head of the respective vectors as projected on sagittal, frontal, and transverse planes. The original field component measurements obtained by a SQUID system in a shielded enclosure had a bandwidth of 0.2-100 Hz with a 60 Hz notch and were signal averaged for 30 cardiac cycles to reduce noise.

illustrate how the vectors change magnitude and rotate in space by showing the locus traced through a cardiac cycle by the point of the heart vectors as projected on the yz, zx, and xy-planes. These diagrams are known as the vector magnetocardiogram and vector electrocardiogram. The approximate 90° angle between magnetic and electric vectors during the early portion of the QRS complex has been noted by B. Denis and D. Matelin [50] for a number of normal subjects. From detailed analysis on one subject, J. P. Wikswo, Jr. and W. M. Fairbank [61] have found the angle to vary from 40° to 160° during ventricular depolarization.

The analysis of normal and abnormal MCG's and comparisons with ECG records is now being carried out by several groups, and it is too early to reach firm conclusions. The preliminary results show a close similarity in MCG and ECG records for normal subjects. On the other hand, abnormalities appear differently emphasized in the MCG and ECG. Examples of abnormalities have been reported by a number of groups [19, 39, 50, 53, 55], and there is some evidence that the preferential sensitivity of the MCG to near-lying sources plays an important role [39]. Whether the MCG will be clinically useful is a question whose answer must await the completion of these studies of abnormal patients.

One very interesting application of the advantage afforded by SQUID systems for dc measurements has been reported by D. Cohen and L. A. Kaufman [62]. Following an earlier pilot study by D. Cohen, J. C. Norman, F. Molokhia, and W. Hood, Jr. [63], they sought to establish whether a dc current may be produced by the heart when it suffers injury. For example a shift of the ST segment of the ECG toward the R peak of the QRS complex is a clinical indication of an abnormality such as infarction or ischemia; and the shift was commonly attributed to a current of injury that appears only during the ST portion of the cardiac cycle. Experiments on dogs indicated that quite the opposite is true during an occlusion of the coronary artery and that in fact a dc current of injury is set up which is arrested during the ST portion of the cycle. The occlusion was produced by inflating a tube that had been installed previously around the artery. An example of the MCG results illustrated in Fig. 12 shows the SQUID magnetometer output as a sedated dog is wheeled up to the detection coil, held in place for about two cardiac cycles, and wheeled away. During conditions of occlusion the MCG reference line shifts upwards, indicating the presence of a dc current of injury. There is also a downward shift of the ST segment which nearly exactly compensates for the dc effect and thus indicates an interruption of the dc injury current. The ST shift is thus a secondary effect, and the primary effect is actually the dc current of injury, which is not detectable by conventional ECG measurements.

Clinical applications need not rely on having an accurate

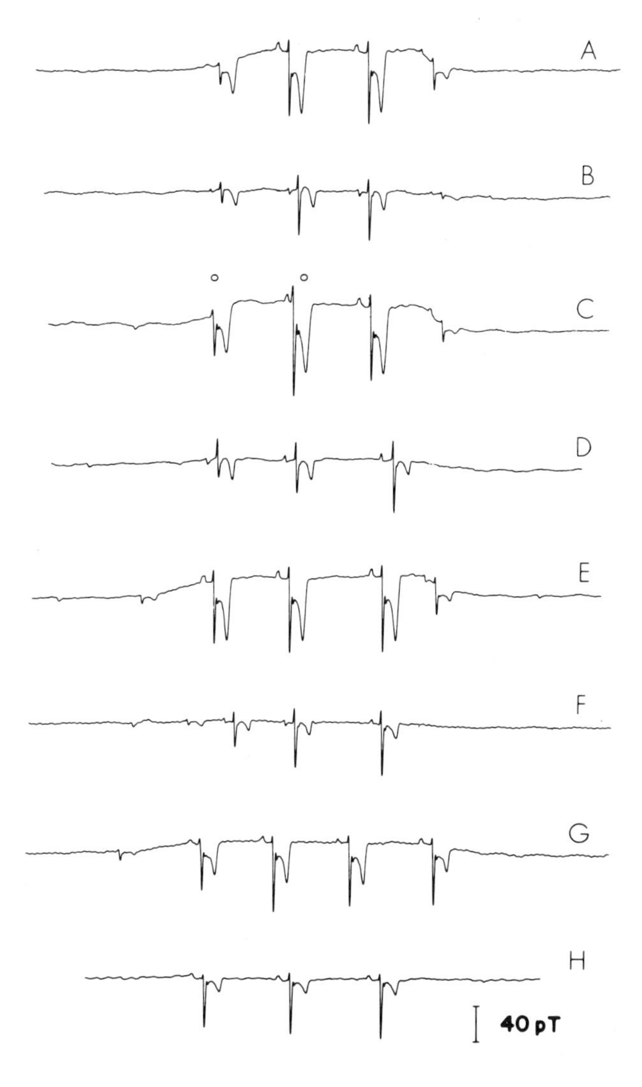

Fig. 12 MCG records for the bandwidth 0-40 Hz during a sequence of artery occlusions and releases in a dog. A: During the first 2-minute trial occlusion. B: After release of the second 2-minute trial occlusion. C: During a 5-minute trial occlusion. D: Just after release of the 5-minute occlusion. E: 15 minutes into a 28-minute occlusion. F: 30 seconds after release of the 28-minute occlusion. G: During the second short occlusion following release of the 28-minute occlusion. H: Immediately after release of this second occlusion. From D. Cohen and L. A. Kaufman, Circulation Res. 36, 414-424 (1975).

electrical model for the heart. The magnetic dipole model for representing the source of observed fields is a traditional first approximation, but in application to the heart is not correct in detail. MCG results over the past several years demonstrate that the field during the QRS complex cannot be attributed entirely to a magnetic dipole because the field amplitude does not vary as the inverse cube of the distance to the heart [19, 39]; the field pattern has improper symmetry [39] and the maps over the chest of the field magnitude do not yield correct numerical agreement with calculations [19, 52]. Generalizing the model to permit the position of the dipole to move during the cardiac cycle also does not yield agreement [19]. Formally, the field distribution outside the torso can be described exactly as the net effect of an infinite sum of magnetic multipole sources at the heart [64]. The experimental task is then one of determining at each instant the parameters associated with each multipole - 3 for the dipole, 5 for the quadrupole, 7 for the octapole, and so on. The advantage of such a description is questionable, especially as the higher multipole terms have no obvious correspondence with the physiology of the heart. Some effort has been invested to test the applicability of magnetic dipole arrays, without remarkable success.

The first physiologically inspired model which was introduced as a source to explain the MCG was the electric dipole, because it had enjoyed considerable success as a first order approximation for describing the ECG. The electric dipole is equivalent to a battery with specified voltage, and it varies in magnitude and orientation through the cardiac cycle. G. M. Baule and R. McFee [6] suggested that the electrical current associated with the dipolar electric field pattern in the surrounding medium (i.e., the volume current) establishes a magnetic field pattern which to a first approximation agrees with features observed during the QRS complex. D. Cohen and L. Chandler [47] subsequently provided additional support for the electric dipole model from MCG mapping data for the magnetic field vector at various locations about the torso, but discrepancies also were observed. An inadequacy for such a simple model should not be surprising, since a number of ECG studies have revealed contributions from strong higher electric multipole moments of the heart [43, 65, 66].

More sophisticated models are based on the current dipole as the elemental unit of field source. It can be viewed as a battery with terminals close together, impressing by biological action a specified current directly from the negative to positive terminal. When surrounded by a homogeneous, conducting medium, an equal backflow current (or volume current) flows in a dipole pattern from positive to negative terminal in response to the current dipole. The magnetic field is thus contributed by the current dipole and the volume current, whereas in the electric dipole model only the volume current contributes. A distribution of current dipoles within a volume is described by the primary current source density $\vec{J}^i(\vec{r})$. A distribution over a surface of current dipoles aligned normal to the surface can thus represent a double layer of charge, so far as representing sources of volume current is concerned. Because of the close correspondence between the current dipole and currents which are biologically pumped across membranes, the current dipole is the basic element in physical models of biological current generators [67]. The application of distributed current dipole models to MCG problems that also include the effect of a realistic torso can be found in the work of B. M. Horacek [68] and B. N. Cuffin [69].

Two routes toward a solution of the inverse problem for more complicated heart models have been explored recently. One is to impose a number of constraints on a model source, which thereby establish a set of parameters to be evaluated from data. An example of this approach for analyzing MCG and ECG data in terms of an array of current dipoles has been given by D. Geselowitz and W. T. Miller [70, 71, 72]. For example, they found that the multipole moments for a model 20 dipole array in an homogeneous conducting sphere (model torso) could be determined with greater accuracy in the presence of noise if electric data are supplemented with magnetic data. The other route is to cast the data into a format that permits a researcher by visual inspection to deduce the source configuration. In this spirit H. Hosaka and D. Cohen [73] have introduced "arrow maps" which display information about the field patterns of the MCG at any chosen portion of the cardiac cycle. The arrows are positioned on a grid, and their length and direction indicate the local gradient in the frontal plane of the field component normal to that plane. Such an off-diagonal gradient can be measured directly with a detection coil having the configuration shown in Fig. 1C. The patterns can be interpreted in terms of a distribution of current dipole sources, and the analysis of such maps for several subjects by H. Hosaka and D. Cohen [73] provide the first determination of distributed generators of the human heart obtained from MCG data.

In parallel with the numerous studies cited above, considerable effort has been devoted toward finding the most advantageous set of field or gradient measurements which yield maximum sensitivity to primary source currents ($\vec{J}^i$) and minimum sensitivity to volume currents (or equivalent secondary currents $\vec{K}^i$), which are affected by neighboring internal organs and the shape of the torso boundary. This study has become known as "lead field theory" by analogy with similar efforts in electrophysiology [65, 67]. The theory has developed on the basis of a reciprocity theorem first applied to biophysics in 1853 by H. Helmholtz [73]. The theorem states that each element of a current source [$\vec{J}^i(\vec{r})$ or $\vec{K}^i(\vec{r})$] contributes to the total flow of current through the leads of a galvanometer as would flow through the element itself (at $\vec{r}$) were the element's equivalent electromotive force applied instead across the galvanometer. The extension of this theorem to ac effects such as the MCG has been made by R. Plonsey [9]. In essence, lead field theory permits the spatial sensitivity of any detection coil to be deduced from the magnetic field pattern which would be produced by that coil if a current were to pass through it. The first application of lead field theory to the MCG was discussed by G. M. Baule and R. McFee [7, 34], who noted that the x, y, or z-component alone of the heart's dipole moment $\vec{m} = \frac{1}{2}\int \vec{r} \times \vec{J}^i dv$ is detected when the coil produces a uniform field in that direction at the heart. The theory has been applied to a number of model situations, including a current dipole in a spheroid [64]. J. Malmivuo [35] has extended the analysis and, assisted by empirical studies, has found a particularly convenient measurement protocol called the "unipositional lead system" whereby the x, y, and z-components of the field directly over the heart are measured, with minimal contributions from the combined effect of internal inhomogeneities and the torso boundary. An example of heart parameters deduced from measurements with the unipositional lead system has been given in Fig. 11. A number of estimates indicate that the effect of the torso boundary on external field patterns could be large [7, 64], but fortunately the low electrical conductivity of the lung tends to confine volume currents near the heart and reduce the influence of internal inhomogeneities [35, 75]. H. Hosaka and D. Cohen [73, 76] and H. Hosaka, D. Cohen, B. N. Cuffin and B. M. Horacek [77] have estimated the effect of the torso boundary on the normal field component for five different torso models. They find that the contribution of the volume current is only ~ 30% of the contribution from a current dipole representing the heart and it is on the basis of this relative insensitivity that the arrow maps were used to determine the source configuration.

In summary, theoretical efforts are continuing in order to isolate the contributions of primary sources to the MCG from those of volume currents. A number of other technical questions are being considered, such as what is the optimal set of input data in the presence of noise for a given source model. And analysis of the relative sensitivity of various detection coil configurations is being extended through use of the lead field theory. Together with these theoretical advances, the ongoing studies of the MCG from large numbers of normal and abnormal patients continue, so that evaluation of the clinical utility of magnetic measurements should be at hand in a matter of several years.

V. FETAL MAGNETOCARDIOGRAM

Measurements of the fetal heart rate after about 18 weeks of gestation is a routine method of determining the condition of the fetus in modern obstetrics. The regularity of the fetal heart rate is an indicator of proper oxygen supply to the fetus. Early detection of irregularities and implementation of standard clinical procedures can reduce intrapartum fetal brain damage and subsequent mental retardation. Unfortunately, when voltages are measured near the abdomen of the mother, detection of the weak fetal electrocardiogram (FECG) is sometimes impossible because it is masked by the much stronger ECG of the mother. Furthermore, between 27 and 34 weeks of gestation the FECG weakens and is often unobserved in cases where it previously could be detected.

The first observation of the fetal magnetocardiogram (FMCG) by V. Kariniemi, J. Ahopelto, P. J. Karp, and T. E. Katila [78] showed that the greater localization of the cardiac field compared with electric potential provided a more effective separation of signals from mother and fetus [79]. An example of the improved record is shown in Fig. 13. These data were taken over the abdomen with the mother prone, using a SQUID gradiometer (Fig. 2) in a wooden cabin, without shielding. The maximum amplitude of the QRS peak of the MFCG amounting to

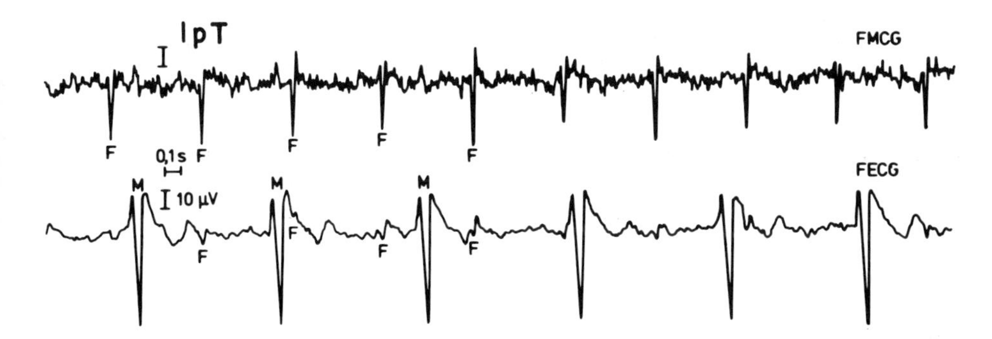

Fig. 13 Fetal magnetocardiogram compared with the fetal electrocardiogram after 29 weeks' gestation recorded by the group at the Department of Technical Physics, Helsinki University of Technology and the First Department of Obstetrics and Gynecology, University Central Hospital. This is a particularly graphic example of the better resolution of the fetal heart signal (F) compared with the mother's signal (M) afforded by magnetic detection.

~ 5 pT is about ten times weaker than the MCG of the mother when measured over the heart. In a study of 35 cases after the 30th week of gestation conducted by K. Hukkinen, T. E. Katila, H. Laine, R. Lukander, P. Mäkipää, and V. Kariniemi [80], fetal heart rate determinations from the FMCG succeeded in 21 cases, and of these the rate could be determined from the MECG in only 12 cases. Of the other 14 cases where the FMCG was not observed, the FECG was obtained in only one case.

One disadvantage of the FMCG is the strong dependence of the signal amplitude on the position of the fetus. Another is the failure so far to detect the FMCG earlier than the 25th week of gestation, whereas the FECG often can be detected more than a month earlier. Thus the FMCG will not replace the FECG, but rather appears to complement it.

VI. MAGNETOMYOGRAM

Human skeletal muscle is striated, and this feature distinguishes it from smooth muscles such as the myocardium. D. Cohen and E. Givler [81] using a SQUID magnetometer in the M.I.T. shielded room discovered that human skeletal muscles at the elbow produce both dc and ac magnetic fields when contracted. The fields are quite large near the skin, having an amplitude of ~ 20 pT which is comparable to the cardiac field near the chest. An example is illustrated in Fig. 14 for fields near the elbow and palm. Also shown is a recording of the ac component taken near the forearm by M. Reite, J. E. Zimmerman, J. Edrich, and J. Zimmerman [82] with a SQUID gradiometer in the N.B.S. eddy current shielded enclosure. These signals have been called the magnetomyogram (MMG) after the electrical counterpart obtained from skin electrodes and known as the electromyogram (EMG). The EMG has been used for a variety of applications, including the study of muscle movement, tension, and human relaxation [83].

A skeletal muscle is comprised of bundles of muscle fibers which are connected to the terminal branches of a controlling nerve cell. An impulse descending from the nerve axon causes all of the muscle fibers in one of these motor units to contract nearly simultaneously. Thus the apparently smooth contraction of a muscle is in actuality the result of each motor unit contracting by a series of rapid twitches. The asynchronous twitches from the different motor units are believed to give rise to the ac voltages in the domain 10-2,000 Hz recorded by the EMG. The spectra of the ac component of the MMG computed by D. Cohen and E. Givler [81] for the muscle just above the elbow has a relatively sharp peak located at about 40 Hz; and the muscle in the palm has a broad peak at about 80 Hz.

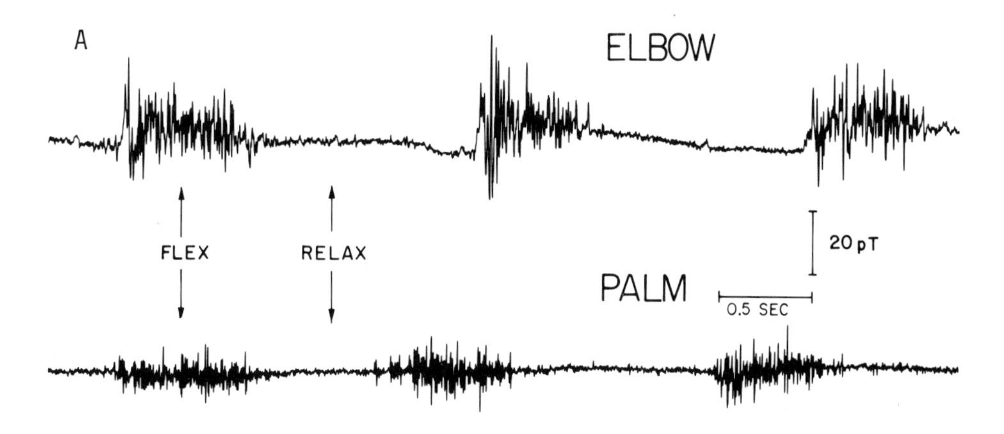

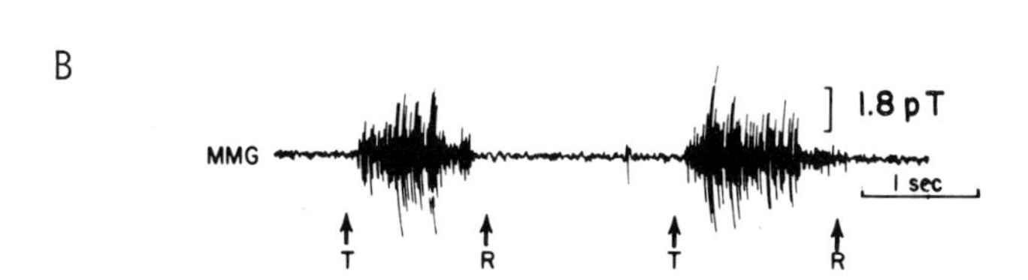

Fig. 14 A: Magnetomyogram obtained in the bandwidth 0-150 Hz near the elbow and palm. From D. Cohen and E. Givler, Appl. Phys. Lett. 21, 114 (1972). B: Magnetomyogram from the forearm with muscle tensed (T) and relaxed (R). From M. Reite, J. E. Zimmerman, J. Edrich, and J. Zimmerman, Electroenceph. Clin. Neurophysiol. 40, 59 (1976).

An interesting feature illustrated in the MMG of Fig. 14 and not yet identified in EMG records is the dc component which appears when a muscle is flexed. The dc currents within the muscle which give rise to this effect diminish slowly when the muscle is allowed to relax. A detailed study of these potentially important features and how they depend upon muscular condition has not yet been carried out. An interpretation of a dc electrical analog which might appear in skin-potential measurements would be complicated by artifacts from changing contact and skin potentials. Thus the MMG offers here a distinct advantage.

VII. MAGNETO-OCULOGRAM

The normal human eyeball maintains a relatively high voltage of ~ 100 mV between the retina and cornea. Electric dipole models have been proposed to describe the current distribution in the surrounding medium in order to explain the changes in skin potential when the eyeball (with dipole) moves. The observed effect is called the electro-oculogram (EOG). The change in dc field associated with the change in current distribution or magneto-oculogram (MOG) has been detected by P. J. Karp, T. E. Katila, P. Mäkipää, and P. Saar [84] using a gradiometer (Fig. 2) in a wooden cabin with no shielding. The baseline of the gradiometer is sufficiently long that the field couples only to the pickup coil. With the subject lying on his side, the axis of the gradiometer is directed normal to the sagittal plane with the pickup coil near the orbit of the eye. When the subject deflects his eye by 75° with the eye's axis remaining in the sagittal plane, a shift in dc field of ~12 pT is observed.

The shift in the MOG on deflecting the eye was influenced by changes in the light adaptive state of the eye. On going from light adapted (photopic) to dark adapted (scotopic) the amplitude decreased by ~ 20-30% over a period of ~ 12 min, then slowly regained its former value. This effect is illustrated in Fig. 15. On going from dark to light adapted vision there is a rapid increase in the shift by 50-100% over a period of 7-12 min, and thereafter a decline to the original value after a total of ~ 30 min. Similar effects are found in the EOG when skin electrodes are placed off the nasal and temporal sides of the orbit [85]. As yet no new feature has been found in the MOG that is not known in the EOG so far as the effect of eye movement is concerned. But the success enjoyed by the EOG in clinical diagnosis of retinal diseases suggests that monitoring changes in the dc field with changing eye condition may turn out to offer new information.

It may be possible to observe magnetically another effect produced not by eye movement, but by change in electrical potential of the cornea. The electroretinogram (ERG) can be

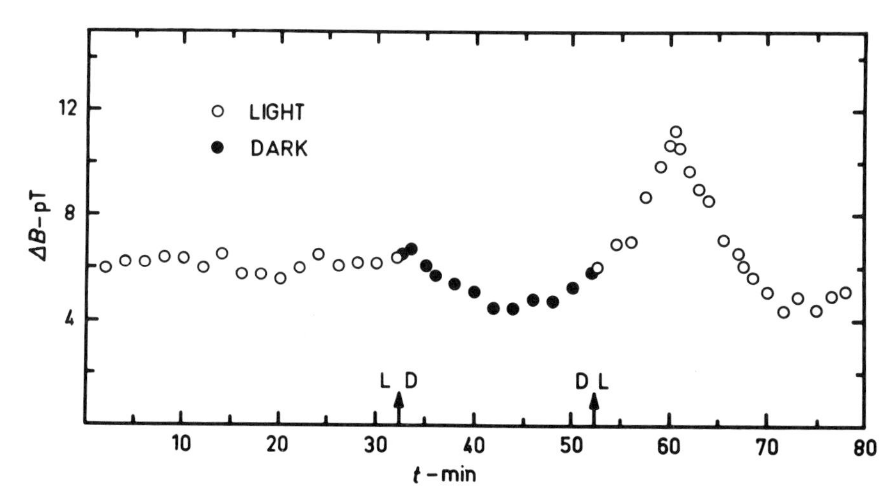

Fig. 15 Magneto-oculogram giving the change in dc field component normal to the sagittal plane when the eye is deflected 75° in the sagittal plane, under conditions of light and dark adaptation. From P. J. Karp, T. E. Katila, P. Mäkipää, and P. Saar, Digest of 11th International Conference on Med. and Bio. Eng., Ottawa, Canada, August 1976.

recorded by placing an electrode on the cornea (usually mounted in a contact lens) and an indifferent electrode elsewhere on the body. The potential difference between these electrodes can be made to vary by as much as ~ 100 μV with a number of different stimulus parameters such as abrupt changes in the level of illumination [86]. While it is now believed that the ERG results largely from current flow in the pigment epithelium of the retina, there is evidence that some effects could be due to receptor activity and excitation of bipolar cells. At present there has been no successful recording of the magnetoretinogram (MRG). However, the successful detection of light adaption changes in the MOG make it likely that the MRG can also be detected. This would be a particularly valuable objective for researchers since the ERG has been useful in diagnosis of visual impairment (e.g., retinitis pigmentosa) and as a basic research tool.

VIII. MAGNETOENCEPHELOGRAM

Neural activity in the human brain produces changes in the distribution of electrical potential over the scalp. Measurement of the temporal variation in voltage between two points resulting from the spontaneous activity of the brain yields a record which is called the electroencephelogram (EEG). The principal uses of the EEG are in the clinical diagnosis of epilepsy and sleep research. Neural activity also produces a magnetic field, and to distinguish it from biomagnetic fields of other origins such as muscle we call it the "neuromagnetic" field. The first observation of a neuromagnetic field was the magnetoencephalogram (MEG) obtained by D. Cohen in 1968 in a magnetically shielded room with an induction coil of 10^6 turns and ferrite rod core [23]. To bring the signal above the thermal noise of the coil, which was reported to have an rms value of ~ 0.3 pT $Hz^{-\frac{1}{2}}$ at 10 Hz, the output was filtered to allow a bandwidth of 5 Hz at a center frequency of 10 Hz and was then signal averaged, using the simultaneously monitored EEG as a reference trigger. Subsequently, D. Cohen [87] demonstrated in the M.I.T. shielded room with a SQUID magnetometer that signal averaging and narrow band operation were not necessary for a good quality MEG. Using a SQUID gradiometer, J. Ahopelto, P. J. Karp, T. E. Katila, R. Lukander, and P. Mäkipää [88] subsequently were also able to observe the MEG in a suburban location without recourse to magnetic shielding.

The amplitudes of the MEG and EEG show marked intersubject variability. The strongest MEG signals are typically ~ 2 pT. Field and voltage variations are found predominantly at low frequency (< 30 Hz), and the power spectra depend upon the state of wakefulness of the subject. The origin of these signals has not been identified nor is it clear to what biological function they relate. The component found in the range 9-12 Hz is called the "alpha rhythm" and is generally dominant. When present in the MEG it can be observed over nearly all the scalp. For reference, we show in Fig. 16A the standard notation for regions on the scalp. D. Cohen [23] found that the general shape of the field pattern of the alpha rhythm is asymmetrical as illustrated in Fig. 16B. The amplitude is suppressed when the subject has his eyes open, as the record in Fig. 17 indicates [87], and is enhanced when the subject is in a restful state [89]. From a series of measurements in the N.B.S. eddy current shielded enclosure, M. Reite, J. E. Zimmerman, J. Edrich, and J. Zimmerman [82] report a strong positive correlation between the amplitudes of the MEG and EEG from one subject to another. They found that the power spectra for the MEG of five subjects has a dominant peak at 11-12 Hz for alpha activity, which is close to the location of the peak in the EEG spectrum. Additional components at lower frequencies were comparatively stronger in the EEG than MEG. The field patterns of these components have not yet been established.

More extensive comparisons between the MEG and EEG from seven normal subjects have been reported by J. R. Hughes, D. E. Hendrix, J. Cohen, F. H. Duffy, C. I. Mayman, M. L. Scholl, and B. N. Cuffin [89]. Measurements were

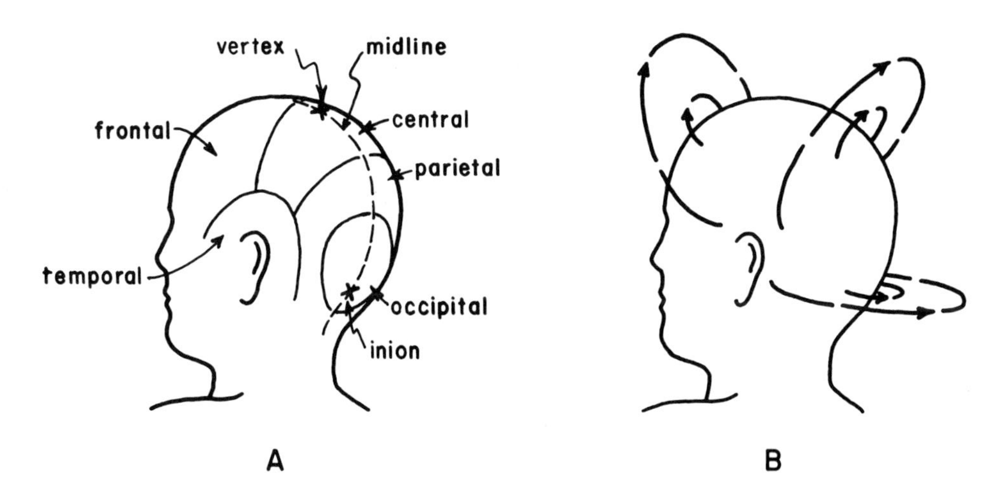

Fig. 16 A: Designations of locations on the scalp, as viewed from the posterior. B: General features of the magnetic field patterns at the alpha frequency as reported by D. Cohen, Science 161, 784 (1968).

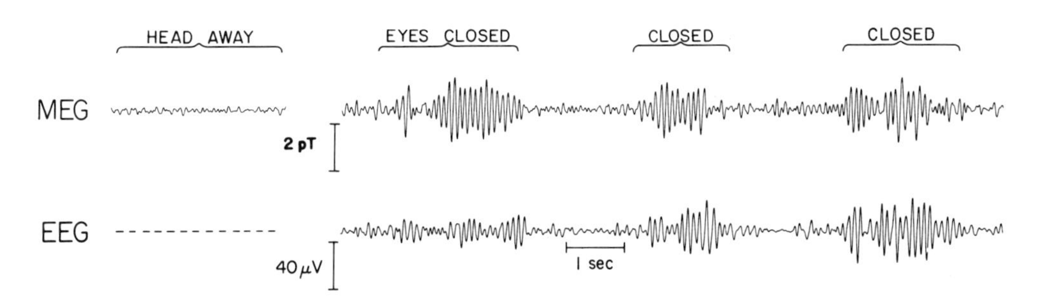

Fig. 17 Simultaneous magnetoencephalogram and bipolar electroencephalogram recordings from the left occipital region for a bandwidth of 4-15 Hz. From D. Cohen, Science 175, 664 (1972).

carried out in the M.I.T. shielded room with a SQUID magnetometer oriented to detect the component of field normal to the scalp. The MEG in the bandwidth 2-45 Hz was typically ~ 1 pT; and the corresponding EEG with one electrode at the vertex and the other at positions including the anterior temporal, mid-temporal, and occipital was ~ 50 μV. They found that when the dominant component of either the MEG or EEG is at the alpha frequency the same frequency component is dominant in the other, and conversely, when the dominant component in one is not at the alpha frequency the frequency for the dominant component of the other cannot be predicted. The alpha frequency signals when dominant in the waking state with eyes closed are strongly correlated in the MEG and EEG, with a direct relationship in phase, depending on the location of the magnetometer. On the left side, EEG and MEG tend to be in phase; and on the right side, out of phase. This relationship was observed on 89% and 69% of the time for the respective sides; the correlation was strongest at the temporal areas, but nonexistent at the occipital areas. Since the EEG tends to have the same phase on left and right sides of the head when the vertex is used as a reference, the observed correlation is consistent with the magnetic field pattern developed earlier by Cohen (Fig. 16B). Occasionally the alpha component of the MEG will lead the EEG by 20-40 ms, but more often it lags. At times the MEG will remain in phase with the EEG and then gradually develop an increasing lag. This may evidence a type of traveling wave in the EEG as previously reported by H. Petsche and A. Marko [90]. An example of the correlation between simultaneous EEG recordings at different locations and also the EEG and MEG is illustrated in Fig. 18A for a wakeful condition. But this correlation is lost during sleep, as shown in Fig. 18B. It is apparent that a close correlation still exists between signals at different electrode positions, but not between the EEG and MEG.

To appreciate another variability of behavior it is important, as emphasized by Hughes et al. [89] to note one feature about the EEG reported by D. Lehmann [91]. During sleep, there is a tendency for scalp electrodes to show left-right (bilateral) symmetry with signals in-phase referenced to the vortex. But during wakefulness the maximum EEG signal strength may move slowly about the head, moving from the right occipital area to left occipital area, then to left temporal, left frontal, right frontal, and so on. However, synchronism remains among the continuing signals in all areas. Apparently there is a correlation between EEG and MEG activity only during this progression during wakefulness, but not during the comparatively steady distribution that characterizes sleep.

Quite a different picture emerges when other frequencies of the EEG and MEG are considered. Often, a half dozen or more

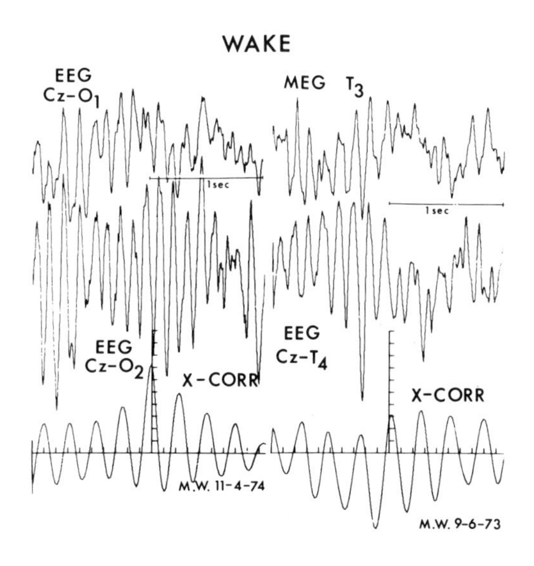

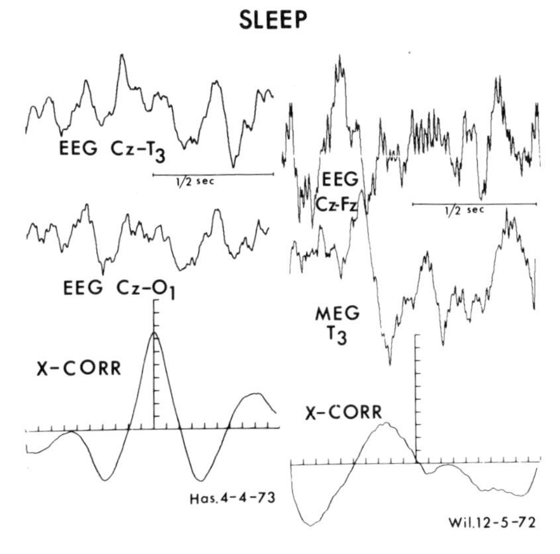

Fig. 18 Three EEG traces at different locations and an MEG trace for the wakeful (A) and sleeping (B) states of subjects. The respective cross-correlograms of the two EEG records and of the EEG and MEG records are shown below with units of 0.1 on the vertical axis and a 1-sec span from left to right on the horizontal axis for A and 0.5 sec for B. From J. R. Hughes, D. E. Hendrix, J. Cohen, F. R. Duffy, C. I. Mayman, M. L. Scholl, and B. N. Cuffin, Electroenceph. Clin. Neurophysiol. 40, 261 (1976).

components can be found in each record. There is a weak suggestion that if the dominant component of the EEG is at 6 or 8 Hz the dominant component of the MEG will be at the same frequency, but this is often not the case. Frequency components of the theta rhythm (4-5 Hz) are often prominent in the MEG during both waking and sleep record, even when the EEG fails to show this activity. No correlation was found by Hughes et al. [89] between low frequency components of the EEG and MEG. And when a subject hyperventilates, diffuse slow waves (~ 1-2 Hz) may appear in only the MEG or only the EEG or both. The sleep record shows a similar apparent independence for slow waves (~ 1-2 Hz).

A sleep "spindle" is a burst of a high frequency signal (~ 14 Hz) having a duration of 1-2 s which is occasionally seen in the EEG. They are infrequent in the MEG and may be entirely absent [89]. "Vertex sharp waves" are another phenomenon observed in the EEG, consisting of one isolated cycle at a frequency of ~ 3 Hz at perhaps twice the average EEG amplitude. It has been seen on the MEG by Hughes et al [89], but with a much more variable amplitude than in the EEG.

It is clear from these important results that a close, but not complete association exists between the generating sources for the alpha rhythm observed in the EEG and MEG. The sources for signals at other frequencies appear either to be disassociated or preferentially oriented so that their magnetic fields are largely parallel to the scalp and hence remain undetected. To check this, measurements of the parallel field would be useful. So little information is available about magnetic field patterns that it is premature to speculate about the nature of the primary and secondary current sources. It is perhaps disappointing that as yet no new feature not previously seen in the EEG has been discovered in the MEG.

IX. VISUALLY EVOKED FIELD

One of the most recently discovered biomagnetic phenomena is the neuromagnetic field produced by the brain when one of the senses is stimulated. Observation of the response evoked by a visual stimulus was first reported by D. Cohen [5], who employed a SQUID magnetometer in the M.I.T. shielded room. The visually evoked magnetic response or visually evoked field (VEF) produced by a briefly flashed light is approximately 10 times smaller than the spontaneous MEG obtained with the subject's eyes closed and therefore is the weakest biomagnetic field yet detected. Signal averaging with repeated stimuli is necessary to bring the typical field amplitude of ~ 0.3 pT above the background noise and spontaneous activity of the brain.

An example of the VEF obtained by M. Reite, J. E. Zimmerman, J. Edrich, and J. Zimmerman [82] for a flash stimulus produced by a strobe lamp with the subject in the N.B.S. eddy current enclosure is shown in Fig. 19. The VEF averaged over 100 responses is qualitatively similar to the visually evoked potential (VEP) obtained from a skin electrode placed under the magnetic field detector near the vertex, referenced to combined mastoid electrodes. In this example the maximum field directed outward from the scalp occurs simultaneously with the maximum positive skin potential. T. Teyler, N. Cuffin, and D. Cohen [92] have reported similar but more extensive measurements in the MIT shielded room. They find that the VEF and VEP amplitudes increase with the luminance of the stimulus, but a tendency toward saturation in the VEP at the highest luminance and decrease in delay for appearance of the first and second peaks are not observed in the VEF data. In studies of the position dependence of the field normal to the scalp, the VEF was not found near the inion low on the midline at the back of the head; but it did occur 45° from the midline in the transverse plane. A reported VEF 135° from the midline near the anterior temple differs qualitatively from the signal illustrated in Fig. 19; and since the location is sufficiently close to the eye it may have been affected by eye movement artifacts (the magneto-oculogram as suggested by T. E. Katila and co-workers).

A technique to observe the VEF in a normal laboratory setting without need for shielding has been described by D. Brenner, S. J. Williamson, and L. Kaufman [93]. The versatility of this technique makes a wide class of magnetic evoked response recordings generally available for psychological and neurological studies. The detection coil is an adjustable second-derivative gradiometer as described in Section III, and the relatively high noise at low frequencies (see Fig. 5) is avoided by using bandpass filters and detecting responses above 8 Hz [38]. This method has indicated the steady state ac response of the brain to a periodic visual stimulus, whereas the studies cited previously detect the transient response to an impulse stimulus. In the first reported steady state measurements, the subject observed an oscilloscope screen (as in Fig. 4) on which a bar grating pattern was flashed on and off in square wave fashion at 10 Hz. Detailed mapping studies of the field normal to the scalp showed that the VEF is localized near the visual cortex, above the inion in the posterior occipital region. A significant response could be found only over a ~4 cm region of the midline and ~10 cm laterally to the side of the midline. This localizability of the VEF near the visual cortex is evidence that the field is produced by currents in the primary projection areas. By contrast, the VEP is more diffuse and can be detected nearly everywhere on the

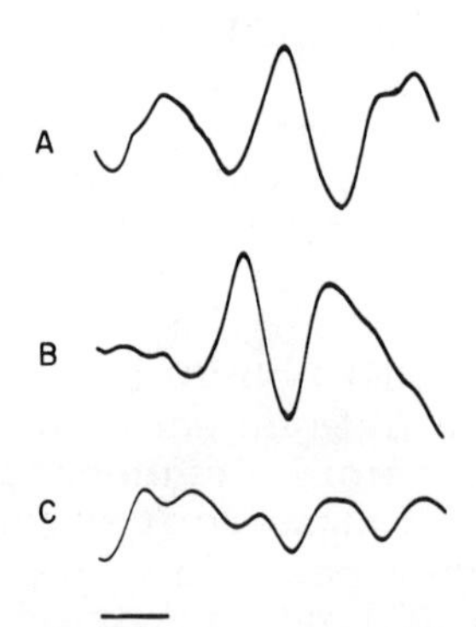

Fig. 19 A: Magnetic evoked response averaged over 100 flashes, with upward deflection indicating a field directed into the scalp. B: The simultaneously averaged voltage evoked response, with upward deflection indicating positive potential at the field location. C: Magnetic noise level indicated by the signal when averaged over 100 flashes with the subject absent. The field sensitivity for A and C are identical. The time calibration at the lower left is 50 msec. From M. Reite, J. E. Zimmerman, J. Edrich, and J. Zimmerman, Electroenceph. Clin. Neurophysiol. 40, 59 (1976).

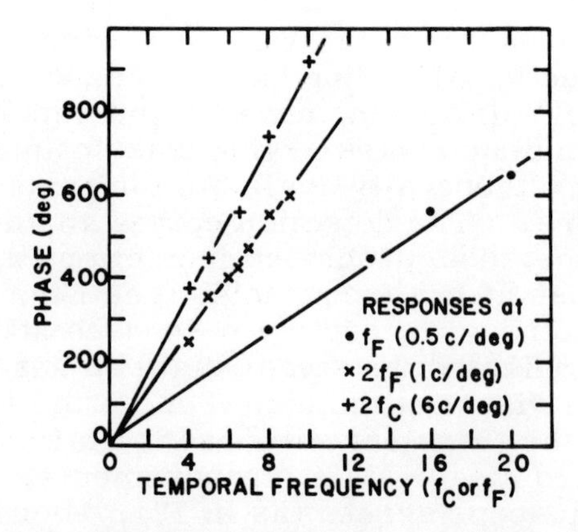

Fig. 20 Phase of magnetic response averaged at the first harmonic (f_F) or second harmonic ($2f_F$ and $2f_C$) relative to the onset of visual stimulus for various temporal frequencies of visual stimulus as indicated on the horizontal axis. The responses for representative spatial frequencies of the visual pattern are shown. From S. J. Williamson, L. Kaufman, and D. Brenner, submitted for publication.

scalp, with large amplitudes near the vertex as well as at the occipital pole near the inion.

Recently S. J. Williamson, L. Kaufman, and D. Brenner [94] have demonstrated that the VEF provides fundamentally new information about the visual system not previously found in the VEP. A direct relationship was found between features of the neuromagnetic response and human reaction time. In separate experiments two different stimuli were used, but both involve a retinally vertical grating formed on an oscilloscope screen by a sinusoidal variation in luminance across the screen in the horizontal direction. For a "flicker" stimulus the pattern was presented for a brief time, then removed for the same duration with the screen maintaining the same spatially average luminance throughout. When the stimulus is presented at a frequency f_F, the response was detected at f_F or harmonics ($2f_F$, $3f_F$, etc.). A "contrast reversal" stimulus differs from the flicker stimulus in that the grating pattern is not suppressed during the last half of the cycle, but instead is shifted horizontally so that the bars fall in between their original locations. When this stimulus is presented at a frequency f_c, no response is detected at f_c (as would be expected if the brain does not respond asymmetrically to pattern shifts); but nevertheless a response is found at $2f_c$.

The phase of the maximum outward response field relative to a reference point on the temporal stimulus cycle is found to depend on the stimulus temporal frequency as illustrated in Fig. 20. For convenience the reference point was chosen as the onset of uniform luminance (flicker stimulus) or horizontal shift of the grating pattern (contrast reversal stimulus). Figure 20 shows, for three representative responses from one subject, that the observed phase increases linearly with the temporal frequency of the stimulus. This proportionality implies that the time interval between peak response field and the reference point of the stimulus cycle is independent of stimulus frequency and therefore is an intrinsic characteristic of the human visual system. This time interval is called the "latency" of the steady state response. Using grating patterns with different uniform spacings between bars, it was found that the latency for the responses when f_F or f_c is below 13 Hz varies in the same way with the spatial frequency of the pattern, as given by the number of bars per degree of visual angle (c/deg). This pattern-specific feature is illustrated in Fig. 21 for responses at $2f_c$. The data for a number of different stimulus frequencies form a universal curve, with latency increasing with spatial frequency. Several characteristics of this curve have been interpreted in terms of the effects of specialized perception channels in the visual system [94]. Another important result is evident when the data are compared with measurements of human reaction times as determined by B. Breitmeyer [95] in psychophysical measurements. Observers briefly presented with a grating pattern on an oscilloscope screen, have reaction times which vary with the spatial frequency of the pattern. The two solid curves in Fig. 21 show the reaction times for two observers. It is apparent that the reaction time differs from the latency by a constant time interval, independent of spatial frequency. One can infer from this that the peak neuromagnetic field is associated with the aggregate activity of many neurons in the visual cortex which signal the motor area to generate a reaction to the stimulus. The motor activity adds a constant to the response time that is independent of pattern features. The pattern specificity of the latency may arise from differences in neural conduction speeds or possibly from different processing times at various levels in the visual system.

These examples indicate that a detailed understanding of physiological current sources is not always a prerequisite for inferring general features of biological activity from biomagnetic data, for virtually nothing is known about the sources of the neuromagnetic fields which are undoubtedly very complicated. This area of study and the search for magnetic responses from other sensory stimuli (aural evoked fields, tactile evoked fields, etc.) hold promise that the application of magnetic monitoring techniques to brain research will continue to be active and fruitful in the coming years.

X. EXPECTATIONS

Biomagnetic studies have been carried out on a large number of effects since the first observation of the cardiac field by G. M. Baule and R. McFee in 1963 and the introduction of effective magnetic shielding and pioneering applications of the SQUID magnetometer by D. Cohen and coworkers. In some magnetic studies distinct advantages over voltage measurements are apparent, as in the fetal magnetocardiogram, magnetic evoked response, and dc magnetocardiogram and magnetomyogram. For others the differences between magnetic and voltage records suggest new and interesting effects are present but their significance remains to be proven. Magnetic studies continue to reveal new phenomena that indicate internal activity of interest to biologists, physicians, and psychologists that cannot otherwise be studied in intact organisms.

This review has not discussed a number of additional dc phenomena associated with the human body discovered by D. Cohen. One for example is the dc field near the abdomen [1, 96]. This has an unknown origin but the field component normal to the torso can be as large as 50 pT. Cohen has also found a dc field which is produced in the chest by a reflex current after a person drinks cold water. With an initial magnitude at

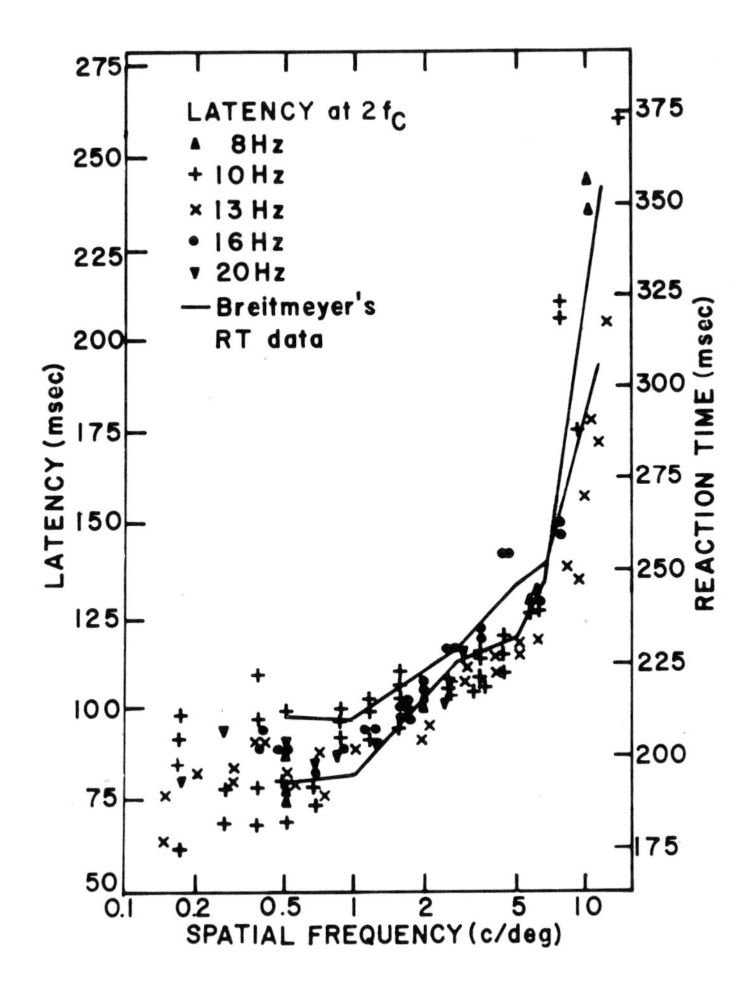

Fig. 21 Latency of the second harmonic magnetic response for a contrast reversal stimulus, as deduced from data at a number of stimulus frequencies indicated in the legend. The variation with spatial frequency of the stimulus pattern is identical with that observed in the reaction time measurements reported by B. Breitmeyer. From S. J. Williamson, L. Kaufman and D. Brenner, submitted for publication.

times exceeding 200 pT this field steadily diminishes to undetectable levels within 5 or 10 minutes after ingestion. Abnormalities from the same stimulus have also been observed in the ECG [43]. The use of magnetic shielding, while not necessary in these particular studies, is at present essential for the convenient study of the weaker dc phenomena, whose study promises to yield much significant information.

It is our opinion that the growth of the discipline of biomagnetism will accelerate in the coming years now that there is active participation of more biologically-oriented scientists in what had become largely a low temperature physicist's domain. Further stimulation comes from technical advances which have enabled commercial SQUID second-derivative gradiometers to be offered with a low intrinsic noise level of 40 fT $Hz^{-\frac{1}{2}}$ down to 1 Hz outdoors and 10 Hz in a laboratory. Asymmetric detection coil configurations show promise of reducing this level by as much as a factor of four [36]. And the working frequency range without recourse to shielding may be extended to lower frequencies with the anticipated development of buckout or feedback techniques to reduce the large contributions from low frequency background fields. Thus sensitivity rivaling the best yet achieved in a shielded enclosure is forseeable for a wide frequency range, except possibly at the very lowest frequencies. A broad class of high sensitivity biomagnetic measurements will be possible in a normal laboratory setting, and this availability undoubtedly will lead to many novel and significant applications.

ACKNOWLEDGEMENTS

In preparing this manuscript we have benefitted from discussions with numerous colleagues. In particular we wish to thank D. Cohen, D. Crum, T. Katila, J. Wikswo, Jr., and J. E. Zimmerman.

REFERENCES

1. D. Cohen, Science 180, 745 (1973).
2. J. P. Wikswo, Jr., J. E. Opfer and W. M. Fairbank, A.I.P. Conference Proc. 18, 1335 (1974).
3. D. Cohen, IEEE Trans. Magn. MAG-6, 344 (1970).
4. D. Cohen, Physics Today 28, 34 (1975).
5. D. Cohen, IEEE Trans. Magn. MAG-11, 694 (1975).
6. G. M. Baule and R. McFee, Am. Heart J. 66, 95 (1963).
7. G. M. Baule and R. McFee, J. Appl. Phys. 36, 2066 (1965).
8. F. Grynszpan, Ph.D. dissertation, University of Philadelphia, 1971, unpublished.
9. R. Plonsey, IEEE Trans. Biomed. Eng. BME-19, 239 (1972).

10. S. Rush, IEEE Trans. Biomed. Engn. BME-22, 157 (1975).
11. J. A. Stratton, Electromagnetic Theory (McGraw-Hill, New York, 1941).
12. B. M. Horacek and H. J. van Eck Ritsema, Proc. International Symposium on Electrical Field of the Heart (North-Holland, Amsterdam, 1971).
13. D. B. Geselowitz, Biomed. Sci. Instrum. 1, 325 (1963).
14. R. Plonsey, CRC Critical Reviews in Bioengineering 1, 1 (1971).
15. H. P. Schwan and C. F. Kay, Circulation Res. 5, 439 (1957).
16. D. B. Geselowitz, Biophysical J. 7, 1 (1967).
17. D. B. Geselowitz, IEEE Trans. Magn. MAG-6, 346 (1970).
18. O. D. Kellogg, Foundations of Potential Theory (Springer-Verlag, N.Y., 1922), p. 221.
19. J. P. Wikswo, Jr., Ph.D. thesis, Department of Physics, Stanford University, 1975, unpublished.
20. Yu. D. Safonov, V. M. Provotorov, V. M. Lube, and L. I. Yakimenkov, Bull. Exp. Biology and Medicine 64, 1022 (1967).
21. D. Cohen, Science 156, 652 (1967).
22. G. M. Baule, Trans. N.Y. Acad. Sci. Series II 27, 689 (1965), and N.Y. State J. Medicine 67, 3095 (1967).
23. D. Cohen, Science 161, 784 (1968).
24. J. E. Zimmerman, P. Thiene, and J. T. Harding, J. Appl. Phys. 41, 1572 (1970).
25. J. E. Zimmerman, Cryogenics 12, 19 (1972).
26. R. P. Giffard, R. A. Webb and J. C. Wheatley, J. Low Temp. Phys. 6, 533 (1972).
27. L. D. Jackel and R. A. Buhrman, J. Low Temp. Phys. 19, 201 (1975).
28. S. H. E. Corp., San Diego, CA. 92121.
29. A. C. Fraser-Smith and J. L. Buxton, J. Geophys. Res. 80, 3141 (1975).
30. J. E. Zimmerman, J. Appl. Phys. 42, 4483 (1971).
31. J. E. Zimmerman and N. V. Frederick, Appl. Phys. Letters 19, 16 (1971).
32. A. Rosen, G. T. Inouye, A. L. Morse, and D. L. Judge, J. Appl. Phys. 42, 3682 (1971).
33. J. E. Opfer, Y. K. Yeo, J. M. Pierce, and L. H. Rorden, IEEE Trans. Magn. 9, 536 (1974).
34. G. M. Baule and R. McFee, Am. Heart J. 79, 223 (1970).
35. J. Malmivuo, Acta Polytechnica Scandinavica, Electrical Engineering Series No. 39, Helsinki, The Finnish Academy of Technical Sciences (1976).
36. J. E. Zimmerman, 1976 Appl. Superconductivity Conf., Stanford, Calif., August 1976; also J. Appl. Phys. in press.
37. D. Brenner, S. J. Williamson, and L. Kaufman, Low Temperature Physics-LT14, M. Krusius and M. Vuorio, Eds. (North Holland, Amsterdam, 1975), Vol. 4, p. 266.

38. D. Brenner, Ph.D. Thesis, Department of Physics, New York University, 1976, unpublished.
39. M. Saarinen, P. J. Karp, T. E. Katila, and P. Siltanen, Cardiovascular Research 8, 820 (1974).
40. D. Brenner, L. Kaufman, and S. J. Williamson, 1976 Appl. Superconductivity Conf., Stanford, Calif., August 1976; and IEEE Trans. Magn., in press.
41. D. Cohen, Rev. de Phys. Appliquee 5, 53 (1970).
42. D. Cohen, E. A. Edelsack, and J. E. Zimmerman, Appl. Phys. Letters 16, 278 (1970).
43. R. McFee and G. M. Baule, Proc. IEEE 60, 290 (1972).
44. W. Einthoven, G. Fahr. and A. de Waart, Pflügers. Arch. Physiol. 150, 275 (1903); Engl. transl.: H. E. Hoff and P. Sekelj, Am. Heart J. 40, 163 (1950).
45. A. D. Waller, J. Physiol. 8, 229 (1887).
46. R. A. Stratbucker, C. M. Hyde, and S. E. Wixson, IEEE Trans. Biomed. Elec. 10, 145 (1963).
47. D. Cohen and L. Chandler, Circulation 39, 395 (1969).
48. A. Rosen, G. T. Inouye, and A. L. Morse, J. Appl. Phys. 43, 1908 (1972).
49. D. Matelin, Thesis, Université Scientifique et Medicale de Grenoble, Grenoble, 1974, unpublished.
50. B. Denis and D. Matelin, to be published.
51. D. Cohen and D. McCaughan, Am. J. Cardiol. 29, 678 (1972).
52. A. Rosen and G. T. Inouye, IEEE Trans. Bio. Med. Eng. 22, 167 (1975).
53. D. Cohen, H. Hosaka, E. Lepeschkin, H. D. Levine, B. F. Massell, D. McCaughan, and G. Myers in Ref. 54.
54. Report of the Low-Field Group: The Magnetocardiogram, Francis Bitter National Magnet Laboratory, M.I.T., December 1975.
55. D. Cohen, E. Lepeschkin, H. Hosaka, B. F. Massell, and G. Myers, J. of Electrocardiology, Suppl., October 1976.
56. M. Saarinen, P. Siltanen, J. Ahopelto, and T. E. Katila, Proc. Nordic Meeting on Med. and Bio. Engr., A. Vusitulo and N. Saranummi, Eds., III, 56.1-56.3 (1975).
57. W. M. Wynn, C. P. Frahm, P. J. Carroll, R. H. Clark, J. Wellhoner, and M. J. Wynn, IEEE Trans. Magn. MAG-11, 701 (1975).
58. E. Frank, Circulation 13, 737 (1956).
59. J. P. Wikswo, Jr., J. A. Malmivuo, W. H. Barry, G. E. Crawford, W. M. Fairbank, R. D. Giffard, D. C. Harrison, and R. H. Roy, Proc. San Diego Biomed. Symp. 14, 359 (1975).
60. J. P. Wikswo, Jr., J. A. Malmivuo, and W. M. Fairbank, "Magnetocardiography: Theory, Instrumentation, and Results", Advances in Cardiovascular Physics, Vol. III: Diagnostics, D. N. Ghista, E. van Vollenhoven, and W. Yang, Eds. (Delft University Press, in preparation).

61. J. P. Wikswo, Jr., and W. M. Fairbank, 1976 Appl. Superconductivity Conf., Stanford, Calif., August 1976; and IEEE Trans. Magn., in press.
62. D. Cohen and L. A. Kaufman, Circulation Res. 36, 414 (1975).
63. D. Cohen, J. C. Norman, F. Molokha, W. Hood, Jr., Science 172, 1329 (1971).
64. F. Grynszpan and D. B. Geselowitz, Biophys. J. 13, 911 (1973).
65. D. A. Brody and R. C. Arzbaecher, IEEE Trans. Bio. Med. Eng. BME-14, 22 (1967).
66. J. H. Holt, Jr., A. C. L. Barnard, M. S. Lynn, and P. Svendsen, Circulation 15, 687 (1969); J. H. Holt, Jr., A. C. L. Barnard, M. S. Lynn, Circulation 15, 697 (1969); J. H. Holt, Jr., A. C. L. Barnard, M. S. Lynn, and J. O. Kramer, Jr., Circulation 15, 711 (1969).
67. R. Plonsey, Bioelectric Phenomena (McGraw Hill, N.Y., 1969).
68. B. M. Horacek, IEEE Trans. Magn. MAG-9, 440 (1973).
69. B. N. Cuffin, Ph.D. Thesis, Dept. Electrical Eng., The Pennsylvania State University, 1974, unpublished.
70. D. B. Geselowitz, J. Franklin Inst. 296, 379 (1973).
71. D. B. Geselowitz and W. T. Miller, IEEE Trans. Magn. MAG-9, 392 (1973).
72. W. T. Miller and D. B. Geselowitz, Ann. Biomed. Eng., in press.
73. H. Hosaka and D. Cohen, J. of Electrocardiology, Suppl., October 1976.
74. H. Helmholtz, Ann. Phys. und Chem. 29, 211 (1853).
75. G. M. Baule, Ph.D. Thesis, Dept. Electrical Eng., Syracuse University, 1964, unpublished; additional features of lead field theory are described in Adv. Cardiol. 10, 304 (1974).
76. D. Cohen and H. Hosaka, J. of Electrocardiology, Suppl., Oct. 1976.
77. H. Hosaka, D. Cohen, B. N. Cuffin, and B. M. Horacek, J. of Electrocardiology, Suppl., Oct. 1976.
78. V. Kariniemi, J. Ahopelto, P. J. Karp, and T. E. Katila, J. Perinat. Med. 2, 214 (1974).
79. J. Ahopelto, K. Hukkinen, T. E. Katila, H. Laine, and V. Kariniemi, Ann. Chir. Gyn. Fenn. 64, 152 (1975).
80. K. Hukkinen, T. E. Katila, H. Laine, R. Lukander, P. Mäkipää, and V. Kariniemi, Am. J. Obstetrics and Gynecology, in press, 1976.
81. D. Cohen and E. Givler, Appl. Phys. Lett. 21, 114 (1972).
82. M. Reite, J. E. Zimmerman, J. Edrich, and J. Zimmerman, Electroenceph. Clin. Neurophysiol. 40, 59 (1976).
83. J. V. Basmajian, Science 176, 603 (1972).
84. P. J. Karp, T. E. Katila, P. Mäkipää, and P. Saar, Proc. Second International Conference on Biophys. and Biotech. in

Finland, Feb. 1976; and Digest of 11th International Conf. on Medical and Biological Engr., Ottawa, Canada, Aug. '76.
85. H. M. Burian and J. H. Jacobson, Eds., Clinical Electroretinography (Pergamon Press, 1966).
86. R. Granit, Receptors and Sensory Perception (Yale University Press, New Haven, 1955).
87. D. Cohen, Science 175, 664 (1972).
88. J. Ahopelto, P. J. Karp, T. E. Katila, R. Lukander, and P. Mäkipää, International Conference on Biomedical Transducers (Biocapt 75, Paris, 1975), Vol. 1, p. 347.
89. J. R. Hughes, D. E. Hendrix, J. Cohen, F. H. Duffy, C. I. Mayman, M. L. Scholl, and B. N. Cuffin, Electroenceph. Clin. Neurophysiol. 40, 261 (1976).
90. H. Petsche und A. Marko, Wien. Z. Nervenheilk 12, 87 (1955).
91. D. Lehmann, in H. Petsche and M. A. B. Brazier, Eds., Synchronization of EEG Activity in Epilepsies (Springer-Verlag, New York, Vienna, 1972), p. 307.
92. T. J. Teyler, B. N. Cuffin and D. Cohen, Life Sci. 15, 683 (1975).
93. D. Brenner, S. J. Williamson, and L. Kaufman, Science 190, 480 (1975).
94. S. J. Williamson, L. Kaufman, and D. Brenner, submitted for publication.
95. B. G. Breitmeyer, Vision Res. 15, 1411 (1975).
96. D. Cohen, J. Appl. Phys. 40, 1046 (1969).

Chapter 9

A PROGRESS REPORT ON COMMERCIAL SUPERCONDUCTING INSTRUMENTS IN THE UNITED STATES

M. B. Simmonds

S. H. E. Corporation

San Diego, California

I. INTRODUCTION

SQUID based instruments have been commercially manufactured in the United States for over six years. However, until about three years ago these systems were only suitable to be used by low temperature physicists and other researchers who were willing to become initiated into the somewhat black art of SQUID operation. The early instruments presented users with a choice between temperamental point-contact junctions and temperature sensitive thin-film bridges. It was often up to the user to piece his system together from various suppliers, and then have to worry about RFI (radio frequency interference), slew rate limitations, mysterious flux jumping, ground loops and a host of other distracting problems. There were no comprehensive research contracts let to industry for the systematic development of SQUID systems, so that out of necessity the development work was done in a piecemeal fashion. The terms "special" and "optional" have sometimes been used as euphemisms for "untried" when soliciting fixed-price contracts. These practices have resulted in some desperately heroic development feats. They have also, on occasion, left the customer with the task of salvaging a less than successful design.

The situation has improved dramatically during the last two or three years. Manufacturers have finally had production experience with enough different kinds of systems that customers can get many types of SQUID instrumentation "off the shelf" or from completely standardized designs with well-documented performance. As an example, it is now quite possible for an exploration geophysicist with no experience in cryogenics to buy a standard three-axis magnetometer which is as straightforward to use and as reliable as any of his other field instruments. He does not need to learn all about SQUIDs in order to assure himself of specifying a useful instrument. His only reminder that the system is "exotic" is that he must replenish the liquid helium periodically.

There are presently two companies in the United States which build superconducting instruments [1]. It is our goal here to describe in general terms what sorts of instruments are currently available and what their critical specifications are. We shall also try to indicate which systems have been engineered to the point of being entirely standard and which ones require a certain level of sophistication on the part of the customer in specifying what he really wants. We shall avoid specific reference to companies or prices, although some product lines may be easily recognized from our descriptions. We have divided the field into the following categories: SQUID Sensors, Laboratory Probes, Geophysical Magnetometers, Magnetic Anomaly Detectors, Biomedical Magnetometers, Sample Measuring Instruments, and Shielded Environments.

II. SQUID SENSORS

The latest generation of sensor can be characterized as a completely sealed four-terminal device. Two terminals are connected to the rf pump circuit and two are connected to the superconducting input circuit. There are no contact adjustments to make and no tiny coupling coils to fabricate. The devices are very resistant to both mechanical and thermal shock. They will operate comfortably through a range of several degrees around 4 K; selected units may be used down to millikelvins.

The general reliability of these sensors is emphasized by the fact that one company found 98% of its SQUIDs still within specifications after one year of use.

The most basic figure of merit that can be applied to these sensors is their energy sensitivity per unit bandwidth. This is defined as $\frac{1}{2}LI_n^2$, where L is the input inductance of the device and I_n is the equivalent input current noise per root Hertz. A routinely obtainable energy sensitivity, referred to the output jack of an operating system, is 5×10^{-29} J/Hz. It is worth digressing to note that standard SQUID sensors are limited entirely, both in regard to energy sensitivity and bandwidth, by the commercial electronics to which they are mated. Energy sensitivities of 6×10^{-30} J/Hz have been demonstrated with these standard sensors, and signal bandwidths of 10 mHz appear feasible without any sensor re-design. Immediate progress could be made in commercial equipment if development money were available.

The noise spectrum of these SQUIDs, when operated in a freely vented helium bath, is white from some cutoff frequency down to about 0.01 Hz. Below this frequency the spectrum has a 1/f character. It has been demonstrated that with some SQUIDs this 1/f knee can be pushed to lower frequency by temperature stabilization. This is also an area where we see good possibilities for advances in the state of the art.

Although we have been discussing SQUID sensors, per se, they are not ordinarily offered as separate items. One of the reasons for this reluctance is that while it is possible to define an intrinsic noise level for a SQUID sensor, the performance of the overall system is critically dependent on achieving a proper match between the SQUID, the rf preamplifier, and the feedback network. Thus the basic commercial package comprises an electronic control unit, an rf head which mounts on top of the dewar, a probe structure which houses the SQUID, and the SQUID itself. We characterize a system such as this as a Laboratory Probe.

III. LABORATORY PROBES

The minimal versions of this instrument provide the user with a fully operating SQUID which may be connected to his own input circuit. If his circuit is a superconducting loop, then he can measure magnetic flux changes in that loop. If the sensor is connected to some electrical device, then he can measure small current changes in the circuit. The ultimate sensitivity of the system in either of these modes is readily calculable from a knowledge of the SQUID's energy sensitivity and its input inductance.

More elaborate, yet still very much standard, versions of this laboratory probe are also manufactured. These probes have resistance and mutual inductance standards mounted in the tip of the probe along with the SQUID. There are also several shielded and filtered lines which run down to the probe tip. All of these items are available for interconnection by using a built-in terminal board. Using this type of probe it is simple to measure voltages or resistances with a resolution better than 10^{-12} volt or 10^{-11} Ω.

The mutual inductance standard which is available in the probe tip may be used in conjunction with a laboratory ac bridge to make measurements of self and mutual inductance to a resolution of 10^{-12} H. A system of this type is frequently used for millikelvin thermometry since the mutual inductance of two concentric coils can be made to depend on the susceptibility of a Curie law salt placed inside them.

Laboratory probes may be specified unsealed for direct immersion in a helium bath, or with a hermetic seal at the bottom so that the SQUID may be mounted in a vacuum can.

None of these probes are very exotic in their design or operation; nevertheless, they are useful. We might view them as cryogenic multimeters. Following this analogy, the SQUID itself would be like a bare galvanometer: not as convenient to use for general measurements.

IV. GEOPHYSICAL MAGNETOMETERS

These instruments are used by geologists and geophysicists to measure ground conductivity tensors. There are various techniques for determining this tensor from magnetic field data, but a common feature of them all is that the depth to which one probes is inversely proportional to the measurement frequency. Since superconducting instruments have a response which is flat down to dc, they offer a great advantage over conventional induction coils. It is especially important in this market for the manufacturer to fully understand the nature and problems of geophysical measurements and to incorporate this knowledge into the design of the system. The customer is frequently not familiar with the special problems of SQUID instruments, and must therefore rely heavily on the experience of the manufacturer. We would like to emphasize to the uninitiated user how important it is to get a standard, well-proven design.

The geophysicist should expect to get a standard system with many or all of the following specifications: a sensitivity of 10^{-10} Gauss/$\sqrt{\text{Hz}}$ down to 0.01 Hz; three identical channels which are orthogonal to better than 10^{-4} radians; an output that is linear over at least a 130 dB dynamic range; availability of electronic roll-off and notch filters; and built-in RFI shielding that is adequate in noisy environments. There are two more performance specifications that require special consideration. They are the maximum slew rate of the phase-lock network, and the roll-off frequency of an eddy current shield around the pickup coils. For a given pickup coil sensitivity these two parameters must be specified high enough and low enough, respectively, that the instrument can track the ever-present magnetic bursts known as sferics. In the past, sensitivity to RFI and sferics have been the most debilitating problems encountered in geophysical SQUID magnetometers.

The dewar must be designed as an integral part of these instruments, both mechanically and electromagnetically. The magnetometer probe should plug tightly into the dewar so that the pickup coils are very well constrained. The best dewars for this application are made of a fiberglass/epoxy composite that is both nonmagnetic and extremely strong. They are most commonly supplied in 10 to 25 liter capacities, but this is one specification that is entirely up to the customer. Instrument operating times of 18 to 24 hours per liter of liquid helium are typical. The dewars are invariably "vapor shielded" and therefore require no liquid nitrogen.

Commercial magnetometers now have specifications that are more than adequate to satisfy geophysicists' needs. The advances that we anticipate are directed more at convenience than performance. Digital conditioning and formatting of the magnetic field data would be one example of these. Remote or automated control of the instrument's tune-up and operation would be another. As more elaborate arrays of magnetometer systems are contemplated, however, these conveniences will become more necessary.

V. MAGNETIC ANOMALY DETECTORS

These are all highly balanced magnetic gradiometers but, beyond that, it is difficult to make generalizations. They have been used for detection and tracking of military targets, location of geological anomalies, and monitoring of geophysical phenomena. Gradiometers are made by taking two identical superconducting loops and connecting them in series opposition so that they show no response to uniform fields. If the loops are normal to a common axis, the configuration measures one of the diagonal elements of the tensor. If the loops lie in a common plane, then it measures an off-diagonal element.

More research and development has probably gone into this category of instrument than any of the others, yet no standard designs are available. The applications are so varied and usually so demanding that, in our experience, each instrument must be a completely custom design. Thus, a much greater responsibility is placed on the customer to understand what he really needs to measure, and what tradeoffs are available. To say that no standard designs exist does not imply that manufacturers are inexperienced. Many special techniques have been successfully developed for use in these instruments. This means that while the designs are custom, they need not be created "from scratch".

It is difficult to appraise the state of the art for this category of instrument since most of them are being evaluated for military applications and their performance is not widely advertised. We shall merely describe qualitatively what has been done in some previous systems. The most ambitious instrument ever delivered had five sensors for measuring all independent elements of the gradient tensor, plus three sensors for monitoring the magnetic field. The gradiometers were individually balanced from outside the dewar by the adjustment of superconducting vanes inside. It was a large, complex instrument, requiring a 90-liter dewar to house it, and a great deal of determination to balance it.

This concept of balance is central to the problem of designing gradiometers. Claims are frequently made that some gradiometers can be balanced to a very impressive level, i.e., one part in 10^7. Actually, this balance is neither very difficult to achieve nor very useful, in itself. The more relevant specifications are: (1) stability of the gradiometer balance against temperature fluctuation in the dewar; (2) stability of the balance against vibration and rotation of the dewar; and (3) stability of the balance against cycling the system to room temperature.

The importance of these specifications can be illustrated by a gedanken experiment. Suppose that a 20-cm baseline gradiometer of moderate sensitivity (10^{-11} gauss/cm-$\sqrt{\text{Hz}}$) is placed in the earth's field and is balanced to 1 part in 10^7. Suppose the coilform upon which the pickup loops are wound is made of pyrex, and in the course of operation the temperature fluctuates spatially by 10^{-3} K. That is, the mass of pyrex inside one coil is 10^{-3} K warmer than the mass inside the other coil. Since pyrex has a temperature-dependent paramagnetism, this process will create a change in gradiometer balance. If we make a few assumptions about the coupling between the coilform and the coil, we can determine that the balance change will only be one part in 10^8. However, this corresponds to a drift that is 50 times larger than the per-root-hertz noise of the system.

Error signals are a product of the uniform applied field and the rejection ratio (or balance) of the pickup coils. When looking at system stability we must consider both terms in the total differential of this product. The basic level of balance required for a given gradiometer depends on its mode of operation. Consider our hypothetical gradiometer once more. If it is to remain stationary in the earth's field, then the criterion for balance is that the largest expected fluctuation in the earth's field should not result in spurious signals that are visible above the system's intrinsic noise. In this case two parts in 10^7 would probably be adequate. On the other hand, if the gradiometer were to be rotated freely, then a physically impossible balance of one part in 10^{10} would be needed to make this rotation "invisible". Our purpose here is not to give a course in gradiometry, but to merely emphasize that a great deal of thought must be given to specification of the instrument as well as to its construction.

Quartz and silicon have both been used successfully as forms for pickup loops. They are both very clean magnetically as well as being dimensionally stable. The silicon offers better thermal conductivity, while the quartz is more readily available and is easier to fabricate. It is very important to provide precisely made grooves or steps to register the superconducting pickup coils, as this gives a better inherent balance. Any mechanism for improving this inherent balance must be extremely rigid, especially if vibration or rotation will be encountered. One of the best techniques is to use a superconducting vane on a carrier which is pressed directly into the coilform. The adjusting mechanism for the vane should provide a fine, calibrated motion. It should then be possible to mechanically disengage the mechanism from the carrier. Proven designs for this, and other balancing schemes, are available from the companies that build these instruments.

Our example of the drift-prone pyrex gradiometer points out the importance of using magnetically clean materials in the vicinity of the pickup coils. This critical volume encompasses not only the lower parts of the probe, but also includes certain parts of the dewar. Since paramagnetic susceptibility has a T^{-1} temperature dependence, construction materials that are cold are more of a problem than those at ambient temperature. In an attempt to reduce this effect, manufacturers have spent considerable time finding alternatives to fiberglass/epoxy as a dewar construction material. Several alternatives have been tried in commercial instruments, including solid fused quartz, polyimid fiber/epoxy and quartz fiber/epoxy composites. There are distinct engineering problems with each of these systems, but for some applications requiring great stability at frequencies below 0.1 Hz they may be required.

The advances we anticipate in gradiometer design are rather interesting. Suppose we add to an already high performance gradiometer, several magnetometers, aspect sensors, thermometers, pressure sensors, and a helium level detector. It should be possible to determine correlations between the gradiometer's residual errors and the outputs of these transducers. The effective balance of the gradiometer could be improved significantly by properly utilizing these correlations. The balance of multi-component gradiometers could also be augmented in this manner. A mini-computer is the obvious choice for implementing the technique.

VI. BIOMEDICAL MAGNETOMETERS

These are instruments that are designed to detect magnetic fields emanating from the human body or other biological specimens. They have primarily been used to examine hearts, lungs, muscles, brains and tumors. The one distinguishing characteristic of these instruments is that in each one a superconducting pickup loop must be mounted very near to the outside surface of the dewar. The present state-of-the-art distance is about 6 mm for a 3 cm diameter loop.

Only a few of these systems have been commercially produced, yet successful, standardized designs are available. These instruments can be tailored for optimum performance in various magnetic environments. Therefore, deciding on a suitable environment for a given experiment or project is one of the first orders of business for the researcher.

It is actually possible to obtain high sensitivity without any magnetic shielding in an average office building by using the right type of instrument. However, with a good shielded enclosure, one could probably gain an additional factor of 5 in signal-to-noise.

If a controlled magnetic environment is deemed necessary for a certain project, there are two basic approaches. The first is to make an enclosure out of high permeability ferromagnetic material. This method is quite expensive. It also takes a great amount of skill and perseverence to get the enclosure working optimally. The large pieces of ferromagnetic material can themselves be noise sources unless the suspension and degaussing systems have been properly constructed. Another difficulty with a ferromagnetic shield is that gradients inside the room are typically 1 μG/cm, or 5000 times as large as the intrinsic gradients in the earth's field. However, this type of shield provides by far the best magnetic isolation available for a room-sized volume.

The second method is to use high conductivity material (such as aluminum) for the walls, and obtain the desired isolation through eddy-current shielding. Such an enclosure can be mounted inside a large set of Helmholtz coils, so that the earth's dc field may be bucked out. It is also possible to use an active feedback arrangement so that low frequency and dc isolation may be increased electronically. A shield like this does not disturb the uniformity of incident fields as much as the high permeability type. Because the field uniformity is preserved, well-balanced gradiometers can be used to full advantage in these eddy-current shields.

Ferromagnetic enclosures are available commercially from several companies that specialize in magnetic shielding. The eddy-current type of enclosure could probably be designed and built by any large metal fabricating shop from simple sketches.

One biomedical instrument, which we feel is a state-of-the-art product, has three SQUID channels and a set of four interchangeable pickup coil arrays. Each of these coilforms has three or four different pickup coils which may be connected, using a shielded switchboard, to the three SQUIDs. Magnetometers, on-diagonal gradiometers, and off-diagonal magnetometers are all available by interconnecting the proper terminals. The basic operating sensitivity of this instrument is 8×10^{-11} Gauss/$\sqrt{Hz}$ with a 2.8 cm diameter pickup coil which is 7 mm from the outside surface of the dewar. All of the pickup loops are mounted on precisely grooved quartz forms. The intrinsic unbalance of the gradiometers was measured to be less than one part in 10^3. No better balance was desired so trimming was not attempted.

Magnetometers and moderately balanced gradiometers perform well in magnetically controlled environments, but they pick up far too much ambient field noise if they are used bare in a typical building. For this application, the highly balanced second-derivative gradiometer is the best available instrument. Pickup coils wound in this configuration respond to d^2B_z/dz^2, with the z-direction typically along the dewar axis. If this coil set is operationally balanced to a few parts in 10^6, the instrument exhibits remarkable immunity to stray magnetic fields. It is the R^{-5} distance dependence of this gradiometer that makes it so selective.

Biomedical gradiometers are often designed with sufficiently large spacing between their pickup coils that only the coil closest to the field source sees the source strongly. Thus, in their response to these local fields, they behave like compensated magnetometers. The one problem with using this compensated magnetometer approach is that a large fraction of the system's sensitivity can be wasted by using up input circuit inductance on the "far" coils which do not see the signal. One technique based on the use of asymmetric coils has already been tried which largely circumvents this problem, and it is certain that other tricks will soon be proposed.

The recent realization that it is possible to study magnetic fields of the brain without a tremendously expensive shielded room has created a flurry of interest in this field.

VII. SAMPLE MEASURING INSTRUMENTS

These instruments are used to measure the magnetic properties of small inorganic and organic samples which are placed inside their sample chambers. They may be divided into two distinct groups, known commercially as rock magnetometers and susceptometers. Most of the other instruments that we have discussed exhibit, by intent, a strong magnetic coupling to their surroundings. Thus, the user must always be concerned with what is taking place magnetically around the laboratory or test site. Sample measuring instruments do not have this complication, since they all use a superconducting shield around the SQUID pickup loops.

The rock magnetometers are specifically designed to measure remanent magnetization (direction and magnitude) of geologic samples as a function of temperature. One version of this instrument has a 3.8 cm diameter room-temperature access into its superconducting shield. The shield has a closed bottom and provides essentially perfect isolation from external fields through its factor of 10^7 attenuation. There are two SQUID channels, one connected to an axial pickup loop, the other connected to a pair of transverse saddle coils on the outside of the access tube. This entire assembly, comprising a thermally insulated access tube, magnetic shield, pickup loops and SQUID sensors, is built as a removable insert for a standard 15 cm neck diameter dewar. A support tower is provided for lowering samples into the pickup coils as well as rotating them.

The basic rock magnetometer makes measurements on room-temperature samples, but an insulated furnace is available which can optically head a sample to 1000°C while it is positioned inside the pickup loops. The sensitivity of the instrument is 10^{-8} emu/$\sqrt{Hz}$, and it has a dynamic range of 100 dB. The magnetometer can be factory-interfaced with a minicomputer for on-line data processing.

The superconducting susceptometer is qualitatively similar to the rock magnetometer except for the addition of a superconducting magnet outside the superconducting shield, and the provisions for maintaining the sample at any temperature between 400 K and 2 K. A standard model of this instrument is available which can apply a 2000 Gauss field to the sample. This field is stabilized to one part in 10^{14} per second by the superconducting shield. The sensitivity of the instrument is guaranteed to be 5×10^{-11} emu/cm^3-$\sqrt{Hz}$ in a measurement volume of 0.4 cm^3. The temperature stability is roughly 0.1%, with a resetability of about 1%.

As with the rock magnetometer, the susceptometer could be mated with data processing equipment. There has been some discussion about using a minicomputer to actually automate the operation of a susceptometer. This is a very attractive idea, since there are a number of processes that must be monitored and coordinated in the course of a complete measurement.

VIII. SHIELDED ENVIRONMENTS

These are the only instruments to be discussed here that are not designed entirely around SQUID sensors. They provide a relatively large diameter room-temperature access into a magnetically shielded environment. The basic technique is to use two nesting dewars, with liquid helium in the annular space between them. A superconducting shield, closed at the bottom, is mounted just inside one of the cold dewar walls. The shield is cooled by thermal conduction through its mounting to the dewar wall. A heater is wound on the shield so that it can be warmed above its transition temperature at will. In one commercial instrument the amount of thermal contact to the helium and the amount of power delivered to the heater were selected so that a complete cycling of the shield could be accomplished in 5 to 10 minutes. This particular instrument has a 15 cm diameter access, is mounted on a portable cart, and may be operated at any angle from vertical to horizontal.

Shielded environments up to 23 cm in diameter are relatively standard, and scaling the instrument up to larger sizes should not present any fundamental problems. It must be remembered, however, that the length and outside diameter of the structure will scale with the diameter of the access. Even a 23 cm system stands about 2.5 m high and weighs 320 Kg.

These instruments may be used as low field environments by placing them in a low field location, such as a shielded room or Helmholtz coil, and cycling the superconducting shield. An extremely stable field as small as a few microgauss may be trapped in this way. They can also be used to trap fields up to about 10 gauss. Attenuation of external fields can be outstanding with this type of instrument. Factors of 10^8 for transverse fields and 10^{13} for axial fields have been claimed.

Some of these shield systems have had superconducting pickup loops and SQUID sensors incorporated into them. This makes them, in effect, giant rock magnetometers.

A number of the shielded environments, especially the larger ones, have been equipped with closed-cycle refrigerators for cooling the thermal shields in the dewars. The refrigerator effectively eliminates all heat flux into the liquid helium, except that which comes down the neck of the dewar. On large systems, this technique can often increase the operation time by a factor of four. One shielded environment which was constructed recently can operate for over 100 days on a single (100 liter) charge of helium if the refrigerator is kept running. These refrigerators have also been incorporated into smaller instruments such as susceptometers and rock magnetometers, but the benefits are not as dramatic.

IX. CONCLUSION

From this rather informal tour of private industry's offerings, it should be obvious that superconducting instruments are no longer a laboratory curiosity. Hundreds of them have been manufactured and sold to a broad cross section of the technical community. More and more of the people who are productively using SQUID systems have only a superficial understanding of how they work, but this is a trend which must naturally continue as these instruments become even more widely used.

REFERENCES

1. S. H. E. Corporation, San Diego, California. Superconducting Technology, Inc., Mountain View, Ca.

Chapter 10

RESISTIVE DEVICES

J. G. Park

Department of Physics, Blackett Laboratory, Imperial College

London SW7 2AZ, United Kingdom

I. INTRODUCTION

Resistive SQUIDs, here called RSQUIDS, may be thought of as SQUIDs in which some of the superconducting material has been replaced by normal metal (Fig. 1). Magnetic flux ϕ_m varying with time, produced by current i_m in a coil inductively coupled to the RSQUID by mutual inductance M, will cause the output voltage v_o to oscillate just as it would for a SQUID, but only if i_m varies sufficiently rapidly. The voltage v_o may be caused to oscillate in another way: if an external current I is passed through the normal metal region N (Fig. 1b) v_o will oscillate at a frequency $f = IR/\phi_o$, where R is the resistance of N.

The RSQUID was not developed from the rf SQUID, but from the voltage-biased weak link by adding rf excitation to it [1] (Fig. 2). Before this excitation has been added, current oscillating at the frequency f, as predicted by Josephson [2], had been detected in point contact weak links [3] as well as in weak links of other kinds. At a frequency f $\sim$ 30 MHz a signal of about 10 μV could be detected using a tuned pickup coil (Fig. 2a), but it was difficult to detect any signal at much lower frequencies. It was soon found, however, that by applying rf excitation at frequency f_r, current oscillating at frequencies f much lower than f_r could be detected. However low the frequency f might be, the amplitude of the voltage across the tuned circuit was modulated at this frequency to a depth $\sim$ 10 μV at 30 MHz, proportionately deeper at higher frequencies of excitation (f_r). It was this amplitude modulated signal which was applied to the detector. The tuned circuit was tuned to the frequency of the rf excitation f_r and did not have to be re-tuned whenever there was a change in f. This

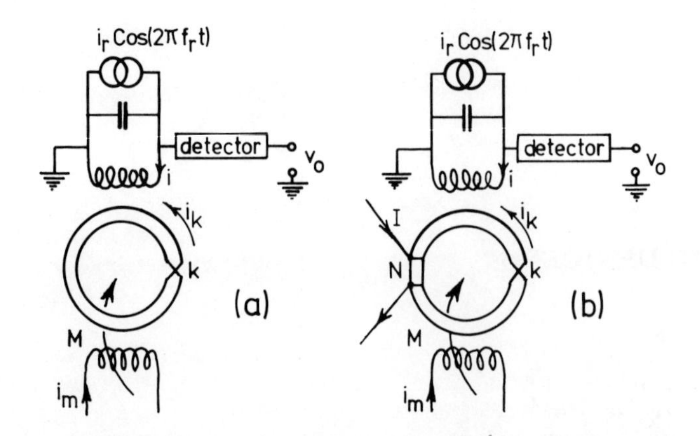

Fig. 1 (a) a SQUID system; (b) an RSQUID system. The SQUID may be imagined derived from the RSQUID by making the normal metal superconducting, every other part remaining unchanged. We call it the 'corresponding SQUID' of the RSQUID. (In order to avoid inessential complications, the effect of rf losses in the normal metal N has been neglected in the text.)

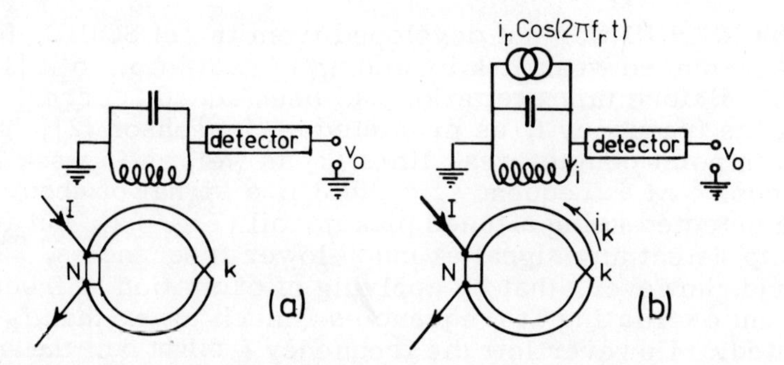

Fig. 2 (a) a weak link biased by a voltage IR; (b) an RSQUID system. It may be regarded as (a), modified by the addition of rf excitation.

method sets an upper rather than a lower limit to the frequency f. Clearly f must be smaller than the driving frequency f_r. The upper limit varies according to circumstance, but is likely to lie in the range 0.1 f_r to 0.01 f_r. Originally [1] the arrangement of Fig. (2b) was called the 'parametric mode'. The alternating voltage induced directly by oscillating current in the weak link, which would have been exceedingly small at low frequencies $f << f_r$, was said to have been amplified by 'parametric up-conversion'. However, in this paper we shall use the first picture, because we find that it provides us with a fruitful point of view: we shall regard the RSQUID as a modified SQUID. In the first part we consider the physical principles of the RSQUID [4]; then, in the second part, we discuss its applications as a noise thermometer, for which it has already been used [5-9]; as a pico-voltmeter, for which it was originally proposed [1] but for which, despite some valuable properties, it has been little used; and, lastly, as a meter of heat current, which we suggest [10] might make RSQUIDs useful for calorimetry.

II. THE 'CORRESPONDING' SQUID

The 'corresponding' SQUID is a SQUID (normally existing only in the imagination) which is identical with the RSQUID in every way, having the same rf excitation and detection circuits and the same kind and strength of weak link, except that the normal metal in the RSQUID is removed and replaced by superconductor. For the corresponding SQUID, as for any other SQUID, there will be a relationship between v_o, i_r and ϕ_m (displayed in Fig. 3), often referred to as the step pattern [11]. Here $\phi_m = Mi_m$: we shall follow the custom of calling ϕ_m the external flux. Any point on this diagram - which we shall call the v_o (i_r) diagram - corresponds to a certain cycle of states through which the SQUID passes repeatedly. The state of the SQUID can be represented by a point on the ϕ(i) diagram which relates the flux in the hole in the SQUID (ϕ) to the value, at any instant, of the current (i) in the coil coupled to it. (For simplicity we imagine that the modulation current i_m passes through the same coil as the rf current - a common if not universally used arrangement.) An example of a ϕ(i) diagram is sketched in Fig. 4. For clarity we have taken for this example a weak link across which the superconducting phase difference is negligible at all currents up to some maximum value - the critical current i_c. Such a model of a weak link is sometimes called a linear weak link, and for it the ϕ(i) diagram consists of straight-line segments each of which corresponds to an integral number of flux quanta. At any point on one of these segments the magnitude of the current in the weak link $|i_k| < i_c$. We use this model for definiteness. Except when we refer to computer calculations which explicitly use this model, all our discussion could be re-phrased and applied

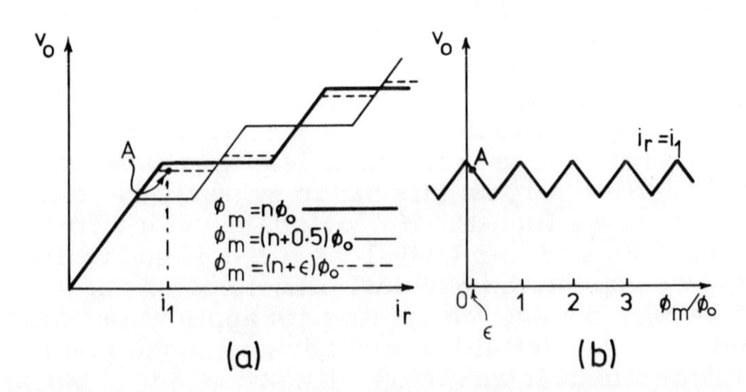

Fig. 3 (a) the 'step pattern' $v_o(i_r)$, somewhat idealized, for a SQUID. The shape of each curve depends on the modulating flux $\phi_m = Mi_m$, but the curves all lie within the envelope formed by those for $\phi_m = 0$, $\phi_o/2$. (b) the 'triangle pattern' $v_o(\phi_m)$ here sketched for $i_r = i_1$.

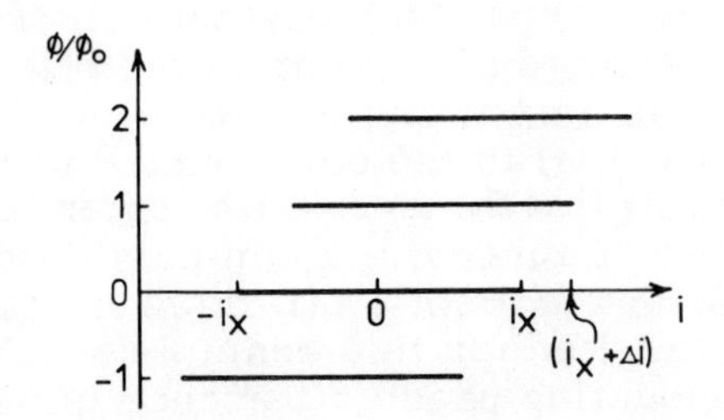

Fig. 4 The flux in a SQUID with a linear weak link. i_c is large enough that $\phi(i)$ is multiple-valued. $M\Delta i$ is the period of $v_o(\phi_m)$.

equally well to any other model of a weak link. Our argument only depends on the existence of a $\phi(i)$ diagram which consists of identical - not necessarily straight - segments, and which has the same symmetry properties as the $\phi(i)$ diagram of Fig. 4.

In order to visualize more clearly the relationship between a point on the $v_o(i_r)$ diagram and the cycle of states on the $\phi(i)$ diagram, consider this description of the cycle when ϕ_m is small ($\epsilon \ll 1$) and i_r is only just sufficient to produce transitions from $n = 0$ to $n = +1$ (or -1), as at A in Fig. 3a. We only consider here a SQUID working in the usual dissipative mode in which transitions are made from one segment of the $\phi(i)$ diagram to the next, and for which it is a necessary condition that $\phi(i)$ be multiple-valued. The cycle will largely be spent building up amplitude on the $n = 0$ segment until i reaches i_x (or $-i_x$). At that point there will be a transition to $n = 1$ (or to $n = -1$), but after a time of order $T_r/2$, where T_r is the period of the rf excitation current, at the moment when $i = -i_x + \Delta i$ (or $i_x - \Delta i$), there will be a transition back again to $n = 0$ where i will oscillate with an amplitude which is reduced at first, but which gradually builds up until $i = i_x$ (or $-i_x$) again, and the process repeats itself. The cycles are in general more complicated than this simple type, but they will always include at least one section which can be described qualitatively as the gradual building up of oscillation amplitude over a number of periods of the rf excitation current, a number varying with the rf level, but whose upper limit is presumably $\sim Q$, the quality factor of the tuned circuit. The period of a cycle (the fundamental period T_f as we shall call it) will clearly be greater than T_r. We must expect $T_f \leqslant QT_r$. (On the steep parts of the steps $T_f = T_r$ probably.) To any point on the $v_o(i_r)$ diagram such as A there corresponds a particular cycle. When $\phi_m = (\epsilon \pm 1)\phi_o$, $(\epsilon \pm 2)\phi_o$..., the cycles are the same as when $\phi_m = \epsilon\phi_o$ but displaced on the $\phi(i)$ diagram by $\pm\phi_o, \pm 2\phi_o$ ..., respectively, along the ϕ-axis, and $\pm\Delta i$, $\pm 2\Delta i$... along the i-axis. The cycle is the same in the sense that the oscillating components of current and magnetic flux in the SQUID are the same. Note that when the behaviour of the SQUID is represented on this particular diagram, the curves $\phi_m = (-\epsilon + \ell)\phi_o$, where ℓ is an integer, are superimposed on the curves $\phi_m = (\epsilon + \ell)\phi_o$, but must be regarded as distinct. It may be found convenient to think of the $v_o(i_r)$ diagram (Fig. 3) as consisting of two sheets, one behind the other. The curves $\phi_m = (\epsilon + \ell)\phi_o$, as ϵ ranges from 0 to 0.5, form one sheet; the curves $\phi_m = (-\epsilon + \ell)\phi_o$ form the other. The two sheets are joined along the edges of the envelope defined by the curves $\phi_m = \ell\phi_o$ and $\phi_m = (\ell + \frac{1}{2})\phi_o$. We shall refer to these latter curves as 'limiting' curves.

With any cycle of states there is associated not only a value of v_o, i_r and ϕ_m, but also a value of the mean phase difference across the weak link Θ_j and mean circulating current I_k. These mean values are averaged over a fundamental period T_f. It is well known that in a superconductor, wherever the current density is negligible, the vector potential $\vec{A}$ and phase θ are connected by the relationship

$$\vec{A}/\phi_o + \nabla\theta/2\pi = 0 . \tag{1}$$

We presume that the superconducting regions in any SQUID considered in this paper are thick enough that a path on which the current density is negligible can be found passing from one side of the weak link to the other. Thus we can write

$$\phi_m - LI_k = \phi_o \Theta_j/2\pi . \tag{2}$$

We have adopted the sign convention for I_k that I_k is positive when, in response to modulation current i_m, it flows in such a direction as to oppose the magnetic flux introduced by i_m. One can show [4], from the symmetry of the $\phi(i)$ diagram, that $I_k = 0$ when $\phi_m = n(\frac{1}{2}\phi_o)$ where n is an integer; that is, whenever the SQUID is represented by a point in the $v_o(i_r)$ diagram on one of the limiting curves. These two curves define an envelope within which all the other curves lie. Hence Θ_j changes by π whenever ϕ_m changes by $\phi_o/2$. We may just as well use Θ_j as ϕ_m as a parameter to label the curves on the $v_o(i_r)$ diagram. The limiting curves ($\Theta_j = 0, \pi$) on which $I_k = 0$, will be the same: the shapes of the other curves will be slightly different when $I_k(\Theta_j)$ is taken into account in Eq. (2), as will a 'triangle pattern' like the one in Fig. 5b as compared with the more usual one plotted against ϕ_m (Fig. 3b). However, the differences will be rather small since the amplitude (I_a) of the oscillatory function $I_k(\Theta_j)$ for a given value of i_r is likely to be small, much smaller at any rate than i_c. (In general, in a more realistic model of the SQUID, when the effect of thermal fluctuations has been taken into account, i_c will not, strictly speaking, represent the critical current of the weak link, but rather the mean current at which transitions take place from one segment of the $\phi(i)$ diagram to the next. It will not cause any confusion in this paper if we continue to use the term 'critical current' for i_c as long as we bear in mind what it really signifies.)

That I_a is likely to be much smaller than i_c can be seen to follow merely from the plausible assumption that $I_k(\Theta_j)$ is a single-valued function, and from the fact that I_k is zero whenever $\Theta_j = n\pi$ and so must therefore be periodic in Θ_j with a period 2π (or less - there may be other zeros; indeed we shall see later that there sometimes are, but we shall ignore their existence for the present since they do not affect the nature of the argument). Consider the $i_k(i)$ curves plotted in Fig. 6b. Note that when $i_m = n\Delta i$ where n is an integer, $\phi_m = n\phi_o$. If we plot $I_k(i_m)$

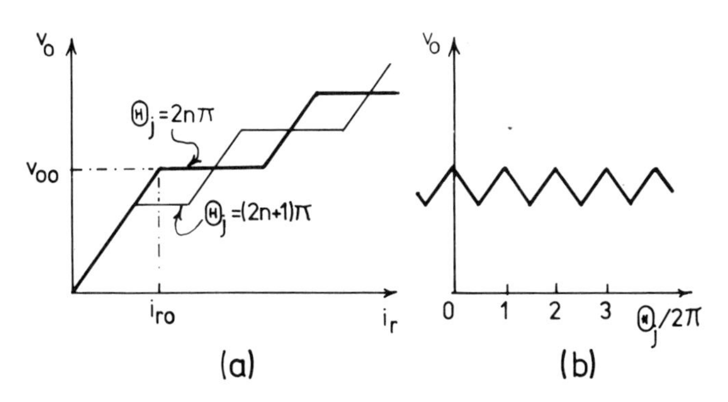

Fig. 5 (a) The envelope of the $v_o(i_r)$ curves. The curves inside the envelope are imagined to be labelled by values of Θ_j rather than ϕ_m as in Fig. 3. (b) $v_o(\Theta_j/2\pi)$ is similar in shape for given i_r to $v_o\ (\phi_m/\phi_o)$ and must coincide with it at the maxima and minima.

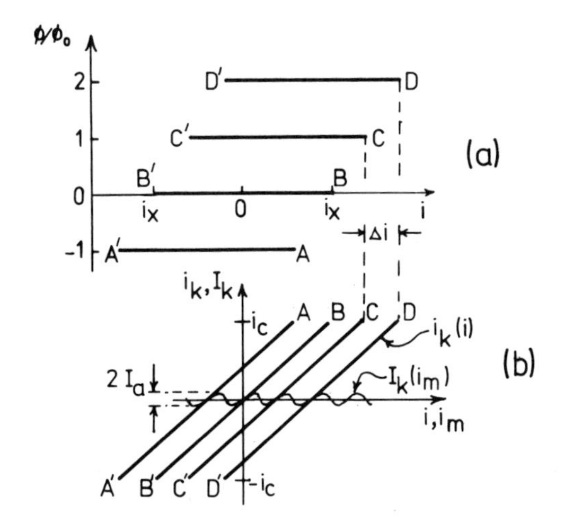

Fig. 6 (a) $\phi(i)$ plotted as in Fig. 4; (b) the current in the weak link $i_k(i)$ using the same i scale as in (a). On the segments AA', BB', CC'.... $|i_k| < i_c$: $\phi(i)$ and $i_k(i)$ can be obtained from the equation $\phi = Mi - Li_k = n\phi_o$ where $n = -1, 0, 1....$ Corresponding points where $i_k = \pm\ i_c$ are labelled with the same letter. Also sketched is a conjecture for the shape of $I_k(i_m)$.

using the same scale for i and i_m on Fig. 6b we can see that since $I_k = 0$ when $i_m = n\phi_o/2M \equiv n\Delta i/2$, $I_k(i_m)$ must have a form like that sketched on the figure. Here we have assumed that $dI_k/di_m \leq di_k/di$ when i, $i_m = n\Delta i$, and that $I_k(i_m)$ is single-valued. Both of these assumptions seem to us plausible: we have not attempted to prove them. Notice that if we make them, it follows directly by geometry that $LI_a/\phi_o < 1$, whereas for $\phi(i)$ to be multiple-valued, as it must be in the 'dissipative' mode, $2Li_c/\phi_o > 1$ (for the linear weak link model).

These are our reasons for expecting that $I_a \ll i_c$. They seem to be confirmed by our calculations mentioned below, at least in the range of parameters we have investigated.

III. THE RSQUID AND ITS 'CORRESPONDING' SQUID

The connection between an RSQUID and its corresponding SQUID can now be seen. I_k is the same function of Θ_j for both of them; and so is v_o. The difference in phase θ_n across the normal metal changes at a rate given by the equation

$$\dot{\theta}_n - \dot{\theta}_s = (2\pi/\phi_o)V,$$

where V is the instantaneous value of the potential difference across it, and θ_s is the phase difference that would exist across the same region if it were superconducting. This result can be derived [4] (as can Josephson's equation relating to the voltage across a weak link to the rate of change of phase difference across it by postulating the existence of such a quantity as the phase θ, which in a superconductor obeys Eq. (1) when the current density is negligible, and combining Eq. (1) with Maxwell's equations. At sufficiently high frequencies, where the skin depth δ in the normal metal is much less than the width w, one can show that the rf potential difference is proportional to $w/\delta \exp(-w/\delta)$. We shall assume that for an RSQUID the rf voltage across the normal metal will have a value so small as to be negligible. If the normal metal is brass with a resistivity $\rho \sim 10^{-7}$ ohm-m, $f_r = 15$ MHz, and if we require $\delta/w < 0.1$ for the rf voltage to be negligible, we need $w > 200\ \mu$m. If the metal is fairly pure copper, with $\rho \sim 10^{-10}$ ohm-m, we would require $w > 6\ \mu$m to satisfy the same criterion.

This is why we can assert that the cycles for an RSQUID and its corresponding SQUID are the same for a given value of Θ_j. Throughout a cycle, Θ_j is not affected by the presence of the normal metal since the rf potential difference is negligible. However, Θ_j will change gradually in the presence of direct or low frequency current through the normal metal. Not only will low frequency external current I produce a potential difference V, but so will the circulating current I_k, even when I = 0. These voltages, together with the Johnson noise voltage $v_n(t)$, will cause Θ_j, and the cycle (and all the other quantities such as the detector output v_o which depend on Θ_j) to vary with time. Most of these results can be summarized in the equation:

$$\dot{\phi}_m - L\dot{I}_k - (I_k + I)R - v_n = (\phi_o/2\pi)\dot{\Theta}_j. \tag{3}$$

It is interesting to compare this equation with the one we obtain for the SQUID if we differentiate Eq. (2):

$$\dot{\phi}_m - L\dot{I}_k = (\phi_o/2\pi)\dot{\Theta}_j \, . \tag{4}$$

For completeness we might have to add another term to the left-hand side of Eq. (3) to take account of the thermal EMF that would be produced by any stray heat current. We shall be considering the effect of a thermal EMF later, but it will be convenient to neglect it for the time being and assume that heat currents are absent.

The commonly observed forms of behaviour of an RSQUID can be derived directly from Eq. (3) when the term I_kR is negligible. If we neglect I_kR, and I is zero, Eqs. (3) and (4) become identical except for the noise voltage v_n. Hence it may be deduced that the response of an RSQUID to magnetic modulation is the same as the response of a SQUID except for the effect of noise, provided that i_m oscillates with sufficient amplitude at high enough frequency. This is what is generally observed. $v_o(i_m)$ has the same period ($\Delta i = \phi_o/M$) for an RSQUID as it does for the corresponding SQUID. When the triangle pattern $v_o(i_m)$ is displayed in the conventional manner on an oscilloscope (deriving the x-deflection from the current i_m alternating at a frequency of order 1 kHz, and the y-deflection from v_o) a triangular wave pattern is observed, as though a piece of the function $v_o(\phi_m)$ were observed through a window. However, because of Johnson noise the pattern appears to vibrate in a random fashion behind the window along the x-axis, instead of being stationary (as it would be for a SQUID).

If both I_k terms in Eq. (3) are neglected and the modulation current i_m is removed:

$$IR + v_n = (-\phi_o/2\pi)\Theta_j \, , \tag{5}$$

which, if I is a direct current, corresponds to oscillation in Θ_j and v_o, with a frequency modulated by noise about the mean (IR/ϕ_o), as observed in practice.

Throughout this paper we assume (implicitly) that the RSQUID and its corresponding SQUID are toroidal. If they are not, $\Delta i = \alpha\phi_o/M$, where [4] $\alpha > 1$. This fact can be taken into account by writing $\alpha\phi_o$ instead of ϕ_o on the right-hand side of Eq. (3). Notice that in the absence of modulation current i_m a current I produces oscillations at a frequency $f = \alpha^{-1} IR/\phi_o$.

The effect of I_k will be discussed later. Notice first that Eq. (3) is the same relationship we would obtain for the RSQUID in the absence of rf excitation if I_k and Θ_j were instantaneous rather than average values, and if the weak link had been modified so that the current-phase relationship was $I_k(\Theta_j)$ instead of $i_k(\theta_j)$. In this sense Fig. 7 may be regarded as an equivalent circuit, identical in type to the original circuit in Fig. 2a except that the method of detection by which $v_o(\Theta_j)$ is obtained is not shown. This method can be used (unlike the direct detection circuit of Fig. 2a) down to very low frequencies whose limit is determined by thermal fluctuations rather than the smallness of the induced EMF. If $I_k(\Theta_j)$ is sinusoidal, the analysis of the behavior of a Josephson junction in series with an inductor and resistor, biased by a source current I, may be applied to the RSQUID.

If $I_k(\Theta_j)$ is known, the response of the RSQUID to modulation current, whether external current I or flux modulation current i_m, may be deduced by applying Eq. (3). The response of the RSQUID may be deduced as long as Θ_j changes slowly enough that its variation may be treated as quasistatic, the cycle of states depending only on the value of Θ_j, and being the same at a given value of Θ_j as the cycle executed by the corresponding SQUID at the same Θ_j when Θ_j is constant in time. Our analysis, and the results obtained by applying Eq. (3), are only valid when this condition is satisfied. Clearly our analysis can only be valid when Θ_j oscillates at a frequency small in comparison with T_f^{-1}. Since T_f varies with the conditions under which the RSQUID is operated, so too will the frequency of oscillation of Θ_j above which our analysis becomes invalid. From what we have already said about T_f, it would seem likely that our treatment will only remain valid as long as f remains much less than a frequency which, depending on the level of rf excitation, varies from f_r to about f_r/Q.

In general we are not at present likely to know $I_k(\Theta_j)$ unless it has been determined experimentally. However, even without detailed knowledge of the form of $I_k(\Theta_j)$ there are several qualitative results we may deduce from Eq. (3) by applying it to some already published data.

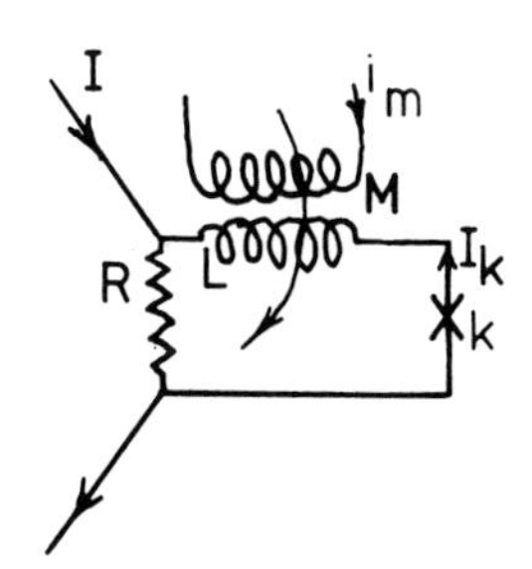

Fig. 7 Equivalent circuit of an RSQUID for direct or low frequency modulation current I and i_m.

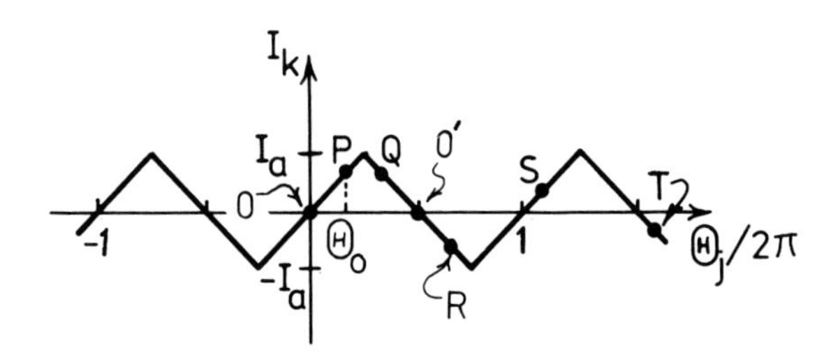

Fig. 8 A possible form for $I_k(\Theta_j)$.

IV. BEHAVIOR WHEN MODULATION CURRENTS I AND i_m ARE ABSENT

A. Stable and Unstable Equilibrium

Let us assume $v_n = 0$ and neglect the effect of noise in Eq. (3). In the absence of any modulation current, i.e., if $i_m = 0$ and $I = 0$, $\dot{\Theta}_j$ will be zero also if $I_k = 0$: the detector output will be constant in time. We may say then that the RSQUID is in equilibrium. To see how equilibrium is approached from a condition in which I_k is at first not zero, let us consider a particular form that $I_k(\Theta_j)$ might have. We shall refer to a point on the $v_o(i_r)$ or $I_k(\Theta_j)$ diagrams as the 'state point', sometimes just 'point'. Each such point represents a particular cycle around the $\phi(i)$ diagram. Suppose $I_k(\Theta_j)$ were triangular as indicated in Fig. 8 and that at $t = 0$, $\Theta_j = \Theta_o$. The RSQUID is represented by the state point P at first. It lies on the branch OP on which $I_k = I_a\Theta_j/\pi$ and varies with time according to the equation, obtained directly from Eq. (3): $\tau\dot{\Theta}_j + \Theta_j = 0$. Hence $\Theta_j = \Theta_o \exp(-t/\tau)$ where $\tau = [L/R + \phi_o/(2I_aR)]$, i.e.,

$$\tau \approx \phi_o/(2I_aR) \qquad \text{if } 2LI_a/\phi_o \ll 1 \,. \tag{6}$$

This inequality is likely to be satisfied for reasons given above. Typical values of L and I_a are 10^{-10} H and 10^{-7} A, respectively, so that a typical value of $2LI_a/\phi_o \sim 0.01$. The RSQUID behaves as though its inductance and the time constant L/R were both increased by a factor $\phi_o/(2I_aL)$. In general, whatever the form of $I_k(\Theta_j)$, $\dot{\Theta}_j$ is negative when I_k is positive, so that at points such as P, Q and S the state point moves to the left. Likewise, at points such as R and T, where I_k is negative, the state point moves to the right. In either case it moves with a speed proportional to the distance of the state point above or below the axis. It follows that points such as 0 (at which $I_k = 0$ and $dI_k/d\Theta_j > 0$), are points of stable equilibrium. On the other hand, points such as 0' (where $I_k = 0$ but $dI_k/d\Theta_j < 0$) are points of unstable equilibrium in the sense that however close to 0' the state point may be initially, the point will eventually move away from rather than towards 0'. We shall not discuss the possibility here that 0' may be stable too if $I_k(\Theta_j)$ is so steep that $(2\pi L/\phi_o)(dI_k/d\Theta_j) < 1$.

Except for the explicit form of $\Theta_j(t)$, and the expression for τ, the same conclusions will be arrived at for any other form of $I_k(\Theta_j)$. I_a will have to be re-defined (particularly if $I_k(\Theta_j)$ is asymmetric (as encountered in Section IVC), but it will generally have the same order of magnitude as the maximum value of I_k. Likewise, although there will be some characteristic time $\tau \sim \phi_o/(2I_aR)$, in general Θ_j will not return to equilibrium in the simple exponential manner derived above, except for small displacements.

B. Deviations from the Standard Behavior

The standard relationships normally observed ($f = IR/\phi_o$, when $i_m = 0$; $\Delta i = \phi_o/M$, when $I = 0$) are likely to be modified when one cycle of oscillation in v_o takes place in a time comparable with τ. For these relationships to hold one must have $f^{-1} << \tau$ in one case and $\phi_o/\dot{\phi}_m << \tau$ in the other. These conditions can be combined in the simple statement that I_k can be neglected (a sufficient condition) provided that the voltage $(IR + \dot{\phi}_m) >> I_a R$. Harding and Zimmerman [12] observed that a rate-dependent effect on Δi could not be observed for their RSQUID under magnetic modulation, even when the time taken for one period of oscillation $\phi_o/\dot{\phi}_m$ was as large as 30L/R. Their observation can be understood if we suppose that $2LI_a/\phi_o < 0.06$ in their experiments. From Section II we can see that this is not unlikely.

When $i_m = 0$ and an external current is applied which is slightly greater than I_a, the output voltage v_o oscillates at a frequency f lower than IR/ϕ_o. If $I_k(\Theta_j)$ is known to be sinusoidal, the form of the oscillations in v_o with time can be found analytically [13] as a function of I/I_a. When L can be neglected it is found that $f = (IR/\phi_o)(1 - I_a^2/I^2)^{\frac{1}{2}}$. If $I < I_a$, then v_o will not oscillate, since all the current will pass through the weak link. However, under these conditions, as we shall see later in Section IVD, the value of v_o will vary with the value of I even though it does not oscillate.

C. The Form of $I_k(\Theta_j)$

We now consider the evidence [17] that $I_k(\Theta_j)$ does not always have the form sketched in Fig. 8, but that in a wide range of circumstances there are extra zeros in I_k corresponding to stable equilibrium at values of Θ_j intermediate between $2n\pi$ and $(2n+1)\pi$. When, at a particular value of i_r there exists one of these extra zeros, the zeros of I_k when $\Theta_j = 2n\pi$ and $(2n+1)\pi$ are unstable, i.e., $dI_k/d\Theta_j < 0$ there. This is what we find in the computer calculations described below. We suggest that these results reveal a general kind of behaviour common to all RSQUIDs. We find that the existence of these extra zeros enables us to understand a result obtained by Zimmerman et al. [14], who measured, among many other things, $v_o(i_r)$ when i_m and I were both zero and i_r was increased slowly.

The result is shown in Fig. 9. As can be seen, the curve follows the $\Theta_j = 0, \pi$ lines of the corresponding SQUID on the rising parts of the steps, but in the region of a step such as aa' crosses

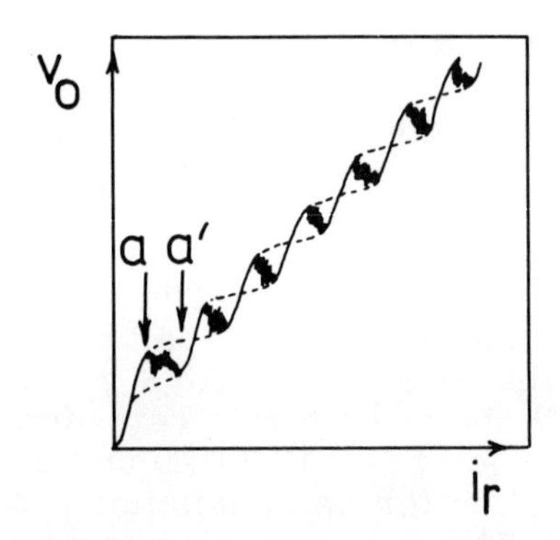

Fig. 9 $v_o(i_r)$ as observed in an experiment [14] in which the amplitude i_r was swept slowly upwards. On applying external direct current I, the state point oscillated between the dashed curves - which were thus determined experimentally.

over from one line to the other, following a noisy path. The two lines $\Theta_j = 0, \pi$, shown as dashed lines in the figure, which limit the envelope on the $v_o(i_r)$ diagram, can be determined with the apparatus used for such an experiment merely by passing a current I through the normal metal: v_o then oscillates up and down at any particular value of i_r so that as the value of i_r is increased the state point sweeps out the area inside the envelope.

In our computer calculations of $I_k(\Theta_j)$ we have taken for our model a SQUID incorporating a linear weak link. This model is not a particularly realistic one. For instance, it does not take any account of thermal fluctuations. But the qualitative features we are considering are probably independent of the model used for the weak link. Between transitions, using this model we had a simple linear circuit, an LCR circuit of quality factor Q coupled to a superconducting loop (coupling constant K) excited by a source of current alternating at the resonant frequency. The current and voltage in any part on the circuit could be expressed as the sum of a transient term, and a steady state term. In the calculation the values of the currents and voltages were found at the moment i_k reached i_c, and then the phase and amplitude of the transient were obtained after each transition had taken place. This procedure, starting from prescribed but usually arbitrary initial conditions, was continued until a pattern repeating itself sufficiently closely was found. Some results of such a calculation when Q = 50 and K = 0.3 are shown in Fig. 10. Here v_o is the quantity that would

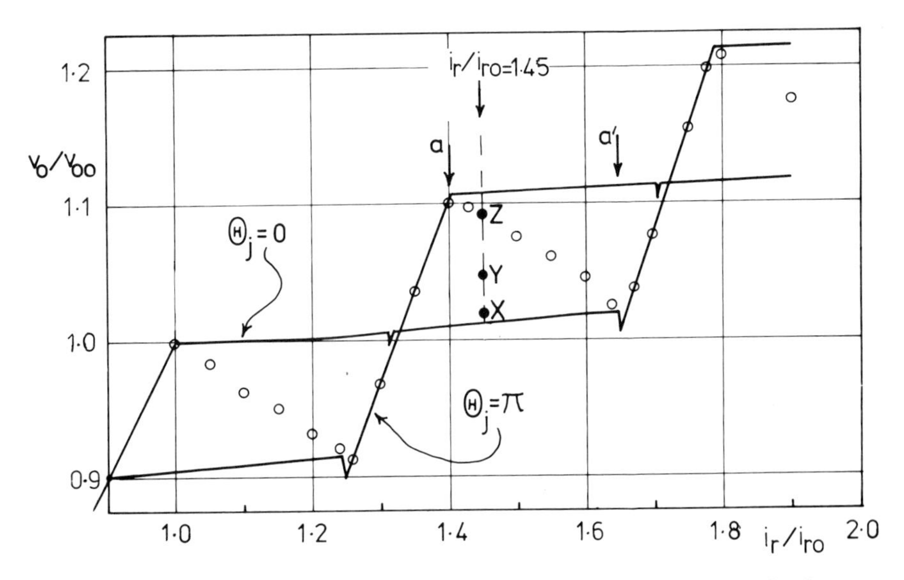

Fig. 10 $v_o(i_r)$ obtained for a linear weak link SQUID by calculation. (The tuned circuit quality factor Q and K, the coupling constant between coil and SQUID, were 50 and 0.3 respectively in this example.) v_o and i_r are normalized to the corner of the first step for $\Theta_j = 0$, where $v_o = v_{oo}$ and $i_r = i_{ro}$ as indicated in Fig. 5.

be measured by an ideal peak detector. (In this example values for T_f as high as 140 T_r were found. $I_k(\Theta_j)$ for three values of i_r on the second step are shown in Fig. 11. The range aa' in Fig. 10 corresponds to a range of i_r/i_{ro} between 1.4 and 1.65. Inside regions such as aa' (as when $i_r/i_{ro} = 1.45$), points like Z are the points of stable equilibrium. On the other hand, just below a (e.g., when $i_r/i_{ro} = 1.35$) points such as 0' ($\Theta_j = (2n+1)\pi$) are stable; while just above a' (e.g., when $i_r/i_o = 1.75$) points like 0 ($\Theta_j = 2n\pi$) are stable. In Fig. 10 we have plotted the calculated points, within the envelope, at which $I_k = 0$. It will be seen that lines through these points cross over from one side of the envelope to the other just as the curve shown in Fig. 9 does. The experimental curve can be understood if one supposes that the sweep rate is slow enough for the state point to move along the locus of equilibrium points, but that there are fluctuations about these points due to Johnson noise. From Fig. 11 we are led to suggest that the fluctuations are much larger at points inside the envelope than they are about points on its boundary (the $\Theta_j = 0, \pi$ curves) because $dI_k/d\Theta_j$, which determines the amplitude of fluctuations, is much smaller there. This can be seen from the results of the next section.

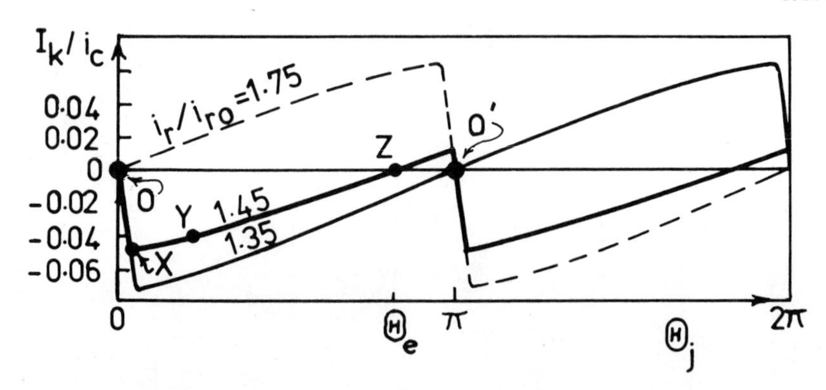

Fig. 11 $I_k(\Theta_j)$ for three values of i_r; obtained from the same computer calculation as Fig. 10.

D. Fluctuations about Equilibrium

The power spectrum of fluctuations in Θ_j about an equilibrium point $\Theta_j = \Theta_e$ such as Z in Fig. 11 when $i_r/i_{ro} = 1.45$ (or, to take another example, about 0 when $i_r/i_{ro} = 1.75$) can readily be calculated as long as the rms displacement of Θ_j from Θ_e is sufficiently small that $I_k(\Theta_j)$ may be treated as linear. We can then write $I_k = I_a \Theta_j/\pi$ as we did before in connection with Fig. 8 - but with a slight change in the definition of I_a. We can find the fluctuation spectrum of Θ_j from the equation $\tau\dot{\Theta}_j + \Theta_j + v_n\pi/(I_aR) = 0$, by exactly the same method one would use in order to find the noise current in a circuit consisting of a resistance and inductance. Thus if we write $\Theta = \Theta_j - \Theta_e$:

$$\langle\Theta^2(\omega)\rangle = (2/\pi)\,[kT/(I_a^2R)]\,(1+\omega^2\tau^2)^{-1} \quad \mathrm{Hz}^{-1}, \tag{7}$$

and

$$\delta\Theta \equiv \langle\Theta^2\rangle^{\frac{1}{2}} = [2\pi kT/(I_a\phi_o)]^{\frac{1}{2}} \tag{8}$$

E. Behavior when I is Small

Lines can be drawn on the $v_o(i_r)$ diagram which are the locus of other values of I_k beside zero. When I is small, the effect of I is merely to displace the state point until $I_k = -I$. When, for example, $i_r/i_{ro} = 1.45$ in an RSQUID, for which the model SQUID of Figs. 10 and 11 serves as the 'corresponding' SQUID, a current $I = 0.04\ i_c$ will displace the equilibrium point from Z to Y, Θ_j and v_o taking up new values appropriate to these points. If, however, I exceeds $0.048\ i_c$ - a current of the same magnitude but opposite sign to the value of I_k at X - Θ_j must decrease monotonically with time. The state point on Fig. 10 will then oscillate and so likewise will v_o.

V. EXPERIMENTS WITH EXTERNAL CURRENT I (ac OR dc)

Let us outline some experiments which illustrate these ideas. Our experiments were done on an RSQUID of toroidal geometry shown in Fig. 12. A ring of copper was pressed between discs of niobium. A toroidal coil carried both the rf excitation current and the modulation current i_m when required. A screw of niobium was adjusted in situ at low temperature to make the weak link. The ring was covered on top and bottom surfaces with a thin layer of Pb-Sn soft solder, which is superconducting at the temperature of our experiments (2-4 K), in order to make the surfaces equipotential and to minimize the contact resistance. Since the surfaces were equipotential, the configuration of the normal metal region was well enough defined that we could do some experiments comparing the behaviour of our RSQUID in practice with theoretical predictions. We were able for example to confirm, with accuracy of order 1%, that under magnetic modulation Δi was the same for the RSQUID as for the corresponding SQUID. Δi was also independent of the frequency of alternating modulation current down to 1 Hz even though for this RSQUID $R/(2\pi L) = 3$ Hz. In order to make these measurements the amplitude of modulation was maintained large enough that $\dot{\phi}_m \gg I_a R$. Hence there was no reason to expect to find any frequency dependence of Δi. It seemed sufficiently remarkable, however, to be worth checking that it was so. This was done for this RSQUID, and also for one of the same shape and size in which all the parts were made from copper except for the niobium screw and two patches of soft solder, one patch making contact with the thread, the other with the point of the niobium screw (Fig. 13).

When i_m was zero and a direct external current I was applied, v_o oscillated and could readily be observed on an oscilloscope. The frequency fluctuated as a result of Johnson noise. In order to measure the average frequency we used a simple if rather slow method which was sufficiently accurate for our purpose. The time base was triggered externally by a stable oscillator and I adjusted to successive values at which, as judged by eye, the picture appeared to be fluctuating about a fixed point rather than one which drifted to right or left. The result of one such experiment is shown in Fig. 14. Except for the points close

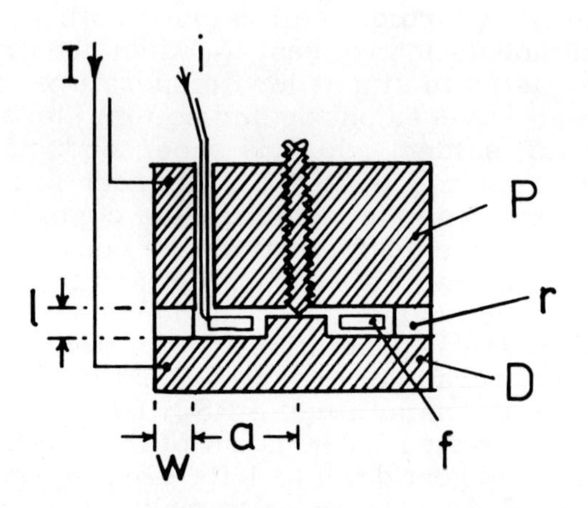

Fig. 12 The toroidal RSQUID used in our experiments. A copper ring (r) of thickness ℓ, width w and inner radius a is pressed between discs of niobium (P and D). The RSQUID is excited by rf current in the toroidal coil f.

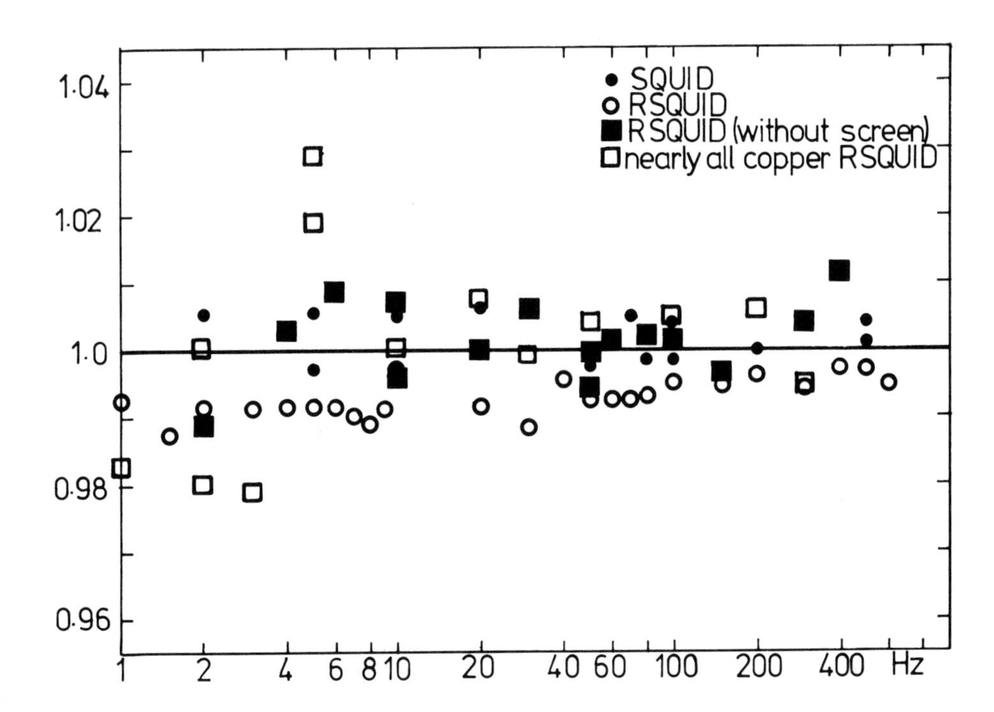

Fig. 13 The period Δi of the 'triangle pattern' (Fig. 3b - but regarded as a function of modulation current i_m rather than external flux ϕ_m) measured for a toroidal RSQUID in different circumstances. Δi_o is the value of Δi for the corresponding SQUID. The measurements were made from oscilloscope photographs by the use of modulation current i_m alternating at a frequency f_m. A corresponding SQUID was made by substituting for the copper ring a superconducting ring of the same dimensions. Δi_o was determined by taking the mean of a number of readings at $f_m \sim 1$ kHz. Measurements were made of Δi as a function of frequency f_m: (a) for the corresponding SQUID - as an indication of the intrinsic uncertainty in a single measurement made by the method we used. (There is an additional uncertainty in the measurements on the RSQUIDs due to Johnson noise); (b) for the RSQUID, with and without a loosely fitting lead screen; (c) for an RSQUID of the same dimensions all made from copper except for two patches of soft solder and a niobium screw.

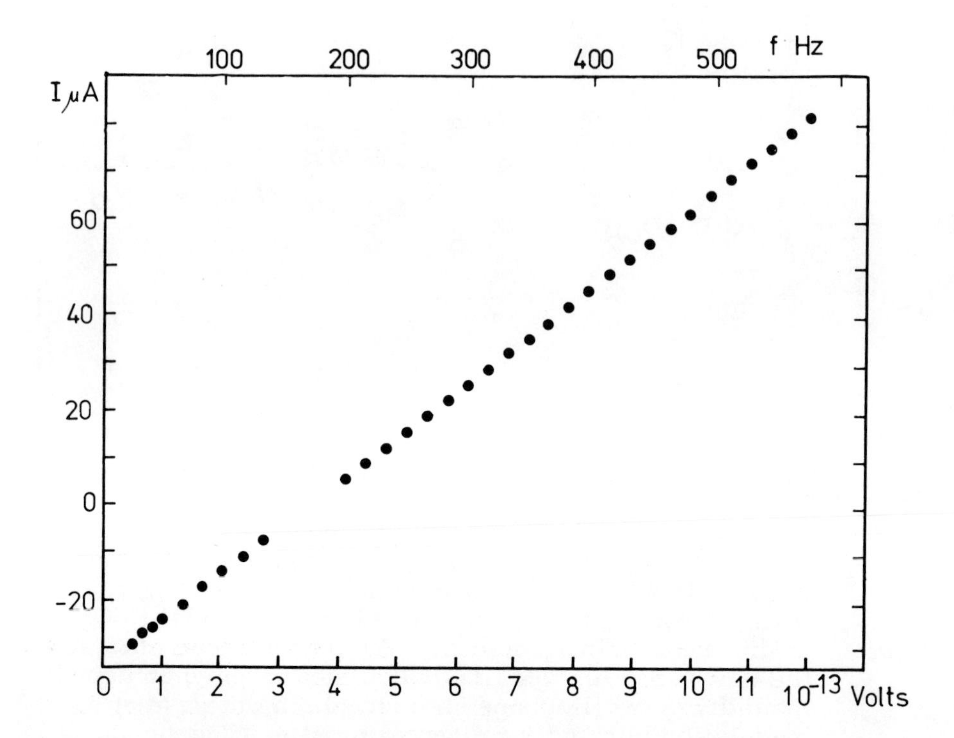

Fig. 14 The frequency of oscillation f for the RSQUID of Fig. 12 as a function of external current I; measured by the stationary pattern method described in the text. A thermal EMF of 3.5 10^{-13} V was present. The voltage values have been obtained merely by multiplying f by ϕ_o. A resistance of 1.04$_6$ 10^{-8} ohm was deduced for the copper ring (ℓ = 0.05 in, w = 0.125 in, a = 0.375 in).

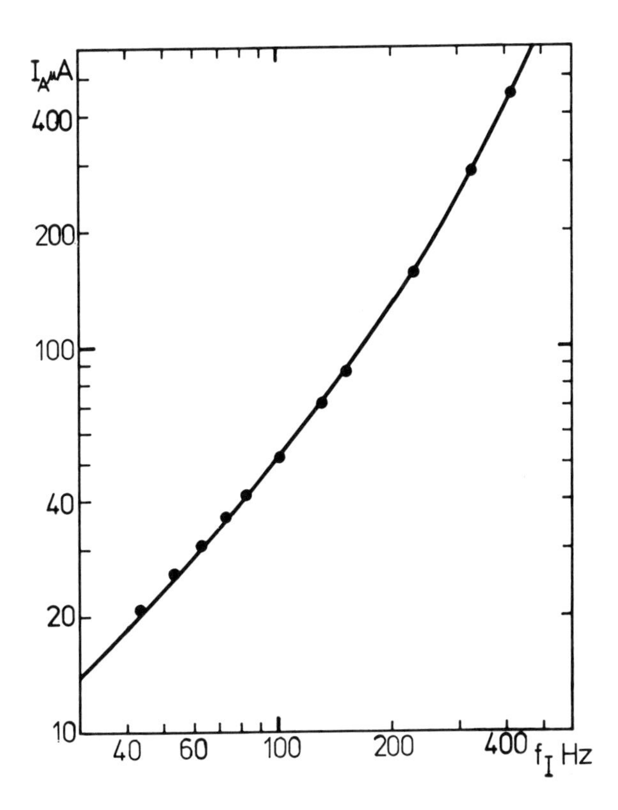

Fig. 15 The amplitude I_A of external current required, for the same toroidal RSQUID, to produce a given amplitude of alternating voltage V_A across the normal metal, plotted as a function of frequency f_I. The conductance I_A/V_A rises with frequency above f_o (where the skin depth $\delta = w$) because of the skin effect which causes the external current to concentrate and run down the outside surface of the ring. When $f \gg f_o$,

$$I_A \approx (V_A/R)(\delta/w)\exp(w/\delta) \equiv (V_A/R)(f_o/f_I)^{\frac{1}{2}} \exp(f_I/f_o)^{\frac{1}{2}}.$$

The curve drawn is the theoretical curve calculated using the values of R, a, w, and ℓ given in the previous caption. From them a value f_o = 47 Hz can be deduced. It has been assumed that contact resistance can be neglected and that the tinned surfaces of the ring can be treated as equipotential.

together on the left, the time base was triggered at 60 ms intervals. Thus the points corresponded to frequencies which were multiples of 16.67 Hz. The few points close together on the left corresponded to frequencies at 8.33 Hz intervals. A rough determination of the frequency of one of the latter points enabled the frequency of all the points to be determined precisely. As can be seen from the figure, there was a thermal EMF due, one presumes, to heat currents in the liquid helium surrounding the RSQUID which caused v_o to oscillate even when I was zero.

The geometry of this RSQUID is well enough defined that the effect of external current I can be predicted even when it is alternating. In this case the ac skin effect observed when the frequency f_I becomes sufficiently high, produces a decrease in the resistance $R_A = V_A/I_A$, where V_A and I_A are the amplitudes of the alternating voltage across, and current through, the normal metal. If ℓ is the thickness of the ring parallel to the direction of current flow $V_A = E(a) \cdot \ell$, where E(a) is the amplitude of the electric field E in this direction on the inside surface of the ring. In this geometry the current is distributed uniformly around the ring but varies radially, as the electric field does also. Thus the current produces no magnetic field inside the ring. Inside the ring, therefore, and in particular on its inside surface, the rate of change of vector potential $dA/dt = 0$: hence in the ring $E(r) = \nabla\phi_\rho$ when $r = a$ (where ϕ_ρ is the scalar potential), but not otherwise. $E(a)/I_A$ may be calculated analytically and the result expressed in terms of Bessel functions. The result of such a calculation was used to calculate the theoretical curve in Fig. 15. The high frequency behaviour may be derived more directly when the skin depth δ is much less than w, for we can then use solutions which are strictly speaking only valid for a plane parallel slab in a magnetic field oscillating parallel to its plane.

In order to test the prediction of our analysis (and also to confirm that it was reasonable to use our idealized picture of the normal metal ring in which the top and bottom surfaces are equipotential and the contact resistance at the copper-solder boundary is small enough to be neglected) we measured the conductance R_A^{-1} by measuring as a function of f_I the current I_A required, in combination with a direct current I, to produce a given degree of frequency modulation in a voltage which when $I_A = 0$ oscillated with a frequency $f = IR/\phi_o$. The degree of modulation chosen was that which was just sufficient to reduce the amplitude of the fundamental component at f to zero. When this condition was satisfied V_A had been adjusted to the value $V_A = 2.405\, f\phi_o$.

The presence of the fundamental at f was observed by applying the output voltage v_o to a phase sensitive detector (PSD) whose reference voltage was set to a frequency of $f + \delta f$, where $\delta f \sim 1$ Hz. When some of the fundamental was present, the output of this detector underwent oscillations at δf, together with some random phase modulation due to Johnson noise. I_A was adjusted until the amplitude of oscillations at δf was zero. The results are plotted in Fig. 15. Above the frequency f_o, the frequency at which $\delta = w$, the amplitude of I_A rises exponentially as $(f_o/f_I)^{\frac{1}{2}} \exp (f_I/f_o)^{\frac{1}{2}}$, corresponding to the decrease in R_A with increasing frequency. In this experiment we were essentially measuring the ac resistance of a hollow tube of inner diameter 2a and wall thickness w by the use of an ac voltmeter inside the tube. The more familiar skin effect, a resistance rising with increasing frequency as $f_I^{\frac{1}{2}}$ when $f_I >> f_o$, would be observed if the voltmeter were connected to the outside of the tube (Fig. 16).

VI. APPLICATIONS OF RSQUIDs

A. Types of RSQUIDs

Some of the different forms of RSQUID that have been used are shown in Fig. 17. All of them use a niobium point contact as weak link and usually brass, copper or bronze for the normal metal. They are all of a similar size, as can be gauged from the niobium screw used to make the weak link, which in each case has a diameter of about 1 mm. Except for (d) which is toroidal, like the RSQUID shown earlier in Fig. 12, these are all for the most part made from rectangular blocks. Superconducting parts,

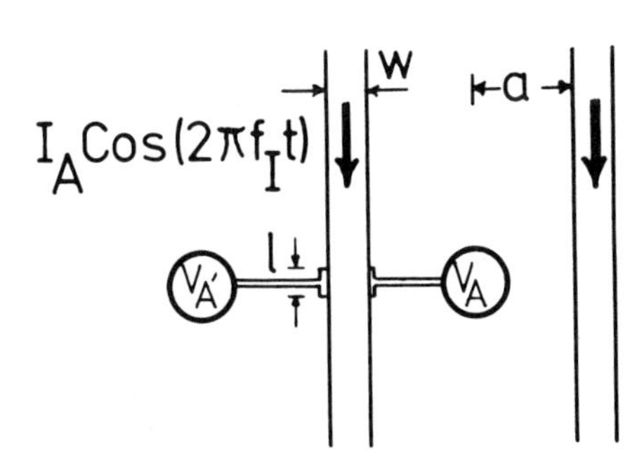

Fig. 16 An ac voltmeter inside a long copper tube of the same cross section as the ring in the toroidal RSQUID of Fig. 12 would measure the same amplitude V_A as the RSQUID fed with external alternating current $I_A \cos (2\pi f_I t)$ if this current were passed down the tube.

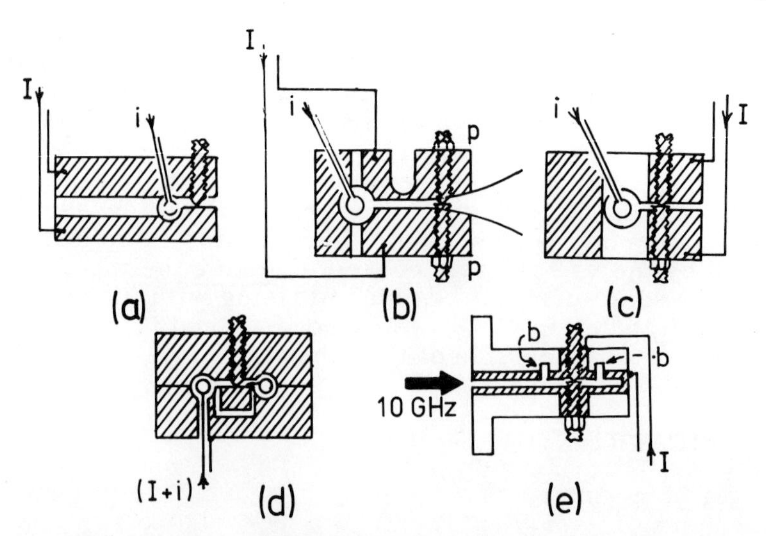

Fig. 17 Some forms of RSQUID that have been used, superconducting parts shaded. All joints are brazed or soldered except for the pressure contacts: N to S in (a), S to S in (d). These RSQUIDs were used by (a) Harding and Zimmerman [12] (flux modulation); (b) Fife and Gygax [15] (microwaves incident on the weak link, via a tapered wave guide, for mixing experiments - not discussed in this paper); (c) Kamper and Zimmerman [6] (noise thermometry); (d) Kamper et al. [7] (noise thermometry); (e) Soulen and Finnegan [9] (noise thermometry).

shown shaded, are made from niobium except in the microwave RSQUID (Fig. 17e) where, except for the two screws, they are made from Babbit metal. The weak link has to be adjusted at low temperature, usually by direct adjustment of the screw. In (b), however, it is adjusted by squeezing the two arms of the block at the two points pp. It may be that an RSQUID can be made along the lines of (c) which is sufficiently compensated for differential thermal contraction that it can be pre-adjusted at room temperature.

In order to minimize electrical contact resistance the normal metal is almost always soldered or brazed to a superconducting metal. It may be soldered or brazed to the superconductor directly as in all but (a) in Fig. 17, or tinned with soft solder and a superconducting connection between the solder and niobium made by a press-contact (as in Fig. 12). When the resistance R is sufficiently high, it may be that a direct pressure contact between normal and superconducting metals is adequate. In (e) the RSQUID is excited by microwaves at 10 GHz: bb is a circular groove, centered on the point contact, in which the Babbit metal is cut away so that current has to pass through the normal metal.

For reasons given earlier our treatment of RSQUIDS here is limited to devices in which the superconductor is thick in comparison with any of the lengths characteristic of superconductivity (penetration depth or coherence length).

B. Picovoltmeters

Although their use as picovoltmeters was the application for which RSQUIDs were first proposed, there has been little development of them for this purpose, unless one regards the RSQUID in the noise thermometer discussed below as a picovoltmeter. They have been used so little partly because picovoltmeters based on the response of SQUIDs to magnetic flux are just as sensitive and convenient, but also because SQUID systems are more readily available and many people have had experience in using them. Moreover, the SQUID can be adjusted once and for all (under favourable circumstances) and does not require any system in the cryostat for adjusting the weak link.

An RSQUID picovoltmeter will have properties that might make it more useful than a SQUID picovoltmeter in some applications. The voltage is applied directly or through a series resistance R_s (Fig. 18) to the normal region of an RSQUID. (The possibility also exists for applying alternating voltage to a coil which is magnetically coupled to the RSQUID, but we shall not consider that possibility here.) The voltage v_o oscillates at a frequency $f = IR/\phi_o$, where $I = V_x/(R + R_s)$, provided $I \gg I_a$. By I_a we mean here the current below which v_o ceases to oscillate.

It is difficult at first sight to see how negative feedback of the conventional sort could be applied to an RSQUID picovoltmeter because f becomes a very nonlinear function of I when $I \sim I_a$ and is zero when $I < I_a$.

There are two ways in which one might circumvent this difficulty and make negative feedback possible. First, one might derive the feedback from v_o directly. As we saw above, if v_E is the value of v_o at the equilibrium point when I is zero, $(v_o - v_E)$ is proportional to $-I$ and hence to $-V_x$ when I is sufficiently small $(I \ll I_a)$. A feedback current proportional to $(v_o - v_E)$ could then

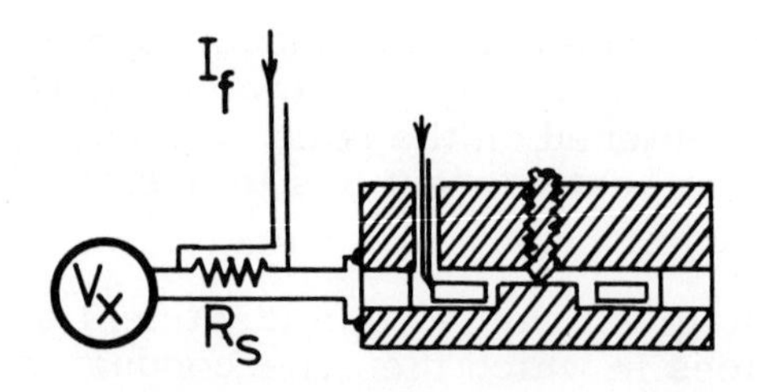

Fig. 18 Toroidal RSQUID picovoltmeter.

be applied to R_s (or even to R if low rather than high input impedance were required).

Second, one might use more conventional methods which require an alternating modulation current i_m, using the modulation to reduce the mean value of I_a. One can see that this is a possibility by working out the mean current in a weak link for which $I_k = I_a \sin(\Theta_j)$ when a triangular wave modulation current of peak to peak amplitude $2\Delta i_m$ is applied to it. At a sufficiently high frequency, $f_m \gg f_o$, the swing in Θ_j is the same as it would be for the corresponding SQUID, i.e., $\Delta\Theta_j = 2\pi M \Delta i_m/\phi_o$. Hence the mean value is given by

$$\bar{I}_k(\Theta_j) = I_k(\Theta_j)(\sin \Delta\Theta_j/\Delta\Theta_j) . \tag{9}$$

The output of a PSD, with v_o for its input, and a voltage alternating at the same frequency as the modulation current for its reference signal, can be fed back to the RF coil, as in a SQUID magnetometer. The feedback current at a given value of I rises linearly with time at a rate proportional to IR just as though the system were an integrator. A current proportional to the PSD output simultaneously fed back to R_s then produces negative feedback of the usual sort.

These suggestions are speculative and have not been tried out as far as I know.

The ultimate sensitivity of an RSQUID picovoltmeter, like that of one based on a SQUID, depends on Johnson noise in the resistance of the measuring circuit. In many applications, however, the ultimate sensitivity is determined by fluctuations produced by other extraneous causes such as magnetically induced rf fields and voltages of low frequency. It is thus worth taking note of the self-screening property of the toroidal RSQUID as shown by the response to alternating external current investigated in the experiments described above (Section V). Not only is the RSQUID screened from alternating current induced in the measuring circuit at frequencies much above f_o or below $L_m/(R_s + R)$, where L_m is the inductance of the measuring circuit, but so also is the measuring circuit screened from the rf excitation current and modulation current inside the RSQUID. Moreover, an RSQUID unlike a SQUID is insensitive to the direct effect of a steady magnetic field, provided the field is not large enough to affect the weak link itself. In some applications the stray field produced at the SQUID in a picovoltmeter by high current leads or large magnetic fields is difficult to eliminate.

C. The RSQUID Noise Thermometer

This device [5-9, 19] makes use of the fluctuations in the frequency with which v_o oscillates in the presence of an external current I as a result of Johnson noise. The frequency spectrum of $v_o(f)$ includes a line at IR/ϕ_o whose width δf can be written $\delta f = 2\pi/t_n$, where

$$t_n = \phi_o^2/(2kTR) . \qquad (10)$$

Rather than measure the width of this line it is found more convenient to make measurements with a frequency counter, which counts the number of cycles undergone by v_o in a certain time τ_g, referred to as the gate time. The variance of a large number (N) of frequency measurements $(f_1, \ldots f_i, \ldots f_N)$ is given by

$$\sigma_f^2 = \langle (f_i - f_o)^2 \rangle = (t_n \tau_g)^{-1} + (12\tau_g^2)^{-1} , \qquad (11)$$

so that the absolute temperature can be obtained directly if R is known, from the relationship

$$T = (\phi_o^2 \tau_g)/(2kR) \; [\sigma_f^2 - (12\,\tau_g^2)^{-1}] . \qquad (12)$$

In both Eqs. (11) and (12) the term containing $12\tau_g^2$ takes account of the fact that the counter can only count a whole number of cycles. That the factor 12 should appear in these equations rather than 6 has been demonstrated recently [19]. Values of σ_f^2, each obtained from N measurements f_i would themselves have a statistical distribution about their mean whose variance would be $2\sigma_f^4$. Hence, one can estimate the rms uncertainty in the temperature calculated from Eq. (12):

$$[\delta T/T) = (2/N)^{\frac{1}{2}} [1 + t_n/(12\tau_g)] \qquad (13)$$

One is generally more interested in the time t_m taken to make a measurement of temperature of given accuracy than the number of values of f_i that are required (particularly if one's results are fed directly to a computer, as they have been in the work done so far on this thermometer). If we write $\tau_g = \gamma t_n$, the relationship between the measuring time t_m and the uncertainty in temperature can be expressed as

$$t_m = 2\gamma t_n [1 + 1/(12\gamma)]^2/(\delta T/T)^2 . \quad (14)$$

According to this relationship, t_m is minimized, for a given value of $(\delta T/T)$, when $\gamma = 1/6$ and then has the value

$$t_{min} = \frac{2}{3} t_n/(\delta T/T)^2 . \quad (15)$$

Thus for a 1% measurement of temperature with R = 10 $\mu\Omega$, one needs, according to this, a measuring time $t_{min} = 2.7\ 10^4 t_n$: 100 sec at 4.2 K, 1 hour at 0.12 K. These times can be reduced if R can be increased. R is limited in the RSQUID of Fig. 17d by the bandwidth Δf permissible for the system that follows the detector: for a typical 30 MHz system it is limited to about 5 kHz if the signal-to-noise ratio of the oscillating signal is to be high enough to avoid counting errors (14 dB is said to be needed) [9]. Presumably Δf must be smaller than this. In that case t_n must be greater than 1 ms and a 1% measurement will take at least 30 min. However, to make a measurement in this time at 2 mK (and thus equal in that respect the performance already achieved with a SQUID noise thermometer) [16] would require an RSQUID with R $\sim 10^{-2}$ Ω. To work with such a high resistance will probably only be possible with a microwave RSQUID; one of these with R $\approx 3\ 10^{-6}$ Ω has already been constructed and successfully operated [9] (see Fig. 17e). With such an RSQUID the signal-to-noise ratio is greater than that for a 30 MHz RSQUID, so that a greater bandwidth Δf can be used.

In order to compare the measuring times required by the two types of noise thermometer it is worth noting that the time constant L/R of the measuring circuit used with the SQUID is equivalent to t_n for the RSQUID system in the sense that a 1% measurement requires $t_m \sim 2.10^4$ L/R in one case and $\sim 2.10^4 t_n$ in the other. Note that t_m is independent of temperature for one (the SQUID) and inversely proportional to temperature for the other (the RSQUID). If t_m is the same for the two types of system at some low temperature, then, as the temperature is raised, t_m will remain constant for the SQUID system, but become shorter for the RSQUID system.

At present, for general purposes, when a noise thermometer is required, a SQUID system is more likely to be chosen, particularly when a small sensor is required or when extremely low temperatures (~2 mK) are to be measured. A commercial instrument is already available. Two properties of an RSQUID system might make it more attractive in the future. These are: (a) the rather direct connection between the absolute temperature T and the measured quantity σ_f^2 (or, more likely $\langle (f_i - f_{i+1})^2/2 \rangle$, which is less sensitive to long-term drifts and very low frequency amplifier noise)[6]; (b) the possibility of the rapid measurement of moderately low temperatures ($T \gtrsim 0.1$ K, say) using $R \sim 10^{-2}\,\Omega$ and the larger Δf available with a 10 GHz RSQUID.

Note that the RSQUIDs used for noise thermometry have been operated in the non-dissipative mode: this does not affect the validity of our remarks in this section.

D. Heat Current Measurement

Lastly we should like to mention a recent suggestion we have made for using an RSQUID as a detector of heat current [10]. It should be possible to determine the heat capacity of a specimen merely by measuring the amount of heat entering or leaving it when its temperature changes from one value to another; and also by the use of ac heating and the properties of thermal waves to make measurements of thermal diffusivity.

The principle is a simple one. A heat current $\dot{Q}$ through the normal metal will produce a thermal EMF $SW\dot{Q}$, where W is the thermal resistance of the normal metal, which will cause the output voltage v_o to oscillate at a frequency

$$f = SW\dot{Q}/\phi_o \, . \tag{16}$$

If all the heat current passes through the normal metal, $\dot{Q}$ can be determined by measuring f. This effect, when caused unintentionally by stray heat current, is generally regarded as a nuisance. Here we examine the possibility of putting it to good use.

We imagine the toroidal RSQUID of Fig. 12, whose proportions have been modified in the manner indicated in Fig. 19. X is a specimen attached to the lower disc D. The lower disc D and normal metal ring N are now much reduced in thickness and diameter so that their heat capacity, which will have to be corrected for, is made small in comparison with that of X. By placing the RSQUID in vacuum and suitably designing the clamp, any heat gained or lost by the specimen X, together with the disc, must pass through N. The rate $\dot{Q}$ at which heat flows to P from a specimen X (together with the disc D) may now be found by measuring f. The frequency f cannot exceed about 100 kHz in SQUID systems at present available, and the maximum value of $S \sim 10\,\mu$ V/K; hence the maximum temperature difference which

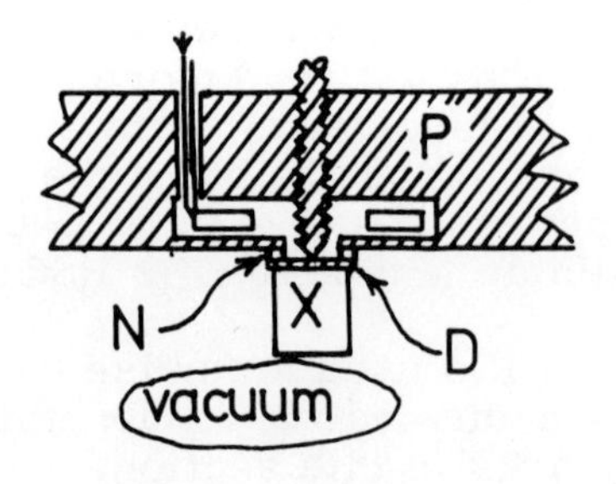

Fig. 19 Form of RSQUID proposed [10] for heat capacity measurement.

can be measured is $\delta T = f\phi_0/S \sim 20\ \mu K$. If the interfacial thermal resistances are sufficiently small one can therefore use a system of this sort to measure heat currents under what for most practical purposes may be regarded as isothermal conditions. One might use it, for example, in order to make a measurement of the heat capacity C of specimen X at temperature T. It should only be necessary to count the number of oscillations in v_0 that took place as the temperature of P changed from T to $T + \delta T$, for then $C = Q/\delta T = (\phi_0 n)/(SW\delta T)$.

It is clearly vital if this scheme is to be useful that the thermal conductance of the weak link be small. It is likely that it will be smaller than that of the ring by at least as large a factor as the normal state electrical conductance ($\sim 10^{-6}$) and can thus reasonably be neglected. (In saying this we are presuming that the Wiedemann-Franz law may be applied to the weak link in the normal state, if only approximately.)

. It is also desirable that the frequency f, for a given value of Q, is not too sensitive to non-uniformity in the heat current density. It can be shown that f is less sensitive to non-uniformity in heat current the more uniform the thickness and composition of the ring can be made. The reason for this result is that if the heat current is not uniform, electric currents flow in the normal metal to maintain the superconducting surfaces equipotential while at the same time being distributed in such a way as to produce zero net current through the normal metal.

The sensitivity will be determined by the resistance (R $\mu\Omega$) of N and its thermopower S. If we write $S = \sigma T \ \mu VK^{-1}$ and assume that the Wiedemann-Franz law holds for the alloy used for N we have

$$n/Q = 2 \times 10^{3} \ \sigma R \ erg^{-1} .$$

Suppose we use CuFe (0.1 at.%) as the normal metal, a material for which $\gamma \sim 3.10^{-4}$ JK^{-2} cm^{-3}, $\rho \sim 1 \ \mu\Omega$ cm, and σ varies from 2.9 at 1 K to 2.0 at 6 K. If the CuFe is formed into a ring 3.10^{-3} cm thick, 1 cm in diameter, 1 mm wide (a volume $\sim 10^{-3}$ cm^{3}), it will have a heat capacity at 1 K of 3 ergs K^{-1}, a resistance $R = 10^{-2} \mu\Omega$, and a value for n/Q of 50 ergs^{-1}.

The ultimate resolution is determined by Johnson noise. If a frequency measurement is made with gate time τ_g the rms uncertainty in f can be obtained by applying Eq. (11). The characteristic time t_n for a ring of $10^{-2} \ \mu\Omega$ at 1 K, is 15.5 s and for this the uncertainty in Q after a measurement made in a time 10 s would be 6.10^{-3} ergs.

Since in this method the amount of heat that has flowed is registered as a number which can be counted up to large values if necessary, it is possible that the correction for the heat capacity of D and N may be made with unusually high precision. In order to gain some idea of the numbers one might have to deal with, consider a specimen X for which the heat capacity of N would be about 1%, such as a 10^{-1} cm^{3} sample (mass 0.9 gm) of CuFe of the same concentration as the ring. It would have a heat capacity at 1 K of 300 ergs cm^{-3} K^{-1}, corresponding to $n/\delta T$ = 15000 cycles K^{-1}.

If the feedback techniques discussed in Section VIB are feasible for picovoltmeters, they also can be used for heat current measurement. Electric current fed back to R_S in a picovoltmeter would be fed back to R (the normal metal) in a measurement of $\dot{Q}$. Note that when feedback current i_f is taken to the rf coil only, it will be proportional to the number of cycles since i_f was last set to zero. It can thus be used as an alternative method for measuring n.

We are at present constructing an apparatus to try out these ideas in practice for measuring the specific heats of dilute alloys containing magnetic impurities. At AERE, Harwell, they are hoping to apply these methods to the actinide elements whose specific heat is difficult to measure at low temperature owing to the self-heating produced by radioactive decay. Using our method it should be possible to balance out the self-heating merely by passing a constant direct current I through the normal metal in the RSQUID.

We are considering also using the RSQUID as a detector of unbalance in a thermal bridge, analogues of the Wheatstone and other bridges; dc bridges for the measurement of thermal conductivity; ac bridges for the measurement of thermal diffusivity. The dc bridge would, however, have to be balanced by using two sources of heat, one of which was variable; the ac bridge would need two sources likewise, one of them being variable in phase as well as amplitude.

REFERENCES

1. Zimmerman, J. E. and Silver, A. H., J. Appl. Phys. 39, 2679 (1968).
2. Josephson, B. D., Phys. Letters 1, 251 (1962).
3. Zimmerman, J. E., Cowen, J. A. and Silver, A. H., Appl. Phys. Letters 9, 353 (1966).
4. Park, J. G., J. Phys. F: Metal Phys. 4, 2239 (1974).
5. Kamper, R. A., Proc. Conf. Superconducting Devices, Charlottesville, Virginia (1967).
6. Kamper, R. A. and Zimmerman, J. E., J. Appl. Phys. 42, 132 (1971).
7. Kamper, R. A., Siegwarth, J. D., Radebaugh, R. and Zimmerman, J. E., Proc. IEEE 59, 1368 (1971).
8. Soulen, R. J. and Marshak, H., Proc. Conf. Appl. Superconductivity, p. 588 (1972).
9. Soulen, R. J. and Finnegan, T. F., Revue de Phys. Appl. 9, 305 (1974).
10. Park, J. G., to be presented at the IC SQUID Conf., W. Berlin, September 1976.
11. Giffard, R. P., Webb, R. A. and Wheatley, J. C., J. Low Temp. Phys. 6, 533 (1972).
12. Harding, J. T. and Zimmerman, J. E., J. Appl. Phys. 41, 1581 (1970).
13. Sullivan, D. B., Peterson, L., Kose, V. E. and Zimmerman, J. E., J. Appl. Phys. 41, 4865 (1970).
14. Zimmerman, J. E., Thiene, P. and Harding, J. T., J. Appl. Phys. 41, 1572 (1970).
15. Fife, A. A. and Gygax, S., J. Appl. Phys. 43, 2391 (1972).
16. Webb, R. A., Giffard, R. P. and Wheatley, J. C., J. Low Temp. Phys. 13, 383 (1973).
17. Park, J. G. and Kendall, J. P., Proc. Int. Conf. Low Temp. Phys. LT14, Helsinki 3, 164 (1975).
18. Park, J. G., Farrell, D. E. and Kendall, J. P., J. Phys. F: Metal Phys. 3, 2169 (1973).
19. Kamper, R. A., in this volume (Chap. 5).

Chapter 11

"HOT SUPERCONDUCTORS": THE PHYSICS AND APPLICATIONS OF NONEQUILIBRIUM SUPERCONDUCTIVITY*

J.-J. Chang and D.J. Scalapino†

Department of Physics, University of California

Santa Barbara, California 93106

I. INTRODUCTION

In recent years there has been increasing interest in the nonequilibrium properties of superconducting systems. In a broad sense, of course, all transport properties involve nonequilbirium (or metastable) states, and hence all of applied superconductivity depends upon these phenomena. However, we know from the fluctuation dissipation relation that linear transport coefficients can be determined by an analysis of equilibrium fluctuations. Thus, for some time efforts focused upon the nearly equilibrium situation and the problems of equilibrium fluctuations. Current interest in nonequilibrium superconductivity is associated with superconductors which are strongly driven so that the quasiparticles, phonons, and the pair field may be significantly different from their equilibrium values. It is this type of system, the "hot superconductor", which will be discussed in this paper.

Before beginning, though, let's say why we believe this is an area worth studying. First, there are already a variety of observed phenomena which appear to be associated with hot superconductors. Wyatt, et al. [1] and Dayem and Wiegand [2] observed enchanced critical currents in superconducting microbridges irradiated by microwaves. Tredwell and Jacobsen [3] observed similar but larger enhancements using high frequency phonons. Klapwijk and Mooij [4] have reported that microwaves can induce superconductivity 16 mK above the equilibrium T_c in an Al film. Figure 1 from Ref. [4] shows plots of critical current versus temperature for a thin Al strip in the absence (I_c) and in the presence (I_p) of 3 GHz radiation. Eliashberg [5] suggested that this effect would occur because of the redistribution of quasiparticles by an external drive which pumps quasi particles from low energy states near the gap to higher energy states. The low-lying quasiparticles are much more effective in

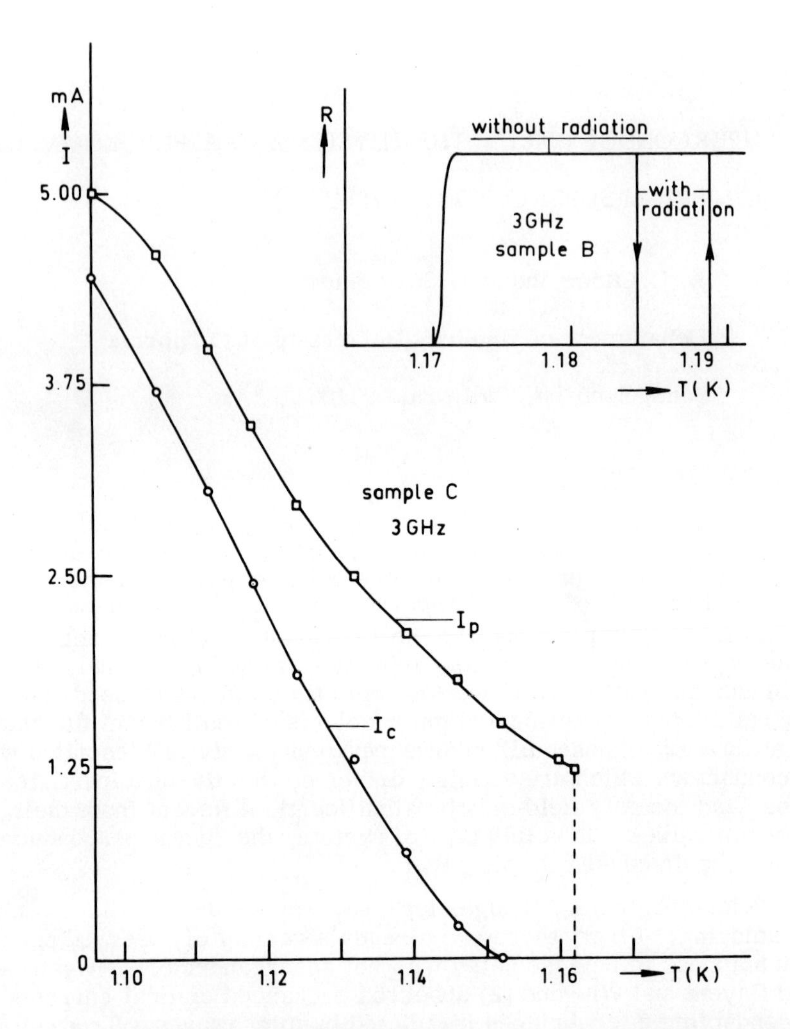

Fig. 1 Critical current versus temperature in the absence of radiation (circles) and in the presence of 3 GHz radiation of constant power (square). Inset shows transition with and without radiation (From Ref. [4]).

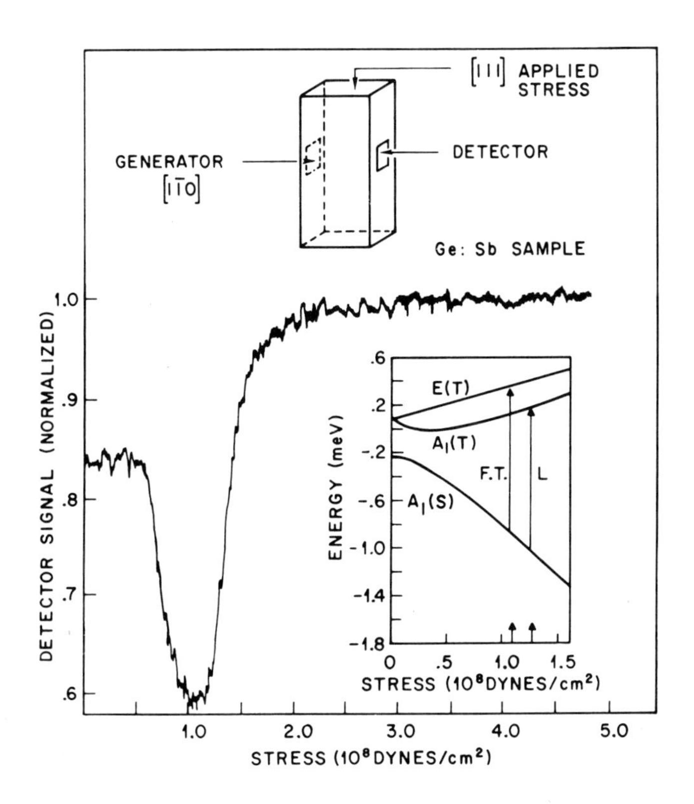

Fig. 2 Detector signal as a function of stress in the experiment to detect phonons emitted from a tunnel junction. For detail see text and Ref. [7].

inhibiting pairing correlations than higher energy quasiparticles. The ultimate limit of this type of pairing enhancement is an intriguing question.

Another type of phenomena involves the energy spectrum of phonons emitted by thin superconducting films driven by quasiparticle injection [6, 7] or heat [8]. Figure 2, taken from Ref. [7], shows in the upper left-hand corner an experimental configuration in which two tunnel junctions are separated by an Sb doped Ge crystal. The spectrum of the phonons emitted from one of the tunnel junctions is then analyzed using the strain-dependent absorption characteristics of the hydogenic-like levels of Sb in Ge and the other tunnel junction as a phonon detector. The observed spectrum has a large peak at an energy $\leq 2\Delta$ where Δ is the equilibrium gap of the emitting tunnel junction. Power levels of order milliwatts were produced. This spectrum can be tuned by applying a magnetic field to vary the gap. Its energy spectrum above 2Δ appeared to be insensitive to bias voltages up to $10\Delta/e$.

Testardi [9] observed a resistive state in a superconducting film irradiated by a laser which occurred at a temperature below the usual T_c and could not be explained simply in terms of heating. This stimulated a series of investigations on superconducting thin films driven out of equilibrium by laser irradiation. It now appears [10, 11] that above a critical intensity, laser irradiation can induce a driven mixed state. Figure 3 shows the normalized change in microwave reflectivity at 70

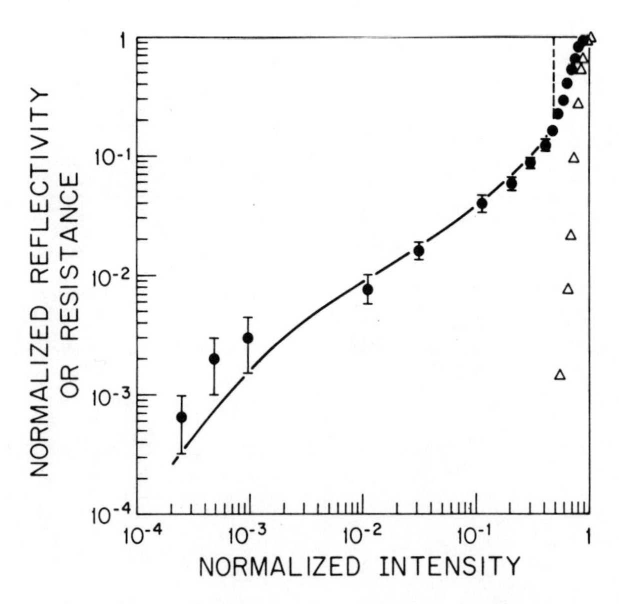

Fig. 3 Microwave reflectivity (0) and dc resistance (Δ) of a Sn film as a function of laser intensity (From Ref. [10]).

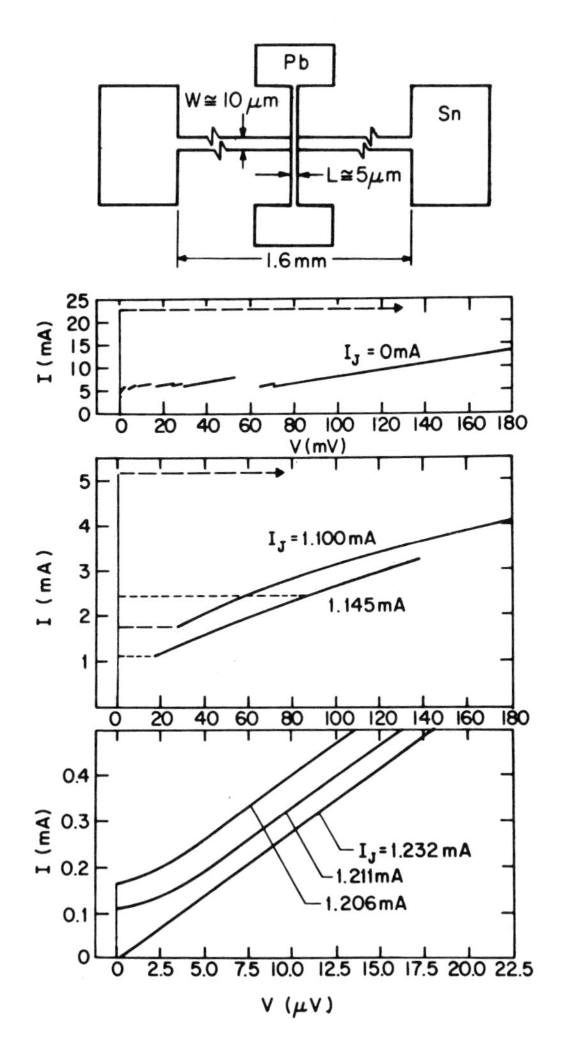

Fig. 4 The quasiparticle injection weak-link configuration and the I(V) characteristics of the driven weak-link for various current injections (From Ref. [18]).

GHz of an Sn film (plotted as the solid points) as a function of incident light intensity. The points are experimental, and the solid line is a fit to a theoretical model of Owen and Scalapino [12]. This model leads to a first-order-like transition to the normal state (dashed line) if the gap is assumed to remain uniform in space. Change and Scalapino [13] showed that, in fact, before the first-order-like phase transition occurs, this model exhibits an instability towards a spatially inhomoneneous state. In Fig. 3, the microwave data shows a gradual transition to the normal state beyond a normalized intensity of about 0.5. In this same region the film has a dc resistance (Δ data points) intermediate between zero and the normal state resistance. This mixed state may consist of a regular spatial array of superconducting regions in a sea of normal regions, or, as suggested by Sai-Halasz et al. [10], the system may fluctuate in time between the normal and superconducting state. Such behavior would be analogous to hydrodynamic instabilities such as the Bernard instability [14] where slightly above a critical Reynold's number stationary spatial variations appear and, as the Reynold's number is further increased, oscillatory and eventually turbulent behavior is produced.

The second reason that makes the problems of hot superconductors attractive is that they form an excellent testing ground for theoretical ideas concerning systems far from equilibrium. The superconductor is a rich system having quasiparticles, phonons, and the pair field, each of which can be monitored. In particular, tunneling provides a probe of both the quasiparticles and the pair field. Here the dc bias provides a direct and convenient way to spectroscopically sweep energy. Furthermore, the theory of equilibrium superconductivity is well understood, and the strength and frequency dependence of the electron-phonon interaction is known from tunneling measurements [15]. Thus the theorist is dealing with a well characterized system in which a vairety of probes are available to tell him if he is on the right track in developing a physical understanding of and a mathematical framework for many body systems far from equilibrium.

Finally, what about applications? Dynes and Narayanamurti [7] used Sn tunnel junctions to generate milliwatts of phonon power over a narrow band near 2Δ. The phonon spectrum emitted by a driven thin film near 2Δ tunes with the gap. Thus hot superconductors can serve as tunable high frequency phonon generators.

Optical irradiation [16, 17] and quasiparticle injection [17, 18] have been used to tune the critical current of weak links. The insert in Fig. 4 shows a quasiparticle-injection weak-link configuration used by Wong, et al. [18]. By injecting quasiparticles from the Pb strip into the Sn film a local nonequilibrium region is created in the Sn bridge. I-V characteristics for various current injections are plotted in Fig. 4 and clearly show the change from a strong hysteretic to "ideal" weak link characteristic as the injection current is turned on. To what extent hot superconductors will provide an analogue of certain doped semiconducting devices, remains to be seen.

The ability to stabilize the superconducting state at temperatures above T_c has been demonstrated. How much above T_c remains to be determined. In the same vein, we expect that the microwave surface impedance will decrease when the system is driven out of equilibrium by a strong microwave field provided the drive frequency is less than 2Δ.

With this as the motivation, let's outline what approach to nonequilibrium superconductivity we will take and what aspects we will treat. First, it is important to realize that, contrary to problems of thermal equilibrium, there will not be a unique state of the system depending upon a few state variables such as temperature, pressure, magnetic field, etc. Rather, there will exist a variety of possible nonequilibrium states which depend not only on how far the system is driven out of equilibrium, but also upon the specific nature of the external source which drives the superconductors. The systems can also exhibit hysteretic behavior.

There are several levels on which the problem of hot superconductors can be approached. In principle, just as in equilibrium, the physical properties of interst can be determined from electron and phonon Green's functions. Because of the nonequilibrium aspect of the problem, one must be careful to construct the correct analytic continuation of the Matsubara frequencies as Gor'kov and Eliashberg [19] have discussed, or to use the Keldysh nonequilbirium formalism [20]. This type of approach is the most basic and is probably essential just below T_c where collective modes of the pair field occur [21-23]. However, in this paper we will not use Green's function techniques. Rather, we will restrict our discussion to a class of problems in which we believe a simpler, Golden Rule Boltzmann equation approach is suitable. This will allow us to concentrate our calculational efforts on the detailed energy distributions of the excitations which we believe are important in determining the physical properties of hot superconductors. In particular we will not use a relaxation approximation but rather the exact solutions of the nonlinear Boltzmann equations.

The class of problems which we will focus on involve a thin superconducting film which is driven in a spatially uniform steady state with a change in magnitude of the gap. The nonequilibrium state will be characterized by the energy distributions of quasiparticles and phonons and the magnitude of the gap. We will assume that the elementary excitations have the same dispersion relations and coherence factors as in the equilibrium state except that the magnitude of the gap will be determined from the BCS gap equations with the nonequilibrium quasiparticle distribution replacing the usual Fermi factor. The distributions will be determined from the Boltzmann equations. Formulated in this manner, the problem is similar to that of hot electrons in a semiconductor. However, in the superconducting case the gap must be calculated self-consistently. Furthermore, in a superconductor, quasiparticle recombination occurs with phonon emission, while in semiconductors hot electrons and holes recombine with the emission of photons. One could characterize our approaches as "quasiparticle engineering" [24], so it seems appropriate for this conference.

In Section II we will discuss the various relaxation processes which can occur in a superconductor which is driven out of equilibrium. Experimental and theoretical results for these relaxation times for superconductors near thermal equilibrium will be given. The relative sizes of these relaxation times as well as their temperature and energy dependence will be reviewed. Proceeding from this, the Boltzmann equations for the quasiparticle and phonon distributions first derived by Bardeen, Rickayzen, and Tewordt [25] will be given.

Before discussing solutions of the Boltzmann equations for some particular cases, it is useful to develop a feeling for the magnitudes of various fluxes and distributions associated with hot superconductors. We treat this in Section III using the macroscopic Rothwarf-Taylor equations [26]. We note under what conditions these macroscopic equations follow from the Boltzmann equation. In the last Section, IV, solutions of the kinetic equations for the quasiparticle and phonon distributions will be given for two examples: (1) phonon generation from driven superconducting films, and (2) the enhancement of the gap by microwave irradiation. We will emphasize the physical properties of these nonequilibrium states.

II. RELAXATION PROCESSES AND THE KINETIC EQUATIONS

In general we must deal with electron-phonon, electron-electron and electron-impurity scattering in a metal. Here we will discuss the electron-phonon scattering because of the essential role it plays in the nonequilibrium properties of most superconducting materials. We assume that the impurities are nonmagnetic and view their effect as simply averaging quantities over the Fermi surface as in Anderson's treatment of the dirty superconductor [27]. This together with the usual slow variation of the quantities of interest with respect to $|\vec{p} - \vec{p}_F|$ allows us to integrate out the momentum dependence and deal with just the frequency (or energy) variables. Electron-electron scattering can be important at both high energies $\gtrsim \sqrt{\omega_D \mu}$ and at low energies $\lesssim \omega_D^2/\mu$. Here ω_D is the Debye energy and μ is the Fermi energy. On the high energy side, the relaxation of optically excited electrons involves the electron-electron interaction. However, this relaxation occurs on a much shorter time scale than will be of interest here. The important relaxation process which determines the quasiparticle and phonon distirbutions occur on an energy scale set by kT_c. For most materials with the exception [28] of Al and possible Zn, kT_c is greater than ω_D^2/μ so that the electron phonon interaction is dominant.

The basic electron-phonon processes are schematically illustrated in Fig. 5. Looking at the top diagram from left to right we have a quasiparticle scattering with the emission of a phonon; while viewing it from right to left (reverse arrows) we have a quasiparticle scattering due to absorption of a phonon. Likewise, the lower diagram viewed from

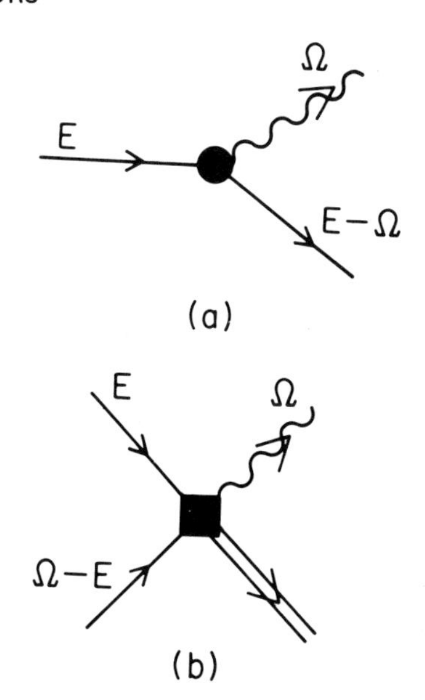

Fig. 5 Basic electron phonon processes: (a) scattering process and (b) recombination process or phonon pair breaking process.

left to right shows two quasiparticles recombining to form a pair with the emission of a phonon. This same figure viewed right to left (reverse arrows) can also be interpreted as a phonon being absorbed with the creation of two quasiparticles from the pair condensate.

Before discussing some theoretical and experimental results for these lifetimes let's make some simple estimates. Consider the scattering rate of a quasiparticle of energy ω (we measure quasiparticle energy relative to the Fermi energy) in a normal metal at zero temperature. A simple Golden Rule calculation gives

$$\frac{1}{\tau_o} = \frac{2\pi}{\hbar} |g|^2 N(0) \left(\frac{\omega}{\omega_D}\right)^3 \tag{1}$$

Here g is the effective electron-phonon coupling, N(0) is the single spin density of states at the Fermi energy. The factor $N(0)\left(\frac{\omega}{\omega_D}\right)^3$ is just the joint final density of states for a scattered quasiparticle and a phonon with total energy ω. We are interested in states of energy $\omega \sim kT_c$. Substituting this into Eq. (1) and rearranging it gives

$$\frac{1}{\tau_o} = \left(\frac{\pi kT_c}{\hbar}\right)\left(\frac{2\,N(0)\,|g|^2}{\omega_D}\right)\left(\frac{kT_c}{\hbar\omega_D}\right)^2 \tag{2}$$

The dimensionless coupling strength of the electron-phonon interaction $\lambda = 2N(0)|g|^2/\omega_D$ is of order unity so that the basic scattering rate is set by the product of the first and last terms. For example, consider Al with $T_c \sim 1K$ and $\omega_D \sim 400$ K. Then $\frac{1}{\tau_o} \sim \left(\frac{\pi kT_c}{\hbar}\right)\left(\frac{kT_c}{\hbar\omega_D}\right)^2 \sim 5 \times 10^{11}\left(\frac{1}{400}\right)^2 \sim 10^6$ so that $(\tau_o)_{Al} \sim 10^{-6}$ sec. While for Pb with $T_c \sim 7$ K and $\omega_D \sim 100$ K, we have

$$\frac{1}{\tau_o} \sim \left(3 \times 10^{12}\right)\left(\frac{7}{100}\right)^2 \sim 10^{10}$$

or $(\tau_o)_{Pb} \sim 10^{-10}$ sec. Thus a wide variation in the basic electron scattering times in metals is possible.

Next, in the same way, let's estimate a phonon lifetime. Consider a phonon of energy Ω in a normal metal at low temperature. Viewing the upper drawing of Fig. 5 from right to left (and twisting the incoming quasiparticle line so that it becomes an outgoing hole) we have a phonon absorbed creating a final state electron and hold pair with energy Ω. The joint density of states for the particle and hole varies as $N(0)\Omega/\mu$, with μ the Fermi energy, so that the Golden Rule gives for the rate of phonon absorption

$$\frac{1}{\tau_o^{Ph}} = \frac{2\pi}{\hbar} N(0) |g|^2 \frac{\Omega}{\mu} \tag{3}$$

For $\Omega = kT_c$ this gives*

$$\frac{1}{\tau_o^{Ph}} = \frac{\pi kT_c}{\hbar} \frac{2N(0)|g|^2}{\omega_D} \frac{\omega_D}{\mu}$$

$$\simeq \frac{\pi kT_c}{\hbar} \frac{\omega_D}{\mu} \tag{4}$$

For Al, $(\tau_o^{Ph})_{Al} \sim 10^{-10}$ while for Pb, $(\tau_o^{Ph})_{Pb} \sim 10^{-10}$. The increase in T_c for Pb relative to Al tends to be compensated by the decrease in ω_D for Pb relative to Al. Because of this, the basic phonon lifetimes do not show the wide range in variation that the electron lifetimes show.

---

*Actually $\lambda \simeq 1.5$ for Pb and $\lambda \simeq .4$ so that there is a factor of order 4 difference even in this rough estimate. As one can see from Table I a more accurate calculation shows that τ_o^{Ph} for Pb is about a factor of 8 shorter than τ_o^{Ph} for Al.

In a superconductor, as we have noted, a variety of relaxation processes can occur. The quasiparticles can scatter or recombine to form pairs, and, in addition, if the quasiparticles are not equally distributed with respect to having p greater or less than p_F, branch mixing relaxation first discussed by Clarke and Tinkham [29] can occur. In addition, there are phonon scattering and pair breaking lifetimes. Recently, Kaplan et al. [30] carried out detailed calculations of the frequency and temperature dependence of these various lifetimes. Here we want to discuss the basic ingredients of these calculations because the same procedure leads to the kinetic equations. In addition, we will review some of the results obtained for the energy and temperature dependences of the lifetimes which will provide insight into the physics of nonequilibrium problems.

For this purpose consider the scattering rate of a quasiparticle of energy ω with the emission of a phonon in a superconductor at temperature T (see the upper diagram of Fig. 5). The scattering rate for this process obtained in Ref. [30] from a reduction of the Eliashberg equations is

$$\frac{1}{\tau_s(\omega)} = \frac{2\pi}{\hbar}\frac{1}{Z_1}\frac{1}{1-f(\omega)}\int_0^{\omega-\Delta} d\Omega\,\alpha^2(\Omega)F(\Omega) \times \mathrm{Re}\frac{\omega-\Delta}{\sqrt{(\omega-\Omega)^2-\Delta^2}}(n(\Omega)+1)(1-f(\omega-\Omega)) \times\left(1-\frac{\Delta^2}{\omega(\omega-\Omega)}\right) \tag{5}$$

This expression has a simple Golden Rule explanation: $\alpha^2(\Omega)$ is the average of the square of the electron-phonon matrix element, $F(\Omega)$ is the final state phonon density of states for a phonon of energy Ω, $\mathrm{Re}\left[(\omega-\Omega)/\sqrt{(\omega-\Omega)^2-\Delta^2}\right]$ is the final state quasiparticle density of states for a quasiparticle of energy $\omega-\Omega$, $(1-\Delta^2/\omega(\omega-\Omega))$ is the electron-phonon scattering coherence factor, $(n(\Omega)+1)$ is the Bose factor for emission, and $(1-f(\omega-\Omega))$ is the usual Fermi factor giving the probability that the final quasiparticle state is empty. The factor Z_1^{-1} renormalizes the electron-phonon interaction calculated for Bloch states to the phonon dressed quasiparticle states (for low frequencies $Z_1 \sim 1+\lambda$). The factor $(1-f(\omega))^{-1}$ represents an enhancement of the decay rate due to the Pauli-principle blocking of the back scattering of other quasiparticles into the occupied quasiparticle state of energy ω.

In thermal equilibrium $n(\Omega)$ and $f(\omega-\Omega)$ are just the usual Bose and Fermi distributions. Thus, all that remains to be determined is $\alpha^2(\Omega)F(\Omega)$. This quantity can be obtained from electron tunneling measurements. At sufficiently low frequencies $\alpha^2(\Omega)$ approaches a constant, and the phonon density of states varies as Ω^2. For the region of energies of order kT_c that we will be interested in, the simple form $\alpha^2(\Omega)F(\Omega) = b\Omega^2$ is quite adequate for all the materials except for Pb and Hg where it can lead to errors of order a factor of 2*. The quadratic form is fit to the tunneling data to obtain b. The parameter τ_0 listed in Table I is a characteristic quasiparticle relaxation time (essentially, Eq. (2)).

$$\tau_0^{-1} = \frac{\pi kT_c}{\hbar} \frac{2b(kT_c)^2}{Z_1(0)} \tag{6}$$

Scattering lifetimes for quasiparticles of several different energies are plotted versus T/T_c in Fig. 6. A quasiparticle at the gap edge, $\omega = \Delta(T)$, cannot emit a phonon and scatter because it is in the lowest energy quasiparticle state. Thus it can only be scattered by thermal phonons, so as the temperature is lowered and the thermal phonon poplulation decreases, τ_s increases. At low temperatures

$$\frac{\tau_s(\Delta, T)}{\tau_0} = \frac{1}{\Gamma(7/2)\,\zeta\,(7/2)} \sqrt{\frac{2\,\Delta(0)}{kT_c}} \left(\frac{T_c}{T}\right)^{7/2} \tag{7}$$

For quasiparticles with energies $\omega > \Delta(T)$, spontaneous phonon emission sets a limit to the scattering lifetime τ_S. This effect is clearly seen in Fig. 6 for the cases $\omega = 1.5\,\Delta(0)$ and $\omega = 2\Delta(0)$.

Similar calculations, using $\alpha^2(\Omega)$, were carried out to determine the lifetime $\tau_r(\omega, T)$ for a quasiparticle to recombine with another quasiparticle to form a pair (lower diagram of Fig. 5). This recombination process is a binary reaction in which one quasiparticle combines with another so that, in thermal equilibrium, at low temperatures τ_r increases exponentially as

$$\frac{\tau_r(\Delta, T)}{\tau_0} \simeq \frac{1}{\sqrt{\pi}} \left(\frac{k\,T_c}{2\Delta(0)}\right)^{5/2} \left(\frac{T_c}{T}\right)^{1/2} e^{\Delta(0)/kT} \tag{8}$$

The exponential growth of τ_r reflects the decrease in the population of quasiparticles as T is lowered. Recombination lifetimes for quasiparticles of various energies are plotted versus T/T_c in Fig. 7. Branch mixing times have been calculated in Ref. [30] but will not be discussed here. Also, Ref. [30] contains various analytic expressions for the lifetimes for particular limits of the ω and T/T_c.

Finally, in Fig. 8, inverse phonon lifetimes for scattering and pair breaking are plotted versus T/T_c. These calculations, carried out

---

*Lifetimes for Pb and Hg calculated using the "exact" experimentally measured $\alpha^2(\Omega)F(\Omega)$ form are given in Ref. [30].

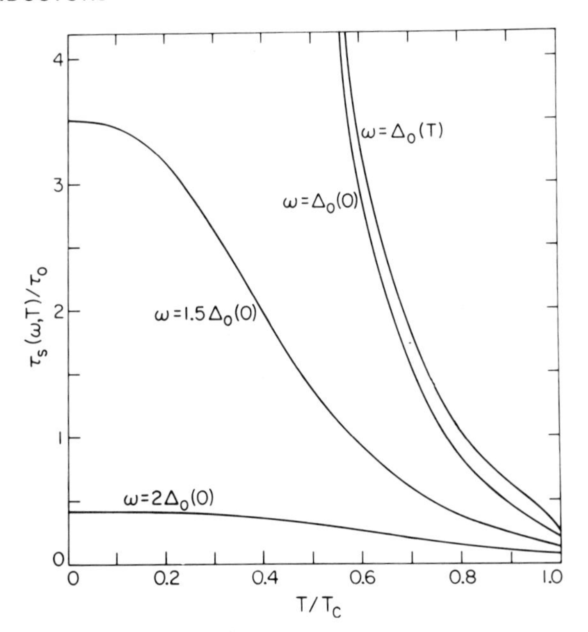

Fig. 6 Temperature and energy dependences of the quasiparticle scattering lifetime.

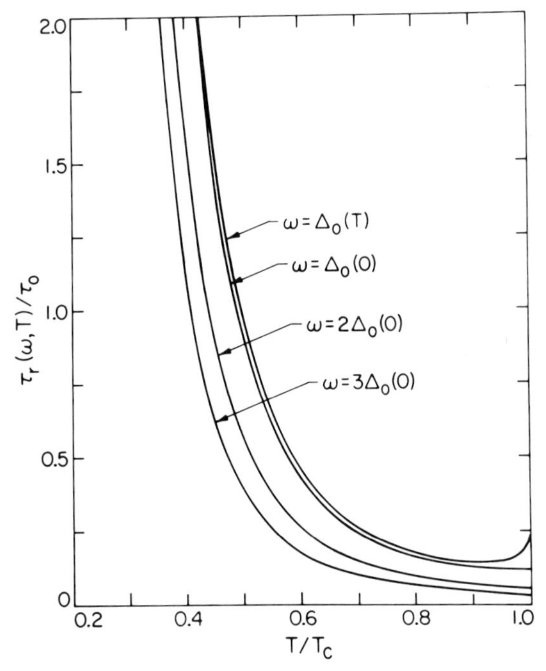

Fig. 7 Temperature and energy dependences of the quasiparticle recombination lifetime.

Table I. Parameters for Superconducting Metals and Alloys

Metal	T_c K	ω_D/k_B K	μ eV	$s \times 10^5$ cm/sec	$N(0) \times 10^{-21}$ States/eV	$N \times 10^{-22}$ ions/cm^3	$\tau_o \times 10^9$ sec	$\tau_o^{ph} \times 10^{10}$ sec
Pb	7.19	105	10.9	1.9	8.63	3.30	0.196	0.34
In	3.40	108	10.0		7.68	3.84	0.799	1.69
Sn	3.75	200	11.6	3.3	8.14	3.70	2.30	1.10
Hg	4.19	71.9	8.29	1.5	7.26	4.07	0.0747	1.35
Tl	2.33	78.5	9.46		11.7	3.50	1.76	2.05
Ta	4.48	240	18.0		40.8	3.52	1.78	0.227
Nb	9.2	275	6.18		31.7	5.57	0.149	0.0417
Al	1.19	428	13.5	6.4	12.2	6.02	438.	2.42
Zn	0.875	327	10.9	4.2	6.64	6.57	780.	23.1
$Pb_{60}Tl_{40}$	6.0						0.0647	
$Pb_{40}Tl_{60}$	4.7						0.118	
$Pb_{60}Bi_{20}Tl_{20}$	7.26						0.0567	
$Pb_{90}Bi_{10}$	7.55-8.05						0.043	

in Ref. [30], are again based upon $\alpha^2(\Omega)F(\Omega) = b\Omega^2$, and they are scaled with respect to a characteristic phonon lifetime $\tau_0{}^{ph}$ (like Eq. (3)) which is given in Table I. The most significant feature of the phonon inverse lifetimes (or scattering rates) is the dominance of the pair breaking process over quasiparticle scattering for phonons with $\Omega \geq 2\Delta$. Phonons with energy greater than 2Δ are far more likely to decay by pair production over almost the entire range of reduced temperatures.

There is an additional relaxation mechanism for phonons which plays an essential role in the nonequilibrium properties of superconducting films. This is the phonon escape time associated with the coupling between the thin film and the bath or substrate. For low frequency phonons, this time depends upon the acoustic matching of the film to the bath or substrate. For the thin films which we will discuss here, an approximate expression for the phonon escape time τ_{es} is:

$$\tau_{es} = \frac{4}{\eta}\frac{d}{s} \quad . \tag{9}$$

Here d is the film thickness, s is the velocity of sound, 4 is a geometrical factor, and η is an escape probability which is of order 0.1 for a metal glass interface and may be as large as 0.5 for a metal He interface. Taking d = 1000 A and s = 5×10^5 cm/sec gives $\tau_{es} \simeq 10^{-10}/\eta$. A convenient parameter is the ratio of the phonon escape time to the zero temperature pair breaking lifetime $\tau_B(2\Delta, 0)$ of a phonon of energy 2Δ. From Fig. 8 we see that $\tau_B(2\Delta, 0) = \tau_0{}^{ph}$, and from Table I we note that $\tau_0{}^{ph}$ is of order 10^{-10} sec. Thus, the ratio $\tau_{es}/\tau_B \sim 1/\eta$ which can vary over a large range (e.g., 1 to 100)*. Formally, setting $\tau_{es}/\tau_B = 0$ corresponds to locking the phonon distribution of the metal to that of the bath, while a τ_{es}/τ_B ratio of 100 corresponds to a weak coupling of the phonons in the metal to the substrate, with the result that highly nonequilibrium phonon distributions can be obtained when the system is driven by an external source.

There are a variety of different experimental checks on quasiparticle and phonon lifetimes. One knows from the success of the theory of strong coupling superconductors that the effects of damping on the quasiparticles and phonons is well described by the Eliashberg forumulation [31]. However, the detailed tunneling studies of the electron spectral weights provide information at higher energies than are of interest to us. Quantities which depend upon the low frequency damping of excitations such as the behavior of the nuclear spin relaxation rate near T_c and the structure of the Josephson current amplitude for bias voltages near $2\Delta/e$ do provide probes for this low energy $\sim \Delta_0$ behavior. However,

*It should be kept in mind that this is at best an order of magnitude estimate. For example, for Pb, $s \sim 1.5 \times 10^5$ cm/sec, $\tau_0{}^{ph} = 3 \times 10^{-11}$ so that $\tau_{es}/\tau_B \sim 10/\eta$ for a 10^3 A Pb film.

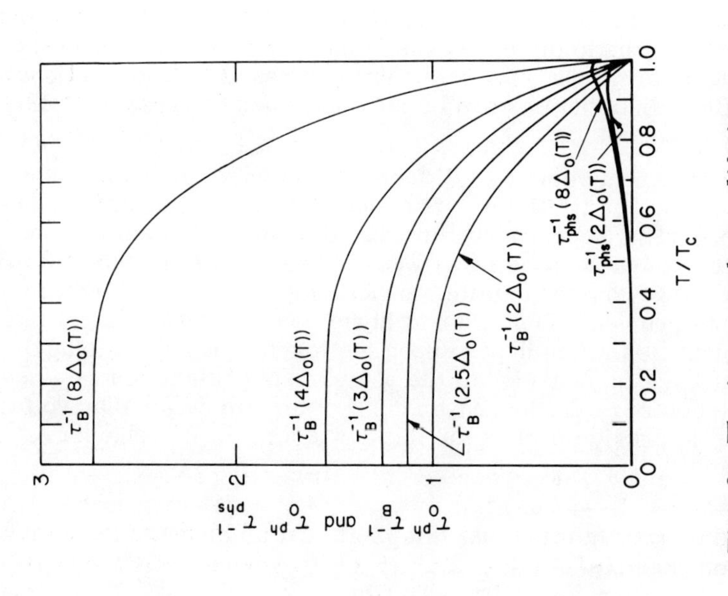

Fig. 8 The inverse phonon lifetimes for scattering and pair breaking processes

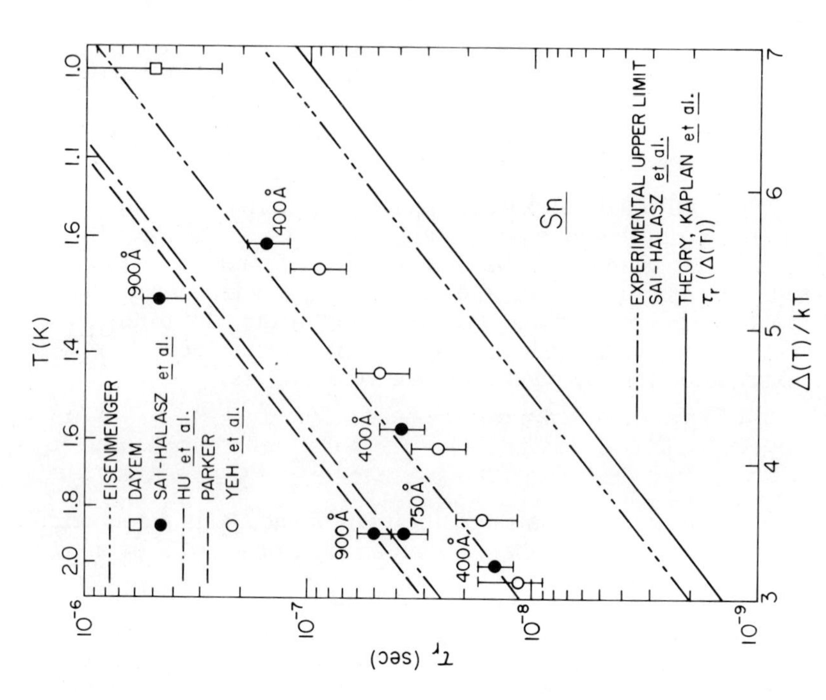

Fig. 9 Low temperature dependence of experimentally determined effective quasiparticle recombination lifetimes for Sn (from Ref. [30]).

these quantities depend only logarithmically on the damping. They have been reviewed in Ref. [30]. Here we will focus on several lifetime measurements.

The recombination lifetimes of quasiparticles were first studied using a double-junction, three-film structure [32]. One junction is used to inject excess quasiparticles into the center film where the increase in the steady state quasiparticle distribution is proportional to the effective recombination time. The second junction monitors this increase. The effective lifetime differs from the recombination lifetime by a factor $(1 + \tau_{es}/\tau_B)$ due to phonon trapping. That is, in a recombination event the emitted phonon has sufficient energy to break a pair. If this happens before the phonon escapes the sample, no net relaxation of the excess quasiparticle density will occur. For a three film Al structure of total thickness ~ 900 Å evaporated on a glass substrate, Gray et al. [33] found an experimental termperature dependence which fits the expected theoretical form. Fitting the low temperature dependence to Eq. (8), an experimental value for τ_0 can be extracted. The phonon trapping factor τ_{es}/τ_B appears to be near unity for this particular configuration so that the effective relaxation times $\tau_0(\mathrm{exp}) = 4 \times 10^{-6}$ sec (in a steady state experiment) and $\tau_0(\mathrm{exp}) = 1.4 \times 10^{-6}$ sec (in a transient experiment) give $\tau_0 \simeq 2 \times 10^{-6}$ sec and $\tau_0 \simeq .7 \times 10^{-6}$ sec respectively. These values are to be compared with our theoretical result of $.4 \times 10^{-6}$ sec [30].

Figure 9 shows the low temperature dependence of a number of different experimentally determined effective quasiparticle recombination lifetimes for Sn. The smaller phonon velocity and the shorter recombination times make the phonon trapping effect much larger for Sn than Al. Most of the data, therefore, give effective recombination times much longer than our theoretical value shown as the solid line. However, Sai-Halasz et al. [10] extrapolated data for various film thicknesses to zero thickness in order to eliminate the phonon-trapping effect. The resulting curve is in quite reasonable agreement with the theoretical predictions.

Quasiparticle lifetimes due to inelastic scattering from phonons can be extracted from a variety of Fermi surface measurements. Table II, taken from Ref. [30], shows experimental τ_0 values obtained from radio frequency size effect (RFSE), surface Landau level resonance (SLLR), and Asbel-Kaner cyclotron resonance (AKCR). The theoretical values are those obtained in Ref. [30] from Eq. (5). As one can see, the agreement for Al and Zn is quite good, but for In it is poor.

From additional data on other systems, from branch relaxation time measurements [29] and from phonon attenuation experiments [34] it appears that the overall temperature dependence of the low energy lifetimes are understood and that the magnitudes agree to within factors of 2 to 10. The energy dependence of the lifetimes of higher energy quasiparticles are known from tunneling to agree well with the theoretical framework of strong coupling superconductivity [15]. Put another way $\alpha^2(\Omega)F(\Omega)$ is obtained from a fit of theory to experiment and damping plays an important role (at higher energies) in the theory. In addition, the frequency

Table II. Comparison of Theory and Experiment for Quasiparticle Lifetimes

	$10^9\ \tau_o$ (sec)	
Metal	Theory	Experiment
In	0.799	3.3[a]
		8.9[b]
Al	438	620[c]
Zn	780	720[d]

a. I.P. Krylov and V.F. Gantmakher, Zh. Eksp. Teor. Fiz. 51, 740 (1966) [Sov. Phys. JETP 24, 492 (1966)].

b. P.M. Snyder, J. Phys. F 1, 363 (1971).

c. R.E. Doezema and T. Wegehaupt, Solid State Commun. 17, 631 (1975)

d. D.M. Brookbanks, J. Phys. F 3, 988 (1973).

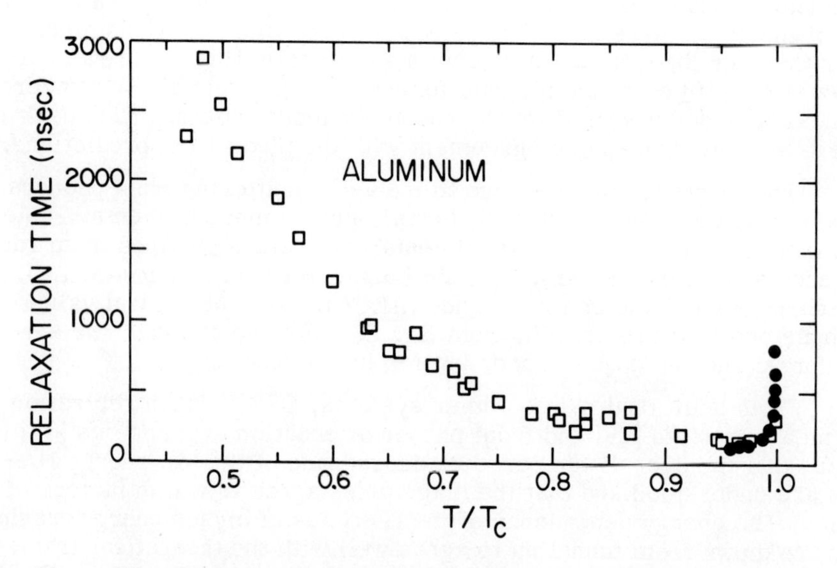

Fig. 10 Temperature dependence of the gap (●) and quasiparticle recombination (□) relaxation times in Al. (I. Schuller, Ph.D. Thesis, unpublished).

dependence of the phonon attenuation has been studied and is in good agreement with theory.

Now, having discussed the relaxation of two components (the quasiparticles and phonons) of the superconductor, we turn briefly to the third component, the condensate. At temperatures above T_c, the fluctuating pair field Δ relaxes according to TDGL with a relaxation time $\tau_{GL} = \pi\hbar/8k(T - T_c)$. However, below T_c the situation becomes much more complex due to the singularities introduced at the gap edge. In addition there are two different modes associated with relaxation of the magnitude and phase of the pair field Δ. According to Schmid and Schön [23] the relaxation times for these modes near T_c both vary as $(T_c/\Delta)\tau_0$, where τ_0 is the electron scattering time. This same form for the branch mixing time was experimentally observed by Clarke [29] and calculated by Tinkham and Clarke [29]. Branch mixing is closely related to the phase relaxation of the order parameter.

Schuller and Gray have studied the relaxation of the order parameter by using the I(V) characteristic to monitor the dynamic change in the gap produced by a laser pulse [22]. These measurements probe the relaxation of the magnitude of the gap. Figure 10 shows a plot of this relaxation time in an Al film versus T/T_c. The solid circles are a result of the laser measurement, and the open squares are steady state quasiparticle lifetime measurements of Gray scaled to overlap in the common temperature range. We see that for $T/T_c \leqslant 0.99$, quasiparticle relaxation determines the time response of the gap. For $0.99 < T/T_c < 1.0$ there is a transition to a collective behavior in which the gap relaxation time diverges as Δ^{-1}.

From this discussion of relaxation times we draw several conclusions. First, except very near T_c, the relaxation of the quasiparticle and phonon distributions determine the rate of approach of the superconductor to equilibrium. Secondly, the escape time of the phonons from the film is such that pair breaking and recombination processes are likely to lead to a nonequilibrium phonon distribution. This is further accented by the sudden increase in the phonon attenuation for $\hbar\Omega > 2\Delta$ due to the onset of pair breaking. Finally, the energy and temperature (for the nonequilibrium case, the quasiparticle density) dependences of the quasiparticle scattering and recombination lifetimes must be carefully treated. This point is illustrated in Fig. 11 where contours of various τ_s/τ_r ratios are plotted in the $(\omega/\Delta(0), T/T_c)$ plane for the thermal equilibrium lifetimes. In the shaded region the recombination lifetime is shorter than the scattering lifetime. In a nonequilibrium situation with excess quasiparticles, the shaded region would extend to $T/T_c = 0$. Thus it is important to treat quasiparticle scattering and recombination on an equal footing.

With these features in mind, we have studied the problem of a thin superconducting film which is driven into a spatially uniform, steady, nonequilibrium state. We have assumed that the nonequilibrium state can be characterized by the steady state energy distri-

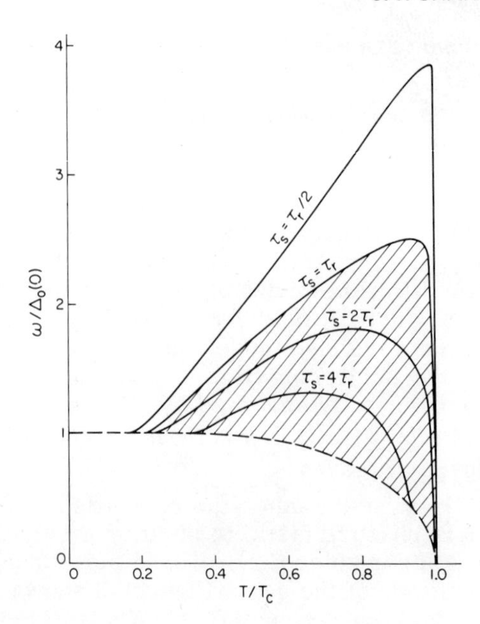

Fig. 11 Comparison of scattering and recombination lifetimes in an equilibrium superconductor (From Ref. [30]).

bution of quasiparticles f(E) and phonons n(Ω) and by the resulting self-consistent gap. The distributions f(E) and n(Ω) are determined from Boltzmann kinetic equations of the type first introduced by Bardeen, Rickayzen, and Tewordt [25] and extended to the strong coupling electron-phonon systems by Prange and Kadanoff [35]. The steady state gap parameter is then calculated from the usual BCS equation [36]

$$\frac{1}{\lambda} = \int_{\Delta}^{\omega_D} dE \; \frac{1 - 2f(E)}{\sqrt{E^2 - \Delta^2}} \tag{10}$$

with f(E) the nonequilibrium quasiparticle distribution.

Eliashberg has obtained a quasiparticle kinetic equation and Eq. (10), proceeding from a Green's function approach [37]. Schmid and Schön generalized this to include time dependent processes [23]. In addition to the usual collision terms, there appear terms in the Boltzmann equation which depend upon whether the phase or magnitude of the gap is changing. We do not believe these terms play a significant role in the problems we will disucss. Only for the tunneling injection, in which a branch imbalance can occur, does the relaxation of the phase of the gap enter. Our present analysis neglects this, and its effect remains to be studied. Rather, we believe that it is very important to treat the dynamics of <u>both</u> the quasiparticles and the <u>phonons</u>. In particular, the energy dependence of the scattering cross section plays an essential role.

For this reason we have concentrated on a detailed analysis of the coupled Boltzmann kinetic equations for f(E) and n(Ω) and the self-consistent BCS-like gap equation. The coupled Boltzmann equations have the following form:

$$\begin{aligned}
\frac{df(E)}{dt} = {} & I_{qp}(E) - \frac{2\pi}{\hbar} \int_0^\infty d\Omega\, \alpha^2(\Omega)\, F(\Omega)\, \rho(E+\Omega) \\
& \times \left(1 - \frac{\Delta^2}{E(E+\Omega)}\right) [f(E)(1 - f(E+\Omega))n(\Omega) - f(E+\Omega) \\
& \times (1 - f(E))(n(\Omega)+1)] - \frac{2\pi}{\hbar} \int_0^{E-\Delta} d\Omega\, \alpha^2(\Omega)\, F(\Omega) \\
& \times \rho(E-\Omega) \left(1 - \frac{\Delta^2}{E(E-\Omega)}\right) [f(E)(1 - f(E-\Omega))(n(\Omega)+1) \\
& - (1 - f(E))\, f(E-\Omega) n(\Omega)] - \frac{2\pi}{\hbar} \int_{E+\Delta}^\infty d\Omega\, \alpha^2(\Omega) F(\Omega) \\
& \times \rho(\Omega - E) \left(1 + \frac{\Delta^2}{E(\Omega - E)}\right) [f(E)\, f(\Omega - E)(n(\Omega)+1) \\
& - (1 - f(E))(1 - f(\Omega - E)) n(\Omega)]
\end{aligned} \tag{11}$$

and

$$\begin{aligned}
\frac{dn(\Omega)}{dt} = {} & I_{ph}(\Omega) - \frac{8\pi}{\hbar} \frac{N(0)}{N} \int_\Delta^\infty dE \int_\Delta^\infty dE'\, \alpha^2(\Omega)\, \rho(E)\, \rho(E') \\
& \left\{ \left(1 - \frac{\Delta^2}{EE'}\right) f(E)(1 - f(E'))n(\Omega) - f(E')(1 - f(E))(n(\Omega)+1) \right. \\
& \delta(E + \Omega - E') - \frac{1}{2}\left(1 + \frac{\Delta^2}{EE'}\right) \Big[(1 - f(E))(1 - f(E'))n(\Omega) \\
& \left. - f(E) f(E')(n(\Omega)+1) \Big]\, \delta(E + E' - \Omega) \right\} - \frac{n(\Omega) - n(\Omega, T)}{\tau_{es}}
\end{aligned} \tag{12}$$

Here $\rho(E)$ is the normalized quasiparticle density of states $\left(E/\sqrt{E^2 - \Delta^2}\right)\theta(E - \Delta)$ and N is the ion density.

In steady state the time derivatives of the distributions vanish and these equations become coupled nonlinear integral equations relating the distributions to the quasiparticle and phonon injection driving terms $I_{qp}(E)$ and $I_{ph}(\Omega)$ to the equilibrium phonon distribution $n(\Omega, T)$ of the bath at ambient temperature T. These driving terms depend upon the nature of the external source, and in the following sections specific examples will be considered.

The various integrals in Eqs. (11) and (12) correspond to the rate of creation or decay of quasiparticles or phonons due to the basic processes illustrated in Fig. 5. For example, the structure of the first term in the second integral of Eq. (11) is similar to the expression for the scattering of a quasiparticle of energy E due to the emission of a phonon. It clearly gives rise to a rate of decrease of f(E) as noted by the minus sign in front of it. The second term in this same integral enters with a positive sign and corresponds to a rate of increase of f(E) due to quasiparticles being scattered into E from lower energy quasiparticle states by the absorption of a phonon. Likewise, the first integral in Eq. (11) contains the contribution to the scattering out of state E due to phonon absorption and scattering into E due to phonon emission. The final integral in Eq. (11) contains the contributions from the quasiparticle recombination and phonon pair breaking processes. In the phonon kinetic equation, the term in the curly brackets with the coherence factor $(1 - \Delta^2/EE')$ is due to scattering processes, while the second term with the coherence factor $(1 + \Delta^2/EE')$ gives the contributions from phonon decay due to pair breaking and phonon creation due to quasiparticle recombination. The last term in Eq. (12) represents phonon exchange with the bath or substrate.

III. MAGNITUDES AND THE ROTHWARF-TAYLOR EQUATIONS

In this section we want to get a feeling for the sizes of the quantities we will be dealing with before getting into the detailed solutions of the Boltzmann equations. We will also introduce a useful set of macroscopic equations for the quasiparticle and phonon densities originally discussed by Rothwarf and Taylor [26].

To set the scale of the density of hot electrons we note that in thermal equilibrium at low reduced temperatures, Eq. (10) gives

$$\frac{\Delta(T)}{\Delta_o} \simeq 1 - 2n \tag{13}$$

Here n is the density of thermally excited quasiparticles measured in units of $4N(0)\Delta_o$.* Suppose the superconductor were at zero temperature,

*This same relationship has been obtained in the low density limits ($n < 0.1$) of both Owen-Scalapino's [12] μ^* model and Parker's [38] T^* model for the nonequilibrium problem.

let's use Eq. (13) to estimate the density of hot electrons necessary to depress the gap by 10%. This means that n = 0.05 and therefore the density is $0.2N(0)\Delta_0$. For Al this is of order 5×10^{17} per cm^3 while for Pb this is of order 3×10^{18} per cm^3.

What sort of external input power is necessary to maintain these excess quasiparticle densities in a thin film? We will estimate this by considering the power output of the phonons emitted from the film into the heat bath. This will be equal to the drive power which must be supplied to keep the system in a steady state. Again we'll treat the ambient bath temperature as being zero.

A density of $0.2N(0)\Delta_0$ quasiparticles recombining in a time τ_r with the emission of phonons of energy $2\Delta_0$ generates phonon radiation of $0.4N(0)\Delta_0{}^2/\tau_r$ watts per unit volume. Now some of these phonons break pairs and are reabsorbed while others escape. The escape depth is approximately given by $\tau_B s\eta/4$. Here τ_B is the phonon pair breaking lifetime, s the velocity of sound, and η the transfer coefficient between the metal and the bath. If the escape depth exceeds the thickness of the film d, then most of the phonons escape, and d replaces the escape depth. Thus the power radiated into the bath per unit area of the film is of order

$$P = \frac{N(0)\,\Delta_o^2}{2\tau_r}\, d_{eff} \tag{14}$$

with d_{eff} the smaller of the two numbers d and $\tau_B s\eta/4$.

Taking $\tau_B \sim \tau_0^{ph}$ and $\eta \sim 0.1$, we find that $\tau_B s\eta/4$ is about 10^3 Å for Al and about 10^2 Å for Pb. Thus an Al film 10^3 Å thick would have limited phonon trapping, while a Pb film of the same thickness and same thermal contact to the bath would have large phonon trapping. The ratio of d to $\tau_B s\eta/4$ is just τ_{es}/τ_B with τ_{es} defined by Eq. (9) in Section II. This parameter plays an important role in determining the structure of the energy distribution of excitations.

To get numbers for the power we will estimate the recombination time τ_r by using a thermal equilibrium value (see Fig. 7) for a thermal density of quasiparticles which reduce the gap by 10%. From the BCS result for $\Delta(T)$ one finds that a 10% reduction of $\Delta(T)$ from $\Delta(0)$ occurs for $T/T_c \sim 0.6$. From Fig. 7 we find that for $T/T_c \sim 0.6$, $\tau_r \sim 0.3\,\tau_0$. Using the times τ_0 listed in Table I it follows that the drive for an Al film is of order a milliwatt/cm^2 while for a Pb film it is of order watts/cm^2.

To conclude our discussion of magnitudes, we will estimate what type of power can be supplied by various driving mechanisms. Consider first the injection of quasiparticles near the gap edge via tunneling. The step in the I(V) characteristic of a typical symmetric SIS tunnel junction corresponds to a current density of order $1A/cm^2$. A strongly coupled junction may have a value for this current density of several hundred A/cm^2. For an Al junction with Δ of order tenths of millivolts the power input with a $10A/cm^2$ current density is milliwatts per cm^2. For a Pb junction Δ is of order millivolts so that for a 10 A/cm^2 current the power density is of order tens of milliwatts. It is therefore difficult to drive a Pb tunnel junction far out of equilibrium.

Next consider a heater mounted on one part of a film. An analysis of the Kapitza-like heat transfer between solids shows that the power transfer per unit area is given by [39]

$$2 \times 10^{10} \frac{\Gamma T^3 \Delta T}{s^2} \text{ watts/cm}^2 \qquad .$$

Here Γ is an angular average of the phonon transmission coefficient between the film and the heater. For $\Gamma \sim 1/2$, T and $\Delta T \sim 1K$, and $s \sim 5 \times 10^5$ cm/sec, this power transfer is equal to 10 m watts/cm^2. Clearly for larger values of T, say 4K, a watt/cm^2 of power could be transferred at a heater film interface with $\Delta T = 1K$.

Finally, another important source of power transfer is microwave radiation. The power transfer per unit area of film is $R_s H^2/2$. The incident power from a plane wave is $\sqrt{\frac{\mu}{\varepsilon}} H^2/2$ with $\sqrt{\frac{\mu}{\varepsilon}} \sim 377\ \Omega$, so that an incident beam of 1 watt/cm^2 will transfer $10^4 R_s / \sqrt{\frac{\mu}{\varepsilon}}$ watts/cm^2 to the film. For a Pb film at 4 K the surface resistance R_s at 10 GHz (x-band) is of order $10^{-5}\ \Omega$ so that the power transfer from a 1 watt/cm^2 beam if of order 1/4 m watt/cm^2. In an x-band resonant cavity with a $Q \sim 10^5$, a watt of power can produce an $H \sim 10^3$ amp turns/m. This would give a power transfer of several watts/cm^2 for $R_s \sim 10^{-5}\ \Omega$.

So far in this section we have proceeded by making order of magnitude estimates of quasiparticle densities and energy fluxes. Some time ago Rothwarf and Taylor [26] introduced a set of phenomenological equations to describe these macroscopic quantities. These equations are useful in providing a more precise framework for determining some of the quantities we have been discussing. We have recently shown [40] under what conditions these equations can be derived from the Boltzmann kinetic equations given in Sec. II. Here we briefly review this.

A quasiparticle density N_{qp} is defined by

$$N_{qp} = \int_{-\infty}^{\infty} d\varepsilon\, 2N(0) f(E) = 4N(0) \int_{\Delta}^{\infty} dE\, \rho(E) f(E) \tag{15}$$

with $E = \sqrt{\varepsilon^2 + \Delta^2}$. N(0) is, as usual, the normal state single spin density of states at the Fermi surface. The factor of 2 takes into account the spin degrees of freedom. Next, a phonon density for phonons of energy $\hbar\Omega \geq 2\Delta$, $N_{ph>}$ is introduced by

$$N_{ph>} = \int_{2\Delta}^{\infty} d\Omega \; NF(\Omega)n(\Omega) \tag{16}$$

Here $F(\Omega)$ is the phonon density of states per ion, and N is the ion density. Now multiplying Eq. (11) by $4N(0)\rho(E)$ and integrating over E one obtains

$$\frac{dN_{qp}}{dt} = \overline{I}_{qp} - 2RN_{qp}^2 + 2\beta N_{ph>} \tag{17}$$

with $\overline{I}_{qp}$ the integrated injection current (per unit volume)

$$\overline{I}_{qp} = \int_{\Delta}^{\infty} 4N(0)\,\rho(E)\,I_{qp}(E)\,dE \tag{18}$$

and

$$R = \frac{4\pi}{\hbar}\frac{N(0)}{N_{qp}^2}\int_{\Delta}^{\infty} dE\,\rho(E)f(E)\int_{\Delta}^{\infty} dE'\rho(E')f(E')$$

$$\times\; \alpha^2 F(E + E')\left(1 + \frac{\Delta^2}{EE'}\right)\left(n(E + E') + 1\right) \tag{19}$$

$$\beta = \frac{4\pi}{\hbar}\frac{N(0)}{N_{ph>}}\int_{2\Delta}^{\infty} d\Omega\,F(\Omega)n(\Omega)\int_{\Delta}^{\Omega-\Delta} dE\rho(E)\rho(\Omega - E)$$

$$\times\; \alpha^2(\Omega)\left(1 + \frac{\Delta^2}{E(\Omega - E)}\right)(1 - f(E))\,(1 - f(\Omega - E)) \tag{20}$$

Note that the scattering terms do not contribute to Eq. (17).

In the same way, if Eq. (12) is multipled by $NF(\Omega)$ and integrated over Ω from a lower limit of 2Δ, we obtain

$$\frac{dN_{ph>}}{dt} = \overline{I}_{ph>} + RN_{qp}^2 - \beta N_{ph>} - \frac{N_{ph>} - N_{ph>}^o}{\tau_{es}} + S \tag{21}$$

Here $\overline{I}_{ph>}$ is the integrated phonon injection current (per unit volume), and $N_{ph>}^o$ is the thermal equilibrium number of phonons at the bath temperature with energies greater than 2Δ. The S term represents a phonon source and arises from the decay of quasiparticles with energy greater than 3Δ.

$$S = \frac{8\pi N(0)}{\hbar} \int_{3\Delta}^{\infty} dE' \int_{\Delta}^{E'-2\Delta} dE\, \alpha^2 F(E' - E)\, \rho(E)\, \rho(E')$$

$$\times \left(1 - \frac{\Delta^2}{EE'}\right) [f(E')(1 - f(E))(n(E' - E) + 1)$$

$$- f(E)(1 - f(E'))n(E' - E)] \tag{22}$$

If the hot quasiparticle disturbance is small at energies greater than 3Δ, S will not play a significant role. However, this contribution can become important when quasiparticles with energy larger than 3Δ are excited. Note that our derivation shows how both R and β depend upon the steady state distributions f(E) and $n(\Omega)$. These parameters will clearly vary for different driving mechanisms as well as different strengths of the drive.

The R-T equations have a simple physical interpretation. The rate (per unit volume) at which two quasiparticles recombine to form a pair is $2RN_{qp}^2$, while the rate at which pairs are dissociated by phonons to form quasiparticles is $2\beta N_{ph>}$. Thus the recombination lifetime $\tau_R = (2RN_{qp})^{-1}$. The coefficients in Eq. (21) are a factor of 2 less because a single phonon is emitted from the recombination of two quasiparticles, and a single phonon is absorbed in a pair breaking process. Here the phonon pair breaking lifetime is $\tau_B = \beta^{-1}$. The fourth term in Eq. (20) is simply the relaxation of the phonon density towards the thermal phonon density $N^{\circ}_{ph>}$ appropriate to a film at the temperature of the heat bath.

As an example of the R-T formalism, let's repeat the calculation of the relationship between the power and N_{qp}. Consider, for example, a tunnel junction biased at a voltage just above $2\Delta/e$. We'll assume the temperature of the bath is sufficiently small compared to T_c that we may take it as zero. Then the steady state R-T equations are

$$\bar{I}_{qp} - 2RN_{qp}^2 + 2\beta N_{ph>} = 0$$

$$RN_{qp}^2 - \beta N_{ph>} - \frac{N_{ph>}}{\tau_{es}} = 0 \tag{23}$$

Rearranging these equations we have

$$\bar{I}_{qp} = \frac{N_{qp}}{\tau_R (1 + \tau_{es}/\tau_B)} \tag{24}$$

with $\tau_R = (2RN_{qp})^{-1}$ and $\tau_B = \beta^{-1}$ the recombination and pair breaking lifetimes, respectively. Now to get the electric current per unit area of film we must multiply the particle injection rate per unit volume $\bar{I}_{qp}$ by the film thickness d and the electron charge e. Then to get the power input per unit area we simply multiply this current by the bias voltage $\sim 2\Delta_o/e$. This gives

$$P/A = \frac{N_{qp} \quad 2\Delta_o d}{\tau_R(1+\tau_{es}/\tau_B)} \tag{25}$$

This reduces to the two limiting cases previously discussed, Eq. (14), in the limits $\tau_{es}/\tau_B \ll 1$ and $\tau_{es}/\tau_B \gg 1$.

IV. SOLUTIONS OF THE BOLTZMANN EQUATIONS

In this section we will discuss the energy distributions obtained from numerical solutions of the Boltzmann equations of Sec. II. Initially we linearized the kinetic equations and studied the deviations in the distributions produced by various sources: phonon injection from a heater, quasiparticle injection by tunneling, microwave and optical radiation [40]. The changes in the tunneling I-V characteristic, the frequency dependent ultrasonic attenuation and microwave absorption coefficients were evaluated using these nonequilibrium distributions. More recently we have investigated the full nonlinear kinetic equations for several cases [41]. Although the numerical procedure for obtaining selfconsistent solutions of the nonlinear equations is more laborious, the results are no more difficult to understand than those obtained by the linear analysis. In addition they also show interesting physical effects not contained in the linear analysis such as shifts of the gap. Furthermore, from an applications point of view, it is likely that devices making use of the unique properties of hot superconductors will operate well away from equilibrium.

For these reasons we will discuss the results we have obtained using the full set of coupled nonlinear kinetic equations. The problem which will be reviewed is the generation of 2Δ phonons from driven superconducting films. We will also make some comments on work in progress on the behavior of thin films irradiated by microwaves with $\hbar\omega < 2\Delta$.

As discussed in Sec. I, a superconducting film pumped by phonons from a heater or a symmetric SIS tunnel junction biased near $2\Delta/e$ is known to generate a band of phonons of energy near 2Δ. The basic physics of these nonequilibrium devices is simple. In the heater case, the injected high energy phonons first break pairs creating quasiparticles. These quasiparticles scatter, relaxing to low energy states near the gap edge where they recombine to form pairs with the emission of phonons with energy $\hbar\Omega \sim 2\Delta$. In the tunneling device, the dominant tunneling process for a bias voltage $V \sim 2\Delta/e$, injects quasiparticles of energy ranging from Δ to $eV-\Delta$. When the excited quasiparticles relax, they recombine generating phonons with energies between $\hbar\Omega = 2\Delta$ and $2(eV-\Delta)$.

For a superconducting film with part of its surface attached to a heater which is at temperature T_H, the quasiparticle injection current is zero and the phonon injection current in Eq. (12) is given by [41]

$$I_{ph}(\Omega) = \frac{\Gamma s}{2d}\left(\frac{1}{e^{\Omega/kT_H}-1} - n(\Omega)\right) \tag{26}$$

Here Γ is an angular average of the phonon transmission coefficient between the film and the heater, s is an average speed of sound in the film and d is the film thickness. A convenient parameter to characterize the strength of this coupling is the dimensionless ratio $(\pi/4)\left(\frac{s\tau_0^{ph}}{d}\right)\Gamma$. Here τ_0^{ph} is the characteristic phonon time discussed in Section II. For a film of thickness a few thousand angstroms $(\pi/4)\left(\frac{s_1\tau_0^{ph}}{d}\right)\Gamma \simeq \Gamma$. We will take $(\pi/4)\left(\frac{s\tau_0^{ph}}{d}\right)\Gamma = 0.01$, a heater temperature $T_H = \Delta_0/k_B$ and assume that the ambient temperature is small compared to the T_c of the film so that T can be set equal to zero. These parameters lead to gap suppressions, for τ_{es} values of interest, which are of order 10% to 20%. While larger values of the gap suppression could be obtained from a tighter acoustic coupling of the film to the heater, this will eventually lead to an inhomogeneous or time dependent mixed state [9, 10, 11, 13]. Here we are considering only the spatially homogeneous nonequilibrium state.

The integral equations for the distributions together with the gap equation were solved numerically for various values of τ_{es}. The steady state quasiparticle distributions are shown in Fig. 12. The curves (a), (b) and (c) correspond to the phonon escape times $\tau_{es} = 3/5\ \tau_B(2\Delta_0)$, $5/4\,\tau_B(2\Delta_0)$ and $3\tau_B(2\Delta_0)$ respectively. Note that the curves stop at different values of E/Δ_0. This is just a reflection of the gap suppression. The ratio of the steady state gap parameter Δ to the zero temperature equilibrium gap Δ_0 has the values 0.92, 0.88, and 0.82 for cases a, b and c respectively. Clearly increasing τ_{es} leads to an increase in the number of excited quasiparticles.

Next we will examine the phonon distributions. The physical situation is clarified if $n(\Omega)$ is multiplied by the phonon density of states factor Ω^2. The resulting distributions are plotted in Fig. 13. The structure at 2Δ arises as expected from the emission of phonons when quasiparticles at the gap edge recombine. Furthermore, phonons with frequency equal to or greater than 2Δ are resonantly trapped by pair breaking processes for large τ_{es}/τ_B values, while phonons with frequencies

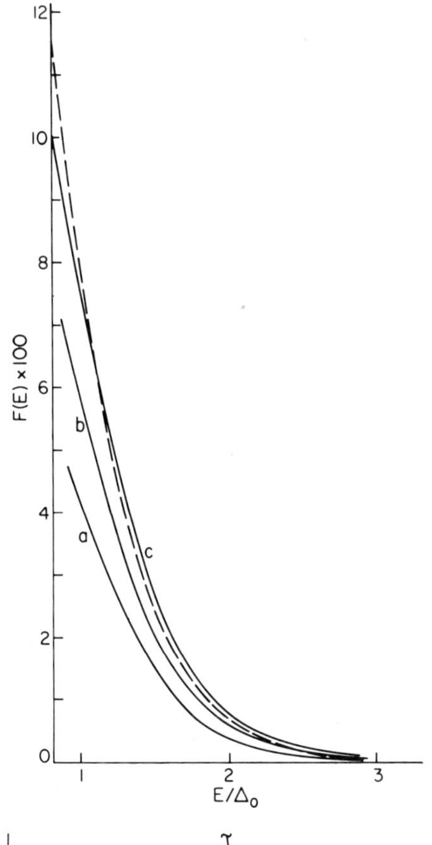

Fig. 12
The steady state quasi-particle distribution in a superconducting thin film driven by a heater of temperature $T_H = \Delta_o$. Curves (a), (b), and (c) correspond to $\tau_{es} = \frac{3}{5}\tau_B, \frac{5}{4}\tau_B$, and $3\tau_B$ respectively. The dashed curve is a fit to the T* model [38] for $\tau_{es} = 3\tau_B$.

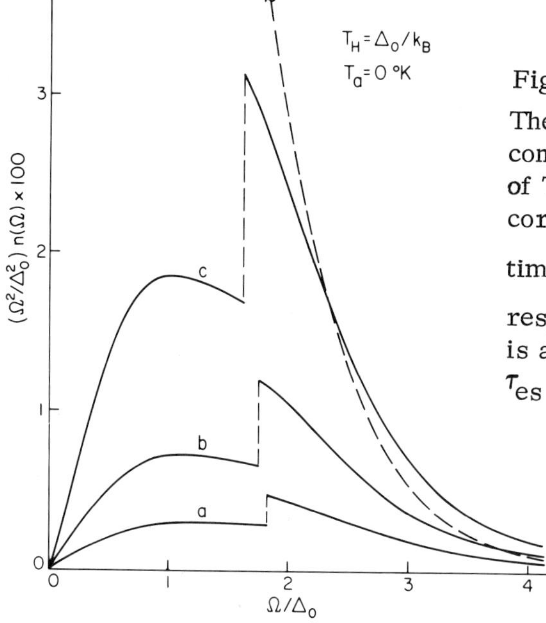

Fig. 13
The phonon spectra for a superconducting film driven by a heater of $T_H = \Delta_o$. Curves (a), (b) and (c) correspond to phonon escape times $\tau_{es} = \frac{3}{5}\tau_B, \frac{5}{4}\tau_B$, and $3\tau_B$ respectively. The dashed curve is a fit to the T* model [38] for $\tau_{es} = 3\tau_B$.

less than 2Δ can only be scattered. Since the scattering cross section is significantly smaller than that for pair production except near T_c (see Fig. 8) these low frequency phonons tend to escape more readily. The increased suppression of the gap as τ_{es} increases, and more quasiparticles are created, is clearly evident.

It is interesting to compare these solutions of the Boltzmann equations with an analytic approximation proposed by Parker [38]. Parker argued that when τ_{es}/τ_B is large, the phonons above 2Δ and the quasiparticles form a strongly coupled system which could come into internal thermal equilibrium at a temperature T^* above the bath temperature. The effective temperature T^* is adjusted to produce the correct number of phonons in the energy region above 2Δ. We will see how well this theory fits our case (c) in which $\tau_{es}/\tau_B = 3$. The dashed curve in Fig. 13 represents a thermal distribution of phonons in the region $\Omega \geq 2\Delta$. The temperature T^* is chosen so that the number of phonons with energies equal to or gerater than 2Δ is the same as that under curve (c) (over the same region $\Omega \geq 2\Delta$). This gives a $T^* = 0.72\ T_c$. In Fig. 12, the dashed line represents a Fermi distribution at this temperature. We see that it gives quite a good fit to the calculated nonequilibrium distribution (curve (c)). The thermal equilibrium BCS gap for a reduced temperature of 0.72 is $0.8\ \Delta_0$, also in close agreement with the value of our numerical solution.

It can be shown that the nonequilibrium distributions produced by optical radiation are similar to these produced by a high temperature heater [40]. Figure 14 shows a plot of the phonon distribution, times the usual phase space factor $(\Omega/\Delta_0)^2$, for $T_H = 1.5\ \Delta_0$ and $\tau_{es}/\tau_B = 2$. The dashed curve shows the spectrum of the drive phonons from the heater. These results for the phonon spectrum from a superconducting thin film under heat pumping are qualitatively different from those calculated in Ref. 7. It will be interesting to measure in more detail the spectral distribution of phonons emitted from <u>thin</u> films.*

Next consider the case of a symmetric SIS tunnel junction biased at a voltage V slightly greater than $2\Delta/e$. In this case the phonon injection current is zero and the quasiparticle injection current is given by

$$I_{qp}(E) = \frac{\sigma}{de^2 N(0)} \{\rho(eV - E)(1 - f(eV - E) - f(E))\,\theta(eV - E - \Delta)$$
$$+\rho(E - eV)(f(E - eV) - f(E))\,\theta(E - eV - \Delta)$$
$$-\rho(E + eV)(f(E) - f(E + eV))\} \tag{27}$$

*It is also of interest to measure the structure of the quasiparticle distribution. The I(V) characteristics for junctions driven out of equilibrium by a variety of sources have been discussed in Refs. [40] and [42].

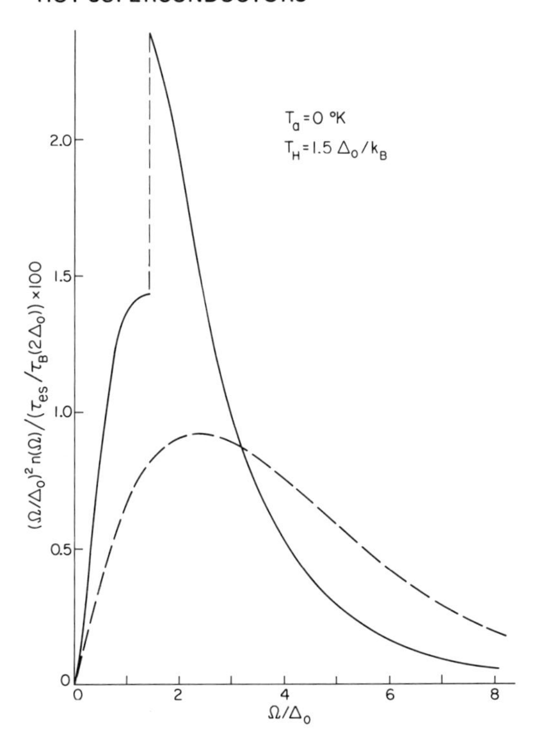

Fig. 14
The phonon emission spectra for a thin film attached to a heater of $T_H = 1.5\,\Delta_o$. The dashed curve is the spectrum of injected phonons.

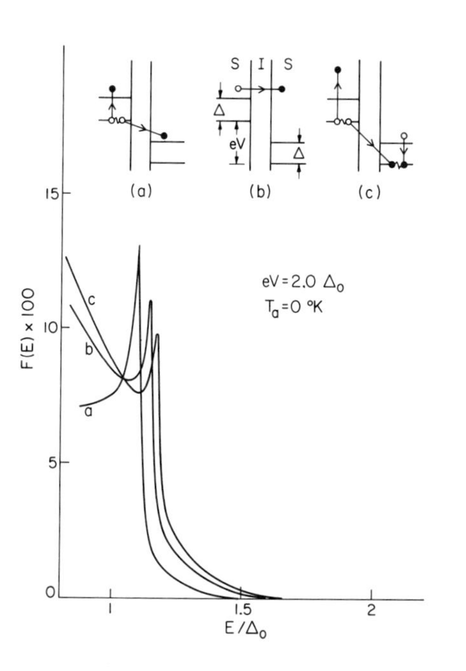

Fig. 15
The steady state quasi-particle distribution in a thin film under quasi-particle injection through an SIS tunnel junction. Curves (a), (b), and (c) correspond to $\tau_{es} = \tau_B$, $4\tau_B$, and $8\tau_B$ respectively. Inset shows the three processes which contribute to the tunneling current.

Here σ is the normal state junction conductance per unit area, d is the junction thickness and N(0) is the single spin Bloch density of states per unit volume. A convenient dimensionless ratio to measure the coupling strength of the current injection is $\frac{\sigma \tau_o}{de^2 N(0)} \left(\frac{T_c}{\Delta_o}\right)^3$.

Here we take this parameter to be 0.01. For an Al tunnel junction with film thicknesses 500 A, this corresponds to a junction conductance σ of order $10^4\ \Omega^{-1}/cm^2$ or, putting it another way, a Josephson current density of order several A/cm^2.*

The three terms in Eq. (27) represent the contributions from processes (a), (b), and (c) shown in the inset of Fig. 15. At bias voltages just above the $2\Delta/e$ step, the current is mainly due to process (a), and quasiparticle states with $p > p_F$ and $p < p_F$ are equally populated. This avoids the branch imbalance problem [29], and we can use the Boltzmann equations introduced in Sec. II. In our numerical analysis, the singularities in $I_{qp}(E)$ which occur at $E = eV \pm \Delta$ have been smoothed out by a gaussian function of half width $0.05\ \Delta_o$. Physically this could arise from lifetime broadening or the inhomogeniety and anisotropy of the gap.

Results for the steady state quasiparticle distributions for different τ_{es}/τ_B ratios are shown in Fig. 15. We have taken $eV = 2\Delta_o$, $d = 500$ Å and a zero bath temperature. Note that because of the suppression of the gap the bais voltage $2\Delta_o/e$ exceeds the $2\Delta/e$ threshold. Curves (a), (b), and (c) correspond to $\tau_{es}/\tau_B = 1$, 4, and 8 respectively. When the phonon escape time increases, the steady state gap decreases from $0.89\ \Delta_o$ for curve (a) to $0.84\ \Delta_o$ for curve (b), and to $0.81\ \Delta_o$ for curve (c), reflecting the increasing quasiparticle density as τ_{es} increases. Gap suppressions of this order of magnitude have been recently observed in Sn tunnel junctions by Yeh and Langenberg [43]. This effect is clearly seen in Fig. 16 from Ref. [43] which shows the I-V curve of a high current density tin-tin oxide-tin junction near the gap edge.

Returning to the quasiparticle distributions, note that as the escape time increases, the peak structure at $eV - \Delta$ shifts to higher energy (remember eV is fixed at $2\Delta_o$) and weakens. The phonon spectral distribution in the thin film is shown in Fig. 17. The sharp peak at $E \sim 2(eV - \Delta)$ in curve (a) reflects the structure in f(E) at $E = eV - \Delta$. These features are weakened as the phonon escape time increases and more phonon-quasiparticle recombination and pair breaking processes

---

*For Sn it would correspond to about $10^3\ A/cm^2$ and for Pb about $5 \times 10^4\ A/cm^2$. Values of several hundred A/cm^2 have been reported for Sn. It does not seem likely that Pb can be driven as hard as the parameter 0.01 would imply. This is because of the short τ_o for Pb and has been discussed already in Section II.

occur. The low frequency phonons peak at eV arises from the phonons emitted as quasiparticles scatter from eV - Δ down to the gap edge. For the example of an Al junction biased at $2\Delta_0$ and carrying a current of A/cm^2, the power output of phonons is of order one-half a milliwatt per cm^2.

We have studied the effects on the phonon emission spectrum due to finite ambient temperature and for an increase of bias voltage eV. We found that the width of the phonon band at $E \geqslant 2\Delta$ increases as T and eV increase. Furthermore, when eV is reduced towards 2Δ, fewer low energy phonons ($\hbar\Omega < 2\Delta$) are emitted. However, the sharp peak and the shoulder structure become more dramatic. The width of the narrow band phonons is limited to approximately $2(eV - \Delta) - 2\Delta = 4(eV/2 - \Delta)$, which can become extremely narrow as eV is turned towards twice the nonequilibrium gap.*

The effect of phonon injection, quasiparticle injection, and optical radiation (also microwave or infrared radiation with $\hbar\Omega > 2\Delta$) is to create more excitations in the superconductor. These additional excitations depress the gap. A different situation however, can come about it the superconductor is driven by microwaves or phonons with $\hbar\Omega < 2\Delta$. In this case pairs cannot be broken, and the quasiparticles are simply redistributed. Roughly speaking it is as if the low lying quasiparticles congregated in the high density of states near the gap edge were shoveled up to an energy $\Delta + \hbar\omega$.

Now, as Eliashberg [5] has pointed out, it is the occupation of the low energy quasiparticle states that most severely interfere with the pairing correlations. Thus, clearing out the low lying quasiparticle states by pushing the excitations into higher states can increase the gap. Furthermore, since the recombination rate varies as the final state phonon phase space, higher energy quasiparticles can recombine more rapidly than lower energy ones (see Fig. 7). To make this last point more specific, suppose microwave radiation with $\hbar\omega = \Delta/2$ is shown on a superconducting film. The phonon phase space varies as the final state phonon energy squared. Thus this factor for the phonon emitted when two quasiparticles at the gap edge recombine varies as $(2\Delta)^2$. However, if these quasiparticles each have energy $\Delta + \hbar\omega = 3\Delta/2$, they would emit a phonon of energy 3Δ, and the final state phase space factor would be $(3\Delta)^2$. This would mean a recombination rate over twice as fast. Since the excited quasiparticles recombine more rapidly, there can actually be a net reduction in the quasiparticle density when microwaves irradiate a thin film. This can also lead to a gap enhancement. As we mentioned in the introduction, the enhancement of the critical current in a weak link exposed to microwave [1, 2] or phonon [3] radiation with $\hbar\omega < 2\Delta$ has been observed as well as a $\sim 2\%$ increase in T_c of an Al film [4].

Here we will discuss the properties of a thin superconducting film irradiated with microwaves. The driving terms in this case have $I_{ph}(\Omega) = 0$ and

* It is also possible that the phonon spectrum emitted by a thick film is narrowed.

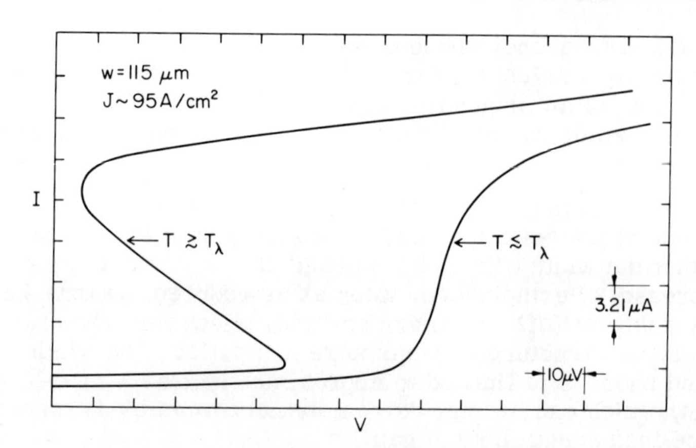

Fig. 16 I(V) characteristics of a tin tunnel junction driven far from equilibrium with the bias voltage $V \sim 2\Delta/e$ (from Ref. [43]).

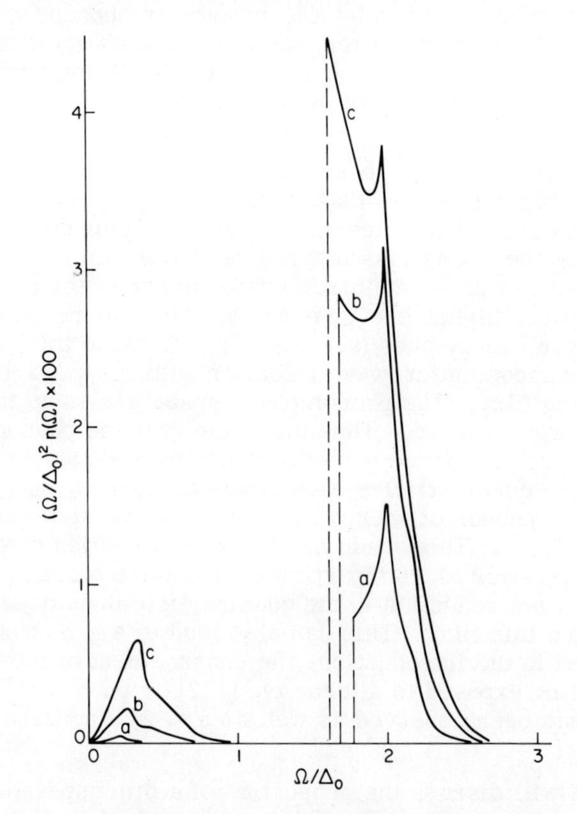

Fig. 17 The phonon spectra correspond to the quasiparticle distribution curves shown in Fig. 16.

$$I_{qp}(E) = \frac{R_N H^2}{2dN(0)\Delta^2} \left\{ \frac{E+\omega}{\sqrt{(E+\omega)^2 - \Delta^2}} \left(1 + \frac{\Delta^2}{E(E+\omega)}\right) (f(E+\omega) - f(E)) \right.$$

$$\left. \times \; \theta(E+\omega-\Delta) + (\omega \rightarrow -\omega) \right\} \tag{28}$$

Here H is the magnetic field and R_N is the normal state surface resistance. In picking a suitable parameter to characterize the driving strength it is important to remember that the superconducting surface impedance changes rapidly with T and ω. For this reason, a useful parameter with which to characterize the coupling is $B = \tau_0 R(\omega, T)H^2/2N(0)\Delta^2 d$, where $R(\omega, T)$ is the thermal equilibrium surface resistance of the superconductor at microwave frequency ω and a temperature T. The terms in Eq. (28) represent the scattering into and out of state E with the adsorption and emission of microwaves. In the following analysis the singularities at the gap edge have been smeared out by averaging over a gaussian smoothing function of width $0.1\,\Delta$. This has the effect of artificially broadening structure but leaving the overall behavior essentially correct.

Since the interest in this case is in the redistribution of the quasiparticles, we take the bath temperature to be $T_c/2$. The microwave frequency is taken to be $\omega = \Delta_0/2\hbar$. Furthermore, we will look at the deviations in the distributions, and contrary to the previous part, look at results obtained from solutions of the linearized Boltzmann equations [40]. The deviations of δf in units of the drive strength B are plotted in Fig. 18 for $\tau_{es}/\tau_B = 0, 1$, and 8. These curves all show a depletion of the quasiparticle distribution just above the gap with a resulting population peak at an energy $E \sim \Delta + \hbar\omega = 3\Delta/2$. As τ_{es}/τ_B increases, the deviations increase in size and show a broader structure.

The deviation in the phonon number distributions $\delta n(\Omega)F(\Omega)$ are shown in Fig. 19 for $\tau_{es}/\tau_B = 1$ and 8. The low energy peaks at $\hbar\Omega \sim \Delta/2$ arise from the scattering of the excited quasiparticles at $3\Delta/2$ down to the gap edge with the emission of a phonon of energy $\Delta/2$. The sharp depletion of phonons starting at 2Δ is due to the fact that in the nonequilibrium state there are fewer quasiparticles in the low energy states near the gap edge than in equilibrium. Thus the equilibrium balance $\hbar\Omega \rightleftarrows e+e$ in this region shifts to $\hbar\Omega \rightarrow e+e$, and the phonon states at 2Δ and slightly above are depleted. When the phonon escape time is larger, the structure is enhanced (note the different scale for curve (d)) with the low energy peak particularly increased.

In Ref. [40] a variety of transport properties of a thin film driven by a strong microwave source were calculated. Figure 20 shows the change in the real part of the surface conductance $\delta\sigma_1(\omega)$ relative to the equilibrium surface conductance $\sigma_1(\omega)$ due to irradiation with a drive field at a frequency $\Delta_0/2$. This is calculated in the linear approximation so that the size of the effect is proportional to B. We estimate that

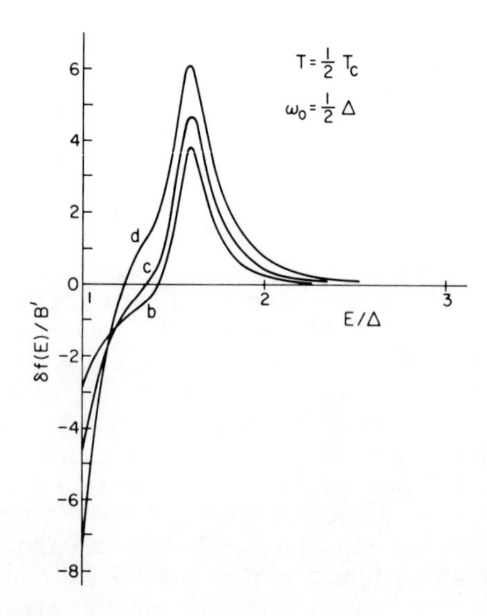

Fig. 18 The energy dependence of $\delta f(E)$ in a thin film under microwave irradiation. Curves (b), (c), and (d) correspond to $\tau_{es} = 0$, τ_B, and $8\tau_B$ respectively. Results are obtained from the linearized coupled kinetic equations.

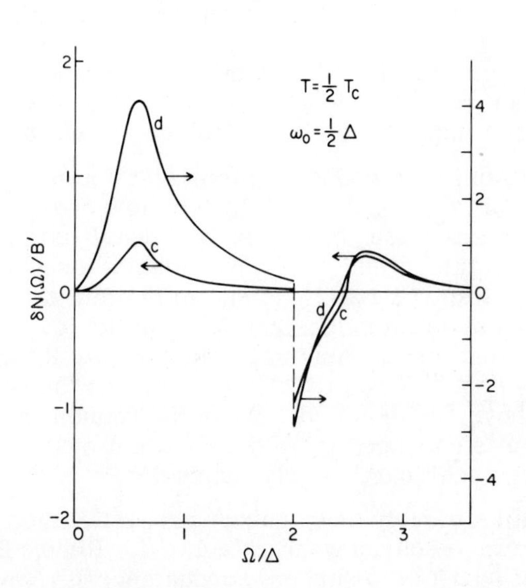

Fig. 19 The change in the phonon spectra in the thin film driven by microwaves. Curves (c) and (d) correspond to those quasiparticle distributions shown in Fig. 18.

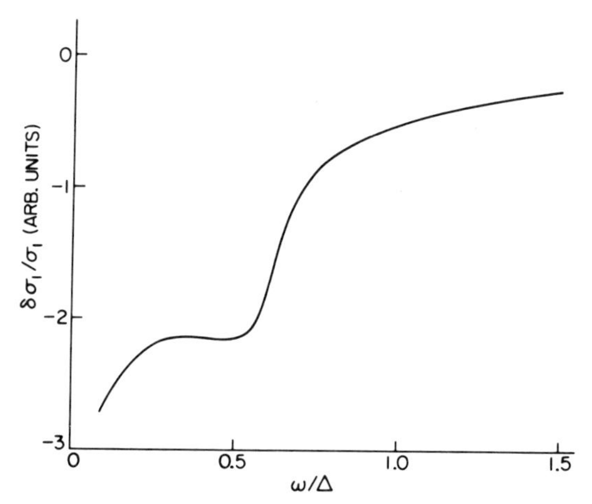

Fig. 20 Change in the real part of the frequency dependent conductivity $\delta\sigma_1(\omega)$ relative to the equilibrium value $\sigma_1(\omega)$ for a thin film irradiated by a microwave source with frequency $\Delta_0/2$.

changes in σ_1 of order 10% should be easily produced. We are presently studying the nonlinear equations in order to determine how large an effect can be produced. We have also carried out preliminary calculations of the gap enhancement using the nonlinear equations. At T/T_c = .9 we have found greater than 10% enhancement with a quite reasonable choice of parameters.

* Work supported by the Office of Naval Research.
† J.S. Guggenheim Foundation Fellow (1976-77).

REFERENCES

1. A.F.G. Wyatt, V.M. Dmitriev, W.S. Moore, and F.W. Sheard, Phys. Rev. Lett. 16, 1166 (1966).
2. A.H. Dayem and J.J. Wiegand, Phys. Rev. 155, 419 (1967).
3. T.C. Tredwell and E.H. Jacobsen, Phys. Rev. Lett. 35, 244 (1975) and Phys. Rev. B 13, 2931 (1976).
4. T.M. Klapwijk and J.E. Mooij, Physica 81B, 132 (1976).
5. G.M. Eliashberg, ZhETF Pis. Red. 11, 186 (1970), [JETP Lett. 11, 114 (1970)].
6. W. Eisenmenger and A.H. Dayem, Phys. Rev. Lett. 18, 125 (1967).
7. R.C. Dynes, V. Narayanamurti, and M. Chin, Phys. Rev. Lett. 26, 181 (1971); R.C. Dynes and V. Narayanamurti, Phys. Rev. B 6, 143 (1972).
8. V. Narayanamurti and R.C. Dynes, Phys. Rev. Lett. 27, 410 (1971).

9. L.R. Testardi, Phys. Rev. B 4, 2189 (1971).
10. G.A. Sai-Halasz, C.C. Chi, A. Denestein, and D.N. Langenberg, Phys. Rev. Lett. 33, 315 (1974).
11. J.N. Lechevet, L. Morelli, L. Leopold, and W.D. Gregory, Bull. Am. Phys. Soc. 19, 438 (1974); P. Hu, R.C. Dynes, and V. Narayanamurti, Phys. Rev. B 10, 2786 (1974).
12. C.S. Owen and D.J. Scalapino, Phys. Rev. Lett. 28, 1559 (1972).
13. J.J. Chang and D.J. Scalapino, Phys. Rev. B 10, 4047 (1974).
14. S. Chandrasekhar, Hydrodynamics and Hydrodynamic Stabaility (Clarendon, Oxford, 1961).
15. W.L. McMillan and J.W. Rowell, Superconductivity, R.D. Parks, ed. (Marcel Dekker, New York, 1969); J.M. Rowell, W. L. McMillan, and R.C. Dynes, "A Tabulation of the Electron-Phonon Interaction in Superconducting Metal and Alloys, Part I", in press.
16. R. Janik, L. Morelli, N.C. Cirillo, Jr., J.N. Lechevet, W.D. Gregory, and W.L. Goodman, Trans. Magnetics MAG-11, 689 (1975).
17. T.W. Wong, J.T. C. Yeh, and D.N. Langenberg, Applied Superconductivity Conf., 1976.
18. T.W. Wong, J.T.C. Yeh, and D.N. Langenberg, Phys. Rev. Lett. 37, 150 (1976).
19. L.P. Gor'kov and G.M. Eliashberg, Zh. Eksp. Teor. Fiz. 55, 2430 (1968) [Sov. Phys.-JETP 28, 1291 (1969)].
20. V. Keldysh, Zh. Eksp. Teor. Fiz. 47, 1515 (1964) [Sov. Phys.-JETP 20, 1018 (1965)].
21. R.V. Carlson and A.M. Goldman, Phys. Rev. Lett. 31, 880 (1973) and 34, 11 (1975).
22. I. Schuller and K.E. Gray, Phys. Rev. Lett. 36, 429 (1976).
23. A. Schmid and G. Schön, J. Low Temp. Phys. 20, 207 (1975).
24. S. Doniach, private communication.
25. J. Bardeen, G. Rickayzen, and L. Tewordt, Phys. Rev. 113, 982 (1959).
26. A. Rothwarf and B.N. Taylor, Phys. Rev. Lett. 19, 27 (1967).
27. P.W. Anderson, J. Phys. Chem. Solids 11, 26 (1959).
28. K.E. Gray, J. Phys. F 1, 290 (1971); D.C. Lankashire, J. Phys. F 2, 107 (1972).
29. J. Clarke, Phys. Rev. Lett. 28, 1363 (1972); M. Tinkham and J. Clarke, Phys. Rev. Lett. 28, 1366 (1972).
30. S.B. Kaplan, C.C. Chi, D.N. Langenberg, J.-J. Chang, S. Jafarey, and D.J. Scalapino, Phys. Rev. B, to be published.
31. G.M. Eliashberg, Zh. Eksp. Teor, Fiz. 38, 966 (1960) [Sov. Phys.-JETP 11, 699 (1960)]. For a review of strong coupling superconductivity see D.J. Scalapino in Superconductivity, R.D. Parks, ed. (Marcel Dekker, New York, 1969), Chap. X.
32. D.M. Ginsberg, Phys. Rev. Lett. 8, 204 (1962).
33. K.E. Gray, A.R. Long, and C.J. Adkins, Phil. Mag. 20, 273 (1967).
34. L.M. Falicov and D.H. Douglass, Progress in Low Temperature

Physics, J.C. Borter, ed. (North Holland Pub. Co., Amsterdam, 1964).
35. R.E. Prange and L.P. Kadanoff, Phys. Rev. 134, A566 (1964).
36. J. Bardeen, L.N. Cooper, and J.R. Schrieffer, Phys. Rev. 108, 1175 (1957).
37. G.M. Eliashberg, Zh. Eksp. Teor. Fiz. 61, 1254 (1971) [Sov. Phys.-JETP 34, 668 (1972)].
38. W.H. Parker, Phys. Rev. B 12, 3667 (1975).
39. W.A. Little, Can. J. Phys. 37, 334 (1959).
40. J.-J. Chang and D.J. Scalapino, Bull. Am. Phys. Soc. 21, 102 (1976) Phys. Rev. B, to be published.
41. J.-J. Chang and D.J. Scalapino, Applied Superconductivity Conference, Stanford, 1976,
42. J.-J. Chang and D.J. Scalapino, Phys. Rev. Lett., 37, 522 (1976).
43. J.T.C. Yeh and D.N. Langenberg, preprint (to be published).

Chapter 12

COMPUTER APPLICATIONS OF JOSEPHSON JUNCTIONS

P. Wolf

IBM Zurich Research Laboratory

8803 Rüschlikon, Switzerland

I. HISTORICAL NOTES

The first superconducting device for digital applications was the cryotron [7]. It did not succeed for a number of reasons including: (a) the mode of switching was based on the slow and power-consuming transition from the superconducting to normal state in metals; (b) the too early attempt to use a highly integrated technology; and (c) the lack of a theory of superconductivity. As an outgrowth of the BCS theory, the Josephson junction was discovered in 1962 (see historical discussion in Chapter I of this Proceedings). Its potential as a switching device which was superior to cryotrons was realized by J. Matisoo [8].

II. THE JOSEPHSON JUNCTION AS A SWITCHING DEVICE

For literature references on Josephson junctions in general, see Ref. 9. In digital circuits tunnel junctions are used as switches between the Josephson state and the resistive state. Switching is accomplished by the interaction of a magnetic field on the Josephson current. The field is generated by currents in one or more control lines sitting on top of the junction [1].

Note: This is an extended abstract of two lectures given at the NATO Institute on Small-Scale Applications of Superconductivity. A list of references is appended to give the reader a quick guide to the literature. For reviews treating digital applications of Josephson junctions see Refs. 1-6.

Tunnel junctions are quite well understood theoretically. This allows good paper designs of circuits. However, owing to the strong nonlinearity of the Josephson effects, analysis must often be carried out numerically by computer simulations. A sufficient junction equivalent circuit is the Stewart-McCumber model [10, 11]. In order to include control by the magnetic field it has to be extended to a distributed model [1]. In applications one usually wants to control a certain Josephson current with a minimal control current. This is accomplished best with junctions having a length of about three Josephson penetration depths.

Due to the high junction capacitance, the McCumber factor β of unloaded tunnel junctions is high and they have a large hysteresis, i.e., they are bistable [10, 11]. The switching time from the Josephson to the resistive state is proportional to the ratio of the junction capacitance per unit area to the Josephson current density [12]. The switching time can be decreased by decreasing the thickness of the tunnel insulator, because then the current density increases exponentially, but capacitance increases only as the reciprocal thickness. Measurements on low current-density junctions show good agreement with theory [12, 13]. However, junctions with high current densities switch so rapidly that at present it is not possible to measure them. For instance, lead-alloy junctions with current densities of 1.5 kA/cm^2 theoretically are expected to switch in 5-10 psec, whereas available equipment allows a resolution of only 30-40 psec [14].

Compared to semiconductor circuits Josephson junctions have two advantages. Not only is the switching speed faster by one or two orders of magnitude, but the power consumption is smaller by more than three orders of magnitude. The latter becomes essential in a high-speed computer. In order to keep propagation delays on interconnection lines between circuits small, which becomes more critical the shorter the circuit switching times, it is necessary to pack the devices densely together. To minimize the heat extraction problem, a low power consumption is of extreme importance.

In circuit applications, an output circuit is connected in parallel to the junction into which part of the junction current will be transferred when the junction goes to the resistive state. The output circuit consists of a superconducting line, which can be the control-line of other junctions and is terminated with a resistor. Two extreme cases are possible: (1) the resistor matches the transmission line, which avoids reflections and gives minimum delay [3]; (2) the resistor is zero so that the line is superconducting throughout. This case has longer delay than the matched case, and is used with short lines or where delay is not critical.

It is more complicated than the matched case [13].

Instead of single Josephson junctions, it may be advantageous to use two- or three-junction interferometers as switching elements. For a review see Refs. 15 and 16.

III. DEVICE FABRICATION

A description of device fabrication can be found in Ref. 17. The various electrodes are evaporated lead with additions of indium and gold [18]. The tunnel insulator is a native oxide formed in an rf plasma discharge [19]. Thicker insulating films are Nb_2O_5 obtained by anodization or evaporated SiO. Pattern definition is accomplished with a photolithographic lift-off process, adapted from semiconductor technology. Junctions with sizes down to about 1 μm have been made [14].

Measured characteristics of Pb-alloy junctions are in reasonably good agreement with theory, with respect to Josephson and single-particle currents and to diffraction patterns [20]. Attempts have been undertaken to fabricate niobium junctions; however, these junctions still have high leakage currents [21].

IV. CIRCUITS

A computer basically consists of logic circuits and memory circuits. Both types of circuits have been built.

A. Logic Circuits

The logic functions AND, OR, etc., can be performed in junctions with two or more control lines by suitable superposition of the input currents. Several circuit approaches have been realized.

1. Latching logic

Here the junction is relatively lightly loaded by the output circuit so that it still shows hysteresis. Accordingly, the junctions stay in the switched state even after the control currents have been removed. In order to reset the junctions the power supply has to be switched off. Basic investigations of such circuits are reported in Refs. 22 and 23. Quite complex circuits have been built, among others a one-bit adder [24], and a four-bit multiplier [25]. The latter circuit consists of 45 junctions, and has a measured four-bit multiply time of 27 nsec. Power consumption is 1.6 mW.

2. Nonlatching logic

Here the circuit is designed in such a way that it resets after the control inputs have been removed; a mode of operation similar to that of semiconductor circuits.

One design approach uses the jump from the resistive to the superconducting state which occurs if the junction voltage is below a value V_{min}. The circuit consists of two series-connected voltage-biased junctions. Such a circuit has been made with a measured switching time of 60 psec and a power consumption of 17 μW [26, 27].

Another approach uses junctions which are so heavily resistively loaded that the junction hysteresis disappears. A circuit with a three-junction interferometer as a switching device had a power consumption of only 40 nanowatts [15, 28].

B. Memory Circuits

The simplest superconducting storage device uses a persistent ring current in a superconducting loop. The first proposal for such a memory cell with Josephson junctions and its potential in a memory system was described in Ref. 2. Later, various cells based on this principle have been fabricated [29, 30, 31]. The smallest made had an area of about 900 $(\mu m)^2$ and a measured writing time of about 80 psec. All these cells have nondestructive read-out by sensing the magnetic field of the ring current with another junction.

In superconducting rings the magnetic flux is quantized. The above-mentioned cells contain so many quanta that they are little affected by quantization. For memory applications it is possible to make use of flux quantization with devices which have overlapping vortex modes. In the overlap regions the device has two stable superconducting states which differ by one flux quantum, and in which one bit of information can be stored [32]. Overlapping vortex modes occur in long junctions as well as in interferometers, and both have been investigated as storage elements, with emphasis on the two-junction interferometer. Two read-modes have been explored, both of which are desctructive. One detects with a Josephson junction the voltage spike which develops over the device when the information is changed [32, 33]. The other makes use of vortex-to-voltage transitions [34]. Recent results are given in Ref. 35.

The above-mentioned results show that promising high-speed circuits with Josephson junctions can be built. For a computer, additional problems have to be solved including reproducible mass fabrication of circuits, a packaging system with high volumetric density in order to keep propagation delays small, and cooling of the system [36].

REFERENCES

1. J. Matisoo, "Josephson-Type Superconductive Tunnel Junctions and Applications", IEEE Trans. Magn. MAG-5, 848 (1969).
2. W. Anacker, "Potential of Superconductive Josephson Tunneling Technology for Ultrahigh Performance Memories and Processes", IEEE Trans. Magn. MAG-5, 968 (1969).
3. W. Anacker, "Josephson Tunneling Devices - A New Technology with Potential for High-Performance Computers", AFIPS Conference Proceedings 41, 1269 (1972).
4. P. Wolf, "Der Josephson-Kontakt", Neue Zuercher Zeitung, March 25, 1974.
5. J. Matisoo, "Josephson Computers", IEEE Trans. Magn. (to be published).
6. W. Anacker, "Josephson Junctions as Computer Elements", in 1976 ESSDERC Proceedings, Conference Series of the Institute of Physics, London (to be published in March 1977).
7. N. S. Prywes, Amplifier and Memory Devices with Films and Diodes, McGraw-Hill, New York, 1965.
8. J. Matisoo, "The Tunneling Cryotron - A Superconductive Logic Element Based on Electron Tunneling", Proc. IEEE 55, 172 (1967).
9. L. Solymar, Superconductive Tunneling and Applications, Chapman and Hall, London, 1972.
10. W. C. Stewart, "Current-Voltage Characteristics of Josephson Junctions", Appl. Phys. Letters 12, 277 (1968).
11. D. E. McCumber, "Effect of AC Impedance on DC Voltage-Current Characteristics of Superconductor Weak-Link Junctions", J. Appl. Phys. 39, 3113 (1968). See Chapter 4 of this Proceedings for Equivalent Circuit Models.
12. W. C. Stewart, "Measurement of Transition Speeds of Josephson Junctions", Appl. Phys. Letters 14, 392 (1969).
13. H. H. Zappe and K. R. Grebe, "Dynamic Behavior of Josephson Tunnel Junctions in the Subnanosecond Range", J. Appl. Phys. 44, 865 (1973).

14. W. Jutzi, Th. O. Mohr, M. Gasser and H. P. Gschwind, "Josephson Junctions with 1 μm Dimensions and with Picosecond Switching Times", Electronic Letters 8, 589 (1972).
15. H. H. Zappe, "Josephson Quantum Interference Computer Devices", IEEE Trans. Magn., Applied Superconductivity Conference August 1976, Stanford, Cal. (to be published).
16. P. Wolf, "SQUIDs as Computer Elements" in SQUID and its Applications, Walter de Gruyter, Berlin (to be published).
17. J. H. Greiner, S. Basavaiah and I. Ames, "Fabrication of Experimental Josephson Tunneling Circuits", J. Vac. Sci. Technol. 11, 81 (1974).
18. S. K. Lahiri, "Metallurgical Considerations with Respect to Electrodes and Interconnection Lines for Josephson Tunneling Circuits", J. Vac. Sci. Technol. 13, 148 (1976).
19. J. H. Greiner, "Josephson Tunneling Barriers by rf Sputter Etching in an Oxygen Plasma", J. Appl. Phys. 42, 5151 (1971).
20. S. Basavaiah and R. F. Broom, "Characteristics of In-Line Josephson Tunneling Gates", IEEE Trans. Magn. MAG-11, 759 (1975).
21. R. F. Broom, R. Jaggi, R. B. Laibowitz, Th. O. Mohr and W. Walter, "Thin-Film Josephson Tunnel Junctions with Niobium Electrodes", Proceedings LT14, Vol. 4, p. 172, North Holland, Amsterdam, 1975.
22. W. H. Henkels, "An Elementary Logic Circuit Employing Superconducting Josephson Tunneling", IEEE Trans. Magn. MAG-10, 860 (1974).
23. D. J. Herrell, "Femto-Joule Josephson Tunneling Logic Gates", IEEE J. Solid-State Circuits SC-9, 277 (1974).
24. D. J. Herrell, "A Josephson Tunneling Logic Adder", IEEE Trans. Magn. MAG-10, 864 (1974).
25. D. J. Herrell, "An Experimental Multiplier Circuit Based on Superconducting Josephson Devices", IEEE J. Solid-State Circuits SC-10, 360 (1975).
26. W. Baechtold, Th. Forster, W. Heuberger and Th. O. Mohr, "Complementary Josephson Junction Circuit", Electronics Letters 11, 203 (1975).
27. W. Baechtold, "A Flip-Flop and Logic Gate with Josephson Junctions", 1975 International Solid-State Circuits Conference, Digest of Technical Papers 18, 164 (1975).
28. H. H. Zappe, "Quantum Interference Josephson Logic Devices", Appl. Phys. Letters 27, 432 (1975).
29. H. H. Zappe, "A Subnanosecond Josephson Tunneling Memory Cell with Nondestructive Readout", IEEE J. Solid-State Circuits SC-10, 12 (1975).
30. R. F. Broom, W. Jutzi and Th. O. Mohr, "A 1.4 Mil^2 Memory Cell with Josephson Junctions", IEEE Trans. Magn. MAG-11, 755 (1975).

31. W. Jutzi, "An Inductively Coupled Memory Cell for NDRO with Two Josephson Junctions", Cryogenics 16, 81 (1976).
32. P. Guéret, "Experimental Observations of Switching Transients Resulting from Single Flux Quantum Transitions in Superconducting Josephson Devices", Appl. Phys. Letters 25, 426 (1974).
33. P. Guéret, "Storage and Detection of a Single Flux Quantum in Josephson Junction Devices", IEEE Trans. Magn. MAG-11, 751 (1975).
34. H. H. Zappe, "A Single Flux Quantum Josephson Memory Cell", Appl. Phys. Letters 25, 424 (1974).
35. P. Guéret, Th. O. Mohr and P. Wolf, "Single Flux Quantum Memory Cells with Josephson Junctions", IEEE Trans. Magn. (to be published).
36. W. Anacker, "Superconducting Tunnel Devices as an Alternative to Semiconductors for Fast Computer Circuits", 1975 International Solid-State Circuits Conference, Digest of Technical Papers 18, 162 (1975).

Chapter 13

PROGRAMS ON SMALL-SCALE SUPERCONDUCTING DEVICES IN CANADA

James A. Blackburn

Physics Department, Wilfrid Laurier University

Waterloo, Ontario, Canada

I. INTRODUCTION

As is pointed out in Dr. Bogner's paper [1], Canada has an established program on large-scale applications of superconductivity, specifically for magnetically levitated trains employing linear induction motor drive. In the area of small-scale applications, the effort is more scattered, and is the result of a relatively small number of investigators (perhaps 20 or 30) working more or less independently of each other. Academic institutions, government laboratories, and a few industrial laboratories are all involved in this work.

The following list includes both current and recently completed studies related to device applications.

A. SQUID Magnetometers

Investigators: M. Burbank, A. Fife, R. Lomnes
Canadian Thin Films, Ltd.
Burnaby, British Columbia

In 1971, Consadori, Fife, Frindt and Gygax reported [2] a technique for constructing rf SQUIDs from single crystals of niobium diselenide. This superconducting layered structure can be cleaved to a thickness of a few hundred angstroms and then scribed to form a microbridge a few microns wide. It has a T_c of about 7.0 K and forms very stable weak links which can be thermally recycled without degradation. Canadian Thin Films has developed this type of SQUID into a complete modular instrument (CTF Series 100) with sensor noise rated at less than 3×10^{-10} gauss rms/$\sqrt{Hz}$.

Recent efforts have been aimed at magnetic gradiometers for geophysical work. A SQUID array with a baseline of about 10 centimeters will be combined with suitable electronics to form an airborne package. Target date for this phase of the project is late 1977.

B. SQUID Resistance Bridge [3]

Investigators: J. A. Rowlands, S. B. Woods
Department of Physics
University of Alberta
Edmonton, Alberta

If two resistors - an unknown R_x and reference R_S - are tied together at one end and connected via a sensitive galvanometer at the opposite ends, and if currents I_S and I_x are supplied to R_S and R_x, respectively, then when the galvanometer is nulled R_S/R_x is equal to I_x/I_S. This principle has been applied in an instrument capable of measuring unknowns in the range 10 $\mu\Omega$ - 1 mΩ to a precision of about one part in 10^7. Operation is at liquid helium temperatures. Measuring currents may be as low as 30 mA and power dissipation is typically a microwatt per resistor. A commercial version of the MacMartin and Kusters [4] dc current comparator (Guildline Instruments Model 9920) was used to measure the current ratios to better than 0.1 parts per million, while a point-contact SQUID and associated electronics (S. H. E. Corp. Model 202A) served as a superconducting galvanometer.

A suitable reference resistor must of course be provided. Some desirable attributes of such a comparison standard would be: (1) R_S nearly independent of measuring current; (2) R_S nearly independent of temperature; (3) small absolute thermoelectric power; and (4) small magnetoresistance. It was discovered that the alloy, Pd - 30 at.% Pt, possessed all of these properties to a high degree. Resistivity of the material is 18.25 $\mu\Omega$ - cm at 4.2 K.

C. Voltage Standards [5]

Investigators: G. H. Wood, A. F. Dunn
Division of Physics
National Research Council of Canada
Ottawa, Ontario

A small group at NRC has established an emf monitoring facility using the well-known interaction of a Josephson device with an oscillating radiation field. The procedure adopted is as follows. A thin film Pb-PbO-Pb junction is irradiated by the output from a crystal stabilized klystron of frequency f and is current-biased at one of the resulting steps in its I-V characteristic. The voltage associated with this step is then compared with the output of the standard cell under test by means of a bridge circuit containing a stabilized current source and precision resistor network. The ratio of the specially wound resistors is fixed at 400:1 and balance is achieved by varying f. Because the equation connecting step voltage and microwave frequency depends only upon the ratio of fundamental constants (e/h), long-term drifting of the national standard cells can be evaluated from the variations in f required to null the bridge. During the past two years efforts have been concentrated mainly on reducing thermal emf's and on improving shielding. Sensitivity is currently about 0.1 ppm.

D. Thin Film Pressure Sensor [6]

Investigators: P. W. Wright, J. P. Franck
Department of Physics
University of Alberta
Edmonton, Alberta

The object of this study was to examine the possible application of tunneling devices to direct pressure sensing. Junctions were fabricated from aluminum - aluminum oxide - lead/indium alloy thin films deposited in a standard crossed geometry. An ac modulation bridge designed by Adler and Jackson [7] was used to record junction resistance as a function of dc bias and test cell pressure. It was discovered that at 30 K and for pressures up to about 4 kilobar the value of R at zero bias depended upon P in the following manner:

$$\log R = 1.740 - 0.0552\, P + 0.00282\, P^2$$

where P is expressed in kbar and R in ohms. This calibration turned out to be quite accurate over a temperature range of 7 K to 77 K. In addition, reproducibility of better than 0.5% was found for repeated pressure cycling.

E. Sliding Superconducting Contacts [8]

Investigators: B. L. Blackford, C. J. Purcell, G. Stroink
Physics Department
Dalhousie University
Halifax, Nova Scotia

Work is continuing world-wide in the development of mechanical devices partially composed of superconducting elements - for example, rotatable sample holders, motors, generators, and so forth. In such apparatus there is an obvious need to make electrical contact to the rotor; however, noise and dissipation in the brushes can be severe problems. This project has revealed the attractive properties of solder coated phosphor-bronze wires. Lossless and noiseless ($< 10^{-8}$ volt) sliding electrical contact was made to a spinning superconducting cylinder at slip velocities of 0.5 cm/sec and for currents below 100 mA. Contact pressure was 0.5 N. Performance was found to degrade with higher forces, currents, and velocities.

F. Other Projects

Here we mention additional projects of more limited scope or of less direct applicability to superconducting devices.

Meincke and Moore [9] (Erindale College, University of Toronto) have designed and constructed a low mass point-contact driven by an improved differential screw. This has been used in both single particle and Josephson tunneling experiments and has proved to be versatile and stable. Smith (University of Waterloo) and Blackburn (Wilfrid Laurier University) have done extensive experiments and computer simulations on relatively high inductance (microhenry) superconducting loops interrupted by thin film Josephson junctions [10]. The detailed dynamics of magnetic flux entry have been revealed and it is anticipated that this work will have implications for memory devices employing weakly closed rings. Nerenberg (University of Western Ontario) and Blackburn (Wilfrid Laurier University) have recently completed theoretical studies of self-resonance in circular and rectangular planar Josephson junctions [11]. The dc current steps and their magnetic field dependences have been investigated; the results may be useful in assessing the behaviour of devices employing this type of tunneling junction. Dmitrevsky and Habib [12] (Electrical Engineering, University of Toronto) have done some theoretical work on voltage-biased Josephson mixers in the small signal regime. A practical detector was also planned for use in a nearby radio telescope. However, shortcomings of the radio antenna and fiscal limitations seem to have halted this plan.

REFERENCES

1. G. Bogner, Review of Large Scale Applications, this proceedings.
2. F. Consadori, A. A. Fife, R. F. Frindt, and S. Gygax, Appl. Phys. Letters 18, 233 (1971).
3. J. A. Rowlands and S. B. Woods, Rev. Sci. Instr. 47, 795 (1976).
4. M. P. MacMartin and N. L. Kusters, IEEE Transactions on Instrumentation and Measurement 15, 212 (1966).

5. G. H. Wood, A. F. Dunn, and L. A. Nadon, IEEE Transactions on Instrumentation and Measurement 23, 275 (1974).
6. P. W. Wright and J. P. Franck, Rev. Sci. Instr. 46, 1474 (1975).
7. J. G. Adler and J. E. Jackson, Rev. Sci. Instr. 37, 1049 (1966).
8. B. L. Blackford, C. J. Purcell, and G. Stroink, Cryogenics 15, 283 (1975).
9. S. E. Moore and P. P. M. Meincke, J. Appl. Phys. 44, 3734 (1973).
10. H. J. T. Smith and J. A. Blackburn, Phys. Rev. B 12, 940 (1975).
11. M. A. H. Nerenberg and J. A. Blackburn, Phys. Rev. B 9, 3735 (1974); M. A. H. Nerenberg, P. A. Forsyth, Jr., and J. A. Blackburn, J. Appl. Phys. 47, 4148 (1976).
12. S. E.-D. E.-S. Habib and S. Dmitrevsky, J. Appl. Phys. 46, 900 (1975).

Chapter 14

PROGRAMS ON SMALL-SCALE SUPERCONDUCTING DEVICES IN FRANCE

R. Adde

Institut d'Electronique Fondamentale, Université de Paris-Sud 91405 - Orsay, France

We give a listing of the laboratories, programs and researchers working on small-scale superconducting devices in France.

Fontenay-aux-Roses:

Laboratoire de Génie Electrique de Paris (T. Pech) and Laboratoire Central des Industries Electriques (F. Delahaye): 33 avenue du Général Leclerc, 92260, Fontenay-aux-Roses.

Maintenance of the volt, tunnel junction fabrication and optimization for voltage maintenance (supported by Bureau National de Metrologie).

Laboratoire de Génie Electrique de Paris (J. Baixeras):

Studies of lamellar eutectics Pb-Sn.

Grenoble:

Laboratoire d'Electronique et de Technologie de l'Informatique, Centre d'Etudes Nucléaires, Av. des Martyrs, 38041, Grenoble.

(a) Laboratoire de microélectronique Physique (D. Zenatti):

Magnetometry: Realization of a 300 MHz squid magnetometer (in collaboration with the group of M. Sauzade, Orsay). Studies of high T_c squids for an autonomous magnetometer in a cryogenerator.
Logic: Realization of elementary gate circuits. Project: Realization of a system with several memory units (in collaboration with the group of D. Randet).
Junction fabrication: programs in collaboration with several groups.

ENS (A. Libchaber): long tunnel junctions.
Univ. Orsay (M. Sauzade): 9 GHz squids
Univ. Orsay (R. Adde): microbridges, small tunnel junctions for detection & mixing.

(b) Laboratoire d'Electronique Physique Appliquee: (D. Randet):

Evaluation of JJ performance compared with other high speed memories. Logic (project): realization of a system with several memory units (in collaboration with the group of D. Zenatti).

Nice:

Laboratoire de Physique de la Matière Condensée, Parc Valrose, 06034, Nice.

Proximity effects in Pb-Sn-Pb Josephson structures (A. Gilabert). Non-equilibrium conditions in superconductors irradiated by laser light. Application to the fabrication of superconducting weak links (Ostrowski).

Orsay:

Institut d'Electronique Fondamentale, Bât. 220, Université Paris-Sud, 91405, Orsay.

Superconducting cavities (A. Septier, N. T. Viet): Development of high Q superconducting cavities and superconducting oscillators. Development of a 9 GHz superconducting oscillator for laser frequency metrology (supported by the Bureau National de Métrologie).

Magnetometry (M. Sauzade): Realization and optimization of a 300 MHz squid magnetometer (in collaboration with the group of D. Zenatti). Applications to magnetobiology. Development of a 9 GHz ac squid magnetometer.

High frequency detection and mixing (R. Adde, G. Vernet): Studies on high frequency and noise properties of Josephson junctions. Development of a 300-1000 GHz Josephson receiver (oscillator-mixer). Application to radioastronomy. Development of a Josephson harmonic mixer (x 400) for FIR laser frequency metrology (supported by Bureau National de Mesures). Fabrication and properties of Nb microbridges (in collaboration with the group of D. Zenatti, Grenoble).

Paris:

Ecole Normale Supérieure, Groupe de Physique des Solides, 24, rue Lhomond, 75231, Paris (A. Libchaber).

Nonlinear effects in the Riedel anomaly. Nonequilibrium effects in superconductors by microwave pumping in the FIR.

Ecole Supérieure de Physique et de Chimie, 10, rue Vauquelin, 75005, Paris (J. Lewiner):

Coupling of Josephson junctions at microwave frequencies.

Rennes:

Institut National des Sciences Appliquées, 35031, Rennes (J. Rosenblatt):

Phase transition to a coherent state in three-dimensional granular superconductors, possibility of super-radiance. Microbridge arrays.

Chapter 15

PROGRAMS ON SMALL-SCALE SUPERCONDUCTING DEVICES IN GERMANY

S. N. Erné

Physikalisch-Technische Bundesanstalt

3300 Braunschweig, Germany

First I wish to point out that in Germany the important superconductivity industrial activities are not centered on small-scale applications of superconductivity: the main interest of the German industry concerns large-scale application of superconductivity, i.e., power machines, cables for energy transfer or high field magnets.

Here we define "small-scale applications of superconductivity" as:

(a) small-scale application of bulk superconductivity
(b) applications of "weak" superconductivity

The first topic includes superconducting resonators for elementary particle accelerators, superconducting lenses for electron microscopy, cryo-current-comparators, etc. The second topic includes SQUIDS, with their various applications, and Josephson junctions.

Applications of Bulk Superconductivity

1. A group at Siemens AG under Dr. Pfister is working in cooperation with a group of the University of Karlsruhe under Dr. Halbritter to produce resonators of high quality factor Q of the order of 10^{10}. They use niobium and Nb_3Sn. Such resonators are very important for high efficiency accelerators.

2. A group under Mrs. Dr. Dietrich at Siemens AG in Munich is working on applying superconducting magnetic lenses to

electron microscopy. The advantages of such a system are well known: one can obtain by small dimensions higher field and field gradients - in other words, better optical properties.

3. Using the field shaping properties of superconducting shields, a group of the Phys.-Techn. Bundesanstalt is developing cryo-current-comparators. A cryo-current-comparator is a device which compares two currents. I can explain the working principle of the comparator by a simple configuration: we use as comparator a superconducting cylinder. If we have a current-carrying conductor in the inside of the cylinder, as a consequence of the magnetic properties of the superconducting material, the field distribution outside is independent of the position of the conductors. Therefore, if we have two conductors, the condition for equal currents is the vanishing of the field outside of the cylinder, without regard for the geometry of the system. Based on this principle, this group has built devices which permit comparison of currents with a relative accuracy better then one part per 10^{11} (the limitation is given by the resolution of the magnetic field detector).

To use such devices we need a detector for the zero flux condition or balance condition. The SQUID magnetometer is the ideal device for this application. It is very sensitive and works at the same temperature as the comparator.

Application of SQUIDs and Josephson Junctions

Only isolated activities in various universities comprising groups of only one or two scientists.

A group at the University of Münster under Prof. Heiden is working on a high slew rate magnetometer for study of fast magnetic phenomena in superconductors. They have developed a 10 GHz SQUID with a closed loop bandwidth in the MHz region.

Mr. Ludwig at the Max Planck Institute for Metal Research in Stuttgart is using a commercial SQUID for studies of flux creep in Type II superconductors. Results will be reported at the IC SQUID conference next October in Berlin.

A group of the Bavarian Academy of Sciences in Garching is using SQUIDs for noise thermometry (contact: Mr. Eska). In the last conference of the German Physical Society they reported the results of comparison between noise thermometry and gamma anisotropy experiments. They use the same sample as a noise source and a gamma radiation source. The two thermometers agree within the measurement accuracy. A group in the

Max Planck Institute for Physics and Astrophysics in Munich (contact: Dr. Kadlec) is working in application of Josephson junctions as wide-band infrared detectors.

Finally, the activities of the Physikalisch-Technische Bundesanstalt include perhaps fifteen scientists who are working in theory and application of superconducting devices. Up to now the main application of small-scale superconductivity has been metrology, so that the largest group is working in this field at the PTB. A list of some of the research is given below:

1. Theory of SQUID, flux dynamics, emission spectra of resistive SQUIDs, models and simulation for heterodyning and for magnetometers in the various working modes like hysteretic, non-hysteretic or at high frequency.

2. Theory of Josephson junctions, study of the interaction with microwave fields, analysis of the effect of the single current terms of the current phase relationship on the IV characteristics.

3. Development of rf SQUID magnetometers.

4. Development of dc SQUID magnetometers.

5. Josephson voltage standard.

6. Cryo-current-comparator as ratio standard.

7. Current measurement with current comparator and SQUID.

8. Susceptibility measurement.

9. Noise thermometry for temperatures below 4 K.

10. Medical applications in prenatal medicine.

Not all the fields have the same state of development. applications are in an early stage of development, whereas the topics at the beginning of this list have been pursued for several years.

Chapter 16

PROGRAMS ON SMALL-SCALE SUPERCONDUCTING DEVICES IN ITALY

M. Cerdonio*

Gruppo Nazionale Struttura della Materia and

Physics Department, University of Rome, Rome, Italy

Research on fundamental and applied aspects of small-scale superconducting devices in Italy is carried out at 4 universities, 4 government laboratories and 1 industrial laboratory. Financing of the university laboratories comes through Gruppo Nazionale Struttura della Materia, the national agency for research on the structure of matter, from funds administrated by Consiglio Nazionale delle Ricerche (CNR), the national research council. CNR finances directly its own two laboratories in which superconductivity research is present; the leading research program of one laboratory (Arco Felice) is on cybernetics, and of the other (Rome) is on electronic properties of solid state. The other two government laboratories where small-scale superconductivity is pursued are CNEN at Frascati (Rome), the major laboratory of the Italian nuclear energy agency, and Istituto Elettrotecnico Nazionale "G. Ferraris" in Turin, the Italian metrology laboratory. The industrial laboratory is SNAM-Progetti at Monterotondo (Rome), which is part of the holding company of the major Italian gasoline company.

Overall, more than thirty research scientists or staff are involved at present, to which some twenty students, temporary, visiting and part-time people should be added.

The scientific research spans the more fundamental problems, such as light-induced Josephson currents [1] and fluxon propagation [2], to fully developed instrumentation, such as high-resolution susceptometers [3] (now in routine use for biochemical

* Present Address: Faculty of Science, Free University of Trento, Povo, Trento, Italy

studies) and strain transducers [4] to be used in gravitational wave experiments. There also are studies on properties and fabrication techniques of weak links [5], and development of superconducting cavities for which $Q \approx 10^9$ and fields of 10 MV/m have already been achieved. Metrology is also pursued with voltage [6] and frequency standards and current comparators. A recent development [7] has been a superconducting bolometer with this interesting performance: a low NEP $\approx 10^{-13}$ watt/$\sqrt{Hz}$ together with a fast response time $\tau \approx 1$ μs.

Table I shows the distribution of these activities in various laboratories. A list of recent references is also included [8-14].

Table I: Small-Scale Superconducting Efforts in Italy

Laboratory or University	Number of research staff	Topic
Catania (+ CNR)	2	Josephson junctions with Formvar barrier.
Genova (+ CNR)	3	Bolometers, use of SQUIDs for transport properties.
Genova (+ INFN)	3	Superconducting cavities for accelerators and frequency standards.
Rome (+ CNR)	7	Susceptometers for biophysical chemistry, strain transducers for gravitational wave antennas.
SNAM-Progetti (Monterotondo)	2	Susceptometers for biophysical chemistry.
CNR - Arco Felice (Naples)	2	Light-sensitive Josephson junctions, large Josephson junctions.
Salerno (+ CNR)	5	Fluxon propagation in long Josephson junctions, proximity effects.
CNEN-Frascati	2	Josephson junctions, SQUID technology.
CNR - Rome	2	Weak links: fabrication and properties.
IEN - Turin	4	Frequency and voltage standards, current comparator.

REFERENCES

1. A. Barone, G. Paternó, M. Russo and R. Vaglio, Phys. Letters 53A, 393 (1975).
2. G. Costabile and R. D. Parmentier, Proc. LT14 (Helsinki, 1975).
3. M. Cerdonio, C. Cosmelli, C. Gramaccioni, C. Messana and G. L. Romani, Rev. Sci. Instr. 42, 1 (1976).
4. L. Adami, M. Cerdonio, R. F. Ricci and G. L. Romani, Appl. Phys. Letters
5. P. Carelli and I. Modena, Phys. Letters
6. D. Andreone, E. Arri and G. Marullo, Alta Frequenza
7. G. Gallinaro and R. Varone, Cryogenics, 5 (1975).
8. E. P. Balsamo, G. Paternó, A. Barone, P. Rissmann and M. Russo, "Temperature Dependence of the Maximum (dc) Josephson Current", Phys. Rev. B10, 1881 (1974).
9. A. Barone, W. T. Johnson and R. Vaglio, "Current Flow in Large Josephson Junctions", J. Appl. Phys. 46, 3628 (1975).
10. R. D. Parmentier, G. Costabile, P. Rissmann and E. P. Balsamo, "Temperature Dependence of the Maximum Josephson Current in Nb-NbO_x-Sn Junctions", J. Low Temp. Phys.
11. M. Cerdonio and C. Messana, "High Resolution Superconducting Magnetometer", IEEE Transactions, Magnetics 11, 778 (1975).
12. M. Cerdonio, F. Mogno and G. L. Romani, "Vibrating Sample Superconducting Magnetometer", Proc. LT14, 4 (Helsinki, 1975).
13. M. Cerdonio, G. L. Romani and S. Pace, "SQUID Operation with Ferromagnetic Core Superconducting dc Transformers", Cryogenics, 5 (1975).
14. G. Paternó, P. Rissmann and R. Vaglio, "Temperature Dependence of the Maximum Josephson Current in Nb-NbO_x-Pb Junctions", J. Appl. Phys. 46, 1415 (1975).

Chapter 17

PROGRAMS ON SMALL-SCALE SUPERCONDUCTING DEVICES IN THE NETHERLANDS

R. de Bruyn Ouboter

Kamerlingh Onnes Laboratory

Leiden, The Netherlands

We give a listing of the laboratories, programs and researchers working on small-scale superconducting devices in The Netherlands.

Kamerlingh Onnes Laboratory, Rÿksuniversiteit, Leiden

Thermoelectric effects in superconducting point contacts. Flux penetration in a many point contact device.

R. de Bruyn Ouboter, A. A. J. Matsinger

Afdeling der Technische Natuurkunde van de Technische Hogeschool te Delft

Detection of sub-millimeter and millimeter radiation with point contacts and microbridges, coupling problems. Fabrication of Nb-, Nb_3Sn-variable thickness microbridges. Behaviour of microbridges. Ginzburg-Landau calculations on the critical current $I_c(T)$ and the supercurrent $I_s(\Delta\phi)$. Nonequilibrium aspects, phase slip, increase of the gap and the critical temperature with microwave radiation.

J. E. Mooij, T. M. Klapwijk, G. M. Daalmans.

Department of Space Research, Rÿksuniversiteit, Groningen

Detection of submillimeter and millimeter radiation from astronomical sources. Wideband and heterodyne detection.

H. Tolner

Afdeling der Technische Natuurkunde van de Technische Hogeschool Twente, Enschede

Biomedical investigations with SQUIDS.

J. J. Walter-Peters

Afdeling Experimentele Natuurkunde der Katholieke Universiteit, Nijmegen

Superconducting galvanometer.

P. Wyder, H. van Kempen

Natuurkundig Laboratorium der N. V. Philips' Gloeilampenfabrieken, Eindhoven

Some research on Josephson junctions.

Chapter 18

PROGRAMS ON SMALL-SCALE SUPERCONDUCTING DEVICES IN THE UNITED KINGDOM

J. G. Park

Department of Physics, Blackett Laboratory, Imperial College

London SW7 2BZ, United Kingdom

Table I is a summary of information gleaned during the last three weeks of August, 1976. Although it may not include every research group involved in Small-Scale Superconducting Devices, it is believed to be sufficiently complete to show what is being done and where in the United Kingdom. An attempt has been made to include an indication of the stage reached by the work. The extent to which it is dependent on funds outside those available within the various establishments themselves is also indicated.

Included is the development of devices, and also any work that is being done using them for making measurements of physical quantities such as electrical resistivity ρ or magnetic susceptibility χ; or that is aimed at increasing our understanding of their underlying physical principles. The use of SQUIDs is not yet very widespread, so that I have not had to omit a mention of any experiments using them that are known to me, or that have been brought to my attention. As yet there are few applications long enough established in any one place to have become routine. The SQUID used in the maintenance of the Josephson voltage standard at the National Physical Laboratory (NPL) is one of the few examples of a device whose use has become a routine matter.

In the table, the names of the principal investigators have been given in abbreviated form, together with the name of the establishment in which their laboratory is to be found. Smith + 2 + S means that one research student and two people other than research students are working on the project in addition to Smith. Names and addresses, referred to by the number on the left-hand side of the table, are given in full in an appendix. In the right-

Table I

Research on Small-Scale Superconducting Devices in the United Kingdom

	Devices	Status		Project
SQUIDs, thin film				
3	Aplin + s (Bristol)	1	S	SQUID Parametric Amplifier (gravity waves)
25	Goodall (Appleton Lab, Slough)	1	S	35 GHz Parametric Amplifier, unbiased array of microbridges (Radio-Astronomy)
18	Donaldson + 3 + 2s (Strathclyde)	0	S	DC SQUID (integrated magnetometers-medicine, etc.)
SQUIDs, not (knowingly) thin film				
10	Meredith + s (Lancaster)	0		UHF SQUID
14	Unvala + s (Imperial College)	1	S	500 MHz SQUID (4 GHz eventually)
31	Good + 2 (Cryogenic Consultants)	0		500 MHz SQUID
21	Richards + Clark + s (Sussex)	1	S*	DC, UHF, 10 GHz SQUIDs (Magnetometers - NMR?)
27	Petley + 3 (NPL)	4		10 GHz SQUID (Attenuator Calibration)
Other Weak Link Devices				
4	Beck + s (Engineering, Cambridge)	3		Detection and mixing, of mm microwaves
6	Waldram (Cambridge)	0		Detection and mixing (point contacts)
28	Blaney + 2 (NPL)	1	E	Detection and mixing (radio-astronomy)
Physical principles of weak links and RF SQUIDS				
7	Halse + s (Kent)	4		$i_2 \cos\theta$ term in $i(\theta)$ from I-V curves near T_c
22	Richards, Clark + s (Sussex)	1	S_a	$i(\theta, t)$; fluctuations and SQUIDs
23	Richards, Clark + s (Sussex	1	S_a	Frequency and phase locking in arrays
5	Waldram + s (Cambridge)	3		$i(\theta)$; microbridges of varying length
Magnetometers				
1	Vinen, Muirhead (Birmingham)	3		Magnetic thermometer for C_p, K_{th}
		0		χ of metals
15	Walmsley + s (Coleraine)	1		Magneto-cardiograms

Table I (continued)

	Devices	Status		Project
Magnetometers (continued)				
16	Cooke, Swithenby, Wells (Oxford)	3	S?	χ of rare earth orthoferrites, etc.
17	Aitken, Walton (Archaeology, Oxford)	2	N	Archaeological specimens
18	Donaldson + 3 + 2s (Strathclyde)	0	S	Environmental and medical applications
19	Finn, Kiymac (Sussex)	4	S	χ of Pt-Mn (1-4 K, 0-15 G)
20	Brewer, Truscott, Betts (Sussex)	2	S	T for superfluid ^{3}He experiments
21	Richards, Clark + s (Sussex)	1	S*	for NMR?
31	Good + 2 (Cryogenic Consultants)	4		Variable temperature (low field)
		1		Variable temperature (high field)
Galvanometers and Picovoltmeters				
2	Gugan + s (Bristol)	0		High resolution ρ(T)-alkali metals
3	Aplin + s (Bristol)	1	S	Gravity wave detector
5	Waldram + s (Cambridge)	3		i(θ); microbridges of varying length
8	Guenault, Pickett + s (Lancaster)	3	S	Thermopower (current in SC loop)
9	Guenault, Pickett + s (Lancaster)	3	S	ρ in magnetic field (SC Int. state)
11	Greig (Leeds)	1	S?	Thermopower, ρ, K_{th}
12	Caplin, Park + s (Imperial Coll.)	2	S	High resolution ρ(T)
13	Park + s (Imperial College)	1	S_a	RSQUID for heat current (C_p, K_{th})
24	Lee, Charlesworth + 1 (AERE)	0		RSQUID for C_p of Actinides
26	Gallop + 1 (NPL)	3		Current comparator (voltage standard)
29	Gallop + 1 (NPL)	4		Galvo' (Josephson voltage standard)
30	McDonald (Oxford Instruments)	1		Systems (e.g., magnetic thermometer)

KEY: S = SRC; S* = SRC, NRDC, Paul Instrument Fund
S_a = SRC (grant applied for); E = ESRO; N = NRC (Canada)
i(θ) = current-phase relationship measured; SC = superconductivity
C_p = specific heat; ρ = resistance or resistivity; χ = magnetic susceptibility;
K_{th} = thermal conductivity

CASE = Co-operative Award in Science and Engineering (work must involve collaboration with industry)
NPL = National Physical Laboratory
AERE = Atomic Energy Research Establishment
SRC = Science Research Council
NRDC = National Research and Development Council
ESRO = European Scientific Research Organization
NRC = National Research Council

hand column of the table a brief indication of the nature of the work is given. In the middle column the stage reached by the research is indicated in a crude but direct way by a number between 0 and 4, according to the following scheme: 0, initial plans are being made; 1, apparatus is being built; 2, the apparatus has been built; 3, results have been or are being obtained; 4, the experiment has been completed. The following letter (usually an S) indicates the source of financial support if any, other than the resources of the establishment itself, has been obtained. Thus, S = Science Research Council (SRC). Other abbreviations are given at the end of the table.

For some years there has been much skepticism about the possibility of anyone but a specialist, highly trained in the arts of getting SQUIDs to work, applying superconducting devices to measurements. This feeling is now less widespread in the UK. The frequency with which the figure 0 or 1 occurs in the middle column of Table I gives some idea of the level of current interest in possible applications of SQUIDs. These are projects in the planning or design stage. Weak links for rf applications are still matters for specialists; but for people who are not specialists in superconductivity or electrical engineering, and who want to use SQUIDs to measure quantities too minute to be measured before – because the magnetic properties are too weak perhaps, or because the sample is too small – the SQUID is beginning to move out of the category of a novel scientific toy. The idea that a SQUID is a device which enables measurements of previously unheard-of sensitivity to be made under conditions unattainable in practice in any real experiment dies hard. It is beginning to die partly because of what one hears that often people have been able to do, and because more people are using these techniques. The existence of local expertise, or lack of it, is an important factor in deciding whether a good idea is followed up or not. It is still true that a newcomer to the subject might be forgiven if he were to feel, on the basis of what he reads in the literature, that it would be better if more emphasis were now given to the problems using SQUIDs for making measurements under realistic conditions; and if a little less attention were lavished on the factors that govern the ultimate sensitivity under ideal conditions.

The wide range of topics is evident from the list of applications in Table I, and these seem to be a good balance between the amount of effort going into applications of the present standard SQUID (driven at about 20 MHz) and on the development of SQUIDs driven at higher frequencies. The latter should eventually extend the range of possible measurements, not only because the signal-to-noise ratio will be raised, but also because the speed of response will be increased.

Appendix

Names and Addresses of Researchers Listed in Table I

1	Prof. W. F. Vinen and Dr. C. M. Muirhead	Dept. of Physics, University of Birmingham, Birmingham B15 2TT.
2	Dr. D. Gugan and s.	Dept. of Physics, University of Bristol, Tyndall Avenue, Bristol BS8 1YL
3	Dr. P. Aplin and s.	Dept. of Physics, University of Bristol, Tyndall Avenue, Bristol BS8 1YL
4	Prof. A. H. Beck and s.	University of Cambridge, Dept. of Engineering, Trumpington St., Cambridge, CB2 1PZ
5 6	Dr. J. R. Waldram and s.	University of Cambridge, Dept. of Physics, Madingley Road, Cambridge, CB3 0HE
7	Dr. M. R. Halse and s.	The Physics Laboratory, The University, Canterbury, Kent
8)) 9)	Dr. A. M. Guenault and)) Dr. G. R. Pickett and s.)	Dept. of Physics, University of Lancaster, Lancaster LA1 4YB
10	Dr. D. J. Meredith and s.	Dept. of Physics, University of Lancaster, Lancaster LA1 4YB
11	Dr. D. Greig	Dept. of Physics, University of Leeds, Leeds, Yorkshire LS2 9JT
12	Dr. A. D. Caplin and) Dr. J. G. Park and s.)	Blackett Laboratory, Imperial College, Dept. of Physics, Prince Consort Road London SW7 2BZ
13	Dr. J. G. Park and s (CASE)	As above
14	Dr. B. A. Unvala and postdoc	Blackett Laboratory, Imperial College, Dept. of Metallurgy and Material Science, Prince Consert Road, London SW7 2BZ
15	Dr. D. G. Walmsley and s.	New University of Ulster, Coleraine, Northern Ireland
16	Dr. A. H. Cooke and Dr. S. J. Swithenby and M. J. Wells	University of Oxford, Dept. of Physics Clarendon Laboratory, Parks Road, Oxford OX1 3PU
17	Dr. M. J. Aitken and Dr. D. Walton (of McMaster University)	Research Laboratory for Archaeology, Keble Road, Oxford
18	Dr. G. B. Donaldson and 1 staff and 2 postdocs and 2 s.	University of Strathclyde, Dept. of Applied Physics, Glasgow, Scotland

Appendix (continued)

Names and Addresses of Researchers Listed in Table I

No.	Names	Address
19	Dr. C. B. P. Finn and) Dr. K. Kiymac)	School of Mathematical and Physical Sciences, University of Sussex, Physics Division, Falmer, Brighton, BN1 9QH, Sussex
20	Prof. D. F. Brewer) Dr. W. S. Truscott) Dr. D. S. Betts)	As above (Sussex)
21	Dr. T. D. Clark) Dr. M. G. Richards and s) (CASE))	As above (Sussex)
22	Dr. T. D. Clark) Dr. M. G. Richards and s.)	As above (Sussex)
23	Dr. T. D. Clark) Dr. M. G. Richards and s.)	As above (Sussex)
24	Dr. J. A. Lee and) Dr. J. P. Charlesworth) and 1 staff)	AERE, Chemistry Division, Harwell, Didcot, Oxon.
25	F. Goodall	Appleton Laboratory, Slough, Bucks.
26	J. C. Gallop and assistant	National Physical Laboratory, Teddington, Middlesex
27	Dr. B. W. Petley and) K. Morris and R. W. Yell) and R. N. Clarke)	As above (National Physical Laboratory)
28	Dr. T. G. Blaney and) assistants)	As above (National Physical Laboratory)
29	J. C. Gallop and A. Hartland	As above (National Physical Laboratory)
30	P. MacDonald	The Oxford Instrument Company, Osney Mead, Oxford OX2 ODX
31	Dr. J. A. Good and) Dr. N. Lindsay and) E. White)	Cryogenics Consultants, 231, The Vale, London W3

PROGRAMS ON SMALL-SCALE SUPERCONDUCTING DEVICES IN THE UNITED STATES

R. Brandt

Office of Naval Research, Pasadena, California 91106

E. Edelsack

Office of Naval Research, Arlington, Virginia 22217

I. SUMMARY

A survey of major U.S. programs on small-scale applications of superconductivity is presented. Programs are classified and tabulated into six categories. Over forty-five programs are reported. For each program, the organization, key person, brief description, source and amount of funding are given. Brief texts accompany each of the six summary tables.

II. INTRODUCTION

As a guide in our compilation of material for this survey, we adopted an arbitrary operational definition of "small-scale application of superconductivity". If a research or development program in the U.S. attempted to exploit a unique property of superconducting materials and the resultant device or application consumed or produced no more than a few watts of power, it was defined as a "small-scale application" and included in this paper. The programs are either currently active or have been recently completed. Programs completed prior to 1976 are excluded.

Information was obtained by several means. First attempted were various computer researches: (1) files of the National Technical Information Service, Springfield, Virginia, yielded reports of the ongoing or recently completed research and development programs supported by the Federal Government via contracts and grants; (2) published journal articles on file at the Cryogenic Data Center of the National Bureau of Standards at Boulder, Colorado; and (3) documentation describing ongoing, in-house research performed at the laboratories of the National Bureau of Standards at Gaithersburg, Maryland and Boulder, Colorado. A second method involved detailed review of the files of individual Federal agencies supporting programs on applications of superconductivity. This procedure was particularly helpful in the case of the Engineering and Material Sciences Divisions of the National Science Foundation. A third procedure was to contact key researchers, soliciting information in their areas of expertise. The category "Geophysical Applications" was compiled primarily by this method. A fourth and most successful method of obtaining information was by direct contact with individual investigators. Much of the information in the category "Device Properties" was obtained in this manner.

The compiled data were impressive in their diversity - diversity of types of applications, diversity of the backgrounds and training of the persons involved in these efforts, and diversity of the sources of funding. For example, types of applications ran the gamut from new techniques for defining the standard volt to measurements of geomagnetic anomalies deep in the ocean. The diverse backgrounds of persons involved in these applications ranged from radio astronomers measuring millimeter-wave radiation from extraterrestrial sources to biomedical reseachers interested in magnetic signals from the human heart and brain. The diversity of Federal sponsors ranged from the Department of Defense, spending several million dollars per year on a broad range of programs involving small-scale applications of superconductivity, to the National Aeronautics and Space Administration with interests restricted to a few special superconductive applications. This diversity of sponsor is also evident in the U.S. industrial sector where IBM has a very large effort devoted to computer applications of superconductivity, while small companies are engaged in programs involving a single specialized application. In addition to diversity, we were impressed by the large number of persons engaged in work on small-scale superconductive applications. We have attempted to include all existing programs in the U.S. If some programs have unintentionally been omitted, we welcome pertinent information to include in an updated survey.

The data were divided into six broad categories listed in Table I. Five categories are based on specific applications: (1) biomedical, (2) metrological, (3) geophysical, (4) detection and generation of electromagnetic radiation, and (5) digital processing. The sixth category, called device properties, includes research programs which appear to be directly related to small-scale superconductive devices. Theoretical and experimental research which was related to superconductivity more generally was excluded. A total of 47 programs is reported. For each program that was separable, we reported the organization, the principal contact person (for cooperative efforts, the principal collaborators are noted), a brief description of the program with the important application areas underlined, and the source and amount of funding. For each category there is a summary table listing the pertinent programs and an accompanying brief interpretive and descriptive text. Preceding the six summary tables is a short list of abbreviations used in the tables. These are primarily abbreviations of funding agencies (see Table II).

With regard to funding information, in several instances the level of funding is uncertain, usually for one of two reasons: corporate funding is often considered proprietary information or the work reported may be part of a larger program such that the fiscal data on the small-scale superconducting device effort are not readily available. Where the data appear incorrect or have significantly changed, we solicit correct information for inclusion in an updated survey.

Table I

Small-Scale Applications of Superconductivity

BIOMEDICAL
METROLOGICAL
GEOPHYSICAL
ELECTROMAGNETIC DETECTION AND GENERATION
DIGITAL PROCESSING
DEVICE PROPERTIES

Table II

Abbreviations Used in Summary Tables

ERDA	Energy Research and Development Administration, Germantown, MD 20767
NASA	National Aeronautics and Space Administration, Washington, DC 20546
NIH	National Institutes of Health, Bethesda, MD 20014
NSF	National Science Foundation, Washington, DC 20550
NYHRC	New York Health Research Council, New York, NY
USGS	U.S. Geological Service, National Center, Reston, VA 22090
AFOSR	Air Force Office of Scientific Research, Bolling AFB, Washington, DC 20332
ARO	Army Research Office, Research Triangle Park, North Carolina 27709
ARPA	Advanced Research Projects Agency, Department of Defense, Arlington, VA 22209
NAVELEX	Naval Electronics Systems Command, Washington, DC 20360
ONR	Office of Naval Research, Arlington, VA 22217
DoD	Other DoD Agencies
my	man-years ($50 - 100 K/yr)
JJ	Josephson junction
G	1 gauss = 10^{-4} Tesla

III. BIOMEDICAL

The use of SQUID magnetometers and gradiometers to measure the minute magnetic fields produced by various organs of the human body has probably created more interest in and out of the scientific community than all other small-scale superconducting device applications combined. Measurements of the magnetic signals from various parts of the human body can yield new information about the organs which generate the electric currents, not available to surface electrodes, and also about organs which contain foreign ferromagnetic particles. Signals measured from the heart are called magnetocardiograms (MCG), those from the brain are called magnetoencephalograms (MEG), and those due to eye movements are called magnetooculograms (MOG). Magnetic signals have also been detected from blood flow, injured tissues and fetuses in utero. Measurements of magnetite dust in the lung have two potential biomedical applications: (1) as a deliberately inhaled harmless tracer for pulmonary diagnosis, and (2) for assessment of the amount of asbestos accumulated in the lungs of heavily-exposed workers, since most asbestos (which is clinically harmful) occurs with adhered magnetite.

The magnetic signal levels vary from 10^{-7} G down to 10^{-10} G or less at frequencies from 1 to 100 Hz. The peak magnetic field of the human heart a few centimeters from the chest wall is about 10^{-7} G and that of the brain just outside the scalp is about 10^{-9} G, except for a few individuals with large alpha-rhythm currents. The magnetic fields of interest decrease as the cube of the distance from the source, while the gradients decrease as the fourth power and second-derivatives as the fifth power of the distance. Thus second-derivative gradiometers are able to measure the very weak magnetic fields of the brain without any magnetic shielding as long as the sources of interference are at least several meters away.

Three different techniques have evolved, all yielding comparable results but with somewhat different projections for the future. Cohen et al. pioneered the use of SQUID magnetometers for these applications. They employ a magnetometer in a sophisticated expensive multi-layer magnetically-shielded enclosure about 2.5 meters in internal diameter. This system approaches 10^{-10} G/$\sqrt{\text{Hz}}$ in sensitivity. Wikswo et al. use a gradiometer in a magnetic shield that is less expensive and probably less sensitive than Cohen's system. Most of these measurements are concerned with vector magnetocardiography (VMCG). The magnetometer flux transformer is designed to allow sequential measurement of three orthogonal VMCG components. Williamson et al. use a second-derivative gradiometer with no shielding and are performing some excellent studies of the magnetic fields of

Table III

Biomedical Programs

Organization	Contact	Program Description	Funding ($K) and Source
National Magnet Laboratory Mass. Inst. of Technology Cambridge, MA 02139	D. Cohen	Magnetocardiography, magnetoencephelography, magnetic measurements of lungs and total body.	200/NSF, NIH and ARPA
National Bureau of Standards Cryogenics Division Boulder, CO 80302	J. Zimmerman	Magnetic measurements of the human heart and brain.	15/internal
Stanford University Department of Physics Stanford, CA 94305	J. P. Wikswo	Vector magnetocardiography; magnetic measurements of blood flow	50/NSF
New York University Departments of Physics and Psychology New York, NY 10003	S. J. Williamson L. Kaufman D. Brenner	Measurements of visually evoked magnetic fields from the human brain.	25/ONR 25/NSF 10/NYHRC
Oakland University Department of Physics Rochester, MI 48063	N. Tepley	Magnetic measurements of blood flow.	11/Michigan Heart Assoc.
TRW Redondo Beach CA 90278	A. Rosen	Magnetocardiography; diamagnetic susceptibility measurements of carcinogenic tissue	0.5 my/ internal

the brain in the range of 10^{-9} G or less. Recently the NYU group have reported interesting results involving changes in the latency of the neuromagnetic response of the human visual cortex. They have shown that these variations are correlated with reaction time.

IV. METROLOGY

JJ technology offers the metrologist possibilities for developing dc, rf, and microwave systems for more accurate and wider range of measurements than can presently be performed. Presently the use of a superconducting primary voltage standard is probably the single most important application in this area.

The U.S. legal volt is routinely maintained at NBS/ Gaithersburg by measurment of 2e/h via the ac Josephson effect. An all-cryogenic 0.01-ppm Josephson Effect Voltage Standard is presently under development. The new standard will incorporate major improvements, including: (1) a single microstripline-coupled Josephson tunnel junction; (a) a cryogenic voltage divider; and (3) a cryogenic null detector. A portable 1-ppm Josephson Effect Voltage Standard is also under development. The 2e/h voltage maintenance system now in use at NBS consists of a superconducting Josephson device (fabricated at NBS), a klystron microwave generation system and a room temperature dc voltage comparator. The JJ is irradiated with microwaves and the voltage produced ($V = nh\nu/2e$) is directly related to the fundamental constants e and h; n is an integer and ν the microwave frequency. The small voltage ($\approx$10 mV) is stepped up by a voltage comparator and compared to a standard cell emf (1 V), which in turn is used to calibrate other standard cells. The other cells provide the reference standard between 2e/h measurements. The 2e/h measurements are presently difficult to make, limited mainly by room temperature instrumentation. An increase in accuracy and ease-of-use will result if all critical components of the system can be operated at liquid helium temperatures.

In the area of superconductive thermometric fixed points, R. Soulen et al. are engaged in research aimed at extending the International Practical Temperature Scale (IPTS-68) below its present limit of 13.8 K. NBS has developed and sold some 60 calibration sets which provide precise, reproducible superconductive thermometric fixed points. Each set provides five fixed points from 0.5 K to 7.2 K with a precision of $\pm$ 1 mK (Cd - 0.5 K, Zn - 0.8 K, Al - 1.2 K, In - 3.4 K and Pb - 7.2 K). With the goal of extending the range of available thermometric fixed points, the reproducibility and width of the following superconductive transitions are under study: W (0.015 K), Be (0.024 K), Ir (0.100 K), $AuAl_2$ (0.158 K), $AuIn_2$ (0.208 K), Nb (9.3 K), V_3Ga (14 K), and Nb_3Sn (18 K).

Table IV

Metrology Programs

Organization	Contact	Program Description	Funding ($K) and Source
National Bureau of Standards Cryogenics Division Boulder, CO 80302	D. McDonald	Rf power and attenuation standards.	80/DoD 80/internal
National Bureau of Standards Electrical Measurements Division Gaithersburg, MD 20234	R. Dziuba T. Finnegan	Maintenance of US legal volt via ac Josephson effect measurements.	30/internal
National Bureau of Standards Electrical Measurements Division Gaithersburg, MD 20234	R. Soulen	Precision (± 0.1%) absolute noise thermometry from 11 mK to 10 K using Nb point contact JJs.	50/internal
National Bureau of Standards Electrical Measurements Division Gaithersburg, MD 20234	J. Schooley R. Soulen	Superconductive thermometric fixed points below 13.8 K.	15/internal
University of Utah Department of Physics Salt Lake City, UT 84112	G. Symko	Nuclear magnetic thermometry from 1 K to sub-milliKelvins using SQUID magnetometer.	40/NSF
National Bureau of Standards Time and Frequency Division Boulder, CO 80302	S. Stein A. Risley	Spectrally pure superconducting Nb-cavity oscillator for frequency synthesis and multiplication.	40/internal
Stanford University Department of Physics Stanford, CA 94305	J. Turneaure	Superconducting cavity stabilized oscillator achieves 6 x 10 (-16) stability for 10 to 1000 sec.	50/ONR

Symko et al. have developed a magnetic thermometer for temperature measurements from 1 K down to the sub-milliKelvin range. They measure the static magnetization of a nuclear paramagnet using a SQUID magnetometer and then infer the temperature using Curie's law. Such a thermometer combines the high resolution of a SQUID magnetometer in measuring static nuclear magnetization with the flexibility for internal calibration. Symko has carefully selected the nuclear paramagnets he uses in order to avoid the problems of impurity electronic contributions which could obscure the static nuclear magnetization measurements.

McDonald et al. have made rf attenuation measurements using an L-band SQUID system. A recently redesigned system has resulted in a significantly simpler geometry which provides an adjustable coupling for precise matching to the electronics. Attenuation measurements with this technique rely heavily on proper signal processing in room temperature components and require careful analysis of the sources of possible error.

Stein et al. are developing a 9 GHz parametric oscillator incorporating a superconducting cavity for use as a source in a JJ harmonic generator. Harmonic orders up to 400 are to be generated for frequency comparison with far-infrared laser signals.

R. Soulen, following upon the early research of R. Kamper and J. Zimmerman, has used a SQUID as a low-noise parametric amplifier to extend the range of noise thermometry down to 10 mK. Soulen applies a dc bias to a resistive SQUID and then measures the random frequency modulation of its self-oscillation caused by thermal noise. From these measurements he is able to determine the absolute temperature since the theory of frequency modulation establishes a relationship between noise power and effective bandwidth which eliminates the uncertainty in measuring the latter quantity.

V. GEOPHYSICAL

The use of superconducting devices in geophysical studies appears to be growing slowly. In the light of various potential applications of superconductive systems in geophysics, this is surprising. The paucity of entries in Table V may, in part, be due to our inability to locate investigators. For example, the measurement of rock magnetism using superconductive magnetometers is in progress at several institutions. One U.S. company has built and sold over twelve superconductive systems for rock magnetism measurements. We have not attempted to contact the users of these systems. Research using superconductive

Table V

Geophysical Programs

Organization	Contact	Program Description	Funding ($K) and Source
National Bureau of Standards Cryogenics Division Boulder, CO 80302	J. Zimmerman W. Campbell[a]	Ultra-low frequency geomagnetic field-test measurements with superconductive magnetometers. Comparison with conventional magnetometers.	15/internal
Stanford University Electronics Laboratories Stanford, CA 94305	A. Fraser-Smith J. Buxton[b]	Superconductive magnetometer measurements of 0.1 - 14 Hz geomagnetic background demonstrate superiority over conventional systems. Future work planned.	40/ARPA
Naval Research Laboratory Communication Sciences Div. Washington, DC 20375	J. Davis R. Dinger M. Nisenoff	Superconductive magnetometer measurements of 45-75 Hz natural and manmade signals at and below sea level. Work continuing.	200/NAVELEX
Physical Dynamics, Inc. P.O. Box 556 La Jolla, CA 92037	G. Gillespie W. Podney J. Buxton	Measurements of 5 x 10 (-4) to 20 Hz geomagnetic gradient noise. Superconductive gradiometer measurements of ocean-wave generated electromagnetic fields to be made.	400/ARPA
University of California Department of Physics Berkeley, CA 94720	J. Clarke	Superconductive magnetometer and gradiometer measurements of geomagnetic gradient activity and magnetic activity precursor to earthquakes.	20/USGS 20/ERDA
University of California Dept. of Mat. Sci. & Eng. Berkeley, CA 94720	H. Morrison W. Dolan A. Dey	Airborne superconducting coil system developed for prospecting via earth conductivity measurements.	400/AMAX. EXPLORATION INC.
Naval Coastal Systems Laboratory Panama City, FL 32401	K. Allen R. Clark J. Titus M. Wynn	Airborne superconductive magnetic sensor system for dipole tracking.	10my/DoD

a. - U.S. Geological Survey, Denver, Colorado 80225.
b. - Stanford Research Institute, Menlo Park, California 94025.

gravity wave detectors appears to be outside the scope of geophysical applications and thus was omitted. There are three programs in this area: (1) W. Fairbank at Stanford University; (2) W. Hamilton at Louisiana State University; and (3) G. Dick at the California Institute of Technology. There are probably a few efforts involving the use of superconductive magnetic gradiometers for prospecting and related sub-surface studies, but we are not aware of any work in progress at this time.

For those interested in reviewing the literature of this field a few recent representative publications are listed below:

1. Zimmerman, J. E. and Campbell, W. H., "Tests of Cryogenic SQUID for Geomagnetic Field Measurements", Geophysics 40, 269 (1975).
2. Buxton, J. L. and Fraser-Smith, A. C., "A Superconducting System for High Sensitivity Measurement of Pc 1 Geomagnetic Pulsations", IEEE Trans. Geosci. Elect. 12, 109 (1974).
3. Fraser-Smith, A. C. and Buxton, J. L., "Superconducting Magnetometer Measurements of Geomagnetic Activity in the 0.1 to 14 Hz Frequency Range", J. Geophys. Res. 80, 3141 (1975).
4. Frederick, W. D., Stanley, W. D., Zimmerman, J. E. and Dinger, R. J., "An Application of Superconducting Quantum Interference Magnetometers to Geophysical Prospecting", IEEE Trans. Geosci. Elec. 12, 102 (1974).

VI. DETECTION AND GENERATION

The use of superconductivity in the detection and generation of electromagnetic radiation has attracted considerable interest and attention, as is evidenced by the breadth of the program described in Table VI. In the detector application, JJ s are expected to respond sensitively to frequencies far higher than do conventional semiconductor devices. An upper frequency limit can be calculated for Josephson detectors based on the energy gap, yielding about 1400 GHz for niobium, but there is evidence, for example, from the frequency comparison work of McDonald et al. of the National Bureau of Standards (see section on Metrology) that Josephson devices may provide a useful response to frequencies much higher than the gap limit.

To construct a competitive detector of high sensitivity many practical engineering problems must be solved. In past years the emphasis of Josephson detector work has been on the demonstration and gross characterization of various detection mechanisms, such as video response, mixing and parametric amplification. It is a sign of growing maturity of this field that in recent years there has been an increasing amount of work

Table VI

Electromagnetic Detection and Generation Programs

Organization	Contact	Program Description	Funding ($K) and Source
Aerospace Corporation P.O. Box 92957 Los Angeles, CA 90009	A. Silver H. Kanter	Parametric amplification studies at 9 GHz (degenerate mode) and 90 GHz (preliminary measurements). Super-Schottky mixer work at 9 GHz (13 K noise temperature and 9 dB conversion loss) and higher frequencies.	110/ONR and DoD 170/internal
University of California Department of Physics Berkeley, CA 94720	R. Chiao	Series array of 80 unbiased JJs demonstrates 12 dB parametric gain at 10 GHz in doubly degenerate mode. Continuing theoretical and experimental studies.	50/NSF 50/NASA
University of California Department of Physics Berkeley, CA 94720	J. Clarke P. Richards	Superconductive bolometer built with NEP of 3 x 10 (-15) W/$\sqrt{Hz}$ and 50 ms response time.	30+/ERDA
University of California Department of Physics Berkeley, CA 94720	P. Richards	36 GHz JJ mixer attains 50 K noise temperature with conversion gain. Preliminary measurements at 144 GHz. Parametric amplification and coupling to JJ arrays at 36 GHz studied.	55/ONR
University of California Dept. of Elec. Eng. Berkeley, CA 94720	T. Van Duzer	Silicon-membrane JJs built for video detector and mixer evaluation. JJ array and super-Schottky diode studies planned.	25/ARO
Case Western Reserve Univ. Dept. of Elec. Eng. Cleveland, OH 44106	E. Thompson	100 GHz JJ detector and FM demodulator with internal oscillator locked to external signal to be built.	30/NSF
Louisiana State Univ. Department of Physics Baton Rouge, LA 70803	W. Hamilton	Detectivity of JJ video detector with magnetic field coupling measured from 22 GHz to 1 THz.	30/AFOSR

Table VI (continued)

Electromagnetic Detection and Generation Programs

Organization	Contact	Program Description	Funding ($K) and Source
National Bureau of Standards Cryogenics Division Boulder, CO 80302	D. McDonald J. Edrich[a]	300 GHz receiver containing 57 K noise temperature JJ mixer with 12 dB conversion loss attains 850 K overall noise temperature.[a] Theoretical analysis of high frequency detection mechanisms in progress.	10/NSF 15/ONR 35/internal
Naval Research Laboratory Communication Sciences Div. Washington, DC 20375	J. Davis	ELF H-field detection system using 3-axis SQUID magnetometer and 90-day dewar under test. Scalar output processing developed.	210/NAVELEX
Naval Research Laboratory Communication Sciences Div. Washington, DC 20375	J. Davis	HF antenna array elements using SQUIDs	90/internal
State University of New York Department of Physics Stony Brook, NY 11794	J. Lukens	1-10 GHz tunable coherent source using large JJ arrays and stripline coupling under study.	150/ONR
Research Advisory Institute 218 Monarch Bay South Laguna, CA 92677	J. Mercereau	Use of deformable superconducting cavities for conversion of mechanical and electromagnetic energy under experimental and theoretical study. Prototype construction planned.	100/ARPA
University of Rochester Dept. of Elec. Eng.	C. Stancampiano	Analytical study of detection processes for rf-driven resistively shunted JJ. Comparison with experimental data on injection-locked oscillators.	20/internal
University of Texas Department of Physics	B. Ulrich	JJ video detector used for astrophysical studies at 1 mm. Josephson parametric amplification and mixing under theoretical and experimental study.	10/NASA
Westinghouse Research Labs. Pittsburgh, PA 15235	M. Janocko	Small JJ arrays as microwave sources under study.	33/ONR

devoted to the construction of practical detectors designed for maximum coupling efficiency and sensitivity.

Three excellent examples of this trend are the work of Silver et al. on super-Schottky mixers, the work of Richards et al. on Josephson mixers, and the work of McDonald and Edrich on a Josephson heterodyne receiver. The measurements on super-Schottky mixers were conducted at 9 GHz, the Josephson mixer results were obtained at 36 GHz, and the Josephson heterodyne receiver operated in the neighborhood of 300 GHz. All three devices performed very favorably compared to competing devices at the same frequency. In the first two cases, work is continuing to extend this performance advantage into the millimeter range.

Work on parametric amplification has also become more quantitative. Chiao and coworkers have measured 12 dB gain at 10 GHz for a doubly degenerate mode; work is continuing to characterize this process more fully and possibly extend these results to higher frequency. Several parametric modes have been examined by Silver and Kanter, and detailed performance characteristics are being obtained at 9 GHz; preliminary experiments are in progress at 90 GHz. Other workers are also investigating various aspects of Josephson parametric amplification in order to establish some measure of expected performance.

The superconductive bolometer of Clarke et al. is now a highly refined device capable of sensitive broadband detection at reasonable speeds. Such devices, as well as Josephson video detectors, are ideal for certain applications in physics and astronomy.

All this activity on detectors has stimulated an increased interest in the use of Josephson junction arrays to improve impedance matching and possibly improve signal-to-noise ratio through coherent signal addition. Some of this work on arrays is reported in the section on Device Properties when not specifically oriented to detectors or sources. However, a notable example listed in Table VI is the work of Lukens et al. on large arrays with the objective of producing a tunable source of microwaves from approximately 1 - 10 GHz at a power level approaching 10^{-7} W.

Josephson junctions can also be used as low-power injection-locked oscillators. Stancampiano and Shapiro have been investigating this application primarily using simulation and analytical techniques; additional experimental tests are planned. In a related matter, Thompson has been studying the possibility of building an FM demodulator in which the instantaneous frequency of an external signal is determined by measuring the voltage induced across a JJ detector; the junction is biased to oscillate internally at approximately the signal frequency or some multiple thereof.

Finally, Mercereau and coworkers are attempting to demonstrate the conversion of electromagnetic and mechanical energy using deformable superconducting resonators. In the generator mode, external work performed on the resonator would be converted to output radiation. This process can only occur using superconducting resonators where the effective photon lifetime is much greater than the mechanical period, so that the small power gain produced over one mechanical cycle can be accumulated coherently over a very large number of cycles. A test device, including appropriate switches for transferring electromagnetic power, has been designed, and experimental work is progressing.

Total funding for work on superconductive detectors and generators is substantial, amounting to approximately $1200K. Of this total amount, about 70% is provided by the Department of Defense with the Office of Naval Research the largest contributor supplying 25% of the total. Other large sponsors are the Naval Electronic Systems Command, the Aerospace Corporation, and the Defense Advanced Research Projects Agency.

VII. DIGITAL PROCESSING

Unlike the other application areas discussed here, work on digital applications of superconductive devices is performed largely in industrial laboratories rather than universities or government laboratories. Also, the work of one organization, namely, the International Business Machines Corporation (IBM), is clearly dominant. The total investment by the listed organizations is very large, probably comparable to the expenditures in all other application areas combined. However, the exact magnitudes of these industrial efforts are considered proprietary. The estimates given in Table VII correspond to frequently rumored numbers, although guesses vary by at least an order of magnitude depending in part on the definitions used for man-year of effort.

The reason for this substantial investment is the promise of super-fast computers. In order to achieve greater speeds, it is necessary to minimize signal propagation delays, which requires that computer elements be placed very close together. However, it is no longer possible to increase the packing density of semiconductor circuits because the heat generated by the semiconductor devices is too large. It has been determined that the heat dissipated by a Josephson logic gate is approximately two orders of magnitude smaller than by semiconductor competitors,

Table VII

Programs on Digital Processing

Organization	Contact	Program Description	Funding ($K) and Source
Bell Laboratories 600 Mountain Avenue Murray Hill, NJ 07974	T. Fulton	JJ logic and memory circuit investigation. Analysis of flux shuttle concept.	less than 10 my/internal
University of California Dept. of Elec. Eng. Berkeley, CA 94720	T. Van Duzer	Fabrication of tunnel junctions with barriers of semimetals, silicon membranes and stabilized oxides. Investigation of JJ logic circuits.	35/NSF 40/DoD
Hewlett-Packard Corp. Palo Alto, CA 94300	D. Rose	Investigation of JJ digital circuit applications and development of commercial JJ voltage standard.	Approx. 3 my/internal
IBM Research Center Yorktown Heights NY 10598	J. Matisoo	Evaluation of JJ logic circuits, memory cells, adders and shift registers. Fabrication yield reliability, reproducibility and stability of tunnel JJs under study.	Approx. 50 my/internal
National Bureau of Standards Cryogenics Division Boulder, CO 80302	D. McDonald	Fabrication of stable lead alloy tunnel junctions. Analysis of digital circuit application: picosecond pulsers, A/D converters, three-terminal devices, and transmission lines.	175/internal 15/ONR
Sperry Research Center Sudbury, MA 01776	K. Kroger	Memory and logic circuits using niobium and lead alloy tunnel JJs under study.	Approx. 3 my/internal

permitting higher packing density and greater resultant speed.

The progress achieved by IBM was recently described at the Applied Superconductivity Conference (ASC) held in August 1976. They have succeeded in fabricating reliable Josephson tunnel junctions using a simultaneous evaporation and sputtering technique and have constructed and evaluated a variety of logic and memory circuits. These circuits have been extensively simulated by computer, and the model predictions have been largely verified experimentally. Stripline techniques have been devised for circuit fabrication, and switching times of several hundred picoseconds have been measured for Josephson logic circuits. Both latching and nonlatching logic have been studied. Several memory concepts are also under investigation, including finite loops containing circulating currents and small SQUID loops which accommodate a single flux quantum.

A general exploratory program is also underway at the Bell Laboratories, although at a much lower level than the IBM effort. Work was reported at the ASC on a successful demonstration of a 5-element Josephson logic circuit; no attempt was made to achieve the high speeds inherently possible.

An effort is starting at the National Bureau of Standards to investigate a variety of digital applications. The work to date has been largely analytical. Fabrication and testing is planned.

We found only one university program in this area at the University of California in Berkeley. T. Van Duzer and co-workers have demonstrated junction switching times and logic gate times comparable to those reported by IBM. Instead of using oxide tunnel junctions, they have used semiconductor-barrier junctions which have controllable and reliable properties.

Modest industrial efforts have also been started at the Sperry Research Center and at Hewlett-Packard Corporation.

VIII. DEVICE PROPERTIES

Programs in this category, summarized in Table VIII, are more closely concerned with the physics of Josephson-effect devices. Because the work is of a more general character, it is often related to more than one of the application areas discussed in this paper.

The work on junction fabrication and characterization has as its goal the realization of new devices with improved properties. Clarke, Buhrman, Tinkham, Deaver, and others are addressing

Table VIII

Programs on Device Properties

Organization	Contact	Program Description	Funding ($K) and Source
University of California Department of Physics Berkeley, CA 94720	J. Clarke	Measurement of: (1) JJ properties under microwave irradiation; (2) properties of superconductor/normal-metal boundaries; and (3) LF SQUID noise including theory.	125/ERDA 20/USGS
University of California Department of Physics Santa Barbara, CA 93106	D. Scalapino	Superconductive nonequilibrium phenomena calculations: quasiparticle and phonon energy distributions; microwave and phonon enhancement of T_c, etc.	30/ONR
California Institute of Technology Department of Physics Pasadena, CA 91125	J. Mercereau	Experimental and theoretical study of coupling between JJ array elements in terms of element spacing, material, frequency, temperature, etc.	60/ONR
Catholic University Department of Physics Washington, DC 20017	R. Peters	Experiments on nonequilibrium effects in superconducting devices.	19/ONR
Cornell University Department of Applied Physics Ithaca, NY 14850	R. Buhrman	SQUID noise measurements and theory. Heating effects in microwave irradiated thin-film JJs.	47/ONR
Naval Research Laboratory Material Sciences Division Washington, DC 20375	M. Nisenoff	Niobium microwave excited SQUIDs evaluated. Niobium thin-film JJs fabricated.	95/internal
Stanford University Department of Applied Physics Stanford, CA 94305	M. Beasley	Investigation of weak-link and tunnel JJs fabricated from high T_c materials.	20/DoD

Table VIII (continued)

Programs on Device Properties

Organization	Contact	Program Description	Funding ($K) and Source
University of Pennsylvania Department of Physics Philadelphia, PA 19174	D. Langenberg	Theoretical and experimental study of nonequilibrium effects in JJs. Excess quasiparticle production via tunnel barrier injection and laser irradiation.	30/ONR
Science Applications 1205 Prospect Street La Jolla, CA 92037	D. Rogovin	Theoretical analysis of quantum wave function phase locking in coupling between JJ array elements.	25/ONR
Stanford University Department of Physics Stanford, CA 94305	R. Giffard	SQUID noise fully characterized. Study of LF ultra-low-noise amplifiers using SQUIDs with tuned input circuits.	40/ONR
University of Wisconsin Dept. of Elec. Eng. Madison, WI 53706	J. Nordman	Niobium-based semiconductor-barrier tunnel JJs studied for switching applications.	35/NSF
Yale University Dept. of Applied Science New Haven, CT 06520	D. Prober	dc SQUIDs using alloy thin-film micro-bridge JJs evaluated.	15/NSF
Harvard University Department of Physics Cambridge, MA 02138	M. Tinkham	Fabrication of variable thickness micro-bridge JJs. Evaluation of microwave and laser response. Study of fundamental device limits.	40/NSF 50/DoD

the problem of removing heat from the region of the junction in order to improve the high frequency response. Different techniques are being explored for adding heat-sink material in proximity to the junctions. Nordman et al. are concerned with achieving desirable junction properties for switching applications through a variety of approaches, including semiconductor-barrier junctions.

Rogovin's theoretical work on arrays addresses the fundamental question of the physical basis of array coupling via phase-locking of individual quantum wavefunctions, while Mercereau et al. are attempting to deduce these coupling mechanisms from a systematic experimental study of array properties.

The behavior of Josephson-effect devices under nonequilibrium conditions is a subject of growing interest, both from academic and practical viewpoints. Techniques for producing nonequilibrium distributions of quasiparticles and phonons by tunneling and laser illumination are being investigated, and theoretical descriptions of these phenomena are being developed. These efforts are well represented by the work of Scalapino, Langenberg, and co-workers.

SQUIDs continue to receive attention. Clarke et al. have been concerned with the origin and magnitude of low-frequency (1/f) noise, while Buhrman et al. have developed a theoretical explanation for SQUID noise at higher frequencies (white noise region). Giffard is examining the possibility of ultra-low-noise amplification using a SQUID with a tuned input circuit. Nisenoff and Wolfe have been investigating SQUID operation at microwave excitation frequencies, while Prober will be studying alternative methods for achieving enhanced sensitivity using dc SQUIDs.

Research on device properties is funded by a number of agencies. Of the total funding in this category of \$650K, more than 40% is provided by the Office of Naval Research. Other major sponsors are the National Science Foundation and the Energy Research and Development Administration.

IX. TRENDS

Attempts to predict the future of some segment of science and technology often tend to reflect the professional interests, prejudices and limited knowledge of the self-proclaimed "prophets". Bearing in mind this admonition, we will in this last section discuss a few of the possible future directions of some of those areas of small-scale applications of superconductivity (SSAS) reviewed in this paper.

During 1975 six Federal agencies supported a total of about $3.5 - 4 million in basic and applied research and exploratory development in the area of SSAS via grants, contracts and in-house laboratory projects. This figure is up significantly from what we estimate it was in 1972, indicating that the field is experiencing a healthy growth. The largest single sponsor is the Department of Defense and, within the military, the U.S. Navy supports over 75% of the total DoD effort in SSAS. In the industrial sector IBM is by far the largest single sponsor (see Table VII). Estimating a man-year at $50,000, the annual IBM expenditure in this area is about $2.5 million. Including the other industrial activities listed in Table VII, the total investment by U.S. industry in SSAS is around $3.5 million. Thus the dollar value of the total U.S. effort in SSAS, government and private industry, in 1975 was $7 - 8 million. We believe this to be a conservative estimate. We do not have reliable data to estimate the annual rate of growth.

There are a few factors which could significantly influence the rate of future growth. Discounting such "breakthroughs" as room temperature (or even liquid-nitrogen temperature) superconducting materials, the present efforts to increase the transition temperature a few degrees above the present record of 23 K would appear to have little effect on either the near or far-term future growth of SSAS. One factor which could have a significant effect on the rate of future growth is the commercial availability of an inexpensive, reliable, miniature, portable, low-maintenance cryogenic refrigerator operating at about 4-5 K, with a cooling capacity of a few tenths of a watt and requiring low power input. The ability to eliminate the need for liquid helium and the periodic transfer of helium would, we estimate, act as an important stimulus to the future increased growth of SSAS, particularly in the areas of biomedical, geophysical and military applications. Within the area of military applications, surveillance, communication, digital processing and weaponry could be strong candidates for SSAS employing miniature closed cycle cryogenic refrigerator systems. Many of the potential military applications require the refrigerator to be reasonably free of mechanical and magnetic noise at the cooling head so as not to interfere with the long time routine operation of a single SQUID or array of SQUIDs. In these applications it is a strong requirement that the refrigerator be sufficiently simple in operation and rugged in construction so as to be operable in field conditions, away from a laboratory, by personnel with little formal training in cryogenic engineering. To our knowledge there is no commercial refrigerator presently available which meets all or even most of these constraints without extensive modification. Severe as these requirements may appear to be, they may be realizable in the next five years. Our optimism stems in part from the results of the preliminary research of J. Zimmerman of NBS/Boulder who has built a laboratory model low-power Stirling-cycle refrigerator which has maintained a temperature of about 16 K for more than 2000 hours. Most recently Zimmerman has succeeded in operating the unit down to 13 K. Also, he has successfully demonstrated the operation of an rf-biased niobium-tin SQUID at 16 K. It will take a considerable development effort to integrate these devices into a single system which is available as an off-the-shelf item. However, we are optimistic that such systems will be available in the future provided industry becomes convinced of the commercial potential of miniature cryogenic refrigerator systems.

An alternate less desirable possible future system which would still have appeal for several civilian and military applications is the use of an integral refrigerator to reduce the evaporation rate of a liquid helium cryostat so that the interval between transfers is significantly extended. This has the advantage that the refrigerator can be turned off for those periods of time when it interferes with the operation of the SQUID. This system still retains the problems attendant with the transportation, supply and transfer of liquid helium.

In terms of the present state-of-the-art in SSAS and possible future trends in specific areas, the following comments are included:

1. Considerable progress has been made in recent years in designing and constructing practical Josephson-effect detectors which have certain performance advantages over competitive detectors in the same region of the electromagnetic spectrum. For instance, in the microwave region the super-Schottky mixer of Silver et al. has a lower noise temperature than the best competitive mixers. At 36 GHz the Josephson mixer of Richards et al. is characterized by a noise temperature about four times lower than the best competitor and also produces conversion gain. The 300 GHz receiver of McDonald and Edrich, which utilizes a Josephson mixer and a maser IF amplifier, also constitutes a substantial improvement. The work on Josephson parametric amplifiers is at a somewhat less advanced stage, but the measurements made to date suggest that these devices may also have certain performance and possibly cost advantages over other paramps.

It is likely that this work on quantitative characterization of Josephson-effect detectors will continue for the next several years in order that the performance limits can be established in detail. These Josephson low-noise receiver components will then be available for use in a variety of systems applications. It is expected that the first use of this new technology will be in radio astronomy and other space-to-ground and space-to-space applications.

2. Work on Josephson sources is at an exploratory stage, and it is expected that research activity in this area will increase. Possible applications are injection-locked local oscillators and rapidly tunable narrowband sources where only low power levels are required. The problems of extracting radiation from Josephson junctions are poorly understood and must be investigated.

3. Arrays of Josephson junctions may be useful for both detection and generation applications. However, the coupling mechanisms in arrays are not yet fully understood. The phenomena of frequency locking and phase locking must be investigated in detail. It is expected that research interest in this subject will remain at a high level.

4. Fundamental to a proper understanding of junction arrays is a thorough understanding of the physics of an individual Josephson device. Of particular practical importance are the distinctions between weak links and tunnel junctions, specifically, in terms of coupling with an external radiation field and the potentialities for coupling in an array. The role of heating in junctions will also be investigated further, and possible improvements using variable thickness bridges will be established.

Another physics matter receiving considerable attention at the present time pertains to nonequilibrium effects. Such effects must occur, to a degree, in all operating devices, but the research is concerned with studying large departures from equilibrium produced by particle injection or through interaction with external radiation. It is possible that this work may result in new, useful device configurations.

5. Work on digital processing applications is particularly demanding in terms of technology requirements, since large numbers of devices and circuits must be made reliably, reproducibly and economically. These technology matters will continue to be dominant in the near future, but we can also expect a greater diversity of applications being explored as more groups become interested in this extremely promising field. As one example, ultrafast analog-to-digital converters using Josephson logic circuits appear to be promising possibility. As the capabilities of this new digital technology become known more widely, it is likely that systems requirements will provide direction for further research and development.

6. SQUID magnetometers are by far the most sensitive magnetic sensors available today and for the foreseeable future. In the laboratory, flux sensitivities of 10^{-19} W or less can be achieved. For example, Wikswo (see Table III) routinely makes measurements down to 50 pT, which is some six orders of

magnitude smaller than the earth's magnetic field and about three orders smaller than the magnetic noise in an average laboratory. The extreme sensitivity and broad band frequency response of SQUIDs assures their continued and growing future use. There are two U.S. companies producing superconductive magnetometer systems for both laboratory and field use. What presently appears to limit the growth in some application areas is the need to use liquid helium. Once this hurdle can be overcome, the biomedical, geophysical and military applications of superconducting magnetometry should experience marked growth.

7. In electrical metrology, the extraordinary sensitivity of SQUIDs from dc through the microwave range assures their future in precise measurements and standards. Thus for the maintenance of the legal volt, precise measurements of direct current, voltage, rf power and attenuation, the future of SSAS is good. For example, routine voltage measurements down to 10^{-15} volts are readily attainable with a SQUID voltmeter. One property of SQUIDs that makes them highly desirable for rf measurements is their linearity. This property will become increasingly important with the trend towards techniques for precise rf measurements which can overcome the present-day dependence on carefully made and maintained artifact standards.

Recent improvements in noise thermometry have significantly reduced the integration time required for a given measurement precision. The fact that 0.1 percent precision noise thermometry measurements can now be routinely made over the range from 10 mK to 10 K gives some indication of the bright future of this area of SSAS.

New applications of superconductivity involving the precise measurements of selected electrical and thermometric quantities as well as the development of precision devices and components appears promising. What appears particularly interesting for the future is the blending of cryoelectric techniques with superconducting devices. Thus an entire system can be at cryogenic temperatures. This offers the possibility of solving metrology problems which are not readily soluble by any other means. While the development of practical superconducting metrological systems is on the increase and will continue to grow, the number of potential future users appears to be relatively small and thus the commercial market for these applications may remain limited.

8. The biomedical applications are among the most interesting and also among the most difficult to assess in terms of future trends. As a research tool in physiology, biology and medicine, there seems little doubt that SSAS will continue to grow. The research activities of Williamson, Cohen, and Wikswo are excellent examples of the present-day endeavors in this field.

As their research results become more widely known and as the commercial availability of rugged, reliable, simply-operated SQUID magnetometer systems increases, more workers will certainly enter the field. Also an encouraging note is the growth of the number of Federal agencies supporting research in MCG and MEG.

In terms of clinical applications it is premature to make any assessments as to the future. As a diagnostic tool to detect abnormal clinical conditions, there appears some promise. Cohen et al. found that MCG may yield clinical information for a number of abnormalities not seen on the ECG. Further work is needed in this area to assess clinical utility. Such is also the case for magnetic measurements of blood flow and the detection and localization of neoplastic tissues. The ease with which clinical measurements can be made will significantly determine the extent of future clinical use. Here the second derivative gradiometer technique of Williamson et al. or a sensitive differential magnetometer system appears to have good potential for routine hospital use in both MEG and MCG. The elaborate shield of Cohen is not needed for routine clinical uses. A simple shield the size of a telephone booth is probably feasible for most measurements. The majority of magnetocardiograms reported to date have been obtained by measuring a single component of the magnetic field near the chest wall. These measurements do not permit a complete assessment of the spatial and temporal variations of the magnetic field of the heart. Wikswo's vector magnetocardiographic technique determines the three orthogonal components of the heart's magnetic field and offers much promise in terms of further research and clinical applications.

ACKNOWLEDGEMENTS

The authors wish to express their appreciation for assistance provided by the following persons: Neil A. Olien, Chief of the Cryogenic Data Center, NBS, who supplied extensive computer compilations of pertinent publications; Drs. James Zimmerman, NBS, and David Cohen, MIT, who provided material concerning biomedical applications; and Prof. A. C. Fraser-Smith of Stanford, who prepared an informative report on geophysical applications of superconductivity.

Part II
Machines

Chapter 20

LARGE-SCALE APPLICATIONS OF SUPERCONDUCTIVITY

G. Bogner

Research Laboratories of Siemens

AG Erlangen, Germany

I. INTRODUCTION

Research and development in the field of applied superconductivity has been carried out for more than 15 years. Many large scale applications have been proposed in this period, but until now superconductivity has found access only to a few areas which in addition are not part of classical electrical engineering, e.g. magnets for scientific research or High Energy Physics. On the one hand, this fact is not very satisfying for physicists and engineers who are aware of the great principal potential of superconductivity and are working strenuously in this field. On the other hand, this situation can be simply explained and there is no reason to sink into despair. Without any doubt superconductivity will lead to unconventional solutions to problems in classical power engineering and to new solutions to problems which are impossible by means of conventional electrical engineering. The question as to when this application will occur, however, remains a difficult one to answer.

The reasons for the delay in the industrial application of superconductivity are manifold.

The limited performance of existing superconducting materials, especially their low transition temperatures, makes the application of a low efficient cooling technique necessary but at present the problems of the compressors involved in this technique have not been satisfactorily solved. The low temperature environment, where superconducting devices have to be operated, also causes numerous problems for the non-superconducting components such as structural and insulating materials. The complexity of the new technique has two consequences: (1) Long research and development periods until a new product of high performance and reliability can be offered (a similar situation is to be observed in other branches of modern physics and electrical engineering, e.g. controlled fusion). (2) The new superconducting technology will be limited to apparatus with high power rating.

Competition from existing conventional methods must be considered. Superconductivity will only be applied if it offers a significantly better performance and/or costs than the alternative methods. A typical example of this is in the case of magnetic separation.

Progress in conventional nonsuperconducting devices have resulted in an increase of the upper power ratings. In some fields of power engineering, as for example power generation and transmission, conventional techniques have been and are continuing to be improved, in a manner which 10 years ago had been thought to be impossible. This progress together with the forecasted decrease in annual growth of electrical energy consumption will lead to a considerable time shift in the application of superconductivity in power generation and transmission. For certain systems, superconductivity might be the only possible practical and economic solution, but at present it has not been applied since other system components have yet to be developed. Typical examples for this case are controlled thermonuclear fusion and MHD.

In some cases superconductivity is being introduced into areas with traditionally long innovation periods, e.g., a new ground transportation system. The application of a magnetically levitated train also depends on political and social aspects. For instance in Europe such a system is only feasible if it includes several countries, not only Germany where levitated trains are being primarily developed at present.

Reliability and safety are of great importance for large scale applications. The benefit of superconducting devices in capital costs and efficiency can be quickly cancelled if their outage time is longer than that of conventional devices. The reliability cannot be proved to the users from small model testing. Large demonstration devices have to be constructed which are very expensive and need considerable time for their realization.

The following paper treats large scale applications which have been proposed so far and which seem from a today's point of view (1976) to have good chances for realization. It is the aim of this review to offer a review of the background, the basic principles, the state of the art, future programs and possible benefits and economic aspects of the large scale application of superconductivity. For more details the special literature given in the reference list, and the excellent book on Superconducting Machines and Devices, edited 1974 by S. Foner and B.B. Schwartz is recommended. The latter is a compendium of review papers on large scale applications of superconductivity presented by outstanding specialists in this field at the NATO Advanced Study Institute in Northern Italy in September 1973.

Predictions about the first and ultimately the most successful areas of application are very subjective. At present it seems that superconducting machinery for ship propulsion and large power generation, magnetically levitated trains and magnetic separation have the best chance of being commercially exploited before the year 2000. The industrial application of superconducting cables and large magnet systems for fusion power reactors will most probably be developed in the period beyond the year 2000.

II. SUPERCONDUCTING MATERIALS AND MAGNETS

A. Introduction

Magnetic fields for electrotechnical applications are in general produced by means of copper or aluminum coils with iron-cores. The maximum economically obtainable flux densities of such conventional electromagnets are restricted to values of about 1 to 2 Tesla, due to the saturation induction of iron, which is approximately 2 T. Special electromagnets with pole pieces which concentrate the magnetic flux can produce flux-densities up to 7T, but only in very small volumes. Another disadvantage of conventional magnet technology is the limited current carrying capacity of copper and aluminium conductors, which in classical coils is not higher than 20 A/mm^2 even with forced water cooling.

As far as research in high intensity magnetic fields is concerned, where economics do not play the dominant rôle, the limits of iron core magnets can be surmounted by the use of axially water cooled magnets with disc-shaped conductors or of helical coil design, so called Bitter-type magnets. With these magnet types magnetic flux densities of up to 25 T have been produced [1] in cores with diameters of a few centimeters. The power consumption of these magnets can reach several Megawatts. The power is totally transformed into heat which has to be removed from the winding by cooling-water with a flow rate of several cubic meters per minute. It can easily be understood that such magnet installations require a large expenditure on power supplies and cooling systems both of which demand a great deal of space. Such high magnetic field installations are therefore restricted to a few laboratories in the world, the most famous one being, without doubt, the Francis Bitter National Magnet Laboratory at MIT, USA. Other high field laboratories have been developed in Braunschweig, Germany; Grenoble, France; Leiden, Netherlands; Canberra, Australia and Lebedev Institute, Moscow.

If the duration of an experiment in a high magnetic field can be limited to definite time intervals with a special duty cycle, the energy loss can be considerably reduced by the use of pulsed magnetic fields which are synchronously adjusted to the experiment. For instance, in large scale experiments in plasma physics for controlled fusion, the magnetic confinement fields are produced using pulsed techniques. Flux-densities of 3 to 4 T for a time interval of several seconds are produced by pulsed water-cooled copper magnets [2]. For short pulses ranging between micro- and milliseconds peak fields of up to 100 T have been reached. Peak fields in the range of 100 to 1400 T with pulse durations

$< 10^{-5}$ s have been produced by field compression utilizing explosive methods, with which the experimental system is destroyed at each pulse [1].

Cryogenic magnets also produce high magnetic fields with comparatively low Joule losses. The interest in these magnets is based on the considerable reduction of the resistivity of ultra-pure metals at cryogenic temperatures. High purity aluminium in form of tapes is especially useful as a conductor at cryogenic temperatures. It is commercially available with resistivity ratio $r > 10,000$. Since at temperatures of about 20 K the real efficiency of refrigerators is about 1.5%, a considerable reduction in power consumption can be achieved in comparison with ambient temperature magnets of equal geometry and flux-density. With magnets of larger size, problems arise from the poor mechanical strength of ultra-pure materials. This makes reinforcements by high strength materials necessary. The deformation of the aluminium has to be kept within the elastic range, otherwide a considerable increase in electrical resistivity occurs. For larger coils, considerable refrigeration power is also required, if such magnets are to be operated continuously. Helium, hydrogen, neon and nitrogen cooled cryogenic magnets, with bore diameters of a few centimeters, have been operated continusously or in pulsed modes up to flux-densities of 25 T [1,3].

The use of superconductors in coils producing magnetic fields has been already discussed by the discoverer of superconductivity, Kammerlingh-Onnes, in 1911. The superconductors available at that time, however, had very small critical magnetic fields which excluded their technical application. The discovery of type II superconductors at the end of the fifties, especially the high-field superconductors with their high current densities at elevated magnetic fields, introduced the possibility for the development of powerful superconducting magnets. Superconducting magnets are the best or practically the only possible solution if very high magnetic fields have to be produced within limited available space. They are also necessary for the economic production of high fields within large volumes. Therefore, at present, the application of superconducting magnets to high energy physics, fusion projects, levitated trains, ore separation, energy storage, electrical motors and generators is being seriously considered.

Before discussing the technology of superconducting magnets and their applications, we give a short review of the characteristics of the presently available high-field superconductors and those under development.

B. High-Field Superconductors

1. Superconducting materials

At the present time NbTi is the most frequently used material for the construction of superconducting magnet windings. It has a critical temperature of about 9.5 K, an upper critical field of 13 T at 4.2 K and the current density lies between 2×10^5 and 3×10^5 A/cm^2 at 5 T and 4.2 K. This superconductor is suitable for the construction of magnets with maximum flux-densities of 8 T to 9 T at 4.2 K. Because of its comparatively low critical temperature its critical data are significantly reduced if it is operated at temperatures markedly above 4.2 K. As one can derive from Fig. 1, an increase in temperature from 4.2 K to 5.5 K leads at 5 T to a reduction of the critical current by almost a factor of two. This fact has two consequences: for economic reasons, the operation temperatures on NbTi-high field magnets should be lower than 5.5 K, the safety margin against random temperature rises is very low with NbTi.

More attractive than NbTi, especially because of its higher transition temperature of 18.4 K and a higher critical magnetic field ($B_{c2} \approx$ 22 T at 4.2 K) is the A-15 compound Nb_3Sn, which is also commercially available. Disadvantages are its brittleness which leads to difficulties in handling and its comparatively high cost. This material can be used for magnetic flux densities up to 16 T at 4.2 K or for lower fields at higher operating temperatures. This results in improvements in refrigerator efficiencies and leads to elevated redundancy in operation. Figure 2 shows the temperature dependence of the critical current of Nb_3Sn. The A-15 compound V_3Ga ($T_c \approx$ 15 K, $B_{c2} \approx$ 23 T) is another possible commercial high field superconductor, which is even more expensive than Nb_3Sn. Its application is therefore restricted to the field range of 16 T to 18 T at 4.2 K, where its current carrying capacity is higher than that of Nb_3Sn. The mechanical properties unfortunately are similar to those of Nb_3Sn.

Besides the above mentioned commercially available A-15 superconductors, another series of high-field superconductors of this class is under development. The most important of these are $Nb_3(Al_{75}Ge_{25})$ with an upper critical flux density of about 40 T at 4.2 K and a transition temperature T_c of about 21 K [4], and Nb_3Ge, with a T_c of 23.2 and an upper critical field of 36 T at 4.2 K [5]. Nb_3Ge is produced by a sputtering or chemical vapor deposition process. It is the superconductor with the highest known transition temperature. A transition temperature of about 25 K for heat treated Nb_3Ge was reported by A. I. Broginski at the 1976 Applied Superconductivity Conference in Stanford, 17-20 August 1976. Figure 3, which compares the critical data of NbTi, Nb_3Sn and Nb_3Ge, demonstrates the high potential of Nb_3Ge as high field superconductor [6]. Unfortunately, the production of commercially useful wire from these A-15 compounds is even more complicated than that of Nb_3Sn or V_3Ga. These difficulties will probably result in very expensive conductors made from these materials.

Recently, a new class of superconducting materials with the general formula $M_yMo_6X_8$ has been found. M stands for a variety of different metals (e.g., Pg, Sn, Cu, Yb, La, Gd, Eu, Mn, Cr, Fe) and X

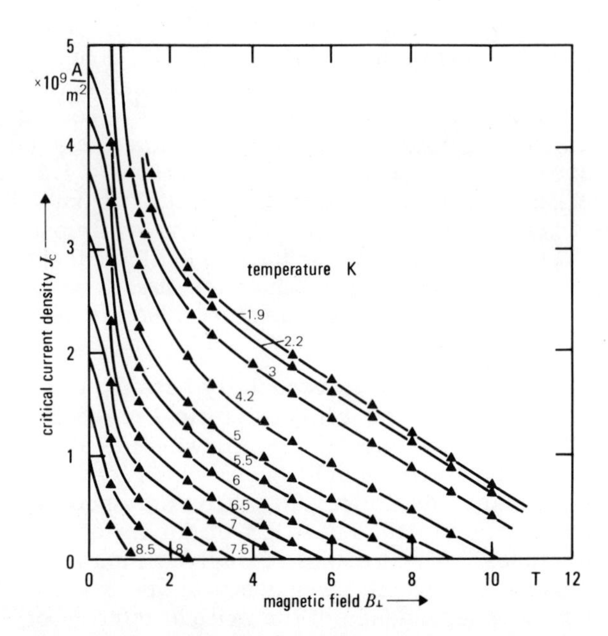

Fig. 1 Critical current density of Nb_xTi versus transverse magnetic field in the temperature range of 1.9 K to 8.5 K [3].

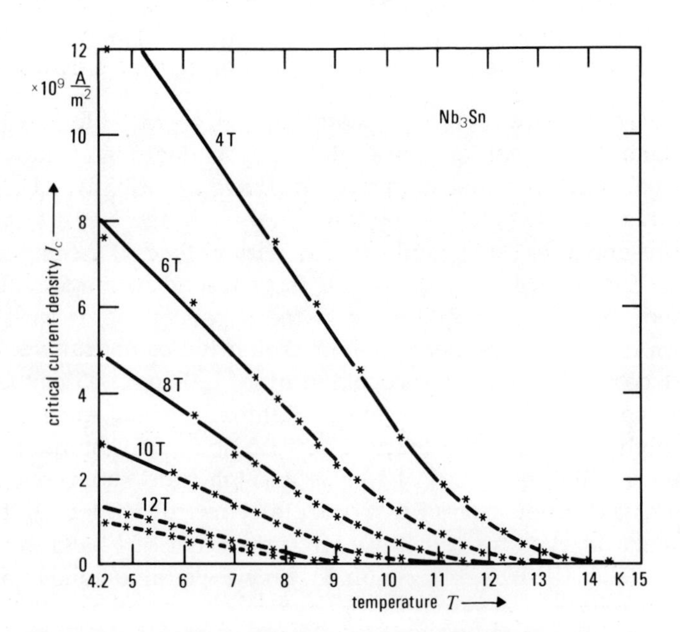

Fig. 2 Critical current densities, J_c, vs temperature at various magnetic fields for a Nb_3Sn multifilamentary wire [12].

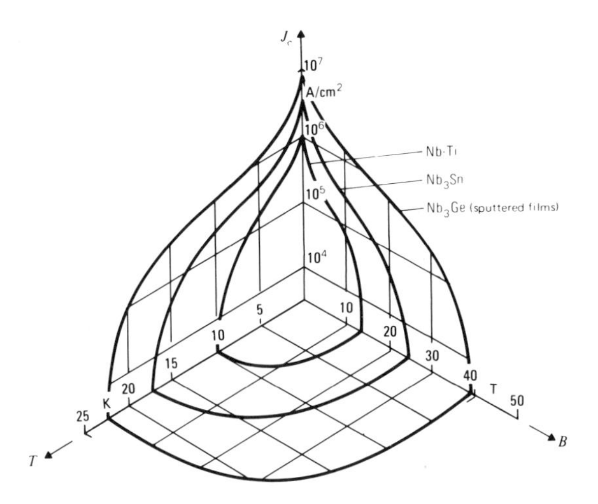

Fig. 3 Current-field-temperature characteristics of high field superconductors [6].

is either Sulphur, Selenium or Tellurium. The highest T_c value in this class is about 15 K and is found for $PbMo_6S_8$. Critical fields of the order of 60 T have been found in this material, which can be increased up to 70 T [7]. In principle it should be possible to fabricate wires from this material. However, these materials are more brittle than the A-15 compounds. Another disadvantage is the small current density currently attained with these materials [8]. It is therefore very doubtful that this material will be fabricated in the form of usable conductors in the near future.

Finally, the class of quasi two-dimensional layer-compound superconductors of the type $NbSe_2$ should be mentioned. In a TaS_2 layer compound, intercalated with pyridine, an upper critical field (parallel to the layers) of about 20 T was found [9]. The low T_c of about 3.5 K for this material, however, makes it unattractive for technical applications.

C. Stabilized High Field Superconductors

High field superconductors which are operated at very high current densities in magnet windings have high electrical resistivities in their normal state. This, in combination with high currents, would lead to a local thermal destruction of the high-field superconductor if a normal resistance region appears in it. Normal resistance regions in high-field superconductors can be caused by "flux jumps", by friction during conductor movement, by external heat pulses, or by insufficient cooling. To prevent damage of the superconductor during a resistive event mentioned above, superconductors are "stabilized" for technical applications. They are combined with high-conductivity normal-conducting materials such as copper or aluminium which have to be in excellent electrical contact with the superconductors. We distinguish between three types of stabilization[10]:

1. Cryostatic stabilization

In this case the maximum possible operating current can safely flow in the stabilizing material when the superconductor is driven normal by an internal or external disturbance. The ratio of the cross-sectional areas of superconductor to normal metal is chosen to satisfy the following conditions: At a given conductivity of the normal material and at a given minimum heat flux from the free conductor surface to the coolant, the composite conductor can recover to its superconducting state. The current returns into the superconducting wire when the disturbance has passed and the superconductor has returned to the resisanceless state. Fully-stabilized conductors in general have normal metal to superconductor ratios of 10 and greater. They are especially used in large magents with large amounts of stored energy, where safety plays a dominant role and the overall current-density is of less importance. If copper is used as stabilizing material it often can be designed to act simultaneously as a reinforcing material. This extra strength is necessary in large magnet. systems in order to withstand the large magnetic forces. The cryostatic stabilization is effective against internal as well as external disturbances (flux jumps, conductor movements, external heat pulses). Even in the case of a momentarily insufficient cooling of conductor sections, it will prevent conductor damage.

2. Adiabatic stabilization

For adiabatic stabilization, the wire is made thin. As the geometrical dimensions of the superconductor wire perpendicular to the field direction is decreased, the heat per unit volume which is dissipated during a magnetic instability is reduced to a level at which it can be absorbed by the heat capacity of the conductor material itself. In the case of a superconducting wire, no flux-jumps (normal region in the superconductor) should occur, if the wire diameter is reduced below a critical value. For NbTi with a critical current density of 2×10^5 A/cm^2 at 5 T this critical wire diameter is found to be about 50 μm.

3. Dynamic stability

In this case the formation of normal region in the superconductor during magnetic instabilities is prevented by adding a high conductivity normal conductor. The normal conductor magnetically damps the growth of the disturbance, so that the heat produced in the superconductor can be carried away by thermal conduction.

It should be pointed out that the last two stabilization methods only prevent the formation of those normal regions in the superconductor which are initiated by internal disturbances (magnetic instabilities). Normal conducting regions in the superconducotr which are caused by external disturbances (e.g., moving superconductor) cannot be overcome by these methods. In practical conductor configurations, which will be discussed in the following paragraphs, the last two methods can be used to protect the conductor from damage when disturbances (also external) occur, provided the current is switched off quickly enough.

D. Conductors for dc and ac Magnets

Today NbTi is the most frequently used high field superconductor. It is manufactured almost exclusively as multifilament conductor where a large number (up to several 10,000) NbTi-filaments with diameters down to a few micrometers are embedded in a matrix of copper. Figure 4 shows a NbTi-multifilament conductor manufactured by Vacuumschmelze Hanau. With these conductors, in general, both the adiabatic as well as the dynamic stabilization criteria are fulfilled. They are called intrinsically stable superconductors. The manufacturing process is comparatively simple. One starts with an assembly of NbTi- and copper rods in a copper tube which is first reduced in size by a coextrusion process. During the further manufacturing process, heat treatments and cold work by drawing down alternate with each other. This creates the desired defect structure (pinning centers) in the NbTi (dislocation walls with a density of 10^{12} cm^{-2} and normal conducting α - Ti precipitates with a density of about 10^{10} cm^{-2}) which is necessary for high current densities [11].

Stabilized Nb_3Sn and V_3Ga conductors were fabricated by vapor deposition or diffusion processes until recently in form of thin layers on tape substrates. This was necessary because of the high brittleness of these materials which prohibits any coextrusion process. Stabilization was done by subsequent addition of copper tapes which were soldered onto the conductors.

For about two years Nb_3Sn and V_3Ga have also been commercially available in the form of multifilament conductors [12, 13, 14, 15]. The fabrication methods, however, are more complicated and expensive than with NbTi multifilament conductors. The fabrication method of the A-15 multifilament conductors is based on a solid-state diffusion process. The formation of the brittle compounds is postponed until the desired geometrical configuration for the conductor (such as multifilamentary wire) is achieved. Figure 5 illustrates two possible ways of producing Nb_3Sn-multifilament conductors. A billet consisting of Nb rods in a Cu-Sn alloy (Sn content $\leq$ 13.5%) is extruded down to the final wire size (Fig. 5a). Since the Cu-Sn matrix work hardens appreciably due to the drawing operations, it has to be annealed several times during the extrusion process. The composite is then heat treated (20 to 100 hrs at 650 to 750° C) and Nb_3Sn layers (thickness: 1 - 3 μm) are produced at the interfaces of Nb and the Cu-Sn matrix. Another possible method of manufacturing Nb_3Sn-multifilament conductors is shown in Fig. 5b. An ingot of Nb rods in a copper matrix is drawn down to the final geometrical size, coated on its surface with Sn and then heat treated. Sn diffuses into the copper matrix and again Nb_3Sn layers are produced at the Nb cores. This

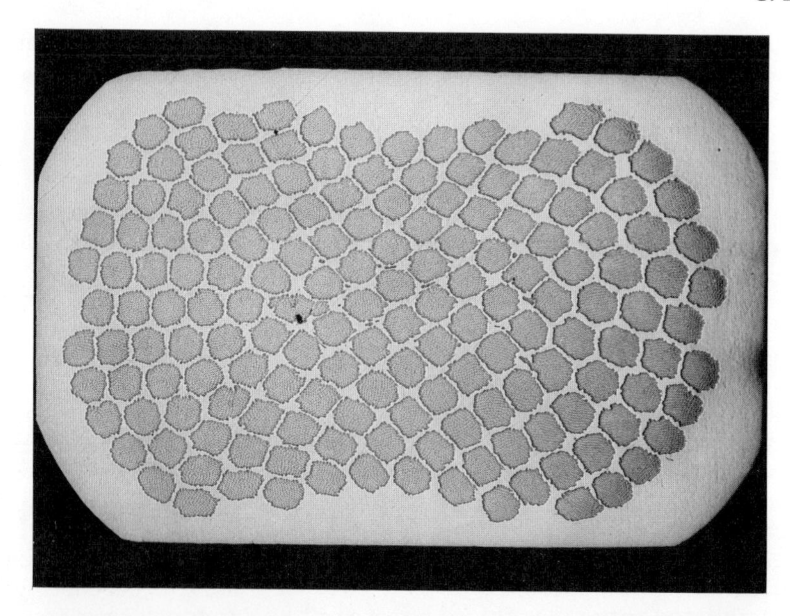

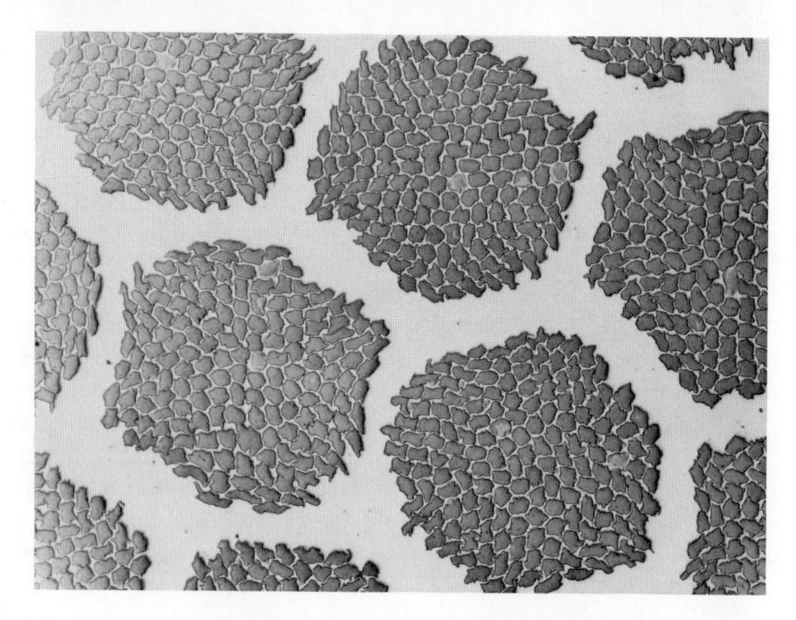

Fig. 4 NbTi-multifilament conductor with copper matrix
Conductor cross-section 9.2 mm x 5.7 mm
163 x 163 NbTi filaments with 25 μm diameter
Current carrying capacity: 20,000 A at 5 T and 4.2 K
(a) total view; (b) sectional view
(Conductor manufactured by Vacuumschmelze, Hanau).

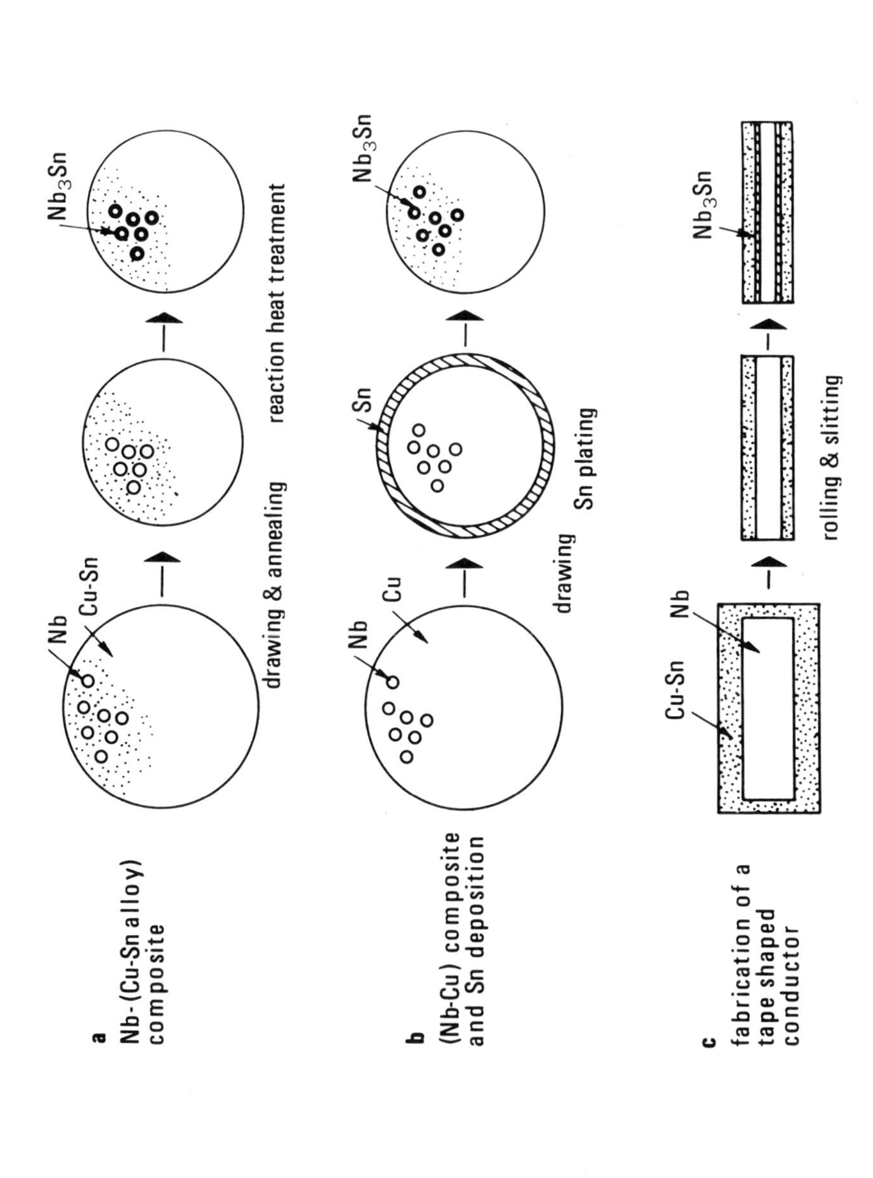

Fig. 5 Schematic diagrams of Nb_3Sn composite superconductor fabrication processes using the solid state diffusion process [12].

fabrication process is limited to small conductor diameters because of the Kirkendall-Effect.

Equivalent processes are applied to produce V_3Ga multifilament conductors. In this case V rods and Cu-Ga alloys with Ga contents of up to 20% are applied at reaction temperatures of 600 - 700° C. Figure 5c illustrates that the solid-state diffusion process can be also used to fabricate tape-shaped Nb_3Sn or V_3Ga conductors. Figure 6 shows a Nb_3Sn multifilament conductor which contains 1615 filaments of 5 μm diameter with a 1 μm thick Nb_3Sn-layer. Conductors with more than 600,000 Nb_3Sn-filaments have already been fabricated [14].

Because of the brittleness of the A-15 compounds, two problems arise with multifilament conductors. First, they have to be handled very carefully when winding them into a coil. Each conductor geometry has a critical bending diameter. If during the construction of a magnet, the wire is wound in a diameter lower than the critical value, a significant reduction in critical current of the conductor will occur. In the case of Nb_3Sn multifilament conductors, the critical diameter is given by the relation $D/A = 250$, where A is the largest distance between a filament and the unstressed neutral line. Figure 7 shows the dependence of the critical current on applied bending diameter for Nb_3Sn-multifilament conductors as well as for V_3Ga conductors [16, 17]. The bending diameter is not so critical with flat multifilament conductor tapes, or with cables made of several thinner conductors where internal gliding between individual conductors is possible.

There is also great interest in the mechanical behavior of these ductors at low temperatures. Because they are used for the high current densities in high flux-density magnets, they are in general exposed to higher forces than NbTi-conductors. It was demonstrated that for a Nb_3Sn-multifilament conductor, which was additionally stabilized with copper per, the critical stress loads at 4.2 K are 300 MN/m^2 to 350 MN/m^2 corresponding to a strain of about 0.5% [18] (see also [45].

Another problem of these conductors is stabilization. Because the matrices of the metal alloys have high resistivity, it is necessary to apply additional copper to the conductors. In the case of Nb_3Sn a large number of superconductors are contained in a high purity Cu-matrix consisting of a bundle of Nb_3Sn filaments in Cu-Sn-alloy surrounded by a Ta diffusion barrier. Similar stabilizing methods have been applied to V_3Ga. Figure 8 shows a section of a Nb_3Sn-multifilament conductor where another stabilizing method is used. Cu-rods enclosed by a Ta layer are embedded into the Cu-Sn matrix.

From the brief descriptions above, it is clear, that the fabrication processes of the A-15 compound multifilament conductors are complicated and therefore lead to comparatively expensive conductors. Due to the high cross-sectional portion of normal conducting materials (Cu-Sn alloy, Cu, and Nb and Ta) in these conductors, the overall current densities are only a small fraction of the high current densities of the A-15 compounds. For intrinsically stable conductors overall

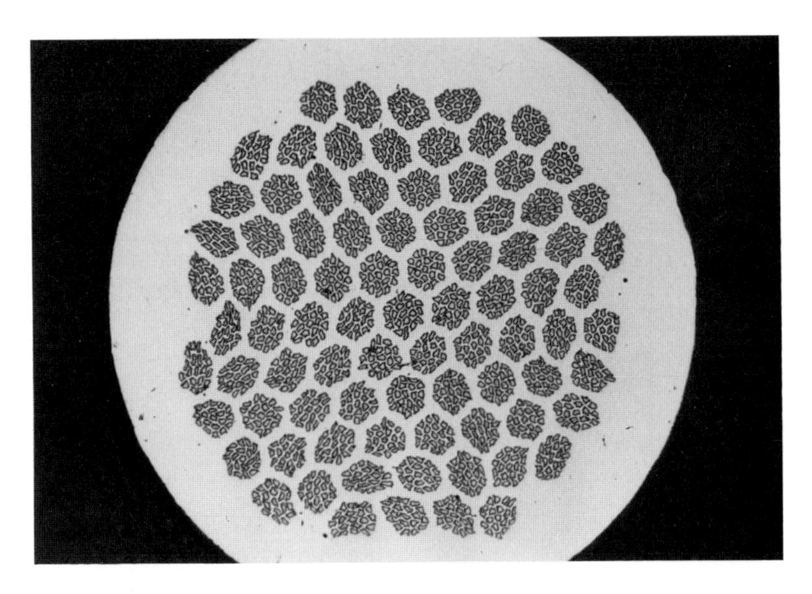

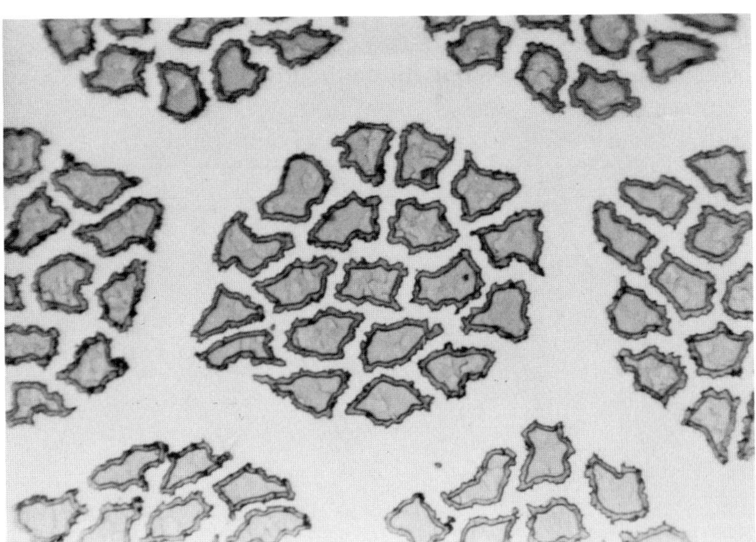

Fig. 6 Nb_3Sn-multifilament conductor, Cu-Sn-matrix.
Conductor diameter 1 mm.
85 x 19 filaments, filament diameter $\approx 10\ \mu m$
Thickness of Nb_3Sn layer $\approx 1\ \mu m$
Current carrying capacity: 300 A at 10 T.
(a) Total view; (b) Sectional view.
(Conductor manufactured by Vacuumschmelze Hanau).

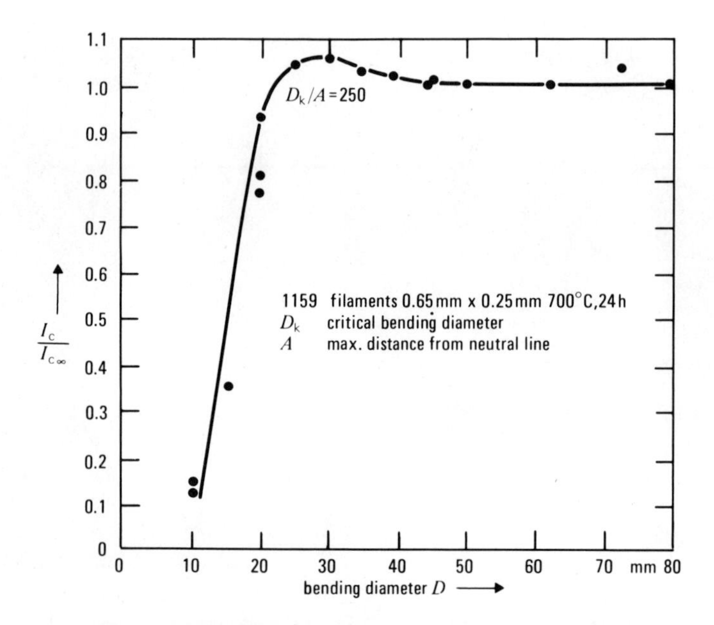

Fig. 7a Critical current of a Nb_3Sn-filament conductor as a function of the bending diameter [16].

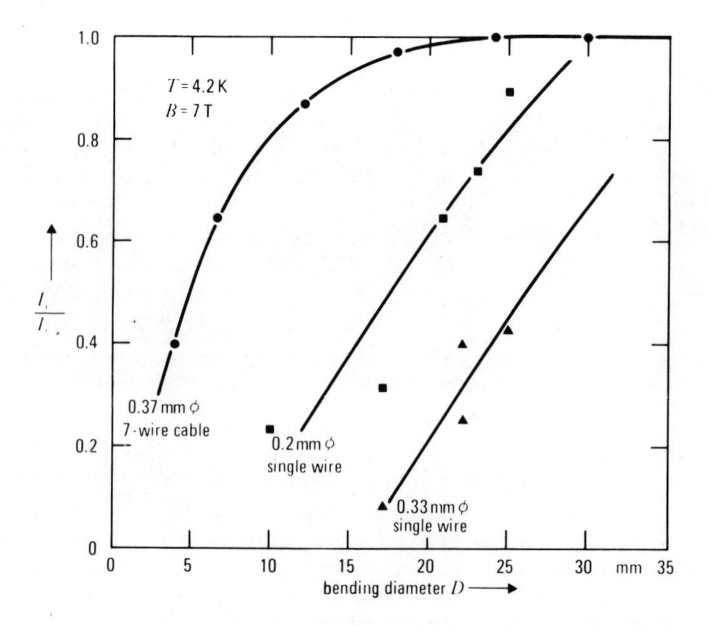

Fig. 7b Critical current degradation as a function of bending diameter for V_3Ga conductors [17].

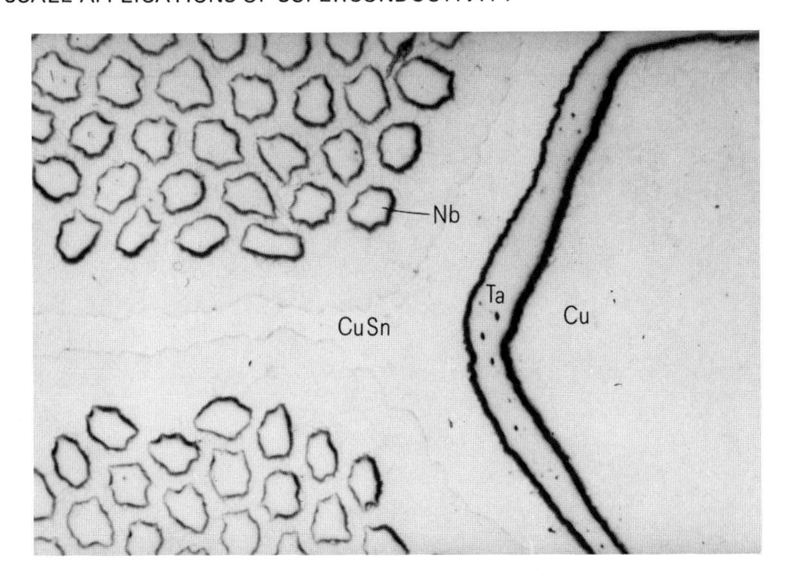

Fig. 8 Multifilament Nb_3Sn-conductor (sectional view) internally stabilized with copper enclosed by a Ta layer. Filament diameter 5 μm. (Conductor manufactured by Vacuumschmelze Hanau).

current densities of 1.5×10^5 A/cm^2 at 5 T and 5×10^4 A/cm^2 at 10 T can be expected.

When high field multifilament superconductors are used in windings for alternating magnetic fields or alternating currents, hysteresis losses occur in the single filaments. It can easily be shown that the magnitude of these losses is proportional to the diameter of the single filaments [10]. Thus for ac applications multifilament conductors should have as small superconducting filament diameters as possible. Diameters in the neighborhood of 1 μm are the limit for fabrication technology. Typical commercial ac conductors have filament diameters of about 5 μm. The hysteresis losses per unit volume of a NbTi-multifilament conductor with filament diameters of 5 μm are approximately 30 mWs per cubic-millimeter when cycled at field amplitudes of 5 T.

The effect of subdividing the superconductors into fine filaments is unfortunately cancelled by the coupling of the individual filaments by the normal conducting metal matrix. Alternating magnetic fields perpendicular to the conductor axis induce voltages between the filaments. This drives large shielding currents across the ohmic matrix and results in eddy current losses in the latter which can be much higher than the hysteresis losses of the filaments. Eddy current losses are also produced in the matrix layer which in general encloses the filament bundle of the presently available composite conductors. Additional negative effects of the shielding currents are flux jumps and residual fields

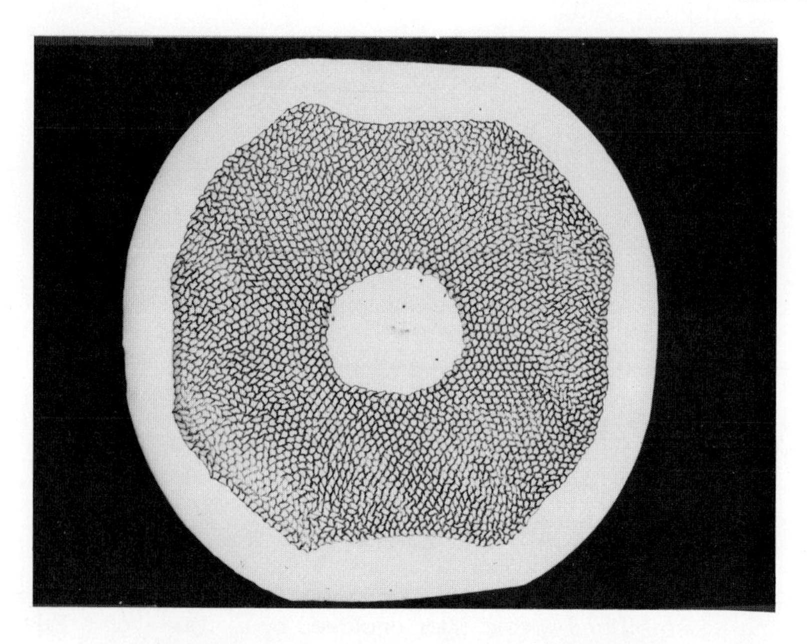

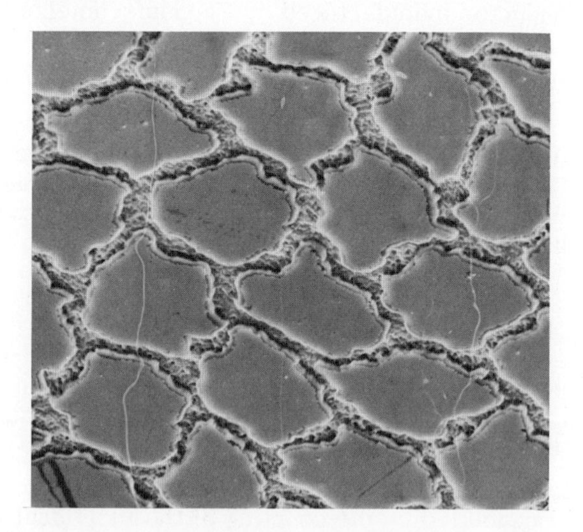

Fig. 9 NbTi-multifilament conductor with two component matrix (CuNi/Cu). Conductor diameter 0.85 mm, 2600 filaments, filament diameter 10 μm. Single filament enclosed by a CuNi-layer and embedded in Cu matrix.
Current carrying capacity: 7000 A at 5 T and 4.2 K
(a) total view; (b) sectional view
(Conductor manufactured by Vacuumschmelze Hanau).

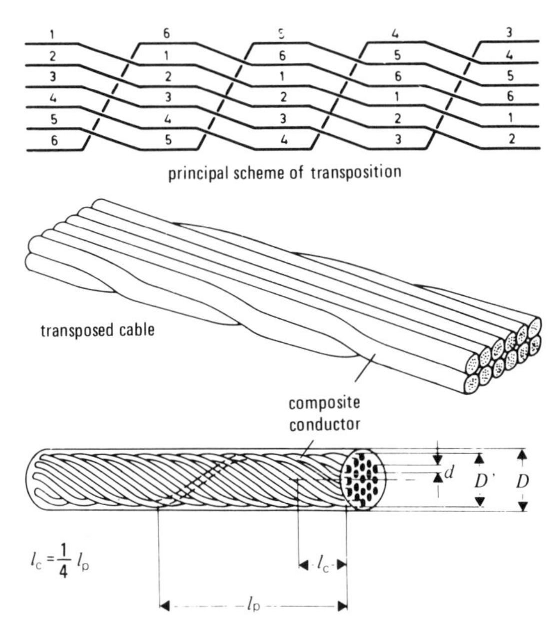

Fig. 10 Fully transposed cable. A single composite conductor (strand) with twisted filaments is shown at the bottom [3].

with large time constants. The losses and other effects can be largely reduced or cancelled if the filaments are electrically decoupled by twisting the conductors along their axis [10]. Present advanced multifilament conductors are twisted for dc applications also because even in dc magnets the field has to be varied (e.g., during charging).

The length of the twist-pitch is limited by mechanical constraints to ⩾ 10 D, where D is the conductor diameter. Accordingly at higher field sweep rates or at larger conductor diameters twisting becomes more and more ineffective for decoupling with a pure copper matrix. Therefore matrix materials with high resistivities are used for fast ac applications. In the case of NbTi multifilament conductors a cupro-nickel alloy (Cu-Ni) is often used, which has a resistivity of about 1000 times that of copper. However, it has been found experimentally that NbTi multifilament conductors with pure CuNi-matrix behave in a very unstable fashion. Therefore two component matrices are frequently used, where e.g., NbTi-Cu multicore subconductors surrounded by a CuNi-mantle are embedded in a copper matrix. Several combinations of two component matrices are possible and have already been realized in commercial conductors. Figure 9 shows a Nb Ti multifilament conductor with a two component matrix where the single NbTi filaments are surrounded by a CuNi-mantle and embedded in the Cu-matrix. A-15 compount multifilament conductors have originally high resistivity alloy matrices.

Very recently it has been recognized that longitudinal ac fields (parallel to the axis of a twisted multifilament conductor) give rise to a new kind of filament coupling by inducing counter-flowing currents in the inner and outer concentric parts of the composite, thereby crossing radially the normal conducting matrix. The corresponding extra losses are found to have the order of magnitude of the hysteresis losses. These losses can be markedly reduced by applying an alternating twist [19]. Longitudinal ac fields can play a role in many magnet applications, e.g., fusion magnets.

Another source of instabilities or losses is the self-field of multifilament conductors, for which twisting the conductor is not effective (the radial positions of the filaments are unaffected by twisting). The filaments remain electrically coupled with respect to the self-field in spite of the twist. In the case of conductors with larger diameters, say $\phi > 5$ mm, (also valid for rectangular conductors with large dimensions) this leads to self-field losses which can be higher than the hysteretic losses.

Because of these losses and also of the limited maximum attainable twist pitch length of multifilament conductors of larger size one is obliged to construct high current capacity ac conductors in the form of fully transposed conductors which are composed of a multiple of small conductors with a short enough twist pitch and small self field losses.

For dc applications, compact twisted multicore conductors with copper matrix is the proper solution for both high and low current capacity conductors. For ac applications small current capacity conductors ($\leq$ 1000 A at 5 T) have to be twisted and should have a two component matrix in the case of higher frequencies. High current ac conductors have to be assembled from fully transposed smaller twisted multifilament conductors. The various types of fully transposed conductors are: single layer cables, braids and Roebel bars. These transposed conductors are either impregnated with thermoplastic resins, with the individual strands fully insulated, or impregnated with highly resistive alloys (SnAg, InTl) with the individual conductors uninsulated. The conductors receive the mechanical rigidness from the impregnation which is necessary for a precise winding and to prevent movement of single strands under magnetic forces. In general the conductors are mechanically compacted to achieve a high packing factor (80 to 85% have been achieved), and are also geometrically calibrated which is absolutely necessary for the use in magnets with highly precise field configurations. With the metal impregnation, good heat conduction and better magnetic stability can be obtained. On the other hand, additional losses are produced by coupling the individual strands. To keep these losses to a minimum, the metal impregnation is carried out in such a way that high resistivity reaction layers are formed around each individual strand.

The methods described above are not sufficient for the application of high field superconductors at power frequencies (50 Hz or 60 Hz) and at high magnetic field amplitudes. Under such conditions the hysteresis and eddy current losses would be intolerably high. Therefore the application of high-field superconductors is restricted to frequencies in the neighborhood of 1 Hz at high-field amplitudes, or at power frequencies to small field amplitudes of about 0.1 T. These limitations strongly restrict the technical application of high field superconductors. Pure Nb is the superconductor with the smallest hysteretic-losses at power frequencies, but its application is also limited to magnetic flux densities of about 0.1 T. This level is interesting for the development of superconducting ac cables (see Section V).

E. Irradiation Effects in Composite Superconductors

Superconductors used for magnets which are placed in particle accelerators or in fusion reactors will eventually be exposed to high irradiation doses. In the case of fusion reactors, the irradiation consists of fast neutrons ($E > 0.1$ MeV). In the case of accelerators, it is composed of charged particles and fast neutrons (secondary particles). The irradiation with charged or neutral particles produces lattice damage in the superconductor as well as in the stabilizing matrix material of the composite. These damages influence the properties of the latter in a characteristic way.

Since the generation of lattice defects is most effectively achieved with neutron irradiation, we will mainly restrict the discussion to results of fast neutron irradiation experiments. These experiments are also qualitatively representative for results with charged particles. We further limit the discussion, with one exception, to irradiation experiments at low temperatures ($T \leq 15$ K). Only low temperature irradiation realizes the conductor operating conditions in magnets.

Irradiation of high purity, low resistivity stabilizing materials, e.g., copper or aluminium, leads to a considerable increase in their resistivity, resulting in less stability of the composite conductor. The irradiation of copper (RRR = 714) and aluminium (RRR > 1000) at 5 K with fast neutrons at a dose rate of 10^{18} n/cm^2, increases the resistivity by a factor of 20 (Cu) and of about 150 (Al) [20]. The increase of resistivity can be explained by the formation of Frenkel pairs. In addition to the increase in resistivity, the mechanical properties of the stabilizing materials are changed, i.e., the yield and tensile strength are increased, indicating embrittlement. After annealing for a short time at ambient temperature, the conductivity and mechanical behavior can be recovered very easily to values near the unradiated value (95 -98%). The irradiation effects on cupro-nickel which is also used as a component in composite conductors can be neglected.

In the case of a high field superconductor all of the critical data (critical temperature T_c, critical current density j_c and upper critical field B_{c2}) are influenced by irradiation:

The critical temperature T_c is in general unchanged at low irradiation dose rates and decreases with increasing dose rates. A typical example is given in Fig. 11 in which the dependence of T_c on the neu-

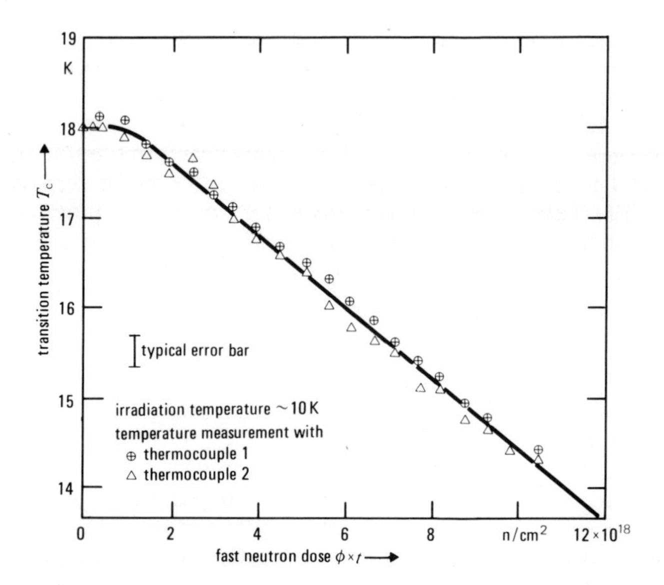

Fig. 11 Transition temperature T_c versus fast neutron dose ϕt (E > 0.1 MeV) for a Nb_3Sn diffusion wire [21].

tron dose rate is depicted for a Nb_3Sn-diffusion wire [21]. No change in T_c is observed up to a dose rate of about 10^{18} n/cm^2. At higher dose rates a remarkable decrease in T_c occurs. For V_3Ga (multifilament conductor) irradiation with 50 MeV deuterons a steep decrease of T_c results for dose rates above 10^{17} n/cm^2 [22]. The T_c decrease is caused by radiation induced disorder in the A-15-lattice. Similar measurements on the T_c of NbTi have not been reported in the literature. It is anticipated (by extrapolation of critical current density results) that for neutron dose rates of up to 4×10^{18} n/cm^2, no significant change in T_c will occur.

The behavior of the critical current density under irradiation strongly depends on the kind and density of pinning centers introduced during manufacturing and also of the field region under consideration. If the original pinning structure of an irradiated high field superconductor was not optimized, i.e., the superconductor has not achieved its maximum possible current density, then irradiation results in an enhancement of the current density. If the superconductor is optimized with respect to its current density (= structure of pinning centers), and this is generally the case with commercially available conductors, then irradiation results in a decrease of the current density. For NbTi multifilament wires the reductions are moderate, e.g., 15 % at a dose rate of 4×10^8 n/cm^2 [23]. In contrast the current density of optimized A-15 superconductors (Nb_3Sn and V_3Ga) is markedly decreased when irradiated with higher dose rates. Reductions to 50% of the original value have been reported for V_3Ga at 4 K and a 50 MeV deuteron irradiation dose rate of about 3×10^{17} cm^{-2} [22]. The critical current density of optimized Nb_3Sn conductors seems to be less sensitive to irradiation than V_3Ga. With neutron irradiation at 330 K a very rapid decrease of I_c (4 T) was observed for dose rates greater than 2 - 3 x 10^{18} n/cm^2 [24]. Similar results are expected for irradiation at He-temperatures. The effect of irradiation on the critical current density of optimized high field superconductors can be explained by the generation of homogeneously distributed small pins. The pins do not contribute to the current density but weaken the pinning force of the pinning centers introduced during manufacturing.

At high irradiation dose rates a reduction of the upper critical field Bc_2 is also expected. It is concluded that for Nb_3Sn a linear decrease occurs at dose rates > 10^{18} n/cm^2 corresponding to the reduction in T_c (viz. $dBc_2\ (T=0)/dT_c \approx 1$ T/K) [25].

On the basis of reactor studies carried out so far, one expects that for a 20 year magnet lifetime, the shielded magnets of fusion reactors are exposed to fast neutron dose rates of 10^{17} - 10^{18} cm^{-2}. For large proton accelerator magnets (1000 GeV) average dose rates per year of about 2×10^{10} rad are expected [26] which is less than the values given for fusion magnets.

From the results obtained so far, it follows that irradiation induced changes in critical parameters of composite conductors used in shielded fusion reactor magnets or accelerator magnets are not a critical limitation in the case of NbTi and Nb_3Sn. It is also anticipated that irradiation does rates of 10^{17} - 10^{18} cm^{-2} will have a minor influence on the mechanical properties of structural materials (e.g., stainless steels) and on the mechanical and electrical characteristics of pure or glass fibre reinforced epoxy resins used in magnets.

F. General Design Aspects of Superconducting Magnets

1. Intrinsically stable coils

Small and medium sized magnets are primarily designed with regard to high overall current densities, e.g. 100 - 400 A/mm^2 at 5 T. This is necessary from an economic point of view, but often also is useful when limited space is available or when field configurations with high precision are required. High current densities can only be obtained by using multifilament conductors with a low cross-sectional ratio of normal conducting matrix material to superconductor, i.e., with intrinsically stable superconductors. One very important prerequisite for obtaining current densities near the critical ones is the exclusion of any conductor movement, otherwise a severe degradation in current density is expected. To prevent such movements, the impregnation of superconducting coils with epoxy resins was introduced several years ago. This impregnation also plays an important role with respect to a homogeneous support and the transfer of the high mechanical loads in superconducting windings caused by the high flux and current densities.

It has been found however, that the impregnation of coils with epoxy resins cannot totally overcome, or may even be the cause of an undesired effect, called "training". This effect is particularly strong in coils with straight-sided configurations, where in the energized coil there is no support of the tension such as hoop stress, as is the case of simple solenoids. Training means, that in general magnets do not reach their design field or current at the first energizing but are training up to their final value which optimally is the short sample critical current. The short sample critical currents are measured at short conductor samples in external magnetic fields in a liquid helium bath. The training involves a number of premature normal transitions (in extreme cases amounting to several hundred), each time releasing a large amount of energy into the helium bath, which is evaporated. Figure 12 shows the behavior of a dipole magnet. The design field of 4.5 T was attained after about 100 quenches (transitions to the normal state) which is an economically intolerable result [27]. Many reasons have been quoted for the appearance of this phenomenon, including:

A not compact enough arrangement of the winding sections which gives rise to movements and friction heating

Strain introduced to the impregnant during manufacture of the coils. In particular strain introduced during the curing of the resin or clamping of the coil structure

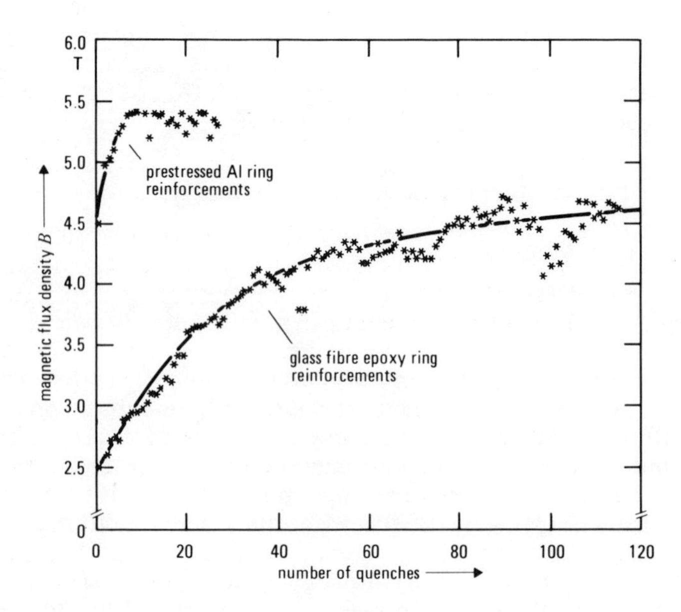

Fig. 12 Training behavior of a dipole magnet before and after being armed with shrink fitted aluminum rings [27].

Strain energy stored in the impregnating epoxy resin during cool down, or under magnetic loads which are released by local cracking. This causes a significant local temperature rise above T_c [10, 28]

Serrated yielding of the superconductor or plastic deformation of the normal metal matrix. Both are linked with the dissipation of energy and a temperature rise [29, 30, 31].

We consider the last reason for the training phenomenon. It has been found by experiments [32] that serrated yielding of the superconductor and plastic deformation of the matrix most probably is not a cause for the training phenomenon. Serrated yielding in NbTi-multicore wires occurs at strain levels larger than 1%, a value which is not expected in superconducting magnets with NbTi-conductors. Distinct training behavior, however, was found in NbTi-specimens without copper matrix, starting at strains far below 1%. This leads to the conclusion that one cause of training may be found in the superconducting material itself [32].

To study the influence of the epoxy resin impregnation on the training effect, a series of investigations was carried out in several laboratories around the world. Several potting methods were studied including vacuum impregnation, wet winding, application of high pressures, different filling materials and curing processes. A wide choice of epoxy resins or other impregnating materials were considered, for example transformer oil or paraffin wax [28].

When using epoxy resins as impregnate the following principles should be considered to prevent significant current degredation or training:

The thermal contraction coefficients of the resins should be matched to that of the conductor as far as possible by using inorganic fillers. Resin layers in the coil interior which are too thick should be avoided because of the increased tendency to crack.

If vacuum impregnation does not allow the use of fillers because the viscosity is too high, one must make sure that only extraordinary thin layers of the impregnant exist in the interior of the coil winding, because pure resins store the highest strain energies.

It has been found that using conventional epoxy resin types applied normally at ambient temperature (also with straight sided coils), quite satisfactory results can be obtained. Operating currents of 80% to 90% of the critical current can be obtained after only a few training steps, when applying the principles mentioned above. The choice of epoxy resin or the hardener is then of less importance. However, at present, impregnating with epoxy resins leads to a complex manufacturing procedure without a sufficient degree of safety for the realization of a low strain impregnant.

For this reason the use of impregnants with low tensile strength, such as oil or wax has been suggested [10]. These materials should be incapable of storing sufficient strain energy to cause significant degradation or training. It has been shown in the case of a quadrupole-prototype, that wax impregnated coils with proper mechanical support at the coil boundaries, achieve operating currents of 85% to 90% of the critical current with negligible training [28].

The importance of an extremely rigid external support compressing the whole coil arrangement was demonstrated in the case of the dipole of Fig. 12. After replacing the glass fiber bandages with which the dipole was first equipped by prestressed Al-alloy rings, the training of the coil could be reduced to 7 steps (upper left part of Fig. 12) [27].

The explanations above show that there really exist a number of reasons which can lead to degradation or training in intrinsically stabilized coils (especially in the case of straight-sided configurations). They may occur alone or in parallel, but fortunately can be overcome by the methods previously suggested. At present it cannot be proved whether training in coils can also be initiated by intrinsic mechanical instabilities in the superconductors themselves. Since these instabilities are closely linked with the metallurgical structure, which is responsible for obtaining high current densities, this source of training would be difficult to eliminate.

Special manufacturing problems arise with magnets, where highly precise field configurations are necessary, e.g., beam guiding or accelerator magnets for high energy physics. With conventional magnets these field configurations are achieved with specially formed iron pole pieces. With superconducting magnets iron is only used as a magnetic shielding means. The desired field configuration has to be accomplished by the winding arrangement itself. Cosine distributions around a circular aperture or intersecting circle or ellipse configurations with constant current density are used. These configurations are approximated by blocks and layers of conductors. The positions of the blocks or layers are computer calculated. Field tolerances of $\Delta B/B<10^{-3}$ need accuracies in the positioning of coil sectors, conductor layers and poles during construction which are better than 0.1 mm. It has been demonstrated with prototype magnets, that this accuracy can be achieved.

For maintaining this accuracy during longterm static or dynamic operation, the winding configurations have to be supported by strong bandages which can be metallic or of fiber-reinforced resins. Often the helium cold shielding iron cylinder is used for mechanical support. The space between the iron cylinder and the windings is bridged with filled resins. Alternatively, shrinking techniques are applied. If forces have to be transferred from the helium-temperature to ambient temperature environment, glass fiber reinforced resin supports are used. They combine high mechanical strength with low heat conductivity. The mechanical and thermal behavior of resins filled with powders or reinforced with fibers has been studied in detail during the past years. Figure 13 is a summary of the mechanical and thermal properties of a glass fiber re-

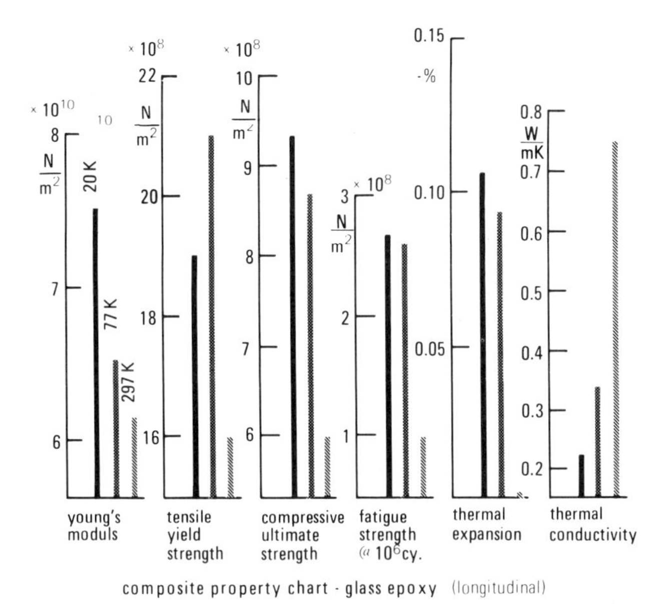

Fig. 13 Summary of mechanical and thermal properties data for glass-epoxy composite [33].

inforced epoxy resin (mechanical load parallel to the fibers) [33].

Until now, the cooling of intrinsically stable magnets has been performed by immersing them into a liquid He-bath at 4.2 K and 1 bar. Coils with thin winding packages are cooled only from the surface. Coils with thicker winding packages or ac coils with continuous internal heat generation have to be provided with internal cooling. This can be performed by cooling channels between individual winding sections filled with He or by heat drains of high conductivity metals which are in contact with the helium bath. Cooling channels in fully impregnated coils can be fabricated by incorporating Teflon rods which don't react with the resin and can be removed after curing.

Recently it was demonstrated very successfully that intrinsically stable high current density coils can also be indirectly cooled by forced flowing supercritical He [34]. In the case mentioned, an eight meter long intrinsic stable solenoid, consisting of 16 coils, producing a flux density of 5 T in a bore of 12 cm is cooled via copper tubes through which supercritical He is circulated. The copper pipes are spirally wound onto the solenoid surfaces and fixed to them by the resin impregnation. Forced cooling with one phase He was also applied to intrinsically stable magnets which were used for levitation experiments. In this case only one cooling loop is used. The helium vessel itself serves as the main cooling channel in which the winding with its own, small, uniformly distributed cooling channels is embedded. More details are found in Section III.

2. Fully or cryostatically stabilized coils

As already mentioned large superconducting magnets are, for safety reasons, constructed with fully stabilized conductors. High current densities are of less importance. Current densities of 10 to 50 A/mm^2 at flux densities of 2 T to 5 T are usual. Impregnation is not absolutely necessary since conductor movement can be overcome by full stabilization. Large fully stabilized coils realized up to now are in general composed from several disc-shaped coils wound from compact rectangular conductors. The assembly is immersed in a liquid He-bath. Each individual conductor is in direct contact with the coolant by cooling channels filled with liquid He.

There is a strong tendency, however, especially with respect to large MHD- and fusion reactor magnets, to construct large magnets with fully stabilized hollow conductors, cooled by a forced flow of supercritical He. This technique is similar to that used in conventional water-cooled copper magnets. The most important advantage of forced cooling is the possibility of having a better mechanical construction. Coils with hollow conductors can be entirely potted with an adequate resin after winding. This results in a much more compact and stronger structure than with bath-cooled unpotted magnets. Better mechanical quality as well as an improvement in electrical coil insulation is to be expected. This fact becomes increasingly important for larger sized coils, where quite high voltages can occur during emergencies. Often better cryogenic safety is quoted as a further important advantage, since the amount of He stored in the magnet system is smaller than in a bath-cooled magnet. In addition, hollow conductors can withstand a much higher pressure than large cryogenic vessels. Finally, force-cooled magnets can be operated in any position.

A disadvantage of all types of forced cooling with supercritical He is the fact, that temperature gradients are built up across the coils. The outlet-temperature is higher than the inlet temperature leading to a non-optimal utilization of the superconductor. To prevent too high a temperature rise due to the isenthalpic expansion of the helium in a larger coil with long conductor length, the coil has to be subdivided into sections. The sections are connected in series by intermediate heat-exchangers which cool down the circulating He every time to its initial temperature (see Fig. 14a) [35]. He-pumps can be used instead of coupling the refrigerator directly to the magnet. With pumps it is preferable to have only modest pressure drops. Therefore, in this case the coil has also to be subdivided into several sections, this time cooled in parallel, Fig. 14b.

It is apparent that the forced cooling requires additional equipment in comparison to a liquid bath. It requires heat exchangers, additional transfer lines, pumps, insulating joints vacuum tight at high pressures. This leads to a less effective and less reliable cooling cycle. Sufficient reliability can probably be achieved, when more experience is gained.

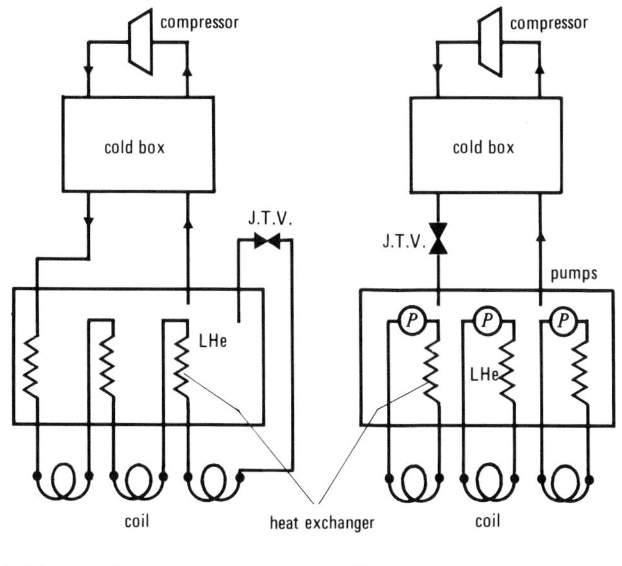

Fig. 14 Forced cooled coils.

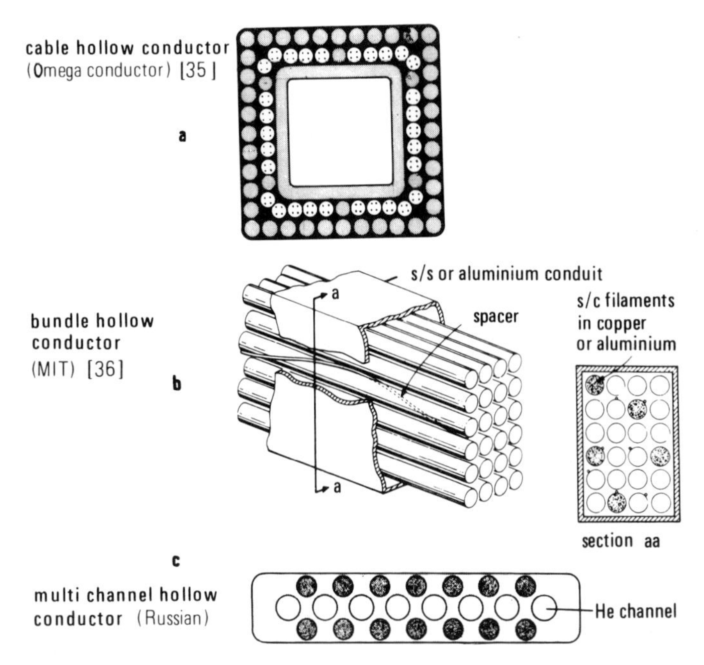

Fig. 15 Hollow conductors for forced flow helium cooling.

Until now, fully stabilized NbTi hollow conductors have been fabricated chiefly by a coextrusion process, or by cabling and soft soldering multicore wires on a central copper tube. Figure 15a shows such a cabled conductor. The disadvantage of this type of conductor is its small ratio between the wet perimeter and conductor cross section. This leads to comparatively high helium mass flow rates in order to achieve satsifactory stability. This disadvantage can be eliminated by the application of bundle hollow conductors. In this type of conductor a bundle of current-carrying twisted strands is contained within a tubular conduit, resulting in a considerable increase in heat transfer area between conductor and intersticially flowing He. Figure 15b shows such a bundle hollow conductor [36]. Computer calculations and experiments have proved that with this type of conductor cryostatic stability can be achieved with a modest fluid velocity [37].

An interesting type of hollow superconductor is shown by Fig. 15c. It is a NbTi-Cu composite conductor with nine cooling channels. This conductor was developed in Russia and is used in the T-7-prototype Tokamak toroidal magnet system at the Kurchatov Institute Moscow [38]. It is fabricated by assembling individual NbTi-Cu multicore conductors and Cu-tubes by adding additional Cu in a electrolytic plating process. The final conductor geometry is achieved by mechanical calibrating.

Finally it was suggested [39] to use two-phase helium instead of supercritical He for forced cooling in hollow conductors. With two-phase He a much better cooling capacity can be attained than with one phase supercritical He. The potential of this kind of cooling will be further investigated.

In large magnets the cross-section of the stabilizing material is not sufficiently large to support the internal magnetic forces. Additional high strength material, e.g., stainless steel, is used for backing the conductor or as an external support structure. For very large magnets (fusion, energy storage) strengthening using stainless steel structures would become too expensive. Therefore the use of external concrete or even rocks for supporting the enormous mechanical loads has been suggested [40, 41]. The load transfer from the cold to the warm parts is again accomplished by fiber reinforced resin structures.

3. Current leads and coil protection

Current leads and coil protection systems are two important components of superconducting magnet systems. Their importance is increasing as superconducting magnet systems become larger and larger. Within this paper only a brief summary of the most important facts will be given.

Large superconducting magnets planned for the future will work at currents > 10,000 A. These currents will have to be fed into the He-cryostat with a minimum heat input to the He. The technique which can accomplish this is well known and proved. Helium gas cooled normal metal (in general copper) leads are used. If a good heat transfer from the normal metal surface to the cold helium gas is guaranteed, the losses can be reduced to about 1.2 mW per Amp and lead. To achieve a good heat transfer to the cooling He-gas, normal metal leads, subdivided into a multiple of subconductors offering a large surface to the helium gas are used. Figure 16 shows the optimized heat dissipation fo gas cooled current leads as a function of the resistivity of the copper used [42]. Another important aspect of magnet current leads is their dielectric strength. This is of special importance in energy storage coils with short charging- and discharging times, where voltages up to 100 kV can occur across the terminals. Even in dc-magnets comparatively high voltages can arise during emergencies. Here the experience gained in the development of terminals for superconducting cables can be used.

Superconducting magnets store large amounts of energy per unit volume. During a quench of the magnet this energy will be released. To prevent damage of the conductor by over-heating, insulation breakdown or excessive boil-off, protective circuits are connected to the superconducing magnets. Such a circuit is schematically shown in Fig. 17. The field energy is dissipated by an external ohmic shunt and is partially inductively coupled into auxiallieary parts such as the cryostat walls, iron yokes etc. The value of the shunt resistance is determined by the desired decay time or the maximum permissible voltage across the terminals, typically 1000 V. The discharge of the energy into the shunt is initiated by a detection circuit. The voltages across identical coil sections are controlled by this circuit (bridge). If an ohmic voltage occurs across one section because of a normal conductor region, the corresponding signal is used to disconnect the power supply from the magnet by opening a switch and is used to extract the energy into the shunt. Ninety to nearly hundred percent of the stored energy can be extracted from the magnet in this way.

In the case of a fault in the safety circuit for a bath cooled magnet, all the liquid He will eventually evaporate. Safety valves and rupture discs are installed to prevent the damage of the cryostat. In forced cooled magnets, only a limited amount of He is used and the cooling tubes are designed to withstand high pressures ($\geq$ 100 bar). In fully stabilized magnets the total stored energy could be absorbed by the heat capacity of the conductors alone without increasing the temperature above 80 K or so.

G. Superconducting Magnets for Laboratory Application

Laboratory scale superconducting magnets are in general not considered to be among large scale applications. However, they can be considered as belonging to this group because of the large number of them being operated for the past years in many hundreds of laboratories all over the world. The relevant industry is able to deliver complete magnet facilities including all auxiliary equipment such as cryostats, power

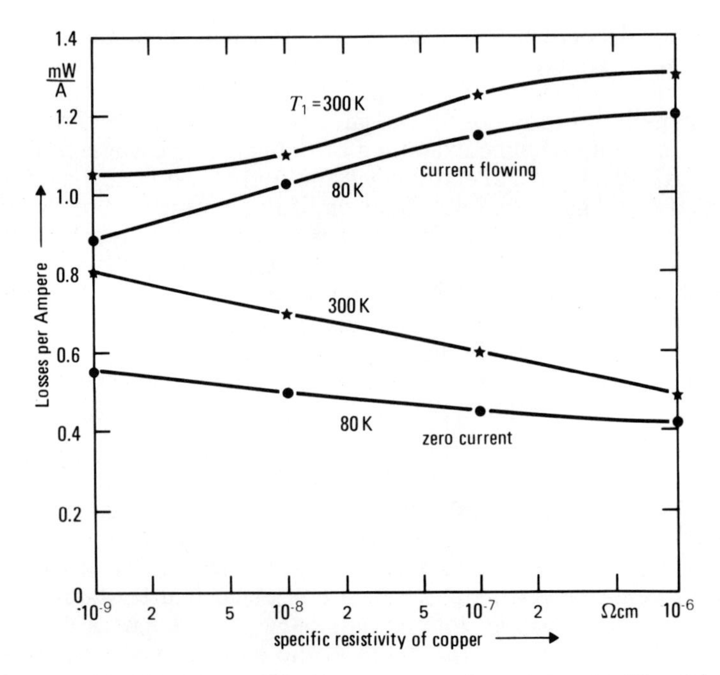

Fig. 16 Optimized heat dissipation as a function of ρ(0) with and without current [42].

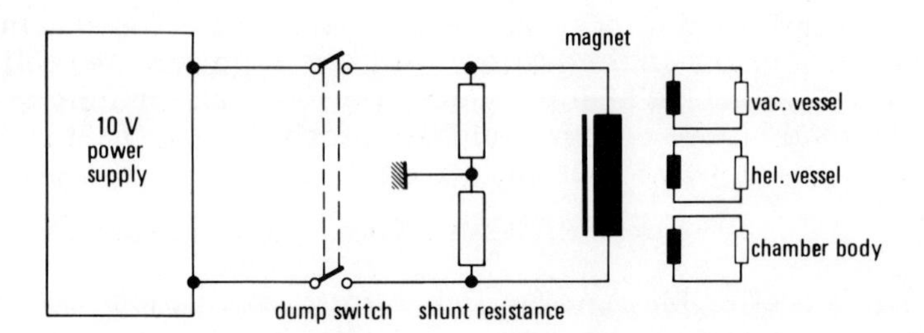

Fig. 17 Coil protective circuits. The superconducting coil is connected to a low value resistance for fast energy extraction. If the coil consists of two or more sections, the voltage difference between sections in case of a quench is used to open the switch connecting the magnet to the power supply [3].

supplies, automatic He-refilling systems and safety systems.

Economical magnets (simple solenoids, split coils, or more complicated arrangements, e.g., for nuclear magnetic resonance experiments) with intrinsically stable NbTi-multifilament conductors are available which generate flux-densities of up to 8.5 T at 4.2 K in volumes of several cubic centimeters. Fields of up to 10.5 T can be achieved with NbTi-coils when operated at 1.8 K, a temperature which can easily be attained by lowering the pressure of a helium-bath.

Magnets for fields higher than 10.5 T must be built with Nb_3Sn or V_3Ga conductors. Until now commercial Nb_3Sn-magnets have been constructed exclusively with tape shaped conductors with thin layers of the superconductor. The highest field reached in a solenoid with a bore of 25 mm ø and a length of 260 mm was 15.8 T at 4.2 K and 16.5 T at 3.0 K [43]. Recently a superconducting magnet (solenoid) installed at the Japanese National Research Institute of Metals has attained a field of 17.5 T. This represents the highest field ever obtained with a superconducting magnet.

The 17.5 T magnet consists of two sections: an outer section constructed with Nb_3Sn-tapes and an inner section of V_3Ga-tapes. The magnet has an outer diameter of 40 cm, a height of 65 cm and an inner clear bore of 3 cm. Design and fabrication of the coil was a joint venture between the US company Intermagnetics General and the Japanese Company Vacuum Metallurgical Corporation [44].

Superconducting magnets with tape-shaped conductors mainly suffer from the following disadvantages:

Large hysteresis losses and instabilities in magnet end parts. Essentially the field components perpendicular to the tape width results in limited field charging or changing speeds.

Comparatively high residual field amplitudes. This is due to the large persistent shielding currents in tapes when the widths are exposed to perpendicular field components.

The above mentioned disadvantages initiated the development of A-15 multifilament conductors. The first test coils of laboratory scale size have been constructed and successfully tested. There are two possible ways to build a coil with A-15 multifilament conductors:

The first uses an unreacted conductor, i.e., a conductor consisting of a bronze matrix plus pure Nb filaments or pure V-filaments which is wound into a coil. The system is then heat-treated and reacted for several hours at about 700 °C. This coil fabrication technique puts special requirements on the auxiliary materials (e.g., electrical insulation).

The second uses a prereacted conductor, i.e., in which the diffusion process and the creation of the A-15 layers around the Nb- or V-filaments have already been prepared before winding the coil. In this case similar winding techniques as with NbTi-multifilament conductors can be used, but with additional precautions due to the great brittleness of the A-15 layers.

For larger coils the use of prereacted conductors is preferred since heat treatment of large structures at temperatures of about 700 °C also lead to mechanical problems, besides the insulation problems already mentioned.

In what follows we report the test results of two selected coils of laboratory scale size built with Nb_3Sn-multifilament conductors. These types of coils are very interesting with respect to magnets for laboratory applications. They were developed within programs aiming for large magnets with A-15 multifilament conductors, e.g., for large fusion magnets or for field windings of large turbo-generators.

The first magnet was built by Siemens AG with prereacted rectangular Nb_3Sn conductors which were stabilized additionally with copper. It was operated in the background field of a NbTi multifilament coil and attained 13 T in a free bore of 55 mm without any current degradation. The cooresponding stored energy of the total magnet system was about 270 kJ. With a further Nb_3Sn multifilament conductor coil insert a flux density of 14.5 T was reached in a free bore of 11.5 mm [45]. The conductors used in this experiment were fabricated by Vacuumschmelze Hanau.

The second coil was built by the Lawrence Livermore Lab with prereacted, rectangular internally copper stabilized Nb_3Sn-multifilamentary conductors, manufactured by Airco. This solenoid, with a comparatively large inner diameter of 27 cm, attained a maximum field of 10 T, when operated in the 5 T background field generated by a surrounding NbTi-coil. Maximum field and current of the coil correspond to the short sample value of the conductor used [46]. The magnet was built and investigated within a program for the development of large fusion magnets.

With present commercially available A-15 superconductors, the economically attainable upper field limit of superconducting magnets lies in the neighborhood of about 18 T. It is anticipated that this limit will soon be approached by laboratory scale coils wound from available A-15-multifilament conductors.

To attain even higher fields with reasonable power consumption combinations of normal conducting and superconducting wire are used in so-called hybrid magnet systems. Hybrid magnets consist of a large bore superconducting magnet enclosing high current density "resistive" magnets. A number of hybrid magnets have been constructed and more or less successfully operated. Some examples of hybrid magnets are:

McGill University, Montreal. The magnet consisted of three sections; an outer section of NbTi multifilament conductor, an intermediate section coil of Nb_3Sn-tape and an inner section cryoresistive coil of high purity Al-tape cooled at 8 - 19 K. The magnet was planned to generate 25 T (10 T by the resistive coil added to 15 T by the superconducting coils) [3]. It did not reach its design value because of force problems.

MIT, Cambridge USA. A magnet was constructed which consists of two watercooled copper coils surrounded by a superconducting coil built from NbTi multifilament wire. The system was designed to generate 22 T in a bore of 3.8 cm ∅ (6 T of field was contributed by the superconducting coil) with a power consumption of 4.4 MW for the water cooled coils [3]. The system suffered from degradation of the NbTi-coil. An improved 25 T system is under construction for the University of Nijmegen, Netherlands.

Kurchatov-Institute, Moscow. The magnet which was jointly developed with the Efremov Scientific Research Institute of Electrophysical Apparatus, Leningrad, consists of two coaxially arranged watercooled normal resistive coils (chromium bronze and hard copper conductor) which are enclosed by a superconducting magnet built of multifilamentary electroplated strips of NbTi and NbZr. The magnet was successfully tested during 1974 and generated a field of 25 T (the normal conducting coil produced 18.4 T and the superconducting coil produced 6.6 T) [47].

H. Magnets for High Energy Physics

High energy physics is the area where the most powerful superconducting magnets have been applied until now. For a long time its needs lead to the advancement of the technology of applied superconductivity for energy technology and other industrial applications. The reasons for the use of superconducting magnets are mainly economic and new experiments.

The power required for large superconducting magnets or magnet arrangements is at least an order of magnitude smaller than for conventional systems, leading to much smaller operation costs. The large European bubble chamber magnet provided with conventional coils would need a power of 70 MW. The superconducting version consumes less than 1 MW. The possibility of higher magnetic fields leads to a drastic reduction in the size of magnet facilities and the necessary buildings.

New kinds of investigations have required advanced superconducting magnets, e.g., experiments with shortlived particles. Because of the necessity of short distances between particle source and analyzer, superconducting magnets with a high bending or focusing power and simultaneously short length were required.

A series of large superconducting particle detector magnets are in successful operation at different laboratories around the world. Table 1 gives the main parameters of some important devices. The magnets are fully stabilized and, with the exception of the Omega magnet, bath cooled. The Omega-magnet is an iron magnet excited by two superconducting coils constructed with hollow conductors cooled by forced flowing supercritical He. The CERN bubble chamber magnet (Fig. 18) with a stored energy of almost 1 GJ is the largest existing magnet. The DESY-

Table 1 Parameters of big particle detector magnets in operation.

	Brookhaven Nat. Lab.	Argonne Nat. Lab.	Nat. Acc. Lab.	CERN	CERN Omega	DESY Colliding Beam
	Bubble chamber				Spark chamber	
Magnetic field (Tesla)	2.8	1.8	3.0	3.5	1.8	2.1
Current KA	5.6	2	5.0	5.7	5.0	1.2
Inside/outside diameters of winding (m)	2.44/2.76	4.78/5.28	4.27/5.08	4.72/6.0	3.58/4.90	1.5/1.61 1.73
Length of winding (m) (including gap)	2.4	3.04	2.89	4.4	4.8	0.9
Stored energy MJ	64	80	400	830	50	4.5

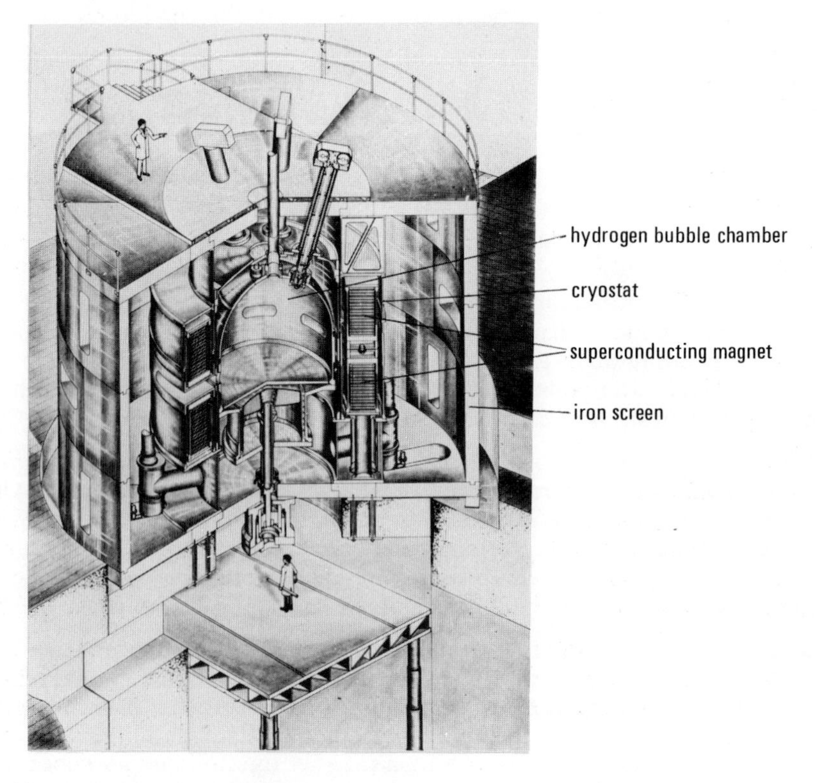

Fig. 18 Large European hydrogen bubble chamber with superconducting magnet (CERN).

magnet, a detector magnet for an electron colliding beam experiment, was constructed by Siemens. These large magnets represent an important step in the direction of even larger and more complex magnets which will be required for plasma research facilities.

With respect to colliding beam experiments there is growing interest in a new type of detector magnet. Large solenoids with diameters greater than 1 meter, lengths of some meters and flux-densities of about 1.5 T are required for detecting and analysing charged particles which are created from the collision of two particle beams. To be able to investigate particles outside the magnet, magnets with an extremely thin winding thickness are necessary (thinner than one radiation length) to allow the particles to pass through the magnet wall. To achieve a very thin solenoid the current density in the winding must be very high i.e., about 10^9 A/m^2 .

Two prototype magnets with half a meter length and one meter diameter have already been constructed and successfully tested at current densities in excess of 10^9 A/m^2 [48]. The coils consist of two layers of intrinsically stable NbTi-multifilament wires (copper to superconductor ratio: 1.8:1 and 1:1) wound on a cylindrical Al-former. The windings are cooled with forced circulated, two-phase He flowing through Al tubes which were directly wound on the superconducting layer. The windings, Al-tubes and Al coil former are cast in epoxy to form a single, unified magnet and cryogenic system with a thickness of about 2 cm, corresponding to a radiation length of about 1/4.

A number of superconducting dc beam line magnets have already been in operation for many thousands of hours in various high-energy physics laboratories. They demonstrate the economics and reliability of these types of magnets. Only a few examples are described in the following sections.

During 1973, a bending magnet which provides a deflection of 8°, was installed in the beam line to the large European bubble chamber BEBC at CERN. The magnet which is of the intersecting ellipse type is 2.4 m long and has a bending power of 5.6 Tm [49].

At BNL (Brookhaven National Laboratory) an 8° bending magnet system, installed in an extracted proton beam from the AGS to the 2.1 m bubble chamber, has been in operation since the end of 1973. It consists of two identical magnets (B_{max} = 4 T, magnetic length: 4 m) with an iron yoke and a rectangular aperture with a separate sextupole correcting coil. The two magnets are identical to better than 10^{-4} over the useful aperture and over the entire field range [50].

During the summer of 1976 four dipole magnets will be installed in a beam line which must supply particles of momentum up to 30 GeV/c from the AGS (Alternating Gradient Synchrotron) to a large volume multiparticle spectrometer. This beam transport system will be the most advanced realized to date. The four magnets (with approximated cos θ current distribution) are each 2.5 m long and produce a flux density of

Fig. 19 Beam guiding dc dipole magnet.
Magnet bore 250 mm. Magnet length 2.5 m.
Operating field 4 T.
(Photo: Brookhaven National Laboratory).

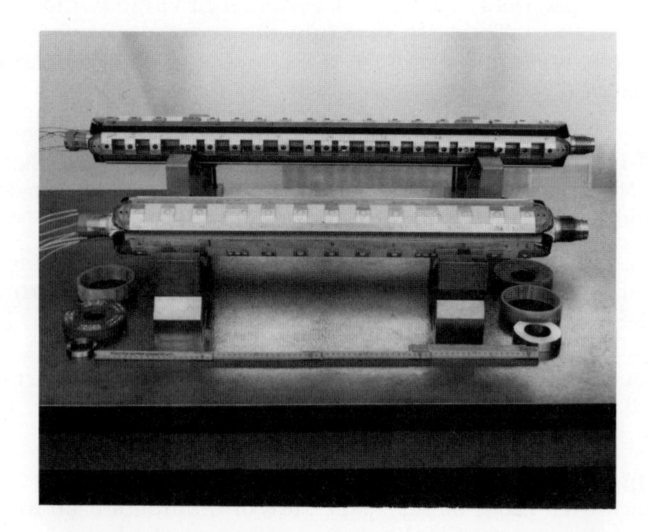

Fig. 20 Superconducting windings for two dc quadrupoles in the CERN-SPS experimental area manufactured by Gesellschaft für Kernforschung, Karlsruhe.

4 T in a warm bore of 20 cm. All four magnets have been constructed and the first one was successfully tested [51]. Figure 19 shows one of these magnets during installation in its cryostat. A dc quadrupole doublet is installed at Saclay in a pion beam line of the Saturne Synchrotron. The quadrupoles have identical lengths of 0.68 m and large apertures of 26 cm and 36 cm. Peak focusing gradients are 35 T/m.

A similar project is under way at IEKP (Institut für Experimentelle Kernphysik) Karlsruhe. Recently a quadrupole doublet for focusing a secondary hyperon beam at CERN was completed and successfully tested. The magnets have a length of 1.4 m and 1.1 m respectively and field gradients of 156 T/m in a cold aperture of 36 mm [52]. Figure 20 shows the windings of the two quadrupoles.

Finally, with respect to dc beam magnets some other interesting developments should be briefly mentioned. First is the trapping of fields, e.g., dipole fields in assemblies of superconducting cylinders. The desired magnetic field configuration is applied when the cylinders are normal. The cylinders are then cooled to their superconducting state and the external field is switched off, creating permanent currents in the cylinders which maintain the original field pattern. This method of producing magnetic field configurations offers a way of saving investment costs for dc beam magnets.

For saving energy, in times of expensive power, the replacement of resistive copper coils in iron magnets by superconducting windings will play a major role in the future. Such an energy saving magnet with a flux density of 1.5 T is under construction at CEN/Saclay [50]. Another interesting beam line magnet assembly has already been mentioned when discussing various cooling methods of magnets, the SIN (Schweizer Institut für Nuklearforschung, Villigen) superconducting muon channel solenoid [34].

Besides dc magnets, pulsed magnets are of great interest for application in proton synchrotron accelerators, in accelerating storage rings or as energy storage systems. At the beginning of the seventies a number or projects were started in several high energy physics laboratories to develop pulsed magnets for accelerators. The goal of these programs was the realization of proton accelerators with energies in the neighborhood of 1000 GeV, about 2 to 3 times the envisaged energy of accelerators with conventional magnets under construction at that time at CERN Geneva and NAL, Batavia USA. Because of the high magnetic fields attainable with superconducting magnets, the size (diameter) of these super-accelerators would not have to be greater than that of existing conventional accelerator magnet rings (300 GeV CERN; 400 GeV NAL).

In Europe, three high energy physics laboratories (Rutherford Lab. Didcot England, CEN/Saclay France and IEKP Karlsruhe Germany) have had close collaboration (GESSS) during the past years, aimed at the development of superconducting synchrotron dipole magnets with high field homogeneity and low losses at pulsed mode operation. This work was directed towards the possible construction of a large superconducting proton accelerator at CERN. It was performed in close contact with the corresponding national industries. At present, this work has been reduced since the superconducting accelerator at CERN has no chance of being realized in the near future.

However, a series of pulsed field magnets has been constructed at these laboratories, which fulfill accelerator requirements. The latest and most advanced dipole magnets existing at present are: AC 5 (Rutherford) [53], ALEC (CEN/Saclay) [54] and D 2a (IEKP) [27]. These magnets have several design features in common, such as cold iron yokes or epoxy resin impregnated coils. Other important characteristics, such as the type of transposed conductor, the current distribution, and the cooling and mechanical support are different. All magnets show a field accuracy of 10^{-3} or better, reach peak fields between 4.5 T and 5.5 T and have overall losses less than 20 W per meter magnet length at field rise times of several seconds. Some of these magnets have already been pulsed for several times 10^5 without showing field degradation or fatigue. Table 2 gives the main parameters of the three GESSS magnets.

At the present time the Fermi-Lab-Energy Doubler, which would be a full scale accelerator, has great chances for realization. It is planned to install a second ring with superconducting magnets in the tunnel of the existing conventional 400 GeV-machine, thus approaching an energy of about 1000 GeV. This ring would require some 744 dipoles (flux density 4.5 T) 6 meters long, and 240 quadrupoles of varying length from 1.5 to 2 meters. Within the Energy Doubler test program, a series of 6 m long dipole magnets and some quadrupoles have been built and tested [55] with a design field of 4.5 T in a cold elliptical bore. Warm iron and special counter flow cryostats (subcooled He at the winding, surrounded by two phase He) are special design-features of these dipoles.

At the Lawrence Berkeley Laboratory a small experimental superconducting accelerator ring (ESCAR) is under construction and is planned for first operation during 1977. Its maximum energy will be 4.2 GeV (at a pulse rate of 6/min) and it will consist of 24 dipoles of 1 m length and 32 quadrupoles [56]. At the Argonne National Laboratory a superconducting storage ring (96 dipoles, 64 quadrupoles) is planned in combination with the existing 12.5 GeV Zero Gradient Synchrotron [57].

Nearly every high energy accelerator center is considering a superconducting colliding beam facility. Colliding beam machines can generate high center-of-mass energies with relatively modest laboratory energies, corresponding to primary particle energies which can never be achieved with single beam accelerators. Due to the relatively long acceleration periods (e.g., 2 minutes) associated with proton machines, these are particularly attractive candidates for superconducting magnets.

At the Brookhaven National Laboratory a 2 x 200 GeV intersecting proton storage accelerator (ISA or ISABELLE) is planned which would be equipped with superconducting magnets. Its design is based on separate superconducting bending and focusing magnets distributed around

Table 2. Main features of pulsed dipoles built and tested by GESSS.

Laboratory	RHEL	CEN – Saclay	IEKP Karlsruhe
Dipole Designation	AC 5	ALEC	D2 a
Central field design value (Tesla)	4.5	5	4.5
Magnet Parameters			
Cold bore diameter (m)	0.100	0.110	0.08
Outside Coil diameter (m)	0.181	0.216	0.152
Coil length (overall) (m)	~1.000	~1.500	0.172
Iron bore diameter (m)	–	0.280	0.172
Iron outside diameter (m)	0.560	0.600	0.38 × 0.3
Iron length (m)	1.000	1.800	1.46
Current density in winding at 5 Tesla $A\,cm^{-2}$	1.55×10^4 (3660 A)	$\approx 1.3 \times 10^4$ (2000 A)	2.1×10^4 (1576 A)
Stored Energy Ws	1.1×10^5	3.4×10^5	1.7×10^5
Concuctor			
Type	cable	cable	cable
Number of strands	15	42	12
Number of filaments × fil.size	8917 × 7.1 μm	1045 × 10 μm	1000 × 12 μm
Matrix	Cu/Cu Ni	Cu	Cu
Strand insulation	Copper oxyd	Copper oxyd	In/Sn
Twist pitch (mm)	—	6	4
Transposition pitch (mm)	—	—	35
Performance			
Max. Quench Field (Tesla)	5.2	>5	5.3
Losses at design field 0.1 Hz (Joule/cycle)	120	>400	500
Effective magnetic length (m)	0.78	1.50	1.24
Homogenity of field — design value		$\leq 4 \times 10^{-4}$ 70 % aperture	
Homogenity of field — experim.value	10^{-3} for $d<80$ mm		3×10^{-3} for $d<60$ mm

two ring structures, one above the other in the same cryostat and each 2700 meters long. The design provides the use of 4.25 m long 4 T dipole magnets and 1.3 m long quadrupoles. Four full sized prototype dipole magnets have been built so far and two of them successfully tested. Figure 21 shows one of these prototypes without cryostat [51]. For operation at higher temperatures and even higher magnetic fields, the first one meter dipole was wound with prereacted Nb_3Sn multifilament conductor and successfully tested. All other magnets described above were built with NbTi-multifilament conductors.

It should be pointed out once again that the projects briefly described above are only a selection of various existing projects which could not all be mentioned because of lack of space. Finally a striking point should also be mentioned, namely that the majority of recent superconducting magnet projects in high energy physics are carried out by the big laboratories themselves. Industry mostly participates as the conductor manufacturer. In the future, care should be taken so that research laboratories do not themselves build thousands of meters of superconducting magnet length necessary for accelerators.

I. Superconducting Magnets for Fusion Reactors and MHD Generators

1. Fusion reactors

The growing future energy demands, and the continual depletion of conventional energy resources, require new sources of energy. Besides solar energy, which is only adequate for special geographical regions, only nuclear energy, fission (fast breeder reactors) and fusion can meet the long term needs. It is the present understanding that both sources are complementary rather than alternatives. However, fusion reactors are said to offer some long term advantages, e.g., lower radioactivity than with fission and uncontrolled nuclear runaway cannot occur.

Two ways of achieving thermonuclear fusion are being investigated at present, laser induced pellet implosion and magnetic confinement of the plasma. Only the latter is treated here. Three possible magnetic confinement configurations are considered, magnetic mirror, theta pinch and tokamak. Figure 22 shows the principle coil configurations of these confinement types. Early studies of different fusion reactor types have already shown, that, with the exception of the theta pinch concept the magnetic fields of all magnetic confinement configurations have to be generated by superconducting magnets. Conventional magnet systems would need a power consumption which is about equivalent to the electrical power output of the reactor [40]. These studies and new conceptual designs of power reactors have shown, that fusion power reactors with an electrical output of about 2000 MW (the smallest economic plant size) require superconducting magnet coils with about 20 m bore diameter [58]. The stored energy of the whole magnet system exceeds 200 GJ, which is more than two orders of magnitude greater than the stored energy of existing superconducting magnet systems (BEBC-magnet CERN stores about 1 GJ).

Fig. 21 Prototype dipole magnet for the Intersecting Storage Accelerator "ISABELLE".
Magnet bore 120 mm. Magnet length 4.15 m.
Operating field 4 T.
(Photo: Brookhaven National Laboratory).

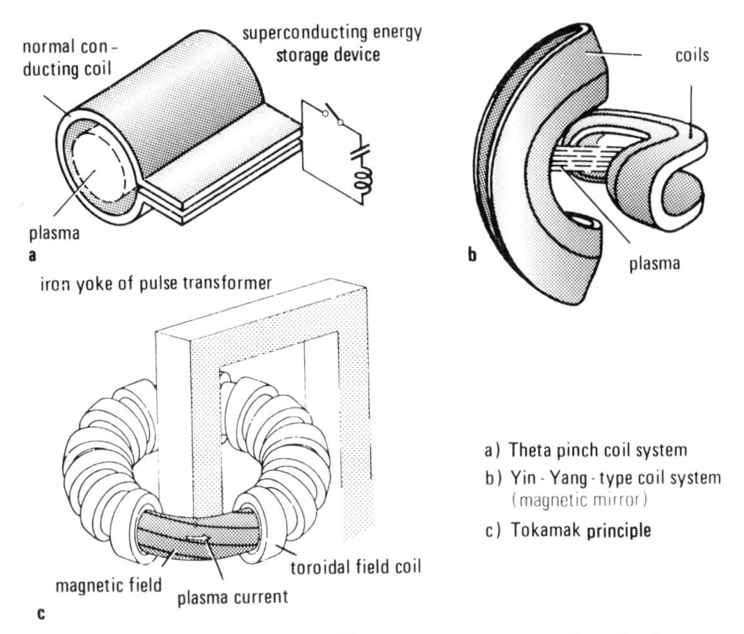

Fig. 22 Magnetic confinement systems for fusion.

Figures 23 a and b are artists' views of a tokamak and mirror fusion reactor and give an impression of the geometrical dimensions of such reactors [59].

Today's experimental plasma reactor facilities are equipped with pulsed water cooled copper coils. This will remain true for plasma facilities which will be constructed in the near future until about 1985. However, fusion experiments with experimental power reactors, planned for the second half of the eighties and in the nineties require superconducting magnet systems for economic reasons. The first operation of a demonstration commerical power plant with a thermal power of about 5000 MW is planned in the US for the end of this century.

The technology necessary for the large magnet systems of fusion power reactors is in principle available. But a series of engineering problems associated with the large size and complex configurations of these magnet systems have to be solved. The problems include:

Tremendous forces and bending moments act on the coil systems, both in the steady state and in the dynamic state.

The coil must be designed with sufficient access to the blanket.

Techniques for on-site fabrication must be developed.

Maintenance after most of the machine has become radioactive must be possible.

High current carrying conductors also for pulsed operation must be developed at low costs.

Power supplies for large pulsed coils which need peak instantaneous powers of several GVA must be developed.

Storage and transfer of the large energies with high electrical efficiency must be solved.

Therefore it is absolutely necessary to start extensive development programs in order to solve these problems and to be able to provide large reliable superconducting magnet systems. Otherwise there will be the danger of a mismatch between supply and demand.

What has been done until now and what are the future programs needed in order to keep pace with the rest of the fusion program? Two "open" confinement systems with superconducting mirror coils have been realized and tested some years ago. The baseball II coil has a stored energy of 10 MJ in a neutral beam injection experiment and was tested at the Lawrence Livermore Laboratory [60]. The mirror quadrupole coils in the IMP (Injection Mirror Plasma) experiment has been tested at ORNL (Oak Ridge National Laboratory) and has a total stored energy of 2.4 MJ [61].

With respect to closed confinement systems, outstanding experiments were made with levitated superconducting rings buried within the plasma [62,63]. These experiments were of principal interest for study-

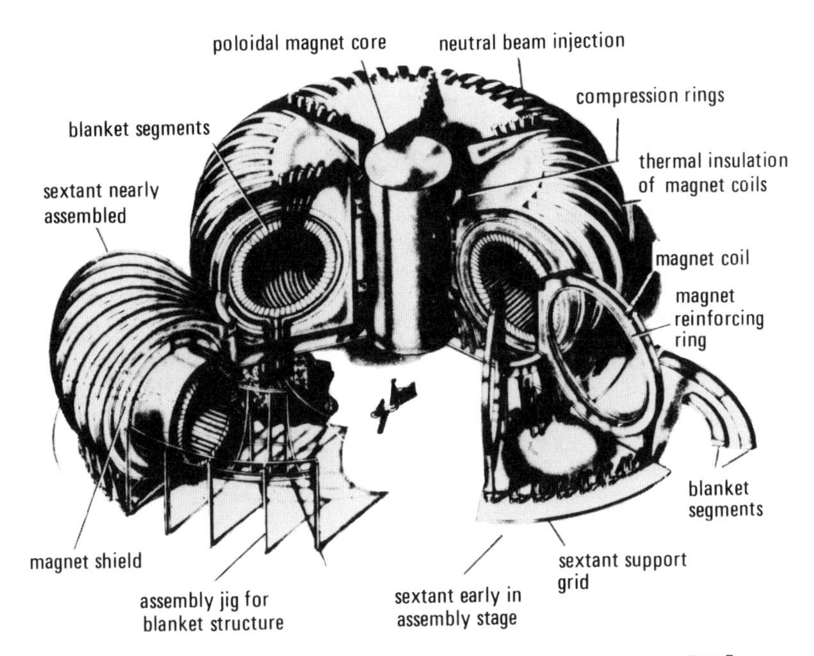

Fig. 23a ORNL tokamak fusion reactor magnet [59].

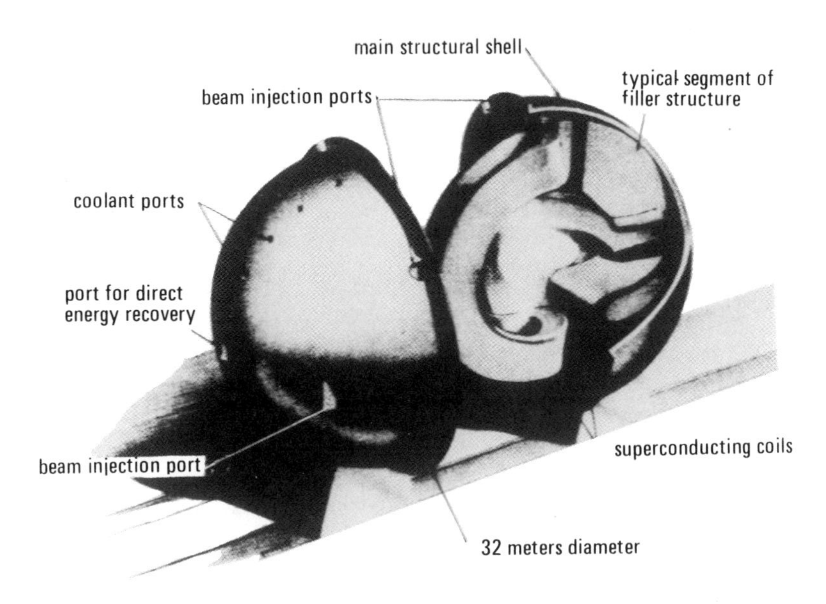

Fig. 23b LLL Mirror fusion reactor magnet [59].

ing magnetic confinement of plasmas but are of less importance with respect to confinement systems of actual reactors.

Presently the closed Tokamak confinement system is considered to be the most promising one. Figure 24 shows the different coil systems required for confinement and operation of the plasma in a Tokamak reactor. A steady helical magnetic field (B_{res}) is produced by the superposition of the fields of the toroidal field (B_T) coils and of the plasma current (B_p). The latter is induced by a flux change parallel to the axis of the toroid and is generated by pulsing transformer coils (Tr). For compensating the field asymmetry due to the bending of the plasma axis vertical field coils V are required. To protect the plasma from an influx of impurities (erosion from the wall of the plasma tube) divertor coils D are needed. While the main toroidal coils are dc operated, all others are operated in pulsed mode (field rise about 1 T/s) and will probably have to be superconducting. The divertor coils are a possible exception.

A superconducting toroidal system is the NASA Lewis bumby torus magnet facility. Twelve superconducting coils with inner diameters of 22 cm form a toroid with a major radius of 0.76 m and generate a toroidal field of 3 T [64]. A prototype disc coil for a superconducting toroid with a warm bore of 1 m, a maximum field of 6.4 T and a stored energy of 1 MJ was built by Siemens, Germany for IPP Garching [65].

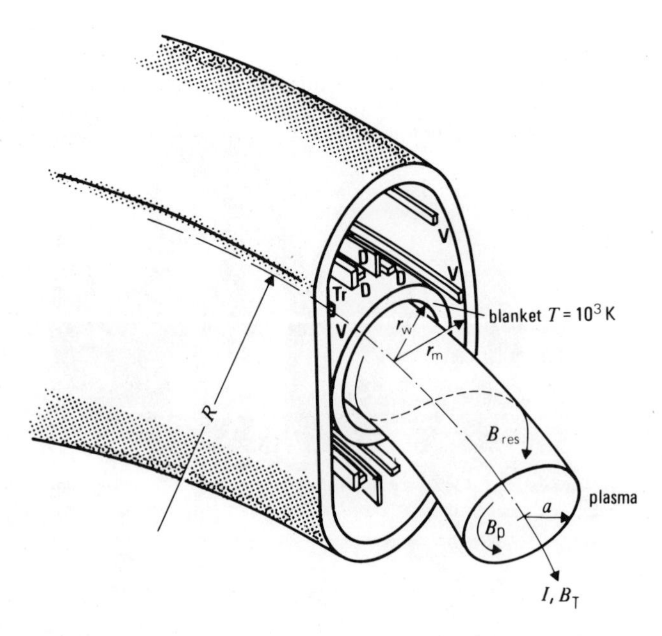

Fig. 24 Confinement of the plasma in a Tokamak reactor and arrangement of the magnet coils [79].

An outstanding superconducting confinement system will be completed during 1976. It is the toroid for the tokamak T-7 experiment at the Kurchatov Institute, Moscow. This magnet system consists of 8 sections, each section containing six double-pancake coils. The coils are wound with a rectangular fully stabilized NbTi-Cu conductor which contains cooling ducts through which supercritical He will be circulated. The coils have an inner diameter of about 0.8 m and are arranged at a toroid major diameter of 2.5 m. The coils are embedded in a structure of Al-alloy for mechanical support. They will produce a field on the toroid axis of 3 T and will store an energy of about 20 MJ. First plasma experiments are foreseen for 1977 [66].

Development on the third possible confinement system, the θ-pinch, was also carried out at several laboratories including Los Alamos, US, Efremov Institute (Leningrad) and IEKP (Karlsruhe) especially investigating energy storage by superconducting coils and superconducting homopolar generators. Within this research, superconducting power switches were developed which are able to handle power pulses on the order of 100 MW at currents of larger than 1 kA [67,68,69].

At the present time the US apparently has the most advanced and detailed fusion program, including major sized superconducting test facilities. This program is sponsored by ERDA (Energy Research and Development Administration) and was started in 1974. The tasks have been distributed among national laboratories and large research universities such as Princeton and MIT. Conceptual design contracts for full-size Experimental Power Reactors (EPR) have been placed with several industrial firms. The program is planned up to the year 2000 and includes two experimental power reactors with thermal powers of 100 MW and 500 MW (1985 - 1997) and a demonstration commercial power plant with a thermal power of 5000 MW in about the year 2000. The EPRs and the DEMO will be equipped with large superconducting magnet systems [70,71].

In the US tokamak and mirror reactors are considered as the most promising candidates for the successful realization of commerical fusion reactors. Therefore the development of this reactor types is favored. Table 3 and Fig. 25 illustrate the present trend of the development of large superconducting magnet arrangements for both reactor types. Before starting the construction of the first Experimental Power Reactor, the feasibility of the D-T burn must have been demonstrated. This will be the goal of the Princeton Toroidal Fusion Test Reactor (TFTR) which uses conventional watercooled pulsed copper magnets and will be in operation between 1980 and 1985.

To prove the feasibility and to provide the technology of such large superconducting magnet systems, two subsize coil systems are to be developed by 1981 or 1982 respectively. The first toroid with six NbTi coils of about 3 m diameter, will be a scaled-down simulation of the magneto-mechanical environment of a Tokamak Experimental Power Reactor. Responsibility for this development is with ORNL. The second magnet

Table 3. US-Tokamak Fusion Power Program (1976) including superconducting magnets (ERDA-DCTR).

Name of Device	ORNL Large Coil Project	Experimental Power Reactor No. 1 EPR # 1	Experimental Power Reactor No. 2 EPR # 2	Princeton Reference Design Model DEMO
Years of Operation	Starting 1980	1985-1989	1989-1997	1997 +
Power Level	–	100 MW_{th}	500 MW_{th}	5000 MW_{th}
Purpose	Simulate the magneto-mechanical environment of an EPR. Demonstrate performance of s.c. magnets	Engineering Power Producing Experiment	Demonstrate Utility Operation	Demonstration of a Commercial Power Plant
Diameter of S.C. Toroidal Field Coils	3m	≈ 7m	≈ 11m	≈ 22m
Plant Cost	28 M$	500 M$?	?	>1500 M$

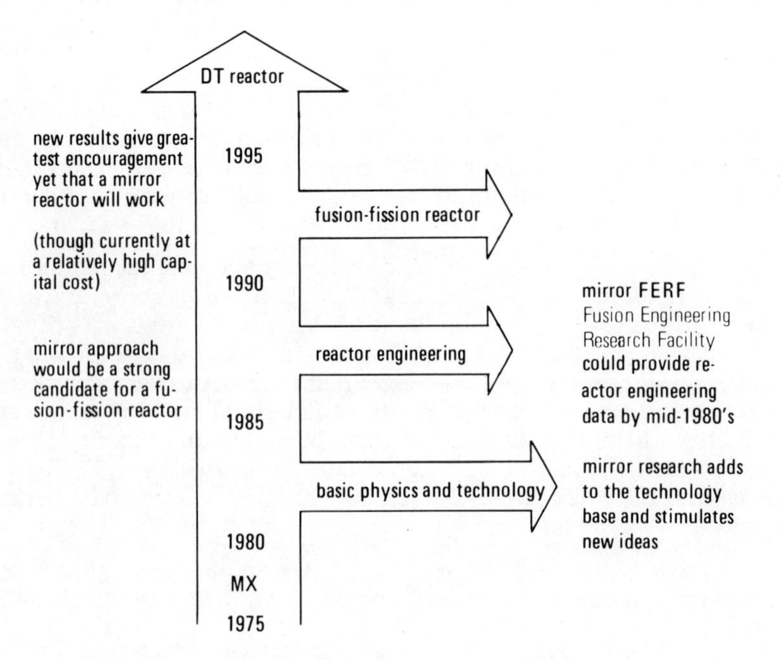

Fig. 25 Development tendencies for Mirror-Type-Reactors including superconducting coils (Lawrence Livermore Laboratory).

which is as yet only in the proposal stage, is a mirror coil system, called MX. It will incorporate two large Yin-Yang NbTi coils having a peak field of 7.5 T and a stored energy of 500 MJ. If approved, this project will be carried out by the Lawrence Livermore Laboratory (LLL), which is also responsible for the development of Nb_3Sn filamentary conductors for fusion applications, e.g., for magnet systems with flux densities > 8 T.

At Los Alamos Scientific Labs (LASL) a program for the development of a magnetic energy transfer system is being carried out with respect to theta-pinch machines. The corresponding Fusion Test Reactor (FTR) is based on a stored energy of about 400 MJ divided into 800 units each of 500 KJ. They are assembled in 100 modular-plastic cryostats, each containing 8 storage coils with their switches. LASL intends to build one module till 1980 but containing only three 500 KJ coils. Another object of their investigations is the use of superconducting homopolar generators for energy storage. Since theta-pinch plasma physics is not looking too promising it is uncertain whether the program will be continued in the originally planned form.

In Europe a similar but less ambitious program is under discussion with Euratom. The aim of these discussions is the realization of a toroidal demonstration magnet system within the next 8 years. At the present time, the kind of demonstration magnet system has not yet been decided. It might be a complete toroid of coils with about 3 m bore each, plus pulsed poloidal windings, or sections of medium sized coils (6 m ∅), or a single large coil.

In Germany, a collaborative program exists between IEKP-Karlsruhe and IPP Garching. With respect to the technology of superconducting fusion magnets, a toroidal apparatus consisting of six D-shaped coils with diameters of about 0.5 m and a stored energy of 6 MJ will be built at IEKP during the next few years. The very advanced conductor for this toroid was developed by Vacuumschmelze Hanau. With this experimental apparatus (called TESPE) the main probelms for developing a superconducting toroidal magnet system will be investigated on a small scale [72]. The work on energy storage will also be continued at IEKP-Karlsruhe.

The investigations in Japan are restricted to toroidal confinement systems. Because of the outstanding importance of fusion, the Japan Atomic Energy Commission decided in 1975 to promote fusion research as a National Project till 1980. The main objective of this program is the construction of JT-60, a large Tokamak system, but with conventional pulsed coils. The program for superconducting magnet systems for fusion application has not finally been decided but the first budget has been approved for FY 1976. The longterm Japanese fusion technology program provides for an EPR in 1990 with a thermal output of 100 MW and a Demonstration Commercial Power Plant in 2000 with a thermal output of 2000 MW [73]. For both plants, superconducting magnet systems are indispensible.

Economic evaluations show that the capital costs of large fusion power plants are comparable with those of present modern nuclear plants. Figure 26 compares the capital costs per kilowatt of a tokamak reactor plant design (Princeton Fusion Power Plant with an electrical power rating of 2030 MW) with those of all nuclear power plants operating in the US prior to 1975. Based on 1974 dollars and including an interest rate of 8% over a 3 year construction period the capital costs are about 730 $/kW [74]. The estimated cost of energy at the busbar for the fusion reactor is about 0.018 $/kWh and is favorable when compared to present day nuclear plants. The fuel cost for a fusion reactor is small when compared to all other fuels. Since it is to be expected that the costs of fossil and nuclear fuels will increase at rates higher than the normal inflation rate, the energy cost at the busbar will become even more attractive for fusion reactors.

2. MHD generators

The main advantage of MHD power plants (combined steam and MHD electrical generating plants) is the expected high overall efficiency (50-60%) which makes MHD especially attractive for energy conservation. These high efficiencies can only be achieved when superconducting magnets are used for the generation of the magnetic field of the MHD-part. Studies have proved that superconducting magnets are the only economic solution of MHD power plants, independent of whether they are planned as peak or as base load plants [75]. Three possible loop systems are possible: liquid metal, nobel gas closed cycle and combustion gas open-cycle MHD generators. Most effort was and is concentrated on the last one.

The MHD channel cross-section of this type of generator increases towards the exit end and requires a larger warm aperture of the magnet at this site. Tapered dipole magnets with saddle shaped coils are most suitable to meet the field requirements of the hot plasma. Figure 27 shows schematically the shape of such a MHD magnet and its typical field distribution along the channel axis together with the principle of the generator channel [76]. Conceptual studies have shown that the peak channel flux-density should be near 6 T [77]. Therefore NbTi in the form of cryostatically (fully) stabilized multifilament conductors has been suggested for most magnet designs. The application of Nb_3Sn for operation at higher temperatures or fields when it will be available at a reasonable price were also considered.

The interest in MHD generators in the western world had its ups and downs. At the beginning of the seventies MHD research was reduced almost to zero. This was not a questioning of MHD superconducting magnets. It was evident in principle, that with the appropriate engineering effort even very large dipole magnets necessary for large base load MHD plants could be realized. There exist other MHD problems which were assumed to be unsolvable, e.g., the corrosion of channel wall materials for electrodes and insulators. On the contrary, large activities in the development of MHD-generators could be observed in Japan and USSR, including the development of corresponding superconducting magnets. In

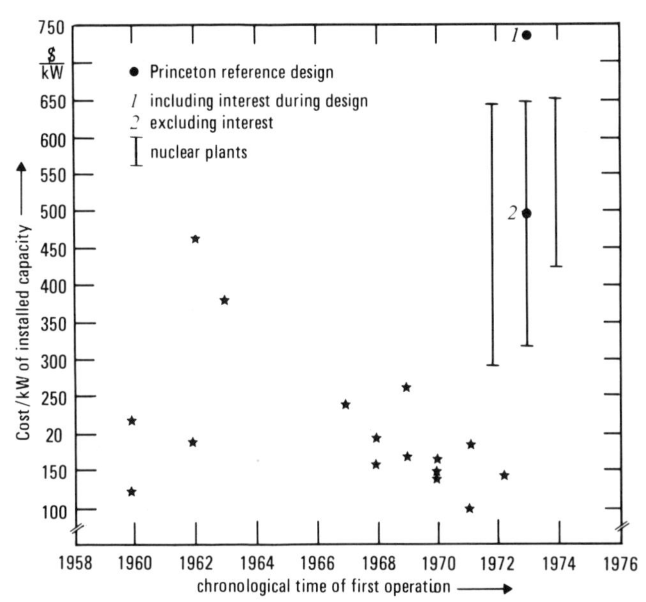

Fig. 26 Cost per kilowatt of installed capacity for all commercial nuclear plants operating in the United States prior to 1975 and for the Princeton Fusion Power Plant Design [74].

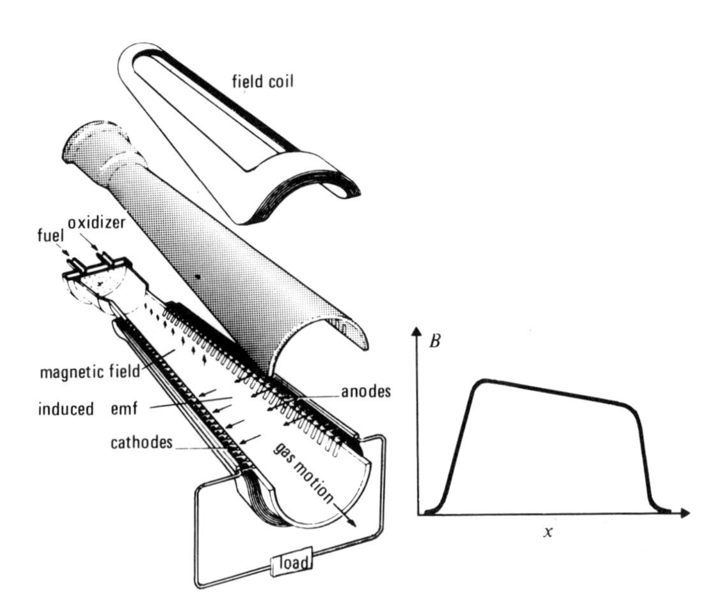

Fig. 27 General shape of superconducting magnet and axial field distribution for a MHD generator [76].

Japan this development was embedded in the largest national program of MDH energy conversion which was finished in 1975. It led to the construction of the largest superconducting dipole magnet which has been operated to date [78]. Its main data are given in Table 4, which is a compilation of superconducting MHD magnets constructed so far [79].

The "energy crisis" and new discoveries with respect to the lifetime channel wall materials have stimulated new interest in MHD-power generation, especially in the US, where a national program for MHD development has been approved. This program is planned to lead to a 500 MW_{el} base load demonstration MHD-system until the end of the eighties. Investigations on open cycle systems with coal combustion are especially emphasized, because it is hoped that this system will be economically most competitive. In addition the US has recently made an agreement with the USSR on a joint MHD development program. As part of this agreement, the Argonne National Lab is responsible for designing and constructing a superconducting dipole magnet for use in the second loop of an existing conventional 20 MW-MHD-facility (called U 25) at the High Temperature Institute, Academy of Sciences, Moscow. The magnet system, to be delivered to Moscow, is expected for early 1977, and will include all important components; magnet, cryostat, liquefier, power supply, He transfer and low pressure storage system and control system. As a consequence of supplying the magnet system, the US MHD-research group, including MIT, will be able to test its MHD channel designs in the U25-facility. No equivalent facility is presently available in the US. The main parameters of the magnet are also given in Table 4. The saddle shaped dipole, which will yield a tapered on-axis field is wound of rectangular, cryostatically stabilized NbTi-Cu multifilament conductors, which are cooled by a liquid He-bath. No magnetic iron is provided [80].

Although much progress with respect to channel materials, air preheater construction and seed recovery systems has been made, the future of commercial MHD-generators is not without doubt. Most optimism comes from the USSR. They plan to build a commercial base load plant in the early 1980's at the HTI Moscow, with a power output in the range of 500 to 1000 MW [81]. Conceptual studies on large MHD power plants (e.g. 600 MW_{el}) carried out in the US showed that large sized dipole magnets are necessary for such plants. An active length of nearly 20 m, inlet and outlet diameters of the warm bore of 2.5 m and 4.8 m respectively and a total weight of 1700 tons will be required [77].

J. Superconducting Magnets for Inductive Energy Storage

The direct storage of electrical energy, without any conversion into other kinds of energy, offers the advantages of momentary access to electrical energy, short discharge times and high efficiencies. Until now electrical energy storage was primarily in capacitor banks in pulsed systems. Since the energy density (stored energy per volume unit) available in ordinary capacitors is comparatively low (≤ 1 Ws/cm^3) the storage of large amounts of energy requires uneconomically large capa-

Table 4. Compilation of superconducting magnets built for MHD power generation [79].

Laboratory	Used in MHD test facility	Warm bore, cm	Hom field length, cm	Central field, T	Year of first test	Stored energy, E_s, kJ
HTI, USSR	shock tube	3.5 × 7	15	1.5	1966	ß
Avco, USA	–	–	120	3.7	1966	3900
ETL, Japan	Mark IV	17 ϕ	20	2.4	1968	350
Hitachi, Japan	–	25 ϕ	60	4.7	1969	4500
HTI; USSR	–	–	20	3.1	1970	120
Gardner, USA KFA, Germany	Argas	22.5 × 29.5	100	3.3	1970	6000
ETL/Mitshu-bishi, Japan	–	21.4 ϕ	~ 40	1.9	1970	150
MCA, USA	–	18 ϕ	90	5	1972	920
Ferranti Pack, Canada	–	27.7 ϕ	105	4.5	1973	1400
ETL/Hitachi Japan	Mark V	39 × 130	120	4.55	1973	60000
ANL/MIT/HTI USA-USSR	U 25	40 ϕ inlet 60 ϕ outlet	250	5 inlet 3 outlet	1977	20000

citor banks. The application of capacitive energy storages is therefore restricted to a maximum stored energy of about 10 MJ and discharge times shorter than about 1 ms.

For larger energies and/or longer discharge times superconducting inductive energy storage systems, with an energy density of about 10 Ws/cm^3 at 5 T are considered. They are of particular interest with respect to large pulsed magnet systems in High Energy Physics (synchrotron accelerators) and in fusion technology (θ - pinch, pulsed coils of tokamak confinement systems). Energies on the order of Gigajoules are stored in large proton accelerators envisaged for the future. Even higher energies, in the range of 10 to 50 GJ, would be stored in the poloidal field windings of tokamaks (see Section II H). Peak instantaneous powers of several 100 MVA or even several GVA are required depending on the rise time of the field pulses, which are in both cases (HEP and fusion) about 5 s to 10 s. This peak power cannot be taken from the power system without serious disturbances.

The principle of the operation of a superconducting inductive pulse generator is shown in Fig. 28. A superconducting coil is charged at a low power, then a superconducting resistive switch is closed and a persistent current circulates through the coil and the switch. Fast opening of the superconducting switch (i.e. fast transition to the normal state) generates a high voltage across it which transfers the energy to an inductive load. A disadvantage of a circuit with a resistive transfer

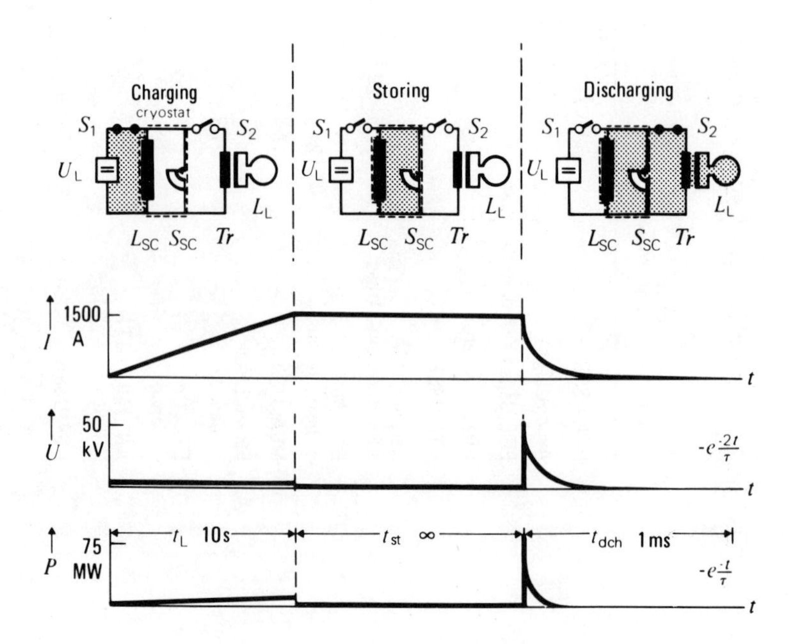

Fig. 28 Operation modes of a superconducting energy storage [69].

element (superconducting switch) is that the maximum transferred energy amounts to only 25% of the stored energy. Half of the stored energy is dissipated in the switch or its shunt resistance and 25% remains in the storage coil. The system is also nonrepetitive. Energy storage systems used in fusion reactors or as power supplies for accelerators must be repetitive and must have a high efficiency of energy transfer. In the case of fusion reactors this efficiency is of crucial importance for the total efficiency of the reactor and therefore for its competitive position.

To increase the energy transfer to the load the circuit must include a reactive element, either a capacitor or a mutual inductance. In practice these reactive elements are replaced by equivalent transfer circuits, e.g., superconducting homopolar generators, rotating electromagnetic converters, or electronic circuits (dc-ac-dc-converter) [79]. An advantage of the circuit with resistive transfer element is the fact that the energy can be transferred in the time of ms. It is therefore still of interest where the efficiency does not play a dominant role or for experimental equipment.

As already mentioned in Section II H, experimental work on superconducting energy storage for fusion is carried out at LASL [67]. Tests with 300 KJ storage coils and either vacuum or superconducting switches have been conducted. At IEKP Karlsruhe a 15 KJ-storage coil (ESPE 1) was operated in a stepwise discharge by a small thyristor switched capacitor achieving an energy transfer efficiency of 80% at a discharge time of 0.5 s. Another pulse generator (ESPE 2) has just been put into operation at this institute. It has a stored energy of 220 KJ, maximum pulse power of 60 MW at about 50 kV [69].

Finally, the application of superconducting pulse generators as power supplies for high power lasers should be mentioned. For this purpose, a 600 KWs energy storage system was built by the French company CGE. It was discharged in 12 ms corresponding to a pulse power of 50 MW and a maximum induced voltage of 100 kV [82]. This kind of development has a military application, therefore only a few details are known.

Besides the possible applications of superconducting coils as pulse generators, their use as energy storage systems in utility systems where very large energies and comparatively long charging and discharging times are required has also been discussed very seriously. In particular two laboratories have been concerned with this topic for some years, the Engineering College, University of Wisconsin [83] and the Los Alamos Scientific Lab (LASL) [84], both having close collaboration with US utility companies [41].

Increasing fuel costs, increasing capital costs of nuclear power plants and decreasing load factors make energy storage by superconducting magnets an attractive component in future utility systems for load leveling, peak shaving and increased system stability. The load factor (ratio of average power to peak power) of present utility systems approaches 50%, i.e., the installed generating capacity must be roughly twice the capacity required to meet the average load. Large scale storage systems for electrical energy are a means for power-leveling, i.e., for increasing the load factor and thereby utilizing more of the comparatively inexpensive off-peak power. Consequently this would also lead to an intensified use of the most efficient base load generators running continuously at optimum load. This form of operation would replace old inefficient coal, oil and gas fired generators, and use more effectively nuclear power plants, thus reducing the number of conventional plants. Finally it would lead to a smaller number of generators necessary to meet the power demand. In the past, energy storage has been used on a limited scale in power leveling and was restricted to pumped hydroelectric plants. Their application is limited by geographical constraints.

The need for energy storage systems will increase in the future as electric power systems grow, i.e., as the unit power outputs of future nuclear plants (fast breeder reactors) and generators are increased. Superconducting magnetic energy storage (SMES) would be in many ways an ideal energy storage system for electrical utilities. It consists in principle of a dc storage coil, and an interface to the three-phase power system, which is a thyristorized converter of the Graetz bridge type. Such types of converters have already been used successfully with dc power transmission over long distances, satisfying the high reliability requirements of the utilities.

The main advantages mentioned for superconducting magnetic storage systems are, very high efficiency, i.e., about 90% in comparison with 65% to 70% of pumped hydro storage, and near independence of geographical areas. Superconducting storage can be placed near the load centers, with little environmental impact.

Further advantages lie in the characteristics of the converter system which is the interface to the power system. These can provide power control in a wide range, with a response-time to the systems demands of a few milliseconds. This feature can significantly improve system stability and automatic generation control.

Energy storage capacities of about 10^4 MJ to 4×10^7 MJ have been considered in the studies mentioned above [83, 84, 41]. The latter approximately equals the storage capacity of the largest existing pumped hydro installation at Ludington US. Figure 29 gives a schematic survey of the possible installation of a 10 GWh superconducting energy storage system for load leveling [41]. The coil type chosen as storage is a solenoid with a thin winding. It has been shown that for energy stored per unit volume, solenoids are more effective by about a factor of two than toroidal coils. The stray field has to be compensated, or the coils have to to be positioned in such a way that the stray field causes no disturbance. The main problems of such storage coils arise from their extremely large geometrical size, which is about an order of magnitude bigger than that of magnets needed for fusion power plants. The stored energy is about two orders of magnitude larger. These larger dimensions together with the high fields (e.g., 5 T) lead to tremendous axial and radial forces.

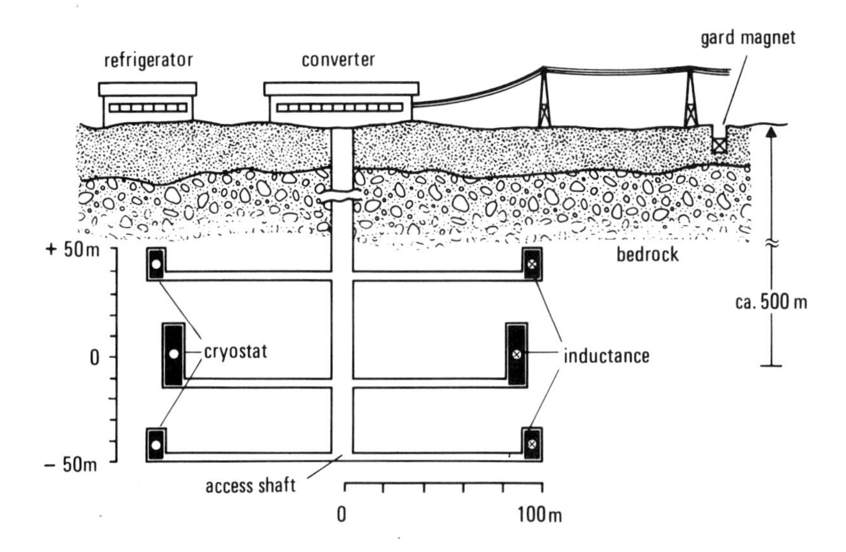

Fig. 29 10 GWh storage unit [41].

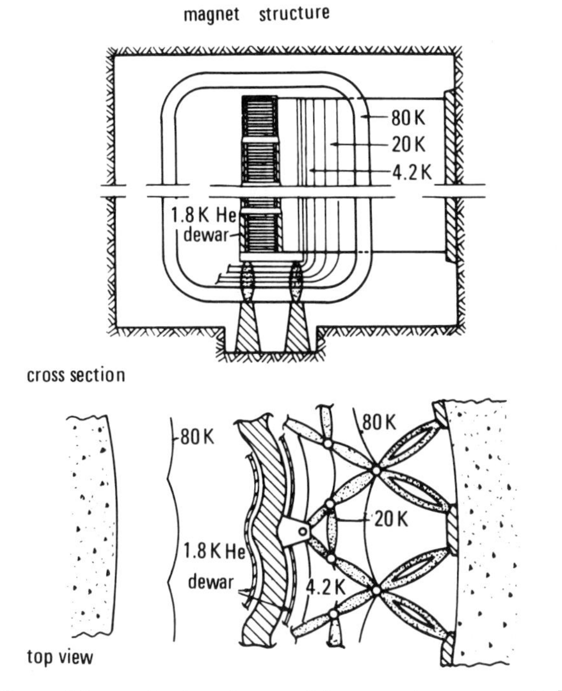

Fig. 30 Magnet structure of large storage unit [83].

In Figure 29, the complete magnet system is mounted in narrow underground circular tunnels in bedrock. This bedrock, which should have a high mechanical strength and high modulus serves as a warm support for the radial as well as for the axial forces of the magnet system. It has been shown that a cold stainless steel support structure would be about four times more expensive than the warm support. This would lead to a totally uneconomic storage system. Since the circular tunnels also form the outer wall of the vacuum chamber they have to be coated with vacuum tight liners. The mean depth of the system is about 500 m to reduce stray fields at the ground surface.

The coil of Fig. 29 has been divided into three sections for supporting the axial forces by the rock. Radial and axial forces are transmitted from the cold coils to the rock surface of the tunnel walls by epoxy fiberglass bumpers, Fig. 30. The Wisconsin study [41] suggested a rippled, with cast Al, fully stabilized NbTi/Cu multifilament superconductor reinforced with stainless steel. It would operate at a current of 157000 Amps at 5 T wound into a single layer coil. The ripple is used to resist circumferential load and to allow for compensation of the radial expansion of the rockwalls and the compression of the epoxy struts. The same is true for the rippled dewar walls and shields.

Because of the excellent cooling conditions of superfluid He, (high heat conductivity, no bubbles) bath cooling at a temperature of 1.8 K is suggested. The saving in NbTi/Cu conductor more than compensates the additional expenditure for refrigeration. The power consumption for refrigeration of a 10,000 MWh storage unit referred to a daily cycle is about 10% of the delivered energy (equal to 75% of the stored energy). The losses of the converters are about 2% of the power handled.

Table 5 [84] gives a comparison of estimated costs of delivered energy for various energy storage systems. The costs for the SMES's are based on superconductor prices considerably lower than present day costs, because mass production was assumed. Table 5 indicates that only for very large storage systems (e.g., > 5000 MWh), SMES will be competitive with other alternatives for load leveling. For smaller units the only economical application will be for peak shaving and system stability.

The contribution of the single components of a very large superconducting storage unit (4×10^7 MJ) with warm reinforcement to the total costs is approximately the following: superconductor 20%; dewar including support structure 62%; refrigerator 2% and converter 6%. About 10% of total costs have to be added for control, computer and personnel facilities.

Besides feasibility studies, experimental work with respect to superconducting energy storage in utility systems is also being carried out. After testing smaller storage units (in the range of 100 KJ), a 100 MJ energy storage system with SCR converter is in its final design stage at the LASL [85]. This project will adress all important problems associated with such storage units except the use of bedrock as mechanical structure. Figure 31 shows a drawing of the planned experimental facility.

Table 5 A comparison of delivered energy, in mills/kWh, for various energy storage systems based on off peak energy costs of 10 mills/kWh [84].

Operating time hours/yr	2000	3000
Pumped-hydro, 0.67 eff Capital cost = $220/kW	33	29
Gas turbine Capital ccst = $150/kW Fuel cost = 300¢/$10^6$ Btu	52	49
Batteries 10h/day discharge, 0.7 eff Capital cost = $15/kW		
5 year life	30	26
3 year life	45	38
SMES solenoid warm reinforcement 3.9×10^7 MJ	30	26
SMES solenoid cold reinforcement 3×10^4 MJ	60	51

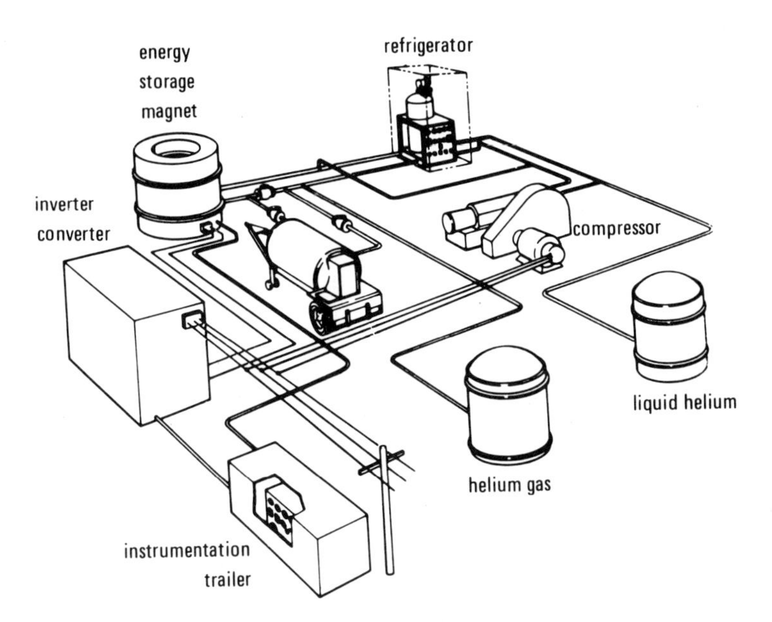

Fig. 31 Installation of 100 MJ energy storage coil including inverter converter, instrumentation trailer, refrigerator and other equipment [84].

K. Superconducting Magnets for Magnetic Separation

In a magnetic separation process, ferromagnetic or paramagnetic particles are separated from a large mass stream of background material by magnetic forces. The magnetic force on the particles is generated by the application of a magnetizing field H_0 and a gradient dH/dx and is given by $F_m = V \cdot M\ (dH/dx)$, where V is the volume of the particle and M its magnetization in the magnetic field H_0. For a reasonable efficiency of the separation process, the magnetic force has to be great enough as to overcompensate opposing forces, which are the gravitational and hydrodynamic drag (viscous) forces.

Magnetic separation already has a wide spread use in the removal of so-called "tramp iron" from materials or fluid streams in industrial processes. Magnetic separators are also widely used in the food industry for cleaning food, e.g., flour from iron pieces (bolts etc.). Of great importance are the magnetic separators used to improve low grade mineral ores. The reduction to the metal requires a certain percentage of mineral in the ore. This is not always found in nature. In this process the ore is pulverized, slurried in water and the magnetic component is separated from the background material (e.g., silicate gangue). Besides enrichment of valuable minerals, magnetic separation is also used for purification, e.g., removal of iron strained TiO_2 from clay used for paper coating.

Separators frequently used in the past were of the drum-type with comparatively small magnetic fields and field gradients suitable only for enrichment of minerals with large particle size (> 10 μm) and of great magnetic moment (e.g., magnetite Fe_3O_4). For the benification of materials with smaller particle size and of weak magnetic properties, so-called high gradient magnetic separators (HGMS) have been developed. In these separators higher field gradients are created by the use of sharp pointed elements, grooved plates and spheres of steel.

Two types of high gradient separators can be distinguished [86]. The first is a trapping type separator with some sort of magnetized structure generating the field gradient. The second is a deflection type separator by which the magnetic particles are selectively deflected in a flowing stream.

A special type of a HGMS has been developed by scientists of the National Magnet Lab at MIT [87]. In this device the particles are captured on a magnetized stainless steel wool matrix. Figure 32 is a schematic drawing of this high gradient matrix separator. Within this matrix which is magnetized to a saturation maximum, field gradients of the order of 0.1 $T\mu m^{-1}$ can be achieved along the edges of the steel wool strands. This steel wool matrix separator is inherently suited to trap particles in the size range between about 0.5 μm to 300 μm. The magnetic trapping is most effective when the particle size and matrix wire diameter are of the same order of magnitude.

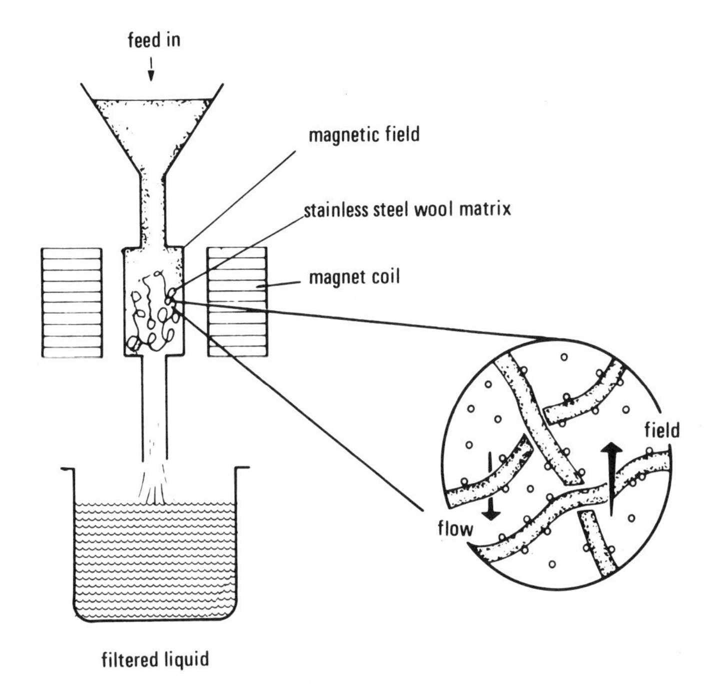

Fig. 32 Schematic diagram of a high gradient magnetic separator (HGMS) [87].

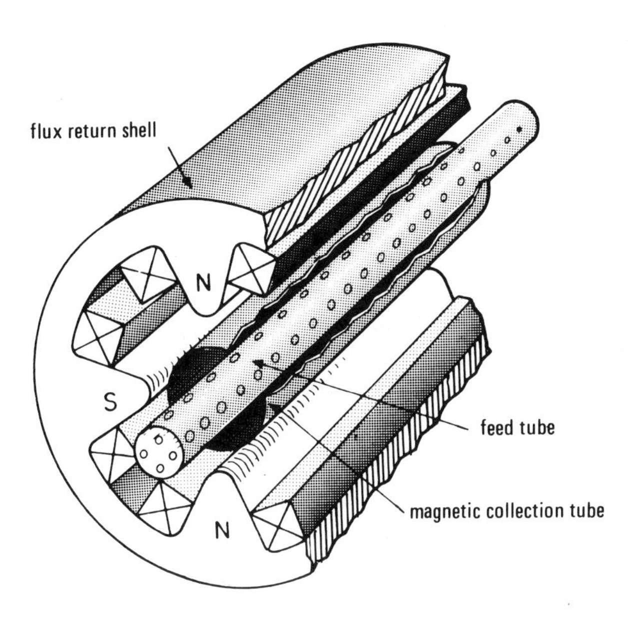

Fig. 33 Quadrupole magnetic separator [86].

Figure 33 shows a deflection type separator which consists of a quadrupole magnet along whose axis a pair of concentric tubes are aligned connected by holes in the wall of the inner pipe [86]. Magnetic particles carried in a slurry in the inner pipe diffuse through the holes into the outer pipe under the influence of the magnetic gradient. The field gradients obtainable with this device are smaller than with the matrix-type so that it is more suited to larger or more strongly magnetic particles. An advantage of this deflection separator is the possibility of continuous operation without additional means.

The trapping type separator requires either periodic flushing or provision for continuously moving the matrix in and out of the magnetic field in order to strip off the trapped particles. This can be accomplished by a rotary matrix, a so-called carousel matrix.

The separation devices discussed have been applied in industrial processes with conventional electromagnets generating maximum fields of 1 - 2 Tesla. At inductions of this magnitude, the separator matrix will be saturated, consequently an increase in magnetic field above two Teslas will not lead to higher field gradients.

From the above argument, the question arises, what benefits can be achieved in magnetic separation by the application of superconducting magnets? Possible improvements are discussed below.

In the case of paramagnetic particles, the magnetic moment will be enchanced linearily with increasing field, so that only faintly paramagnetic small size particles can be separated with considerable efficiency. With open gradient separators, e.g., with a quadrupole magnetic field, both the magnetizing and the field-gradient can be considerably increased by the use of superconducting windings. The flotation process which is normally used for the benefication of weakly magnetic fine grained minerals (e.g., copper and nickel ores) might be replaced by magnetic separation. Also an extension in the use of magnetic separation with weakly magnetic iron ores, i.e., hematite (Fe_2O_3) and geothite (FeOOH) might be expected.

The efficiency of a matrix separation process, which is determined by two quantities, grade and recover, can also be markedly improved. The recovery (ratio of the amount of magnetic material recovered in the magnetic component to amount of magnetic material in the feeds) increases with increasing field and decreases with increasing speed of the slurry (highter field more trapping, higher velocity less trapping). The grade (percentage of magnetic material in the magnetic component) varies little with field, but increases with the velocity of the slurry (less mechanical trapping at high velocities) [88]. With superconducting magnets high fields and high velocities can be used, resulting in high recovery and high grade.

A typical ore processing plant might handle 20,000 tons of material per day. A large steel mill may discharge around half a million cubic meters of contaminated water per day. Such large plants would require large separator volumes of the order of cubicmeters surrounded by large magnets. These devices may be constructed with normal magnets but the energy consumption would be enormous. The use of superconducting windings would drastically reduce the power consumption leading to low operational costs.

Besides ore separation and clay purification already mentioned, there is also interest in other applications of magnetic separation. One example is coal cleaning instead of costly coal gasification. The principal virtue of gasification is the removal of sulphur and ash. Approximately half othe sulphur in coal is pyritic and can be magnetically separated, while half is organic and cannot. For removing all of the sulfur by magnetic separation, means have to be developed to convert the organic sulphur to pyritic sulphur. Investigations with respect to magnetic coal cleaning have been carried out at MIT [89].

Another example is the purification of contaminated water systems. Water purification includes the removal of suspended solids (turbidity), bacteria, color particles, dissolved heavy metal constituents and ions (e.g., orthophosphate ion which is a nutrient for algal growth) [90, 91]. To remove the contaminants a seed, finely devided iron oxide (magnetite Fe_3O_4 : 100 to 1000 ppm), and flocculant (Fe(III) or Al(III) ions) have to be added to the water system. The iron oxide is coagulated with the contaminants into flocs which are then collected in the separator. Very fast flow rates are possible because of the large open volume of magnetic separators compared with mechanical filters. Recycling of the magnetite seed is possible. The insoluble precipitates are adsorbed in a tight lattice around the magnetite particles. They can be stripped by the use of a second magnetic separator and acid cleaning [91]. The application of magnetic separation in chemical processes, biochemistry and physical chemistry has been also suggested [92].

A series of companies and laboratories which are specialists in the field of high gradient magnetic separation with conventional magnets are seriously considering the application of superconducting magnets for this pupose. The principal features of high gradient separation can be investigated with the aid of conventional magnets and the results can be extrapolated to higher field values. Until now experiments have been carried out principally with conventional equipment. Experimental devices with superconducting magnets have been operated at Magnetic Corporation of America (MCA) [93], English Clays Lovering Pochin, Cornwall, England [94] and at Friedrich Krupp GmbH, Essen, Germany. In Germany Klöckner-Humboldt-Deutz, Köln, in collaboration with the Institut für Experimentelle Kernphysik, Karlsruhe is developing a superconducting magnet system.

Because of the anticipated advantages presented above, it seems that superconducting magnets may have their first large scale industrial application in the field of magnetic separation. However, in the future it has to be proved whether the benefits, i.e., the magnetic separation of fine grained minerals instead of flotation, the higher efficiencies and the energy saving will cancel the larger efforts necessary with a cryogenic installation. Preliminary design studies indicate that for large units a superconducting system is less expensive in capital costs and significantly less expensive to operate than a separator with conventional magnets [93]. On the basis of over-simplified estimations and comparisons [95] it was concluded that superconducting magnets will only be accepted by industry when energy costs have increased five to ten-fold.

III. LEVITATED VEHICLES WITH SUPERCONDUCTING MAGNETS

A. Introduction

Magnetic levitation and propulsion offers the possiblity of increasing the speed of tracked vehicles to 500 km/h for mass transportation. Intensive research and development work has been carried out in the US and is still underway in Japan, Canada, Great Britain and West Germany. Many ideas, proposals and experiments concerning either components or complete vehicles have been published, see the review [96].

The area of application of such a new high speed ground transport system would be medium distance traffic, i.e., with distances of 200 - 2000 km, where especially the following problems exist:

Increasing requirements due to economic growth, increasing division of labor and further development of tourism resulting in increasing requirements.

Environmental concern for medium distance travel. Air traffic with its high number of take-offs, landings and waiting-operations represents a remarkable part of the environmental impact (noise, air pollution) and any increase will be faced with fierce resistance.

Increasing demands of the public and government. The criteria of safety, time spent, punctuality and comfort are of increasing interest for the user. Their realization with conventional means becomes more and more problematic.

Fuel conservation. The progressive shortage of oil will result in a restriction of its use (e.g., to the chemical industry or intercontinental air traffic).

The technical and economical upper speed limit of the conventional wheel-rail system is supposed today to be at about 350 km/h. This speed limit is determined by wear phenomena, especially in two critical areas.

Wear is a problem in the small contact area between rail and wheel in which the exchange of supporting, guiding, propelling and drag forces and the return current transmission occurs.

Wear is also a problem in the contact area between trolley wire and trolley arm which has to transmit the whole propulsion power (which is for instance in the case of a Trans European Express (TEE) locomotive 7.4 MW continuously).

It is the technical aim of the new developments to avoid such highly concentrated mechanical loads and to find new ways for the transmission of the propulsion power. With respect to levitation and guidance of fast tracked vehicles two electromagnetic systems are under investigation at present, the electromagnetic system and the electrodynamic system. The first is based on attractive forces between conventional copper-iron electromagnets and iron rails. Its main disadvantages are a comparatively small clearance between magnet core and rail (e.g., 15 mm) and the necessity of active control of the clearance distance because of the unstable stability of this system. The electrodynamic system is based on the repulsive forces occurring between moving ironless superconducting magnets and stationary sheets, loops or coils of copper or aluminium at ambient temperature. In the following, we restrict ourselves mainly to the electrodynamic system.

B. Basic Features of the Electrodynamic System

The electrodynamic system is characterized by the following features:

Vehicles are lifted and guided statically stable by repulsive forces without mechanical contact.

During operation clear air-gaps of more than 100 mm can be realized.

The levitation and guidance systems represent, in the most simple case, progressive and undamped springs with a resonant frequency of about 1.5 Hz. Additional damping is required.

At low velocities and at standstill wheels are used for support and guidance.

The generation of lift and guidance forces results in drag forces which have to be overcome by the propulsion system.

The superconducting levitation and guidance magnets on board have to be cooled to liquid He-temperatures. They operate in persistant mode (short circuited) and need no power supply on board.

The propulsion can be accomplished in a contactless manner by several phase windings laid into the track, which are sectionally supplied with power and which interact with the superconducting magnets on board.

The propulsion and guidance forces can be combined in an advantageous manner.

C. Principle of the Electrodynamic Levitation System

If an excited magnet moves above and along an electrically conducting plate at a constant distance from the plate surface, the following forces are acting (see Fig. 34 [97]):

(1) The lift force, F_z. It increases proportionally to the square of the velocity and approaches saturation at velocities higher than 120 km/h.

(2) The drag-force, F_x. It varies linearly with the velocity at low speeds and has a maximum at about 20 km/h. It decreases at higher speeds in a hyperbolic manner.

(3) The unstable lateral force, F_y. It arises from a lateral displacement of the magnet from its central position with respect to the plate.

With a simple electrodynamic arrangement (normal flux system) a ratio

$$F_z/F_x \approx 35$$

can be technically achieved at 500 km/h. The lateral force and the reduction of the lift force by lateral displacement can be limited to some percent of the lift force by an adequate choice of sheet magnet width and lift height.

The characteristic speed-dependence can be simply derived from the behavior of the induced currents I_2 (see Fig. 35). These currents increase linearly with increasing velocity, as long as the reactive component of the plate resistivity can be neglected. At medium or high velocities the reactive resistivity dominates. Induced voltage and impedance grow in the same manner, consequently the induced current becomes saturated. The lift force is proportional to the stored magnetic energy of the arrangement and the drag losses are equivalent to the ohmic losses in the sheet.

D. Various Lift- and Guidance Systems

Various possibilities are given for the arrangement of magnets and sheets. Single magnets are positioned above a simple plane plate (called "normal flux systems") or double magnets are excited in opposition and enclose the reaction sheet (called "null-flux system"). The functions of lift and guidance can be accomplished separately or in combination by a suitable arrangement of the reaction sheets. Figure 36 depicts various possibilities of magnet and strip combinations. Table 6 is a summary comparing the characteristic features of normal- and null-flux systems with reaction sheets and coils [98]. The main advantage of the null-flux system is its smaller specific drag force (= drag force per lifted mass). On the other hand this system suffers from a comparatively high number of ampere turns (10^6 At) and large internal magnetic forces. These require large efforts in mechanical reinforcement and lead to high weight vehicles. These disadvantages led to the decision not to investigate the null-flux system with homogeneous sheets in larger experimental systems.

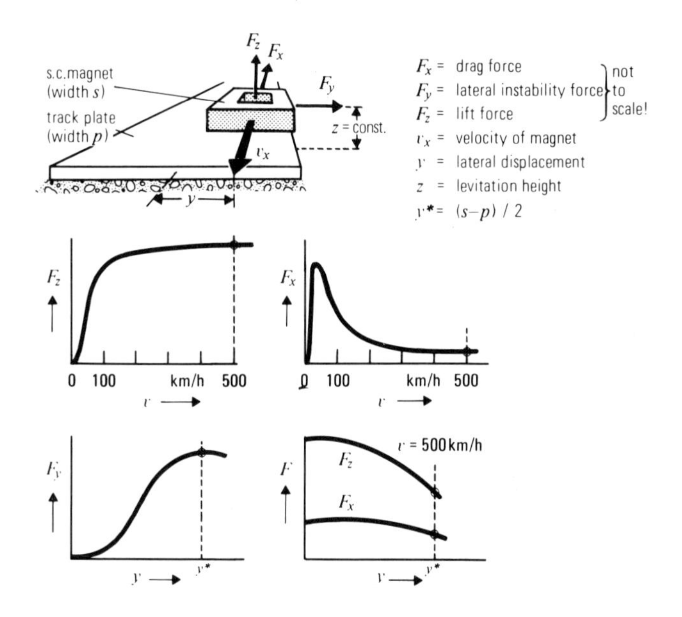

Fig. 34 Electrodynamic levitation: Basic behavior [97].

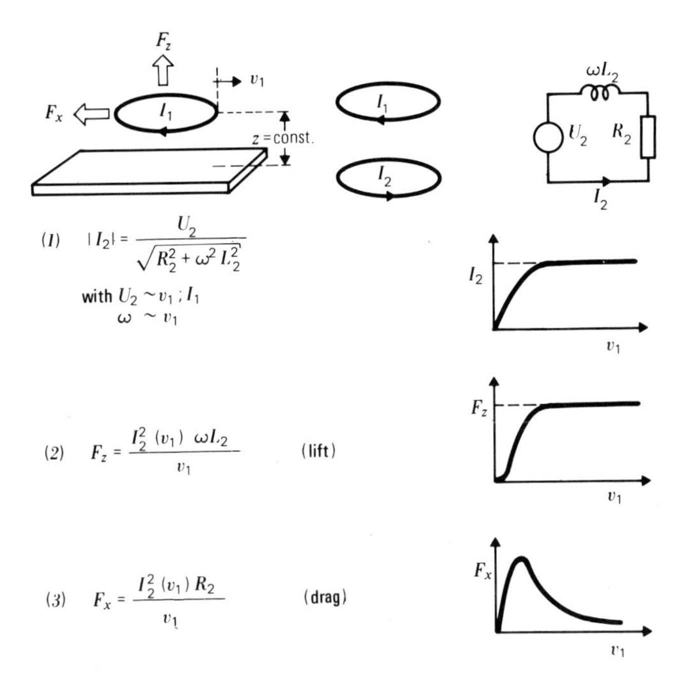

Fig. 35 Electrodynamic levitation: Simplified explanation of the speed dependence of lift- and drag forces.

Table 6 Characteristic features of normal and null-flux systems with reaction sheets [98].

		Normal Flux	Null-Flux
Action of force		unidirectional	symmetrical (bidirectional)
Typical number of Ampere turns per magnet	A	$\leq 0.5 \times 10^6$	10^6
Favourable levitation hight or displacement	mm	250	20 to 60
Specific drag losses	W/N	3 to 4	1.5 to 3
Track sheet: thickness	mm	>15	8 to 16 *)
Specific resistivity	Ωm	$<3.3 \times 10^{-8}$	$<3.3 \times 10^{-8}$
Spring constant		small	large
Typical eigenfrequency	Hz	1.5	1.8
Self damping		small	small
Magnet structure		normal	more complex
Magnet weight		small	large
Thermal losses		small	large
Track structure		comparatively simple	more complicated
Preferable application		lift	guidance

*) has to be optimized from case to case!

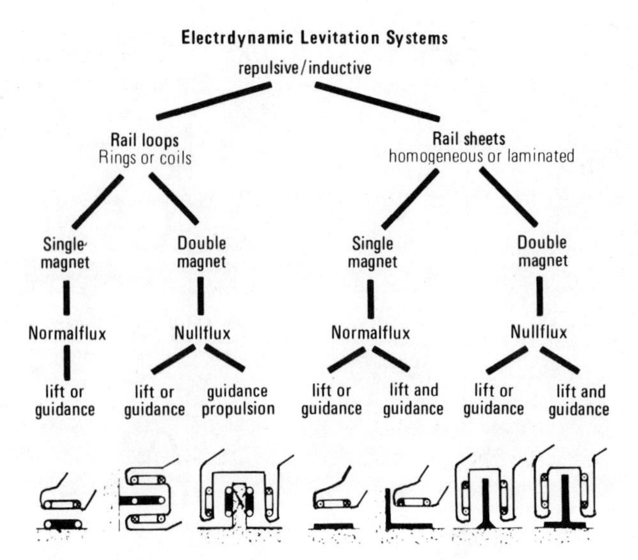

Fig. 36 Basic types of electrodynamic levitation systems with reaction sheets [98].

Instead of homogeneous reaction sheets or strips, discrete loops in the form of short circuited coils can be used. Contradictory opinions exist on the suitability of discrete loops for lift. Most research teams prefer sheets because of the anticipated smaller expenditure for fabrication and material. Another advantage is seen in the large variety of possible sheet arrangements which can be very simply achieved and used for combined lift and guidance. Homogeneous sheets generate about double the lift-force as discrete loops if the same amount (weight) of aluminium is used. The use of discrete loops results in smaller specific drag losses as can be seen from Table 7 [99].

With respect to losses and adaption, discrete loops promise advantages for guidance. Therefore present design concepts foresee the additional use of the magnetic excitation of the superconducting pole magnets of a linear synchronous propulsion motor (see Section III F) for guidance [100]. This can be realized by laying into the track a second winding layer within the area of the propulsion winding. In principle this so-called difference-flux winding consists of a series of successive loops in the form of an eight, laid transverse to the direction of motion. At a symmetric position of the propulsion magnets relative to the difference-flux windings, the voltages induced in the left and right part of the latter are cancelled. In the case of a lateral displacement however differential voltages and currents are generated in the loops resulting in repulsive guidance forces. Figure 37 shows schematically the principle of this guidance system.

Electrodynamic guidance systems of this kind are being developed as horizontal versions in Germany and Canada. In Japan they are being developed as vertical systems with loops vertically mounted at the sides of a central track part. The combination of linear synchronous propulsion and difference flux guidance, called CSPG (combined system for propulsion and guidance), is provided for in all the projects underway.

E. Damping

From a static point of view, electrodynamic levitation systems are self-stabilizing, i.e., at slow changes a new stable position is achieved without any regulation. However, the self-damping of the system is not large enough to compensate for dynamic instabilities caused by wind forces or track-irregularities (e.g., sheet joints). To achieve proper ride comfort the vertical acceleration of the passenger compartment must be less than $0.1\ m/s^2$ in the critical frequency range between about 0.5 Hz and 20 Hz. This can be realized if additional active, normal-conducting, primary damping windings are used. They must be mounted into the airgap between the lift magnets and the track-sheet, and a secondary damping system must be placed between the passenger compartment and the magnet carriers.

Proposal have also been made which suggest only an active primary damping of the magnet system without a secondary damping of the compartment which is rigidly connected to the magnet-carriers [98].

Table 7 Homogeneous rail sheets and discrete coils. Comparison of lift- and drag-forces (Normal flux system) [99].

Cross section of rail coils	Weight of material in the rails	v = 500 km/h h = 250 mm		v = 100 km/h h = 215 mm	
		lift force	specific losses	lift force	specific losses
mm²	t/km	t	kW/t	t	kW/t
Cu					
50 × 50	180	9.4	37	8.2	35
70 × 70	355	10.3	23	11.1	21
100 × 100	700	10.8	14	13.2	13
Al					
50 × 50	55	8.4	63	5.3	59
70 × 70	107	9.7	38	8.3	36
100 × 100	212	10.4	23	11.0	22
homogeneous Al-sheets 20 × 800	173	20	38	20	33

Superconducting magnets: length = 2m; width = 0.3m; current = 800kA
Discret coils: length = 1m; width = 0.3m

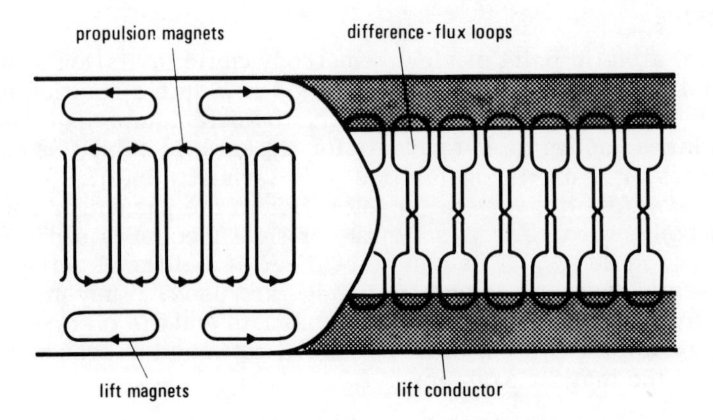

Fig. 37 Principle of the difference-flux guidance system.

F. Propulsion Systems

Because of environmental factors, only electromagnetic systems are considered for the propulsion of magnetically levitated vehicles. In principle there is the choice between linear induction motors = LIM (Types A) and linear synchronous motors = LSM (Types S) as is depicted in Fig. 38 [101]. (It should be mentioned that the necessary travelling magnetic wave field can also be generated by rotation of charged superconducting windings (A3) [102]). In addition two alternatives are given for each case, if the propulsion energy is transmitted to the vehicle (active vehicle: primary windings on the vehicle) or to the track (active track: secondary windings on the vehicle). Ironless linear synchronous motors with active track (Fig. 38, version S2 and S3) seem to be the most promising solution for electrodynamically levitated vehicles. They have the following main advantages.

The propulsion energy (e.g., 18 MVA in the case of a 120 ton vehicle) does not have to be picked up by the vehicle. The only way to accomplish this would be by brushes sliding on current rails. This solution suffers from wear problems which, for high speeds, have not yet been satisfactorily solved.

The clearances achievable between the primary windings (track-windings) and the vehicle "secondary" (superconducting pole windings in in the vehicle) have about the same size as the clear gaps of the electrodynamic lift- and guidance-system, i.e., 100-250 mm. In contrast the air gaps of linear induction motors are limited to clearances of $\leq$ 20 mm in order to attain reasonable reactive power consumption and efficiencies.

The vehicles have a considerably smaller weight than those with conventional linear induction motors. Their electrical equipment is much simpler. Besides the motor units active vehicles need on-board-converter systems with full power rating for thrust control.

Weight and over all economy considerations have led to the decision of developing linear synchronous motors with active tracks for the next generation of vehicles, i.e., revenue vehicles with speeds up to 500 km/h.

Figure 39 depicts the basic mode of operation of a LSM with active track [101]. The propulsion energy is successively fed into about 5 km long track sections by power sub-stations. Meander shaped cables with insulated strands are laid into these track sections forming a multiphase winding which, when excited, generates a traveling electromagnetic wave. The synchronous speed v_s of this wave, which is equivalent to the velocity of the vehicle, is determined by the exciting frequency and the pole pitch of the winding. To allow for any speed up to the maximum value in arbitrary track sections, the feeding sub-stations must work as frequency converters, providing an excitation current with variable frequency.

The propulsion force F_x and a vertical force component F_z, which supports the lift system, are caused by the interaction between

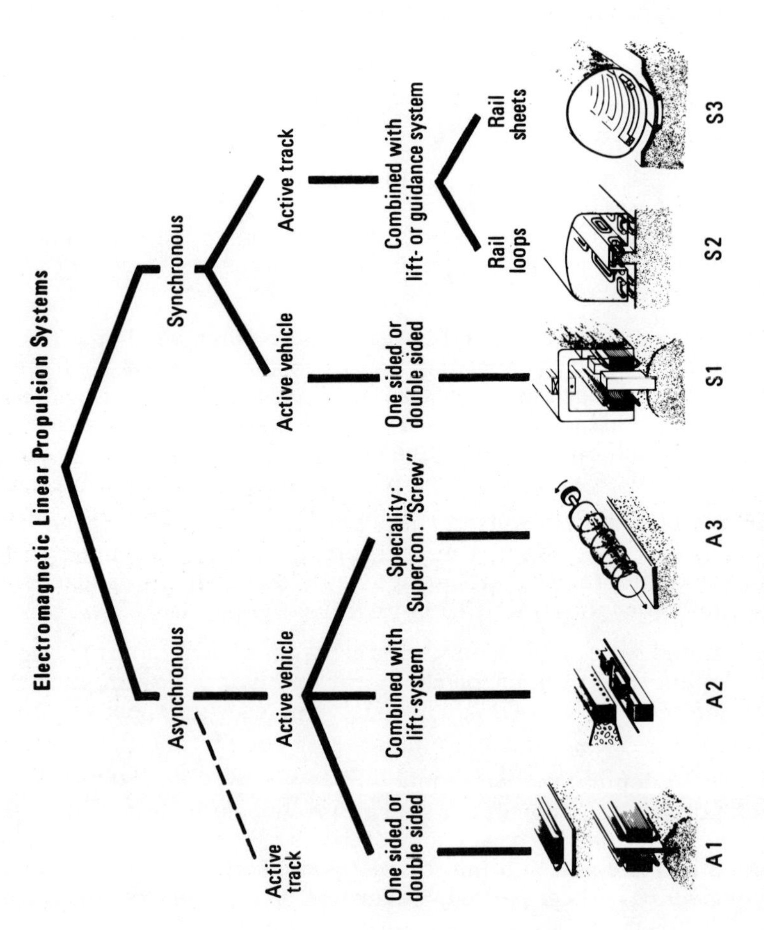

Fig. 38 Basic types of linear electromagnetic propulsion systems [101].

the traveling wave and the superconducting pole magnets installed in the bottom of the vehicle. It is seen from the diagram of Fig. 38 that both force components can be adjusted independently of each other by the current I_1 of the track winding and the relative position of the vehicle magnets to the traveling wave as measured by the slip angle λ. With this propulsion system, the thrust and velocity of the vehicle can be regulated within the range of operation. In addition a contribution to the static lift force is achieved when track windings and pole magnets are horizontally arranged (case of Fig. 39). If a vertical arrangement of both is chosen, the additional force contributes to the lateral guidance force. The dynamic behavior of the vehicle can be substantially improved in the case of a horizontal arrangement by adequately controlling the slip angle λ and hence the force F_z. It has been shown that high motor efficiencies can be expected, e.g., $\eta_s = 0.88$ with a $\cos \varphi = 0.93$ at the optimum working point of $\lambda = 123^\circ$ [103].

The economic realization of the LSM is based on the use of superconducting pole windings in the vehicle. The high magnetic fields generated by these windings reduce the field and therefore the conductor requirements for the primary windings in the track to a reasonable level.

G. On-Board Cooling Systems

In the past, mobile application of superconducting magnets has been the exception. Even under stationary conditions there is not much experience with the combined operation of several distributed individual magnets. There are several possibilities for the supply of cryogenics to the superconducting magnets distributed in the lower part of a vehicle.

One can use sealed-off cryostats. Each magnet cryostat is filled with an amount of liquid He sufficient for a one day operation and is then tightly closed. During operation the temperature and pressure of the coolant are increasing. In a nightly maintenance period the coolant is brought back to its initial state by recondensing or exchange of the fluid. No on-board distribution, liquification or recovery is needed [104].

A second method is exchangeable on-board storage tanks. Groups of magnet units are cooled via storage tanks, which are periodically replaced by full tanks. The evaporated He is compressed by on-board compressors and stored in exchangeable high pressure cylinders.

Thirdly, one could carry on-board refrigerators. The magnets are coupled to refrigerators and cooled in closed cycles. There is the choice between one refrigerator per magnet, per magnet group, per coach or per train.

It is anticipated that for a complete transportation system, the supply of magnet groups by active on-board refrigerators is the most promising solution.

Figure 40 shows a corresponding proposal made by Linde AG, Germany which has the following features:

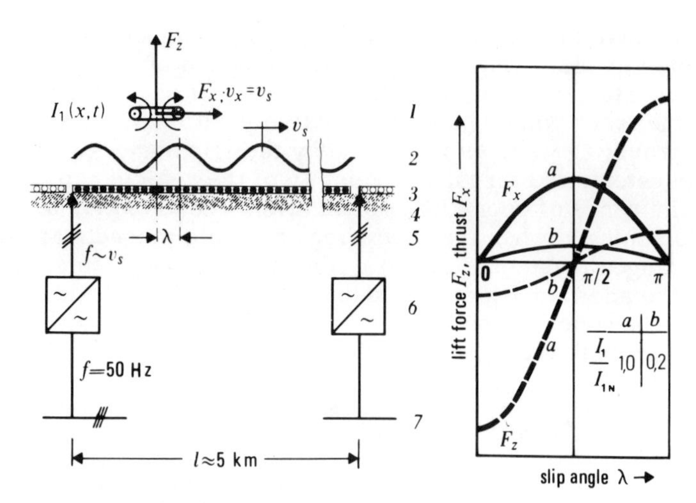

Fig. 39 Principle of an ironless linear synchronous motor with active track [101].

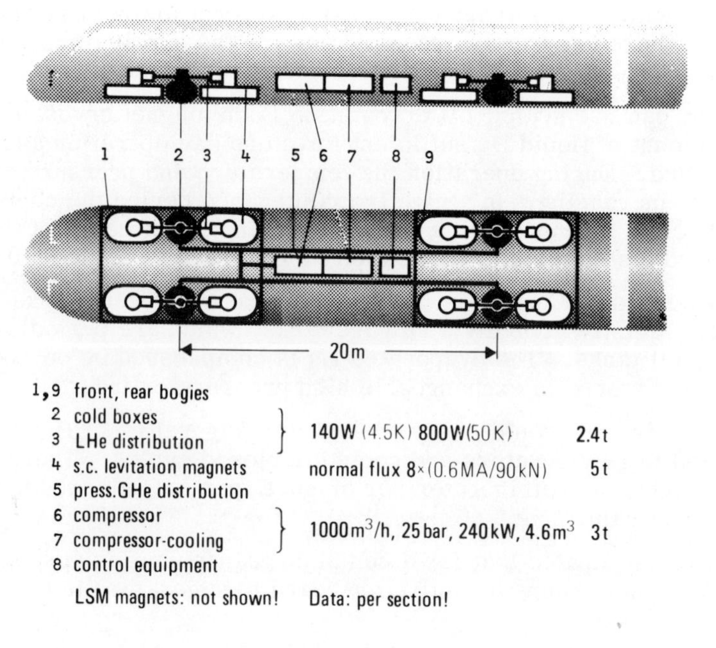

Fig. 40 70t-Section of ED Revenue Vehicle: Cryogenic system proposed by Linde A.G. (1974; data per section) [97].

There is one cold box for two magnets one on either side of a bogie, so liquid He is to be transferred through short, stiff pipes only.

The compressor, compressor-cooling and control equipment make up a central unit on the coach (or coach frame) for all cold boxes of one coach.

Only pressurized gaseous He from the compressors and warm return gas from the cold boxes need be conducted per considerable lengths and across the gap between bogies and coach.

The most important candidate for further development in such a system is the compressor set. In existing facilities it represents the item of biggest dimensions, weight and noise. Currently, a stationary facility for 140 W at 4.5 K and 800 W at 50 K would have an overall weight of more than 10 tons, whereas 5 to 6 tons seem to be achievable for mobile application within several years. A compact light-weight rotary compressor is under development at Linde.

H. Magnetic Shielding of the Passengers

The problem or necessity of shielding the passengers from the fringe field of the levitation and propulsion magnets has not yet been determined. At present no general standards exist for the maximum tolerable magnetic field levels and exposure time. It is anticipated that field levels on the order of 10^{-2} T are harmless to humans permanently exposed to them. Immediately above the levitation and propulsion magnets these levels are exceeded. The level of magnetic fields in the passenger compartment of levitated vehicles depends very strongly on the vehicle design, i.e., the distance between magnets and compartment, since magnetic field strength decreases rapidly with the distance from the magnets. Therefore designs with secondary suspension between compartment and magnet frame are favorable with respect to the magnetic field level to which passengers are exposed. If the field level is intolerably high, additional shielding coils must be mounted between the compartment and the magnets. From the above, it becomes clear that detailed investigations of biological effects of magnetic fields are necessary in order to develop future standards.

I. Electrodynamic Levitation Projects

1. FRG – the Erlangen test carrier and track

Besides studies on the technical and economical limits of the conventional wheel-rail system two magnetic levitation projects are being carried out in the FRG by two different industrial groups under the common sponsorship of the German federal government. One project deals with the electromagnetic system (see Section III A) and is conducted by the Arbeitsgemeinschaft Transrapid EMS (a company which was founded by the Federal Department of Research and Technology and the two companies Krauss Maffei and Messerschmidt-Bölkow-Blohm).

The second project is investigating the electrodynamic system. In this project, the three electrical companies AEG-Telefunken, Brown Boverie & Cie and Siemens in cooperation with Linde AG have combined their efforts as has been outlined earlier [105].

As a means of testing subsystems of realistic dimensions for several hours continuously it was decided in 1972 to build a test carrier (EET = Erlanger Erprobungsträger) running on a circular track. The schematic drawing Fig. 41 together with the data from Table 8 summarize the technical features of the EET. The vehicle has 16 guiding and supporting aircraft wheels for landing and starting. The carrier is operated at 45° banking angle against centrifugal forces. It is propelled by an on board linear induction motor (LIM), generating a thrust of 45 kN maximum at low speeds, decreasing to about 22 kN at 200 km/h. The electrical energy for the double sided motor (maximum power consumption 5 MVA) is picked up from 6 current rails and conditioned by a static converter. A linear induction motor instead of the favorable lineac synchronous motor with active track was used since more experience was available with the LIM. This promised an earlier beginning of large scale tests with the Erlangen test carrier.

Eight superconducting magnets were originally planned to lift and guide the carrier when at operational speeds (> 80 km/h). The magnets which are identical both for lift and guidance are assembled on two magnet platforms together with the cryogenic supply systems and the necessary measuring devices which form widely autonomous subunits. Due to large scale linear synchronous motor (LSM) experiments which were planned in the meantime, the EET program concerning lift and guidance has been cut down to the study of lift only. Guidance will be provided by eight aircraft wheels, although all eight magnets have been constructed and successfully tested in the stationary mode. Figures 42 and 43 show an aerial photograph of the test track and the Erlangen test carrier.

The drawing of Fig. 41 does not show the stationary equipment. These include the control room with telemetric and data handling system, the energy supply, and the stationary cryogenic supply system, which consists of a 300 W refrigerator in combination with spread out distribution including a double set of liquid helium pumps. Forced-flow cooling was provided for the magnets to prevent liquid He level problems due to centrifugal forces and to allow uniform cooling independent of their working position. Figure 44 gives a cross-sectional view of the magnet. Subcooled helium (4.5 K, 2 bar) is forced to flow through the winding (see spacer plates in Fig. 44) once around a "race track". It is conducted to the next magnet, after part of this stream has been split off and expanded in two-phase flow for cooling a first copper shield at 4.5 K and a second copper shield with gaseous helium at about 50 K. Table 9 gives the main data of the magnet [106]. During operation of the test carrier the magnets are operated in the persistent mode by a mechanical switch, which in the closed state and at 500 A has a resistance of 10^{-8} Ω.

The cooling of the magnets on board of the EET is provided by a

Table 8 Erlangen test facility: main data.

Track, stationary	
Diameter, central line	280 m
Banking angle	45°
Track element, concrete	144 blocks
Track element, shape	double C
Track element, length	6 m
Reaction plates for lift	Aluminium
and guide: width; thickness	600 mm, 20 mm
LIM reaction plates: height,	800 mm;
thickness	11 mm
Crane for 45° load movement	20 t

Test carrier	
Total final mass	18 t
Length: frame, overall	10 m, 12 m
Rated speed	200 km/h
Minimum operational lift gap	100 mm
Rolling mode lift	8 tyres
Rolling mode guide	8 tyres
Gauge, same as magnets	2,6 m
Rated thrust of LIM	22 kN
Converter system	0-105 Hz; 5 MVA
Current rails for LIM supply	3 rails, 3 kV, 1000 A
Current rails, auxiliary	3 rails, 380 V, 200 A

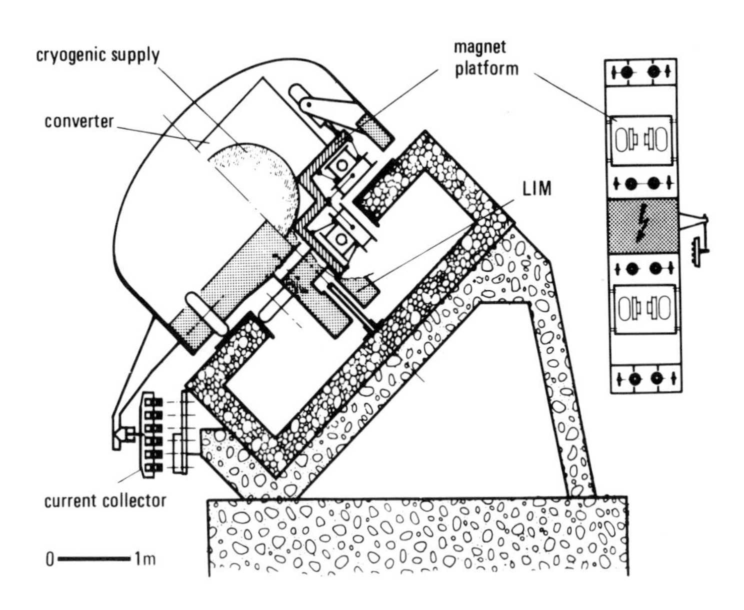

Fig. 41 Test carrier EET: schematic drawing.

Fig. 42 The test facility at Erlangen 1975: aerial photo.

Fig. 43 The Erlangen Test Carrier EET passing the stop point (1974).

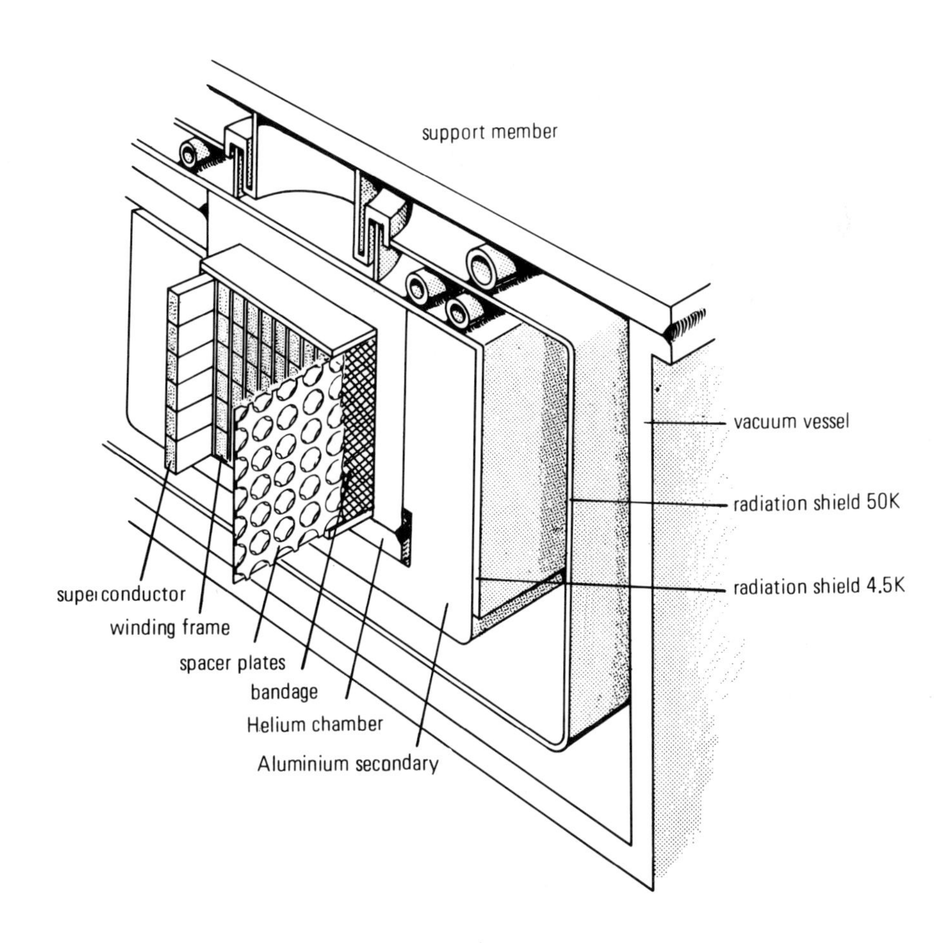

Fig. 44 EET Magnet: Cross-section [106].

modified tank system developed by Linde as an intermediate step towards active facilities (See Fig. 45) [107].

A small compressor set generates He gas pressure (2 bar) in a 150 1 liquid He tank. It forces the liquid to pass the superconducting windings of a series of magnets, after being subcooled to one phase state in a heat exchanger. A Joule-Thomson valve changes the state of the coolant into two phase flow allowing for an efficient cooling of primary thermal shields in the magnets. Much of the gas enthalpy is made use of by a set of heat exchangers before the gaseous He feeds the room temperature compressors. Part of the gas is branched off at about 50 K for thermal shielding of magnets and connection pipes.

The Erlangen test carrier has been operated since March 1974 in the rolling mode, i.e., on its air craft wheels and without superconducting magnets. In March 1975 it achieved a velocity of 230 km/h in this mode, which was the mechanical limit of the track when loaded with a 12 ton vehicle. The EET equipped with four lift magnets and then weighing 16 tons was successfully electrodynamically levitated for the first time in March 1976. A clearance of 100 mm between the track surface (Al-sheet) and the bottom of the cryostats was achieved at speeds up to 120 km/h. Figure 45 indicates the speed range where the EET was levitated and the dependence of total load, lift and drag forces on speed.

In parallel with the lift-experiments, investigations on the vehicle dynamics and design studies on a commercial vehicle are being carried out. In addition, experiments with LSM-propulsion systems are in preparation [108].

A test-rig with a full size water cooled magnet for investigating the effectiveness of various primary damping methods yielded the first results. The additional ampere-turns required under each lift magnet for an adequate ride comfortwill not be more than about 3% of the ampere-turns in the superconducting magnets. This leads to a control power of about 0.4 kVA per ton levitated.

The next steps in this program are corresponding measurements with a special superconducting magnet to check the combined function of the normal control coils and a superconducting magnet in persistent mode, and the application of the developed control strategy and damper layout to an EET magnet platform on the EET.

With respect to LSM-propulsion, a rotating drum test-rig with a diameter of 5.8 m is under construction. The conversion of the 280 m-circular track to LSM-propulsion is in preparation for 1977. Corresponding superconducting pole magnets for the EET are under construction.

At present, a 120 ton, 200 seat commercial vehicle is under design in the project group Maglev (AEG, BBC and Siemens) at Erlangen. It consists of two equal sections coupled together resulting in a total length of 50 m. The principle design features can be seen from Fig. 47. The propulsion of the vehicle will be accomplished by a linear synchronous motor (active track type) with a thrust of 87 kN at a speed of 500 km/h,

Table 9 EET magnet: design data [106]

Conductor (VAC)	
Material	NbTi 50, Cu (VACRYFLUX)
Cross-section	2,45 mm × 1.4 mm
Critical current (3.5 T/4.2 K)	1000 A
Cu:NbTi	5:1
Twist length	50 mm
Filaments	300 × 50 μm

Magnet	
Rated lift force	60 kN
Rated current (3.4 T/5 K)	500 A
Ampereturns	515 kA
Effective current density	81 A/mm^2
Maximum flux density	3.4 T
Stored energy	120 kWs
Total mass	540 kg
Mass at 4.3-5 K	200 kg
Winding cross-section	83 mm × 76 mm
Winding length	1.0 m × 0.3 m
Cryostat length	1.4 m × 0.6 m
Height of cryostat body	0.24 m

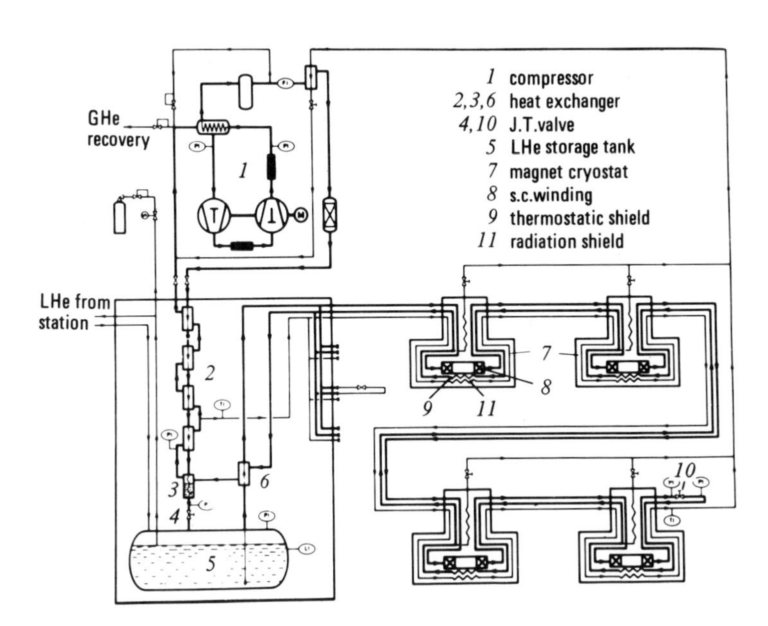

Fig. 45 EET-on-board cryogenic system (Linde A.G.) Flow diagram.

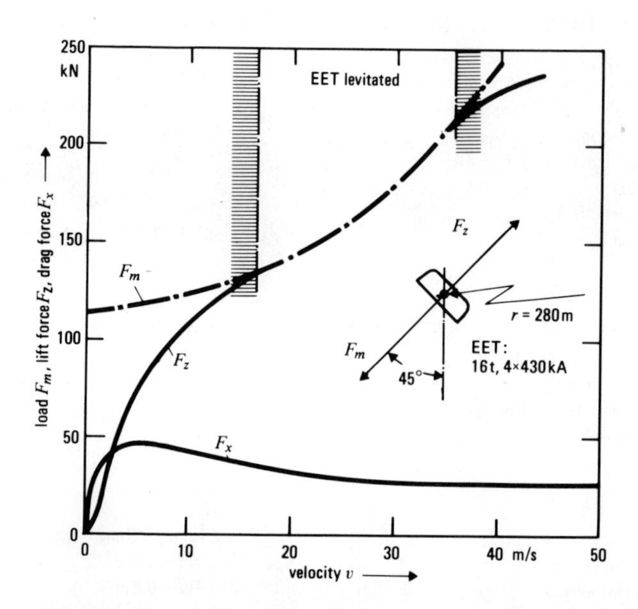

Fig. 46 First EET levitation experiment (March 1976): total load, lift and drag forces [108].

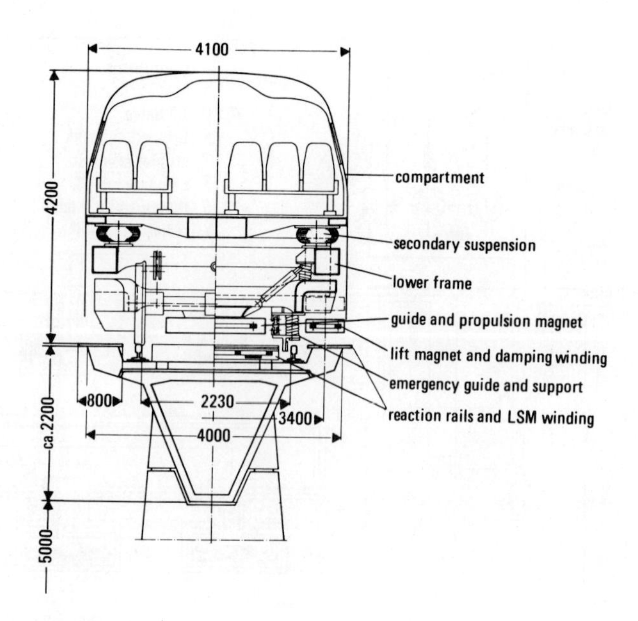

Fig. 47 Conceptual EDS vehicle: Cross-section (dimensions are in mm) [108].

which corresponds to an input power of 18 MVA. The lift with a free air gap of 100 mm will be generated by 10 pairs of superconducting magnets in normal flux operation and in interaction with aluminium sheets 0.8 m wide and 0.02 m thick and mounted on the track. The magnet windings are 1.5 m long and 0.3 m wide carrying a total current of 0.5 MA. Guidance of the vehicle will be with a difference flux system (figure-of-8-loops) laid into the track and interacting with the propulsion magnets. The superconducting magnet windings are planned to be bath-cooled by liquid He which is continuously refilled and recovered by a decentralized refrigerator system on-board of each vehicle section (see also Fig. 40).

The program described above will be critically assessed at the end of 1977. At this time a major comparison of the results of the two German magnetic levitation projects will be carried out. This will lead to a decision as to which of the two levitation systems (EM system or ED-system) will be developed and sponsored afterwards. A commercial vehicle of the selected type is planned to be ready for testing in the early eighties. By that time a 60 km long national test track (or sections of it) is to be constructed in the Donauried north of Augsburg.

2. The Japanese National Railway magnetic levitation project

This project was initiated officially in 1970. Its original background was the high rate (30% per year) of increase of passengers on the Shinkansen (New Tokaido Line). It reached an average of 300,000 passengers per day in 1972. (The Shinkansen runs at a maximum speed of 210 km/h between Tokyo, Osaka and Hanaka, which is about 1000 km south from Tokyo). It is expected that a saturation of the passenger capacity of the line, especially between Tokyo and Osaka, will occur by 1980.

Japan had to deal with the problem of increased passenger capacity. In addition, the problems arose during the past years of environmental impact, especially noise and vibrations caused by a conventional high speed train like Shinkansen in heavily populated areas. As in Germany, the electromagnetic and the electrodynamic system are being investigated but with distinct emphasis on the ED-system. The project is carried out by the JNR in close collaboration with the electrical companies Fuji, Hitachi, Mitshubishi, and Toshiba. The development program is based on discrete coils in the track and superconducting normal flux-coils in the vehicle charged with a current of 250 KAT (AT = ampere turn) for levitation. For propulsion and guidance, as in the German project a combined system is provided, i.e., using superconducting field windings on-board in combination with an active track for LSM-propulsion and in combination with difference flux windings for guidance. A current of 500 KAT is provided for the propulsion and guidance magnets [109, 110]. Table 10 gives the preliminary specifications which were set for the system and component evaluation.

During the course of the program a number of light-weight magnets with flat cryostats were constructed in order to study their mechanical, electrical and thermal behavior. The type of cooling envisaged is bath cooling in sealed off cryostats or in cryostats coupled to an on-board refrigerator system. With respect to active on-board refrigerator systems development was carried out on a movable vane type rotary compressor and the slash plate type helium compressor.

In order to study the behavior of LSM-propulsion, experiments were conducted with a circular test track 20 m in diameter, and a model train with 8 vehicles (equipped with ferrite magnets). It was observed that the propulsion control system showed excellent characteristics [109].

A number of simulating rotary disc- or drum-type experiments were carried out for investigating lift- and drag-forces of normal flux and track coil systems and in order to compare the experimental results with theory. These types of experiments also included the problem of damping. In 1973 a test facility had been completed at the JNR Technical Laboratory where the combined system of LSM-propulsion and difference-flux guidance (CSPG) was studied [111]. The experimental results were in good agreement with theory and confirmed the possibility of realizing a CSPG.

In addition to the development work described so far, wind tunnel experiments with respect to optimizing pitching moment and rolling, and investigations of biological effects in magnetic field environments have been carried out.

In order to study actual levitation characteristics and to demonstrate the feasibility of the technique, two vehicles were built and operated in 1972. Vehicle No. 1 weighing 2 tons was propelled by a one-phase LSM utilizing the fringe field of two superconducting magnets (for lift) and ran over a track of total length of 220 m. Vehicle No. 2, called ML-100, was built and operated in order to commemorate the centenary of Japanese railway in 1972 and for demonstration purposes. Figures 48 and 49 show a photograph and a cross-section of ML-100. The vehicle has a weight of 3.5 tons. It was propelled by a linear induction motor with active track up to a speed of 60 km/h over a distance of 480 m. It was levitated by four superconducting magnets with 250 KAT each.

Recently the test facility of ML-100 has been modified. In addition to the normal flux levitation system a combined LSM-propulsion and difference flux guidance system was installed [112]. The modification was carried out in order to attain know-how. This information can be fed back to improve a 7-km experimental line which is now under construction at a site in Kyushu, Miyazaki Prefecture.

The test line is elevated in structure, almost straight and even and has an inverted T-guideway. The horizontal portions of the guideway are equipped with the lift coils, the vertical part comprises vertically arranged windings for a combined LSM-propulsion and difference flux guidance system. Corresponding horizontal and vertical superconducting coils are installed in the vehicle. Figure 50 depicts a cross section of the guide way and the test vehicle. The major specifications of the experimental unit are given in Table 11.

Table 10 Preliminary specifications for the Japanese magnetic levitation system [109].

Maximum number of cars per train	16
Maximum speed	550 km/h
Maximum acceleration	3 km/h/s $\hat{=}$ 0.1 g
Deceleration (normal brake)	5 km/h/s
Deceleration (emergency brake)	10 km/h/s
Take-off speed	below 100 km/hr
Levitation height	250 mm
Accuracy of track	± 5 mm/10 m
Hours of operation	18 hr (6:00 AM to midnight)
Number of superconducting magnets	
levitation	4 × 2 rows/car
guidance and propulsion	4 × 2 rows/car
Car:	
weight	30 tons
dimensions	25 m × 3.4 m × 3.4 m (length × height × width)
Propulsion	active track LSM

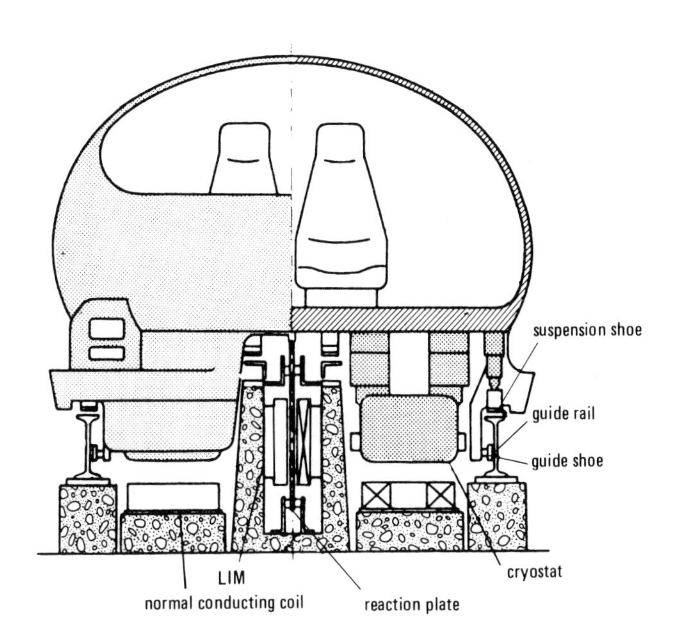

Fig. 48 Cross-sectional view of the ML-100 [109].

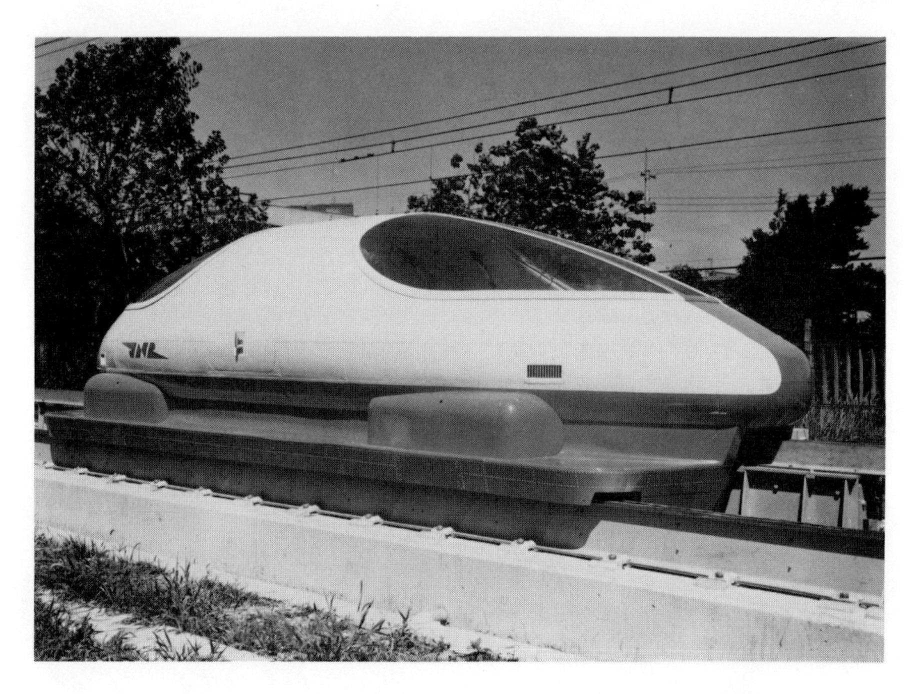

Fig. 49 Japanese test vehicle ML-100 [109].

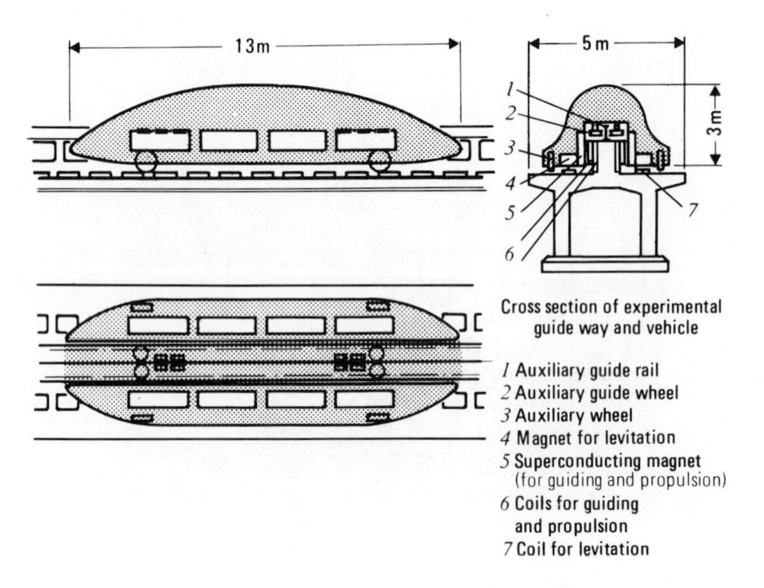

Fig. 50 Japanese Test Vehicle for the 7 km experimental line (State 7/75).

Table 11 Major specifications of the Japanese 7 km experimental line [112].

Maximum speed:	500 km/h (5 s continuously)
Clearance:	100 mm
Vehicle LXWXH:	13 m × 4 m × 3 m
Weight:	10-tons
Auxiliary support guide device:	wheels
Emergency brake device:	type friction
Levitation:	Inductive repulsion type by means of superconducting magnet and difference-flux type guide system
Propulsion:	Combined guide and LSM propulsion system
Guideway:	Inverted-T type (central convex portion is demountable) entire-line elevated guideway

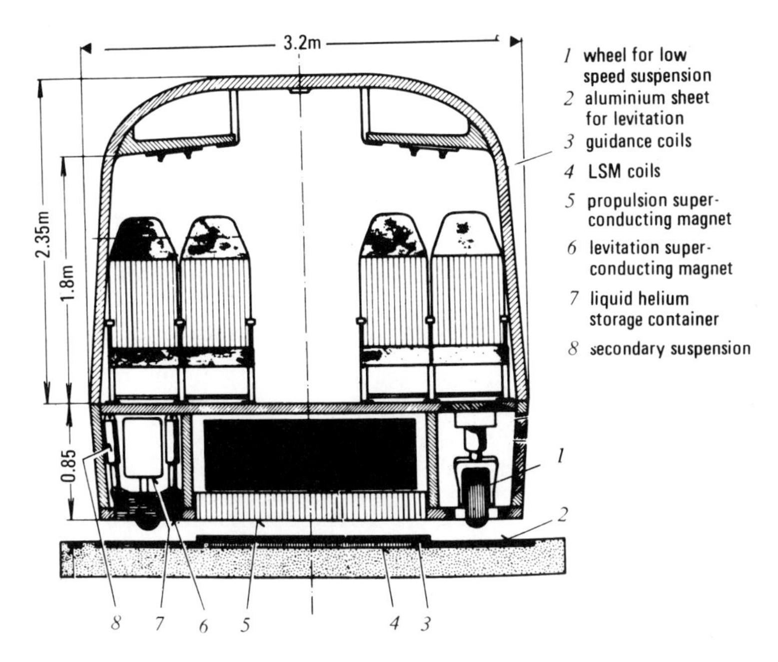

Fig. 51 Cross-section of proposed Canadian superconductive Maglev vehicle and its guideway [113].

The construction of the 7 km line is planned in two steps. The construction of the first track section has already started and will be completed in spring 1977. The completion of the test vehicle which is half the size of the final system is expected by the same date. Travelling experiments on the first track section with reduced speeds are foreseen for the end of 1977 [112]. If the results with the 7 km test facility lead to a go-ahead decision, a test line extending more than 25 km will be constructed to repeat full-scale operational tests.

3. The Canadian Maglev-project

The Canadian Maglev group, an interdisciplinary team of scientists and engineers from Queens University, Kingston, the University of Toronto and McGill University, Montreal was set up under the Administration of the Canadian Institute of Guided Ground Transport at Queen University. Its aim is the development of an electrodynamic levitation system with linear synchronous motor propulsion for high speed inter-city transportation along the Toronto-Ottawa-Montreal corridor for the 1990's.

Single 100-passenger vehicles are provided, weighing 30 tons and levitated with a clearance of 15 cm by eight 385 KAT, 100x30 cm superconducting magnets. The magnets interact with two 80x1 cm aluminium strips laid in parallel along an elevated flat-topped guideway. The aluminium sheet thickness is graded from 1 cm at high speed to 3 cm in low speed sections to maintain total drag (magnetic and aerodynamic) almost independent of speed. For linear synchronous propulsion the vehicle is equipped with fifty superconducting magnets 40 cm x 150 cm, carrying 500 KAT each. The track is comprised of split three phase windings, energized by variable frequency current source converters in 5 km sections. The propulsion magnets simultaneously interact with flat difference flux loops overlying the LSM windings causing the main portion of lateral guidance, which has a stiffness of 1000 kN/m. Figure 51 shows the cross-section of the proposed Canadian superconductive Maglev vehicle and its guideway. The conceptual design provides an on-board cryogenic system with sealed off cryostats. The maximum temperature and pressure in these cryostats will rise to 13 K and 20 bar respectively. Therefore the use of Nb_3Sn-magnets is indispensible [113, 114].

During the course of the program, theoretical and modelling experimental studies of all important components and preliminary economic analysis have been carried out. For full scale tests, a 7.6 m diameter wheel rotating around a vertical axis at peripheral speeds up to 100 km/h has been developed and put into service. Guideway components can be attached to the vertical rim. Vehicle-borne components are mounted in a stationary harness and six component balance for positional adjustment and for the measurement of forces and torques. Experiments on combined propulsion and guidance are under way with half-scale superconducting magnets [114].

4. Work on magnetic levitation in Great Britain

In England two levitation projects are under way. One is based on the repulsive electrodynamical system. The other is a very new system which is called "Mixed System".

The Wolfson project, University of Warwick, is concerned with the construction of a 550 m long test track, for exploring electrodynamic levitation. The system under investigation combines the functions of levitation, guidance and linear propulsion. The vertical lift and the lateral guidance forces of a 150 kg, 2.5 m long test vehicle are generated by two superconducting magnets moving above two aluminium strips which are arranged in parallel along the guideway. Within the gap, between the two levitating strips, active track windings are laid which also interact with the superconducting magnets and provide the vehicle thrust. Most of the magnetic flux of the superconducting coils is used for propulsion and only a small portion for levitation and guidance. Besides a linear synchronous motor, a linear commutator machine with active track is also being investigated [115].

Work on a new levitation system called the "Mixed System" was reported on very recently [116]. It is carried out jointly by the Rutherford and Culham Laboratories. The levitation system is based on the fact that with arrangements of superconducting coils and superconducting magnetic flux screens or with split superconducting constant flux coils magnetic field configurations can be attained which allow a stable suspension of an iron body. Fig. 52 shows one of three experimental arrangements, with which the new levitation principle was verified at the Rutherford Lab. A superconducting magnetic flux screen is concentrically positioned inside a short superconducting solenoid providing a radial stabilizing force for the iron disk in addition to the original existing axial stabilizing force generated by the solenoid. It has been suggested by the two laboratories that such a scheme can be used for contactless levitation in high speed ground transportation. A practical arrangement would use a passive iron track with superconducting screens and/or coils mounted on the vehicle. The following advantages of the mixed system are proposed.

One obtains the full lift forces without any vehicle movement.

The magnetic drag losses are small compared with ED-system. The system is not based on induced currents and only parasitic eddy currents are generated.

Large clearance gaps between vehicle and track can be obtained.

The levitation is stable in all directions.

At present the development work seems to be in the stage of theoretical considerations and model-experiments. So far, no practical coil-track arrangement has been suggested.

5. The US program on magnetic levitation

In 1966 Powell and Danby, of Brookhaven National Lab [117] sug-

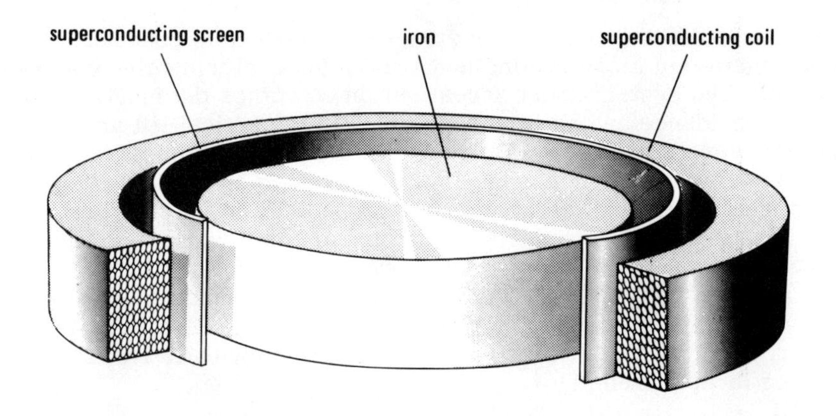

Fig. 52 Iron disc suspended inside a superconducting magnetic flux screen (Rutherford-Culham Labs).

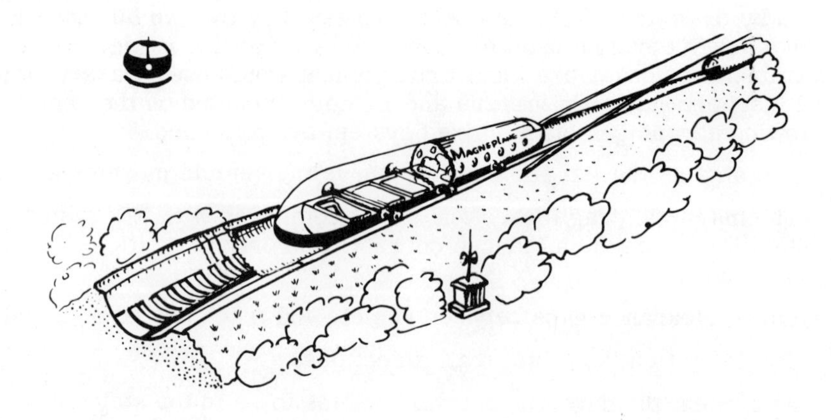

Fig. 53 Principal concept of the MIT – MAGNEPLANE [119].

gested that superconducting magnets in combination with discrete loops in the guideway could be used to levitate and guide fast tracked vehicles. This suggestion initiated the above mentioned research and development programs in a number of industrialized countries. Before 1971 theoretical and experimental work was done by Stanford Research Institute (SRI) and by Ford Motor Company. These studies provided a detailed mathematical formulation and experimental verification of the forces on a rectangular coil operated at an arbitrary height and velocity above a continuous conducting sheet of finite thickness, as well as over an L-shaped guideway. Since the beginning of 1971, the work has been sponsored by the US Department of Transportation (DOT) and the National Science Foundation (NSF) [118].

In the years following, intensive work on dynamic vehicle stability and ride quality was carried out by SRI and Ford, including work on a test sled vehicle running at low speeds. The program was refunded by DOT in the middle of 1974 aiming for a conceptual design and study of a passenger carrying system and the design and testing of a high speed (500 km/h) rocket propelled test sled to collect data relevant to a commercial vehicle conceptual design.

The commercial vehicle with an empty weight of 30 tons, passenger capacity of 80 seats is propelled along an inverted T guideway by two noise-suppressed ducted fans driven by regenerative gas turbines. Lift and guidance is accomplished by eight superconducting normal flux magnets (0.5 x 1.5 m, 350 KAT) and L-formed Al-sheets in the guideway resulting in a levitation height of 30 cm.

In addition to the ducted fan propulsion system, LIM and LSM propulsion systems were also considered seriously. Furthermore, a very advanced propulsion system was suggested. A rotating superconducting paddle wheel or helix [102] (see also Fig. 38) is installed in the vehicle and generates thrust and lift when interacting with the Al sheets in the guideway. Elimination of the levitation magnets should be possible with this kind of propulsion scheme.

During January 1975 the program described above was cancelled by the executive branch of the US Government. As a result, the high speed rocket propelled test sled will not be constructed. In parallel with the work of SRI and Ford Company, outstanding work on electrodynamic levitation was carried out at MIT, sponsored by NSF. Kolm and Thornton have proposed the bending of the continuous track sheet into a semicircular arc surrounding the lower third of the vehicle's circumference [119]. This concept allows self banking of the vehicle and thus eliminates the need for an articulated suspension or an unrealistically smooth guideway which is accurately banked for curve travelling. Figure 53 shows the concept of the MIT-Magneplane. The feasibility of a LSM-propulsion system was demonstrated using a 1/25 scale model developed by MIT and Raytheon and running along a 120 m guideway [120]. Pancake shaped superconducting coils provided both the levitation forces and the field for the LSM. The MIT-Raytheon project was also terminated early in 1975.

IV. ELECTRICAL MACHINES

A. Introduction

Electrical machines have been the subject of steady development since the formulation of the dynamo-electrical principle and the construction of the first electrical dynamo more than a hundred years ago. Today, electrical machines have reached a degree of quality and performance which is rarely encountered in other fields of engineering. It is obvious that at this stage of development further improvements are limited to small steps. In some cases the limits have been reached already, or are within sight.

The advent of superconductors, however, has opened up new prospects. The use of high-field superconductors in electrical machines promises to be a large step forward, particularly as far as the limit ratings, specific output and efficiency are concerned. At the present time, the application of superconductors is restricted to the field windings of dc and synchronous machines. This section is therefore limited to these machine types. In view of the cryogenic equipment required, application of superconductors is recommendable for large machines only.

B. Limits of Conventional Machines

The power of an electrical machine with a cylindrical air gap (see Fig. 54) is expressed by the well-known equation

$$P = CD^2 \ell n \tag{1}$$

The Esson (output) coefficient is expressed by

$$C \approx 0.12\, A_{eff} B_r \quad \text{in kVA min/m}^3 \quad , \tag{2}$$

where D is the armature diameter in m (air-gap diameter), ℓ is the length of active part of armature in m, n is the speed in rev. per min., A_{eff} is the effective linear current density of air-gap periphery in kA/m, and B_r is the radial component of air-gap flux density in T.

The quantity of C also depends on the field pattern to a certain extent. Only the fundamental-wave amplitude of B_r is normally substituted in Eq. (2). For synchronous machines, the power equation yields the apparent power (in kVA), and the active power (in kW) for dc machines. As indicated by Eq. (1), the Esson coefficient is a measure of the performance of a machine with given dimensions, and for the mechanical and electrical stressing of the machine components.

The obtainable air-gap flux density of conventional machines, which are excited through a magnetic core with a copper winding, is limited to values slightly over 1 T. This is due to the necessary slotting of the rotor and stator cores and the saturation induction of the cores,

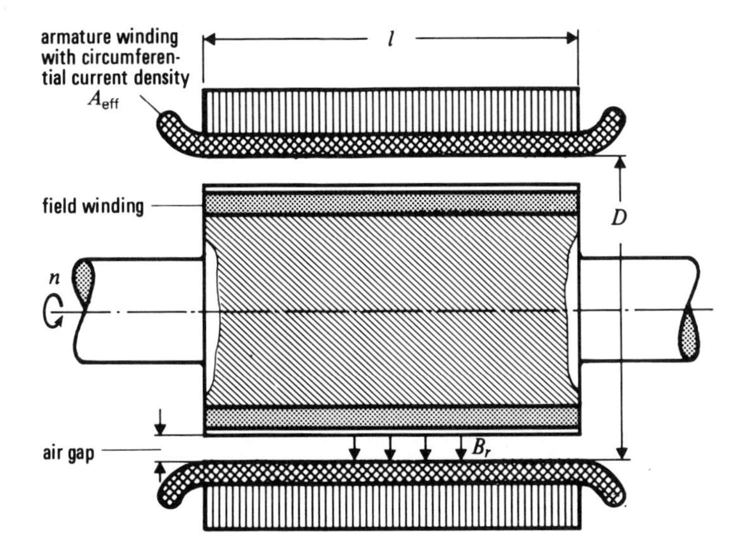

Fig. 54 Schematic view of a rotating electrical machine.

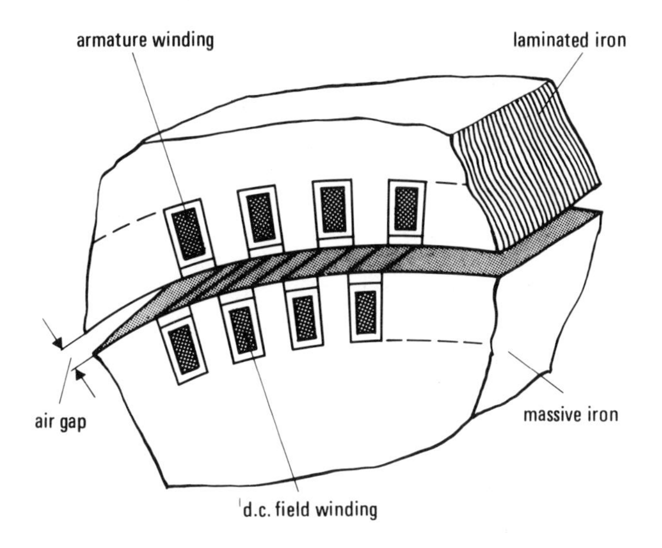

Fig. 55 Conventional electrical machine with iron.

which is approximately 2 T. The magnitude of the technically obtainable linear current density is limited by the slot cross-sections available, the conductor insulation and the current density attainable with copper conductors (up to about 20 A/mm^2). Figure 55 gives an impression of the space available for copper conductors in a conventional machine with iron.

The Esson coefficient of the largest machine built so far, the four-pole 1500-MVA turbo-generator of the Biblis nuclear power station [121], is $C \approx 40$ kVa min/m^3. Its effective linear current density of air-gap periphery is of the order of 275 kA/m and its air-gap flux density around 1.2 T. With smaller machines, the values of A_{eff} and B_r are limited to lower values.

According to Eq. (1), an increase of the machine limit rating would, in principle, be possible by enlarging the armature volume or by increasing the speed. These possibilities are, however, subject to limitations which are determined by the mechanical properties of the materials (centrifugal forces), the maximum permissible distance between the bearings (vibrations), and transportation problems. These limits have nearly been reached by the large machines built recently. Thus, an increase of the machine power will only be economically possible if the linear current density of the armature and the flux density are increased without having to accept a reduction in efficiency. Subject to certain conditions, the use of stabilized high-field superconductors offers promising possibilities.

C. Superconducting Machines: General Remarks

When exposed to high magnetic field amplitudes at power frequency, stabilized superconductors experience intolerably high losses. The use of superconductors is therefore limited to electrical machines with dc field windings, i.e., dc and synchronous machines. Copper conductors still have to be used for the ac armature windings of these machines. The armature of superconducting homopolar machines is also made of normal conductors. Although the armature carries pure direct current the current collection from a rotating armature cannot be governed in a suitable manner at the extremely low temperature involved. Despite this limitation, the use of high-field superconductors on the machines mentioned can bring about considerable improvements, as explained below.

Field windings made of stabilized high-field superconductors are capable of producing current densities which are one to two orders of magnitude higher than those of effectively cooled copper windings (taking the slot cross-section as a reference quantity). Superconducting field windings thus allow the generation of a high air-gap flux density without having to use iron in the field winding and armature. Magnetic iron is only required for the screening of the machines against stray fields. The omission of magnetic iron in the machine has two important consequences. First, the air-gap flux density is no longer limited by the saturation induction of the iron. Thus a considerably higher flux density is obtainable than in conventional machines. Second, almost the entire armature circumference becomes available for the installation of copper conductors. This provides a considerable increase of the armature linear current density.

The two factors together provide a considerable increase of the Esson coefficient, Eq. (2). A correspondingly high increase of the specific output (in kW/kg) and of the limit rating results. Higher efficiencies are to be expected because of the small losses of superconducting windings. The small losses are largely due to the thermal insulation, the connecting high-current leads and eventual conductor joints.

In order to achieve low volume and low weight, field windings of intrinsically stable superconductors have to be used which allow the necessary high current densities. In the case of dc machines (especially large homopolar machines) current densities of about 100 A mm^{-2} or even less are sufficient. For large synchronous generators densities of > 100 A mm^{-2} are indispensible. Besides dc loadings, superconducting field windings have also to resist to a certain extent alternating fields and currents. These ac loadings occur due to variations of the exciting field (within seconds) and as a result of armature reaction, particularly during transient conditions, such as load changes and short circuits.

In the case of dc machines, armature reaction is negligible or zero. Therefore, the design of the field winding has to take into account only changes of field excitation. If the variations necessary are weak and slow (e.g., motors for naval propulsion), intrinsically stable NbTi-multifilament superconductors with copper matrix may be used. If rapid changes of the excitation have to be applied (e.g., generators for naval propulsion), multifilament conductors with a two component matrix and/or transposed conductors in the form of Roebel bars is a proper solution. Superconducting windings for dc machines are either medium sized (heteropolar machines) or of the simple solenoid type (homopolar machines). Therefore the designer of superconducting dc machines can rely a great deal on the experience gained from the construction of superconducting magnets [122].

In contrast to dc machines in the case of large synchronous machines, the armature reaction has a considerable effect on the field winding under actual conditions. To keep the losses due to unbalanced loading and transient phenomena within acceptable limits and to reach the required high field winding current densities, the armature and field winding have to be "decoupled" by means of damping cylinders. Some of the main problems encountered in the development of large synchronous generators are the accurate prediction of the alternating fields and currents affecting the superconductors, the theoretical and experimental determination of the resulting losses and current degradation, and the design of the superconductors selected to cope with these conditions.

The results obtained so far show that interlaying of damping cylinders may reduce the armature reaction to such a small level that the conductor for the field winding also has to be selected mainly with respect to changes of the field-excitation. The field windings of large synchronous machinery are however of rather complicated shape, i.e., dipoles or quadrupoles, and simultaneously of enormous size. In addition, they have to be rotated at high speeds (e.g., 3600 r.p.m.). It is obvious that these conditions create new problems which have not yet been solved within the framework of the existing technology of ordinary superconducting magnets.

D. DC Machines

At the present time the use of dc machines is limited largely to motors where a good degree of speed control is necessary, for example, in high-power high-torque motors (typically 5000 kW at 100 rev/min) for steel mill drives where precise speed control is absolutely essential. There are a vast number of small dc motors employed in industry but because of refrigerator requirements these small motors are not of interest for application of superconductivity. Only in a few isolated instances are dc generators employed. Traditional dc machines are of the heteropolar type.

1. Heteropolar machines

It seems reasonable first to study the benefits which may be gained by the use of high field superconductors in heteropolar machines, especially motors.

Figure 56 shows the schematic of a heteropolar machine. Its main components are the stationary dc field winding (in the example a dipole field), the rotating armature and the commutator. Although a dc current is supplied to the armature terminations, the current flowing through the armature windings is alternating. The actual current amplitude and direction in the armature conductor under consideration depends on the relative position of this conductor to the direction of the field generated by the field winding. The dc to ac conversion is accomplished by the commutator which is fed with current via brushes and acts as a mechanical dc to ac inverter.

In reality, this commutator consists of a plurality of copper segments (in the example only two) separated by insulating pieces and connected to the terminals of a plurality of armature windings. The voltage U_s between adjacent commutator segments is not allowed to exceed 35 V, otherwise brush sparking will occur. The maximum power rating of a heteropolar machine is realted to this voltage U_s by Eq. (3), and is limited to approximately 15 MW [123].

$$P_{max} = \frac{1}{2} U_s A \pi D \alpha \qquad (3)$$

where

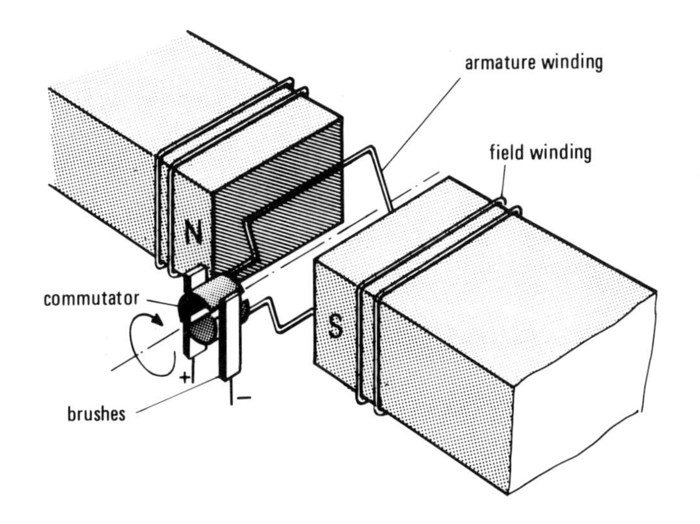

Fig. 56 Schematic of a heteropolar dc machine.

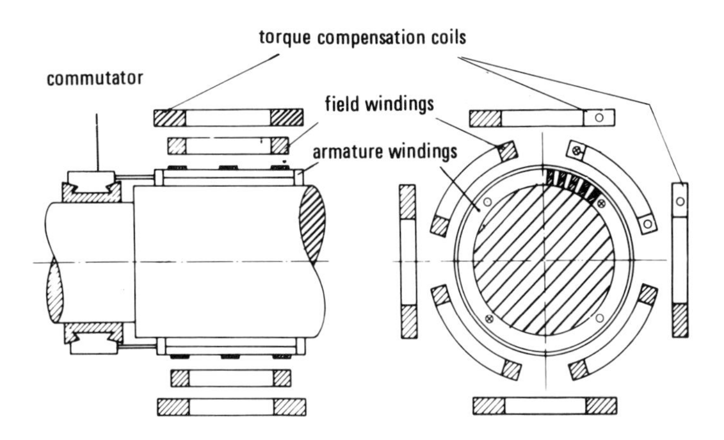

Fig. 57 Heteropolar dc machine with superconducting field windings.

$$U_s = 2\pi D \ell B_r n \tag{4}$$

Again A is the linear current density at the armature periphery and α is the pole arc to pole pitch ratio. The above equations are valid for single-turn lap windings.

Figure 57 shows the principle of a 4 pole superconducting heteropolar machine. The design is very similar to that of a conventional machine. The main difference is that the normal conducting multipole dc field winding is replaced by superconducting windings. The rotating armature winding is constructed with copper conductors as in the case of conventional machines. This is absolutely necessary since the armature windings carry an alternating current. In order to prevent excessive eddy current losses in the rotor body, it has to be constructed of a high strength plastic material, e.g., fiber reinforced epoxy resins, instead of iron. The omission of iron leads to high magnetic flux cutting the copper conductors of the armature winding. Twisted cables instead of compact conductor bars have to be used for the armature winding to eliminate too high losses.

In the design of Fig. 57 normal conducting torque compensation coils are employed. They are excited by the armature current providing cancellation of the torque acting on the superconducting field windings. If the compensating coils are omitted, the superconducting field windings are loaded by the torque. This solution is also possible. Replacing the copper field windings by superconducting ones and increasing the air gap flux density does not elevate the upper limit of the power rating. However, the maximum power of superconducting heteropolar machines will be available at lower rotational speeds.

What else can be expected from the application of superconducting field windings? Within the limit rating, the application of superconducting windings offers a considerable increase in the specific machine output (in kW/kg), owing primarily to maximum air gap flux densities of 3 T or more. Such flux densities can be produced with NbTi windings without serious problems. According to the equations above, an increase of B_r yields a corresponding decrease of $D\ell$ (with n and U_s being kept constant) and, therefore, with A constant, a corresponding reduction of the armature volume and mass. A further reduction in mass is anticipated according to the low density of the plastic material used for the rotor body. In addition, by employing a nonmagnetic rotor body, additional space will be available to increase the armature conductor volume, resulting in lower ohmic losses for the same total armature current. With these possible savings in eddy current and ohmic losses, a total gain in efficiency of between 2% and 4% seems to be attainable in such machines [124].

It seems that development work on large superconducting heteropolar dc machines has so far been carried out only in USSR and in the FRG at Siemens. In the USSR a small 3 kW motor was constructed in 1970 by the All Union Research Institute for Electrical Machinery Industry, Leningrad. A 200 kW motor is currently under construction which serves as a model for larger machines for the steel industry [125]. A 10 MW unit is under design [126].

At Siemens, a detailed design study supported by experimental investigations of single components was carried out during the past years.* Figure 58 shows the basic features of the four pole machine under consideration, the output power of which was determined to be 1.7 MW at 260 r.p.m. Both the field windings and the commutating pole coils are superconducting and are installed within the same cryostat. Compared to the field windings, the commutating pole coils are very small. Because of the increase in air-gap flux densities and the omission of iron in the rotor, the armature current fields will have less effect on commutation than in conventional machines. A special feature of the design presented is that the field windings are subjected to the full torque reaction. The torque has to be transmitted by support members (of low heat conduction) through the cryostat to the outer machine foundations. The design study proved that the total energy consumption of such a solution can be considerably reduced, compared to the other known design principle which uses additional normal-conducting coils for compensating the torque reaction [127].

The study led to the following main results.

Although the active volume of the machine is only about 50% that of a corresponding conventional 8-pole machine, the total volume including refrigerator is about equal.

The maximum benefit in efficiency is about 1.5% (96% compared with 94.5%). This efficiency decreases very rapidly for applications involving frequent and very fast torque reversals, as in the case of motors for rolling mill drives. (The increased losses occur in the field winding during field changes of the commutating pole coils.)

The weight of the superconducting version, including the refrigerator, is only about half that of a conventional machine. It will increase if a large portion of iron is necessary for shielding the magnetic stray field.

The cost (based on today's prices and including all peripheral components) is considerably higher than that of a conventional motor.

The results mentioned above indicated that superconducting heteropolar machines with power ratings of the order of 2 MW will not replace conventional machinery for applications such as motors for rolling mill drives or ship propulsion units. It is clear that the advantages of superconducting heteropolar machines concerning volume, weight and efficiency, will be greater with increasing power rating and size. But even for larger machinery (e.g., 10 MW), it remains to be seen whether the advantages are great enough to replace conventional heteropolar machines in future.

* The work has been supported by the Federal Department of Research and Technology of the FRG.

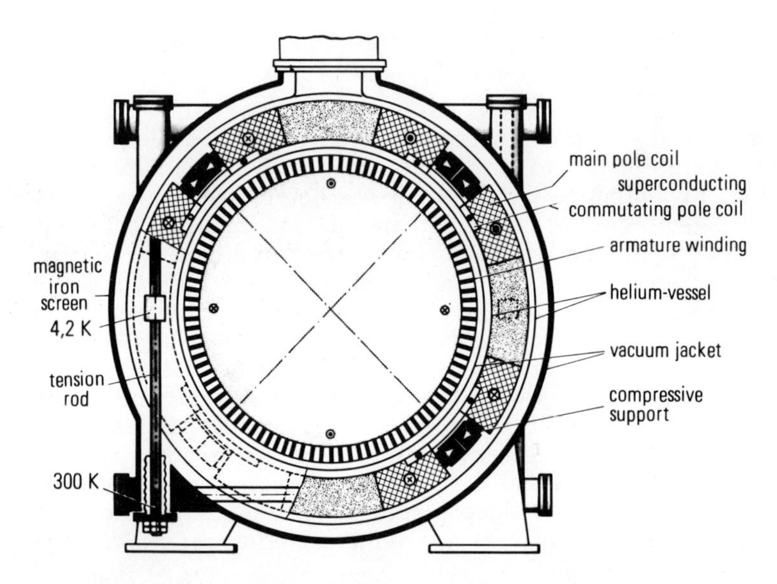

Fig. 58 Basic features of a superconducting heteropolar dc machine, power rating 1.7 MW at 260 r.p.m. [124].

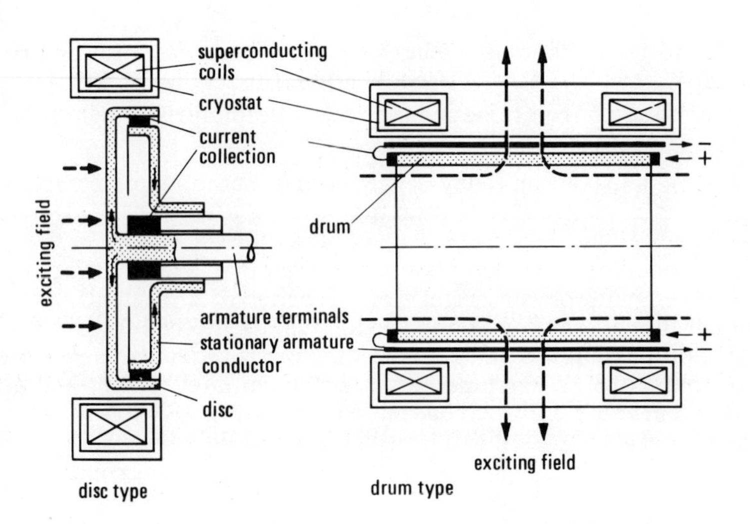

Fig. 59 Principle of superconducting homopolar machines [128].

2. Homopolar machines

a. Basic design features

In the case of dc machines, the main interest today is focused on homopolar machines. These machines have a disc-type or drum-type armature. They do not have a commutator and feature a comparatively simple construction. Figure 59 is a schematic view of a disc- and drum type machine with normal conducting armatures and superconducting field windings [128]. The machines exhibit no armature reaction (since the excitation flux and the armature flux are in quadrature) and the field coils do not experience any torque reaction. As a rough guide, the drum type machine would be employed when the ratings are in excess of about 5 MW. The current collecting system which accomplishes the current transmission between rotating and stationary parts is of special importance and suffers from a number of problems, as we will see in what follows.

The working principle of this machine is based on an arrangement which is known as Faraday's disc. If such a disc, or disc segment, is rotated in a magnetic field with rotationally symmetrical field lines (axis of rotation parallel to field lines), a pure direct voltage free from harmonics is generated. This is in contrast to heteropolar machines.

Since a solid disc or a drum acts like a single armature conductor, the voltages attainable with such an arrangement are comparatively low. The equation for the voltage U between the inner and outer contact of a rotating disc or between the left and right side contact of a rotating drum is

$$U = \phi n \quad . \tag{5}$$

ϕ is the total flux penetrating the disc or mantle of the drum between the inner and outer contact or the left and right side contact, and n is the rotational speed in rotations per second. Assuming a disc with a diameter of 3 m, a homogeneous magnetic flux-density of 1.5 T, and a speed of 100 r.p.m. the resulting voltage is about 18 V (minor slipring diameter neglected).

The inherently low voltage and high currents of homopolar machines were the main reasons that this type of machine has never been widely employed in industry. Superconducting field coils improve the situation because the flux densities are increased by at least a factor of two. This improvement alone is not sufficient. It is necessary to elevate the machine voltage by a series connection of several armature elements which are insulated from each other. Figure 60 shows three ways which have been proposed in order to achieve higher voltages, (a) the segmentation of a disc, and series connection of (b) several discs or (c) several drums [128, 129]. However, whether the voltage is 10 V or 1 kV, the problem of current collection from slip rings remains. It becomes obvious from Fig. 60 that the use of several armature conductor elements

Table 12 Properties of current collection systems.

Type of contact	Maximum current density Am^{-2}	Maximum peripheral speed ms^{-2}	Voltage drop order of magnitude V	Wear per 1000 h mm	Coefficient of friction (brush pressure Ncm^{-2})
Conventional carbon-copper solid brushes	3×10^5	about 40	pos: 0.15 neg: 0.14	pos: 3 neg: 3	0.15 (~3)
Metal plated carbon fibre brushes	1.5×10^6	about 120	pos: 0.35 neg: 0.15	pos: 10 neg: 1	0.4-0.5 (~1)
Liquid metals	$\geqslant 5 \times 10^6$	$\geqslant 120$	0.002-0.005 at $5 \times 10^6 Am^{-2}$ (neglecting MHD effects)	Purification and degassing necessary	friction loss per sliding area $1 MW m^{-2}$ at $150 ms^{-1}$ NaK liquidmetal

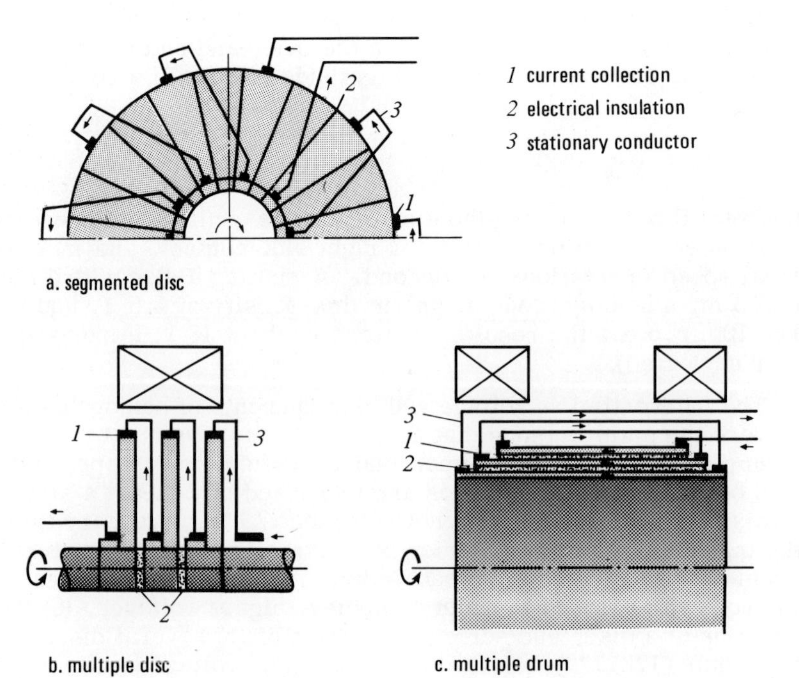

Fig. 60 Homopolar machines with multiple armature conductors.

requires a proportional increase in the number of rotating current contacts.

The use of ordinary carbon brushes is insufficient to attain, with superconducting homopolar machines, linear armature current densities which are usual in air cooled conventional machines, viz. 30 to 100 kA m^{-1}. Indeed, for many of the more important applications, superconducting homopolar machines would not be economic with conventional brushes. Therefore within the existing homopolar machine programs, the main attention is focused on the development of new highly efficient and reliable current collection systems. In contrast the development of superconducting field windings for homopolar machines is of secondary importance, since the necessary technology is principally available.

b. Current collection systems

The requirements which have to be accomplished by these contact systems are high current densities up to high slipring speeds to achieve sufficiently high armature current densities, low electrical and mechanical losses in order to attain high efficiencies, a low wear rate to achieve long life time, and reliable behavior to guarantee reliable operation of the machine.

At present, two current collection systems are under special consideration, metal plated carbon fiber brushes and liquid metal contacts. The first solution is proposed by the IRD (International Research and Development) Company in England [130, 131, 132]. Liquid metal contacts have been mainly investigated in the US [133, 134] and for more than 15 years by Prof. Klaudy, Graz, Austria [135, 136]. Table 12 summarizes the characteristic properties of carbon fiber brushes and liquid metals and compares them with the corresponding data of conventional carbon-copper brushes. It is seen that the properties of carbon fiber brushes are significantly superior to those of conventional brushes and that the liquid metal contacts have excellent electrical properties, i.e., high current densities at high speeds and low voltage drops. The use of liquid metals, however, is related with some problems which do not exist with fiber brushes.

Before discussing liquid metals, a short explanation of the basic features of the metalized carbon fiber brushes is given. These brushes consist of a plurality of fine carbon fibers with diameters of $\leq 10\ \mu m$, plated with a layer of Ag about $1\ \mu m$ in thickness. The thin metallic layer has two tasks, namely to improve the electrical and thermal conductivity along the fiber and to prevent the formation of heavy oxide layers on the copper slip rings by mechanical cleaning. The high current-densities available with these brushes are explained by a comparatively large contact area accomplished by punctual contacts of the large number of single fibers.

Whereas this new type of carbon brush can be handled like a conventional component for designing electrical machinery, liquid metals introduce a new technology and increase the complexity of the machine

Table 13 Physical properties of liquid metals for sliprings.

Liquid metal	Melting point	Density	Viscosity	$\rho^{0.9}\eta^{0.1}$	Surface tension	Resistivity
	°C	10^3kg/m^3	10^{-3}kg/ms	kg/m$^{2.8}$s$^{0.1}$	N/m	$10^{-6}\Omega$m
Mercury	−38.9	13.1	1.20	2600	0.48	1.04
Gallium	29.8	6.09	≈2.0	1370	0.36	0.26
GaIn-76	15.7	6.30	1.50	1300	?	0.29
Sodium	97.8	0.93	0.68*	220	0.195	0.096
NaK-78	−12.5	0.85	0.53*	200	0.115	0.33

The data normally refer to the melting point. * Temperature 100°C

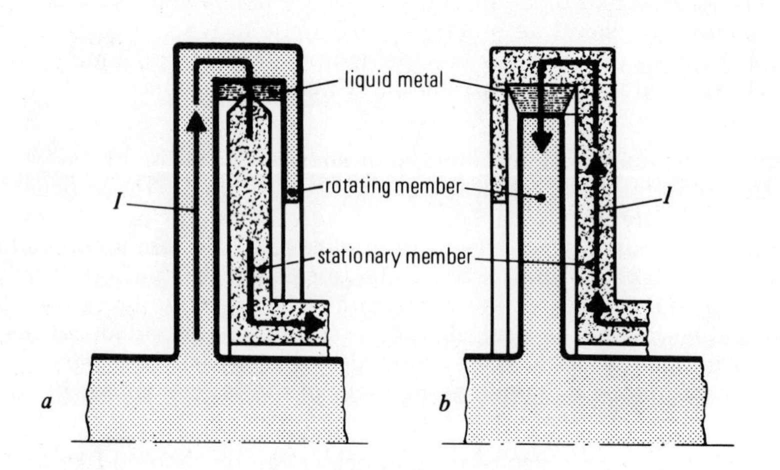

Fig. 61 Schematic of rotating liquid metal contacts [124].

design. Not only must an optimal liquid metal be selected, but also complex support systems must be considered as we will see in the following.

Figure 61 shows the two basic contact configurations. In both cases, the stability of the liquid metal, within the gap between the rotor and the stator, depends on the action of the centrifugal forces. Depending on the viscosity of the liquid and its wall friction, this confinement principle will work only for speeds of the rotating contact member, which exceed a certain minimum value. For speeds below this minimum value or in the case of standstill conditions, the contacts must be able to operate with a liquid sump.

When designing a homopolar machine with high efficiency and performance, it is of primary importance to reduce the losses of the liquid metal contacts to a minimum. It is also important to develop a stable confinement for the liquid metal contacts, in the presence of disturbing magnetic fields. In a proper design, hydrodynamic or viscous losses represent the main contribution to the total losses of the liquid metal contacts. The losses are due to shear in the liquid metal caused by relative motion of the inner and outer portions of the collectors, and are proportional to the density and viscosity of the liquid ($\sim \rho^{0.9} \times \eta^{0.1}$), the cube of the rotor surface speed and the wetted area. The viscous losses can be kept to a minimum by using a liquid metal with a low value ($\rho^{0.9} \times \eta^{0.1}$). As can be seen from Table 13 the eutectic NaK 78 yields the lowest value. At a velocity of 150 ms^{-1}, a speed which generally is assumed to be the upper limit, viscous losses of about 1 MWm^{-2} are to be expected with NaK. Besides NaK, which is relatively difficult to handle, GaIn is being considered as a suitable contact liquid. GaIn is not dangerous, but in contrast to NaK, it has the disadvantage of forming alloys or amalgams with most of the commonly used contact materials.

If there are axial fields acting on a liquid metal ring loaded with a radial current I, circumferentially directed Lorentz body forces are created in the liquid metal. These Lorentz body forces influence the fluid power losses in two ways. First, they distort the fluid's velocity profile and therefore change the hydrodynamic losses, and second by electrodynamic effects. It was proved for either a motor or generator, that the Lorentz body forces increase the total power loss of a liquid metal current collector [137]. The additional losses caused by an axial magnetic field are called MHD-losses.

Due to the velocity profile of the liquid across the gap between stator and rotor, considerable eddy current losses are generated by magnetic fields with a component normal to the surface of the collector rotor. Finally, there are also magnetic forces which tend to remove the liquid metal from the current collectors. The load current produces a magnetice field which runs circumferentially around the axis of the machine (see also Fig. 62). This field interacts with the load current flowing across the gap. It generates a force which tries to expel the fluid laterally out of the collector gap.

Besides continuous liquid metal contacts extending around the total slipring circumference, discrete localized liquid metal contacts are under investigation. One of these is "liquid metal rolling contacts" [138]. Here relatively small diameter rotating wheels replace the large stationary inner contact member of Fig.61. The surface velocity of the wheels is adjusted to be the same as that of the large rotating ring, thus considerably reducing hydrodynamic losses. Another concept provides the use of so-called hybrid collectors or "floating pads". It is a cross between a solid brush and a liquid metal annulus collector. It consists of hydrostatically positioned pads which utilize liquid metal for hydrostatic support as well as current transfer [134].

From the above, it becomes clear, that in ironless superconducting machines with their high flux densities the liquid metal contact zones have to be magnetically shielded. Fields below 0.2 T are required to restrict magnetic losses and disturbing forces acting on the liquid metal confinement.

Besides magnetic effects, other disturbing phenomena are the formation of aerosol and of gas layers at the surface of the rotating contact member. Gas entrainement in the liquid metals is caused by the cover gases (dry nitrogen) which surround the contact zones under a slightly elevated pressure in order to prevent oxidation and contamination.

In order to guarantee the correct operation of a liquid metal current collector, the liquid metal has to be continuously purified and cooled. This is accomplished by a closed liquid metal loop including the current collector and a number of other components, sump tank, liquid metal pump purifier and cooler, flowmeter, etc. Figure 62 shows the principle

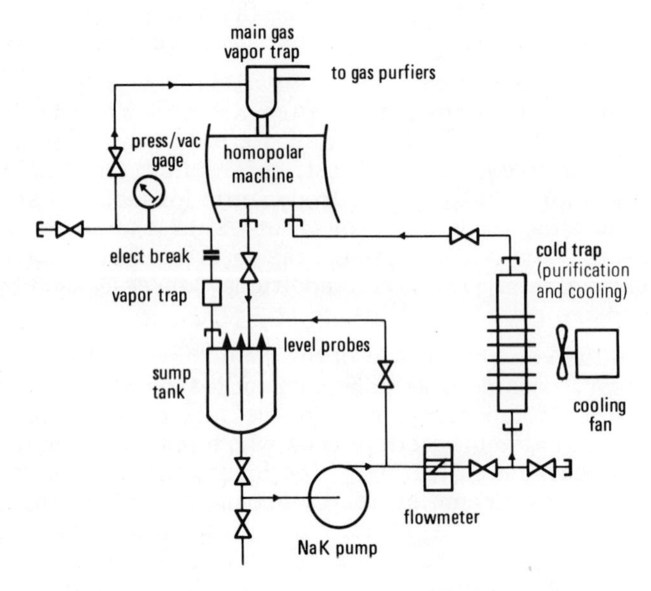

Fig. 62 Purification and cooling loop for a liquid NaK current collector [134].

of such a liquid metal loop. This loop is rather expensive since its components are not ordinary engineering products.

As mentioned above superconducting homopolar machines are in general designed with multiple conductor armatures requiring several current collectors at different electrical potentials. One-loop systems which can service multiple current collection sites are indispensable. Techniques for electrically isolating several flowing liquid streams from each other are under development. In addition to the liquid metal loops, recirculation and purification systems for the cover gas are necessary.

Although the technique of liquid metal current collection systems is without any doubt complex and difficult it is anticipated by experts that all problems can be satisfactorily solved. Results with NaK-prototype loops, continuously running more than 9000 h seem to prove this optimism [134].

c. Applications of homopolar machines

Assuming that current collecting systems with high current densities, low losses, high reliability and long life time are available, superconducting homopolar machines with ratings of several MW will be superior to conventional dc machinery with respect to specific volume, weight, efficiency and capital cost. The fact that they can be produced with ratings in excess of 200 MW (no limitation by commutation the armature current as in the case of heteropolar machines), make dc machines available for a number of applications for the first time.

Superconducting homopolar motors may be designed to have extremely high torques ($M = IN\Phi/(2\pi)$, where I is the current and Φ is the flux per armature conductor and N is the number of armature conductors) and with good speed control. This makes them attractive for steel mill drives, mine winders, paper mill drives and other industrial drives, e.g., auxiliary drives in power stations (large boiler feed pumps and air fans).

Large superconducting dc generators may replace ac generator-transformer-rectifier sets for supplying high dc power directly to aluminium smelters. The total installed capacity for the production of dc power for aluminium smelting is in excess of 30,000 MW and growing at about 10% per annum. The use of superconducting dc generators for the production of chlorine has also been proposed [139]. The installed electrical dc power required for this purpose is at present about 5,000 MW and is increasing at about 10% per year.

With regard to superconducting homopolar motors and generators the major development is, however, directed towards marine propulsion systems for which low mass, low volume and high efficiencies are important requirements. Although, at present, the development is mainly restricted to naval vessels, the civil marine field will benefit significantly from the results which will emerge. Studies on the prospects for the application of homopolar machines to commercial ships indicate that there is at least good qualitative evidence to suggest that the benefits will be substantial [131].

The principal scheme of an electrical ship propulsion system is given in Fig. 63. A high- or medium speed prime mover (steam turbines, gas turbines or diesel engines) with normally constant speed, drives a superconducting dc generator which feeds a high torque low speed dc motor directly coupled to the propeller. Propeller speed variation or reversion can be accomplished by controlling the generator field, whereas the motor field remains constant. A drive of this type operates continuously and provides excellent maneuverability for ships.

The generator-motor set replaces the gearboxes (including reversible gearboxes) and controllable pitch propellers of conventional propulsion systems. It is anticipated that the power transmission efficiency of a superconducting system (generator-motor set) can reach about 98% compared to an upper value of about 96% for mechanical gears. Besides benefits in efficiency, the flexibility and reliability offered by superconducting propulsion systems is particularly useful for commercial ships. Because power is transmitted through cables it is now possible to design the motor and generator locations in a manner such as to achieve the optimal use of the ship's loading space. The remote location of the prime mover-generator part from the motor can lead to an increase in load capacity of 10% in volume limited ships [131]. Moreover, propeller motors of high ratings can be coupled with several generators and prime movers to produce the drive power according to the actual need at optimal prime mover operating conditions (fuel consumption and lifetime). There are many other ad-

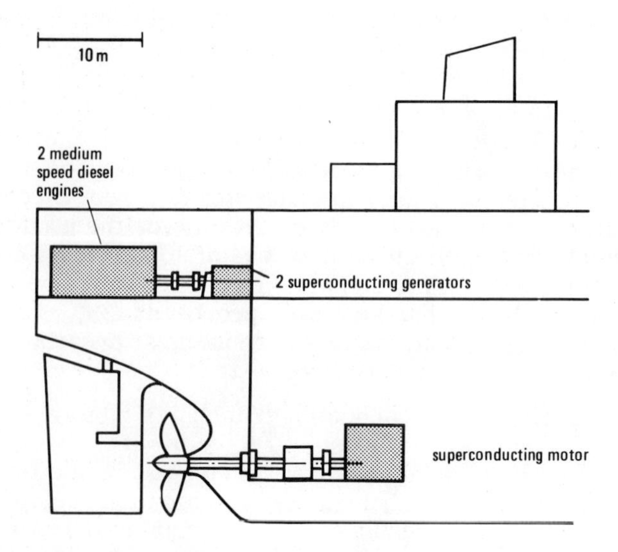

Fig. 63 Principal arrangement of an electrical ship propulsion system [131].

vantages of superconducting dc propulsion systems which are treated in detail in the literature [131, 140].

Experts in the field of applied superconductivity are of the opinion that dc machinery for ship propulsion may be the first large scale application of superconductivity in electrical power engineering. Although the need for high speed commercial ships with large propeller powers (≥ 30 MW) has been markedly reduced in the past due to energy saving policies (needed power about proportional to cube of speed), it is forecasted by ship builders that the reduction will be only momentary.

d. Experimental work on homopolar machines

Until now the development of superconducting homopolar machines has been carried out mainly in Great Britain and in the US. The world's first superconducting dc motor was built in 1966 by IRD (International Research and Development Co.) England. It had a rating of 37.5 kW at 2000 r.p.m. This motor was followed by a 2400 kW prototype motor with a speed of 200 r.p.m. which was successfully tested in 1971. It used a segmented type disc-armature and conventional brushes and operated at 430 V and 5.8 kA [139]. The field coils consisted of fully stabilized NbTi multifilament conductors. Thus the machine was neither optimized with respect to weight nor to efficiency. Recently land tests were completed by the British company on a 1 MW motor-generator set for a naval ship propulsion system. The motor was equipped with conventional brushes. The 370 V, 1500 r.p.m. generator was equipped with the new metal plated carbon fiber brushes. Both machines contain segmented disc armatures. The field windings are built of intrinsically stable NbTi-multifilament conductors.

In the US homopolar machines, exclusively with liquid metal collectors (NaK), have been developed with respect to naval ship propulsion systems. Five years ago, General Electric Co built and tested a 150 kW (10 V, 15 kA, 3600 r.p.m.) dc generator with a multiple-disc armature [141]. The Naval Ship Research and Development Center (NSRDC) Annapolis built and successfully tested a 300 kW ship propulsion motor, which had ingenious design features. The superconducting coil and the dewar, which has a very simple cylindrical form, are located inside the drum armature. A ferromagnetic screen enclosing the rotating armature provides an appropriate field shape [142]. NaK was used as current collector system. A 300 kW homopolar generator is currently under test. Laboratory tests of the integrated system (motor and generator) with a gas turbine are scheduled for the fall of 1976. Shipboard tests on a test-craft are provided for fiscal year 1977. This advanced machine concept is also the basis for the development of a homopolar 30 MW, 180 r.p.m. ship propulsion system which is underway at the Garrett Corporation and General Electric for the US Navy. Such propulsion systems are of interest for hydrofoils and Small Waterplane Area Twin-Hulled (SWATCH) crafts. In the course of this development work prototype motors and generators with powers of 2200 kW (100 V, 1200 r.p.m.) are now under construction by both companies. Recently GE published some data on two 2200 kW motors which they are constructing under the Navy contract [143]. The motors which will be delivered in June 1977 and in March 1978 respectively, have a weight of 3200 kg, i.e., only 30% of the weight of a conventional dc motor. Their volume will be only 45% the volume of a conventional machine. The new superconducting motors are expected to be so reliable that they can be hermetically sealed in a steel enclosure aboard ship to reduce safety hazards and protect the liquid NaK from humidity and salt water.

With regard to liquid metal contacts extensive work has also been carried out at Westinghouse [134]. This work is being performed for the development of a special non-superconducting homopolar propulsion system (SEGMAG) for advanced concept vehicles for both terrain and marine environments.

At present in the USSR development work on superconducting homopolar machines is directed towards power supply schemes for aluminium smelters, i.e., large dc generators [144]. A prototype drum type machine with NaK current collectors and a power of 600 kW at 30 V and 3000 r.p.m. is under construction. A design study on a 60 V, 160 kA generator for aluminium electrolysis is under way.

In France the Laboratoire Central des Industries Electriques has been concerned with a low speed motor for naval propulsion. A 60 kW prototype motor, with liquid metal contacts in the form of a "flooded rotor" was built and tested [145]. In Japan a 3000 kW dc generator (150 V, 20 kA, 1000 r.p.m.) with a single disc armature and conventional solid copper-carbon brushes was built for a copper refining plant [146]. In both countries the construction of new homopolar machines is not contemplated for the near future.

In the FRG design studies on dc motors for ship propulsion are being conducted. Investigations on liquid metal contacts have been carried out at Siemens.* Figure 64 shows the basic concept of the motor under consideration [124]. It is a multiple drum type machine with a cylindrical superconducting field winding placed within the armature and arranged symmetrically to the latter (similar to the machine type of NSRDC, mentioned above). It must be emphasized, however, that the efforts on homopolar machines in Germany are small compared to those in the US or Great Britain.

E. Synchronous Machines

At the present stage of development, the three-phase synchronous machine is the preferred type for the application of superconducting windings. It is a favorable condition that the rotating winding revolves synchronously with the induced field of the three-phase armature winding. Therefore, under ideal conditions, i.e. with steady-state symmetrical loading (neglecting the effect of harmonics), the field on the superconductors remains unvariable with respect to time, despite the armature reaction. Figure 65 shows the schematic of an ac synchronous machine

---

* The work has been supported by the Federal Department of Research and Technology of the FRG.

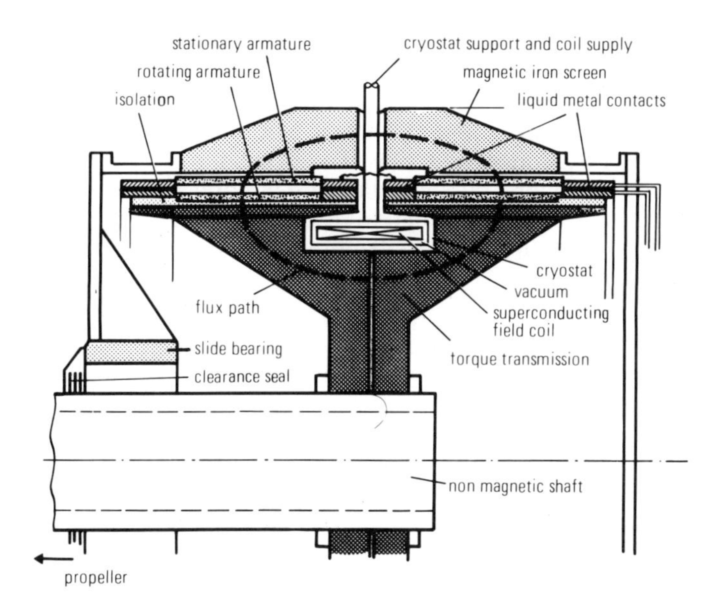

Fig. 64 Schematic of a superconducting homopolar motor for ship propulsion [124].

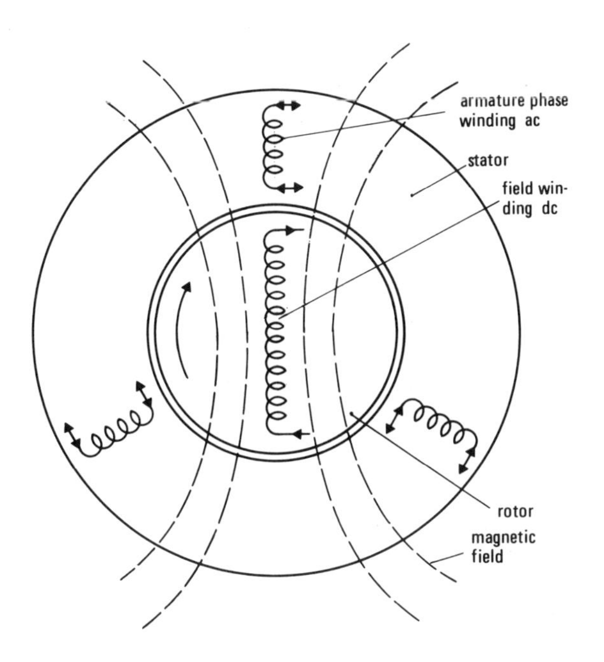

Fig. 65 Schematic for an ac synchronous machine [147].

[147]. In the case of a superconducting machine the dc field winding is superconducting. The armature windings, however, are normal conductors.

Development work on synchronous machines with superconducting windings is primarily concerned with large turbo-generators for central power stations. However, superconducting windings may also be advantageous on smaller machines in order to improve the power-to-size ratio and efficiency, e.g. for ship's propulsion machinery, marine and air-craft generators.

1. Technical limits of conventional turbogenerators

Developments in large nuclear power plants have led to a rapid rise in the unit ratings of turbogenerators over the past ten years. At the present time the largest unit ratings on record are about 1200 MVA for 2-pole machines with speeds of 3000 min^{-1} or 3600 min^{-1} and about 1600 MVA for 4-pole machines with speeds of 1500 min^{-1} or 1800 min^{-1}. Further improvements in the water cooling of the rotor and stator would make it possible to raise the outputs of 2-pole machines to about 2000 MVA and those of 4-pole machines to over 3000 MVA.

The maximum possible outputs for a given type of machine are determined by electrical and mechanical limits [148]. As conventional generators are scaled to larger sizes the machine reactances have to be kept low enough for transient stability by increasing the air gap. The wide air gap requires increased magnetomotive force in the field winding. This corresponds to an increase in exciter power and a decrease in machine efficiency. In the case of 2-pole machines, mechanical limits are caused by centrifugal forces, bending stresses, bending vibrations and torsional vibrations. For reasons associated with material strength and the required operating stability, these parameters limit the maximum rotor diameter (to 1300 mm at 3000 min^{-1}) as well as the ratio of rotor length to diameter. Present-day machines have a maximum rotor-length-to-diameter ratio of 7:1. In the case of 4-pole machines, the maximum rotor dimensions are limited primarily by the means available for manufacture and transport. Owing to the low speed, mechanical stresses are only of secondary importance here. These limiting factors also apply to superconducting machines.

For instance, the water-cooled rotor of the 4-pole, 1500 MVA generator for the nuclear power plant at Biblis in Germany has an overall length of about 15 m, an active length of 7.6 m, a maximum diameter of 1.8 m and a total mass of 200 tons. If the demand for electric power continues to grow at a high rate, and if the unit ratings of turbogenerator sets are increased accordingly, it is quite possible that the output limits of conventional machines will be reached within a matter of about 15 to 20 years.

2. Potential advantages of superconducting generators

The potential advantages of superconducting generators have been generally identified as follows.

A reduction in size and weight by at least a factor of two.

The potential for higher ratings.

Higher efficiency compared with conventional machines.

A reduction in capital plant cost.

A reduction in capitalized cost of losses.

An improvement in power system stability.

As we will see later, the various estimates for the economic breakeven-point for superconducting generators vary between 500 MVA and 1500 MVA.

3. Basic construction of superconducting generators

As already mentioned, the construction principles of superconducting generators are the same as those of conventional machines. A superconducting field winding rotates inside a stationary armature winding, which, because of the high ac losses of high field superconductors, contains conventional (water cooled) copper conductors. Both field winding and armature winding include no magnetic iron. Iron is used only as a magnetic shield to minimize the alternating magnetic fields external to the machine.

In the past a number of small experimental generators were built which had rotating armatures and stationary field windings, since a rotating superconducting field winding requires quite new cryogenic solutions. This solution requires the transfer of the armature power by sliding contacts. However, because of the difficulty of transferring a large amount of power through sliding contacts (even with liquid metal contacts), it has always been clear that any alternators intended for central station use would have to have rotating field windings. The development of superconducting generators in the power range of 1 to 10 MVA, e.g. for aircrafts, is also based on rotating superconducting field windings [149].

a. Rotor

In contrast to a conventional generator where the rotor is a solid forged cylinder, the rotor of a superconducting generator is basically an evacuated hollow cylinder in which the superconducting field winding (a dipole or quadrupole) is suspended. This suspension has to fulfill the following important requirements.

It must transmit large mechanical torques from the warm rotor structure to the cold superconducting field winding.

It must minimize the heat inleak from ambient temperature to the low temperature region.

The air gap between field- and armature winding must be small in order to optimally utilize the superconductor.

It must accommodate differential thermal contraction between the cryogenic and ambient temperature components.

It must fit force transmission without introduction of critical vibrations.

The suspension structures proposed so far differ from design to design but they can be reduced to two basic structures which are shown in Figs. 66 and 67. In Fig. 66, concept a, the mechanical torque to the s.c. field winding is transmitted from the driving shaft via the ambient temperature rotor cylinder (= vacuum case) by several radial supports which are themally isolated via long stainless steel cylinders [150]. The heat inleak by conduction is additionally reduced by fixing an extra cooled thermal radiation shield (50 to 100 K) to one of the torque transmitting cylinders. Thermal shrinkages in the radial direction are compensated by elastic deformation of the cylinders and in the axial direction by sliding of the radial supports.

In Fig. 67, concept b [151], the mechanical torque is transmitted from the drive side of the rotor to the field winding via a stainless steel tube. The tube's extensions are cooled by counterflowing He after having passed the field winding. This cooling guarantees minimal thermal conduction losses through the suspension structure. The differential thermal contraction in the axial direction is compensated by axial slide supports. Radial contraction offers no problem, since elastic deformation of the thermal radiation shield, which is thermally coupled to the cold He stream of the torque tube extension, is possible.

Both designs contain electromagnetic damper shields which have to reduce the alternating magnetic field and current amplitudes acting on the field winding due to phase current unbalance, stator winding harmonics and transient operation. There is a striking difference in both designs with respect to the electromagnetic shields. Concept (a) contains two electromagnetic shields, the outer ambient temperature rotor case and the cold thermal radiation shield which acts simultaneously as an electromagnetic shield by choosing copper with a proper thickness as shield material. Concept (b) contains only one electromagnetic shield at room temperature, namely the outer rotor case.

With only one electromagnetic shield the following problem arises. To provide effective shielding of the superconducting field winding, the electromagnetic shield time constant needs to be relatively long. But the selection of a long shield time constant will diminish the effective damping of rotor swings provided by the electromagnetic shield in the case of significant changes in the transmission network. In contrast the double electromagnetic shield version (warm and cold shield) accomplishes sufficient protection of the field winding from alternating magnetic fields and currents as well as effective damping of rotor swings. A compromise between both damping concepts mentioned above is the use of a composite damper shield at room temperature, where the outer stainless steel rotor cylinder is combined with a copper or alu-

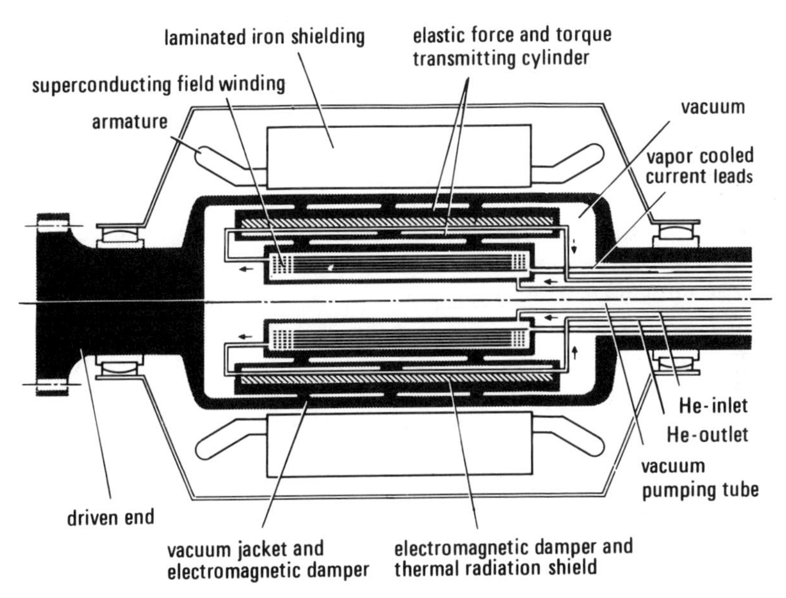

Fig. 66 Schematic of a superconducting generator.

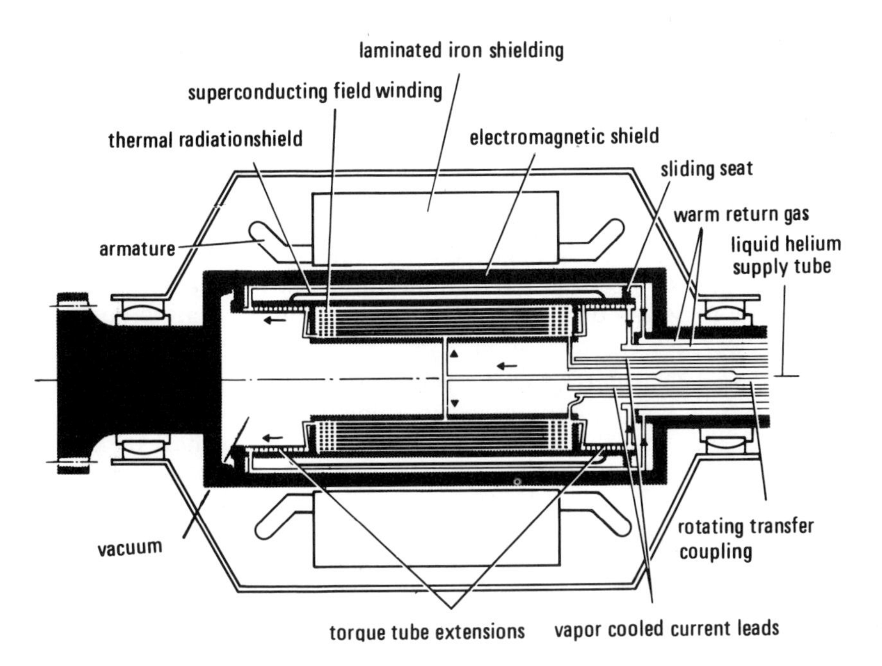

Fig. 67 Schematic of a superconducting generator.

minium cylinder. Most of presently known rotor designs exhibit the principal features of concept (b) namely torque transmission from the drive side by a torque tube cooled at its extensions and the use of only one (composite) electromagnetic damper shield at ambient temperature [152, 153, 154].

During the subtransient period of a short circuit condition, the electromagnetic shields experience large induced currents and pulse torques. The amplitude of these torques can be several times the steady state value of the torque acting on the field winding. This together with radially directed forces leads to severe internal stresses in the cylindrical electromagnetic shield.

These stresses must be taken into account during design. Since the shielding effects of the electromagnetic dampers are based on large induced currents, appropriate cooling has to be provided for them. In order to relieve some of the stresses that occur during a fault, a free spinning electromagnetic shield which surrounds the rotor was suggested [155]. In all rotor designs, the cold helium enters at the exciter end and cools the field winding and then radiation shields, torque transmitting members and current leads and leaves the rotor again at the exciter side.

A question of great importance is that of the appropriate materials to be selected for the rotor. Non magnetic steels or titanium alloys of high strength and toughness must be used for the housing of the field winding, the torque transmission members and the rotor body. Materials that can be considered include austenitic steels such as the precipitation hardened steel alloy A 286 (X5NiCr Ti26 15) or CrNi(Mn)MoN steels and special titanium-aluminium alloys, e.g., Ti-6Al-4V or Ti-5Al-Sn. The housing of the field winding as well as the torque transmission must have excellent mechanical characteristics both at ambient and low temperature since they are exposed to temperature gradients and have to withstand overspeed tests at room temperature. A serious problem that still has to be solved in the future is the manufacture of suitably large forgings from materials of this kind and in the desired shape.

The vacuum required in the rotor for thermal insulation is about 10^{-5} mbar. This can be maintained in a permanently sealed vacuum type rotor, if only small He leaks are assumed. The small leaks can be compensated for by means of gaseous or solid absorption agents (cryosorption, cryotrapping) [156].

If larger helium leaks have to be overcome, continuous vacuum pumping has to be provided. This can be done by means of rotating seals at the exciter end as described later. Both rotos with permanently sealed and continuously pumped vacuum have been proposed. It has also been suggested that the vacuum jacket be made stationary and rotating seals be employed to seal it off from the rotor shaft [157]. It is the general view that for large turbogenerators only the solution with a rotating vacuum jacket can be considered.

b. Superconducting field winding

The design of the excitation winding depends on the electrical and mechanical properties of the superconductors and their loss behavior under the influence of unavoidable armature reaction. Other factors must be considered such as the space required for the electrical insulation and cooling ducts, the minimum dimensions of the supporting structural members necessary to take up the considerable electromagnetic (internal) and mechanical forces (torques and centrifugal forces); and the practical manufacture of the winding, including perfect mechanical fixing within the rotor slot.

The coil types which have to be used for the field windings are dipoles or quadrupoles with approximated sinusoidal current distributions around the cylindrical coil former. The useful flux in the armature has to be generated by the fringe field of these air core coils which rapidly decreases with increasing distance. In order to produce magnetic flux densities at the armature windings which are markedly higher than in conventional machines, the maximum field at the superconducting winding has to have field values of up to 7.5 T depending on the individual rotor design and the generator size. Current densities in the stabilized superconductors of about 200 A mm^{-2} are required at these flux densities to attain the necessary magnetomotive force.

To meet the above requirements the use of intrinsically stable NbTi-multifilament conductors seems to be just sufficient. A prerequisite for this, however, is that even in the worst case of a short circuit the temperature at any point in the field winding should not considerably exceed 4.2 K. For attaining this condition the helium temperature at the rotor entrance must be lowered to about 3.0 K. Because of their higher critical parameters, Nb_3Sn-multifilament conductors would offer a larger safety margin especially for transient conditions. It would, therefore, be more suitable for reliable field windings than NbTi conductors. This is particularly true for generators with very high power ratings ($>$ 2000 MVA). However, appropriate winding techniques have to be developed to deal with the difficult mechanical handling of Nb_3Sn conductors.

In view of the restrictions imposed with regard to the permissible voltage on changes in the exciting field, and also for reasons connected with the actual winding process, it is desirable to have conductor currents as high as possible. In accordance with the capacity of present-day excitation systems, conductor currents of 10,000 A or more are being considered for the field windings of large superconducting generators.

For a proper choice of the conductor type and the cooling means for the field winding, knowledge of the amplitudes and the effects of the ac loads acting on the conductors of the field winding due to armature reaction is of great importance. The ac amplitudes depend on the degree of attenuation provided by the electromagnetic shields and differ from design to design. It has been proved that even in the worst case of a short circuit, the two-electromagnetic shield concept can provide a reduction to values less than 1% of the dc base load [150]. The conductor losses created by these ac loads in the field winding can be tolerated, if they are only of short duration. However, this is only true if the conductors are operated sufficiently below their critical values. It will therefore be advisable to aim at a rated current in the field winding of about 50% of the critical current at the maximum field occurring at the winding. It has to be pointed out additionally that the ac amplitudes can only be kept to tolerable levels if mechanical vibrations of the electromagnetic shields are largely excluded by a proper design. Such vibrations may occur due to large mechanical loads acting on the shields during a fault. In addition to the effects of the armature reaction the field winding will be exposed to ac fields because of necessary regulations of the excitation field. With respect to the ac field loads reported above a fully transposed conductor in the form of a Roebel bar (see Section II, Fig. 10) has to be foreseen for the field winding. To withstand the high forces and torques without movement and damages the field winding has to be resin impregnated and form fitted into the wedge terminated slots of the cylindrical stainless steel winding formers (see Fig. 68).

At this point it should be mentioned that the dimensions and conditions of the dipole and quadrupole field windings required for large generators, significantly exceed the equivalent parameters of corresponding stationary intrinsically stable magnets realized until now.

4. Cooling system

The main sources for the heat load of the superconducting field winding are heat conduction by the suspension system, joule heating and thermal conduction in the current leads, and ac losses due to the armature reaction. If a cold electromagnetic shield is used, heat is also produced at a temperature level of 50 to 100 K, due to the generation of shielding currents. The magnitude of the losses obviously depends on the special design of the suspension structure and also on the choice of the electromagnetic shielding system. For appropriate designs, the losses can be kept surprisingly low. In the case of a 3000 MVA dipole generator with a warm and a cold electromagnetic shield and with radial torque transmission, the losses occurring during steady state operation (5% unbalanced load assumed) can be kept to about 20 kW at the cold shield (80 K), and about 100 W at the field winding (including the losses of current leads overdesigned for safety reasons). These losses correspond to a refrigerator input power of about 600 kW, which is only 0.02 % of the generator power or about 4% of the total generator loss (windage, stator, and iron shield).

Transient heat loads are generated when rapid field changes appear. These field changes may be caused by field ramping or system faults, such as a short circuit. In the latter case high rates of transient heating, of short duration, occur. The short duration keeps the amount of energy dissipated during one such event within reasonable bounds. Since such an event occurs only rarely, it should be possible to absorb the energy released by the heat capacity of the helium cooling the winding without a critical temperature rise being produced in the field winding. Additional heating during a short circuit has also to be taken into account when designing the cooling means of the electromagnetic shields.

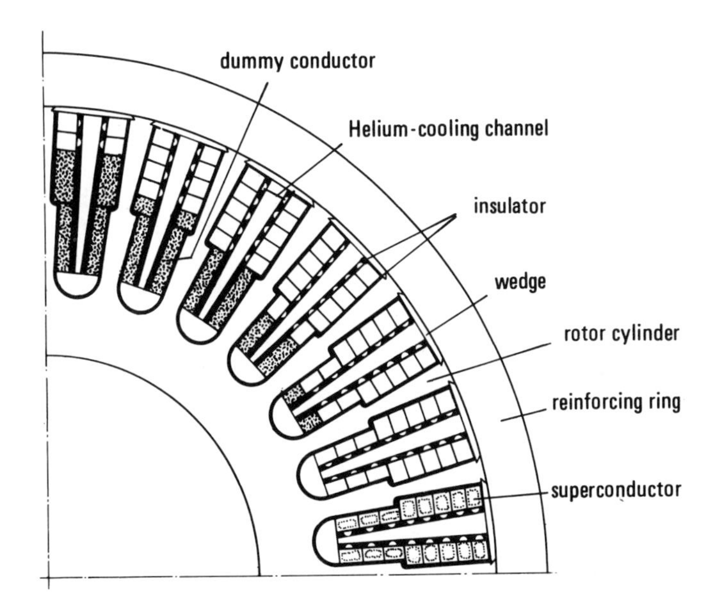

Fig. 68 $\frac{\pi}{2}$ - section of a superconducting dipole-generator field winding.

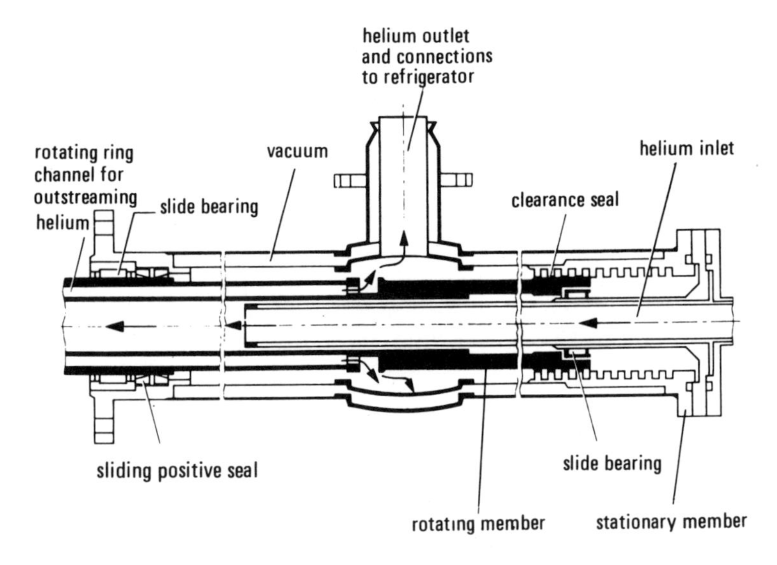

Fig. 69 Schematic of a rotating helium couple.

The losses occurring in the field winding are only a small percentage of the total losses. Refrigerators with capacities necessary to meet the requirements for cooling a large generator are already available. Still many problems exist in connection with the generator cooling and are under investigation within several development programs.

One of these problems is the transfer coupling which has to be accomplished between the stationary refrigerator and the helium supply of a rotating helium system. Such a rotating transfer coupling has the functions of sealing off the helium against outside area, and separating the helium streams of different state (temperature, pressure, gas, liquid) e.g., He-go and return streams. The transfer coupling has to meet the requirements of minimum thermal losses and He leakage, it must permit radial and axial shaft play and have high reliability and availability. It must have a maintenance free time greater than two years.

Several rotating transfer coupling schemes have been developed in the past having similar main features. Figure 69 shows the schematic of a horizontal coupling for a helium go and return stream. The rotating and stationary tubes are fitted into one another, similar to stationary couplings, like a bayonet joint. They are thermally isolated by vacuum jackets within the transition region. The coupling contains clearance seals to separate the cold He inlet stream from the warm return stream. The actual sealing of the rotating member to the exterior is accomplished by a ferrofluidic seal. This seal comprises a magnet in the stator, a magnetic shaft and a magnetic fluid in between. The magnetic fluid is a low vapor-pressure fluid containing colloidal iron oxide particles. It is expected that such a seal can operate for well over two years before replenishment of the carrier fluid is required. Replenishment of the seal fluid can be accomplished during operation. Instead of ferro-fluidic seals gliding face seals, e.g., with carbon filled polyimid rings, were also used successfully as positive seals.

It has been demonstrated by several experiments that the additional losses of properly designed rotating couplings caused by rotation are very small, e.g., $< 1\%$ referred to the amount of liquid He transferred. In long term tests, running times of more than 6500 h have been achieved without any troubles [151]. All experimental helium couplings operated so far work with warm seals. Besides lack of proper cold seals, the warm seals avoid dissipation of friction losses at low temperatures. Cold seals are also under investigation for multiple transfer couplings.

Intensive investigations are being carried out on the thermodynamics of rotating He cooling systems. The processes being considered for cooling the field winding of large generators include, cooling by thermosiphon process [158, 159], cooling by forced flowing He [150], and bath cooling [151]. The classification above was chosen for its simplicity. In reality one effect, namely thermal convection, plays a dominant role in all cooling processes.

It is a common feature of all processes, that the helium in the rotor experiences high acceleration fields. For example, on the periphery of a rotor of 1 m in diameter a rotational speed of 3000 r.p.m. produces a centrifugal acceleration of about 5000 times that of normal gravity. As a result, helium refrigerant fed into the rotor along the rotation axis will experience a centrifugal compression in flowing out to the field winding with pressure increases of up to 20 bars and temperature rises in the region of 20 % of the inlet values. When working with NbTi as a superconductor for the field winding, the inlet temperature of the He must be significantly lowered to compensate for the compression effect.

The existence of the large centrifugal acceleration is the basis for the thermosiphon cooling. The heat is being transported by convectional streams for the winding to a liquid bath on the axis of the rotor. Figure 70 shows the principles of the cooling processes mentioned above. It was shown that similar high heat transfer coefficients can be attained with thermosiphon cooling as with forced cooling [160]. An excellent demonstration of the effectiveness of a bath cooling was given by GE scientists with respect to the cooling system of a 20 MVA model generator [151]. In this case the heat is mainly transported from the interior of the winding to the surface of the boiling bath by heat conduction through a highly conductive coil former.

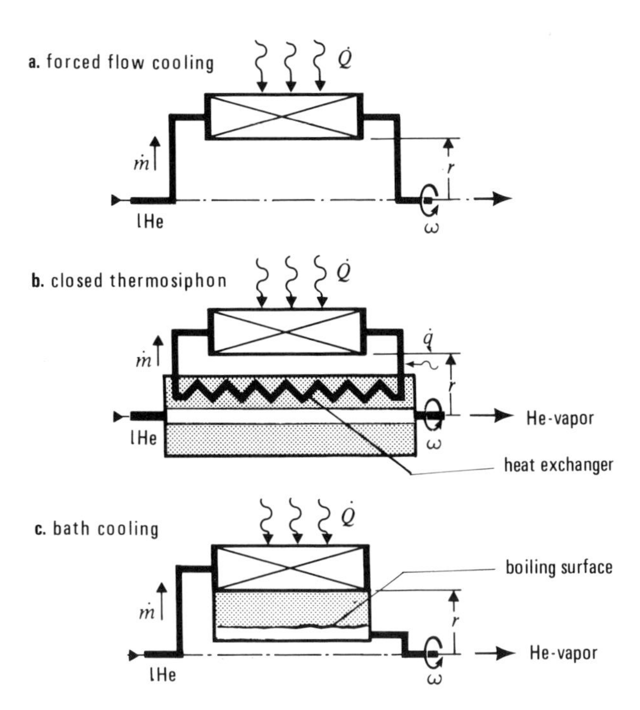

Fig. 70 Different cooling schemes for the superconducting field winding of a synchronous generator.

The level of the boiling surface near the rotor axis can be automatically maintained for various heat loads if the liquid He supply pressure is kept constant.

Besides studying the characteristics of thermosiphon and bath cooling, investigations are under way to determine the influence of the rotation on forced flow cooling. It is theoretically expected that the heat transfer coefficient and the flow resistance are markedly increased by rotation, e.g., by a factor of 2 to 3 at 3000 r.p.m. at a radius of 0.4 m [161]. In the case of thermosiphon cooling the heat transfer coefficient is reduced at high values of $\omega^2 r$ [160].

An important question is the selection of the most appropriate cooling circuit scheme. It can be shown for a superconducting generator that a thermodynamically optimal cryogenic system will consist of a refrigerator with different He streams at different temperature levels. Since the refrigerator input power constitutes only a small fraction of the total generator losses (e.g., 4%), one should attach more importance to reliability and availability than to efficiency. Parallel cooling of torque tubes, radiation shields, and current leads with He vapor leaving the field winding or series cooling of these components leads to more simple designs of rotating He-transfer couplings. With respect to reliability and availability, the use of liquid He for supplying the rotor is proposed. In the case of a failing He liquifier, the repairing time can be easily bridged by a liquid He vessel.

5. Armature winding (stator)

The notable decrease of weight and volume of large superconducting power generators results from an increase of the air gap flux density provided by the superconducting rotor winding and a higher linear armature current density, made possible by the elimination of magnetic stator teeth as they are utilized in conventional machines. In the case of an armature having no magnetic iron, the armature conductors are exposed to the full flux of the rotating winding. In conventional machines most of the flux flows through the iron. Eddy current and circulating current losses in the conductor bars must be reduced to an acceptable low level by subdividing the copper bars into many fine strands, and balancing the transpositions to both radially and circumferentially directed magnetic fields. The armature conductors are cooled by water flowing through ducts built into the conductor bars. The stator winding should preferably be held in non-magnetic cores. Proposals to embed the winding in fiber-reinforced plastic may be difficult to realize because of the mechanical factors involved [162].

In conventional machines, the main component of the reaction torque is taken up by the armature teeth and transferred to the stator frame. However, in the case of superconducting machines having no magnetic iron, the full reaction torque is applied to the armature conductors and must be transmitted through the electrical insulation. Steady-state and transient forces and any resultant vibrations must be successfully accommodated for many years without mechanical failure of the bar or insulation. These forces considerably exceed the forces experienced by the stator bars of conventional machines. Control of the armature forces is one of the major problems occurring with large superconducting generators and is the subject of corresponding development programs.

6. Machine screening

The elimination of magnetic iron in the rotor and armature of a superconducting generator results in relatively large stray fields which must be screened. The methods considered so far [162] to do this include a laminated iron cylinder, a compact iron cylinder, a compact water-cooled cylinder of copper or aluminium and a short-circuited, water-cooled screening winding.

In most machine designs a laminated iron screen is used in the active section. This has the advantage that the useful armature flux is increased for a given excitation winding magnetomotive force and the eddy current losses are relatively low. In the end areas, eddy current screening by means of compact copper components may be provided.

7. Electrical operating behavior and characteristic data

The electrical behavior of a synchronous machine in the power system is determined primarily by its reactances, the synchronous reactance x_d, the transient reactance x_d', and the subtransient reactance x_d''. The synchronous reactance characterizes the steady-state behavior, while the transient and subtransient reactances become effective on load changes and short circuits.

The synchronous reactance is the sum of the leakage reactance of the armature winding and the mutual reactance between armature and field winding. The mutual reactance and thus also the synchronous reactance are much smaller than in the case of conventional generators due to the relatively large air gap and the absence of magnetic iron in the rotor and armature. This condition gives superconducting generators a favorable steady-state behavior for slow load changes thus improving the stability of the power system. The steady-state stability is never reached, i.e., superconducting generators can be operated at their full load ratings to zero power factor ($\cos \phi = 0$) underexited, if no impermissible heating of the stator parts occurs. Changes in machine load produce smaller changes in terminal voltage than with high reactance conventional alternators.

In the case of transient operations (fault conditions), superconducting generators can make an important contribution to the stability improvement in large power systems. Since in a superconducting generator there is some design freedom with respect to the choice of armature current density A and flux density B transient reactances ($x \sim A/B$) can be adjusted within certain limits to the stability requirements of the power system. The small initial (at start of fault) synchronous reactance

Table 14 Comparison of superconducting and conventional turbo-generators.

	1200MVA* 2-pole conventional	1200MVA* 2-pole superconducting	1500MVA** 2-pole conventional	3000MVA** 2-pole superconducting
Rated voltage kV	26	34	27	27
Armature outer diameter m	4.3	3.7	4.3	4.3
Total length m	13	7.2	15	11
Total mass t	630	140	580	480
Synchronous reactance x_d %	181	52	220	90
Transient reactance x'_d %	32	29	31	70
Subtransient reactance x''_d %	26	16	23	55
Efficiency (rated load) %	98.6	99.4	98.75	>99

* Data from Westinghouse [33] ** Data from KWU [30]

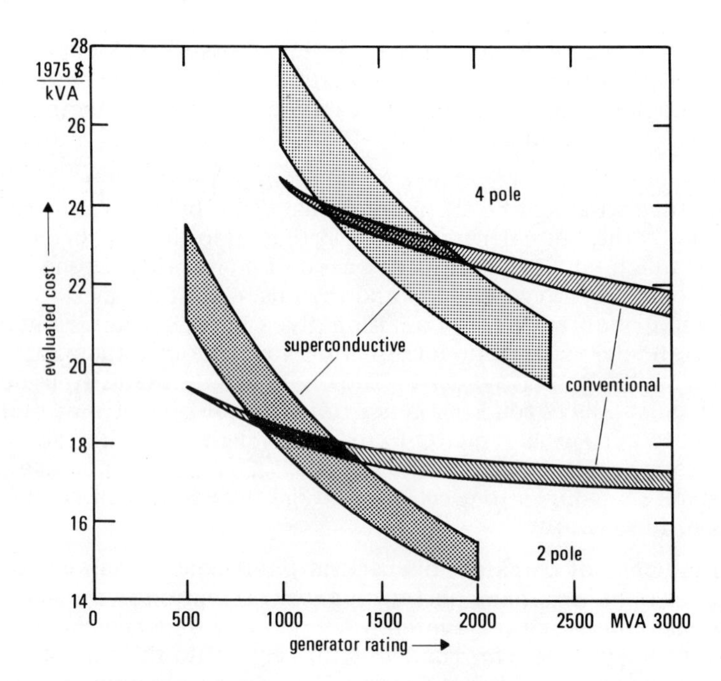

Fig. 71 Comparison of evaluated cost for conventional and superconducting steam turbine-generator sets [163].

also contributes to an improved transient behavior.

Small reactances, however, have a negative effect in that the short circuit currents become higher. This leads to higher losses and to higher mechanical loads in the armature and the electromagnetic shields in the post-fault period. Due to the low resistances of the electromagnetic shields and the excitation circuits, the subtransient and transient time constants in superconducting generators are much larger than in conventional machines. This amplifies the mechanical and thermal stresses in the armature and electromagnetic shields. With appropriate shielding, the additional mechanical stressing of the field winding during a short circuit is negligible. From the above it becomes obvious that the optimal design of a superconducting generator will be a compromise with respect to power density, system stability and acceptable mechanical loads after a short circuit.

Table 14 shows a comparison of the main data of superconducting and conventional two-pole turbine-generators. It reveals that the gain in volume is about a factor of two. The margin attainable in weight is about a factor of two to four and the increase in efficiency about 0.6 to 0.8%. The enhancement of efficiency for four-pole machines is estimated to lie in between 0.3 and 0.5%. The margin in efficiency has mainly to be attributed to the very low losses of the field windings ($\approx$ 0.02% of machine rating) and the lower iron losses, windage losses and bearing losses because of the smaller machine size. The synchronous reactances x_d of the superconducting versions are considerably lower and especially the subtransient reactances x_d'' can also be markedly reduced. The design of the 3000 MVA-generator was rather conservative. The air gap flux density chosen was only 20% higher than in a conventional machine which also leads to comparatively high transient reactances. The list of advantages quoted for superconducting generators is not complete. The specification of all benefits evaluated so far would go beyond the scope of this article.

8. Economic considerations

When calculating the economical gain of large superconducting turbine generators, it is customary not only to consider the saving in capital cost but also to add capitalized benefits due to the better performance of such a generator. As in the case of new large conventional generator installations it is usual to take into account the improvement in efficiency. This leads to a capitalized value of annual fuel saving and to a credit for the additional plant output available (calculated on the incremental cost basis).

Figure 71 is a comparison of estimated costs for conventional and superconducting turbine generator sets [163]. It is based on an improved efficiency and assumes the same high levels of availability and reliability set by conventional generators. On the basis of Fig. 71, the economic cross-over between superconducting and conventional generators seems to be around 1500 MVA. But there also exist more optimistic evaluations which lead to a break-even point of about 500 MVA [152, 162].

With the present level of knowledge, it is not yet possible to make a final statement on the economic benefits of large superconducting generators. Lower capital costs and an improved efficiency may easily be offset by a lower degree of reliability and availability. However, there is no reason to assume that superconducting generators are, a priori, less reliable than conventional ones. To answer these questions, experimental generators with ratings of 100 MVA and more should be built. The operating data obtained from these machines will provide the basis for reliable economic calculations and for availability and reliability forecasts.

In the US the total market for turbine generator sets will grow from about $350 million yearly in 1970 to approximately $1.75 billion in the year 2000 based on 1970 dollars. Over 60 new steam turbine generator units will be needed annualy from 1985 to 1995 in order to supply the capacity requirements called for in the Federal Power Commission (FPC) forecast. The average size turbine generator installed annually will increase to about 1000 MW in 1985 and about 1600 MW in 1995. The latter figure corresponds to maximum generator ratings of 2500 MVA available at this time [163].

In Europe, also, there is a potential market for superconducting generators. It is estimated that from the present day to the year 2000 there could be a need in the power systems of the European Community Member States for about 200 generators of 3000 MVA, if power plants of this size are available [164]. In addition, the demand for generators with ratings of 1000 MVA will run into several hundreds of units.

9. Superconducting turbine-generator projects

Studies on large superconducting turbo-generators with power ratings from 300 MVA to 3000 MVA are under way in practically all industrial nations of the world. Very often these studies include the experimental investigation of single important components such as He-transfer coupling, current leads, cooling circuits, superconductors, winding models, etc. At present, in the US, Westinghouse and General Electric (sponsored by the Electric Power Research Institute EPRI) are carrying out detailed design studies on two-pole 300 MVA generators. These could serve in the near future as demonstration-generators proving availability and reliability of such machines. Rough studies on 1200 MVA machines are also under way. In Western Europe, Alsthom (France) together with Electricité de France, International Research and Development Company, U.K., Brown Boverie and Cie, Switzerland and Kraftwerk Union in cooperation with Siemens AG, Germany are studying large two-pole generators in the power range from 1200 to 3000 MVA. In the USSR, work in this field is carried out under the leadership of the All Union Research Institute for Electrical Machinery Industry, Leningrad. In Japan especially the Fuji-Electric Company and the Mitshubishi Electric Corporation are working in this area.

Besides these studies, a number of model-generators with rotating superconducting NbTi-multifilament field windings and power ratings from 1 MVA to 20 MVA have been successfully operated or are just under construction. First one must mention the 3 MVA-generator built at MIT with which extensive experiments have been carried out. The machine was paralleled to the Cambridge Electric System and operated in the steady state to 3.2 MVA overexcited, and to 2.6 MVA underexcited [165]. Westinghouse has built a 5 MVA generator which was tested under open circuit and short circuit conditions [166]. In the USSR, the All Union Research Institute for Electrical Machinery Industry, Leningrad, has constructed and tested a 1 MVA machine. They are planning by 1980 to build a 20 MVA generator which will be smaller than the 1 MVA machine and which will be operated in the Leningrad power system. For 1987 the aim is to construct a 200 MVA demonstration generator [167]. A model generator with a rating of 6 MVA is under construction at Fuji Electric and Mitshubishi Electric in Japan. A model rotor for this machine has already been successfully tested [168]. The total machine will be completed at the end of 1976. General Electric in the US has a 20 MVA-generator under construction which will be operated at the end of 1977. In contrast to the machines mentioned above, which were operated without real load, the GE machine will be driven by a gas turbine and operated under utility conditions [154].

Besides the most important work on turbine generators, development activities on other types of superconducting synchronous machinery exist. A 400 Hz, 10 MVA light weight generator for airborne applications is under construction at Westinghouse under contract to the US Air Force. The rotor for a 5 MVA machine has already been successfully tested [159]. While the field windings of present machines consist of NbTi-multifilament wires, Nb_3Sn-conductors are foreseen for a new machine. Also new types of ac machines are under consideration. One of these is the so-called dual armature machine, which consists of a stationary and a rotating armature and a torqueless rotating superconducting field winding [147, 169]. This type of machine is especially suited for ship propulsion motors. Attention has also been paid to low speed multipole superconducting synchronous machines, e.g., as motors for ship drives [170]. Finally, due to the success in the development of rotating liquid metal contacts with high current carrying capacities, large turbine-generators with rotating armature are again considered to be feasible [153, 171].

Without any doubt, large superconducting turbine-generators offer the largest economic potential impact when compared to all other machine types, including dc machines. The results of numerous model machines have proved the principal feasibility of this type of machine. However, the significant superiority over conventional generators forcasted by several studies can only be proved by the operation of large superconducting demonstration generators with ratings similar to those of large present day generators.

V. SUPERCONDUCTING CABLES

A. Introduction

The increasing demand for electric energy, reflected in the steady increase in power plant size and in increasing alternator unit ratings, results in a corresponding increase in the transmission system capacity. Overhead transmission lines are still the most economic form of transmission over long distances, even for very large power ratings. However, in the vicinity of built-up areas there is insufficient space available for the large towers of such systems with considerable crossarm overhang. Lack of space and land conservation measures are forcing the utilities to resort to underground power transmission. Conventional oil-filled paper insulated cables for 420 kV can provide a maximum transmission capacity of approximately 500 MVA. The limitation is due to the restricted dissipation of the heat generated in the cable as a result of electric losses to the surrounding earth and to the fact that there is an upper limiting temperature for the cable insulation (approximately 85 °C), which may not be exceeded.

A number of new types of underground transmission systems are being developed to meet future transmission needs. These types include forced cooled cables with oil or water cooling and SF_6 gas spacer cables, of which a few are already in operation. Examples of the application of forced cooling with natural flowing water are three tunnel installations in England and a 16 km long 1 GVA - 420 kV system of the BEWAG in Berlin, Germany. SF_6-gas-spacer cables operated at transmission voltages of 420 or 765 kV can have a transmission capacity of some thousand MVA [172]. A disadvantage of this type of transmission system is the large diameter of the pipe conductors. They generally operate as single-phase conductors, due to the relatively low dielectric strength of the insulating gas of this transmission system. The resultant space requirements are twice that of forced water-cooled oil cables and can rarely be met in heavily built-up areas. SF_6-gas-spacer cables have been employed so far for transmission of energy over relatively short distances and under exceptional circumstances only (e.g., 420 kV feeder from an underground pumped storage power plant in the South of Germany).

In addition to the new types of transmission systems already mentioned, development work is proceeding on liquid nitrogen cooled aluminium or copper conductor cables and on superconducting cables. The latter may be regarded as the ultimate step in the development of forced cooled cables. The development of superconducting cables, which commenced in the early sixties, has concentrated so far on development of the individual cable components (thermal insulation, superconductor, dielectric medium and cooling) and investigation of the properties of these components. Tests on short to medium sized experimental links have however already been carried out or are in preparation.

B. Superconducting Cable Concepts

1. Mechanical construction

A superconducting cable consists in principle of the thermal insulation and the electric conductor system contained therein. So far as mechanical construction is concerned, there are three different cable concepts.

The rigid pipe type cable (Fig. 72a). Thermal insulation and conductors are made of rigid pipes. The advantages of this concept are low thermal losses, simple construction and straightforward electric properties. The main disadvantage lies in the comparatively short manufacturing length, due to difficulties in transportation. This results in a large number of complicated field joints and difficulties in compensating for the longitudinal contraction of the cold cable components. The use of material with a low coefficient of contraction (e.g., Invar $\Delta \ell/\ell \approx$ 0.05%) or of contraction compensating devices, such as corrugated components, is necessary.

The semi-flexible type of cable (Fig. 72b). The thermal insulation is again made of rigid pipes with corrugated components to compensate for thermal contraction. The conductors are flexible, however, and consist either of a corrugated pipe or strips or wires wound helically on a flexible hollow cylindrical support. The main advantage of this type of cable is the fact that the flexible phase conductors can be manufactured in lengths of 200 - 1000 m and transported on drums. The conductors can be pulled into the previously laid rigid thermal insulation in a similar manner to that used for conventional high pressure oil cables in steel pipes. The number of field joints required is considerably reduced compared to the rigid pipe type cable.

The completely flexible type cable (Fig. 72c). Both the phase conductor and the thermal insulation are flexible, the latter being constructed of corrugated metal pipes. Such cables can be manufactured completely in lengths of 200 m and transported on drums. Due to manufacturing difficulties and to limitation of the bending radius, the diameter of currgated pipes is limited to ≤ 300 mm. As a result these cables can be employed for large transmission capacities as single phase cables only, which is an economic disadvantage.

All three concepts are at present under development but most effort is concentrated on the type in Fig. 72b.

2. Conductor configurations

Nowadays, coaxial tubular conductor pairs are used almost exclusively for three-phase cables as depicted in Fig. 73 [173]. The bore of the inner hollow conductor is utilised as a cooling channel for the coolant helium. Each superconducting phase conductor is surrounded by a superconducting shield or return conductor, in which a current flows in opposition to that of the phase conductor. In consequence, the

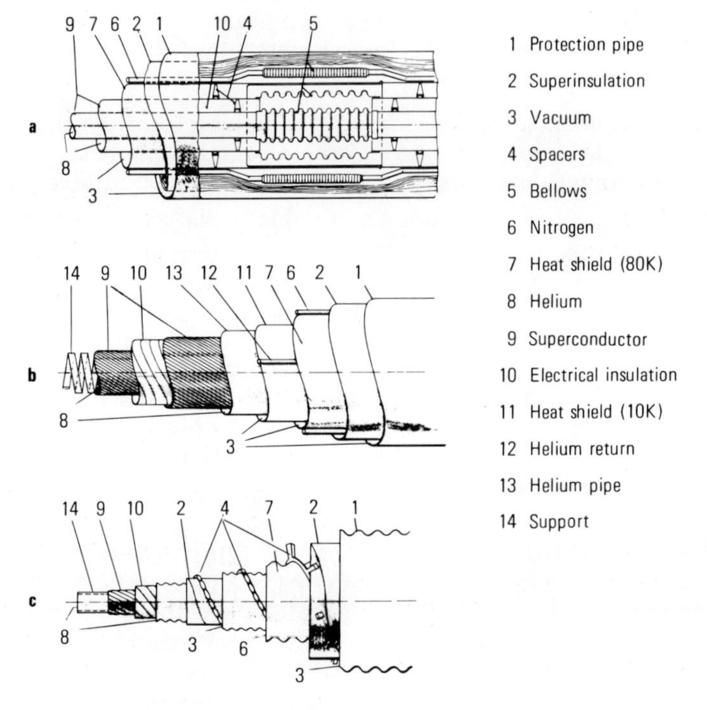

Fig. 72 Mechanical cable designs. (a) Pipe-type (b) Semiflexible (c) Totally flexible [172].

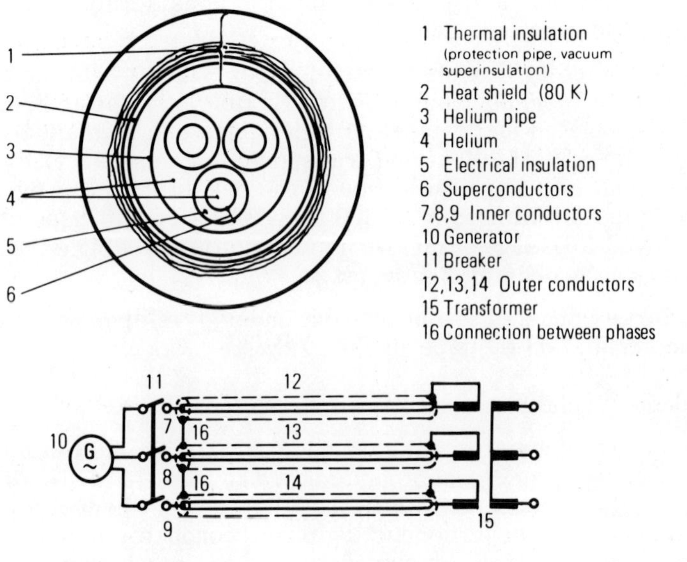

Fig. 73 Superconducting ac cable design with coaxial conductor pairs [173].

electromagnetic field is limited under all operating conditions to the vicinity of the phase conductor and thus no eddy current losses occur in the metal pipes of the thermal insulation. The electrical connection of the cable to achieve this effect is also shown in Fig. 73. The star point of the transformer is brought out and connected to the alternator side via the return conductors. The coaxial conductor arrangement has the added advantage of homogeneous field distribution on the conductor surfaces and freedom from resultant forces due to the symmetrical arrangement.

In the case of direct current cables, hollow conductors are preferred on account of the superior cooling effect. A coaxial arrangement is not absolutely necessary due to the lack of alternating electromagnetic fields in normal operation. The coaxial arrangement has the advantage of freedom from forces when arranged symmetrically, whereas in the case of conductors arranged in parallel considerable repulsion forces can occur at the envisaged large currents (e.g., 50 kA). Figure 74 depicts two proposed cable configurations [174, 175]. In the case of the conductors arranged in parallel, the helium pipes are of magnetic material, which shield the conductors magnetically from each other to avoid the mutual forces. It has also been shown that the coaxial arrangement is also the most favorable with respect to the expenditure for superconductors. Furthermore, compact solid conductor arrangements have been proposed [176]. These have the advantage of high power density. This results in higher magnetic flux which leads to poorer utilisation of the superconductors.

3. Comparison between superconducting direct current and alternating current cables

Superconducting direct current cables offer a number of advantages as a result of their construction. They require a reduced number of conductors (in the ratio 2:3). The choice of the type of superconductors is larger since no provision has to be made for hysteretic losses. Higher current densities and flux densities are also possible. In the case of solid dielectrics, the most critical factor for selection, namely the choice of an extremely low loss factor, can be neglected. As a result kraft paper may be employed.

The advantages described above lead to smaller dimensions for direct current cables for the same transmission capacity. DC cables have negligible electric losses due to considerably lower transmission losses (thermal and electrical) than is the case for three-phase cables. They have about half the diameter. The specific investment costs ($/km. MVA) for direct current cables are lower by a factor of 2 to 3 than those for three-phase cables. The electrical properties of direct current superconducting cables are similar to those of conventional cables. They have no capacitive reactive current losses and there are practically no limits set to the transmission distance.

In spite of the advantages described, the use of superconducting

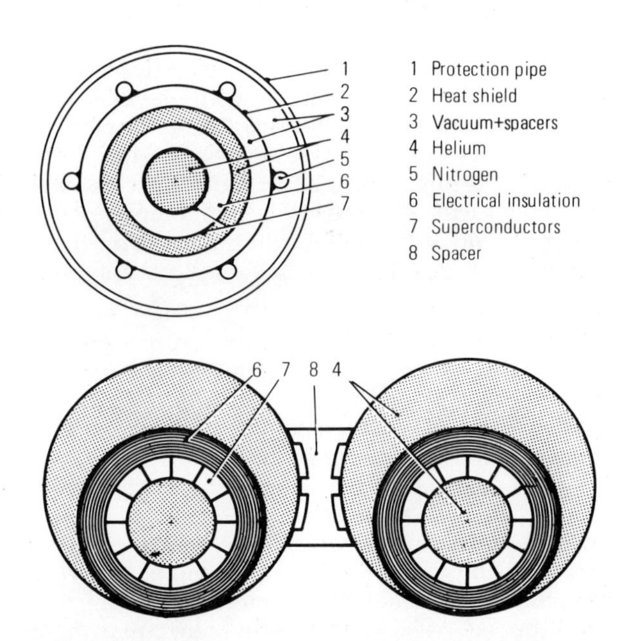

Fig. 74 Cross-sections of superconducting dc cable designs [174, 175].

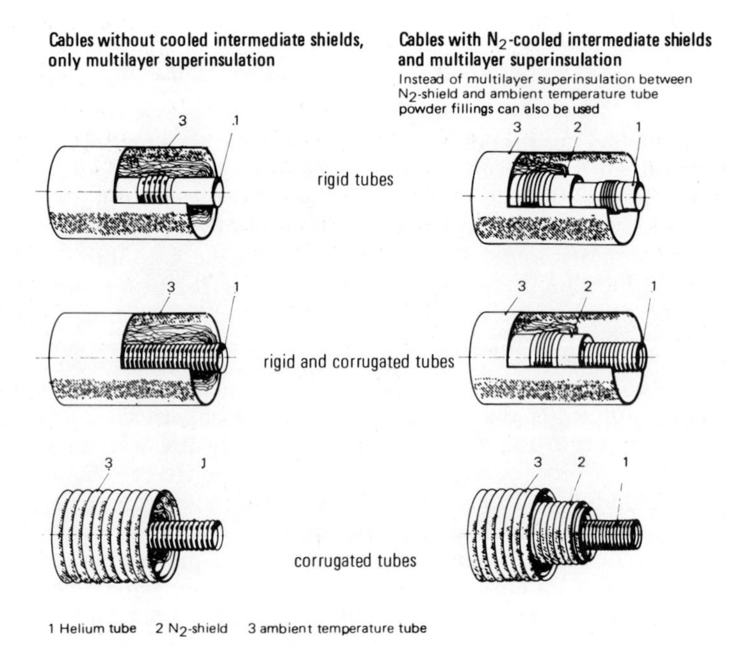

Fig. 75 Constructions of thermal cable insulations [172].

direct current cables is extremely limited by virtue of the considerable investment required for the convertor stations at both ends of the cable (e.g., for 500 kV: 180 DM/kW). For this reason superconducting cables are economic only for transmission distances in excess of 200 km where they face competition from the relatively cheap overhead transmission lines. Three-phase cables are thus the only solution to supply built-up areas over short distances.

In consequence nearly all the cable research projects underway throughout the world are devoted to development of superconducting three-phase cables. Direct current cables have been developed in individual cases for their limiting effect on the short-circuit power of the grid for urban feeders and for network interconnections.

C. Cryogenic Envelope

The thermal insulation (cryogenic envelope) of a cable is assigned the task of reducing the flow of heat from outside to the helium cooled conductor system. It represents, as a number of calculations have shown (c.f. Section V E), one of the most costly elements of the cable. In comparison, less expenditure is required for the helium refrigerator. For this reason it is clear that little is gained by minimizing the thermal losses at the cost of a more expensive cryogenic envelope. There is thus a real need for optimization based on individual requirements.

Nowadays attention is focussed principally on two types of thermal insulation, a flexible cryogenic envelope employing corrugated tubes and a rigid cryogenic envelope employing rigid tubes. As Fig. 75 shows, a combination of both concepts is possible. The space between the outer tube and the helium tube containing the cable conductors is evacuated, as is the case for laboratory cryostats, and it is partially filled with "Superinsulation" (reflecting foils). Often a heat barrier in the form of a nitrogen cooled shield is inserted between the outer and the helium tube. In contrast to the rigid cryogenic envelope the completely flexible cryogenic envelope has the advantage that it requires no compensation for cold contraction of the helium tube and can be wound on drums for outer diameters up to approximately 300 mm. Thus, it is capable of being transported in greater lengths. On the other hand, as a result of the considerable number of mechanical supports required, and on account of the larger surface area, the corrugated tube type of construction has considerably greater thermal losses as shown by the data following.

For the rigid cryogenic envelope the losses are

$$4 - 6\ \mathrm{K}:\ P \leq 100\ \mathrm{mW/m^2}$$
$$77\ \mathrm{K}:\ P \leq 2\ \mathrm{W/m^2}\ ,$$

while for the flexible cryogenic envelope the losses are

$$4 - 6\ \mathrm{K}:\ P \leq 500\ \mathrm{mW/m^2}$$
$$77\ \mathrm{K}:\ P \leq 4\ \mathrm{W/m^2}\ .$$

A number of suggestions have been made to reduce costs. The cryogenic envelope should be constructed without a nitrogen shield using only superinsulation [177]. The thermal losses are then higher by a factor of at least 5 than with a nitrogen shield (based on refrigerator motor rating and for a rigid cryogenic envelope)[172]. In contrast to these, another suggestion [178] envisages six copper shields in the vacuum between the outer and the helium tube cooled by evaporating helium, which is extracted from the helium cooling circuit within the cable junctions. This type of insulation is employed for normal helium cryostats. It is likely to be very expensive for a cable in the form proposed.

Of great importance is the initial realization and maintenance of the insulating vacuum of $\leq 10^{-5}$ mbar. In the case of rigid cryogenic envelopes 10-20 m long, sealed sections can be evacuated in the factory leaving only the connection pieces to be evacuated in the field upon completion of the assembly. It is, however, conceivable that evacuation would be initiated in the field only after the complete assembly of the total length of the cryogenic envelope. This can be effected using vacuum pumping stations at distances of approximately every 1 km.

D. Superconducting Material

1. Direct current superconductors

Superconductors for direct current cables should have high critical current density. This saves material and keeps the cable dimensions small. In addition, a high transition temperature is desirable to allow the cable to be operated at the highest possible temperature. Up to now NbTi and Nb_3Sn were regarded as being most suitable for direct current superconductors [172]. Multifilament conductors are particularly suitable, but the use of strip conductors with thin Nb_3Sn layers has also been proposed.

Recently Nb_3Ge has come to be regarded as a potential candidate as a direct current cable conductor. Its transition temperature of about 23 K to 25 K would allow the operation of superconducting direct durrent cables with subcooled hydrogen in the range of 14 K to 20 K. Besides a marked improvement in the efficiency of the cable cooling, further advantages are the low price and unlimited availability of hydrogen as a coolant in contrast to helium. Of especial interest is the better dielectric strength of liquid H_2 as opposed to liquid helium. At the present moment, however, Nb_3Ge is not available commercially and exists only in the form of laboratory samples generated within numerous research projects.

Similar to superconductors for other power applications, the superconductors for direct current cables must be stabilized with copper or aluminium. Cryostatic stabilization for twice the operating current is sufficient, since the maximum peak short-circuit current amplitude of direct current cables can attain this value only due to the limiting effect of the convertors.

It has been proposed [175], that the superconductor cross section of direct current cables be dimensioned for twice the operating current. In this way the short circuit current can be carried by the superconductor without any temperature or pressure rise in the coolant.

2. Alternating current superconductors

Pure niobium, Nb_3Sn and Nb_3Ge are at present under investigation as to their suitability as conductors for alternating current cables. Niobium has the advantage that it is relatively cheap, easy to handle and has low ac losses. Nb_3Sn has a relatively high transition temperature and consequently facilitates use of a more effective cooling circuit. Its high critical flux density $\mu_0 H_{c2}$ renders it more resistant to short circuits than niobium, which by virtue of its low upper critical flux density ($\mu_0 H_{c2} \approx 0.4$ T at 4.2 K) becomes normally conductive on short circuits (which may lead to currents 15 times the rated current in ac cables).

The use of Nb_3Ge as conductor for a three-phase cable is subject to the same conditions as already described for direct current cable. Here the possibility of improving the dielectric strength as a result of the better properties of liquid hydrogen is of particular importance.

Hysteresis losses, which markedly affect the total losses in superconducting three-phase cables, are exhibited by the aforementioned superconductors at power frequency and at the flux densities which normally occur in cables. It has been known for some time that at power frequencies an extremely smooth, clean and cold-worked niobium surface exhibits very low hysteresis losses. The losses lie well below the limit given for a three-phase ac cable of 10 $\mu W/cm^2$ at 4.2 K and a surface flux-density of 0.1 T [172].

Only surface losses (Meissner condition) occur in the case of niobium at a flux density of 0.1 T, which is typical for a cable, by virtue of its high lower critical flux density $\mu_0 H_{c1}$ (0.14 T at 4.2 K). This lower critical flux density $\mu_0 H_{c1}$ drops to approximately 0.1 T at 6 K. At higher temperatures or at higher flux densities, flux enters the niobium volume. Volume losses occur which are considerably higher than the surface losses. Figure 76 [172] shows the losses of a commercially produced niobium wire as a function of the temperature and of the magnitude of the flux density at its surface. A bend the data curves can be clearly seen. This indicates the transition from surface losses to volume losses. It can further be seen that for a three-phase ac cable with niobium conductors and a maximum surface flux density of 0.1 T if a loss level of ≤ 10 $\mu W/cm^2$ is specified, the maximum operating temperature is approximately 6 K.

The loss level of Nb_3Sn conductors was, until recently, appreciably higher than that of niobium conductors. New production methods now allow production of Nb_3Sn conductors, whose losses are near those of Nb [179]. Nb_3Sn layers, formed in a solid state diffusion process utilizing bronze have remarkably low losses. Figure 77 [180] shows the losses of such a conductor and of a Nb conductor as a function of the magnetic field amplitude. At 0.1 T and 4.2 K, the losses are appreciably less than 10 $\mu W/cm^2$. It has been found that to achieve low losses, the Nb_3Sn

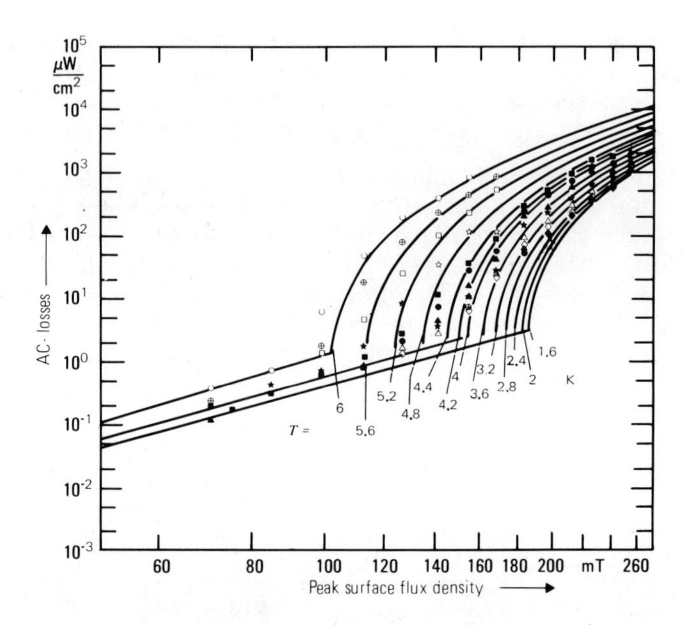

Fig. 76 50 c/s-losses of niobium at various temperatures [172].

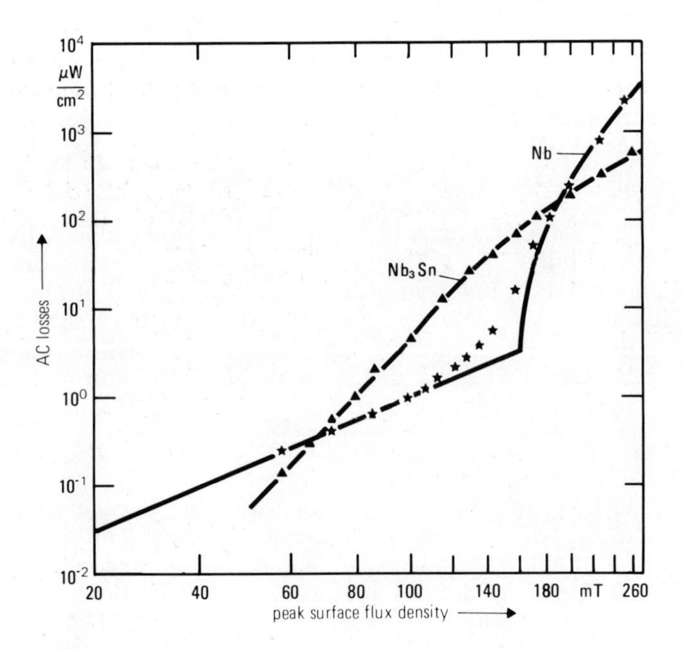

Fig. 77 50 Hz-losses of a niobium and Nb_3Sn conductor at 4.2 K [180].

must have an extremely smooth surface, a high surface shielding field $\mu_0 \Delta H$ [181], and a high critical bulk current density j_c. This is the case for Nb_3Sn formed by a solid state diffusion process. It is characterized by a very fine grain structure in which the grain boundaries form the pinning centers.

Figure 78 shows the losses of a commercial Nb_3Sn sample as a function of temperature and the magnetic flux density [179]. It can be seen that this Nb_3Sn sample can be utilized in three-phase cables up to approximately 8 K with a surface flux density of 0.1 T without the specific losses exceeding 10 $\mu W/cm^2$. Loss measurements were also carried out on strip samples with Nb_3Ge layers produced by chemical vapor deposition [182]. The surfaces of these samples have been both mechanically polished and chemically etched. Losses at 50 Hz and 0.09 T peak flux density have been determined to be as low as 0.2 $\mu W/cm^2$ at 12 K and about 30 $\mu W/m^2$ at 14 K. These low losses make Nb_3Ge an attractive material for superconducting ac cables being operated in the temperature range of 12 K to 14 K.

The relationship between the surface losses of niobium and the surface magnetic field has been theoretically explained by Melville [171]. The temperature relationship of the surface losses is closely approximated when a square law temperature function is assumed for the upper and lower critical fields H_{c1} and H_{c2} [172]. The solid curves of Fig. 76 agree with the theory. The volume losses of niobium are adequately explained by a theory propounded by Dunn and Fournet, Hlawiczka and Mailfert [184] based on a modified Bean-London model. When linear temperature functions are assumed for the surface shielding field ΔH and the critical volume current density j_c [185] their temperature relationship is closely approximated.

The dependence of the losses of Nb_3Sn on the field strength is adequately explained by the theory of Dunn and Fournet, Hlawiczka and Mailfert in the flux range of 70-100 mT approriate to cables. Simultaneously it is confirmed that the lower losses of the Nb_3Sn conductors are not only due to a higher critical bulk current density but also to a relatively high surface shielding flux-density $\mu_0 \Delta H$. The temperature dependence of the losses depicted in Fig. 78 can be most closely approximated, when a square law temperature relationship is assumed for the critical current j_c, and a fourth power law temperature relationship is assumed for the surface field ΔH [179]. In the case of Nb_3Ge the experimentally determined field and temperature dependences of the losses are also in good agreement with the theory mentioned above.

As already mentioned previously, much higher short circuit currents occur in three phase ac cables than in dc cables. The short circuit current, which may reach fifteen times the rated current, flows for approximately 60 - 80 ms in the cable until the current is interrupted. Stabilization of the three-phase superconducting conductors with normal metal of good conductivity, such as Al or Cu, is essential to protect them from destruction by such currents. In order that no eddy current losses occur in normal operation, the stabilizing material of high conductivity should

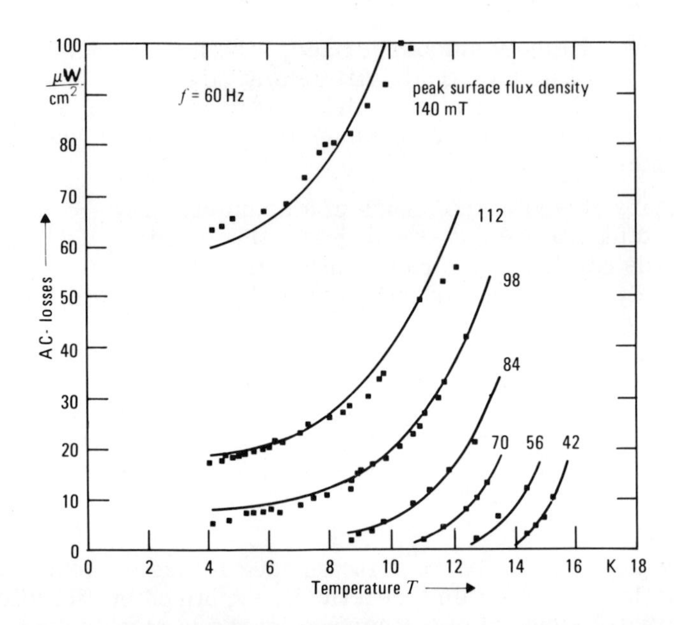

Fig. 78 Temperature dependence of 60-Hz losses of a Nb_3Sn-sample. Solid lines are theoretical calculations [179].

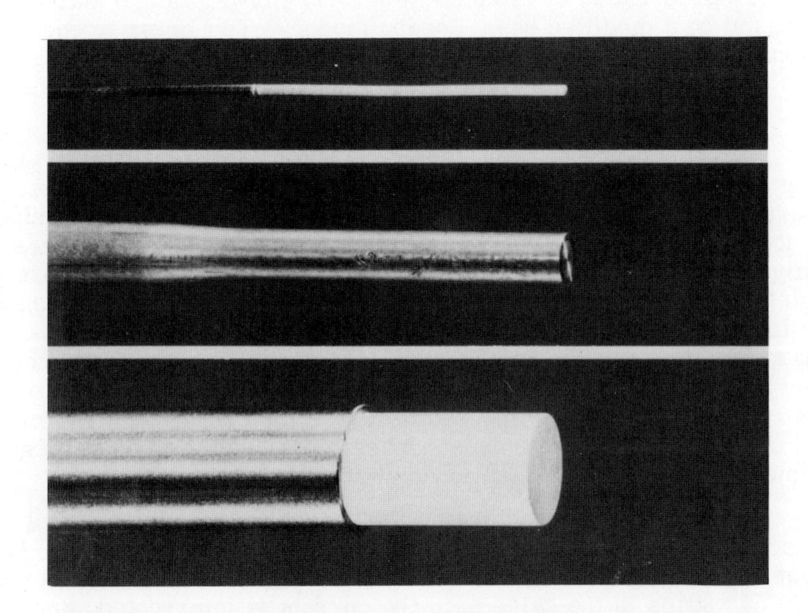

Fig. 79 Aluminium stabilized Nb-conductor for superconducting cables. Three different manufacturing stages (made by Siemens).

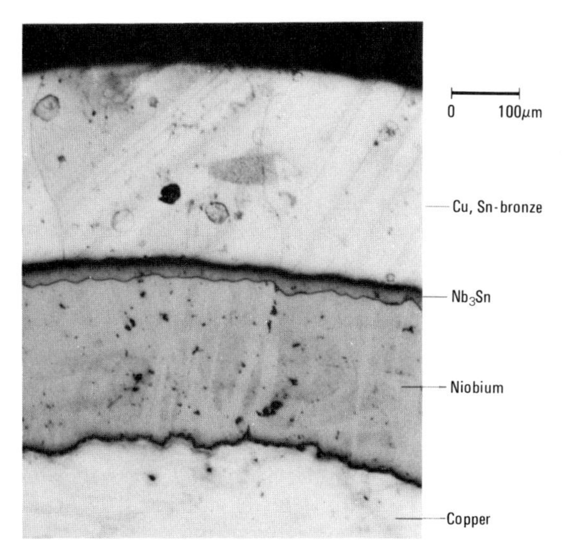

Fig. 80 Cross-section of a Cu-stabilized Nb_3Sn-conductor. Laboratory made sample of Siemens [180].

be fully enclosed by the superconductor,i.e., fully shielded. Figures 79 and 80 depict two conductors, which satisfy this condition, namely a niobium wire with a highly conductive Al core (resistivity ratio ≈ 3000) and a Nb_3Sn wire formed by a solid state diffusion process with a core of highly conductive Cu. The Cu-Sn bronze, which surrounds the superconductor in this case, has to be removed.

It is open to question whether full cryostatic stabilization on short circuit can be achieved for three-phase cables in contrast to dc cables. First, the effective cross section of the highly conductive stabilizing material is limited by the skin effect. It may be possible to obviate this effect by subdividing the stabilizing material into small mutually insulated areas [186]. In the case of Nb_3Sn and Nb_3Ge, the considerable brittleness of these materials is another negative factor. The thickness of the stabilizing material is limited by the bending radius required for cables (see the next section).

Finally, a further unknown is the heat exchange between the conductor and the flowing helium. The short circuit is a transient phenomenon and causes pulse like heating of the superconductor. The temperature excursion of the superconductor for a short circuit is at present neither adequately explained nor measured. Nor is it known whether current reclosure within the period of 0.3 - 0.5 s, usual for conventional cables, can be achieved with three-phase superconducting cables. A positive result in this area has been recently published [182]. Using niobium tube conductors stabilized with copper, it could be shown that superconductivity was restored some tenths of a second after interruption of a short circuit current, whose amplitude was approximately seven times rated current. Further experiments and theoretical consideration of the behavior of three-phase superconductors under short-circuit conditions are essential and are being researched in a number of places.

E. Cable Core

As initially mentioned, both rigid and flexible cable cores have been proposed for the cable projects under investigation. The rigid tube conductors are the easiest to manufacture but have the disadvantage that only short lengths (≤ 20 m) can be transported and that a considerable number of joints is required. A further problem to be solved in conjunction with cable cores is the compensation of thermal contraction without plastic deformation.

In the case of a three-phase cable with rigid hollow conductors, the problem of thermal contraction is solved by means of a three component conductor [187]. Contraction is primarily compensated by a 1.6 mm thick Invar tube (integral contraction 0.04%) on which an 0.8 mm thick copper layer is plated as stabilizing material over which an 0.08 mm thick niobium layer superconductor is deposited. Analogous to the flexible cryogenic envelope using corrugated tubes, flexible 2 component corrugated tube conductors of Nb and copper are under development [188]. In a number of cable projects, a flexible cable core is under development comprising stabilized wire or strip superconductors wound helically on a flexible carrier. Figures 81a and 81b depict such a flexible coaxial phase conductor. The superconductors are wound helically on a flexible tube (PE-helix) to form the phase conductor. A potential smoothing foil and the electrical insulation (wound PE tape) follow, over which a further layer of potential smoothing foil is added. Then a second layer of helically wound superconductors follows forming the current return path, which is protected by a covering of PE tape. This phase conductor with two simple single layers of helically wound wires is comparatively easy to manufacture and makes full use of the material. Compensation of longitudinal contraction on cooling is achieved by correct matching between the pitches of the helical wire layers and the coefficient of contraction of the material employed as a carrier and for electrical insulation.

In the case of coaxial three-phase tubular conductors, the helical arrangement of the individual conductors leads however to a number of problems, which do not occur with rigid tubular conductors, and corrugated tubular conductors. These problems are related to the fact that, as a result of the helical arrangement of the superconductors of the inner and outer conductors, an axial magnetic field is generated within the inner conductor and in the annular gap between the inner and outer conductor. Therefore, the use of normally conductive material must be avoided. Considerable eddy currents with unacceptable losses would occur in the interior of the inner conductor. The magnetic field and the consequent losses in the interior of the inner conductor can be completely eliminated, when the pitch of the inner and of the outer conductor are equal, i.e., $1/P_i = 1/P_e$ where P_i and P_e are the corresponding pitches of the internal and external conductors, and $I_i = -I_e$. The simplest and most

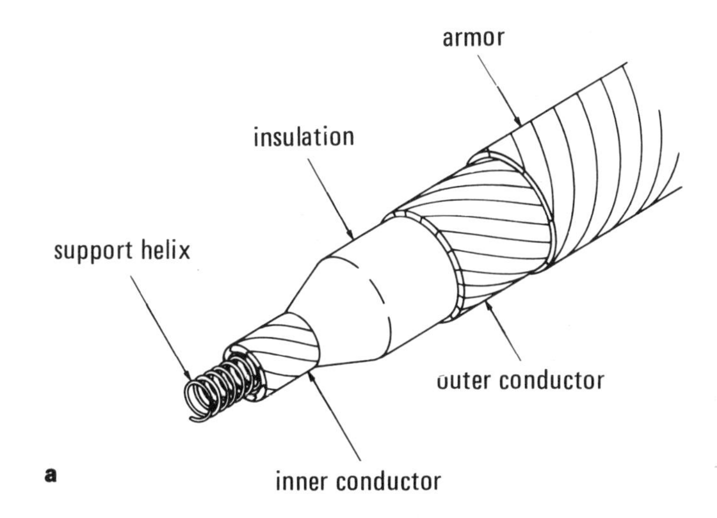

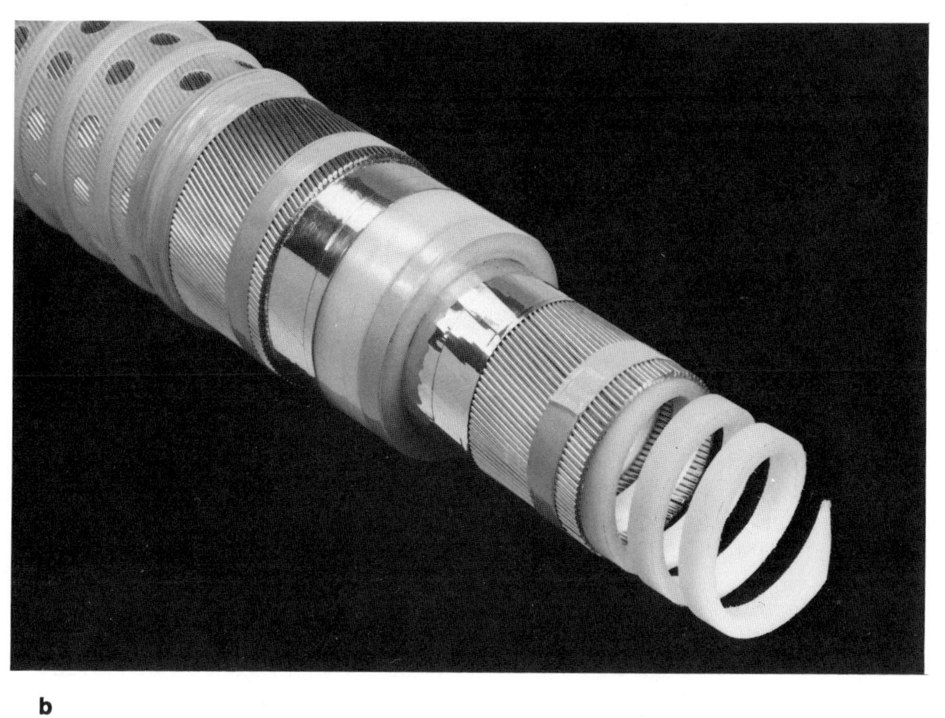

b

Fig. 81 (a) Schematic diagram of a flexible coaxial cable core.
(b) Model of a flexible coaxial cable core (made by Siemens AG).

economical way to avoid losses in the interior of the inner conductor is to utilize a carrier of non-conductive material and to use conductors whose stabilizing material is completely enclosed by the superconductor as shown in Fig. 79.

The helical arrangement of the inner and outer conductors also has the consequence that an axial flux is enclosed by the outer conductor of the coaxial conductor arrangement. This flux induces currents and thus unacceptable losses appear in the normally conductive tubes surrounding the phase conductor. This problem was first reported by Sutton [189]. The losses are especially critical in single core cables. They can be avoided only where the surrounding tubes are provided with internal coatings of superconducting layers or when the axial flux, which is enclosed by the outer conductor, is reduced to zero. This is the case for $I_i = - I_e$ when $d_i^2/P_i = d_e^2/P_e$, where d_i and d_e are the diameters of the internal and external conductors respectively.

The axial flux within the complete conductor has a further undesirable consequence. It induces a voltage in the outer conductor, which can assume considerable dimensions and be the cause of additional expenditure for insulation [190]. The voltage is reduced as the pitch is increased and becomes zero, when the conductors are axially parallel (equivalent to the case of a rigid coaxial tubular conductor pair). Realistic calculations show that conductor configurations are practical in which the pitch is so large, that only relatively small voltages occur along the outer conductor. In cases in which this is not possible, the voltage in the outer conductor can be avoided only where the axial flux is zero.

Finally, it can be shown that for a cable core fixed at both ends, as shown in Fig. 81(a) and 81(b), no axial mechanical forces will occur, if the following equation is valid for the radial contraction of the core,

$$\frac{\Delta r}{r} = \alpha_c \left(1 + \frac{P^2}{(\pi d)^2} \right)$$

where α_c is the integral contraction of the stabilized superconductor. Uniform radial contraction of the cable core, without length shortening, occurs when $P_i/d_i = P_e/d_e$. The radial contraction $\Delta r/r$ of the cable core, as shown in Fig. 81, is determined chiefly by the coefficients of contraction of the carrier and of the electrical insulation.

Cable cores with single layers of helically wound wires present the engineer a problem which has 3 conditions to satisfy of which only one can be satisfied. Which condition should be chosen as a basis for cable design is governed by the cable specification (current, voltage, geometrical dimensions, superconductor material). The condition $1/P_i = 1/P_e$ is the least critical. It can be avoided through the use of stabilized superconductors, in which the stabilizing material is completely enclosed by the superconductor. In the case of coaxial tubular conductors with large diameters, the danger of excessive electrical voltages in the outer conductor can be reduced by the use of a large pitch P. It can be assumed therefore that the essential condition for a cable of the type described is that of stressless contraction.

Another solution to the problem has been proposed by Sutton [189], namely use of double helically wound conductors for the go and return conductors of a coaxial cable as shown in Fig. 82. Both helical windings of each conductor layer have the same pitch but are of opposing direction. As a result of this measure, an axial magnetic field in the cable core is avoided and the cable can be designed to meet the requirements regarding thermal contraction on cooling.

This type of cable core is being investigated in a cable project based on a Nb_3Sn conductor [177]. Stabilization is effected in this cable by means of separate helically wound strip aluminium conductors to take into account the considerable brittleness of Nb_3Sn, the resultant low radius of curvature and to ensure symmetrical current distribution even in the event of short circuit. Two helixes with opposing sense of winding are arranged within the internal conductor and an additional two helixes are arranged external to the outer conductor. The cable conductor thus comprises a total of eight helically wound layers as compared to the two depicted in Fig. 81. It is apparent that such cables are complicated to fabricate and comparatively expensive.

F. Electrical Insulation

The electrical insulation is one of the most important components of superconducting cables. Together with an increase in the dielectric strength of the cable insulation, the economic viability of a cable is increased. The operational reliability of superconducting cables is dependent mainly on the integrity of the insulation. In addition to the dielectric strength, the dielectric losses and the mechanical properties at room and at low temperatures are the most important parameters of the electrical insulation. The dielectric losses are of subsidiary importance in the case of dc cables, due to the absence of an alternating field, and consequently for dc cables result in a greater freedom in the choice of insulating materials. In the case of three-phase cables, the electrical insulation must however satisfy the condition for dissipation factor that $\tan\delta \leq 2 \times 10^{-5}$, in order that the dielectric losses are comparable to other cable losses (conductor, thermal insulation). Basically the dielectrics so far proposed for superconducting cables include [172], (a) vacuum, (b) helium liquid, subcooled or supercritical, and (c) wrapped tape insulation impregnated with helium or in a vacuum.

Vacuum insulation ($\tan\delta = 0$) has been considered particularly in conjunction with rigid tube type cables. The dielectric strength of the insulation is markedly dependent on the necessary spacers, on the geometrical form, the surface quality and the transition of the these to the cathode. In the case of coaxial cables, maximum field strengths of the order of 20 MV/m may be expected at 4.2 K. Major mechanical problems

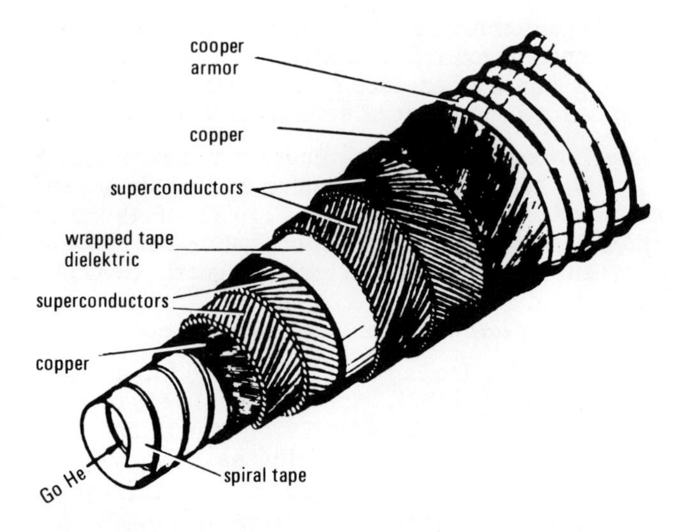

Fig. 82 Flexible cable core with double helix inner and outer conductor layers [215].

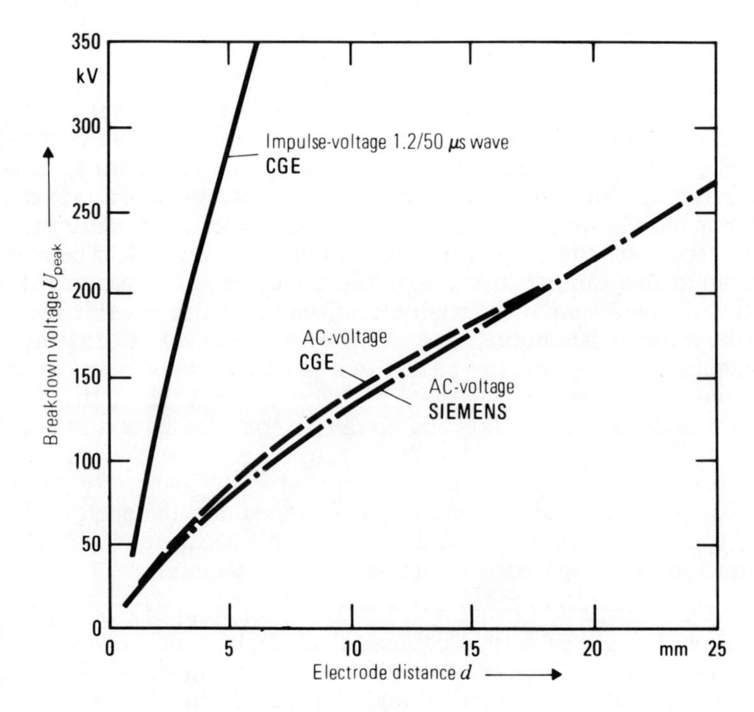

Fig. 83 AC- and impulse-breakdown voltage of liquid He at 4.2 K as a function of electrode distance.

arise from possible differences in the coefficients of expansion of the spacers and of the conductor material and in the difficulty of sealing the insulation to remove vacuum leaks. At present, neither vacuum nor vacuum impregnated foils are under consideration.

In conjunction with rigid tubular type coaxial conductors, helium is also under consideration as an insulating medium. In view of the cable operating conditions (temperature range 4 - 10 K, pressure range 2 to 15 bar), subcooled or supercritical helium appears most suitable. It is recommended that the helium used as an insulating medium be kept separate from the helium coolant. The dielectric strength of helium has been investigated chiefly at 1 bar and 4.2 K. Figure 83 depicts the dependence of the peak breakdown voltage on the electrode gap width for 50 Hz ac and for impulse loads [191, 192]. In the case of larger electrode gap widths, such as are usual for coaxial cables, permissible field strengths of 10 MV/m (peak value) may be assumed for ac loading. In the case of impulse loading, the permissible field strength values may be greater by a factor of at least 3. No discernible difference could be observed between ac and dc loading [193]. Appreciable increase of the dielectric strength (e.g., up to 65%) is obtained with either subcooled helium or with supercritical helium under pressure (e.g., ten bar) [194].

This result can be explained if one assumes that breakdown in helium originates in partial discharges in the gas bubbles in the helium. The breakdown strength of helium gas decreases with increasing temperature or decreasing pressure. Paschen's law (viz. breakdown voltage is proportional to the product of gas density and electrode gap width) is obeyed well for helium gas with densities up to approximately 20 kg/m^3 [195]. Helium gas at atmospheric pressure and temperatures considerably in excess of 4 K, is therefore a relatively bad insulator. This fact leads to complications for those parts of the cable in which helium occurs in gas form, e.g., in the cable terminations cooled by helium gas, which has to be conducted from high voltage to ground potential. The electrode material has little influence on the dielectric strength of helium. Mechanical irregularities of the electrode surface drastically reduce the dielectric strength. Impurities have varying influence. Frozen particles of air have practically no effect, while slight traces of oil drastically reduce the dielectric strength [195].

The dielectric constant ε_r and the dissipation factor $\tan\delta$ of liquid helium at 4.2 K were determined experimentally to be 1.05 and 10^{-6} respectively. Thus liquid helium has negligibly low dielectric losses. To what extent these values are relevant to impure helium still has to be determined.

A coaxial conductor arrangement with liquid helium as a dielectric material requires supports to fix the tubes relative to each other both radially and axially. In the experiments performed, it has been found that the introduction of spacers in liquid helium leads to a marked reduction of the dielectric strength of the latter. This is in agreement with the results obtained with supports in a vacuum insulation. In the case of polyethylene spacers in liquid helium, a reduction to approximately 60% of the value obtained with the liquid alone has been observed [196]. An even more severe reduction (down to approximately 20%) has been recently observed in the case of filled epoxy resin supports in supercritical helium (5 bar) [197]. On the basis of the available results a maximum permissible ac operating field strength of $< 5\,MV/m$ can be taken for helium as a dielectric material with supports. This value is less than the minimum required for an economically viable superconducting cable. Wound insulation with paper tape or synthetic tape is being investigated among the cable projects with flexible cable conductors (refer to Fig. 81). During cable operation, this insulation becomes filled with subcooled or supercritical helium at the butt joints or lapped points so that the foil insulation is impregnated with helium. Synthetic or paper material used for wound foil insulation should have the highest possible intrinsic dielectric strength, adequate mechanical properties at room temperature, and cryogenic temperatures. In the case of ac cables synthetic or paper materials should have low dielectric losses at 4.2 K.

The breakdown strength of foil insulations under continuous load is largely dependent on the state (temperature and pressure) of the surrounding helium. This effect is due to partial discharges in the helium contained in the cavities of the insulation. Pulse breakdown voltage tests are employed in order to be able to make a more quantitative statement regarding the intrinsic dielectric strength of foils. The results of such measurements show that values of over 100 MV/m are obtained for the breakdown strength of standard high density Polyethylene (HDPE) foil, Mylar and Kraft paper. In general homogeneous synthetic foils have a higher dielectric strength than paper like foils of the same material.

All synthetic materials of high inherent dielectric strength and also Kraft paper can be used as insulation for dc cables. For ac cables, only materials with a loss factor for $\tan\delta \leq 2 \times 10^{-5}$ are acceptable. This condition is satisfied only by nonpolar and slightly polar polymeres. Table 15 shows the chain repeat unit and the loss tangent of a number of polymeres, which have been investigated in conjunction with superconducting cables. Since economic factors are of decisive importance for superconducting cables, the price of the foil is given in the table.

A low loss factor, $\tan\delta$ at 4.2 K, is dependent on a low dipole moment of the chain repeat unit. Polytetrafluorethylene PTFE and Polyethylene PE have no permanent dipole moment and have a dissipation factor $\tan\delta < 10^{-5}$. A very low dipole moment of the polymere chain repeat unit is an essential but not the sole condition for a low dissipation factor. The latter is dependent on a number of further parameters, such as the degree of crystallinity of the polymere, the degree of impurities and of additives (e.g., antioxidants, solvents and dyes) or decay products. Unfortunately, up to now, the only polymeres which are available were developed for use at room temperature and not for 4.2 K. PTFE

Table 15. Plastic materials for wrapped tape cable insulations.

Material Trade Name	Ideal Structure	Dissipation Factor tan δ at 4.2K	Remarks	Price per 1000ft of tape 1 inch×5 mil
Polytetrafluorethylene Teflon	$[-CF_2-CF_2-]_n$	5.0×10^{-6}	Nonpolar	$ 150.--
Polyethylene oriented Valeron	$[-CH_2-CH_2-]_n$	7.0×10^{-6}	Nonpolar	1.60
Polypropylene Bicor	$[-CH_2-CH(CH_3)-]_n$	2.0×10^{-5}	Slightly polar	2.50
Polyamide Nylon 11 Rilsan	$[-NH-(CH_2)_{10}-C(=O)-]_n$	2.5×10^{-5}	Slightly polar	7.--
Polysulfone Udel	$[-C_6H_4-O-C_6H_4-C(CH_3)_2-C_6H_4-O-C_6H_4-SO_2-]_n$	3.0×10^{-5}	Slightly polar	13.--
Polycarbonate Makrofol	$[-C_6H_4-C(CH_3)_2-C_6H_4-O-C(=O)-O-]_n$	5.0×10^{-5}	Slightly polar	18.--
Polyimide Kapton	$[-R-N(CO)_2C_6H_2(CO)_2N-]_n$	9.0×10^{-5}	Polar	80.--
Polyester Mylar	$[-(CH_2)_2-O-C(=O)-C_6H_4-C(=O)-O-]_n$	2.0×10^{-4}	Polar	4.--

and Kapton cannot be used in superconducting cables for economic reasons. Mylar can be used only in dc cables on account of its high dissipation factor. Normal polypropylene in the form of solid tapes was found to be mechanically weak and is therefore excluded from further investigations.

As already mentioned, in addition to the electrical properties of polymeres, the mechanical properties are of decisive importance. This was apparent when dealing with the compensation of longitudinal contraction of flexible cable cores when these are cooled down. This process is decisively influenced by the thermal contraction of the electrical insulation. Table 16 gives the electrical and mechanical properties of some polymeres (suitable for ac cables) in tape form as compared to normal paper insulation (Kraft paper) [198].

Polymere foils have an appreciably lower tensile strength and lower E-moduli than Kraft paper which is processed on conventional insulation wrapping machines. Their elongation is also appreciably higher. In order to produce insulation layers with the necessary precision on insulation wrapping machines, polymere tapes must have an E-modulus of at least 700 MN/m^2. This value is just reached by some of the polymeres listed in Table 16.

On transition from ambient temperature to a temperature of 4.2 K the ultimate tensile strength and the E-modules increase greatly. This means that the polymere tapes become embrittled. The maximum permissible elongation is drastically reduced. In order to avoid tensile fracture of the tapes during cooling down, it is necessary that the total thermal contraction between 293 K and 4.2 K be less than the maximum permissible elongation. This is the case for the materials given in Table 16. In order to avoid tearing as a result of local stress concentration, the tapes must exhibit good sliding properties, i.e., the coefficient of friction may not be excessively high. In this respect paper-like plastic foils are superior to solid plastic foils. For this reason mixed insulation of both solid and paper-like foils has been proposed [180].

An important problem of wound foil insulation is posed by partial discharges. These discharges occur in the helium, which impregnates the insulation at the butt joints and overlaps of the tapes. The problem is intensified by the considerable differences of the dielectric constants, of helium and the polymeres. This leads to a stress intensification in the spaces filled by helium. The partial discharges lead to an increase in the dielectric losses and in the course of time may lead to gradual destruction of the insulation and, if intense enough, to electric breakdown. It is therefore a requirement that the inception field strength for partial discharges be considerably higher than the operating field strength. For economic reasons the inception field strength should be at least 10 MV/m. Figure 84 depicts the relationship between field strength and loss factor of PE tape insulation impregnated with helium at 4.2 K [196]. A steep rise in the loss factor occurs at field strengths considerably below the breakdown value and partial discharges can be observed in this region also. The use of higher pressure increases the dielectric strength of the impregnating helium. The higher pressure moves both the inception voltage for partial discharges and the rise of the loss factor to higher field strength values. The highest inception voltages for corona discharge of 16 MV/m at 7 K and 12 bar measured were for laboratory taped windings of Valeron (high density PE oriented and cross laminated, Van Leer Co.) and of Cryovac (high density PE cross linked and oriented) [198]. In the case of model cable cores manufactured on conventional

Table 16. Electrical and mechanical properties of plastic tape candidates for ac cables. Kraft-paper for comparison [198].

	Diel.const. 4.2 K	Tan δ and 60 Hz	Tensile Strength MN/m^2 293 K	Tensile Strength MN/m^2 4.2 K	Yield Strength 0.2 % MN/m^2 293 K	Yield Strength 0.2 % MN/m^2 4.2 K	Tensile Modulus MN/m^2 293 K	Tensile Modulus MN/m^2 4.2 K	Elongation % at 293 K	Elongation % at 4.2 K
Polyethylene high density	2.3	1.5×10^{-5}	50	210	6	200	720	6700	322	3.29
Polyamide Nylon 11 Rilsan	2.4	2.5×10^{-5}	55	150	17	130	730	5240	355	3.13
Polyethylene oriented (Valeron)	2.3	0.7×10^{-5}	40	125	4	125	880	4160	402	3.10
Polysulfon (Udel)	2.5	3.0×10^{-5}	60	120	40	110	1850	4410	64	2.98
Polycarbonate (Macrofol)	2.9	5×10^{-5}	180	340	50	190	3400	4460	33	10.75
Kraft Paper	1.7	3.0×10^{-4}	100	–	80	–	9600	–	1.69	–

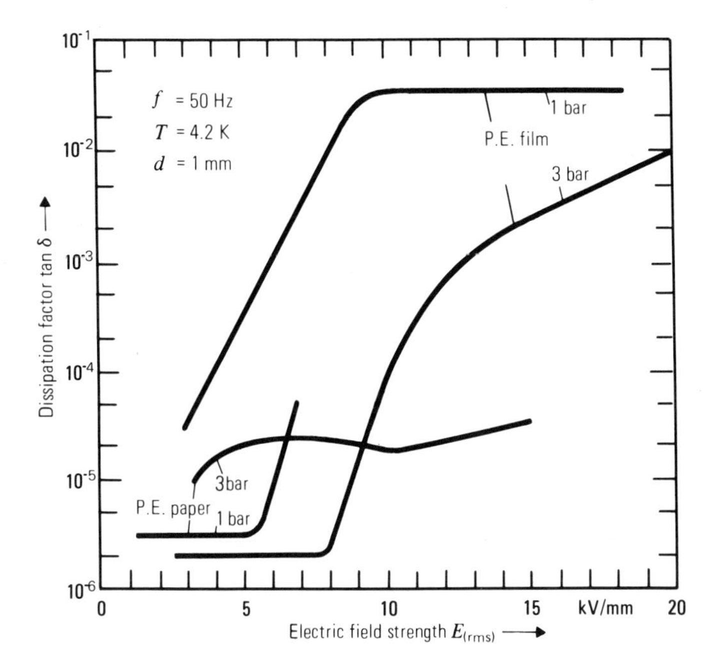

Fig. 84 Dielectric dissipation factor versus the electric field applied to a wrapped tape polyethelene insulation in liquid helium [196].

cable machines lower values have been obtained due, no doubt, to imperfections in the stress cones [177, 180].

In conclusion it may be said that, with respect to their electrical properties, at present high density polyethylene in the form of Valeron and Cryovac is the most promising candidate for wound foil insulation of ac cables. It should be possible to attain operating field strengths of 10 MV/m. Polycarbonate in the commercial form as presently available exhibits the best mechanical properties but still has dielectric losses which are too high. Intensive research must still be carried out in order to achieve a good, reliable and economic insulation for a c cables. Improvement of the performance of polymeres at the low temperatures employed in cables is of special importance. Methods to reach this goal include modifications of the extrusion and polymerisation processes, the compounding of polymeres and a reduction of the concentration of solvents, dyes and other additives. It should further be mentioned that at present little experience is available relating to the long term performance of wound polymere tape insulation subjected to electrical loading. This is a subject which will have to be investigated in the future.

Kraft paper is a possible candidate for flexible dc cables. It has been found that with Kraft paper tape insulation impregnated with helium at temperatures of 4 - 5 K, dc field strengths in excess of 20 MV/m can be permitted [199]. The same statement applies to high density polyethylene tape insulation impregnated with helium [200].

G. Cable Cooling

Cooling of superconducting cables poses a number of problems, which do not occur with other superconducting systems or are of secondary importance. Superconducting cables have relatively large geometrical dimensions, they contain a large volume of stored helium and the cables have difficult access. Many of the cable designs proposed have two coolant circuits, a helium cooling circuit and a nitrogen cooling circuit, which cools a heat barrier to approximately 80 K. It has been proposed that both coolants be employed in single phase form. The nitrogen cooling circuit is relatively free of problems. Operation of a 10 km long 2500 MVA three-phase cable with subcooled nitrogen at inlet conditions of 80 K and 4 bar requires a throughput of approximately 1.6 kg/s to maintain a temperature difference of 4 K between the cable inlet and outlet. By means of a suitable choice of the tube dimensions, the flow losses can be kept below 1% of the dissipated losses to be removed from the cable.

Single phase helium, either subcooled or supercritical has been considered for the helium circuit. The relevant temperature ranges are determined chiefly by the properties of the superconducting materials employed, by the hysteresis losses in the case of ac cables and by the critical current density in the case of dc cables. For this reason, cables employing niobium and NbTi must operate below 6 K, whereas cables with Nb_3Sn are proposed to operate in the range of 6 K to 10 K.

The choice of the most suitable operating pressure is governed by the following considerations.

With respect to avoidance of oscillations, the pressure should be as high as possible. One should operate in a region of the temperature entropy diagram for helium as far away as possible from the critical point and from the transposed critical line (see Fig. 85).

To keep the helium mass flow and as a result the pressure drop as low as possible, the pressure should not be too high. The heat capacity of helium drops at increasing pressure.

A resultant positive Joule-Thompson coefficient should be attained on expansion of the helium along the length of the cable.

The dielectric properties of an electrical insulation employing He or helium impregnated wound foils improve radically with increasing He pressure.

Which of the criteria above is the most decisive can be established only after experimental experience is gained with long sections of test cables. It is however to be expected that the demand for an adequate safety margin, with respect to the electrical insulation, will be a decisive factor in choosing the pressure.

For Nb- and NbTi cables a working pressure of 4 to 6 bar has been proposed and for Nb_3Sn cables 10 to 15 bar (see Fig. 85). The feed pressure drops along the cables can be kept low by virtue of the relatively low mass flow (of the order of some hundred g/s). The main pressure drop occurs in most cable designs in the He return path to the refrigerator. The return path is installed in the vacuum of the cryogenic envelope and on account of space limitations has a relatively low cross section.

Special attention must be paid to transient cooling processes when a cable warms up due to a short circuit or is cooled down or warmed up between ambient and low temperature. These phenomena are complicated by the large ratio of length to cross section and by the limited feed capability from one cable end. It has been shown [172], that the temperature and pressure increases in the He after a short circuit are tolerable. However, a long time is required until temperature and pressure are completely equalized throughout the length of the cable. The additional cooling capacity required for this purpose is included in the safety margin normally provided.

Cooling down and warming-up of cables under the restrictions mentioned above presents a much greater problem. On the one hand, comparatively large masses have to be cooled down, while on the other hand the maximum possible helium throughput is limited by the maximum permissible pressure in the He pipes (e.g., 20 bar). If one assumes that a 10 km cable section has to be cooled down from ambient temperature to 4.2 K by helium gas and liquid under high pressure (20 bar) fed from one end of the cable, a period of some weeks is required for cooling down. The same applies to warming up of longer cable sections.

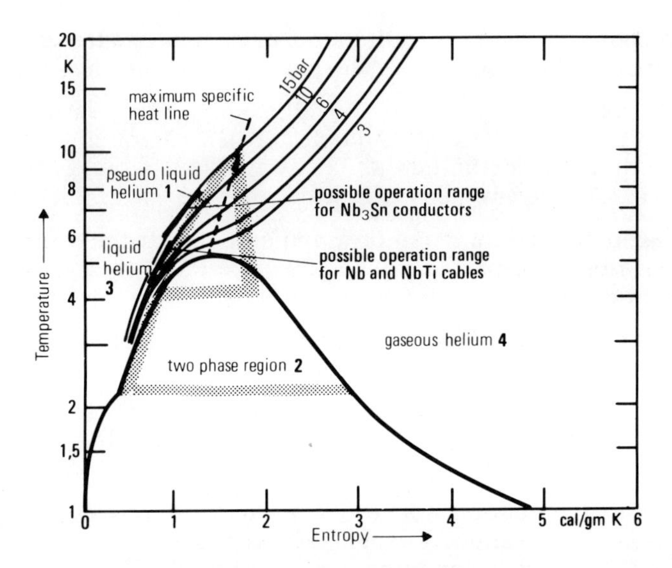

Fig. 85 Temperature-entropy diagram for helium showing the possible regions of cable operation.

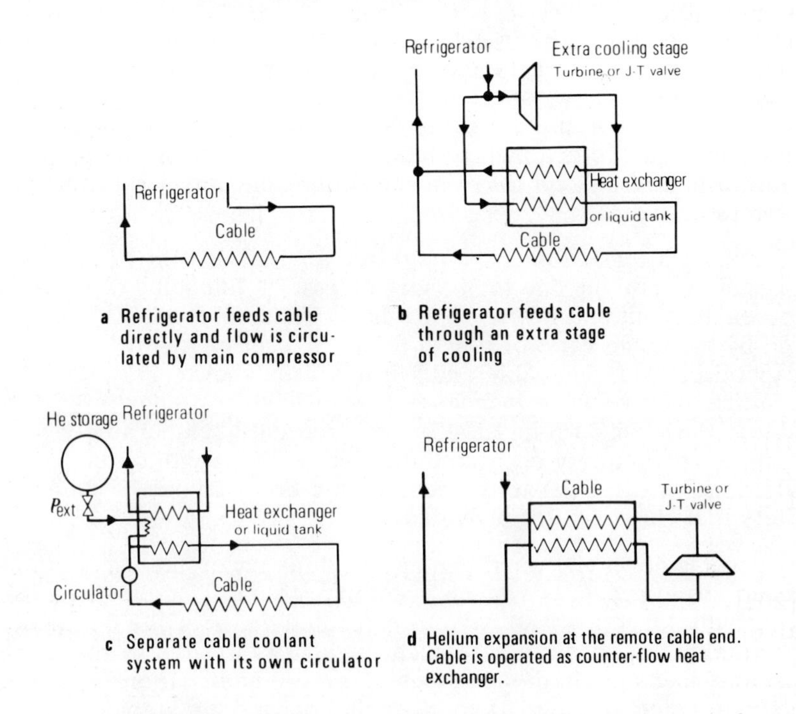

Fig. 86 Possible combination schemes of refrigerators and superconducting cables.

Of special interest is the question of supply of He to the cable during the cooling down phase. Longer cable sections have a liquid helium content of some 100 m^3, which must be delivered in road tankers. In the event of a necessary warming up of the cable, provision must be made to transfer the helium content of the cable back to road tankers and transfer it to other users. Storage of He in gaseous form is impracticable on account of the enormous volumes involved. Removal of the helium contents of cables by means of road tankers necessitates the integration of superconducting cables into a system of other large scale He consumers.

In addition to technical problems, the large helium content of superconducting cables also presents economic questions. Today the helium filling accounts only for some percentage of the cable cost. This would be significantly increased by a higher price for helium. Warnings to the effect that a He shortage might lead to considerable increase in the price of helium in the future cannot be disregarded. Conservation of He and the development of economical processes for production of He are effective and necessary countermeasures.

There are a number of possible ways of combining cable and refrigerator as shown in Fig. 86. In (a) the refrigerator outlet is connected directly to the cable. This combination can be employed for cable operating temperatures > 6 K, that is for cables with Nb_3Sn conductors. In (b) part of the helium is expanded in a turbine or Joule-Thompson valve and coupled to the cable coolant stream via a heat exchanger or a He bath. By this means, temperatures of <6 K may be obtained. This method is suitable for cooling cables with niobium and NbTi conductors. Whereas compressors operating at room temperature are employed to pump He through the cable in the arrangements as shown in (a) and (b), a low-temperature helium pump is employed in (c) with a heat exchanger between the refrigerator circuit and the cable coolant circuit. The view is often expressed that the efficiency of coolant circuits with helium pumps working at low temperature, is less than that of cooling circuits with compressors at room temperature. The disadvantage of a lower efficiency when pumping at low temperatures is largely compensated by the fact that less power is required for pumping at low temperatures. This is a consequence of the greater density and incompressibility of He [201]. The advantage of coolant circuits with pumps lies in the greater flexibility and higher reliability. The arrangement in (c) is suitable for all operating temperatures.

A very effective cooling concept recently proposed for temperatures > 6 K, that is for cables with Nb_3Sn conductors, is shown in Fig. 86(d) [202]. The last expansion stage (expansion turbine or Joule-Thompson valve) of the coolant circuit is connected to the remote end of the cable. The expanded He in the return path is not conducted separately and thermally insulated as generally proposed. It is in heat exchange with the go path. It thus flows back in the helium tube surrounding the cable conductors (Fig. 87) [203]. The cable cryogenic envelope is thus

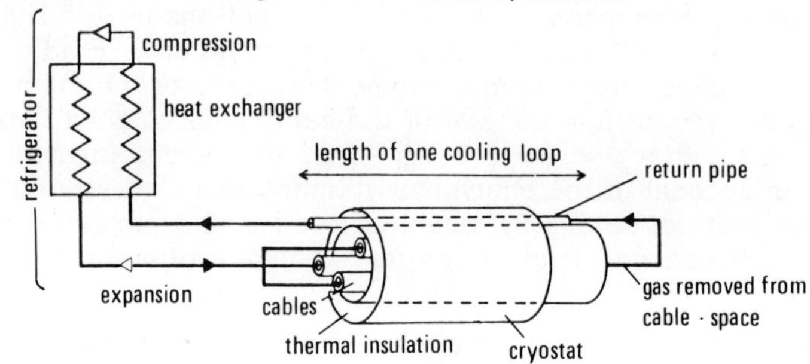

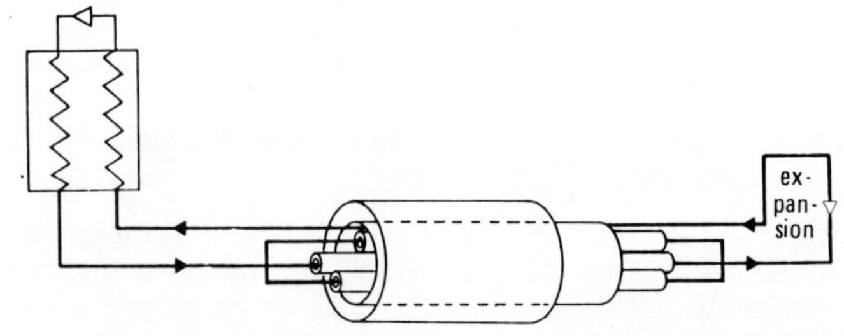

Fig. 87 Combination schemes of refrigerators and cables [203].

utilized as the final heat exchanger stage of the cooling circuit. This cooling principle results in a number of advantages which include a simpler and consequently a cheaper mode for the construction of the cryogenic envelope.

The cable can be cooled from both ends. This leads to a better temperature distribution for a given maximum ΔT.

The helium return path has a larger cross section.

In consequency of the last two points, larger distances between refrigerator stations are made possible. Instead of distance of 10 to 20 km previously for ac cables, distances of up to 60 km should be possible with the new method.

The cooling circuit is more efficient. A compressor input at room temperature of 190 W per Watt of dissipated power at helium temperatures is needed instead of 310 W in the case of conventional circuits.

The total losses (thermal and electrical) of large capacity superconducting ac cables are small compared to other underground transmission systems under development. A specific loss of 200 W/km at 4 to 6 K, and of 2 kW/km at 80 K, occurs in a 110 kV 2500 MVA three-phase cable. If a refrigerator efficiency of 300 is assumed for 4 to 6 K, and of 10 for 80 K, a loss of approximately 80 kW/km is obtained relative to the ambient temperature.

Low losses in conjunction with a high power density ($P > 10,000$ MVA/m^2 referred to the total cable cross section) make superconducting ac cables an attractive proposition. In the case of superconducting dc cables the transmission losses are $\leq 25\%$ of the losses of an ac cable of the same rating [172]. The power density of superconducting dc cables is at least 4 times as high as that of three phase ac cables. The distances between the cooling stations of dc cables can thus be larger than that for ac cables. Whereas distances of 10 to 60 km are considered practicable for ac cables, distances of 20 to 100 km should be possible for dc cables.

Helium refrigerating plants of a capacity suitable for cooling long cable sections (some or several kW capacity at 4 to 10 K) are already available. Such refrigerating systems must, in addition to the required refrigeration capacity, be reliable and be compatible with the requirements for components of a power supply system. The considerations mentioned in conjunction with large superconducting alternators (refer to Section IV) are also applicable. Larger failure rates than those of conventional plants cannot be accepted.

H. Cable Terminations

The cable terminations of superconducting cables are relatively complicated devices and are the most problematic components of superconducting cables. They have a number of functions to fulfill each of which will be dealt with here separately.

Superconducting cables are designed for use in power supply systems in which other important components such as the three-phase windings of alternators, transformers and circuit breakers are of normal conductors. The cable terminations must therefore perform the following functions.

They must be capable of feeding in larger currents from ambient temperature to low temperatures with minimum losses.

They must couple high electric potential to the superconductor.

The injection and removal of the coolant helium must take into account the jump from earth to transmission level potential.

They must transfer He gas, with a very low dielectric strength at room temperature, from a high potential to earth potential.

Figures 88 and 89 depict the principle of construction of a single pole dc cable termination and of a single-phase ac cable termination. In both cable terminations, the problem of current feed-in at minimum loss has been solved by providing current leads of normal conducting material designed as heat exchangers. They are cooled by counter flowing cold He gas, within the area of transition from room temperature to helium temperature. It is thus possible, similar to the method empolyed for current leads of superconducting magnets, to reduce the losses flowing to He temperature (thermal and Joule losses) to 2 to 3 W

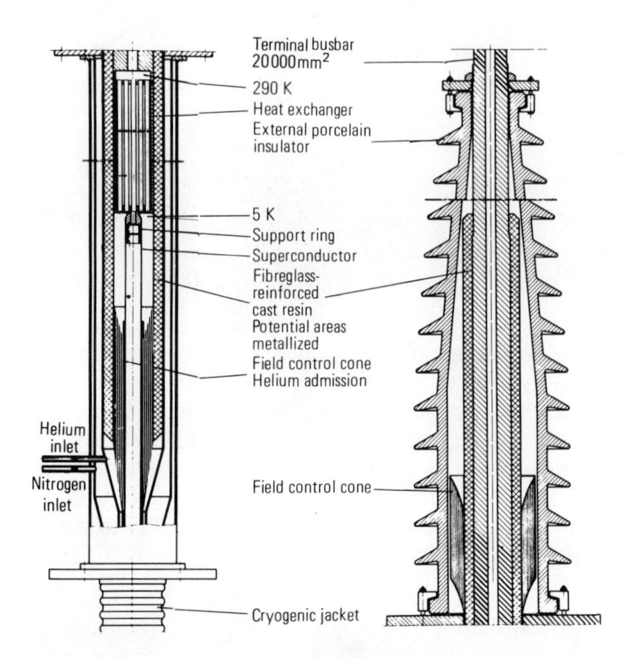

Fig. 88 200-kV-Terminal for a superconducting dc cable (Constructed by AEG, Kabelmetal, Linde).

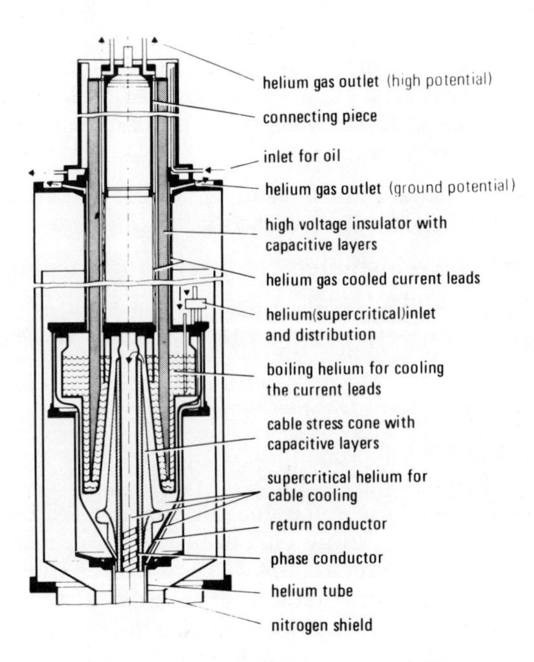

Fig. 89 One phase ac cable termination (Siemens).

per kiloampere feed-in and return current.

To overcome the problem of coupling the high potential, similar methods of construction to those employed in the cable terminations of conventional oil-impregnated paper insulated high power cables have to be used. A longitudinal electrical field occurs in the region of transition from the cable core end to the termination for which the dielectric strength of electrical insulations is drastically reduced. Breakdown values of the order of only 5 - 10% of the values valid for radial fields have been measured at 4.2 K for He impregnated wound foil insulations in longitudinal fields. The reduction is similar to that experienced with oil impregnated paper tape insulations at ambient temperature. In the case of cable terminations for superconducting cable, the situation is aggravated by mechanical stresses in the insulating parts due to differing thermal contractions and temperature gradients. Smooth and continuous modification of the electrical field in the region of transition from the cable core insulation to the insulation of the termination is effected by means of stress cones. It is practically linearised in the case of ac cables by capacitive layers.

By means of the stress cones the field is controlled in the axial channels, in which the liquid He at earth potential is transferred to the inner conductor at high potential. The dielectric strength of He gas, which was used to precool the current leads and which must be brought from a high potential to earth potential at room temperature, can be significantly increased by addition of electrically negative gases or vapors. This insures that the potential drop can be effected over a reasonable distance [172]. Cable terminations as shown in Fig. 88 have been successfully tested in tests lasting up to 10 days at a direct voltage of 200 kV to earth [199]. Tests of an ac termination of the type depicted in Fig. 89 revealed breakdown voltages > 160 kV rms and partial discharge starting voltages of approximately 120 kV rms [204].

I. Superconducting Cable Projects

Theoretical and experimental research on superconducting cable has been and is still being carried out in most industrialized countries. These investigations initially dealt with the development of cable components. Now they have reached the stage of construction and testing of short or medium sized sample or model cables. These experimental cables are used to demonstrate the feasibility and performance of superconducting cables. They supply valuable experience for commerical manufacture of cable conductors and of cryogenic envelopes, for laying, jointing and termination techniques as well as for the cooling of long cable sections.

In Germany, a working group consisting of AEG-Telefunken, Kabelmetal and Linde are developing a dc cable. The cable is a fully flexible type and employs helically wound Nb_3Sn strips as conductor. The electrical insulation comprises wound Kraft paper tape. The working group carried out current and thermo-mechanical tests over two years ago on a 15 m long single phase cable. The cable was bent to form a loop and was short circuited. The current (maximum 30 kA) was induced in the loop by four transformers [205]. At present, voltage tests are being carried out by this working group on a 20 m long experimental cable length fitted with two cable terminations. Direct voltages of 200 kV to earth have been successfully withstood during endurance tests lasting 10 days [199].

A further cable project is being carried out in Germany by Siemens [180, 206, 207]. The aim of this project is the development of a 110 kV 2500 MVA three-phase ac cable of the semi-flexible type (a rigid cryogenic envelope with flexible cores constructed of helically wound conductors). Niobium stabilized by aluminium is under construction for the superconductor as well as Nb_3Sn stabilized by copper (refer to Fig. 79 and Fig. 80). The electrical insulation is wound high density polyethelene (HDPE) tapes. After completion of a phase of intensive development of the individual components, a single phase 35 m long experimental cable length for 110 kV and 10 kA with two cable terminations is under construction (refer to Fig. 90). The cryogenic envelope has been assembled and the flexible cable core with Nb conductors stabilized by Al has been drawn into the envelope. The cable terminations have already successfully passed voltage tests in which voltages up to 160 kV were applied. Both the German cable projects are scheduled to continue to the end of 1977. They have been supported by the Federal Department of Research and Technology of the FRG.

In England, the Central Electricity Research Laboratories (CERL) are engaged in the development of semi-flexible three-phase ac cables [208]. In addition to studies of 400 kV - 5 GVA cables, intensive development of components (Nb/Cu strip conductors, dielectric materials) were carried out. An 8 m long test facility is available. Loss measurements on Nb/Cu conductors and dielectric measurements on 5 m long flexible cable cores with HDPE tape insulation have been carried out. The program is scheduled for completion in Summer 1976 [209].

In France, extensive theoretical and experimental investigations on three-phase ac and dc cables have been carried out by the Compagnie Géneral Electricité (CGE), the Laboratoire Central des Industries Electriques (LCIE), the Electricité de France (EDF) and by Air Liquide. The scope of this work has been drastically reduced and is now limited to investigations of the mechanical and dielectric properties of polymeres.

At the Anstalt für Tieftemperaturforschung (ATF) in Graz, Austria, development of fully flexible superconducting three-phase cables, suggested for the first time by Professor Klaudy in 1965, has been underway for more than ten years. Current tests at low voltages (20 kV) and cooling tests with supercritical helium and liquid nitrogen have been carried out on cables 14 m and 50 m long [210]. A test cable approximately 50 m long (60 kV single phase) with a niobium coated corrugated copper pipe as the conductor will soon be operated under actual grid conditions in parallel to an existing overhead transmission line in one of the Austrian power stations. The associated cable terminations have

Fig. 90 35 m-Test facility, semifelxible. Flexible cable core was just pulled into the rigid cryogenic envelope (Photo Siemens).

been successfully tested at ac voltages of 140 kV rms and 295 kV dc pulse voltage.

In the USSR, system studies and development of components for dc and three-phase ac superconducting cables are being carried out at the Krzhizhanovsky Power Research Institute in Moscow. A 4.2 m long rigid cable model with Nb tubular conductors cooled by supercritical helium went into operation in 1970. It was followed by a 12 m long rigid three-phase cable model for 35 kV and 10 kA with Nb_3Sn conductors (tapes soldered longitudinally to the inner wall of copper tubes). The cable section was provided with cable terminations at each end. At present a rigid test cable 100 m long for 10.5 kV and 10 kA is being assembled with Nb_3Sn conductors employing electrical He insulation. This project is being undertaken jointly by the Krzhizhanovsky Institute and the governmental utility company Mosenergo. The cable is due to be tested under network conditions on a research site at a Mosenergo substation near Moscow during 1976. The next steps provide for a 1 km long test cable (10 kA, 10.5 kV) and a 7-10 km long test cable [211].

In Japan, the Furukawa Electric Company has been in the forefront in developing superconducting cables. Three different types have been investigated.

A 154 kV, 1000 MVA three-phase ac cable with rigid coaxial tubular conductors composed of Nb/NbTi/Cu and rigid thermal insulation, electrical insulation liquid He and Mylar.

A 110 kV, 5000 MVA dc cable employing coaxial tubular conductors composed of NbTi/Cu and a 33 kV, 5000 MVA dc cable employing stranded NbTi/Cu multifilament conductors. Liquid He and nylon respectively were used for the electrical insulation of the dc cables.

Current tests employing 3000 A ac and 5000 A dc have been carried out on a rigid cable model 4 m long with Nb/Cu tubular conductors [212].

At present only paper studies on a 500 kV, 10,000 MVA ac and dc cables are being carried out in Japan. The big national program for high power transmission systems proposed by the Research Committee on New Power Transmission Systems for 1975, in which superconducting cables were to have played a decisive role, has been postponed indefinitely [213].

Numerous development programs are being carried out in the US. Mention is made first of all of the three-phase ac cable project of Brookhaven National Laboratory. This working group is investigating three-phase ac cable designs for 345 kV and 500 kV transmission voltage and capacities of approximately 5000 MVA per circuit. The basic research program of the laboratory is chiefly concentrated on development of wound tape insulation and of Nb_3Sn conductor materials with low ac losses. Using 20 m long test facilities, cryogenic envelopes, and electrical tape insulations, flexible and semi-flexible cable types have been investigated. At present, a three-phase test rig some hundreds of meters long is being constructed at Brookhaven Laboratory, for 69 kV and 3 kA. It is to be parallel to an existing 69 kV overhead transmission line. The cable is of the semi-flexible type (rigid cryogenic envelope and flexible cable core). The latest plans foresee increase of the operating voltage of the test rig to 138 kV [177].

A further three-phase ac cable project is being carried out by the Linde Division of the Union Carbide Corporation using a rigid tubular conductor cable. The electrical conductors are coaxial tubes of a Nb/Cu/Invar composite using supercritical He ($P \leq 6$ bar) as a dielectric medium. They employ spherical radial spacers and have long extended axial supports of filled resins. Most effort is now concentrated on the development of a 138 kV, 3400 MVA three-phase ac cable. A 12.2 m long cryogenic envelope with cable terminations is available for test purposes. Current and voltage experiments up to 16 kA, 265 kV ac and 650 kV impulse voltage (1.5/40 μs) can be carried out on true-to-scale or subscale conductor assemblies [214].

A dc cable project is under way at Los Alamos Scientific Laboratory. Detailed design for semi-flexible cables with multifilament Nb_3Sn conductors (helically wound, operated at 11 - 14 K) for capacities of 1 - 10 GW is in progress [215]. A 20 m long test facility with a cryogenic envelope is available for current tests on Nb_3Sn conductors. Investigations regarding the suitability of Nb_3Ge conductors, manufactured using the CVD process, as a possible conductor material for dc cables are being carried out in addition to those on Nb_3Sn conductors.

Support research programs are being carried out by a number of laboratories for the above mentioned US projects, viz. investigations of dielectric media at low temperatures by the Oak Ridge National Laboratory, dielectric measurements and investigations of Nb_3Sn conductors by Stanford University, investigations of flowing He, development of suitable low temperature measuring techniques, measurement of dielectric losses by the National Bureau of Standards, and the development of Nb_3Ge conductors for ac cables by Westinghouse.

J. The Economics of Superconducting Cables

Before discussing the economics of superconducting cables, some general remarks on integration of superconducting cables in existing networks are called for. Detailed discussion of this theme would go beyond the scope of this paper. Detailed network studies [177, 203, 208] have been carried out with special emphasis on the problem of the maximum short-circuit power and of network stability under contingency conditions. The studies are based on possible areas of application for superconducting cables, viz. as a link between networks, as a link between a large concentration of generators and a network, and as a feeder for metropolitan areas.

It has been shown that compared to other new high power transmission systems, no fundamentally new problems are presented by the introduction of superconducting cables for some or several Gigawatts.

Modifications of conventional transmission system components, such as circuit breakers, surge arresters, and transformers, will be necessary for the large currents which have to be carried at present day operating voltages.

Before comparing different types of superconducting cables quantitatively with each other and with other systems, we note the qualitative dependence of the cost of three-phase ac cables on the transmission voltage, the transmission power and on the maximum permissible operating electrical field strength E utilizing Fig. 91 [200]. R over P (R/P), is the ratio of the external radius of the cable core to the transmitted power P. It is a qualitative measure for the specific cost of the cable. The maximum magnetic field value on the conductor surfaces of the cable cores has been assumed to be 0.07 T (H=40 A mm^{-1}) in Fig. 91. The following facts can be deduced from Fig. 91:

The costs of superconducting three-phase ac cables of equal rating are reduced drastically as the maximum permissible operating field strength E is increased, i.e., as the transmission voltage is raised.

The costs sink rapidly as the power rating is increased for constant physical parameters (E and H) and transmission voltage.

There is an optimum transmission voltage for every power with constant physical parameters.

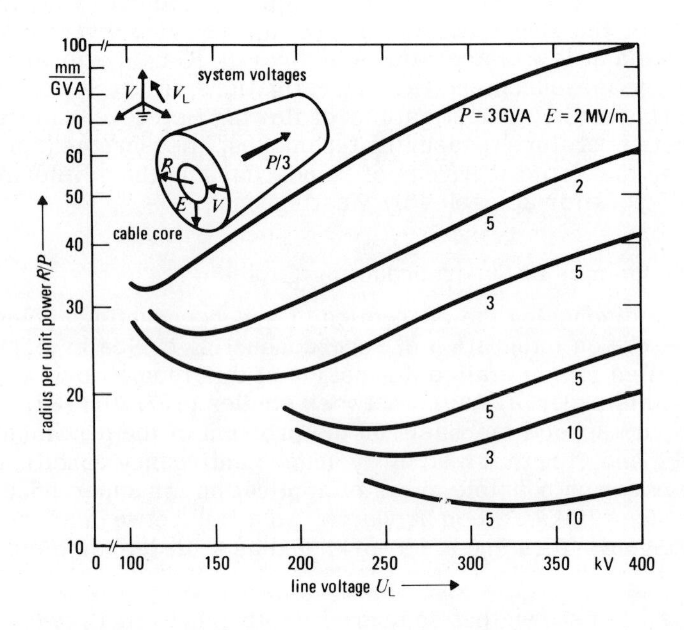

Fig. 91 Cable core radius per unit power R/P for various powers P and electric fields E (ac cable) [200].

Similar conclusions can be drawn from Table 17. The absolute and specific costs of superconducting three-phase ac and dc cables and of a conventional forced water cooled oil-paper cable are compared. Cases 1 and 2 have been reported by Siemens within the framework of a study "High Power Transmission" commissioned by the Federal German Ministry for Research and Technology. The study was carried out by German industry, utilities and research institutes [216]. Cases 3,6 and 7 are derived from reports of the CERL, U.K. [200, 208] and cases 4 and 5 from reports of Brookhaven National Laboratory, US [197, 203]. The operating costs, in which the capitalized loss costs are counted, have been omitted. These are generally of secondary importance for the utilities. The capitalized loss costs of conventional cables are considerably higher than those of superconducting cables, whereas the maintenance costs of the latter are probably much higher. In a number of cases, costs have been given for reactive compensation, in other cases these have been ignored since only shorter cable lengths (< 30 km) were considered or the costs have been included under other items. Different profit margins have probably been assumed from case to case. An operating field strength of 10 MV/m has been taken for all superconducting three-phase ac cables. The specified power ratings and the specific costs are based on a load factor of 1.

Although the same assumptions have not been made for all the examples cited in Table 17, the following conclusions can be drawn.

The American studies report considerably more favorable absolute and specific costs ($/km MVA) than European studies. This discrepancy can only be partly explained by the fact that systems of 2 or 3 cables were considered in the US studies, while the European studies considered only individual cables.

The transmission voltage of superconducting three-phase ac cables must be as high as possible, viz. 345 kV or 400 kV, in order to be able to compete with water cooled oil-paper insulated high power cables.

The break-even point for superconducting three-phase ac cables with a transmission voltage of 400 kV, as compared to forced water cooled oil cables, according to European studies, lies below 5 GVA (within the range of 3 to 5 GVA). It is considerably higher than the figure of 2 GVA reported some years ago. A study commissioned by the EEC [217] comes to the conclusion that forced water cooled paper-oil insulated 400 kV cables are more economical than superconducting three-phase ac cables up to 6 GVA. According to the same source SF_6 gas spacer cables should be more economical than superconducting cables up to 4.5 GVA.

It was established within the framework of a recent study carried out by the ENEL, Italy, for the EEC that in Europe there will probably not be a demand for underground transmission systems with a power per circuit in excess of 2000 MVA until the year 2000 [164]. For the transmission of very large power blocks, e.g., 5000 MVA, redundant transmission circuits must be used to ensure continuity of supply. Similar considerations have led to comparable conclusions in Japan [218].

Table 17 Comparative Costs of Cables

Component	Case 1		Case 2		Case 3		Case 4		Case 5		Case 6		Case 7	
	Superconducting												Conventional forced cooled 2 circuits	
	AC 110 kV 2000 MVA		AC 220 kV 5000 MVA		AC 400 kV 5000 MVA		AC 345 kV 4800 MVA		AC 550 kV 5700 MVA		DC 200 kV 5000 MW		AC 400 kV 5000 MVA	
	$10^3 \frac{\$}{km}$	%	$10^3 \frac{\$}{km}$	%	$10^3 \frac{\$}{km}$	%	$10^3 \frac{\$}{km}$	%	$10^3 \frac{\$}{km}$	%	$10^3 \frac{\$}{km}$	%	$10^3 \frac{\$}{km}$	%
Cryogenic Envelope (water pipes)	770	32	810	26	540	25	585	42	495	24	106	13	440	16
Cable cores	880	36	1350	44	500	23	390	28	990	48	150	17	1080	39
Cooling stations	380	16	420	15	260	12	95	7	121	6	300	34	130	5
Installation + Civil works	210*)	9	210*)	7	320	15	105	8	287	14	116	13	500	18
Jointing	100	4	140	5	240	11	–	–	–	–	106	12	290	11
Terminations	20	1	45	1	–	–	–	–	–	–	–	–	–	–
Helium	50	2	85	2	80	4	31	2	100	5	10	1	–	–
Auxiliaries and miscellaneous	–	–	–	–	180	9	185	13	59	3	86	10	190	7
Reactive Compensation	–	–	–	–	20	1	–	–	–	–	–	–	110	4
Total	2410	100	3060	100	2140	100	1391	100	2052	100	874	100	2740	100
$\$ \cdot km^{-1} \cdot MVA^{-1}$	1200		610		430		290		360		175		550	

*) Not including installation

These new conclusions, which owe their origin in part to the new energy policy formulated since the outbreak of the oil crisis, have led and will lead in the future to a reduction in the planned activity in Europe and Japan in the area of superconducting cables. This has been mentioned in Section VI. In view of the unquestioned long term potential of superconducting cables, it is important that the cable projects in hand be brought to a satisfactory state. This will allow a solid basis for future development work.

The situation in the US seems to be much more favorable than in Europe and Japan. The studies available have calculated more favorable economic break-even points for superconducting cables and foresee in certain cases the necessity of transmitting powers in the region of 5000 MVA to 10,000 MVA in the years 1990 to 2000 [177, 203]. For this reason development is being actively pursued in the US.

K. Future Development of Superconducting Cables

Although in recent years great progress has been made in development of components for superconducting cables, some problems have not been adequately solved. For example, an improvement of the electrical insulation of ac cables to achieve an operating field strength $E > 10$ MV/m would significantly improve the economics. Work is required in this field before further development can proceed. Experiments to determine the behavior under short-circuit conditions are also lacking and should be carried out.

The reliability of superconducting cables is of decisive importance. At present only more or less uncertain concepts exist. Possible faults and the likelihood of their occurrence must be estimated and techniques for their detection and clearance (and where possible prevention) must be developed. For this purpose, tests over long periods of time are required using longer cable lengths and operating under network conditions.

An appreciable advance could be realized through the use of Nb_3Ge as cable conductor. Operating temperatures of 15 K to 20 K, maintained by subcooled liquid hydrogen, would then be possible. As a result better efficiencies of the cooling circuits are possible. Simpler cryogenic envelopes could be employed. Hydrogen has appreciably better dielectric properties than He. It is cheaper and is available in unlimited quantity. Nb_3Ge would consequently lower the economic break-even point of superconducting cables to appreciably lower powers.

REFERENCES

1. H. Brechna, Superconducting Magnet Systems, Springer Verlag, Berlin-Heidelberg-New York, 1973.
2. The JET-Project, EUT-JET-R7, August 1975, published by the Commission of the European Communities, Brüssel, or: B.J.

Green, Europhysics News 6, 6 (1975).

3. H. Brechna, Superconducting Machines and Devices, Large Systems Applications, Nato Advanced Study Institutes Series 1, Plenum Press, New York and London, 139 (1974).
4. S. Foner et al., Phys. Letters 31A, 349 (1970).
5. S. Foner et al., Phys. Letters 47A, 485 (1974).
6. J.R. Gavaler et al., IEEE Transactions on Magnetics MAG 11, 192 (1975).
7. O. Fischer, Europhysics News 7, 1 (1976).
8. O. Fischer, VIth Int. Cryogenic Eng. Conf., Paper No, IP4, Grenoble, 11-14 May 1976.
9. S. Foner and E.J. McNiff, Phys. Letters 45A, 429 (1973).
10. M.N. Wilson et al., J. of Phys. D (Appl. Phys.) 3, 1517 (1970).
11. I. Pfeiffer and H. Hillmann, Acta Met. 16, 1429 (1968).
12. M. Suenaga et al., IEEE Transactions on Magnetics MAG II, 231 (1975).
13. H. Hillmann and E. Springer, Siemens Zeitschrift 49, 739 (1975).
14. E. Gregory, J. of Appl. Polymer Science: Symposium 29, 1 (1976).
15. K. Tachikawa et al., IEEE Transactions on Magnetics MAG 11, 240 (1975).
16. H. Kuckuck et al., Cryogenics 16, 350 (1976), and G. Ziegler et al., Deutsche Phys. Gesellschaft, Frühjahrstagung, Freudenstadt, 5-9 April 1976.
17. Y. Iwasa, IEEE Transactions on Magnetics MAG 11, 266 (1975).
18. D.C. Larbalestier et al., VIth Int. Cryogenic Eng. Conf., Paper No. K4, Grenoble, 11-14 May 1976.
19. G. Ries and K.P. Jüngst, to be published in Cryogenics, 1976.
20. K. Boening et al., Phys. Stat. Solidi 34, 395 (1969).
21. M. Söll et al., Journal of Low Temperature Physics 24, 631 (1976).
22. E. Seibt, IEEE Transactions on Magnetics MAG 11, 174 (1975).
23. M. Söll et al., IEEE Transactions on Magnetics, MAG 11, 178 (1975).
24. D.M. Parkin and A.R. Sweedler, IEEE Transactions on Magnetics MAG 11, 166 (1975).
25. M. Söll et al., 9th Symposium on Fusion Technology, Garmisch-Partenkirchen, 14-18 June 1976.
26. H. Brechna and W. Maurer, Gesellschaft für Kernforschung, Karlsruhe, Report No. 1468, September 1971.
27. P. Turowsky, Proc. Fifth Int. Conf. on Magnet Technology, MT-5, Roma, 541 (1975).
28. P.F. Smith and B. Colyer, Cryogenics 15, 201 (1975).
29. D. Evans, Rutherford Laboratory, Chilton, Didcot, England, Rep. No. RL 73-093 (1973).
30. D.S. Easton and C.C. Koch, Cryogenic Eng. Conf., Paper No. J 5, Kingston, Canada, 22-25 July 1975.
31. J.R. Heim, Fermi Nat. Acc. Lab., Batavia, Illinois, Rep. No. TM 334-B (1974).
32. C. Schmidt, Internal Report of IEKP-Karlsruhe, 75 Karlsruhe, Postfach 3640, FRG (1976).

33. F.R. Fickett, Proc. Fifth Int. Conf. on Magnet Technology MT-5, Roma, 659 (1975).
34. G. Vécsey et al., Proc. Fifth Int. Conf. on Magnet Technology, MT-5, Roma, 110 (1975).
35. M. Morpurgo, Particle Accelerators 1, 255 (1970).
36. M.O. Hoenig and D.B. Montgomery, IEEE Transactions on Magnetics MAG 11, 569 (1975).
37. M.O. Hoenig et al., 6th Int. Cryogenic Eng. Conf., Paper No. I-9, Grenoble, 11-14 May 1976.
38. The author became acquainted with this type of conductor during a visit to the Kurchatov-Institute, Moscow, in September 1975.
39. M. Morpurgo, Paper presented at the Conf. on Technical Applications of Superconductivity, Alushta, USSR, September 16-19 (1975).
40. M.S. Lubell et al., Plasma Physics and Controlled Nuclear Fusion Research III, IAEA-CN-28/K-10, 433 (1971).
41. "Wisconsin Superconductive Energy Storage Project", Final Report to NSF, 1 July 1974.
42. J.M. Lock, Cryogenics 9, 438 (1969).
43. IGC-Newsletter 1, No. 2, 1 (1973).
44. The Japan Economic Journal from 9.3.76.
45. H. Pfister et al., 1976 Applied Superconductivity Conference, Stanford, California, Paper No. W2, 17-20 August 1976.
46. J.P. Zbasnik et al., Cryogenic Engineering Conf., after deadline paper, Kingston, Canada, 22-25 July 1975.
47. P.A. Cheremnykh et al., IEEE Transactions on Magnetics MAG 11, 519 (1975).
48. M.A. Green, 6th Int. Cryogenic Eng. Conf., Paper No. N-4, Grenoble, 11-14, May 1976.
49. G. Kesseler et al., Proc. Int. Conf. on Magnet Technology MT-3, Hamburg, 768 (1970).
50. W. Heinz, Proc. of the Fifth Int. Conf. on Magnet Technology MT-5, Roma, 14 (1975).
51. W.B. Sampson, Brookhaven Nat. Lab., private communication, June 1976.
52. Institut für Experimentelle Kernphysik, Karlsruhe, 31, Jahresbericht 1975, to be published 1976.
53. J.H. Coupland, Proc. of the 5th Int. Conf. on Magnet Technology MT-5, Roma, 535 (1975).
54. G. Bronca et al., Proc. of the 5th Int. Conf. on Magnet Technology MT-5, Roma, 525 (1975).
55. B.P. Strauss et al., IEEE Transactions on Magnetics MAG-11, 451 (1975).
56. T. Elioff et al., IEEE Transactions on Magnetics, MAG-11, 447 (1975).
57. J.R. Purcell et al., IEEE Transactions on Magnetics MAG-11, 455 (1975).
58. W. Heinz, 6th Int. Cryog. Eng. Conf., Paper No. IP6, Grenoble, 11-14 May 1976.

59. A.P. Fraas, ORNL-TM-3096 (1973), R. Moir and C.E. Taylor, URCL-74326, Lawrence Livermore Laboratory (1972).
60. C.D. Henning et al., 8th Int. Conf. of Refrigeration, Int. Inst., Refr. Washington 1971.
61. K.R. Efferson et al., 4th Sympos. on Engin. Problems of Fusion Research, Washington 1971.
62. D.N. Cornish, Culham Lab., Lab. Report No. CLM-P 275, 1 (1971).
63. C.E. Taylor et al., 4th Sympos. on Engin. Problems of Fusion Research, Washington 1971.
64. J.R. Roth et al., Proc. Applied Superconductivity Conference, Annapolis 1972, IEEE Pub. No. 72CH0682-5-TABSC, 361 (1973).
65. G. Bogner, Proc. Appl. Superconductivity Conference, Annapolis 1972, IEEE Pub. No. 72 CH0682-5-TABSC, 214 (1973).
66. Private communication during a visit to the Kurchatov Institute, Moscow, in September 1975.
67. C.E. Swannack et al., IEEE Transactions on Magnetics MAG 11, 504 (1975).
68. Private communication during a visit to the Efremov Institute, Leningrad, in September 1975.
69. A. Ulbricht et al., 6th Int. Cryog. Engin. Conf., Paper No. N11, Grenoble, 11-14 May 1976.
70. USEAC Division of CTR, "Fusion Power by Magnetic Confinement", WASH-1290, UC-20, February 1974.
71. E.J. Ziurys, Proc. of the 5th Int. Conf. on Magnet Technology MT-5, Roma, 296 (1975).
72. C.H. Dustmann et al., 9th Symposium on Fusion Technology, Garmisch-Partenkirchen, Germany, 14-18 June 1976.
73. T. Hiraoka et al., 9th Symposium on Fusion Technology, Garmisch-Partenkirchen, Germany, 14-18 June 1976.
74. J. File, Proc. Fifth Int. Conf. on Magnet Technology MT-5, Roma, 281 (1975).
75. J.B. Heywood and G.J. Womack, Open-Cycle MHD Power Generation (Pergamon Press, 1969).
76. J. Powell, Superconducting Machines and Devices, Large Systems Applications, Nato Advanced Study Institute Series 1 (1974).
77. D.B. Montgomery et al., Proc. 6th Int. Conf. on MHD El Power Generation, Washington, June 1975.
78. Y. Aijama et al., Proc. of the 5th Int. Cryog. Eng. Conf. 1974, Kyoto, 300 (IPC 1975).
79. P. Komarek, Cryogenics 16, 131 (1976).
80. J.R. Purcell et al., Proc. 6th Int. Conf. on MHD El Power Generation, Washington, June 1975.
81. C. Laverick, Applied Superconductivity in the USSR, 31 March 1976.
82. P. Dubois, private communication (1974).
83. R.W. Boom et al., IEEE Transactions on Magnetics MAG 11, 475 (1975).
84. W.V. Hassenzahl, IEEE Transactions on Magnetics, MAG 11, 482 (1975).

85. W.V. Hassenzahl, 9th Progress Report on SMES, LA-6225-PR, UC-95b, issued: February 1976.
86. E. Maxwell, Cryogenics 15, 179 (1975).
87. D. Kelland et al., Superconducting Machines and Devices, Large Systems Applications, Nato Advanced Study Institutes Series 1, Plenum Press, New York-London, 581 (1974).
88. J.A. Oberteuffer, IEEE Transactions on Magnetics MAG 9, 303 (1973).
89. S.C. Trindade and H. Kolm, IEEE Transactions on Magnetics MAG 9, 310 (1973).
90. C. de Latour, IEEE Transactions on Magnetics MAG 9, 314 (1973).
91. C. de Latour and H. Kolm, IEEE Transactions on Magnetics MAG 11, 1570 (1975).
92. H.H. Kolm, IEEE Transactions on Magnetics MAG 11, 1567 (1975).
93. Z.J.J. Stekly, IEEE Transactions on Magnetics MAG 11, 1594 (1975).
94. J.H.P. Watson and D. Hocking, IEEE Transactions on Magnetics MAG 11, 1588 (1975).
95. P.G. Marston, Proceed. 5th Internat. Conf. on Magnet Technology MT-5, Roma, 424 (1975).
96. R.D. Thornton, IEEE Transactions on Magnetics, MAG 11, 981 (1975).
97. C. Albrecht, 14th Internat. Congress of Refrigeration, Moscow, 20-30 September 1975.
98. C. Albrecht, ETZ-A 96, 383 (1975).
99. H. Hieronymus, Technical Report Siemens AG, 22.1.1976.
100. J. Gloel and J. Holtz, Siemens Forschungs- und Entwicklungsberichte 5, 85 (1976).
101. C. Albrecht and G. Bohn, to be published in "Physikalische Blätter" (1976).
102. L.C. Davis and R.H. Borcherts, J. Appl. Phys. 44, 3293 (1973).
103. J. Holtz, ETZ-A 96, 365 (1975).
104. Y. Ishizaki et al., Proc. of the 5th Internat. Cryogenic Eng. Conf., Kyoto 102 (1974).
105. G. Bogner, Superconducting Machines and Devices, Large Systems Applications, Nato Advanced Study Institutes Series 1, Plenum Press, 610 (1974).
106. C. Albrecht et al., Proc. of the 5th Internat. Cryogenic Eng. Conf., Kyoto 28 (1974).
107. St. Asztalos et al., Proc. of the 5th Internat. Cryogenic Eng. Conf., Kyoto, 37 (1974).
108. C. Albrecht, to be published in Proc. of the AMMAC 76, the Institute of Electrical Engineers, London.
109. T. Ohtsuka and Y. Kyotani, IEEE Transactions on Magnetics MAG 11, 608 (1975).
110. Y. Kyotani, Cryogenics 15, 372 (1975).
111. T. Iwahana, IEEE Transactions on Magnetics MAG 11, 1704 (1975).

112. T. Ohtsuka, private communication (1976).
113. D.L. Atherton and A.R. Eastham, IEEE Transactions on Magnetics MAG 11, 627 (1975).
114. D.L. Atherton and A.R. Eastham, Cryogenic Engineering Conf., Paper No. A-1, Kingston, Canada, 22-25 July 1975.
115. R.G. Rhodes and B.E. Mulhall, Cryogenics 15, 403 (1975).
116. G.J. Homer et al., Rutherford Laboratory paper, to be published (1976).
117. J.R. Powell and G.R. Danby, ASME Paper 66 WA/RR 5 (1966); Cryogenics 11, 192 (1971).
118. R.H. Borcherts, Cryogenics 15, 385 (1975).
119. H.H. Kolm and R.D. Thornton, Proc. of the Appl. Superconductivity Conf. Annapolis, IEEE, Pub. No. 72 CH0682-5-TABSC 76, (1972).
120. H.H. Kolm, Proc. of the 5th Internat. Conf. on Magnet Technology MT-5, Roma, 385 (1975).
121. J.S. Joyce et al., IEEE Trans. PAS 93, 210 (1974).
122. G. Bogner, Proc. of the 1972 Applied Superconductivity Conference, IEEE Publ. No. CH0589-5-TABSC, 214 (1972).
123. G. Bogner and D. Kullmann, Siemens Forsch.-u. Entwickl.-Ber., Vol. 4, 305 (1975).
124. G. Bogner and D. Kullmann, Siemens Forsch.-u. Entwickl.-Ber., Vol. 4, 368 (1975).
125. I.A. Glebov, Paper presented at the Conf. on Technical Applications of Superconductivity, Alushta, USSR, 16-19 September 1975.
126. C. Laverick, Review Paper on Applied Superconductivity in the USSR, 31 March 1976.
127. E. Massar, Proceedings of the Autumn-School on the Application of Superconductivity in Electrical Engineering and High Energy Physics, Titisee, 9-12 October 1972.
128. A.D. Appleton, Cryogenics 9, 147 (1969).
129. K.R. Jones, Electr. Rev. 180, 50 (1967).
130. A.D. Appleton, Proc. of the 1972 Applied Superconductivity Conference, IEEE Pub. No. 72 CHO 682-5-TABSC, 16 (1972).
131. A.D. Appleton, IEEE Transactions on Magnetics MAG 11, 633 (1975).
132. A.D. Appleton, Proc. of the Fifth Int. Conf. on Magnet Technology, MT-5, Roma, 447 (1975).
133. R.L. Rhodenizer, Navy Ship Systems Command Report Contract No. N 00024-68-C-5414, 27 February 1970.
134. C.J. Mole et al., APRA-Semi-Annual Technical Rep. for Period Ending 31 May 1975. Contract No. DAHC 15-72-C-02229.
135. P. Klaudy, Electortechn. u. Maschinenbau 78, 128 (1961).
136. P. Klaudy, Archiv für technisches Messen und industrielle Messtechnik 355, R 97 (1965).
137. J.L. Johnson et al., Proceedings of the Holm Seminar on Electrical Contact Phenomena 201 (1973).
138. P. Klaudy, Electrotechnik und Maschinenbau 89, 439 (1972).

139. A.D. Appleton, Superconducting Machines and Devices, Large Systems Applications, Nato Advanced Study Institutes Series, Plenum Press, 219 (1974).
140. W.J. Levedahl, Proc. of the 1972 Applied Superconductivity Conf. IEEE Publ. No. CH0689-5-TABSC, 93 (1972).
141. G.R. Fox and B.D. Hatch, Proc. of the 1972 Applied Superconductivity Conference, IEEE Pub. No. 72CHO 689-5-TABSC, 93, (1972).
142. T.J. Doyle, Adv. in Cryog. Engineering 19, 162 (1974).
143. R.F. Shanahan General Electric Company, Public Information, 9 July 1976.
144. I.A. Glebov, IEEE Transactions on Magnetics MAG 11, 657 (1975).
145. J.P. Chabrerie et al., Proc. of the 1972 Applied Superconductivity Conference, IEEE Pub. No. 72CHO 689-5-TABSC, 93 (1972).
146. M. Yamamoto and M. Yamaguchi, Proc. of the Fifth Intern. Cryog. Eng. Conf., Kyoto, 154 (1974).
147. J.L. Smith Jr. and T.A. Keim, Superconducting Machines and Devices, Large Systems Applications, Nato Advanced Study Institute Series, Plenum Press, 279 (1974).
148. J.S. Joyce et al., Paper presented at the Amer.-Power Conf., Chicago, Ill, 8-10 May 1973.
149. J.H. Parker et al., IEEE Transactions on Magnetics MAG 11, 640 (1975).
150. G. Bogner and D. Kullmann, Siemens Forsch.-u.-Entw.-Ber. 5, 12 (1976).
151. P.A. Rios, et al., Paper submitted to the Conf. on Technical Applications of Superconductivity, Alushta, USSR, 16-19 September 1975.
152. J.L. Smith Jr., Proc. of the Fifth Int. Conf. on Magnet Technology MT-5, Roma, 431 (1975).
153. C.C. Sterrett et al., Paper presented at the ASME-IEEE Joint Power Conference, Portland Oregon, 28 September-2 October 1975.
154. B.B. Gamble et al., Paper submitted to the ASME-IEEE Joint Power Conference, Portland Oregon, 28 September-2 October 1975.
155. J.L. Kirtley Jr. and N. Dagalakis, IEEE Transactions on Magnetics, MAG 11, 650 (1975).
156. G. Klipping, Proc. of 6th Internat. Vacuum Congr., Kyoto, Japan, 25-29 March 1974.
157. J.L. Smith Jr. et al., Proc. of the 1972 Applied Superconductivity Conference, IEEE Publ. No. CH 0689-5-TABSC 145, 1972.
158. A. Hofmann, Sixth Int. Cryogenic Eng. Conf., Paper No. L-2, Grenoble, 11-14 May 1976.
159. R.G. Scurlock, Sixth Int. Cryog. Eng. Conf., Paper No. IP-7, Grenoble, 11-14 May 1976.
160. R.G. Scurlock and G.K. Thornton, Proc. of the Fifth Int. Conf. on Magnet Technology MT-5, Roma, 530, (1975).

161. Y. Mori et al., Int. J. Heat Mass Transfer 11, 1807 (1971).
162. A.D. Appleton and A.F. Anderson, Proc. of the 1972 Appl. Superconductivity Conf., IEEE Publ. No. CH0689-5-TABSC 136 (1972).
163. B.W. Birmingham et al., Proc. of the Fifth Int. Cryog. Eng. Conf., Kyoto, 157 (1975).
164. ENEL – "Study on the future need for large generating units and high power underground cables in the European Community (1980-2000)" Contract No. 082-73-12-ECI. September, 1974.
165. J.L. Smith, Jr., Private communication, April 1976.
166. C.J. Mole et al., IEEE Trans PAS, Conf. Paper C 73259-9 (1973).
167. G. Bogner, Information received during a visit to the All Union Res. Inst. for Electrical Mach. Ind., Leningrad, September 1975.
168. S. Akijama et al., 6th Int. Cryog. Eng. Conf., Paper No. L-7, Grenoble, 11-14 May 1976.
169. C. Pinet, Proc. of the Fifth Int. Conf. on Magnet Technology, MT-5, Roma, 452 (1975).
170. P. Thullen et al., IEEE Transactions on Magnetic MAG 11, 653 (1975).
171. A.D. Appleton et al., Cigre 11-02, 1976-Session August 25-September 2, 1976.
172. G. Bogner, Superconducting Machines and Devices, Large Systems Applications, Nato Advanced Study Institutes Series 1, Plenum Press New York-London, 401 (1974).
173. W. Kafka, Elektr. techn. Zeitschr. ETZ-A 90, 89 (1969).
174. P. Dubois et al., Proc. of the Applied Superconductivity Conference, Annapolis 1972, IEEE Pub. No. 72 CHO 682-5-TABSC, p. 173 (1972).
175. C.N. Carter, Cryogenics 13, 207 (1973).
176. R.L. Garwin and J. Matisoo, Proc. IEEE 55, 538 (1967).
177. E.B. Forsyth, Paper presented at the Conference on Technical Applications of Superconductivity, Alushta, USSR, 16-19 September 1975; BNL 20444.
178. H. Morihara et al., Cryog. Eng. Conf., Atlanta 1973, (Paper not published in Adv. Cryog. Eng., Vol. 19).
179. J.F. Bussiere et al., Appl. Phys. Letters 25, 756 (1974).
180. G. Bogner, Cryogenics 16, 259 (1976).
181. H.A. Ullmaier, Phys. Stat. Sol. 17, 631 (1966).
182. J.D. Thompson et al., 1976 Applied Superconductivity Conference, Paper No. L9, Stanford, 17-20 August 1976.
183. P.H. Melville, J. Phys. C4, 2833 (1971).
184. W.J. Dunn and P. Hlawiczka, Brit. J. Appl. Phys., Ser. 2, 1, 1469 (1968), and G. Fournet and A. Mailfert, J. de Physique 31, 357 (1970).
185. P. Penczynski et al., Cryogenics 14, 503 (1974).
186. P. Penczynski, Deutsche Auslegeschrift No. 2310327, 12 February 1976.
187. W.T. Beall, IEEE Transactions on Magnetics MAG 11, 381 (1975).
188. H. Franke, private communication (1976).

189. J. Sutton, Cryogenics 15, 541 (1975).
190. G. Morgan and E.B. Forsyth, Cryogenic Engineering Conf., Paper No. U-5, Kingston, Ontario, Canada, 22-25 July 1975.
191. G. Matthäus and P. Massek, Siemens Research Labs.
192. B. Fallou et al., CIGRE-Report Nr. 15-04 (1974).
193. R.W. Meyerhoff, Adv. Cryog. Eng. 19, 101 (1974).
194. B. Fallou et al., Conf. Low Temperatures and Electric Power, London 1969, Proc. Int. Inst. Refrig., Comm. 1, Pergamon Press, New York 1970, p. 377.
195. J. Gerhold, Cryogenics 12, 370 (1972).
196. B. Fallou, Proc. of the Fifth Int. Conf. on Magnet Technology, MT-5, Roma, 644 (1975).
197. R.W. Meyerhoff, private communication (1975).
198. E.B. Forsyth et al., Cryogenic Engineering Conf., Paper No. U-7, Kingston, Ontario, Canada, 22-25 July 1975.
199. H. Franke, private communication (1976).
200. D.A. Swift, 14th Internat. Congress of Refrigeration, Moscow, 20-30 September 1975.
201. D.E. Daney, Cryogenic Eng. Conf., Paper No. R-1, Kingston, Ontario, Canada, 22-25 July 1975.
202. J.W. Dean and J.E. Jensen, Cryogenic Eng. Conf., Paper No. R-7, Kingston, Ontario, Canada, 22-25 July 1975.
203. E.B. Forsyth, Paper submitted to the International Journal of Energy (1976).
204. Results obtained by Siemens Research Laboratories.
205. E. Bochenek et al., IEEE Trans. on Magnetics, MAG 11, 366 (1975).
206. G. Bogner, Adv. in Cryog. Eng., 19, 78 (1974).
207. G. Bogner, Cryogenics 15, 79 (1975).
208. B.J. Maddock et al., CIGRE Report 21-05, 25 August-2 September 1976.
209. D.A. Swift, private communication (1976).
210. P.A. Klaudy, Elektrotechnik und Maschinenbau 89, 93 (1972).
211. Y.L. Blinkov, Paper presented at the Conference on Technical Applications of Superconductivity, Alushta, USSR, 16-19 September 1975.
212. Y. Furuto et al., Proc. Fifth Int. Cryogenic Eng. Conf., Kyoto, 180 (1975).
213. Y. Furuto, private communication (1976).
214. R.W. Meyerhoff, Cryogenic Eng. Conf., Paper No. R-8, Kingston, Ontario, Canada, 22-25 July 1975.
215. W.E. Keller and R.D. Taylor, Los Alamos Scientific Lab Progress Report LA-6215-PR, February 1976.
216. Deutsche Systemstudie "Elektrische Hochleistungsübertragung" to be published during 1976.
217. J. Erb et al., Comparison of Advanced High Power Underground Cable Designs, Kernforschungszentrum Karlsruhe, KFK 2207, September 1975.
218. T. Ohtsuka, private communication (1976).

INDEX

PART I – SQUIDs

(See also Index Part II – Machines, P.731)

INDEX – PART II – MACHINES

(See also Index Part I – SQUIDs, P. 719)